Plant Model Greenlab for Botany and Agronomy

Philippe De Reffye • Mengzhen Kang
Marc Jaeger • Baogui Zhang
Editors

Plant Model Greenlab for Botany and Agronomy

Springer

éditions Quæ
Éditions Cirad, Ifremer, INRAE
www.quae.com

Editors
Philippe De Reffye
BIOS UMR Amap
Centre de coopération internationale en recherche agronomique pour le développement (CIRAD)
Montpellier, France

Mengzhen Kang
Institute of Automation Chinese Academy of Sciences
Haidian District, Beijing, China

Marc Jaeger
AMAP GreenLab
Centre de coopération internationale en recherche agronomique pour le développement (CIRAD)
Montpellier, France

Baogui Zhang
College of Land Science and Technology
China Agricultural University (CAU)
Haidian District, Beijing, China

ISBN 978-981-97-7540-8 ISBN 978-981-97-7541-5 (eBook)
https://doi.org/10.1007/978-981-97-7541-5

Translation from the French language edition: "Architecture des plantes et production végétale. Les apports de la modélisation mathématique" by Philippe De Reffye et al., © Editions Quae 2018. Published by Editions Quae. All Rights Reserved.

This Springer imprint is published by the registered company Springer Nature Singapore Pte Ltd.
The registered company address is: 152 Beach Road, #21-01/04 Gateway East, Singapore 189721, Singapore

Preface

I dream of a botany that would be able to determine itself autonomously, according to its own rules, ceasing to lag behind animal or human physiology: taking into account the plant itself, as an original form of life, as a model in terms of autonomy and environmental restoration, it could regain its place at the centre of life sciences. In our world of money, show-off, advertising, noise, pollution and brutality, what better testimony than that of plants, beautiful and useful, discreet and autonomous, silent and totally non-violent?

Translation by F. Varenne from Francis Hallé, *Eloge de la plante – Pour une nouvelle biologie*, Paris, Seuil, 1999.[1]

For some years now, the stimulating books by botanist Francis Hallé have been proving that plants have much to teach us. Fascinated as we are by animals, our brothers in animality, we still very often have a narrow idea of life, of the variety of its forms as well as its behaviour and strategies. It is therefore by no means pointless to try to represent and model plants. It may even be that there is a certain urgency to do so today. Don't we need to represent them, simulate them—as the botanist patiently draws them again—so that finally, more widely, we all learn to see them?

This book of synthesis, written with care and pedagogy by Philippe De Reffye and several of his French and Chinese colleagues, is timely: it takes even further the admirable Copernican gesture of decentring encouraged by Francis Hallé because, in a sense, it gives concrete expression to the expectations of this gesture on the technical, methodological, and epistemological levels. What this book brilliantly shows is that today, perhaps, it is the plant models—foremost among which are those of Reffye and his affiliated teams—that tell us best or least badly how it would be appropriate to model and simulate living things in general. At a time when many modeling and simulation programs with integrative aims are being launched, i.e. these types of research programs aimed at simulating an organ or even an entire organism such as a complete eukaryotic cell, the brain or the human body, and at a time when debates are raging at the international level about the approaches most likely to allow computer modeling that is both biologically relevant and truly

[1] This book has been translated by David Lee: *In Praise of Plants*, Portland, Timber Press, 2002.

predictive, the book by De Reffye and his colleagues may well provide some decisive answers.

But can there really be science behind what may at first glance appear to be simple computer simulations and beautiful images? Until a few years ago, some people might have doubted it. While some simulations may indeed be mere digitisation of scenes for the purpose of shape detection or automatic 3D image synthesis for purely descriptive purposes, it is important to take a closer look. The mathematical models and simulation models of Philippe De Reffye and his teams lead to biologically significant calculations and extremely serious images because they are based on a considerable amount of botanical, agronomic and mathematical knowledge, knowledge that has been patiently accumulated, chosen, criticised, confronted and then patiently interwoven over a long history, the pivotal episodes of which I have tried elsewhere to retrace and interpret[2]. I will mention a few milestones here for the record. When I look back at this story afterwards, with its trial and error and chance encounters, with its opportunities too, and after having discussed it several times with Philippe De Reffye and his colleagues, I cannot help but admire the level of abstraction, efficiency but also, and always, the level of sensitivity to the diversity of real plants that this approach now achieves.

The history of the progressive construction of the GreenLab model—the model presented in this book—has accelerated over the last ten years or so, but it began more than forty years ago now. It is intimately linked to Philippe De Reffye's entire scientific career at CIRAD (International Center in Agricultural Research for Development), to his insatiable vocation as a researcher but also to his happy encounters and early interactions with colleagues such as Hervé Bichat, Alain Coléno, Jean Françon, François Houllier, Daniel Barthélémy and of course with all his doctoral students, including Marc Jaeger, Frédéric Blaise and Paul-Henry Cournède. Based on the essentially conceptual—but also partly institutional—historical investigation I have proposed, we can identify four major periods, each of which goes hand in hand with four major types of architectural models of plants.

The first period is that of the beginnings and dates back to the early 1970s. At that time, Philippe De Reffye was a young agricultural engineer, working at the IFCC (French Coffee and Cocoa Institute, which was later to join CIRAD when it was founded), in Côte d'Ivoire. Within the initial framework of an agronomic problem of forecasting the fruiting of coffee trees, he was led to show, on the one hand, the limits of the biometric models usually used at the time and, on the other hand, the interest in using computers to make a step-by-step representation—a simulation in this sense—certainly very schematic but historically accurate, of the almost complete architecture of the coffee tree. This first computer model for the simulation of the architecture made it possible to trigger stochastic processes during the programme, processes whose parameters were calibrated to the field data. These processes simulated one by one the births of the coffee tree's fruits. This first period therefore saw the first design

[2] See: F. Varenne, *From Models to Simulations*, New York, Routledge, 2018.

of an integrative and agronomy-oriented architectural simulation model for a single species, the coffee tree.

For his 1979 State PhD thesis, a thesis on which unsurprisingly Francis Hallé sat on the jury, Philippe De Reffye proposed a simulation engine of the same type but made "universal", i.e. seeking to simulate all the architectures of vegetative plants in all their diversity, going beyond the case of the coffee tree alone and integrating a large number of the qualitative and quantitative particularities previously listed by botanists. The key idea here was to consider the architecture of a vegetative plant as the result of the dynamics of its meristems, in relation to the particularity of the architectural model specific to its essence. As a consequence of this idea, the simulation engine should focus on the functioning of the meristems.

At the beginning of the 1980s, on Reffye's return to France, at CIRAD, a colleague specialising in combinatorics and discrete mathematics from the University Louis Pasteur in Strasbourg, Jean Françon, amplified the scope of this research by directing it towards fruitful interactions with combinatorics, algorithmics and computer graphics. Françon's students contributed to improving the algorithm, the programming but also the graphic rendering of the simulations carried out by the first De Reffye model. Within the framework of the laboratory then created at the CIRAD of Montpellier, AMAP, the Plant Architecture Modeling Workshop, this third period saw the development of several universal simulation models that were increasingly realistic and better designed from a computer science point of view. However, it soon became apparent that efficient computational processing—favouring sequential calculation—was in contradiction with the biological dynamics of meristems: real meristems operate in parallel and not sequentially. On the other hand, even if they integrated various and heterogeneous knowledge (including that which makes it possible to simulate the mechanical bending of branches or self-pruning) and even if they partly were based on feedback of the structure on the functioning of the plant, the first AMAP simulation models were still considered by many to be simple models for describing structures. In fact, although they simulated development well, they were not able to predict the production of material by photosynthesis and therefore secondary growth in response to given environmental conditions, something which is of great interest to the agronomist.

Within the framework of an new partnership that is still now active with INRA and other institutions such as the University of Montpellier, François Houllier—director of AMAP in the early 2000s—and Philippe De Reffye have resolutely oriented existing structural models towards taking into account physiological and functional knowledge. By integrating eco-physiological knowledge such as the "law of water efficiency" and the "law of temperature response", AMAP tended to develop more and more "structural-functional" models (FSM), which were no longer primarily simulation models. The priority was no longer in the topological and graphical rendering of the architectural structure, but above all in the realism of the production of material in response to environmental conditions. Simulation images were clearly becoming secondary although, by their preserved botanical structural realism and their essential predictive aspect, they still contributed on the one hand to the verification of model hypotheses and on the other hand to the remarkable diffusion (as was the case very early on, as early as the 1980s) of AMAP software towards other disciplines, other than those strictly related to agronomy such as landscaping, urban planning or the analysis and interpretation of satellite images.

From the beginning of the 2000s, we are therefore clearly entering a fourth period, which I have proposed to call the "remathematisation period". This period is illustrated by remarkable progress in terms of the mathematical and computer simplification of algorithms, but also in terms of refocusing botanical driving concepts, in collaboration with researchers from INRIA in France and China, the Institute of Automation, Chinese Academy of Sciences, and the China Agricultural University. The structure previously simulated organ by organ is now calculated phytomer type by phytomer type: it is the quantities and distributions of each type of phytomers that are directly calculated. Their birth is therefore no longer simulated step by step. As we will see in this book, proceeding constructively, the part of the theory of stochastic processes known as "renewal theory" is systematically used in connection with the concepts of order of branching and physiological age, leading to the writing not only of models of stochastic processes but firstly and directly of algebraic equations linking the moments or distributions of the stochastic processes which are involved in each order. Stochastic simulation is therefore less present than in the previous simulation-oriented computer programmes, even if the final variability of the results obtained remains similar. The number of calculation steps is considerably reduced. As it is impossible to express oneself better than the authors themselves on this subject, I will briefly quote here a passage from the end of chapter 5 which seems to me to sum up the content of this recent and formidable advance:

> From GreenLab's botanical automaton, whatever its complexity, we can deduce compact operators in matrix form, which allow us to write recursive algorithms which factor the calculation of development and which directly give the mean and variance of the numbers of phytomers produced per category, without recourse to stochastic simulation. The most important result is that the calculation time depends on the age of the plant and not on the number of elements to be produced, as is the case with conventional plant simulators. The link between the number of phytomers produced and the calculation time is quadratic. Another important result is that the quasi-random mode based on sub-structures gives stochastic simulations of comparable quality to pseudo random simulations for mean and variance with much better performance, because it is based on factorisation in sub-structures. Calculation times are proportional to the size of the collection and to the square of the age of the plant as for the algorithmic mode. Moreover, the quasi-random mode also allows to draw stochastic plants in a more efficient way. This mode is therefore the choice of the Greenlab model.

Thus, after a period that favoured the integration of models of various aspects and/or organs of the plant in the form of an integrative simulation based on a pluri-formalized model, a period of maturity followed that found its way back to elegance and mathematical efficiency. This efficiency is not so much sought-after as such, although the mind can certainly rejoice that a more formally unitary theory has replaced the computational branches of partial models and stochastic processes: as the authors often explain, it is sought after firstly because it allows inverse methods to exist which themselves allow the efficient and rapid identification in the field of the parameters necessary to calibrate the model, and secondly because by simplifying the algorithm, its number of computation steps is considerably reduced. As a result, also, its visualisation in the form of a computer-generated image is made easier. Finally, this simplification is sought after because it also permits the model to be open, to continue to look to the future, i.e. to continue to allow more and more stabilised botanical and agronomic knowledge to be integrated and made compatible with it.

What do we admire in the end in such a particularly realistic plant simulation? It's not so obvious to know, when you think about it. Is it the real plant that one admires in its morphological inventiveness, the simulation being only an evocative representation of it? Or is it the simulation that we admire in its unique ability to replicate and feign this same plant? Or is it not rather the art and technical mastery of the designers of the simulation and the formalisms that we admire? Or finally, isn't it the inventive genius of the "designer" of the real plant, namely the ingenious evolution of species? We suspect it: in the look and in the scientific practice of Philippe De Reffye and his colleagues, it is all of these things at once. It is the living plant first of all, its history, its dynamics, its development and its growth, but also their representations, their formalisations, their simulations.

This is why this book—both technical and synthetic—invites us to new admirations, to an augmented admiration, as we speak of augmented reality. The admiration for plants is always present and even guiding: it opens the way to knowledge, but not just any knowledge, to a knowledge that respects diversity because it is radically multi-type, truly multidisciplinary and not just multifactorial. The collection of data leads here to integrative but significant and measured representations, as well as to operational applications of these measured representations. The scales of the phenomena are not reduced, but respected. The plurality of phenomena is recognised and integrated. What is still ignored is also recognised to a certain extent and taken into account in this sense. This is why, in the end, it seems to me that, despite its undeniable mathematical technicality and its mobilisation of a necessary botanical erudition, this book must find a wide readership: this readership will be amply rewarded if it is true that—which I personally do not doubt—the serious and scientifically fertile practices of integrative modelling and simulation have, beyond a burning topicality and for some of them a somewhat premature visibility, a quite considerable future in the life sciences and even beyond.

Franck Varenne—2022 adapted translation by the author

Rouen, France
November 2022

Franck Varenne

Acknowledgements

The modelling of plant growth and architecture has been the focus of my attention during my research career at CIRAD over more than 40 years. This has often encroached on my family life and it is to my wife Catherine and my children that I express my gratitude in the first place, for having tolerated overtime modelling at home, even during weekends or holidays! To them, a big thank you.

I would like to express my gratitude to the researchers and heads of research institutions who have joined and supported the AMAP and GreenLab projects over the long term.

In the history of the AMAP laboratory, I have a special gratitude for the very first researchers who contributed to its origin, creation and development until today and whose actions and projects have gone beyond the scientific framework to achieve an industrial valorization that has generated significant own resources for the laboratory. Thanks to René Lecoustre and Marc Jaeger, for having made the laboratory a pioneer in the representation of plants and landscapes, from which the company Bionatics was established. Thank you to Joël Sor, Director of Information Systems at CIRAD, for supporting the project since its inception in its hardware, software and commercial components.

Special mention should be made of teacher-researchers who have enriched AMAP-GreenLab projects by contributing their know-how and giving it recognition in their own discipline. In botany: Francis Hallé and Claude Edelin, for the botanical concepts of plant architecture introduced into the model. In agronomy: Michael Dingkuhn, Ep Heuvelink and Jérémie Lecoeur for the concepts of the crop models and their adaptation to the model. In computer science: Jean Francon, Claude Puech, Maurice Goursat, Jean Pierre Quadrat of INRIA, for having formalized the model in the form of grammar, in order to make it more readable to mathematicians.

The interest and support given to the model by mathematicians Philippe Flageolet, André Gagalowitz, Michel Fliess, Michel Sorine from INRIA, and computer scientists David Hill from the University of Clermont Ferrand and Jean Daniel Zucker from the IRD were invaluable encouragement.

My first research in Côte d'Ivoire was supported by successive IFCC directors Marc Belin and Jacques Capot, who gave me their confidence by investing in a Hewlett Packard calculator. Back at CIRAD in Montpellier in the 1980s, my gratitude goes to Hervé Bichat, Director of CIRAD, and Jean Marie Sifferlen, Secretary-General and Director of the Gerdat department, convinced of the quality of the project, they decided to create the AMAP laboratory and supported its development.

In the 1990s, it was thanks to Alain Coleno, director of the plant sector at INRA, Yves Birot, director of the Forestry department at INRA, and Jean Bouchon, unit director at INRA Forêt, that AMAP became a joint CIRAD-INRA laboratory. I would like to thank both of them because they have made it possible to carry out a great deal of work on the architecture and stands of forest trees.

In China, since 1998, at Institute of Automation, Chinese Academy of Sciences (CASIA), the GreenLab project, the successor to the AMAP model, has been created in cooperation with CIRAD, thanks to the support of Professor Songde Ma, Director of CASIA, and Gilles Kahn, Scientific Director of INRIA. This project, led by Professor Baogang Hu, has enabled numerous Franco-Chinese exchanges and has developed more specifically the mathematical aspects on the one hand and agronomic applications in conjunction with the China Agricultural University (CAU).

Without these contributions and their institutional anchoring, the work on modelling would not have been able to benefit from new scientific advances.

Finally Digiplante, the latest addition to the AMAP and GreenLab projects, was first hosted in 2002 in the Metalau project at INRIA Rocquencourt (led by Maurice Goursat). This joint project between INRIA and CIRAD was then transferred to the École Centrale de Paris with the creation of the eponymous laboratory, Digiplante, at the initiative of the director of the MAS laboratory, Christian Saguez. Paul-Henry Cournède, project manager, continues to contribute to the development of the GreenLab model, particularly in terms of formalism and statistical methods. Digiplante has also been a space for numerous Franco-Chinese exchanges, thesis and Post-Docs.

At present, the GreenLab model is being particularly developed and used at CASIA in Beijing, with the support of Professor Wang Fei-Yue. In China, since 2003, PMA symposia on the use of plant architecture in agronomy have been held every three years on the initiative of CASIA.

Finally, we must mention the essential contributions: in France from the ITB, IRD and Irstea institutes, in the Netherlands from Wageningen University, in Canada from UQAM, in Togo from the University of Lomé, and in Côte d'Ivoire from Daloa University.

Of course, I would like to express my sincere and warm thanks to all the authors of the chapters and the contributors to this book for their trust, their judicious remarks and their careful proofreading, which have enabled the book to be considerably improved. Special thanks to Dr. Xiujuan Wang for her tremendous contribution in finalising this book. Finally, to Françoise Réolon and the teams at Quae Publishing, I would like to express my gratitude for all the care taken in the production of this complex work from the beginning to its publication.

Philippe De Reffye

Contents

Contributors

Daniel Barthélémy CIRAD, Montpellier, France

Frédéric Blaise CIRAD, Montpellier, France

Paul-Henry Cournède CentraleSupélec Laboratory of Applied Mathematics and Systems (MAS), Châtenay-Malabry, France

Ep Heuvelink Horticulture and Product Physiology, Wageningen University and Research, PB, Wageningen, The Netherlands

François Houllier Institut Français de Recherche pour l'Exploitation de la Mer (IFREMER), Plouzané, France

Baogang Hu National Laboratory of Pattern Recognition, Institute of Automation, Chinese Academy of Sciences (CASIA), Beijing, China

Jing Hua State Key Laboratory for Multimodal Artificial Intelligence Systems, Institute of Automation, Chinese Academy of Sciences, Beijing, China

Marc Jaeger CIRAD, UMR AMAP, Montpellier, France

AMAP, Univ Montpellier, CIRAD, CNRS, INRAE, IRD, Montpellier, France

Mengzhen Kang State Key Laboratory for Multimodal Artificial Intelligence Systems, Institute of Automation, Chinese Academy of Sciences (CASIA), Beijing, China

School of Artificial Intelligence, University of Chinese Academy of Sciences, Beijing, China

Véronique Letort Mathématiques et Informatique pour la Complexité et les Systèmes, CentraleSupélec-Université Paris-Saclay, Gif-sur-Yvette, France

Yuntao Ma College of Land Science and Technology, China Agricultural University, Beijing, China

Amélie Mathieu AgroParisTech, UMR EGC (1091), Grignon, France

Philippe De Reffye CIRAD, UMR AMAP, Montpellier, France

AMAP, Univ Montpellier, CIRAD, CNRS, INRAE, IRD, Montpellier, France

Sylvie Sabatier UMR-AMAP, CIRAD-CNRS-INRAE-INRIA-IRD-Université Montpellier, Montpellier, France

Baogui Zhang College of Land Science and Technology, China Agricultural University, Beijing, China

Chapter 1
Introduction

Philippe De Reffye and Mengzhen Kang

It's quite wrong to base a theory solely on observable quantities. Because, in reality, things happen in exactly the opposite way. It's only the theory that decides what can be observed.
A. Einstein

1.1 Purpose of this Book

This book deals with mathematical models of plant growth and their applications, which consider architectural development as well as plant growth that includes the production and distribution of synthesized biomass (source-sink relationships). The interaction between plant architecture and growth contributes to the plasticity of plants in response to the environment and to the mode of cultivation. The modeling of plant development (organogenesis) and growth (photosynthesis) requires a multidisciplinary and a multiscale approach (botany, agronomy, applied mathematics, computer science) according to the principles of integrative biology (Chauvet 1993; Chauvet website n.d.). Botanical knowledge on the concepts of herbaceous and woody plant architectures, as well as eco-physiological knowledge used in agronomy, are exposed.

P. De Reffye
CIRAD, UMR AMAP, Montpellier, France

AMAP, Univ Montpellier, CIRAD, CNRS, INRAE, IRD, Montpellier, France

M. Kang (✉)
State Key Laboratory for Multimodal Artificial Intelligence Systems, Institute of Automation, Chinese Academy of Sciences (CASIA), Beijing, China

School of Artificial Intelligence, University of Chinese Academy of Sciences, Beijing, China
e-mail: mengzhen.kang@ia.ac.cn

P. De Reffye et al. (eds.), *Plant Model Greenlab for Botany and Agronomy*,
https://doi.org/10.1007/978-981-97-7541-5_1

1.2 What Is Integrative Biology

In general, such complex systems can be visualized as a graph describing the hierarchy and topology of an evolving anatomical and biological structure in which source-sink actors interact. These are grouped into compartments to define a functional structure associated with the hierarchical structure. This is the case of the growth of a plant's architecture.

The concept was presented by G. Chauvet (Chauvet 1993), who linked different scientific disciplines by defining a level of observation and a time step that bridges them.

Multi-disciplinarity involves a foundation of empirical biological knowledge that needs to be modeled by a mathematical model using physical methods.

Knowledge from different disciplines cannot be directly used. They need to be mathematized and, after being simplified with the agreement of the specialists in each discipline, be able to communicate with each other according to a common framework. This is the parameter identification phase.

The level of observation and the time step are important, as they provide the framework for the mathematical model.

From the level of observation and by successive aggregations (Russian dolls), multiscale structures can appear, without changing the framework of the mathematical model. The functioning of higher-order structures can be predicted from the elementary level of observation.

Without a theoretical model, there are no valid measurements. All we do is to make statistics. Measurements of biological systems are not the parameters themselves; rather, parameters are hidden within the measurements and need to be extracted using inverse mathematical methods. This process is known as parameter calibration. If the mathematical model is suitable, it will accurately adjust the measurements, thereby validating the model.

It is important to study the behavior of the mathematical model through simulation. It enables us to verify whether the observed variations in the functioning of the biological system can be predicted by the model, and to highlight emerging properties. All these conditions are necessary for the model to be considered applicable.

1.3 The Case of GreenLab Model

The particular case of the GreenLab model and its use by the Gloups software (GreenLab Operators Use for Plant Simulation) is taken as a generic example. The methods concerning the modeling, the methodology of measurements, and the estimation of parameters are detailed. Finally, the algorithms used in the numerical simulation and in the two- and three-dimensional (2D and 3D) representation of a computational plant are explained. The aim of the book is to provide a researcher or a team in a multidisciplinary framework with the necessary and sufficient information to implement the use of computational plants in agronomy.

The book is primarily intended for students and researchers in natural and environmental sciences (botanists, geneticists, eco-physiologists, agronomists), computer graphics designers, and finally landscape architects who use 3D computer models of plants. They will find the multidisciplinary knowledge of plant growth and architecture useful and necessary for their research, including the development of simulation software. But the multidisciplinary aspect of the subject is also of interest to scientists from various disciplines (mathematicians, computer scientists, physicists, biologists) who practice modeling, and even to non-specialists who are curious to better observe and understand how the plants that surround them grow on their balconies and in gardens, parks, fields, and forests.

Compared to other international projects in the field of plant sciences, and as pointed out by Winfried Kurth, a well-known German researcher in plant modeling, the GreenLab model presented in this book focuses on the mesoscale represented by the phytomers, the basic components of plant architecture constituted by the internode and its associated organs (leaf, buds, fruits). The GreenLab model does not attempt to model structures and processes at a scale of resolution lower than that of plant organs. GreenLab therefore explicitly excludes plant functioning at the cellular and molecular scale, although it is of great importance in biotechnology today. It is true that current biotechnology research on plant systems at the microscale level has not yet fulfilled its promise of finding applications in agronomy.

For their part, physicists in the nineteenth century knew how to model the laws of thermodynamics, electricity, and optics, while ignoring the atomic structure.

It is a legitimate scientific goal to explore how far one can go with a strictly stochastic model limited to the phytomer level. Obviously, there are many processes for which no consideration of causality at the mesoscale level is necessary. To understand a clockwork mechanism, one does not need to know quantum mechanics! In fact, from a philosophical point of view, the simplicity of the approach is an advantage; "Ockham's Razor" requires that unnecessary assumptions should not be introduced into scientific theories. From this point of view, the GreenLab approach seems well adapted to the needs of agronomy and forestry.

The GreenLab model simply uses knowledge already acquired in the various disciplines. But their integration into a mathematical model is new, and it brings out unexpected and verifiable emergent properties. This is the merit of this book, which is a synthesis between architectural models from botany and crop models from agronomy.

1.4 On the Architecture of Plants

There is a scientific strangeness at the beginning of the twenty-first century: plants are better known and studied at the molecular, cellular, or stand level than at the level of the architecture of the whole plant, as it appears to us, observable and manipulable.

The accumulation of knowledge in molecular biology and plant physiology is not sufficient to account for the functioning of the whole plant system.

It is also illusory to believe that we can understand plant morphogenesis by reducing it to mathematical concepts such as fractals or the golden ratio. It is necessary, according to the methods of integrative biology, to analyze in detail the functioning of the organs, entities of the plant system, to develop validated sub-models. These sub-models, when integrated and consolidated, should reveal at the global level the functioning and construction of the plant, i.e., its morphogenesis.

The first models in plant architecture were conducted within mono-disciplinary frameworks, as outlined in the following.

1.4.1 The Models of Botany

At the scale of the whole plant, the description of the form was traditionally done with very summarized qualitative notions. We speak of crown to characterize the living crown of a tree that carries leaves and of habit to describe its visual aspect (ball, weeping, fastigiate habit, etc.).

In 1978, the botanists F. Hallé and R. Oldeman established a typology of plant development from the description of the functioning of buds (continuous, rhythmic, axial, or terminal flowering): these are the "architectural models". Twenty-six models are sufficient to characterize the architectures of the plant world. According to the botanical tradition, it is the drawing in situ that captures the details of the architecture of the plant. Quantitative criteria do not come into play, but the incomparable quality of the drawings according to a particular technique allows to restore at a glance the stages of the development of the schematized plant. This is the first step in a multidisciplinary modeling of plant growth.

1.4.2 The Models of Agronomy

At the beginning of the twentieth century, Agronomy mainly used statistical methods (experimental designs) to improve cultivation practices and plant selection. Then Agronomy developed crop models to predict the production of plant populations in fields or greenhouses, according to environmental parameters (light, temperature, water, nitrogen, etc.).

The plant is no longer considered individually. What counts here is the plant stand, characterized by its leaf area/m^2 and its production. The harvest index concerns the fraction of the total biomass that corresponds to the useful compartment (leaves, stems, fruits, roots). Agronomists have shown that production is closely related to the energy of incident light (Photosynthetically Active Radiation or PAR), to the leaf area index of the canopy that intercepts this light (LAI), and to a proportionality factor that is the light use efficiency (LUE). The Dutch were the pioneers in this approach to plant production, which was first developed by de Wit in the 1960s (de Wit 1965).

These empirical models, coupled with the water balance (soil water content), give remarkable predictions at the stand level and are widely used in Agronomy, such as Agricultural Production Systems sIMulator (APSIM), a dynamic simulation model for tomato crop growth and development (TOMSIM), Crop environment resource synthesis (CERES), Process-based integrated crop growth and land surface temperature equations (PILOTE), etc.

These models that do not consider the individual plants, despite their good operability, have shown some limitations:

- Lack of prediction of meristem mortality and organ abortion.
- Approximate simulation of leaf area and poor consideration of planting density.
- Difficulty to manage climatic stress (temperature, light, water).
- Lack of consideration of variability.
- Poor communication between different computer modules of the crop model.
- For ornamental plants, no visualization.

To overcome these challenges while maintaining all the advantages of crop models, the introduction of plant architecture is a research direction that deserves to be explored. Plant architecture allows to better consider Botany and Ecology in Agronomy. This approach can obtain more information from experimental designs and provide a deeper understanding of plant functioning. The benefits gained should compensate for the additional measurements needed to incorporate the plant architecture into crop models.

1.4.3 Virtual Plant Models in Computer Sciences

The dream of the computer scientist is to find a universal algorithm that implicitly contains all aspects of plant morphogenesis. What interests computer scientists first of all are the various computer formalisms adapted to this subject (combinatorial trees [Viennot et al. 1989], grammars [Prusinkiewicz et al. 1988; Kurth 1994], fractals [Smith 1984], etc.), which simulate the architecture of plants with a focus on geometric construction. There is a preference for algorithms that drive the drawing of tree structures, with fractals and the golden ratio being frequently cited. The organs represented in these models are often decorative and have no functional role. Virtual plants therefore integrate few botanical and agronomic knowledge.

However, virtual plants produce high-quality synthetic images of plant architectures (Janey 1992; Lintermann and Deussen 1999; Palubicki et al. 2009). What matters is not conformity to the rules of plant architecture (which is only accessible to botanists), or plant production (which is only accessible to agronomists), but the aesthetic aspect of the image. These different approaches are detailed in Chap. 14.

1.4.4 Computational Plant Models in Agronomy and Ecophysiology

These very specific plant models aim to simulate in detail the physiological functioning associated with plant growth and to understand the underlying causality at the cellular or even molecular (hormones) scale (Prusinkiewicz and Runions 2012). Plants capture light with their leaf surfaces, and photosynthesis is simulated based on its physiological mechanisms and organs play a functional role here (Allen et al. 2005; Wernecke et al. 2000; Cieslak et al. 2011; Perttunen et al. 1998). These types of computational plants are based primarily on computer simulation of biological processes. It is most often an academic exercise that capitalizes knowledge to simulate complex plant physiological systems through computer programs.

These computational plants are still unsuited to the practical needs of Agronomy because there is no operational methods to estimate the numerous parameters of the system and to study its behavior and its optimization, like the methods of Physics. Moreover, the lack of genericity (one plant = one program) does not allow us to develop essential and practical tools.

1.5 Reasons for the Difficulties in Using Computational Plants

Plants grow slowly, making growth monitoring a time consuming and labor-intensive process.

The architecture of plants is complex, often voluminous, and its stochastic aspect disturbs the data taking. Mortality processes cause many data to be lost (organ senescence, branch pruning), and do not allow the reconstruction of the missing biomass.

In the absence of a clear mathematical model of architectural development (Franquin 1972), it is not possible to define a system of measurements on plant architecture capable of identifying and estimating the parameters of functioning at the scale of the whole plant. "Without theory, there are no valid measurements," another way of quoting Einstein's citation at the start of this introduction.

The often-observed absence of an inverse method in computational plants to estimate the parameters of the model, does not allow to validate it on a data set.

Moreover, there is no formal proof that the simulation programs of computational plants work properly.

1.6 The GreenLab Model

The GreenLab model presented in this book aims to contribute to computational plants.

Compared to the classical virtual or computational plant models, the GreenLab model has some particularities:

- It integrates knowledge acquired in botany (architectural models), with that acquired in agronomy (crop models).
- It adopts the Bernoulli process as the engine of the stochastic development of plant axes and uses the physiological age as the principle for their differentiation.
- It is limited to the phytomer scale (mesoscale), without taking into consideration the cellular level.
- It aims at a high genericity, treating each plant species (beet, rice, rape, cotton, maple, etc.) as a facet of a single, unified model.
- It uses an empirical approach, relying on experiment and data, and avoids a priori systems that are common in virtual plant models. The latter focus on branching processes, whereas the GreenLab model focuses more on plant axis (development, mortality).
- The model parameters are designed to be derived from data. This implies a strategy of data assimilation and the use of inverse methods, as is a common practice in the physical sciences.

A particularity of the GreenLab model is its explicit mathematization, whose equations drive the functioning. This dynamic model can be calculated through equations (the matrix mode) or simulated by a computer program that builds the plant (the list mode). The numerical results from the simulations are thus validated by the calculation.

Moreover, the mathematical aspect of the model makes it possible to use the tools available in Applied Mathematics (study of the behavior of the model, inverse method, optimization, etc.). Before studying natural plants, computational plants from simulation are of great use in developing data assimilation and inverse methods.

In accordance with the principles of integrative biology, the aim here is to model the functioning of individual organs, and then to study the functioning of the plant structure that assembles them. The computer formalism of the simulation is secondary as long as the results are consistent with the mathematical model.

The GreenLab model is applied in the context of even-aged, mono-specific agronomic plantations, where clones or varieties are grown at homogeneous densities. The same plant is repeated with its inherent variability, which is the case in agriculture. We are far from being able to model and simulate a natural plant scene. It is obvious that we must first place ourselves in stable environmental conditions to know the growth and development of a plant without constraint. This understanding will eventually serve as a reference to measure later the effects of a stress (hydric, etc.).

The four steps that are necessary to simulate the growth and architecture of a plant and a plant stand are exposed in this book.

1.6.1 Mathematical Modeling of Meristem Functioning

The vegetative axes of plants are built by terminal meristems that undergo stages of differentiation called "physiological ages," a concept introduced by Rivals (1965). Each physiological age corresponds to a category of axis that can be identified botanically on a plant. Along these axes, the organs are arranged according to "organic series", a notion due to Buis and Barthou (1984) and which depend on the physiological age and therefore independent of the architectural model of the plant.

The architectural models of plants (Hallé et al. 1978; Hallé and Keller 2019) are simulated using the notion of "botanical automaton" (Chaps. 2 and 4). These automata replicate the functioning of meristems, by building branched axes consisting of a series of phytomers. Time is discretized in development cycles to simulate the stochastic processes of plant development (mortality, branching [Chap. 3]). At each cycle, an active meristem can set up with a certain probability a new phytomer, make a pause, or abort. This approach fully captures the stochastic aspect of plant development. The activity of meristems is continuous in herbaceous plants and shrubs, and becomes rhythmic in trees, which increases the complexity of development (Chaps. 2 and 3).

We will distinguish the simulation aspect, which creates random plant structures using the GreenLab model, and the computation aspect, which directly gives the statistics (means, variances) for the cohorts of organs produced by each plant through their development. The notion of "potential structure" (Chap. 3) is a particularly effective tool to deal with this second aspect.

The architecture of a plant with its branched axes is a materialization of the functioning of meristems in the form of growth trajectories. It implicitly contains the duration of development, and growth monitoring is not necessary, which is a great advantage. The estimation of the development parameters of an observed plant (i.e., the functioning of the meristems) is done according to the "Crown Analysis" (Chap. 3) from measurements made directly on the architecture through targeted sampling. This developmental model has been validated on various cultivated plants such as wheat, coffee, poplar, elm, etc. (Chaps. 3, 9, 10, and 11).

1.6.2 Mathematical Modeling of Organ Functioning

The GreenLab model (Chap. 5) takes up the agronomic laws retained in crop models that concern plant stands, but adapts them to the level of the individual plant and is positioned more particularly at the level of the phytomer and its associated organs. It shares many concepts with the Ecomeristem model (Luquet et al. 2006) and TOMSIM model (Heuvelink 1999).

The growth cycle, during which biomass is produced and allocated to organs, is identified and matched in duration with the development cycle (Chap. 3). This discretizes growth and synchronizes it with development, allowing for the definition of the "demand" of the plant at each cycle. The notion of "potential structure" makes it possible to take into account the stochastic behavior of development and to integrate it into the calculation of demand (Chaps. 3, 5, and 6).

The GreenLab model integrates agronomic knowledge from crop models that focus on plant stands: the notions of common pool, leaf index, harvest index, light interception (Beer Lambert's law), thermal time, climatic efficiency, etc. are adapted here to the level of the individual plant. For a given plant, the notion of "production surface" allows to calculate the biomass produced by photosynthesis at each cycle (Chap. 5) by taking into account its leaf surface area, without the need to calculate the interception of light by the foliage (use of radiosity or ray tracing in radiative transfers).

Characteristics specific to herbaceous plants (basipetal growth) or woody plants (secondary growth) are the subject of dedicated chapters (Chaps. 7 and 8). This growth model has been validated for its genericity on various cultivated plants such as corn, rice, tomato, cotton, teak, maple, etc. (Chaps. 9 and 10).

The estimation of growth parameters for an observed plant (i.e., the source-sink functioning of the organs) is done according to the "organic series analysis" (Chaps. 9, 10, and 11), measurements made directly on the architecture by targeted sampling without the need for continuous growth monitoring, which allows the GreenLab model to be easily operational. The organic series are orthogonal to the cohorts of organs.

1.6.3 Moving from the Individual Plant to the Stand

In order to move from the individual plant to the plant stand and to the notion of plant production, to join the crop models, it is necessary to take into account two additional factors:

- The law of delay in germination: A mono-specific field is composed of cohorts of plants slightly delayed in their germination (Chap. 11).
- The planting density that is not homogeneous: The hazards of sowing, the abortion of seeds, cause differences between plants, which modify the

environment and the space available for the growth and the development of each of them (Chap. 11).

It should be noted that the validation of the GreenLab model is already acquired at the first level of observation, which is that of the crop models. Indeed, if we group organs of the same nature into compartments, GreenLab becomes a classical crop model (like STICS [Brisson et al. 1998], PILOTE [Mailhol et al. 2011], TOMSIM [Heuvelink 1999], etc.) with comparable performances (Chap. 11). With cohorts and organic series, it is the functioning of individual organs that is additionally modeled. This approach provides a more detailed level of information on the plant's functioning, and allows us to reconstruct the architecture of the plants and their growth history.

1.6.4 Agronomic Applications

Chap. 12 presents various applications of computational plants. Estimation of model parameters on crop plants is obviously the most important as it paves the way for new methods in plant breeding and cropping systems where plant architecture can play a central role. These plants should play an important role in data assimilation and phenotyping.

If the visualization of computational plants in GreenLab is not the main objective, it is obvious that the simulation of the architecture of a cultivated plant, whose parameters have been estimated in a plantation, offers an incomparable materialization of the result of eco-physiological functioning. The synthetic image is like a reward for the modeler. Computational plants have benefited from the know-how of virtual plants, in the fields of 3D geometric construction and visual rendering. The botAny and Modeling of Plant Architecture and vegetation (AMAP) and GreenLab models have used the pioneering work developed at the Université Louis-Pasteur (ULP) in Strasbourg, under the direction of the computer scientist Jean Françon.

1.7 To Remember

This book presents a mathematical model of a computational plant, whose formalism has many advantages. It has been developed over a long period of time, thanks to numerous collaborations, during which it has been confronted with the reality of cultivated plants, while striving to meet the four conditions that a Physics model must meet to be practically usable:

- The identification of the parameters of the system and the equation of the mathematical model.
- The study of the behavior of the model and its numerical simulation.
- Parameter estimation (data assimilation and inverse methods).

- Validation of the model on cultivated plants and its applications (in plant breeding, optimization, and control of crop itineraries).

Finally, the theme of modeling and simulation of plant architecture, in addition to its affinities with integrative biology, is also part of the broader context of complex systems related to environmental sciences. We are indebted to Franck Varenne, a historian of science and a specialist in the history of computer simulation, for his insightful discussions on this subject with its particularities in his books (Varenne 2007; Varenne 2019; Varenne et al. 2013). He was ideally suited to write the foreword for this work.

References

Allen M-T, Prusinkiewicz P, DeJong T-M (2005) Using L-systems for modeling source-sink interactions, architecture and physiology of growing trees: the L-PEACH model. New Phytol 166(3):869–880

Brisson N, Mary B, Ripoche D, Jeuffroy MH, Ruget F, Nicoullaud B, Richard G, Beaudoin N, Recous S, Tayot X, Plénet D, Cellier P, Machet J-M, Meynard J-M, Delécolle R (1998) STICS: a generic model for the simulation of crops and their water and nitrogen balances. I. Theory and parameterization applied to wheat and corn. Agronomie 18(5–6):311–346

Buis R, Barthou H (1984) Relations dimensionnelles dans une série organique en croissance chez une plante supérieure. Rev Biomath 85:1–19

Chauvet GA (1993) Hierarchical functional organization of formal biological systems: a dynamical approach. I. The increase of complexity by self-association increases the domain of stability of a biological system. Philos Trans R Soc Lond B Biol Sci 339(1290):425–444

Chauvet Gilbert website (n.d.). https://www.gilbert-chauvet.com/en/index.htm

Cieslak M, Seleznyova AN, Hanan J (2011) A functional–structural kiwifruit vine model integrating architecture, carbon dynamics and effects of the environment. Ann Bot 107(5):747–764

de Wit C-T (1965) Photosynthesis of leaf canopies. Pudoc

Franquin P (1972) Modeles mathematiques de structures chez les vegetaux. II Relations de structures. Cahiers Orstom, sér biol 17:3–21

Hallé F, Keller R (2019) Mais d'ou viennent les plantes? Actes sud

Hallé F, Oldeman R-A-A, Tomlinson P-B (1978) Tropical trees and forests. Springer Verlag, Berlin/Heidelberg/New-York, 441p

Heuvelink E (1999) Evaluation of a dynamic simulation model for tomato crop growth and development. Ann Bot 83(4):413–422

Janey N (1992) Modélisation et synthèse d'images d'arbres et de bassins fluviaux associant méthodes combinatoires et plongement automatique d'arbres et cartes planaires (Doctoral dissertation, Besançon)

Kurth W (1994) Morphological models of plant growth: possibilities and ecological relevance. Ecol Model 75:299–308

Luquet D, Dingkuhn M, Kim H, Tambour L, Clement-Vidal A (2006) EcoMeristem, a model of morphogenesis and competition among sinks in rice. I. Concept, validation and sensitivity analysis. Funct Plant Biol 33:309–323. https://doi.org/10.1071/FP05266

Lintermann B, Deussen O (1999) Interactive structural and geometrical modeling of plants. IEEE Comput Graph Appl 19(1):56–65

Mailhol J-C, Ruelle P, Walser S, Schütze N, Dejean C (2011) Analysis of AET and yield prediction under surface and buried drip irrigation systems using the crop model PILOTE and Hydrus-2D. Agric Water Manag 98:1033–1044

Palubicki W, Horel K, Longay S, Runions A, Lane B, Měch R, Prusinkiewicz P (2009) Self-organizing tree models for image synthesis. ACM Transactions on Graphics (TOG) 28(3):58
Perttunen J, Sievänen R, Nikinmaa E (1998) LIGNUM: a model combining the structure and the functioning of trees. Ecol Model 108(1–3):189–198
Prusinkiewicz P, Runions A (2012) Computational models of plant development and form. New Phytol 193(3):549–569
Prusinkiewicz P, Lindenmayer A, Hanan J (1988) Developmental models of herbaceous plants for computer imagery purposes, proceedings of SIGGRAPH 88. Computer Graphics 22(4): 141–150
Rivals P (1965) Essai sur la croissance des arbres et sur leurs systèmes de floraison (Application aux espèces fruitières). Journal d'Agronomie Tropicale et de Botanique Appliquée 12:655–686
Smith AR (1984) Plants, fractals, and formal languages. In: Proceedings of SIGGRAPH '84 (Minneapolis, Minnesota, July 23–27, 1984). Computer Graphics 18, 3, pp. 1–10
Varenne F (2007) Du modèle à la simulation informatique. Vrin, Paris
Varenne F (2019) From models to simulations (Routledge). Taylor & Francis
Varenne F, Silberstein M, Dutreuil S, Huneman P (2013) Modéliser & simuler. Épistémologies et pratiques de la modélisation et de la simulation, vol 1
Viennot X, Eyrolles G, Janey N, Arques D (1989, July) Combinatorial analysis of ramified patterns and computer imagery of trees. In: *Proceedings of the 16th annual conference on Computer graphics and interactive techniques*, pp 31–40
Wernecke P, Buck-Sorlin G, Diepenbrock W (2000) Combining process- with architectural models: the simulation tool VICA. Systems Analysis ModellingSimulation 39(2):235–277

Chapter 2
Botanical Concepts on the Organization of Plant Architectures

Sylvie Sabatier, Daniel Barthélémy, Philippe De Reffye, and Mengzhen Kang

Abstract The aim is to provide an understanding of the botanical elements (and their combinations) involved in plant architecture and rules for simulating and modeling plant growth and development (the botanical concepts used in the GreenLab model). The structure of a plant is an intricate organization shaped by genetics, modulated by the environment, and resulted from the interaction between development and growth. The phytomer, comprising the internode and node with leaves, flowers, and buds, is the basic unit of plant structure, produced by the terminal meristem of a leafy axis. The meristems' ages (chronological, ontogenic, physiological ages) are crucial for understanding plant development. Thermal time, calculated as the sum of temperatures, is used to track the chronological age of meristems, as it provides a stable measure of their functioning. Plant Structures are organized into architectural models based on meristem functions and growth patterns, applicable to all plant types, and can be represented in chronological or topological modes.

Electronic supplementary material The online version of this chapter (https://doi.org/10.1007/978-981-97-7541-5_2) contains supplementary material, which is available to authorized users.

S. Sabatier
CIRAD, UMR-AMAP, CIRAD-CNRS-INRAE-INRIA-IRD-Université Montpellier, Montpellier, France
e-mail: sylvie.sabatier@cirad.fr

D. Barthélémy
CIRAD, Montpellier, France
e-mail: daniel.barthelemy@cirad.fr

P. De Reffye
CIRAD, UMR AMAP, Montpellier, France

AMAP, Univ Montpellier, CIRAD, CNRS, INRAE, IRD, Montpellier, France

M. Kang (✉)
State Key Laboratory for Multimodal Artificial Intelligence Systems, Institute of Automation, Chinese Academy of Sciences (CASIA), Beijing, China

School of Artificial Intelligence, University of Chinese Academy of Sciences, Beijing, China
e-mail: mengzhen.kang@ia.ac.cn

P. De Reffye et al. (eds.), *Plant Model Greenlab for Botany and Agronomy*,
https://doi.org/10.1007/978-981-97-7541-5_2

Keywords Botany · Phytomer · Growth unit · Chronological and physiological age · Structure · Thermal time

2.1 Introduction

This chapter aims to provide an understanding of the botanical entities (and their combinations) involved in plant architecture and the rules for simulating and modeling plant development (the botanical concepts used in the AMAP model and its successor, the GreenLab model). The illustrations of examples are drawn from the plant simulation software Gloups (de Reffye and Jaeger 2013).

The elementary botanical entity of plant structure is the phytomer, consisting of the internode and the node, which carries leaves, flowers, and axillary buds. The production of phytomers is ensured by the terminal meristem of a leafy axis, an organ that sets up successive phytomers which will give a leafy axis when they elongate. A meristem includes three distinct types of age:

- The chronological age (CA), which refers to the duration of meristem functioning since its creation.
- The ontogenic age (OA), which refers to the age of the plant at the time of meristem creation giving influence on the functioning of the meristem.
- The physiological age (PA), which refers to the degree of differentiation of the phytomers produced by the meristem. This notion is essential to detail the development of a plant.

The structure of a plant is a remarkable organization whose morphogenetic program is genetically determined. Its expression is modulated by the environment. It results from a dual mechanism: development and growth.

The term "development" refers to the kinetics of the placement of leafy axes by meristems, which builds the skeleton of the plant's architecture. Organogenesis is the origin of development.

The term "growth" refers to the dynamics of biomass production by source organs (leaves) and its distribution to sink organs (leaves, internodes, fruits, etc.), which they ensure by expansion. This is defined as "source–sink relationships". Photosynthesis is the origin of growth.

A plant structure includes two aspects:

- The organization of all botanical entities within a plant with their links, without considering the functioning or geometry of the plant, referred to as "topological structure". It can be made into a planar representation. The topological structure also contains hidden information related to the developmental time.
- The three-dimensional (3D) representation of the plant, which uses geometric operators to correctly position and orient the organs, commonly referred to as "architecture". These operators take into account the dimensions of the entities and angles (phyllotaxis, connection, etc.).

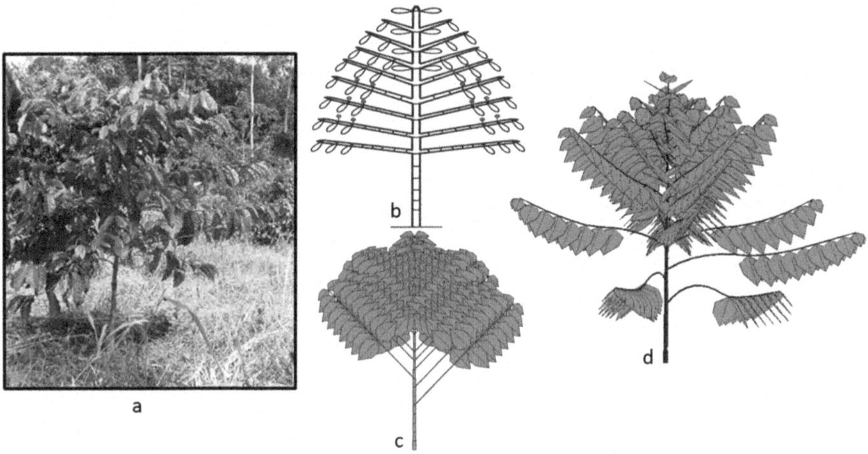

Fig. 2.1 (**a**) Photography of a young *Coffea robusta* (RCI). (**b**) Botanical drawing showing the structure of the plant. (**c**) 2D plant simulation of the structure using the functioning parameters. (**d**) 3D simulation of the plant architecture. (Gloups software)

Architectural analysis (Barthélémy and Caraglio 2007) provides concepts to analyze, simplify, and model the development of a plant during its life. It has been developed and applied to many species.

There is a long tradition among botanists in the art of drawing plants. The botanist F. Hallé, who originated the concept of plant architecture, has created a style of two-dimensional (2D) plant representation. This style aims not for a faithful view of the observed plant but for a schematic planar (even elegant) drawing that allows one to read and clearly reproduce its development pattern and structure. It is the same point of view that is adopted most often in this book with the GreenLab model: the computer simulation of plants in 2D is sufficient and, most importantly, more readable (Fig. 2.1). The three-dimensional (3D) simulation is only necessary for computer graphics applications, especially in urban planning and landscaping, and sometimes in agronomy for light interception computations.

2.2 Botanical Entities of a Plant Structure

Within a plant species, a large variability in shape is observed as a function of age and environmental resources.

A plant structure can be seen as a stack of botanical entities in different levels of organization. Transitions between these levels is done by repetition (Barthélémy 1991). In its topological form, the plant structure displays the branched system and the relationships among each botanical entity, without any concern for geometry, a bit like a subway map that displays the network of lines and stations without considering real distances. It is represented in a planar form with standardized organ dimensions. For greater readability in simulated structures, internodes are assigned colors that correspond to their PA. In addition, a more botanical representation is given by dressing this structure with its organs and more "natural" colors.

2.2.1 *Meristem, Phytomer, and Organ*

The apical meristem of an axis, the place of cell division, initially produces a series of embryonic phytomers that are the basic botanical entities. In a second step, these phytomers are expanded. The result is the formation of a vegetative axis through phytomers series. The embryonic phase is not directly visible because the apical meristem is most often protected inside a bud. This entity is absent in herbaceous plants; it is present in temperate trees as those in the northern hemisphere during the winter (e.g., ash, chestnut tree). In tropical countries, there are trees where the meristem is protected only by embryonic leaves. Many other protective mechanisms for meristems exist, such as by stipules (e.g., coffee) or by secretions. At the time of budburst, new shoots are rapidly expanded by lengthening the organs. As previously written, phytomers consist of a portion of an axis called the internode and, at its end, a node that carries one or more leaves, along with their axillary meristems. These axillary meristems give rise to branches, flowers (Fig. 2.1a) or inflorescences, thorns, twists, etc. The number of leaves per phytomer is generally stable and specific to the species. For example, poplar, maple, and catalpa have one, two and three leaves per node, respectively. The leaves are arranged in a spiral pattern along the axes (phyllotaxis).

Depending on its position within the architecture, the phytomer has botanically established, stable, and specific characteristics for the given species. For modeling and simulation, it should be noted that the phytomer represents the first level of factorization of organ production. To know the number of phytomers resulting from development is also to know the number of organs produced by the meristems.

2.2.2 *Growth Unit and Vegetative Axis*

During its functioning, the meristem emits phytomers one by one giving a continuous aspect to the structure of the stems, or grouped (mainly in trees) giving them a rhythmic aspect. The latter case results in the establishment of a "growth unit" (GU) that corresponds to the portion of the stem produced during an uninterrupted elongation phase (Hallé and Martin 1968). The axis then has two nested organizational levels: the phytomer and the growth unit (Fig. 2.2b).

The result of apical meristem activity is either an axis composed of phytomers for the case of continuous development, or an axis composed of growth units, decomposed into phytomers for the case of rhythmic development (Fig. 2.2a, b). In the latter case, the scale of the axis is double. Depending on the species, the apical

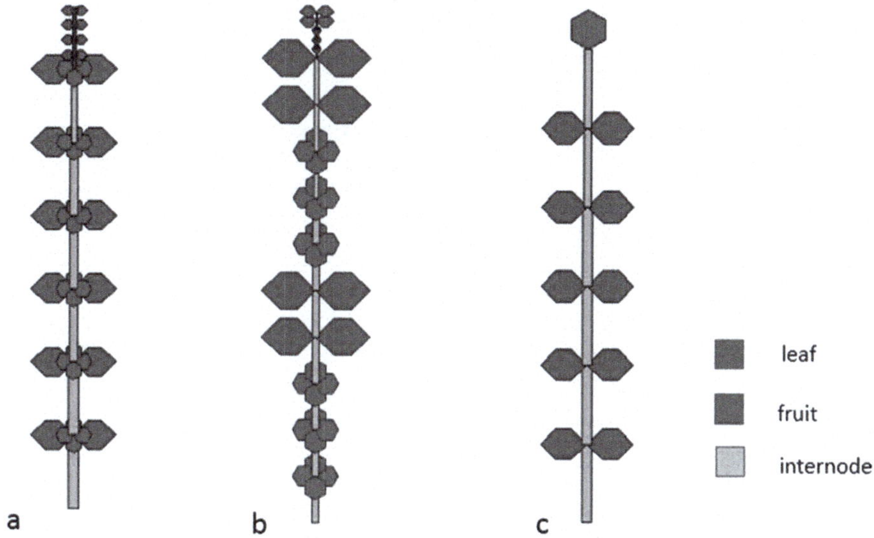

Fig. 2.2 Botanical representation of three types of vegetative axes. (**a**) Axis with indefinite and continuous growth that carries phytomers composed of an internode, two leaves, and four fruits and that ends with the embryonic phytomers (awaiting elongation) resulting from the functioning of the terminal meristem. (**b**) Axis with indefinite and rhythmic growth whose apical meristem produces growth units that contain alternations of two different types of phytomers bearing two leaves or four fruits. (**c**) Axis with definite and continuous growth with transformation of the apical meristem into a fruit

meristem has either indefinite functioning, meaning it can form phytomers throughout the life of the plant, or finite functioning, where the meristem eventually aborts or transforms into an inflorescence, spines, twists, etc. Trees such as cecropia or eucalyptus have continuous and indefinite development, while those such as cherry or poplar have rhythmic and indefinite development. This pattern is called monopodial development, which characterizes these branched structures. In such cases, the mortality of meristems is accidental when it exists.

When development is finite, the mortality or transformation of meristems into inflorescences is programmed. The module corresponds to the portion of the axis set up until the terminal flowering for example (Fig. 2.2c). For many species, the terminal meristem systematically aborts after the new shoot has been placed. Subterminal axillary meristems take over. The axes are then made up of a succession of modules (this is the case for elm, plane tree, birch, apricot, frangipani, etc.). This pattern is called sympodial development. Some tree species first have a monopodial structure in their young phase, followed by a sympodial structure in their mature phase with the appearance of a terminal flowering on the axes, which generates modules with sympodial branches (chestnut, walnut, mango, maple, etc.).

2.2.3 *Branched System, Branching Order, and Morphological Differentiation of the Axes*

Although the architecture of some vascular species (most palm trees, papaya, etc.) is a single stem (or reduced to a single axis during their lifetime), a large number of species show an architecture that becomes more and more complex through the branching process. Generally, the branching is derived from axillary (lateral) meristems, and rarely from the duplication of the apical meristem (as seen in the doum palm). Branches may develop simultaneously with the main stem (immediate branching) or a few weeks to a year later (delayed branching of temperate trees, such as cherry). In old or stressed trees, branching can be delayed for several years due to axis mortality.

Each axis within the branched system is characterized by its order of branching, which represents a simple spatial succession of axes relative to each other. The order 1 is the stem resulting from germination, the order 2 is the carried axis by the order 1, the order axis 3 is the axis carried by the order 2, and so on (Fig. 2.3a, b).

Within the branched system, the axes are also differentiated by their function: some are designed to explore space, others to exploit via photosynthesis, or finally, some are dedicated to reproduction. Fruit trees, such as cherry and apricot trees, have short axes with reduced, few and unbranched internodes that are abundantly flowering and have a short life span, as well as long axes with extended internodes, which contribute to most of the vegetative development of the architecture. The first type of axis, which is short, is called the exploitation axis; it ensures most of the photosynthesis and fruiting. The second type of axis, which is long, is called the exploration axis because it establishes the architecture of the tree within its spatial environment through the branching process (Fig. 2.3c, d).

This differentiation is an endogenous characteristic of the axes. The coffee tree shows vertical or orthotropic axes (the stems) that explore the environment, and horizontal or plagiotropic axes (the branches) that carry reproductive structure. If a branch is cut to make a cutting, it will never give back an orthotropic axis and will produce a creeping plagiotropic architecture. Only cuttings taken from the stems will restore a vertical axis and give back a coffee tree.

2.2.4 *Architectural Unit and Axis Categories*

The first stages of tree development give rise to a system strongly hierarchical around a vertical axis. For each species, an elementary architecture is defined and called an "architectural unit" (AU) that consists of a finite number of axis categories (Barthélémy and Caraglio 2007). Each category groups axes with similar morphological, anatomical, and functional properties (Fig. 2.3a, b), thus defining a given PA.

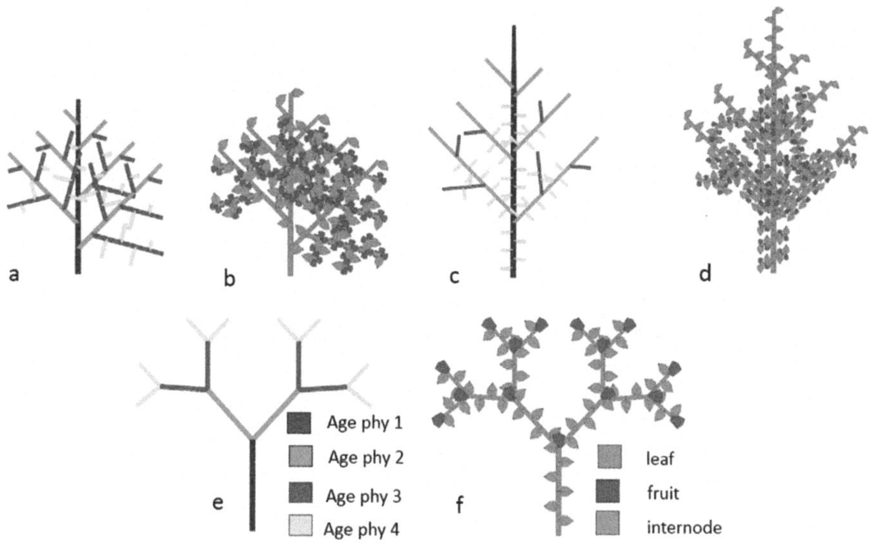

Fig. 2.3 Examples of plant structures. (**a**, **b**) Topological structure and botanical representation of a plant with continuous and monopodial development. There are four categories of axes composed of phytomers: the stem, the main branches, the long branches, and the short branches; the latter have a limited development and bear fruit. Here, there is an agreement between the order of branching and the category of axes. (**c**, **d**) Topological structure and botanical representation of a plant with rhythmic and monopodial development. The axes are made up of growth units (GUs) that are themselves made up of phytomers. Each GU on the stem carries two branches and five short branches; each GU on the main branch carries a long branch and four short branches; each long-branch GU carries a branch and four short branches. Each GU on the long branch carries two short branches. Short branches are not branched. There is no agreement between the branch orders and the axis categories. Short branches have limited development and bear fruit. The GUs of plants carry axes whose vigor decreases from the top to the base. This morphogenetic gradient corresponds to "acrotony". (**e**, **f**) Topological structure and botanical representation of a plant with sympodial development. The axes with definite growth end with a flower. After each flowering, two relay axes are produced

The number of axis categories in an architecture is relatively small and generally less than or equal to 5. Each category of axis corresponds to a degree of differentiation of the building meristem. A complex plant structure is finally broken down into a small number of categories of axes that are arranged according to botanical rules. The grouping of phytomers into categories of branched axes therefore still allows a significant factorization that reveals substructures that repeat themselves many times in the main structure. For example, five categories of axes are listed:

- The trunk, with indeterminate development.
- The long branches carried by the trunk, which have an average life span of 15 years in a dense plantation.
- The branches on the trunk and long branches, with an average life span of eight years.

- Twigs, with an average life span of four years.
- Short branches, which can appear on all categories of axes with a lower PA, and have an average life span of two years.

2.2.5 Reiteration

As the tree develops further, its architecture becomes more complex by duplicating existing structures. This phenomenon of repetition of part or all of the architectural unit is called reiteration (Oldeman 1974); it allows the construction of the crown of the adult tree (Hallé et al. 1978).

During reiteration, apical and axillary meristems are at the same differentiation stage, thus building similar structures (Fig. 2.4). Repetition can affect all branched structures locally and, remarkably, it does not increase, by definition, the number of axis categories.

The repeated structures constitute the highest level of factorization in plant architecture; they constitute the basis for establishing the equations of plant production by meristems. These theoretical structures make it possible to model the development of the plant in a very efficient way.

2.2.6 Sequence of Tree Development

Architectural analysis breaks down the development of a tree into distinct phases and establishes a precise and ordered sequence of morphological events marking the life

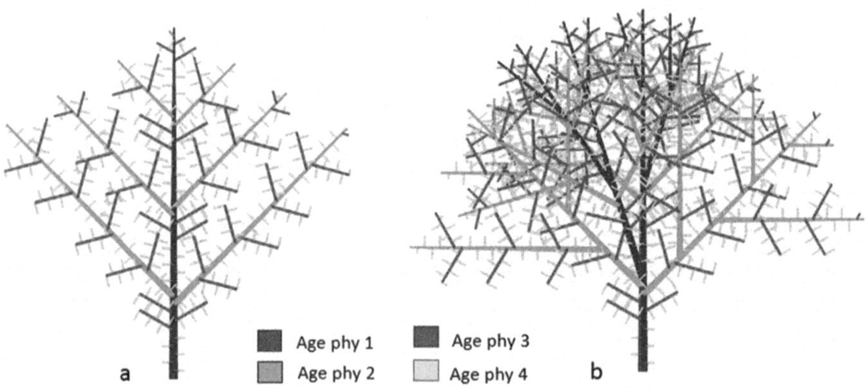

Fig. 2.4 Simulation of the architectural unit of a tree with its reiterations. Here, the architectural unit whose development is rhythmic is composed of four categories of axes with an acrotonic branching. (**a**) Representation of the architectural unit with the four categories of axes. (**b**) The plant structure is composed of repetitions (reiteration) of the architectural unit. The reiteration is carried out on the trunk up to branch order 3 (total reiteration) and on the branches up to branch order 2 (partial reiteration). The whole arrangement is called a "reiterated complex"

of the tree (Barthélémy und Caraglio 2007). From germination, the young tree builds a trunk that branches at varying speeds depending on the species and the surrounding environment. As the tree grows, the trunk shoots elongate and the leafy axes are more and more branched. The architecture of the tree becomes more complex and a specialization is observed among the different axes. The tree expresses its "architectural unit" (AU); its structure is strongly hierarchical around the single trunk. The shape of the crown is often pyramidal. Further development continues in some species such as *Araucaria araucana* (Grosfeld et al. 1999) in accordance with the AU, and maintains a single trunk until its death. Most often, the establishment of the crown in the adult tree corresponds to an enrichment of the branched system by the total and immediate reiteration of the trunk and branches. This stage is identifiable by the structure of its main branches, showing immediate repetitions (forks) with successive lateral branches. The shape of the crown is regular and compact. It becomes rounded at the top. At the later stage, the mature tree corresponds to a reduction in annual shoots and a decrease in branching capacities, with a strong expression of reproductive structures. The tree crown consists of a succession of forks that are closer and closer together over time. The shape of the crown becomes irregular and fragmented into a multitude of small sub-crowns. The tree has reached its maximum height. During the subsequent development, the crown tends to break up due to progressive branch mortality, indicating that the tree is senescent.

At each stage of development, the apical meristems generate GUs of different organizations, which results in a modification of the development parameters. A computer simulation of the development of the architecture must take into account this evolution of the functioning of meristems.

2.2.7 Physiological Age of Meristems and Morphogenetic Gradients

For some continuously developing plants, such as terminal flowering herbaceous plants, the phytomers gradually change morphology from the base to the top of the stems. Along the stem, the internodes lengthen, and the leaves change shape to resemble sepals that, in turn, metamorphose into petals, finally generating a flower. This plant metamorphosis was first described by the writer Goethe (1790), a botanical enthusiast and initiator of plant morphology as a discipline.

During its life, the development of a plant is characterized by a continuous change in the functioning of the meristems, leading to a progressive modification of the architecture. In trees, the architectural analysis of many species has revealed that the axes that constitute an individual's architecture correspond to a succession of entities whose characteristics gradually evolve along the axis. This phenomenon is particularly visible in fruit trees (apple, cherry, etc.). These morphological modifications of the entities reflect the differentiation of the meristems that gave rise to them: we speak of the "PA of the meristem" (Rivals 1965; de Reffye et al. 1991a, b). More generally, at the time of its formation, all the values of the biological characteristics of an elementary entity (phytomer, growth unit, axis), make it possible to define the PA of the meristem that built it. Thus, the short axes of fruit trees are characteristic of a “physiologically aged” state with short, flowering growth units and a short life span. On the other hand, forks (or reiterations) of the trunks characterize vigorous, physiologically young, and perennial productions (Rivals 1965; de Reffye et al. 1991b).

This progressive differentiation is done according to several “morphogenetic gradients” (Barthélémy and Caraglio 2007). Often the most vigorous branches with long life expectancy are located near the top of the growth unit. Below, over a more or less extensive length, a series of twigs whose life span is limited to a few years is observed. This morphogenetic gradient called “acrotony” is observed in many forest and fruit species (maple, poplar, elm, cherry, apricot, etc.). This phenomenon is simulated by prioritizing the categories of axes branched along the GU (Figs. 2.3c and 2.4a).

With an equivalent point of view, Rouane (1977) compared the growth of a cypress branch to a phenomenon of cellular differentiation according to a mechanism close to that which controls morphogenesis, which uses PA as described in de Reffye et al. (1991a). Meristems function for a period of time specific to the physiological age during which they create the same type of phytomere. The cessation of activity of a terminal meristem is caused either by abortion or by a change in physiological age.

In the first case, the death of the terminal meristem can leave the axillary meristems to continue functioning (Fig. 2.5a) or, by dominance, stops all branch development (Fig 2.5b). In the second case, the terminal meristem changes physiological age and continues to develop (Fig. 2.5c). The axis is made up of several parts which show the transition between physiological ages called “drift”. Another gradient bottom up on the stem, name “base effect” follows the evolution of the physiological age of the branches until stabilization (Fig. 2.5d).

2.2.8 *Ontogenic Age of Meristems*

Other information can change the way a plant’s meristems function during the transition and branching processes. This is the ontogenic age, which refers to the age of the plant at the time of meristem creation giving influence on its functioning. Such information can force the synchronization of the meristems functioning when the plant reaches a given CA.

In the case of immediate reiteration, when meristems synchronize according to ontogenic age, the crown of the tree breaks into identical sub-crowns, giving the development a monopodial aspect (Fig. 2.6a). Otherwise, the reiterations resume their development from the beginning of their differentiation. The development takes on a sympodial aspect (Fig. 2.6c).

In annual herbaceous plants, the development time is short, typically lasting only a few months. After an initial purely vegetative development, flowering occurs simultaneously at the top of the axes. It can be seen that the stem and branches

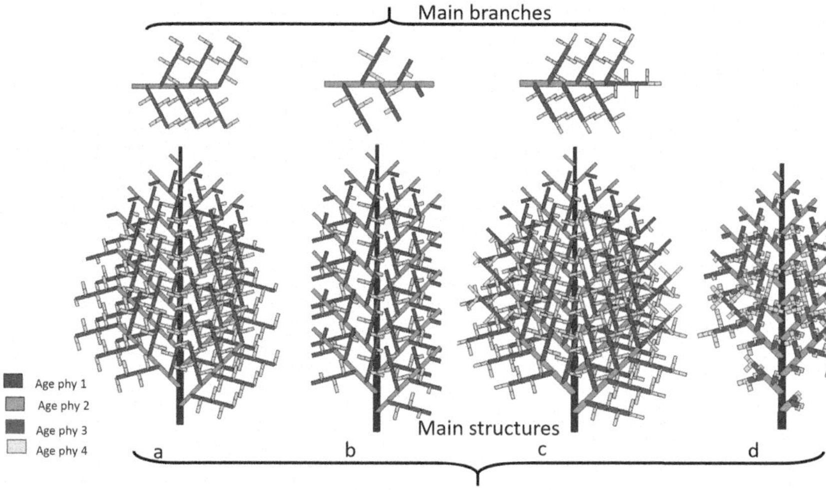

Fig. 2.5 Monopodial development: structures with gradients. Monopodial development structure with aging meristems that undergo physiological age (PA) transitions due to ramification (**a**, **b**, **c**, **d**). The leafy axes are composite via the transition of meristem in higher PA. When meristem has reached the ultimate PA through successive transitions, it dies and the development of the axis is complete

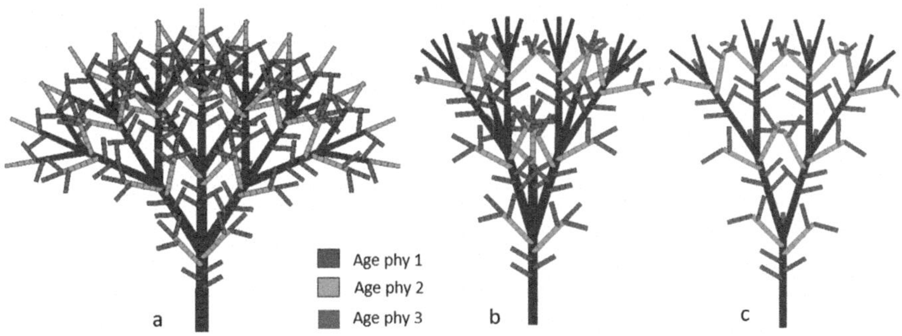

Fig. 2.6 Effect of ontogenic age on the development of immediate reiterations. (**a**) Case of ontogenic age: the transition in physiological ages (PAs) of reiterations is synchronous; development is monopodial. (**c**) The transition in PAs starts from the beginning of axis development. Development is sympodial. (**b**) The architecture is intermediate between the monopodial and the sympodial state. The transition in PAs does not start from the beginning of axis development

synchronize their transformation and flowering occurs at the same time (Fig. 2.7). Initially, the axillary branches develop independently, but when the apical meristem of the main axis turns into a flower, it simultaneously triggers the flowering of all underlying axillary meristems. Axillary meristems therefore have information about the physiological state of the main meristem (i.e., ontogenic age).

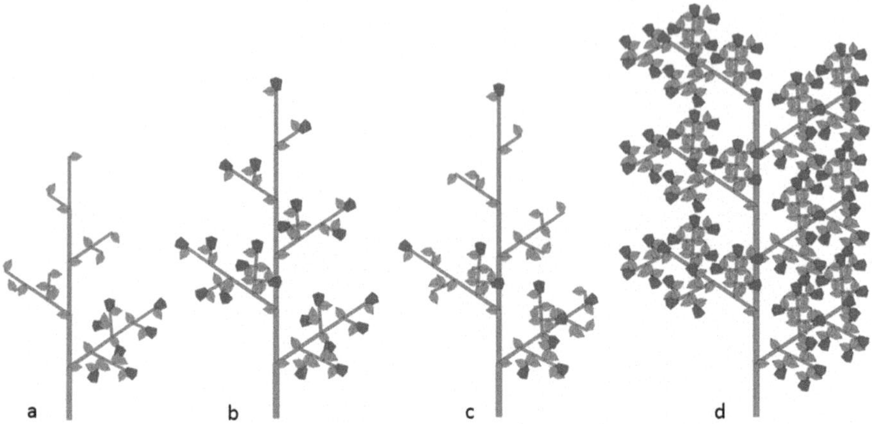

Fig. 2.7 Representation of plant development with flowering governed by ontogenic age: controlled plant growth (**a**, **b**) versus uncontrolled growth (**c**, **d**). The simulated plant structure has four physiological ages (PAs) following the branching order. The axes have a defined development: six development cycles (DCs) for the PA 1, four DCs for the PA 2, three DCs for the PA 3, and two DCs for the physiological age 4. (**a**, **b**) The first branch has completed its development after five DCs, and all its meristems flower simultaneously; the other meristems of the plant continue their functioning. At the age of six DCs, the terminal flowering of the stem triggers the terminal flowering of all developing axes, and the development of the plant is completed. Flowering is synchronous. This case is typical in herbaceous plants. (**c**, **d**) The axes flower according to their PA and stage of development. The development of the plant stops at 13 DCs. Flowering is asynchronous with the extension of branches not blocked by the flowering of the bearing axis. The representation **c** shows the plant at the age of six DCs. At this stage, only the first short axes that have fully developed have flourished. At the 13 DC stage, all the axes have bloomed, and the development of the plant is complete

The ultimate stage of differentiation is the mortality of the axis. The death of a meristem is considered to be the ultimate transition in physiological age in the development model.

2.2.9 *Physiological Age, Reference Axis, and Organization of Plant Structure*

Observation of the constitution of the vegetative axes over long periods shows that meristems can change, in some cases, PAs through successive transitions. Thus, after several decades of functioning, the trunk gradually transforms into a branch, which evolves over a few years into shorter and shorter branches. This phenomenon is particularly visible in conifers, such as cedars and pines, whose tops curve to form a “table” shape, which is essentially a terminal metamorphosis where the trunk transforms into a branch.

The axial ordering in increasing order of all possible PAs of meristems, along with their repetition, constitutes a theoretical axis that compiles the transition stages into a single series, from which a plant is reconstructed. This theoretical axis called the "reference axis" was developed for the first time on Japanese elm tree, the development of which has been studied precisely on individuals aged between one and one hundred years (Chap. 3; de Reffye et al. 1991a) and on Aleppo pine (Caraglio et al. 2007).

At its creation, a meristem is born with a PA that is determined botanically according to its position in the plant structure. By definition, if the stem is born with physiological age 1, the rule is that an axillary meristem has an older or equal physiological age (in the case of reiteration) compared to the apical meristem. All meristems of the same PA produce identical axes and therefore, by branching, they generate identical substructures (Yan et al. 2002, 2003). A large number of copies (instantiations) of structures of a given PA at various stages of development are thus presented in a tree. Their repetition of these structures allows the factorization of development.

The top-down dressing of the main axes by the lateral substructures in increasing CA is analogous to the method of "structured-recursion" developed by Bosanac (1990) in the Tree software. These substructures are embedded in each other, starting from the oldest physiological ones to the youngest (Fig. 2.8). Thus, the plant can be factorized into substructures. Each type of substructure is computed only once, but will be inserted multiple times into the assembly of lower PA structures. The reference axis allows monopodial or sympodial structures to be manipulated at the same level (Fig. 2.9).

The cost of a simulated tree is normally proportional to the number of organs produced for a given age. With the factorization into substructures, this time only depends on its development time (Cournède et al. 2006).

2.3 Concepts of the Chronological Age of Meristems

The previous paragraph describes, from a static point of view, the botanical items and their assemblies that constitute the plant structure. Another essential notion contained within this structure is time, which concerns both the duration of the plant's development and the kinetics of the setting up of phytomers.

2.3.1 Thermal Time, Phyllochron, and Development Cycle

The emission of successive phytomers by the meristems of the axes is done according to a time interval called phyllochron. This interval is called the calendar because it allows the development of the plant to be monitored on a daily basis. But the phyllochron is influenced by environmental factors, especially temperature; the

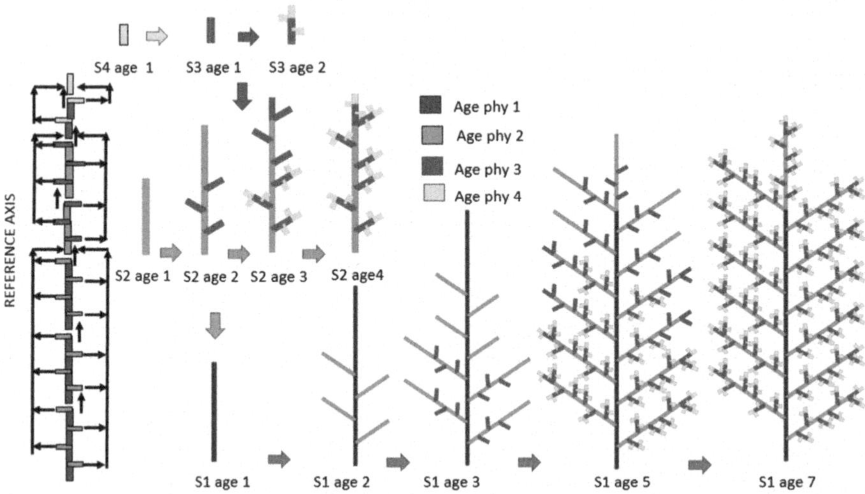

Fig. 2.8 Factoring a plant structure with four physiological ages (PAs) and seven chronological ages (CAs) into substructures, under the dependence of a reference axis that controls the evolution of the PAs of meristems, with transitions either by continuation or by branching. A substructure is characterized by a main axis that has a PA and a CA, on which are substructures that are physiologically older and whose CA increases from the top to the bottom of the main axis branch. Collections of substructures S4, S3, S2, S2, S1 are being developed for the four PAs, starting with the oldest and continuing down to age 1, which is the complete plant structure. When, by transition, the PA of the meristem becomes maximum (4), after running along the reference axis, development stops

axes grow slower when temperatures are low. In agronomic experiments, we observe that the number of phytomers produced by a meristem is proportional to the cumulative average daily temperatures experienced by the plant. This cumulative temperature is called "thermal time". The transformation of calendar time into thermal time regulates the speed of axis development. The thermal time, required for a meristem to form a phytomer is termed the "development cycle (DC)", which is independent of calendar time. At each development cycle, meristem may or may not emit a phytomer. Thus, a stem made up of ten phytomers requires at least ten DCs to be built. The total number of DCs is called "chronological age", which characterizes its stage of development in terms of the number of DCs completed. For agronomic applications, the link between the actual duration of a development cycle (phyllochron) and environmental parameters to monitor development in calendar time (e.g., day-to-day) is explained in Chap. 4.

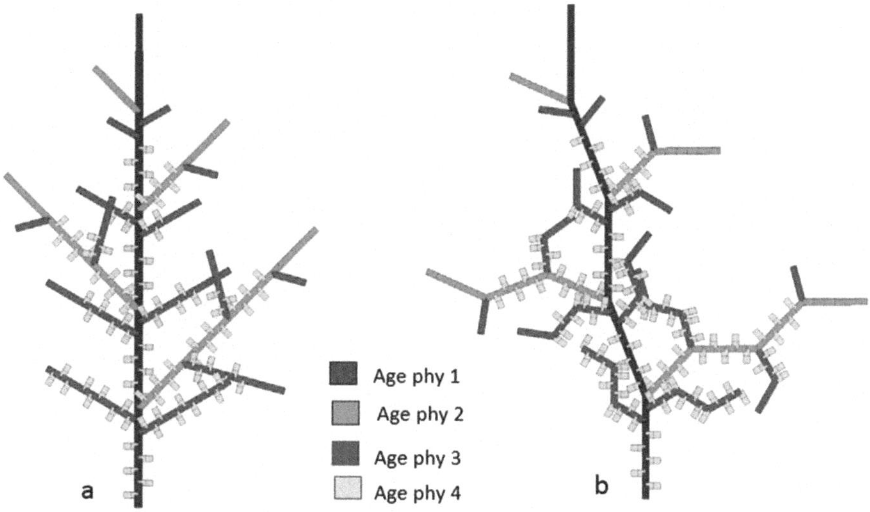

Fig. 2.9 Equivalence between monopodial and sympodial development structures. (**a**) Monopodial architectural unit. The axes have a life span that depends on physiological age (PA). (**b**) Architectural unit of a sympodial type. Meristems die after the growth unit is set up. The axes are extended by subterminal relays, the number of which depends on PA. Here, both structures **a** and **b** produce the same numbers of phytomers types

2.3.2 Organ Cohorts

During the development of the plant, meristems produce at each cycle "cohorts" of organs. These cohorts are characterized by the same type of phytomer (leaf, internode, fruit, etc.), the same date of formation since germination (same ontogenic age), and the same level of differentiation (PA). As these cohorts evolve in DCs, their organs have the same CA, which is counted from the date of their initial creation.

2.3.3 Topological Structure and Chronological Structure

To follow the development of the axes, time is therefore discretized into development cycles (the method is explained in Chap. 3). If the periods of inactivity of the meristems correspond to regular pauses following developmental periods (annual or seasonal growth cessation), they are called structural pauses, as opposed to random, unpredictable pauses. These random pauses occur during the development period and explains, for example, the observed variability in the number of phytomers produced by meristems that are of the same chronological and physiological ages (belonging to the same cohort).

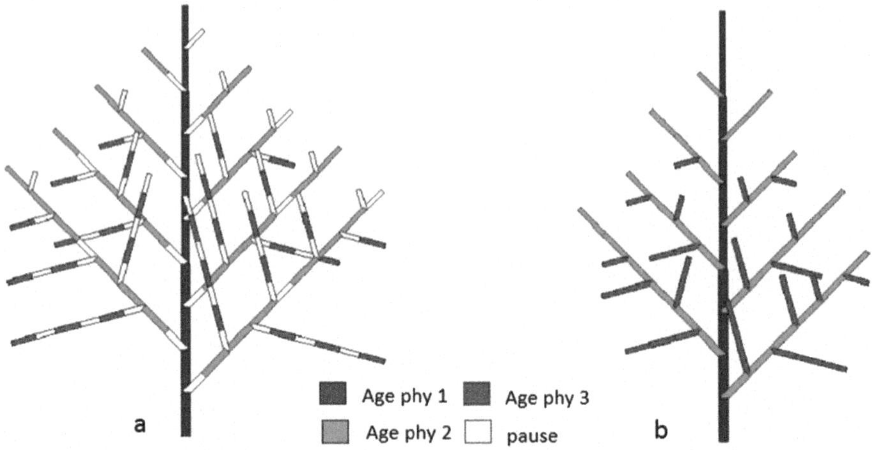

Fig. 2.10 Chronological representations (**a**) and topological representations (**b**) of a continuously growing structure characterized by development rates that are specific to the physiological age (PA) of the axis. In **a**, the pauses (void entities) are represented by white rectangles. There are no pauses on the stem (rhythm $w = 1$). The PA axis 2 has a rhythm $w = 0.75$ (one pause every three cycles). The PA axis 3 has a rhythm $w = 0.5$ (one pause every two cycles). By convention, if $w < 1$, the branched axis starts with a pause before the rhythm pattern is established

In trees with rhythmic development in areas with marked seasonality, apical and axillary buds have synchronized activity. This synchronization is seasonal in nature. During spring or the early rainy season, the buds resume their functioning and give rise to a growth unit. After a more or less long period of activity, the meristems rest until the following spring. These periods of pauses are considered structural. Therefore, all lateral axes of the same PA carried by the stem have potentially the same development, since they generally resume their functioning at the same time. To ensure this synchronization, the initial pauses of the respective buds depend on their positions on the stem, which are related to their formation dates.

Structural pauses correspond to a deterministic development. The duration of these pauses is measured in development cycles. The representation of the simulated structures can then be done in two modes (Figs. 2.10, 2.11, 2.12, 2.13, and 2.14):

- Chronological mode: Phytomers and pauses are displayed to indicate all the details of the development. Pauses, which are void entities, are considered here as botanical entities in the same way as phytomers; they alternate with phytomers to build a vegetative axis. The chronological mode allows a spatial–temporal representation of plant development, which is useful for educational purposes and is called "chronological structure". This mode makes synchronizations between events and the effect of developmental delays clearly visible.
- Topological mode: The topological structure corresponds to the realized phytomers, therefore to the visible part of the plant structure. In this mode, pauses (void entities) are deleted, and the temporal aspect becomes hidden.

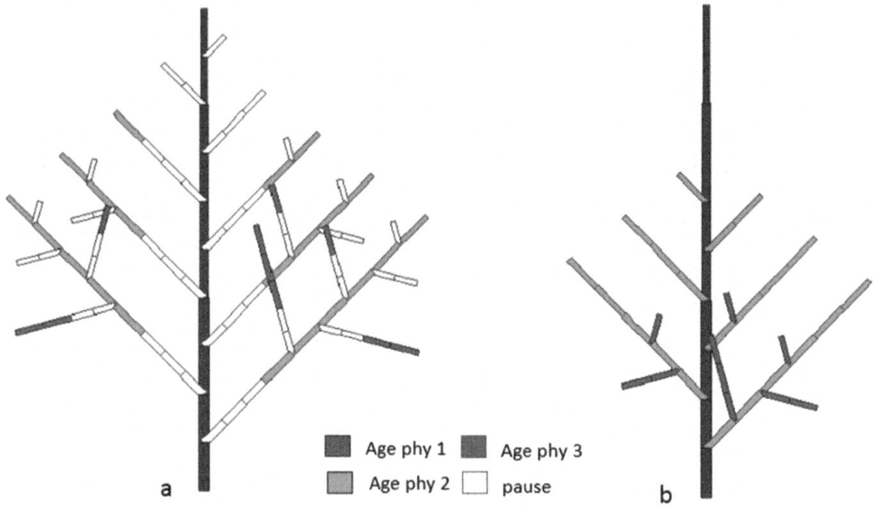

Fig. 2.11 Chronological representations (**a**) and topological representations (**b**) of a structure with continuous growth and delayed branching. In the chronological structure, branches are initiated with a delay of three development cycles (DCs), and these branches in turn bear further branches with a delay of two DCs. These delays are structural (systematic). In contrast, the topological structure gives the appearance of branches on the stems being offset or shifted

The periods of activity and pause of the meristems are of equal importance, from the point of view of development and growth. The leaves remain functional even if no new phytomers are created, and the functioning of the organs (leaves, fruits, rings), i.e., the production of biomass and its distribution, can continue in the plant structure even if development is stopped.

2.3.4 Continuous Development Plants: Rhythm Ratio between the Axes

In plants with continuous and regular, and therefore deterministic development, the emission rates of stem and branch phytomers may be different. Generally, the rhythm of the main stem is used as a reference and is 1 by convention. The ratio between the number of phytomers emitted by the branch's meristem and the number of phytomers emitted by the stem's meristem over the same period allows these variations in development between different axes to be taken into account, called "rhythm ratio" w (Fig. 2.10). The rhythm of a meristem's functioning distributes the pauses along the periodically created axis. The rhythm of development of an axis is the ratio between the number of phytomers produced and the number of developmental cycles (DC) over a given period.

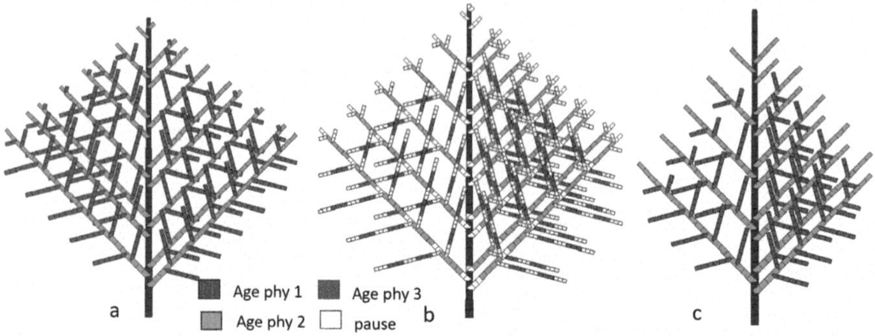

Fig. 2.12 Trees with rhythmic development. (**a**) Immediate branching with chronological and topological representations is identical. The shape is conic. (**b**, **c**) Delayed branching with annual pauses: chronological representations and topological representations of a rhythmic growth structure with delayed branching of an elongation phase. The shape is telescopic because the branches belonging to same growth unit (GU) have the same size. The pauses here are structural because they are regular (in the case of temperate trees such as chestnut, maple, etc.). The annual development time in this example is six DCs. GUs have a decreasing number of phytomers with increasing physiological age (PA) (respectively 5, 4, and 3 for PAs 1, 2, and 3). The durations of the annual pauses complementary to the meristems to synchronize their functioning are therefore respectively 1, 2, and 3 DCs for PAs 1, 2, and 3. In addition, the branching of the axes is deferred by a number of DCs depending on their position so that they can also be synchronized with the terminal meristems when development resumes at the beginning of the season. The topological structure shows that a growth unit has branches of the same size due to the synchronization of their development

2.3.5 Immediate and Delayed Branching

In shrubs, branching can occur either immediate or be delayed by several weeks after the creation of axillary meristems. In the latter case, the meristems undergo pauses counted in DC before being put into functioning. The lateral axes appear as offsets on the supporting axes in the plant structure (Fig. 2.11).

The branching of rhythmically growing temperate tree species usually occurs at the beginning of the following year, so the last formed GU is not branched. As a result, axillary twigs have delayed development (widespread in temperate trees, chestnut, poplar, beech, etc.). In the architecture simulation, taking into account the delays that depend on the position of the twigs on the growth unit allows their elongation to be synchronized (Fig. 2.12b). In some species, twigs develop just after the axillary meristem is placed. This immediate branching is present in many species (alder, elm, beech, cedar, etc.) in the juvenile phase and can disappear later. But the immediate branching is systematic in *Michelia* (a plant species close to *Magnolia*) although it is a rhythmic-growing tree with a synchronous recovering of development for all axes in the following spring (Fig. 2.12a).

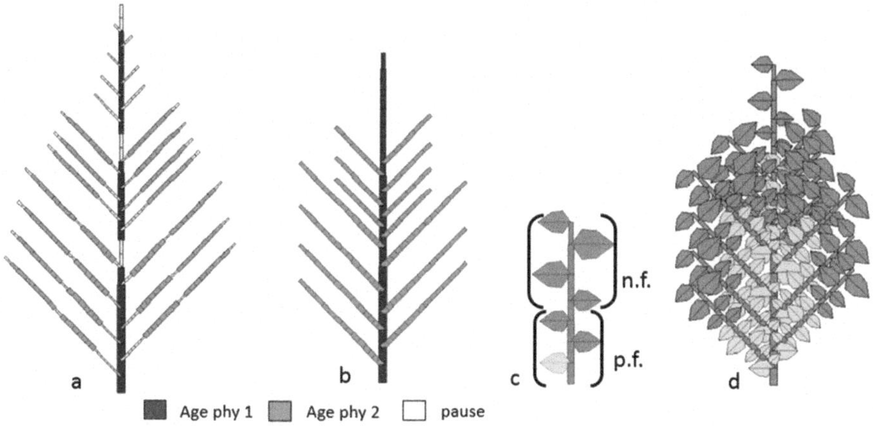

Fig. 2.13 Chronological representations (**a**) and topological representations (**b**) of a rhythmically developing structure with delayed branching of an elongation phase with growth units having a preformed part and a neoformed part. The chronological structure shows in this example the growth units (GUs) formed of nine DCs composed of three preformed phytomers followed by four neoformed phytomers. The rest time between two successive GUs is two pauses. The first three phytomers have the same size because they have expanded simultaneously. The four neoformed phytomers have different sizes because they elongate one by one just after their formation in a variable supply/demand context. The plant at age 1 has a GU (**c**) showing the preformed (p.f.) and the neoformed parts (n.f.). The plant at age 3 has the GUs showing living (green) and dead (yellow) leaves on branches

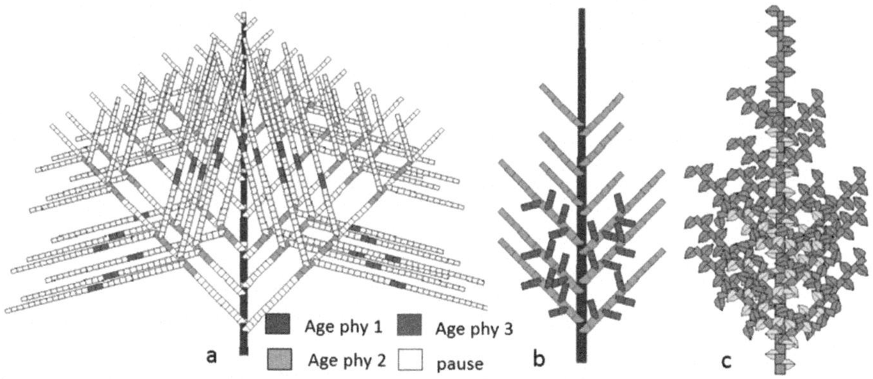

Fig. 2.14 Chronological and topological representations of a structure with rhythmic and polycyclic development, with delayed branching of an elongation phase. Here, the annual shoots develop on 15 DCs. The trunk (physiological age (PA) 1) is polycyclic (three growth units (GUs) per year) and makes GUs with four phytomers per annual shoot. The long branches are polycyclic (two GUs per year) and make a GU with three phytomers per shoot and the twigs make one GU per year (monocyclic) with two phytomers per shoot. (**a**) Simulation of the chronological structure at three years. The branches of the same shoot have the same number of phytomers because their development is synchronous. In this example, the chronological structure contains more pauses than phytomers created. (**b**) Topological representation once the pauses are suppressed. (**c**) Botanical representation of the structure showing living (green) and dead (yellow) leaves. Walnut tree follows this type of development

2.3.6 Synchronous Rhythmic Development Trees: Preformation, Neoformation, and Polycyclism

If the emitted phytomers are formed and stored in the bud in embryonic form during the previous growing season, they are called preformed. They are elongated over a relatively short period of time in spring. The number of preformed phytomers in a growth unit (GU), which is relatively constant, nevertheless decreases often with the increase in the PA of its meristem. In some species (cherry, apricot, poplar, etc.), at the end of the elongation of the preformed part present in the winter bud, new phytomers are created, one by one, by the meristem, which starts to function again. These phytomers are called neoformed. The growth unit then consists of a preformed and a neoformed part. Generally, these two parts are distinguished by their foliar morphology and anatomy (Fig. 2.13). In poplar, the leaves from preformation are small and homogeneous, and those from neoformation reach their maximum size in the middle of the neoformed part. Neoformation occurs from spring to autumn in elm trees or apricots, for example. The number of new neoformed phytomers varies according to age and developmental conditions (fertility and climate) and can exceptionally reach 100 (in apricots, for example).

In some species, meristems establish only one growth unit per year (maple, chestnut, for example). Others show several successive periods of elongation separated by more or less short rest periods; they finally produce an annual shoot consisting of several growth units (Fig. 2.14) that, most often, only branches the following year. This is called polycyclic development (walnut, Aleppo pine). The number of growth units emitted during the year gradually decreases with the increase in the PA of the meristem edifier. The same species can be monocyclic or polycyclic depending on the climatic environment. Scots pine is monocyclic in Finland and polycyclic in southern France.

To simulate tree development, it is therefore necessary to allocate a sufficient number of DCs to the meristems functioning to take into account the periods of structural pauses and meristem activity necessary for annual growth.

2.3.7 Trees with Asynchronous Rhythmic Development

Tropical trees, such as rubber trees, have meristems that do not all work simultaneously, with seasonal recovering of development (called flush). Some meristems remain dormant for varying periods. Branches of the same CA and PA can have a variable number of GUs. The functioning of the meristems becomes desynchronized in this case. In addition, the branches may have a different rhythm ratio compared to the trunk. While temperate trees necessarily make one GU every year or more in case of polycyclism, tropical trees make a variable number of GUs under the influence of stochastic processes. The pauses make void GUs alternating with GUs made of phytomers. For temperate trees, there can be no rhythm ratio between

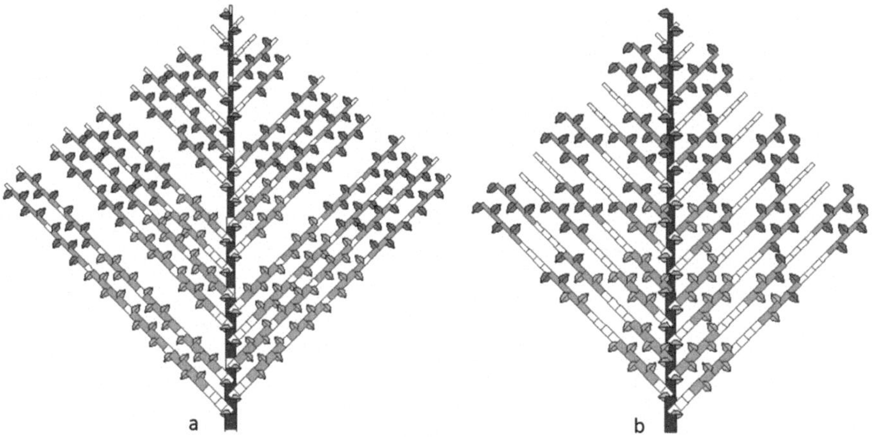

Fig. 2.15 Comparison between the development of temperate and tropical trees. (a) Temperate tree with GU of four phytomers terminated by an annual pause of one DC. (b) Tropical tree with GU of four phytomers and a branch-to-stem rhythm ratio $w = 0.5$. Empty GUs of four DCs alternate with GUs of four phytomers.

trunk and branches, even in case of polycyclism, because of the annual synchronization of development (Fig. 2.15).

2.3.8 Abortion of Meristems and Axis Mortality

Meristems have a life span that depends mainly on their PA. If a PA transition abruptly changes their mode of functioning, abortion puts an end to it. In single-stemmed plants, abortion can ultimately lead to the death of the entire stem. In trees, it is a continuous process that affects meristems according to their chronological, ontogenic, and physiological ages measured in DCs during the development process. Mortality of meristems produces a natural and progressive pruning of the vegetative axes; it increases with PA. Thus, short axes are eliminated quickly, but long axes remain.

In the case of monopodial development, meristem abortion determines the life span of the vegetative axes (Fig. 2.16). Its determinism does not only depend on PA. The environment can produce stresses that significantly increase axis mortality.

In the case of sympodial development, abortion of terminal meristems is systematic at the end of the establishment of growth units (elm, for example). The axes are developed by successive relays provided by subterminal axillary meristems.

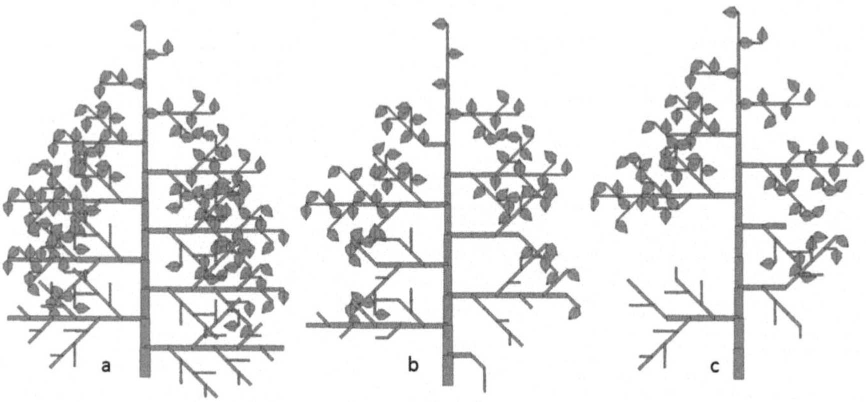

Fig. 2.16 Topological and botanical representations of a structure with continuous development with death process of branches. In (**a**), all branch meristems die simultaneously after axes 2 reaches six DCs that is their life span ontogenic age. (**b**, **c**) Two random simulations where meristems can die independently at any chronological age depending on the rate of reliability

2.4 Deterministic and Stochastic Aspects of Plant Architecture

The observable irregularities in the development of an axis are apparently due to randomness. Regularity of development is the exception in trees (*Cecropia* sp.) and concerns mainly small single-stemmed plants (sunflower, maize, beet) or cultivated and pruned plants (tomato, cucumber). As soon as the plant branches, leafy axes quickly desynchronize in their development in terms of the number of phytomers.

For axes of the same category, the random pauses of the different meristems, over identical development periods, lead to a statistical distribution of the number of phytomers produced. The sequence of activity and rest periods no longer has a periodic character in time, as for rhythm, but becomes aperiodic. The development of the axes then takes on a stochastic aspect.

This phenomenon affects a meristem in all aspects of its functioning, including the placement of phytomers, abortion, and/or the delay in the appearance of a branch. The architecture then loses its regularity, but the similarity between individuals of the same species persists because they are the result of the same random processes. Consideration of stochastic aspects of plant development is essential to define good sampling methods to study plant functioning, both for meristems and source–sink relationships (Chaps. 3 and 6).

The chronological representation mode becomes very useful for visualizing simulated stochastic plants, as pauses are materialized (Figs. 2.10, 2.11, 2.12, 2.13 and 2.14).

2.5 Conclusion

In this chapter, botanical notions of plant architecture strictly limited to the organization of plant structure, i.e., dependencies, by filiation and branching, among botanical entities generated by the functioning of meristems, have been described.

Plant morphogenesis is considered to be the result of two phenomena: development and growth.

Development is the result of the meristems functioning that produce a plant structure whose branched axes are made up of a series of phytomers. The bodies do not play a functional role from a development point of view. Mathematical modeling of development and its simulation is the subject of Chaps. 3 and 4.

Growth is the accumulation of biomass in the plant system through photosynthesis. It is the result of source–sink relationships where organs play their functional roles. It ensures the increase in biomass of the organs produced by development. Mathematical modeling and simulation of growth are discussed in Chaps. 4, 5, 6, and 7.

Three ages characterize a meristem: chronological age, which measures the duration of its functioning since its creation; physiological age, which measures its degree of differentiation; and ontogenic age, which corresponds to its date of creation in the plant structure. Chronological and ontogenic ages are measured in development cycles (DCs), a basic unit of duration that drives development. At each development cycle, a meristem can produce a botanical entity that is either a phytomer or a pause. The cycle is measured in thermal time rather than calendar time, which regulates development.

Depending on the modes of meristem functioning (continuous, rhythmic, polycyclic, definite or indefinite), branching (monopodial, sympodial), and flowering (apical, lateral), plant structures are organized according to a combinatorial set into a limited number of structural patterns called architectural models (Hallé et al. 1978). These models cover all types of plants (Chap. 4). For each structural pattern, there is an appropriate sampling strategy for identifying plant system parameters and appropriate inverse methods (Chaps. 3, 8, and 9). Most herbaceous plants, shrubs, and trees, fit into these structural patterns and can therefore be modeled.

References

Barthélémy D (1991) Levels of organization and repetition phenomena in seeds plants. Acta Biotheor 39:309–323

Barthélémy D, Caraglio Y (2007) Plant architecture: a dynamic, multilevel and comprehensive approach to plant form, structure and ontogeny. Ann Bot 99(3):375–407

Bosanac B (1990) Tree: a knowledge-based system for generating images of trees. Harvard University. http://www.onyxtree.com/about.html

Caraglio Y, Pimont F, Rigolot E (2007) *Pinus halepensis* mill. Architectural analysis for fuel modelling. Proc Int Workshop MEDPINE:43–59

Cournède P-H, Kang M-Z, Mathieu A, Barczi J-F, Yan Y-P, de Reffye P (2006) Structural factorization of plants to compute their functional and architectural growth. Simul-Transact Soc Model Simul Int 82(7):427–438

de Reffye P, Jaeger M (2013) Modèles mathématiques du développement et de la croissance de l'architecture des plantes. Le cas du modèle GreenLab. In: Varenne F, Silberstein M (eds) *Modéliser & simuler : épistémologies et pratiques de la modélisation et de la simulation*, Tome 1, vol 2. Éditions Matériologiques, Paris, pp 625–658, (Sciences &, Philosophie)

de Reffye P, Dinouard P, Barthélémy D (1991a) Modélisation et simulation de l'architecture de l'Orme du Japon *Zelkova serrata* (Thunb.) *Makino* (Ulmaceae) : la notion d'axe de référence. Comptes Rendus du 2e Colloque international sur l'arbre, Montpellier, 10–15 Septembre 1990. Naturalia Monspeliensia, Hors-Série, pp 251–266

de Reffye P, Elguero E, Costes E (1991b) Growth units construction in trees: a stochastic approach. In: 9th seminar of the Theoritical biology school. Solignac, France, 21-23/09/1989, pp 325–342

Grosfeld J, Barthélémy D, Brion C (1999) Architectural variations of *Araucaria araucana* (Molina) K. Koch (Araucariaceae) in its natural habitat. In: Kurmann MH, Hemsley AR (eds) The evolution of plant architecture. Royal Botanic Gardens, Kew, pp 109–122

Hallé F, Martin R (1968) Étude de la croissance rythmique chez Hevea brasiliensis. Adansonia, Série 2(8):475–503

Hallé F, Oldeman R-A-A, Tomlinson P-B (1978) Tropical trees and forests. Springer Verlag, Berlin/Heidelberg/New-York, 441p

Oldeman R-A-A (1974) L'architecture de la forêt guyanaise. Mémoire n. 73. Orstom, Paris

Rivals P (1965) Essai sur la croissance des arbres et sur leurs systèmes de floraison (Application aux espèces fruitières). J d'Agronomie Trop Botanique Appliquée 12:655–686

Rouane P (1977) Un modèle de la ramification et de la croissance végétale en tant qu'image de la différenciation cellulaire. Comptes rendus hebdomadaires des seances. Serie D. Sciences naturelles

(von) Goethe J-W (1790) La métamorphose des plantes. Traduction de Bideau, H., 1975. Éditions Triades, Paris

Yan H-P, Barczi J-F, de Reffye P, Hu B-G, Jaeger M, Leroux J (2002) Fast algorithms of plant computation based on substructure instances. J WSCG 10(1–3):145–153

Yan H-P, de Reffye P, Pan C-H, Hu B-G (2003) Fast construction of plant architectural models based on substructure decomposition. J Comput Sci Technol 18(6):780–787

Chapter 3
The Functioning of Meristems: Equations and Development Operators

Philippe De Reffye, Mengzhen Kang, and Véronique Letort

Abstract The aim of this chapter is to introduce the simulation of plant structures and architectures using the botanical automatons (including deterministic and stochastic development). Deterministic development with constant phyllochron is rare, occurring in some single-stemmed plants, while most branched plants have variable phyllochrons, resulting in diverse phytomer production. The GreenLab model simulates meristem activity stochastically with parameters a, b, and c, representing branching, development, and viability rates. It uses development cycles (DCs) to measure an axis's chronological age, offering both chronological and topological visualization modes. The model produces cohorts of organs and a potential structure that contains the history of development. Monte Carlo simulations are used to estimate axis branching and phytomer distributions, and the crown analysis method infers developmental parameters from plant architecture, aiding in data restoration. The model operates in matrix mode for element production and list mode for creating a dual botanical automaton for three-dimensional (3D) plant simulation in computer graphics.

Electronic supplementary material The online version of this chapter (https://doi.org/10.1007/978-981-97-7541-5_3) contains supplementary material, which is available to authorized users.

P. De Reffye
CIRAD, UMR AMAP, Montpellier, France

AMAP, Univ Montpellier, CIRAD, CNRS, INRAE, IRD, Montpellier, France

M. Kang (✉)
State Key Laboratory for Multimodal Artificial Intelligence Systems, Institute of Automation, Chinese Academy of Sciences (CASIA), Beijing, China

School of Artificial Intelligence, University of Chinese Academy of Sciences, Beijing, China
e-mail: mengzhen.kang@ia.ac.cn

V. Letort
Mathématiques et Informatique pour la Complexité et les Systèmes, CentraleSupélec-Université Paris-Saclay, Gif-sur-Yvette, France
e-mail: veronique.lechevalier@centralesupelec.fr

P. De Reffye et al. (eds.), *Plant Model Greenlab for Botany and Agronomy*,
https://doi.org/10.1007/978-981-97-7541-5_3

Keywords Plant development · Stochastic processes · Development cycle · Chronological · Topological and potential structures · Crown analysis · List and matrix modes

3.1 Origin of Stochastic Development of Leafy Axes

Very few plants have a purely deterministic development. This is typically seen in single-stemmed plants, such as palm trees, certain cultivated varieties of maize, sunflower, or pruned plants such as tomatoes, whose meristems produce new phytomers at regular intervals (i.e., phyllochron is constant) expressed in thermal time (Chap. 2). But often, especially in branched plants, even if expressed in thermal time, the time interval between the emissions of two successive phytomers is variable and follows a certain distribution (i.e., the phyllochron is variable). Thus, due to the random nature of the phyllochron, two similar leafy axes can produce different numbers of phytomers during the same functioning time. An example of this can be found in a coffee tree plant, where a pair of branches sharing the same stem phytomer can yield varying numbers of phytomers.

In accordance with the "Renewal theory", the statistical distribution of the phyllochron is called the inter-arrival law, and the number of phytomers produced during the same period of time by a collection of branches is called the counting law. For example, in *Coffea canephora* trees in Ivory Coast, the meristems of the stems take 16 days on average to create a new phytomer, with variations ranging from 10 to 25 days (Fig. 3.1a), whose mean is $\mu = 15.7$ and the variance $\sigma_2 = 12.6$. This leads to a variable number of phytomers being produced over a 90-day period (Fig. 3.1b). The mean of this law is $M = 5.43$ and the variance is $V = 0.28$.

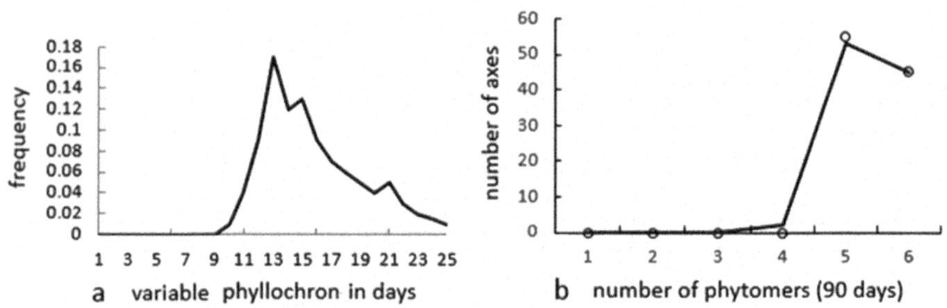

Fig. 3.1 Phyllochron distribution and phytomer production in *Coffea canephora* Stems over a 90-day period. (**a**) The observed distribution of phyllochron in days. (**b**) The number of phytomers produced by 100 stems of the coffee tree during a period of 90 days: observations (symbol "o") and shifted binomial law (solid line) ($K = 6$, $b = 0.29$, shift $= 4$)

3.2 Application of the Renewal Theory to the Functioning of Meristems

The objective of the Renewal theory is to count the occurrences of a given phenomenon over a fixed period of time, knowing that the time between two successive occurrences is independent and follows the same law of variation (inter-arrival law).

Here, the phenomenon of interest is the formation of a phytomer by the apical meristem of a leafy axis. Results from Renewal theory show that the distribution of the number of phytomers produced during a period T (denoted as N_T), converges toward a normal distribution as T increases. This law depends only on the mean μ and the variance σ^2 of the distribution of the inter-renewal times. For sufficiently large values of T, according to renewal theory, the following approximation for the mean M and the variance V of the number of produced phytomers can be made:

$$M \approx \frac{T}{\mu}$$
$$V \approx T\frac{\sigma^2}{\mu^3} \tag{3.1}$$

Moreover, the counting law of the number of produced phytomers tends quickly toward a discretized normal distribution. When the variance of this counting law is lower than its mean, the distribution can be approximated by a binomial law, which tends in turn toward the normal law as T increases.

The mean and variance of a binomial law $B(N, b)$ are written as:

$$M = N \cdot b$$
$$V = N \cdot b \cdot (1 - b) \tag{3.2}$$

Where N is the number of trials and b is the probability of their successes.

By comparing the systems of Eqs. (3.1) and (3.2), it is easy to establish that, in this case:

$$N \approx \frac{T}{\mu\left(1 - \frac{\sigma^2}{\mu^2}\right)}$$
$$b \approx 1 - \frac{\sigma^2}{\mu^2} \tag{3.3}$$

In this way, the binomial law (N, b) can be considered as the distribution of the counting process with a discrete time step called the development cycle (DC). N steps are made, each with a probability b at each DC. Such a process is called the "*Bernoulli* process". The numerical simulation of meristems therefore gives cycles that produce phytomer entities alternately (with a probability b) with void entities (with a probability $1 - b$), which are considered as pauses in the functioning. The use of renewal theory therefore makes it possible to pass from continuous time to discrete time with a good approximation. Assuming that the distribution of the number of phytomers produced by a similar leafy axis over a given period T is approximated with a binomial law $B(N, b)$, the average duration of the DC can be deduced:

$$t_{DC} = \frac{T}{N} \tag{3.4}$$

This model is in good agreement with the use of thermal time in relation to calendar time. The minimum time between the creation of two successive phytomers corresponds to one DC and the number of DCs that count the phytomer entities and pauses is in a linear relationship with thermal time.

Suppose that the renewal time is linearly transformed by a coefficient k, then μ can be expanded by a coefficient k, i.e., $k^*\mu$, and its standard deviation is also multiplied by k. This will have no effect on the value of probability b. Indeed, we have:

$$b = 1 - \frac{k^2\sigma^2}{k^2\mu^2} = 1 - \frac{\sigma^2}{\mu^2} \tag{3.5}$$

The distribution of the number of phytomers produced therefore remains invariant after the linear transformation from calendar time to thermal time. It is a simple expansion of the time scale. Although the duration may vary with temperature conditions, we still reach the same stage of development in terms of the number of DCs and the same distribution of the number of phytomers per axis. Note that the model is applicable only if: $\sigma^2/\mu^2 < 1$.

The case $\sigma^2 = \mu^2$ (which corresponds to Poisson law) would give an infinite number of DCs. Fortunately, in all the analyzed phytomers, the *Bernoulli* process has always adapted well, and the observed values of b typically correspond to $b > 0.7$.

3.3 Numerical Studies on Theoretical Cases

Let us consider two renewal processes, denoted as $(X_1, ..., X_n)$ and $(X^*_1, ..., X^*_n)$, which have the same mean and same standard deviation, with $\mu = 6$ and $\sigma = 2$. We define Nt and N^*t as the associated counting processes (Fig. 3.2).

These two laws can follow a binomial law (for example, $N = 9$, $b = 2/3$), and a geometric law ($N = -1, b = 0.5$) shifted by four DCs (Fig. 3.3a). Both laws have the same mean ($\mu = 6$) and variance ($\sigma^2 = 2$). They correspond to the distribution of the delays in DCs between two successive phytomer formations (i.e., the law of variables $X_1, ..., X_n$ in Fig. 3.1). Both laws tend to give the same counting law.

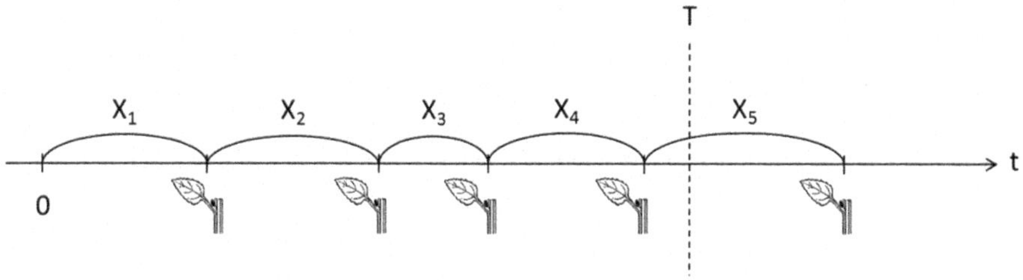

Fig. 3.2 Pattern of the renewal process: the intervals between the formation of two successive phytomers form a sequence of independent and identically distributed random variables ($X_1, \ldots, X_n, \ldots$), each following the same law with a mean μ and a variance σ^2. The associated counting process makes it possible to count the number of phytomers, denoted as N_T, produced over the a given period [0, T]. In the above example, if we look at this period, we obtain $N_T = 4$. The law of this variable N_T can be approximated by either a binomial or normal law

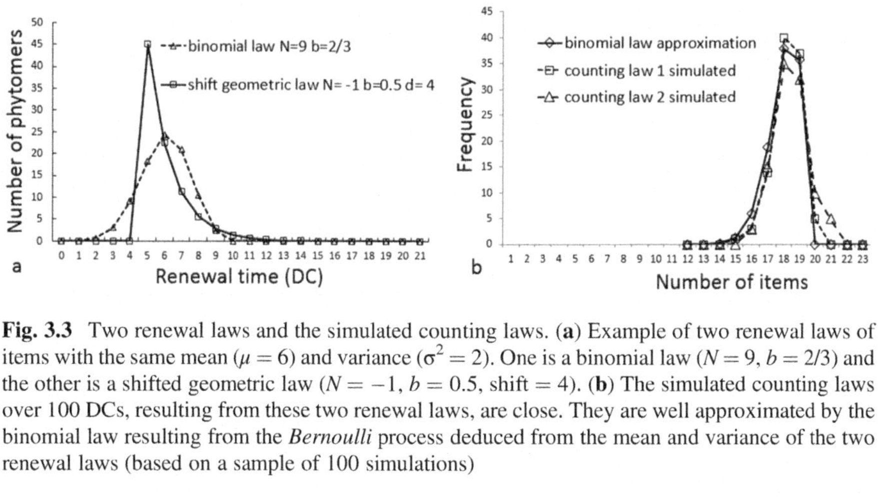

Fig. 3.3 Two renewal laws and the simulated counting laws. (**a**) Example of two renewal laws of items with the same mean ($\mu = 6$) and variance ($\sigma^2 = 2$). One is a binomial law ($N = 9$, $b = 2/3$) and the other is a shifted geometric law ($N = -1$, $b = 0.5$, shift $= 4$). (**b**) The simulated counting laws over 100 DCs, resulting from these two renewal laws, are close. They are well approximated by the binomial law resulting from the *Bernoulli* process deduced from the mean and variance of the two renewal laws (based on a sample of 100 simulations)

From the law of the renewal process, if we consider an observation period of $T = 100$ DCs, then once transformed into development cycles, the distribution of the two associated counting processes, *Nt* and N^*t, can be approximated by the same binomial law (N^*, b^*) with following numerical presentation:

$$b^* \approx 1 - \frac{\sigma^2}{\mu^2} = 0.944$$

$$N^* = \frac{100}{6 \times 0.944} \approx 18.$$

Figure 3.3b shows 100 simulations of the two processes (*X1, ... Xn*) and (X^*1, ... X^*n), over a time step of 100 DCs. The two associated counting processes, *Nt* and

N^*t, are compared with a binomial law with parameters ($N = 18$, $b = 0.944$), corresponding to the binomial approximation of the distribution of Nt and N^*t by the renewal theory. All three laws have similar results.

The approximation of the binomial law due to the renewal theory gives the probability of observing an axis composed of K phytomers for N development cycles, which is equal to:

$$P(X = K) = \binom{N}{K} b^K (1 - b)^{N-K} \tag{3.6}$$

Returning to the case of the coffee tree, where we observe over a period of 90 days an average $M = 5.43$ phytomers created with a variance $V = 0.28$, we deduce the value of the *Bernoulli* parameter b: $b = 1 - 0.28/5.43$, so $b = 0.94$. This value can be computed directly from the phyllochron distribution of the Fig. 3.1a, whose mean is $\mu = 15.7$ and the variance $\sigma_2 = 12.6$. This also gives $b = 1 - \sigma_2/\mu$ $2 = 0.94$. We notice that the inter-arrival law of the phytomers looks like a shifted geometric law (Fig. 3.1b).

In conclusion, the distributions of the number of phytomers produced over the same developmental period by stems of the same nature provide information on how meristems work. Even when ignoring the calendar time required for their creation, the number of phytomers can be related to the number of DCs that generated them, along with an associated probability of functioning. The complex phenomenon of renewal is therefore assimilated with a good approximation to a *Bernoulli* process. Thus, in the rest of the book, the development of the axes will always be modeled by a number N of development cycles (DCs) with a probability b of production of an associated phytomer.

The result of the functioning of a meristem is therefore an axis of development that is composed of a random alternation of phytomers entities with void entities: the representation including void entities is called the chronological representation. By removing the pauses, we obtain the topological representation of the axis, in which we lose the notion of time, and which corresponds to the observable reality.

3.3.1 Continuous Development

In the case of continuous development, phytomers and pauses generated by meristems may alternate in periodic or random patterns, thereby forming a "development axis".

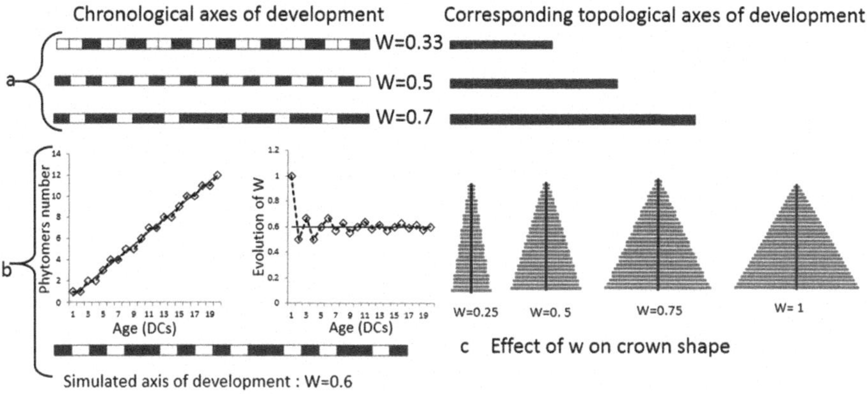

Fig. 3.4 Development axes with different rhythm ratios *w*. (**a**) Periodic development in chronological and topological modes is displayed for rhythm ratios of $w = 0.33$, $w = 0.5$, and $w = 0.7$. (**b**) The evolution of the number of phytomers produced as a function of the development cycles (DCs), for $w = 0.6$. The parameter *w* converges quickly to its nominal value and an algorithm (Bresenham) is used to calculate the alternating pauses and phytomers. (**c**) The value of *w* modifies the crown shapes of plants

3.3.1.1 Periodic Development

Periodic development refers to a deterministic case that within the same architecture, the axes can have different average development speeds. For example, the branches of coffee and cotton grow slower on average than their main stems. The rhythm ratio, denoted as $w = X/N$, is defined as the ratio of the number *X* of phytomers produced for *N* DCs of development (Fig. 3.4). The simulation of the rhythm *w* is performed by a periodic sequence of 0s and 1s, for example, the sequence [001001001...] simulates a rhythm $w = 1/3$.

Assuming $w_1 = 1$ for the rhythm of the stem and $w_2 < 1$ for that of the branches, $w = w_2/w_1$ is the rhythm ratio. This ratio modifies the profiles of the tree crowns (Fig. 3.4).

3.3.1.2 Stochastic Development and *Bernoulli* Process

The development of the axes is simulated by a flowchart using Monte Carlo method (Fig. 3.5a). This method produces a random sequence of phytomers in a proportion *b* (the development rate), and void entities in a proportion $1 - b$ and represent pauses in the meristem's functioning. If *N* is the number of DCs of the axis, there are 2^N possible simulated random axes. Such a stochastic model is called a *Bernoulli* process.

If all the pauses are displayed, the development axes are in chronological mode (Fig. 3.5b). This representation is internal to the model and has a pedagogical value. The realization of a random chronological axis of the development is represented by

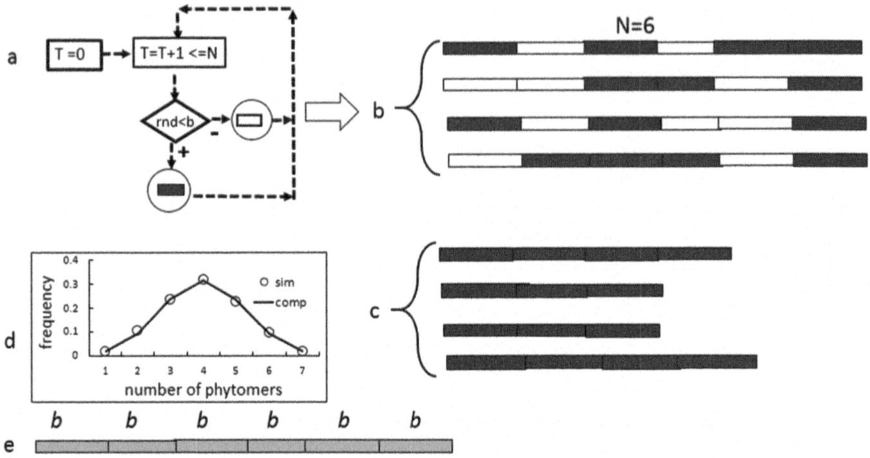

Fig. 3.5 Development axes simulated by a *Bernoulli* process (Monte Carlo algorithm). (**a**) Development flowchart is applied with $N = 6$ and $b = 0.5$. A random number *rnd* is compared to the threshold b. The flowchart produces a phytomer or a pause at each DC in the respective proportions b and $1 - b$. (**b**) Four random simulations are given in chronological mode. (**c**) Simulated axes in topological mode after suppressing pauses. (**d**) The distribution of the number of phytomers per axis is displayed for 500 random simulations (symbol "o" represents the simulations and the curve represents the binomial law). (**e**) The potential development axis with six phytomers, each with a probability of occurrence of b

a random series of 0 and 1, where 0s represent pauses (void entities) and 1s represent phytomers, for example, [101011···].

If pauses are removed, the development axes are then in topological mode (Fig. 3.5c). This representation corresponds to the observations made when counting the phytomers on the leafy axes. In this mode, the time information is no more explicit.

The potential development axis is a chronological axis where all phytomers are displayed with their respective probability of existence (Fig. 3.5e). The distribution of the number of phytomers per axis generated by a *Bernoulli* process at the cycle N with a parameter b follows a binomial law $B(N, b)$ (Fig. 3.5d).

If we denote X as the number of phytomers produced by a simulation on an axis, the probability that this number is K, given by Eq. (3.6). Its mean M and variance V are:

$$M = N \cdot b$$
$$V = N \cdot b(1 - b) \tag{3.7}$$

The mean and variance are linearly linked by the relationship:

$$V = (1 - b) \cdot M \tag{3.8}$$

Note that the model applies with variable probabilities $b(i)$ depending on the cycles. In this case, the respective mean and variance will be:

$$M(N) = \sum_{i=1}^{N} b(i)$$
$$V(N) = \sum_{i=1}^{N} b(i)(1 - b(i)) \tag{3.9}$$

The $b(i)$ can be deduced from the relationship between the means and the measured variances. The first validation of the modeling of the development of leafy axes was carried out on coffee trees (de Reffye 1981).

The deterministic mode of periodic development can be combined with the *Bernoulli* process. In this case, a stochastic pause can only occur outside the periodic pauses induced by the rhythm w (de Reffye 1979, 1981). The *Bernoulli* process and periodic development are thus combined by the scalar product of the development axes obtained in both cases.

3.3.1.3 Viability of Meristems

At a DC, a meristem may abort because of a random event. The viability of a meristem during T DCs is modeled by the survival probability c for each DC. Figure 3.6a shows the simulation flow chart for both the viability and *Bernoulli* process of the meristem. Figure 3.6b–e shows the simulated chronological axes of development, the corresponding topological axes, the distribution of produced phytomers, and the potential axis with associate probabilities.

At the end of T cycles, the probability for an axis to still be alive is $P(T)$:

$$P(T) = c^T \tag{3.10}$$

The probability for a meristem to die at the end of K cycles is then $P(K)$:

$$P(K) = (1 - c)c^{K-1} \tag{3.11}$$

At each DC, part of the meristem population aborts and fixes the sub-distribution of produced phytomers. This is a truncated geometric law whose mean and variance of the number of cycles at mortality are shown to be:

$$M_T = \frac{1 - c^T}{1 - c}$$

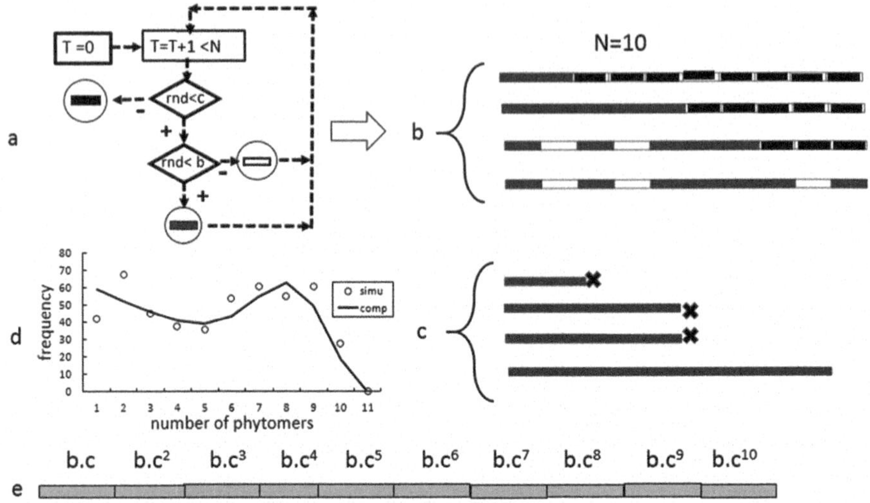

Fig. 3.6 Development of axes simulated by a compound *Bernoulli* process with parameter b and a mortality process with parameter c. (**a**) Development flowchart. The meristems in this example function according to a *Bernoulli* process with a value $b = 0.75$ and a viability rate of $c = 0.97$. Random numbers *rnd* are compared to the probabilities b and c. (**b**) Four random simulations are given in chronological mode (we distinguish void entities from pauses, and black entities indicates meristem abortion). (**c**) Simulated axes in topological mode after suppressing pauses (a cross means a dead axis). (**d**) Distribution of the number of phytomers per stem for a population of 1000 stems over 10 development cycles (DCs) with mortality (the "o" represents the simulations and the curve represents the values of the compound law). In the left part, most axes are dead, and in the right part, the bell shape corresponds to the front wave of the axes still in development. (**e**) The potential development axis produces phytomers with the probability of occurrence as a function of rank k is bc^k

$$V_T = \frac{c}{(1-c)^2}\left(1 - c^T(2T+1)(1-c) - c^{2T+1}\right) \tag{3.12}$$

The overlapping of dead axis distributions by adding the part of living axes generates a compound law. In the case with a development probability b and a viability probability c, it is possible to calculate the probability of observing a dead or living axis composed of K phytomers at the end of T DCs according to the equation:

$$P(K) = \sum_{i=K}^{T-1} (1-c)c^i \binom{i}{K} b^K (1-b)^{i-K} + c^T \binom{T}{K} b^K (1-b)^{T-K} \tag{3.13}$$

The mean and variance of the phytomers in the population distribution are obtained by the equations of the compound laws:

$$M_X = M_T b$$
$$V_X = M_T b(1-b) + b^2 V_T \tag{3.14}$$

When T increases, we can show that the law tends toward a geometric law with a parameter (all axes are dead):

$$c' = \frac{bc}{1-c(1-b)} \tag{3.15}$$

The process creates cohorts of dead axes for which development is definitively stopped.

However, this viability is generally not constant. In cycle T, we can observe $T + 1$ classes of axes, each containing from 0 to T phytomers in the proportions $F(0)$, $F(1)$, . . ., $F(K)$, . . ., $F(T)$, where $F(K)$ represents the proportion of dead axes in the cycle K, for $K < T$, and $1 - F(T)$ the proportion of living axes.

The viabilities for each associated cycle are noted as $c_1, c_2, \ldots, c_K, \ldots, c_T$. The relationships between classes and viabilities are then written as follows:

$$\begin{aligned} F(0) &= 1 - c_1 \\ F(1) &= c_1(1-c_2) \\ F(2) &= c_1 c_2 (1-c_3) \\ F(K) &= \prod_{i=1}^{K} c_i (1-c_{K+1}) \\ F(T) &= \prod_{i=1}^{T} c_i \end{aligned} \tag{3.16}$$

The $F(K)$ values are the observable result of the meristem abortion rates in the cycle K. To obtain their viabilities, it is necessary to inverse the system, which results in:

$$\begin{aligned} c_1 &= 1 - F(0) \\ c_2 &= \frac{1-F(0)-F(1)}{1-F(0)} \\ c_K &= \frac{1-\sum_{i=0}^{K-1} F(i)}{1-\sum_{i=0}^{K-2} F(i)} \end{aligned} \tag{3.17}$$

We assume $Ft(K) = \sum_{i=0}^{K} F(i)$ represents the rate of dead branches in the cycle K.

Then we get:

$$c_1 = 1 - Ft(0)$$
$$c_K = \frac{1 - Ft(K-1)}{1 - Ft(K-2)} \tag{3.18}$$

The average duration of the functioning time in cycles before mortality is given by the equation:

$$T_m = c_1 + c_1 \cdot c_2 + \ldots + c_1 \cdot c_2 \cdot \ldots \cdot c_T \tag{3.19}$$

When c varies, Eq. (3.13) becomes generalized and gives the probability of observing class K.

$$P(X=K) = \sum\nolimits_{i=K}^{T-1} \prod\nolimits_{j=1}^{K} c_j (1 - c_{K+1}) \binom{i}{K} b^K (1-b)^{i-K}$$
$$+ \prod\nolimits_{i=1}^{T} c_i \binom{T}{K} b^K (1-b)^{T-K} \tag{3.20}$$

The mean and variance of the compound law are deduced from this.

In general, the proportion of dead branched axes from the top of the stem to the bottom increases according to a sigmoid law. It can be fitted using a function of the form:

$$S(K) = 1 - \exp(-\beta K^{\alpha}) \tag{3.21}$$

The parameters can be estimated by inverse method. Equation (3.18) gives:

$$c_K = \exp(-\beta((K-1)^{\alpha} - (K-2)^{\alpha})) \tag{3.22}$$

The percentage of survivals at the age K on DC is then written as:

$$F(K) = \exp(-\beta (K-1)^{\alpha}) - \exp(-\beta K^{\alpha}) \tag{3.23}$$

The observation of plants shows the important role played by branch mortality in the architectural development process, particularly in trees whose trunks are pruned due to the fall of dead branches. Mortality is the cause of missing data, since the fresh mass does not exist anymore. The model allows reconstructing it, thus giving access to a reconstruction of the past plant architecture. It is essential to take this phenomenon into account if we want to fully simulate the development of plants.

3.3.1.4 Modality of Branching

A phytomer generally bears one or several axes that can appear either immediately or after a delay. Each axillary meristem follows its own dormancy law according to its position on the axis. This dormancy can be due to structural reasons (in temperate trees, the axillary meristems are dormant on the shoot of the current year) or it can be random.

3.3.1.4.1 Immediate Branching

Many shrubs exhibit immediate branching. It means that axillary meristems can function only at the cycle following the creation of their associated bud. If a branch does not appear at this cycle, it will never do.

The rate of immediate branching to the phytomer of rank i from the base of the axis, denoted as a_i, is defined as the ratio of the number of axillary branches to the number of axillary meristems. In general, the first few phytomers at the base of the stem do not bear branches, and branching gradually appears along the stem until the branching rate stabilizes, a phenomenon known as the base effect.

3.3.1.4.2 Delayed Branching

The branching can be delayed by several DCs according to a specific delay law. If $Fa(i)$ is the proportion of observed branches starting at DC i, then the probabilities of occurrence a_i of twigs at the cycle i are calculated by solving the following system:

$$\begin{aligned} Fa(0) &= a_1 \\ Fa(1) &= (1 - a_1)a_2 \\ Fa(i) &= \prod_1^i (1 - a_i)a_{i+1} \end{aligned} \tag{3.24}$$

Where $Fa(i)$ corresponds to the frequency of appearance of a new axillary branch in the DC i.

By inversion of the model, the a_K values are calculated after estimating the values $\sum_{i=0}^{K} Fa(i)$ to an empirical function according to the equation:

$$1 - a_K = \frac{1 - \sum_{i=0}^{K-1} Fa(i)}{1 - \sum_{i=0}^{K-2} Fa(i)} \tag{3.25}$$

If the values a_K are constant, the distribution of the delay law will follow a geometric law.

In general, the law of delay is complex and in sigmoid form. In coffee tree, the third-order axes, unlike the second-order axes, do not have immediate branching. The values of a_K are variable and are deduced from the observations using Eq. (3.25).

It should be noted that Eq. (3.24) covers all cases. The immediate branching corresponds to: $a_1 = a$ and $a_K = 0$ for all $K > 1$.

There are also plants with delayed branching in shrubs, a phenomenon that is revealed by the observations of shifted plant crowns. The branches are only born on the stem after a certain delay d with $a_d = a$ if $K = d$ and $a_d = 0$ if $K \neq d$. (see Chap. 2, Fig. 2.10)

3.3.1.4.3 Distribution of Branching along the Stem

Assuming that the stem phytomers have only one axillary meristem, the rate (or intensity) of branching a that gives the percentage of branched phytomer is generally insufficient to describe the modalities of branching. Diffuse or grouped branching would be associated to the same rate and cannot be distinguished. Therefore, an additional parameter can be introduced to describe different distribution modes of branches along the stem: the coupling parameter r (Fig. 3.7). This parameter controls to which extent the presence of a branch at one node increases the probability of another branch being present at the subsequent node.

This creates a Markov chain with a probability $r + (1 - r)\cdot a$ of remaining in the branched state and $r + (1 - r)(1 - a)$ in the non-branched state. If $r = 0$, the branching is diffuse; if $r = 1$, the stems are either fully branched or bare in proportion a and $1 - a$. Between the two extreme states, the grouping of branches is more or less accentuated. At the cycle T, by setting $b = 1$ and $c = 1$ for the development parameters, the mean $M(T)$ and the variance $V(T)$ of the number of branches produced by the stem are:

$$\begin{aligned} M(T) &\approx T \cdot a \\ V(T) &\approx T \cdot a(1-a)\left(\frac{1+r}{1-r}\right) \end{aligned} \tag{3.26}$$

The rules for positioning and alternating branches from axillary meristems are generally well described by Markov chains (Guédon et al. 2003), even in the

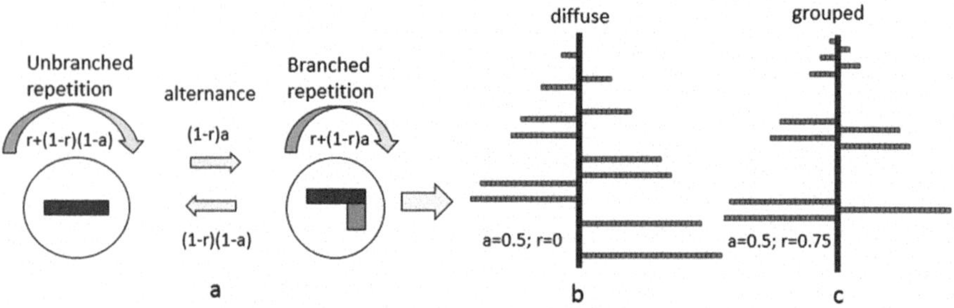

Fig. 3.7 Branching system with two types of branched and unbranched phytomers. (**a**) Schematization of the functioning of the branching by coupling in the form of a Markov chain. (**b**) A simulation with $a = 0.5$, $r = 0$. The coupling is null and the branching is diffuse. (**c**) $a = 0.5$, $r = 0.75$. The coupling is strong and the branching is grouped

complex case of acrotony. In many cases, it is sufficient for the model to simulate the number of axes of different categories, rather than their specific positions on the stem. Thus, the GreenLab model focuses more on calculating the number of axes set up rather than their positioning, which simplifies the calculations. If necessary, the AmapSim simulation software (Barczi et al. 1997, 2008) can more accurately simulate the Markovian aspect of branching.

3.3.1.5 Stochastic Development of Structures

In general, all stochastic processes are combined during plant growth.

Taking the example of a stem of physiological age 1 (PA 1) that carries branches of PA 2, we can distinguish:

- The meristem functioning of the stem is controlled by a *Bernoulli* process with parameter b_1. The stem is often without mortality.
- The meristem functioning of the branch is controlled by a *Bernoulli* process with parameter b_2. Branches have a rhythm ratio w that is generally lower than that of the stem, which is set equal to 1 by convention.
- The branching probability a_2 increases from the base to stabilize. Three parameters are sufficient to characterize it.
- The viability c_2 of branch meristems varies, increasing the mortality rate from the top of the stem in a sigmoid shape. Three parameters are sufficient to characterize it.

Thus, in general, nine parameters are sufficient to simulate the continuous development of a structure with two PAs, which is low considering that it can contain more than a thousand phytomers. These parameters allow to get stochastic structures in development through computer simulations using the Monte Carlo method (Fig. 3.8).

To estimate these parameters from observations on real plants, the method of "crown analysis" was developed. It is presented in the next paragraph.

3.3.1.6 Plant Crown Analysis

The purpose of plant crown analysis, used here as an inverse method, is to extract the *Bernoulli* parameters b_1, b_2 and the rhythm ratio w from a plant architecture. Note that no growth monitoring is required for this process. A crown is considered to consist of a stem that carries branches. In this example, it has two PAs. Absent or dead axes are initially excluded from the analysis.

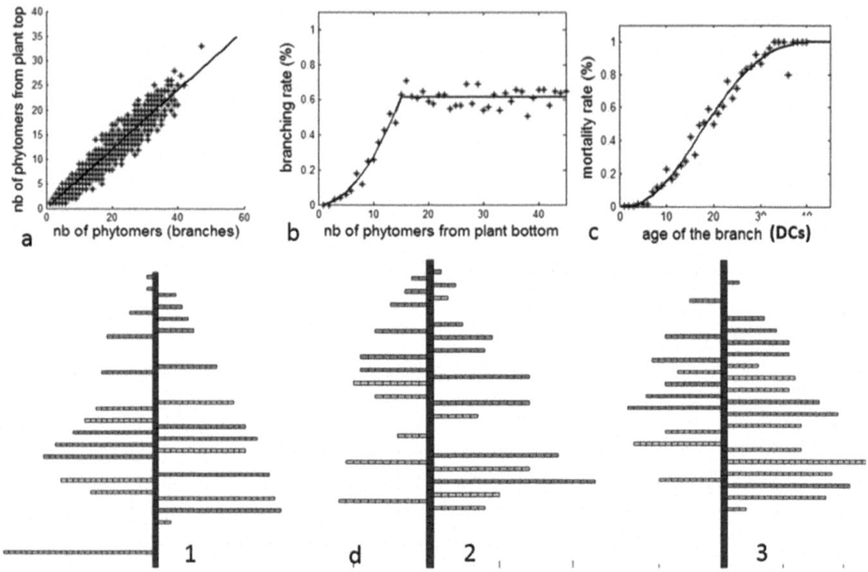

Fig. 3.8 Simulation of a stochastic structure with continuous development with two physiological ages for stem and branches, and parameters for development, viability, and branching (Gloups software). (**a**) Due to the *Bernoulli* process, the living branches at rank *k* below the top have a variable number of phytomers. (**b**) By going up from the bottom to the top, we simulate a branching rate that increases until it stabilizes. (**c**) From the top down, the viability of meristems of branches decreases. The mortality rate of branches increases until the living branches disappear. (**d**) Three random structures aged 40 development cycles (DCs) from 100 simulations illustrate stochastic development (the branches that remain alive are green, the dead ones are gray). (It is not necessary here to specify the numerical values of parameters *b*, *a*, *r*, *c*)

3.3.1.6.1 The Negative Binomial Law and the Crown Analysis

At the rank of the K_1 phytomer below the top of the stem, the chronological age (CA) of a phytomer is distributed according to a negative binomial law $BN(K_1, b_1)$ where b_1 is the parameter of the *Bernoulli* process assigned to the meristem functioning of the stem.

$$P(X = K_1 + x) = \binom{K_1+x}{x} \cdot {b_1}^{K_1} \cdot (1 - b_1)^x \tag{3.27}$$

The negative binomial law (Eq. 3.27) characterizes the number of $K_1 + x$ trails required to obtain the desired number of K_1 successes. Here, the trails correspond to the DCs, the successes (in number K_1) correspond to the creation of the phytomers and the failures (in number x) represent the pauses. The rank therefore characterizes the number of successes (since it counts the number of phytomers formed) and the pauses (hidden) increase the chronological age (CA) of the phytomer that appeared at the rank K_1: this CA obviously has a minimum of K_1 (it takes at least K_1 cycles to

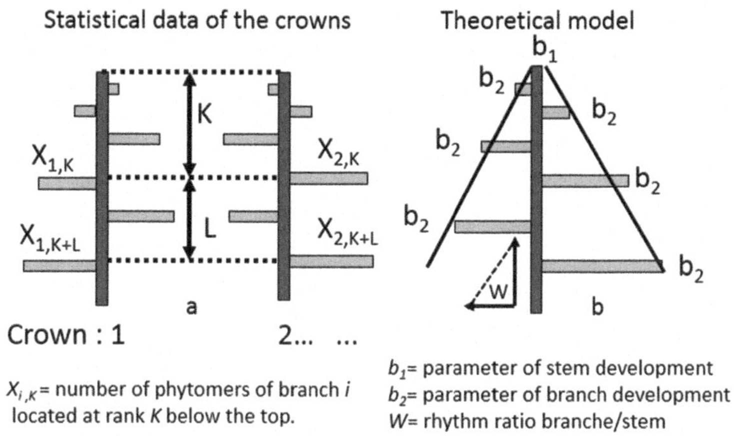

Fig. 3.9 Diagrams of branch identification in crown analysis. The development model depends only on the three parameters b_1, b_2, and w. (**a**) Stochastic variable $X_{i,K}$: the number of phytomers on the branch located at rank K under the top of plant i. The simulation produces random crowns, and the mean and variance of $X_{i,K}$ can be evaluated. (**b**) Theoretical development model. b_1 is the development probability of the stem; b_2 is the development probability of the branch; and w is the rhythm ratio of the stem to the branches

build K_1 phytomers) and is truncated with the tree's age in DCs. Once the CA of the branch K_2 is assigned ($K_2 = K_1 + x$), the number of phytomers on the branch will depend on the *Bernoulli* process with parameters (K_2, b_2). The distribution of the number of phytomers on the branches located at the rank K_1 therefore follows a compound law of binomial laws with parameters (K_1, b_1) and (K_2, b_2). If the rhythm ratio w is not equal to 1, the age of the branch K_2 must be multiplied by w.

3.3.1.6.2 Equations for Plant Crown Analysis

Because of the *Bernoulli* process that discretizes the development, we therefore have a theoretical development model along with statistical data on the crowns. These statistics are expressed as, at the rank K from the top, the percentage of branched nodes, the percentage of dead branches, and the distribution of the number of phytomers in the living branches whose mean and variance are detailed. These statistics depend on the stochastic processes that generate them, and the aim here is to find the relationships that link them to the parameters of the development. These relationships are illustrated in Fig. 3.9. We denote $X_{i,K}$ as the number of phytomers on the branch located at rank K under the top of plant i.

Suppose there are N crowns that have been observed at the rank K on the stem. Only living branches are considered in order to exempt the mortality effect. We can calculate at this rank a mean X_K and a variance V_K of the number of phytomers on these branches, which are expressed according to parameters w, b_1, and b_2. The following equations have been established:

The inter-stem empirical mean is:

$$\overline{X_K} = \frac{\sum_{i=1}^{N} X_{i,K}}{N} \tag{3.28}$$

Its theoretical value is:

$$X_K = w \cdot \frac{K}{b_1} \cdot b_2 \tag{3.29}$$

The inter-stem empirical variance resulting from the composition of stem–branch binomial laws is:

$$\overline{V_K} = \sum_{i=1}^{N} \frac{\left(X_{i,K} - \overline{X_K}\right)^2}{N-1} \tag{3.30}$$

Its theoretical value is:

$$V_K = w \frac{K}{b_1} b_2 (1 - b_2) + (w \cdot b_2)^2 \frac{K(1 - b_1)}{{b_1}^2} \tag{3.31}$$

These two statistics are insufficient for our analysis because there are three parameters to be estimated. The third parameter is provided by the intra-stem variance v_K of pairs of adjacent branches that are branches carried by the nodes with a distant of L ranks (if $L = 0$, the branches are carried by the same node; if $L = 1$, the branches are adjacent).

The intra-stem empirical variance is:

$$\overline{v_K}(L) = \frac{\sum_{i=1}^{N} \left(X_{i,K+L} - X_{i,K}\right)^2}{2N} \tag{3.32}$$

Its theoretical value is:

$$v_K = w \cdot b_2 (1 - b_2) \frac{K + \frac{L}{2}}{b_1} + (w \cdot b_2)^2 \frac{L(1 - b_1)}{2b_1^2} + (w \cdot b_1)^2 \frac{L^2}{2b_2^2} \tag{3.33}$$

By solving the previous equations, we have three expressions that give the solutions for the values of w, b_1, and b_2 at the rank K from the top.

$$
\begin{aligned}
1-b_2 &= \frac{1}{X_K}\left(v_K - \frac{L}{2}\left(\frac{V_K}{K} + L\left(\frac{X_K}{K}\right)^2\right)\right)\\
1-b_1 &= \frac{1}{X_K{}^2}\left(K\left(V_K - v_K + \frac{1}{2}\left(\frac{LX_K}{K}\right)^2\right) + \frac{LV_K}{2}\right)\\
w &= \frac{b_1 X_K}{b_2\, K}
\end{aligned}
\tag{3.34}
$$

In practice, $L = 0$ is chosen for plants that have two axillaries per node such as coffee tree, or $L = 1$ for plants that have a single axillary and adjacent branches such as cotton plant.

In particular, for $L = 0$ (the case of two axillaries/phytomers), Eq. (3.31) is simplified and can be written as:

$$
\begin{aligned}
1-b_2 &= \frac{v_K}{X_K}\\
1-b_1 &= \frac{K(V_K - v_K)}{X_K{}^2}\\
w &= \frac{b_1}{b_2}\frac{X_K}{K}
\end{aligned}
\tag{3.35}
$$

The values of these three parameters can therefore be found using the estimates of X_K, V_K, and v_K calculated from measurements on samples. If these parameters do not depend on the rank K, the values of w, b_1, and b_2 can be refined for greater accuracy by averaging the values obtained over all measured ranks K.

3.3.1.6.3 Incomplete Model Case

If the intra-stem variance cannot be calculated due to the lack of data on adjacent branches, the system is under-determined so an additional constraint must be added. For instance, if this assumption is found in reasonable adequacy with the plant architecture, it can be assumed that $b_1 = b_2$.

Equation (3.29) becomes:

$$X_K = w \cdot K \tag{3.36}$$

Equation (3.31) becomes:

$$V_K = w \cdot K(1-b_1) + (w)^2 K(1-b_1) \tag{3.37}$$

So it gives the *Bernoulli* parameter of the development:

$$1 - b = \frac{\sum_{K=1}^{N} \frac{(X_K - w \cdot \bar{K})^2}{N}}{w(1+w)\sum_{K=1}^{N} K} \tag{3.38}$$

The coefficient w is the rhythm ratio. In that case, by setting $b_1 = b_2 \approx 1$, an average and deterministic model can replace the stochastic one with a good approximation.

3.3.1.6.4 Damping of the *Bernoulli* Parameter

On old coffee trees, we can see that the development probability of branches decreases before the mortality starts. The value of parameter b, calculated by the crown analysis, decreases when the rank K of the branch under the top increases. It is assumed that the parameter b depends on the CA i in DCs of the meristem according to the equation:

$$b(i) = b^i \tag{3.39}$$

Here we assume that whorls are made of m branches with $m \geq 2$.
Following equations have been established.
At the rank K from the top, the average number of DCs in the stem is:

$$N_K = K/b_1 \tag{3.40}$$

The mean of the phytomers in the branch at the rank K from the top is:

$$X_K = \frac{b_2(1 - (b_2)^{w \cdot N_K})}{1 - b_2} \tag{3.41}$$

The average size in phytomers of the branch reaches a limit:

$$X_{Lim} = \frac{b_2}{1 - b_2} \tag{3.42}$$

The intra-stem variance is:

$$v_K = \frac{1 - b_2}{1 + b_2}(X_K + (X_K)^2) \tag{3.43}$$

The inter-stem variance is:

$$V_K = v_K + (w^2 b^{2(w \cdot N_K + 1)}) K \frac{1 - b_1}{b_1} \tag{3.44}$$

Resolution of the system gives:

$$\begin{aligned} C &= \frac{v_K}{X_K + X_K^2} \\ b_2 &= \frac{1 - C}{1 + C} \end{aligned} \tag{3.45}$$

$$w \cdot N_K = \frac{Log\left(1 - \frac{1 - b_2}{b_2}\right)}{Log(b_2)} \tag{3.46}$$

$$1 - b_1 = \frac{K(V_K - v_K)}{\left(w \cdot N_K b^{(w \cdot N_K + 1)}\right)^2} \tag{3.47}$$

$$w = w \cdot N_k \frac{b_1}{K} \tag{3.48}$$

3.3.1.6.5 Profiles of Structures Depending on the Parameter b_2

Figure 3.10 compares the profiles of two random structures: one has a constant parameter b_2 (0.66) and the other has a damping parameter $b_2(i) = 0.95^i$. The first structure has a triangular profile and the second one has an ogive profile.

3.3.1.7 Validation of the Model on Cultivated Plants

3.3.1.7.1 Cotton Plant

In this example, the development of a population of 50 cotton stems was measured over 6 observation dates (de Reffye et al. 1988).

The statistical link between the mean and variance is linear (Fig. 3.11b), which proves the stability of the *Bernoulli* parameter b_1 during the development. We find $V = 0.18{\cdot}M$, which is obtained by Eq. (3.8). Thus, we have $1 - b_1 = 0.18$.

The development probability is therefore $b_1 = 0.82$.

At the last observation stage, the distribution of the number of phytomers per stem is adjusted to a binomial distribution (Fig. 3.11a). The observed mean and

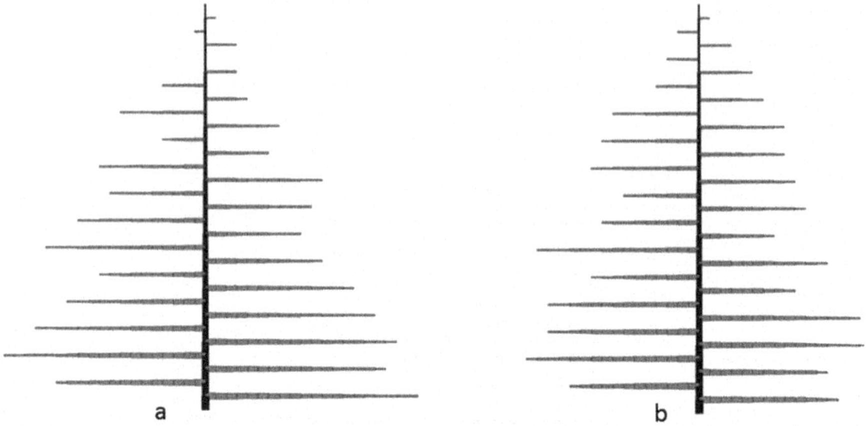

Fig. 3.10 Two structures for plants with an age of 30 development cycles (DCs). (**a**) Tree crown with $w = 1$, $b_1 = 1$, and $b_2 = 0.66$. The form of the planar structure is triangular. (**b**) Tree crown with $w = 1$, $b_l = 1$, and $b_2 = (0.95)^i$. The development of the branches is damped. The form of the planar structure is ogive-shaped

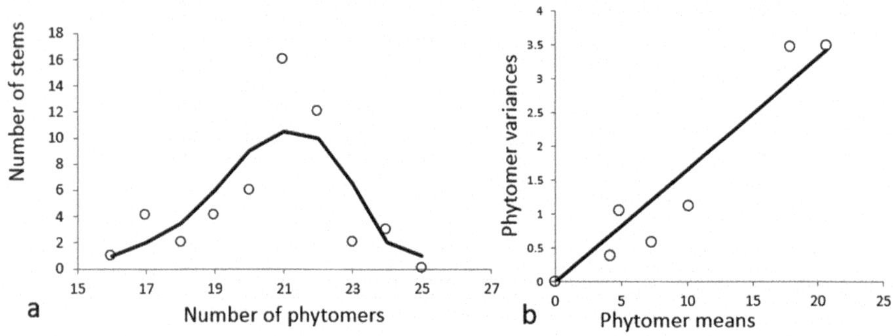

Fig. 3.11 Distribution of phytomers by stem and mean–variance relationship in cotton. (**a**) Fitting the distribution of the number of phytomers observed on 50 cotton stems with a binomial distribution. Since the sample is small, we observe a dispersion of observations (o) around the theoretical values. (**b**) Linear relationship between the mean and the variance of the 50 cotton stems at different development stages. Each of the two figures makes it possible to estimate the *Bernoulli* parameter ($b = 0.82$) (CIRAD-IRCT data, Michel Cognée)

variance are $M = 20.68$ and $V = 3.42$, respectively. The parameters of the binomial law are calculated by:

$$b_1 = 1 - V/M \text{ and } N = M/b_1, \text{ which gives } b_1 = 0.83 \text{ and } N \approx 25.$$

The test for adjusting the distribution to the binomial distribution is statistically significant, and this second method provides another estimate of b_1, close to the first.

The functioning of a population of cotton meristems is therefore characterized by a number of cycles T and a development probability b_1.

Then, by applying the crown analysis (Eq. 3.34), the parameter b_1 is recalculated in a third way, at the same time as the parameters b_2 and w related to the branches. As an example, Table 3.1 gives the number of phytomers of the branches located at the rank 10 and 11 below the top for 50 cotton stems.

The mean of the phytomers of the branches at the rank 10 from the top is:

$$\overline{X} = \frac{\sum_{i=1}^{50} X_{10,i}}{50} = \frac{3+3+4+\ldots}{50} = 3.18$$

The variance of the phytomers of the branches at the rank 10 from the top is:

$$V = \frac{\sum_{i=1}^{50} (X_{10,i} - \bar{X})^2}{50-1} = \frac{3^2+3^2+4^2-50\cdot 3.18^2\ldots}{50-1} = 0.395$$

The variance of the couple of branches at the ranks 10 and 11 from the top is:

$$v = \frac{\sum_{i=1}^{50} (X_{10,i} - X_{11,i})^2}{2\cdot 50} = \frac{(3-4)^2+(3-4)^2\cdot(4-3)^2}{2\cdot 50} = 0.32$$

We find from Eq. (3.34):

$$b_1 = 0.86,\ b_2 = 0.92,\ w = 0.29$$

This analysis can be performed on several ranks on the stem in order to increase the precision of the model (Table 3.2).

This gives finally:

$b_1 = 0.78 \pm 0.08$, $b_2 = 0.88 \pm 0.02$, and $w = 0.23 \pm 0.04$. This third value of b_1 is comparable to the first two calculated ones, which shows the good consistency of the model.

The parameter b_1 of the stem is less precise than the parameter b_2 of the branches, which is normal since there are fewer stems than branches. The developmental rhythm of the branches for this variety is nearly four times slower than that of the stem. Simulations of cotton tree crowns are shown in Fig. 3.12.

This study on cotton led to the conclusion that the stochastic development model of a branched plant based on AMAP-GreenLab model was well adapted to that species.

3.3.1.7.2 Bamboo

A case of compound mortality with development driven by the *Bernoulli* process is illustrated by the distribution of bamboo branch phytomers (Fig. 3.13). Bamboo branches stop their development beyond the rank $K = 20$ from the top of the stem. But some branches abort before reaching maturity. An aged distribution of 22 DCs is

Table 3.1 Number of phytomers per branch at the ranks 10 and 11 from the top for 50 cotton stems

	No. of stem	1	2	3	4	5	6	7	8	9	10	11	12	13	14	15	16	17	18	19	20	21	22	23	24	25
Rank 10	No. of phytomer	3	3	4	3	3	3	3	2	3	4	3	3	4	3	3	3	3	2	3	3	2	3	4	4	4
Rank 11	No. of phytomer	4	4	3	4	3	4	4	4	4	4	3	3	4	3	4	3	3	4	1	3	3	3	4	4	4
	No. of stem	26	27	28	29	30	31	32	33	34	35	36	37	38	39	40	41	42	43	44	45	46	47	48	49	50
Rank 10	No. of phytomer	3	3	3	3	3	3	4	3	3	3	3	3	4	2	2	4	4	3	3	3	3	4	3	5	4
Rank 11	No. of phytomer	4	3	3	4	2	3	4	4	3	3	4	3	5	3	3	4	4	4	3	4	3	4	4	5	4

Table 3.2 Crown analysis results on cotton plant

Rank	b_1 stem	b_2 branch	w rhythm
6	0.93	0.82	0.28
7	0.69	0.89	0.22
8	0.68	0.88	0.22
9	0.72	0.9	0.25
10	0.86	0.92	0.29
11	0.93	0.88	0.34
12	0.72	0.91	0.26
Mean	0.78	0.88	0.27
Standard deviation	0.04	0.01	0.02

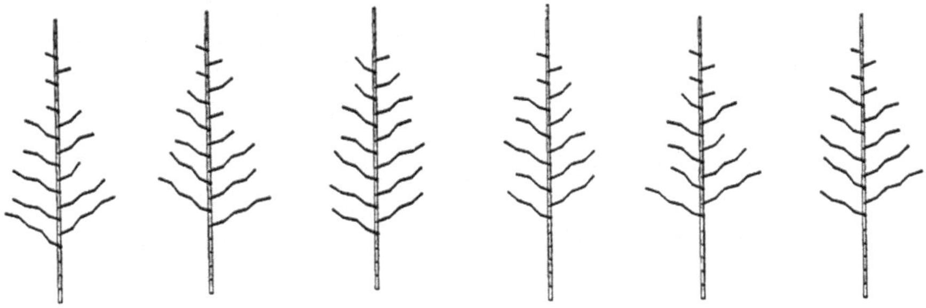

Fig. 3.12 Six stochastic simulations of cotton plant crowns (Amap software)

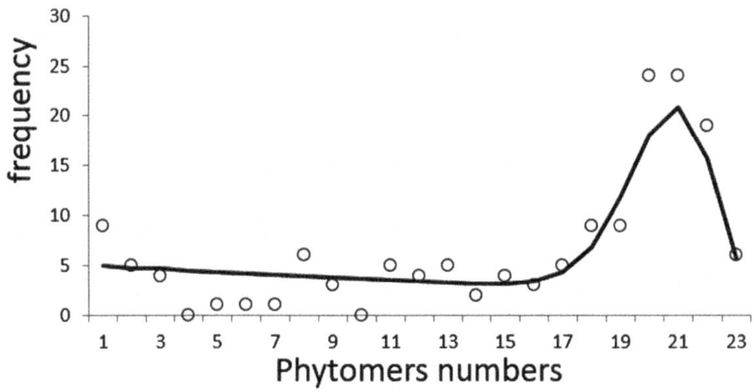

Fig. 3.13 Calculated (line) and observed (○) distribution of the number of phytomers in a sample of 150 bamboo branches

obtained, depending on the composition of a *Bernoulli* process and viability, the parameters of which are constant here (Dabadie et al. 1991).

This case is an experimental illustration of the theoretical process in Fig. 3.6. The computed parameters gives $T = 22$ for the number of DCs for branches, $b = 0.89$ for the probability of the development, and $c = 0.97$ for the viability of the meristem.

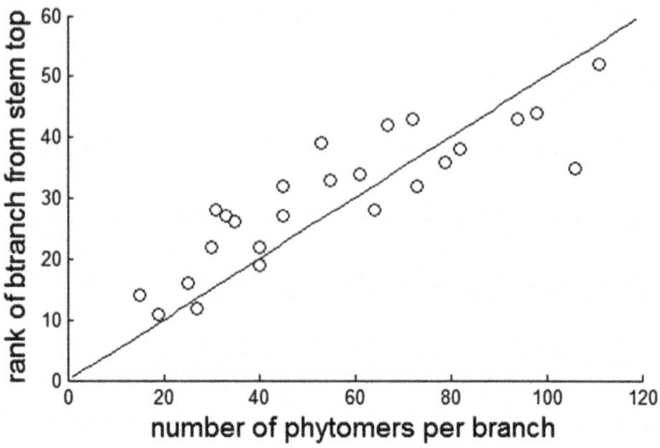

Fig. 3.14 Number of phytomers for the branch according to the rank from the top of the eucalyptus tree. The slope of the line is w assuming $b_1 = b_2$

3.3.1.7.3 Eucalyptus

The eucalyptus data concerned two trees (Diao et al. 2012). The branches were too scattered to calculate an intra-stem variance. In this case, it is necessary to assume $b_1 = b_2$ and apply Eqs. (3.36) and (3.38).

This gives: $w = 0.5$ and $b = 0.48$ (Fig. 3.14).

3.3.1.7.4 Coffee Tree (Robusta)

Coffee tree is an ideal plant to observe the mortality in the development of tree architecture (de Reffye 1979, 1981).

Coffea robusta have an orthotropic stem and two plagiotropic branches per phytomer, so $L = 0$ is placed in Eq. (3.34), which gives Eq. (3.35).

The shape of the tree crown observed in different coffee clones exhibits either triangular or ogive shape. In the latter case, the average size of the branches tends toward a limit value. The mortality of branches during development increases the proportion of dead branches compared to the branches still alive. It is important to note that in the crown analysis, one should not mix living branches with dead ones. Only measurements on living branches can be used in Eqs. (3.34) and (3.35).

The branches die after a variable life span. In general, it is observed from the top to the base of the stem that the rate of dead branches Ft changes according to a sigmoid shape as a function of time (in DC). It is linked to the viability of meristems according to Eq. (3.49).

$$C_1 = 1 - Ft(0)$$
$$C_t = \frac{1 - Ft(t-1)}{1 - Ft(1-2)} \tag{3.49}$$

The relationship between the cycle t and the rank K of phytomer on the stem bearing a branch is written as:

$$t = \frac{K \cdot w}{b_1} \tag{3.50}$$

The observation of the mortality rate of the branches therefore makes it possible to compute the viability of the meristems. We empirically choose a discretized beta law to describe Ft depending on two parameters B_a and B_b.

$$Ft(t) = \frac{\sum_{i=1}^{t} \left(\frac{i}{T}\right)^{B_a} \left(1 - \frac{i}{T}\right)^{B_b}}{\sum_{i=1}^{T} \left(\frac{i}{T}\right)^{B_a} \left(1 - \frac{i}{T}\right)^{B_b}} \tag{3.51}$$

3.3.1.7.4.1 Case of a Clone of Coffea Robusta with a Triangular Shape

In this example, 12 crowns of a 3-year-old *Coffea canephora* tree were detailed to describe the state of their branching at each rank on the stem. We can visualize three crowns found in this collection in Fig. 3.15.

The variability of the number of phytomers per branch according to their rank below the crown of the stem is displayed in Fig. 3.16a. The equations of the crown analysis give the values of the parameters of the functioning of the meristems: $b_1 = 0.85$, $b_2 = 0.79$, $w = 1.1$.

The evolution of the branching rate from the base of the stem is shown in Fig. 3.16b.

The following equation adjusts the data and gives the branch rate a_i according to the rank i from the base:

$$a_i = a_0 \cdot \left(\frac{i}{12}\right)^{aA}, \text{ with } a_0 = 0.88 \text{ and } aA = 2 \text{ if } i \leq 12,\ a_i = a_0 \text{ if } i > 12.$$

At first the branching probability increases parabolically and then stabilizes at the value a_o after the 12th rank on the stem.

Fig. 3.15 Three observed crowns of *Coffea canephora* (S. Akaffou data, Côte d'Ivoire)

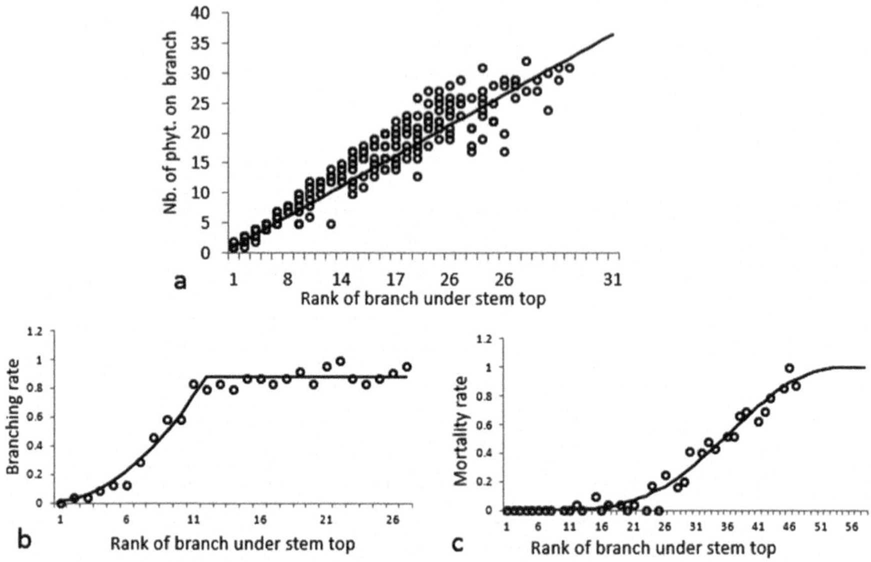

Fig. 3.16 Characteristics of the stochastic functioning (development, mortality) of the crown of a *Coffea canephora* tree (set of 12 crowns according to S. Akaffou [Côte d'Ivoire]). Observed data (○), adjusted by the stochastic model (curve). (**a**) Variability in the number of phytomers of the branches according to the rank below the top of the stem. The equation of the line that adjusts the points is $y = (w{\cdot}b_2/b_1){\cdot}x$. (**b**) Evolution of the branching rate from the base of the stem. This stabilizes at the rank 12. (**c**) Evolution of the mortality rate of branches from the top to the base of the crown

The evolution of sigmoidal mortality is given by the parameters of the beta law $T = 60$, $B_a = 5.45$, and $B_b = 3.58$ that adjust the observations made on the stems for the evolution of the mortality rate of the branches from top to bottom of the stems (Fig. 3.16c). Results in Fig. 3.16, which are based on observations, are very similar to results in Fig. 3.8, which are based on simulations.

3.3.1.7.4.2 Case of a Clone of Coffea Robusta with an Ogive Shape for the Crown

We observe the differences in the evolutions of the means and variances of the orthotropic stems and plagiotropic branches as a function of time measured in months (in humid tropical areas, the chronological and thermal times are linearly linked). We observe the linear development of the stems (Fig. 3.17a) and the damped development of the branches (Fig. 3.17c). The link between the mean and variance is linear for the stems and quadratic for the branches, as stipulated in the model.

A large number of crowns were measured (200 stems) at ranks 10, 15, 20, and 32 below the top of the stem. In cases of damping development, Eqs. (3.41)–(3.45) of the crown analysis must be used.

Table 3.3 shows the evolution as a function of the rank on the stem for the mean X of the number of phytomers of branches, the intra-stem variance v, and the inter-

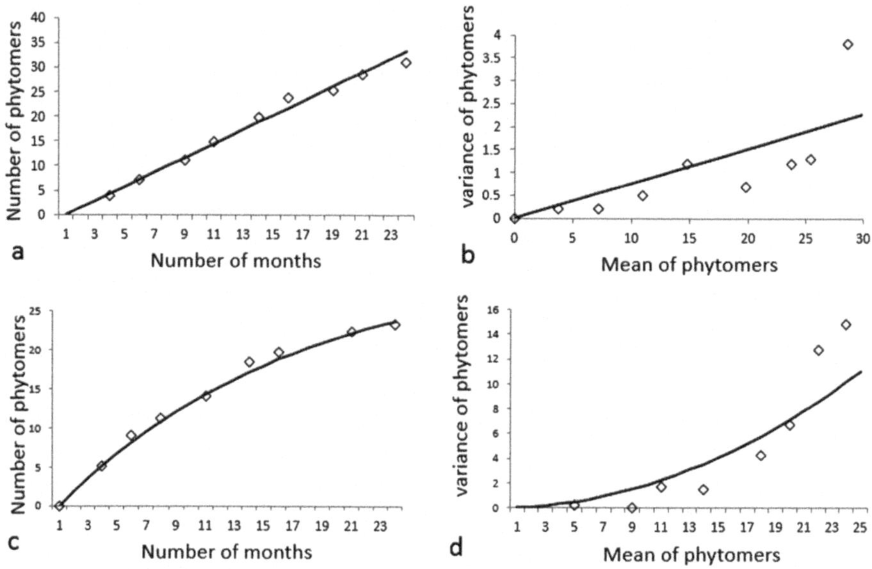

Fig. 3.17 Evolution of the number of phytomers as a function of time measured in months: (**a**) linear development of the stems and (**c**) damping development of the branches. The evolution of the variances is (**b**) linear for stems, and (**d**) quadratic for branches (average of five leafy axes)

Table 3.3 Results of the crown analysis on a Robusta with damping development

Rank	X	v	V	b_1	b_2	w
10	11.8	1.8	2.7	0.9	0.97	1.3
15	16.6	3.5	4.7	0.85	0.969	1.3
20	18.4	6.5	8.6	0.85	0.967	1.3
32	24.9	9.2	10.5	0.9	0.97	1.3

stem variance V. Eqs. (3.41)–(3.45) allow to calculate the *Bernoulli* parameter b_1 for the stems, the damping parameter b_2 for the branches, and the rhythm ratio w at each rank.

The values of the parameters are stable according to the ranks. We retain the average values: $b_1 = 0.9$, $b_2(i) = 0.97^i$, $w = 1.3$. (Note that for this coffee tree we have $w > 1$.)

Similarly, Table 3.4 shows the evolution of branch mortality for the selected ranks.

At rank K of the branches, the average age t in DCs is given by the equation: $t = \frac{K*1.3}{0.9}$

The fitting of the data by a cumulative beta law gives the parameters:

$$T = 60 \text{ DCs}, B_a = 5.24, B_b = 4.0.$$

The simulation of branch development with the composition of the *Bernoulli* process and the death process gives a proportional representation of the distributions of dead and living branches (Fig. 3.18b).

Table 3.4 Rate of branch mortality function of development cycle (DC)

Rank	DCs	% mortality
10	14	0.032
15	22	0.081
20	29	0.32
32	40	0.89
	60	1

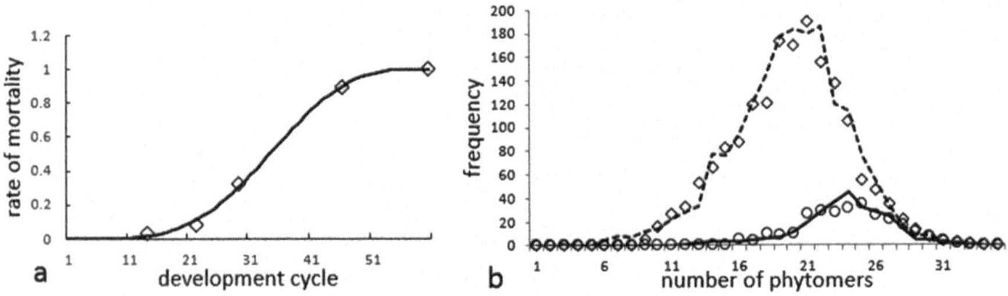

Fig. 3.18 (**a**) Evolution of the mortality rate according to the number of development cycle (DCs). (**b**) Comparison in a Robusta coffee tree between simulated and observed distributions of phytomer numbers for living (solid line and ○) and dead branches (dotted line and ◊). The population represents about 2000 branches of 47 DCs. Taking into account the *Bernoulli* process and the viability of the observed meristems makes it possible to reproduce satisfactorily, by simulation and in good proportion, the distributions of the phytomers for both dead and living axes

3.3.1.7.4.3 *Note on Coffee Tree Development*

It takes six parameters to model the stochastic development of the crowns. As observed, these parameters can be estimated by the inverse method. The coffee tree is the ideal prototype of a computational plant. These parameters characterize the functioning of the crown and therefore the production of phytomers during the development of architecture.

3.3.1.7.4.4 *Stochastic Simulation of the Coffee Tree Architecture*

The previous crown analysis provides the development parameters. Figure 3.19 shows three random structures of the same clone coming from a computer simulation aged of 60 DCs.

3.3.1.7.4.5 *Control of Phytomers Production of the Crown*

The numbers of phytomers of the stems (X_1) and branches (X_2) were counted on ten coffee stems of the same age taken at random.

This gives average values of $X_1 = 62$, $X_2 = 1069$.

A simulation of 100 trees of the same age using the parameters of the crown analysis gives:

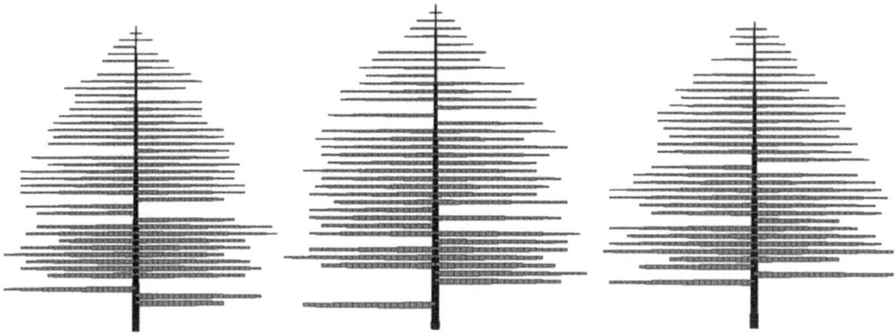

Fig. 3.19 Three stochastic simulations of a robusta clone. The laws of stochastic development of meristems are taken into account. The missing branches are due to either the laws of branching or pruning of mortality (GreenLab software)

$$X_1 = 63 \pm 5, X_2 = 1047 \pm 172.$$

There is a good agreement between the observed and simulated data.

3.3.2 Rhythmic Development: Double-Scale Axis of Development

In trees, development is most often rhythmic and the leafy axes are composed of growth units (GUs), which are composed of phytomers. The extension of these axes comes from the addition of GUs set up by apical meristems during their periods of activity. GUs are characterized by the statistical distribution of their number of phytomers. The leafy axis is of a double scale. The upper scale concerns the mortalities of the successive implementation of the GUs and the lower scale concerns the intra-GU development, i.e., the distribution of phytomers.

The GU can be considered as a kind of "metaphytomer" with its own functioning rules. The duration of a macrocycle is expressed in DC. The number of DCs is constant for macrostates, but the number of microstates that represents the phytomers of GUs can vary according to a distribution. When the formation of microstates stops, extra pauses are added to reach the constant duration of the macrocycle in DCs.

In tropical trees, such as rubber or mahogany trees, the successive production of GUs by the same meristem at the macrocycle level depends on a *Bernoulli* process (parameter b_1, Fig. 3.20). Not all meristems are active at the same time; some are active in proportion b_1 and others are in pause in proportion $1 - b_1$. As as result, some GUs are missing, leading to asynchronous development of the axes. A rhythm ratio w and viability c_1 are also defined at the GU (macrostate) level.

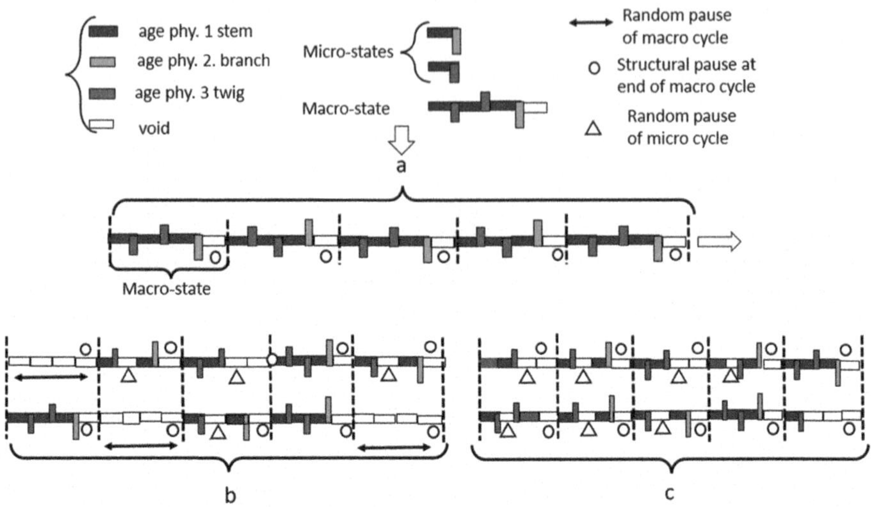

Fig. 3.20 Functioning of the double-scale rhythmic axis. (**a**) Potential axis of rhythmic development with two scales. (**b**) Two random simulations of development axes, stochastics and asynchronous growth. (**c**) Two random simulations of development axes, stochastics and synchronous growth. At the upper scale, for each macrocycle, a new macrostate (growth unit, GU) is set up at the end of the development axis with a probability b_1. If this fails, a void GU entity is obtained that is visible in the chronological representation mode. One or more terminal structural pauses (denoted as o) separate the macrocycles. At the lower scale, each macrocycle is composed of microcycles from a distribution that produces microstates (phytomers) with a probability b_2 according to a *Bernoulli* process ($b_2 = 1$ for the first microstate). In the case $b_1 = b_2 = 1$, we fall back on the deterministic case, and all entities are produced. With $b_1 < 1$, the development of the axes is asynchronous; some GUs are missing (denoted as $\leftrightarrow$); this is the case for tropical trees. With $b_1 = 1$, the development of the axes is synchronous; this is the case for temperate trees. In both cases, some phytomers of GU (denoted as Δ) are missing due to $b_2 < 1$

In temperate trees (e.g., maple), the elongation of the axes is deterministic ($b_1 = 1$, $w = 1$). Each spring, a meristem produces a new shoot annually, unless it aborts. The development of the axes is synchronous, with all meristems functioning at the same time. The stochastic part in the creation of GUs comes only from mortality and branching.

Within the GU, the number of phytomers follows a statistical distribution that is characteristic of the species and the physiological age (PA) of the axis. Each macrocycle consists of DCs that can form phytomers with a probability b_2 (Fig. 3.20). The duration of meristem functioning can vary according to a distribution. The number of phytomers produced by GUs of the same chronological and physiological age follows a compound law (de Reffye et al. 2012).

By fixing $b_1 = b_2 = 1$, we obtain a deterministic structure that contains all possible stochastic realizations (Fig. 3.21) and visualizes the potential structure.

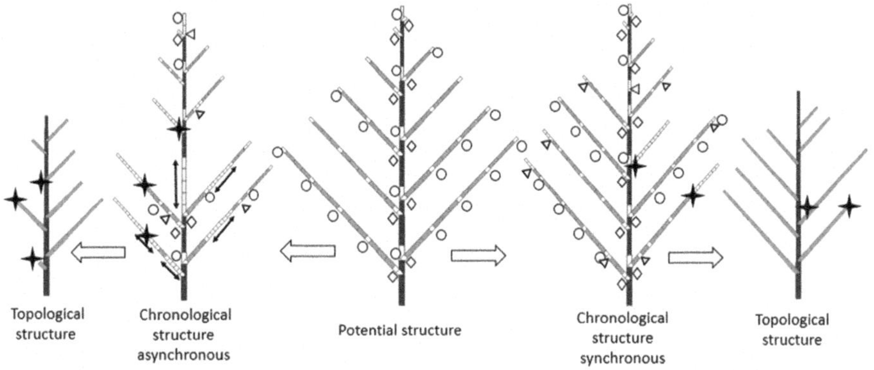

Fig. 3.21 Simulated stochastic structures and the potential structure that contains all possible growth units (GUs) and their feasible macro and microstates. Pauses denoted as o are structural and delimit the GUs. Pauses denoted as ◊ are structural and induce a delay in branching. Pauses denoted as ↔ are random and suppress macrostates. Pauses denoted as Δ are random and suppress microstates. Meristem abortion is denoted as +. When $b_1 = 1$, the elongation of the axes is synchronous (in the case of temperate trees). All GUs of an axis are present unless meristem abortion (denoted as +). When $b_l < 1$, the elongation of the axes is asynchronous (in the case of tropical trees). Some GUs are missing (denoted as ↔). As a result, the plant crowns are more irregular. In both cases, some phytomers (denoted as Δ) are missing within the GUs due to $b_2 < 1$

3.3.2.1 Crown Analysis of the Rubber Tree

Often in the case of tropical trees, the axes are not synchronized and the same meristem can produce a variable number of GUs (up to five) annually. For the same developmental age, the average number of GUs produced and the associated variance characterizes the *Bernoulli* process of parameter b. If the mean–variance relationship is linear, then b is constant. Similarly, the mortality rate of branches as a function of age in macrocycles characterizes the viability of GUs (parameter c). In addition, between the trunk and the branches, the GU development rate of the axes (w) may differ. The study of rubber trees (Costes 1990) made it possible to estimate the values of these parameters in rubber trees.

At the upper scale of the GU series on the axes, the crown analysis for rubber tree applies, which gives for the whole trunk branches: $b_1 = 0.90$, $b_2 = 0.57$, $w = 0.9$. Meristem mortality is zero below five GUs per branch, and beyond that, viability takes a constant value (Fig. 3.22). The newly obtained value of b_2 is close to that obtained by the mean–variance link, which shows the consistency of the model.

3.3.2.2 Crown Analysis of the Symphonia Tree

The *Symphonia* is a tropical tree. Its plagiotropic primary branches consist of asynchronous GUs that carry one or two secondary branches. The *Bernoulli*

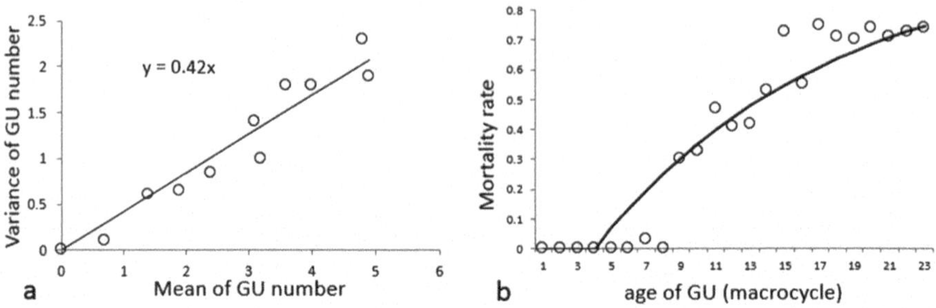

Fig. 3.22 Functioning of asynchronous axes at the growth unit (GU) level: example of rubber tree (from Évelyne Costes 1990). The computed data are indicated by the solid curve, and the observed data are denoted as ○. (**a**) Mean–variance link of the number of GUs per branch as a function of age. The trend is linear in the observed sample. The slope of the line gives the probability of development of the GUs: $b_2 = 0.58$. (**b**) Observed in a sample, the mortality rates of the branches according to their chronological age. Mortality is zero until age 4. Beyond that, the viability takes a constant value; the adjustment gives a viability $c_2 = 0.92$. The cumulative death function is expressed as: $F(i) = 0$, if $i \leq 4$ and $F(i) = 1 - 0.92^{i\text{-}4}$, if $i > 4$

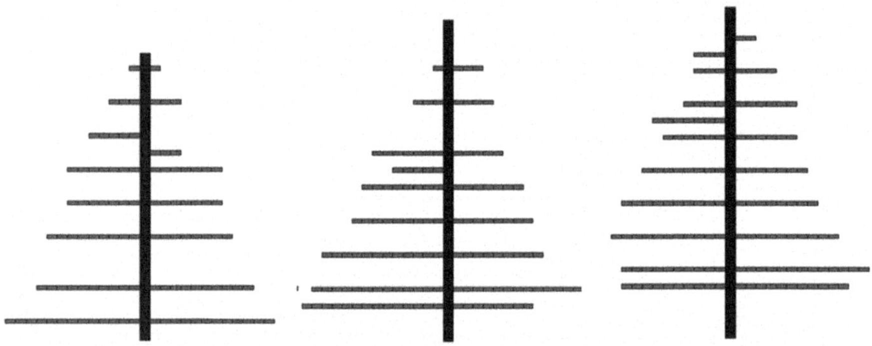

Fig. 3.23 Details of three primary branches measured on a Symphonia tree out of a total of 11. There are two secondary branches per phytomer of main axis. Absences are due to the branching rate (according to Thomas Mangenet's thesis, CIRAD AMAP)

processes and the rhythm ratio are identifiable. Three measured branches are shown in Fig. 3.23 (T. Mangenet: personal communication).

The equations of the crown analysis give the parameter values of *Bernoulli* process: $b_1 = 0.86$, $b_2 = 0.84$, rhythm ratio $w = 0.90$, and branching rate $a = 0.59$.

The stochastic development expressed in GUs of the branches depends here only on four parameters.

3.3.2.3 Statistical Distributions of the Number of Phytomers in GU

In all studied cases, the number of phytomers per GU follows a unimodal or bimodal distribution. The unimodal case corresponds to the elongation of a preformed part of

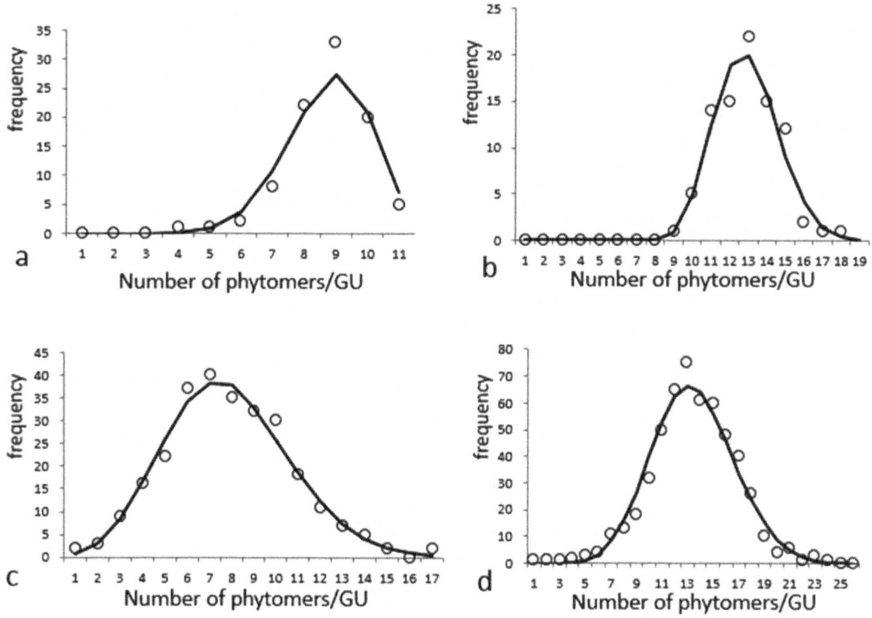

Fig. 3.24 Examples of preformed distributions of phytomers in growth unit (GUs) resulting from the functioning of meristems from the same tree. The computed data are indicated by the solid curve, and the observed data are denoted as ○. These laws are characteristic of a renewal process over a fixed period with a phyllochron (time between the elongation of two successive phytomers) characterized by its mean and variance. All these distributions are adjusted to binomial laws with parameters N, b. The term N refers to the development time of the GU in DCs and the term b refers to the probability of producing a phytomer at each DC (*Bernoulli* process). (**a**) GU of beech, binomial law $B(11, 0.7)$ (Nicolini 1998). (**b**) GU of walnut, binomial law $B(17, 0.63)$ (Sabatier et al. 1998). (**c**) GU of litchi, binomial law $B(17, 0.23)$ (Costes 1988). (**d**) GU of rubber, binomial law $B(23, 0.5)$ (de Reffye et al. 1991)

the GU in the meristem (beech, chestnut, maple tree). The bimodal distributions (poplar, cherry, apricot tree) always come from a preformation and a possible and immediate neoformation, corresponding to the resumption of meristem functioning, after the elongation of the preformed part. The cessation of neoformation development corresponds either to a pause of the meristem in trees with monopodial branches (cherry tree) or to an abortion of the meristem in trees with sub-terminal sympodial branches (apricot tree).

Figure 3.24 shows four unimodal distributions of the number of phytomers corresponding to preformed parts (beech, walnut, rubber, litchi, etc.). These distributions are characteristic of the species in question and adjust well to positive binomial laws.

Figure 3.25 shows bimodal distributions of number of phytomers per GU corresponding to neoformed parts that are added to preformed parts at a certain rate (illustrated for cherry, apricot, poplar, cocoa tree). Meristems have a grouped cessation law of functioning during preformation, but the following neoformation

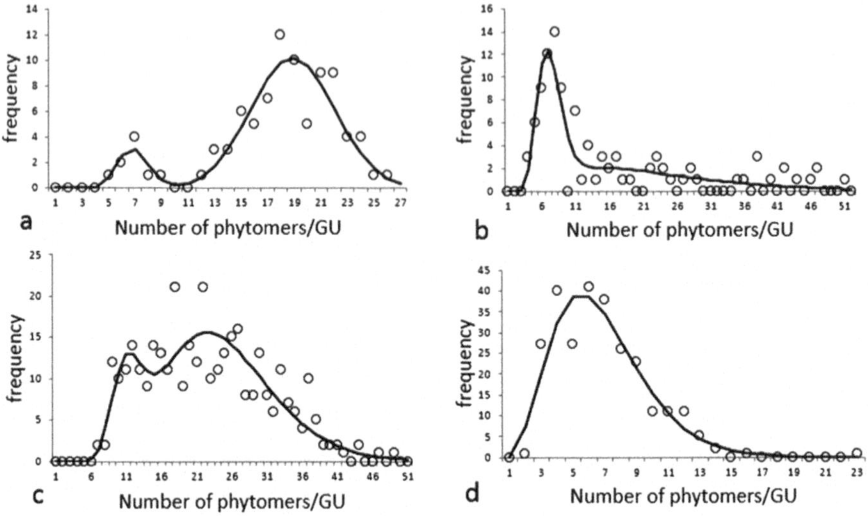

Fig. 3.25 Bimodal distributions of the number of phytomers of pre- and neoformed growth units (GUs) on the branches of four trees species. The computed data are indicated by the solid curve, the observed data are noted ○. (**a**) Cherry GU $B(12, 0.45)$, $a = 0.85$, $B(20, 0.63)$ (Fournier 1989; de Reffye et al. 1991). (**b**) Apricot GU, $B(12, 0.46)$, $a = 0.48$, $BN(2, 0.88)$ (Chavaneau 1989; de Reffye et al. 1991). (**c**) Poplar GU $B(12, 0.68)$, a = 0.86, $BN(5, 0.76)$ (after H. Rey) (de Reffye et al. 1991). (**d**) Cocoa GU, parameters $B(3, 0.72)$, $a = 1$, $BN(2, 0.66)$ (Mialet 1990)

cessation law is scattered; it is always well adjusted to a positive or negative binomial law. These bimodal laws are specific to a tree species and define a signature. Depending on the PA of the meristem, the neoformed part can be always present (the trunk) or absent (twigs). The neoformed proportion decreases with increasing PA.

Generally, if the preformation does not exceed 15 phytomers, the neoformation can exceed 50 phytomers in the apricot tree (Fig. 3.25b). The peak of preformation in the bimodal distribution is most visible in the cocoa tree (Fig. 3.25d), as few GUs produce a neoformed part. The older the PA, the lower the proportion of neoformation. The configuration of GUs with neoformation requires five parameters (N_1, b_1, a, N_2, b_2) that refer to the binomial laws (N, b) of the pre- and neoformed parts and the rate a of the neoformation. We denote B as the case of a positive binomial and BN as the case of a negative binomial. These distributions are characteristic of a tree species.

3.3.2.4 Distributions of Phytomers According to Physiological Age

In the studied trees (elm, poplar, maple, apricot, cherry, etc.), phytomers distributions per GU are good markers of PA. The physiologically oldest (short) shoots only make preformation, whereas the trunk GUs always make neoformation. The

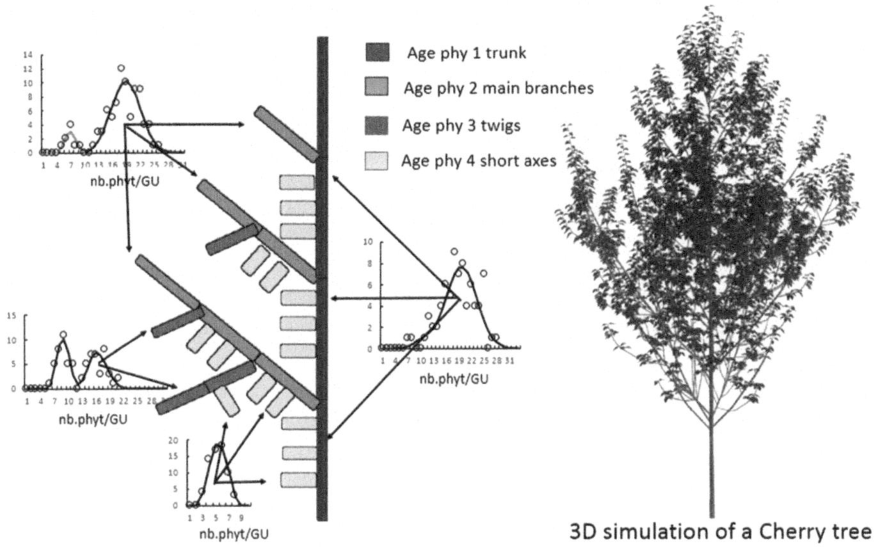

Fig. 3.26 Simulation of the architecture of a cherry tree (Fournier 1989) (Amap software, de Reffye et al. 1991). The distributions of phytomers per growth unit (GU) for each physiological age (PA) are stochastic. Depending on the order of branching or PA, the proportion of neoformed GUs changes. All the GUs in the trunk make neoformation, whereas short axes do not. The main branches of order 2 make neoformation 90% of the time and those of order 3 make 50% of the time

branches express neoformation, according to their vigor and associated with their PA. Generally speaking, as the PA increases, the neoformation decreases, along with other characteristics such as polycyclism and anticipated twigs.

The case of cherry tree is a good example of this behavior. Preformed distributions are stable from one PA to another, while the proportion of GU with the neoformation changes (Fig. 3.26).

3.4 Simulated Structure and Potential Structure

3.4.1 Definition of the Potential Structure

A computer program that runs the GreenLab model allows to simulate plant structures with continuous or rhythmic development. The importance of pauses in the development of the structure, whether of structural origin (rhythms, developmental delays) or stochastic origin, has been highlighted.

The numerical simulation of a random tree builds cohorts of phytomers created by meristems synchronized in their functioning. At each time step, the condition of a meristem is tested against a random variable to determine if it sets up a new phytomer, a pause, an abortion, etc. A pause corresponds to a void botanical entity that cannot branch, but may give a successive phytomer according to the *Bernoulli* process. The abortion of an axillary bud results in the loss of a branch from the structure. The stochastic simulated structure explores only part of the potentially feasible structure, i.e., the structure that would be generated if all meristems functioned normally, without pause (except for structural ones) or abortion. The "potential structure" is a chronological structure that materializes all the possible achievements of the axes generated by the meristems. Theoretically, this structure can be huge, but its elements are countable. A stochastic structure simulated in chronological mode is therefore a particular realization within the potential structure. All possible simulated structures (which can be very numerous) are part of the potential structure. The notion of potential structure is an essential tool for applying statistical methods to the study of plant architecture.

3.4.2 Example of a Potential Structure

Consider a structure with only one PA, where number of axillary buds per phytomer is constant and equals to m, and all meristems are synchronized with a rhythm of 1. The probabilities of meristem functioning are defined: the *Bernoulli* process develops the axes with a parameter b, the meristems keep alive with a viability parameter c, and the axillary buds have an immediate branch rate with parameter a. These functioning probabilities compound to determine the probability of existence for each phytomer within the potential structure. A phytomer can give two types of filiation (Fig. 3.27):

- By apical extension: Along an axis, the probability of existence of a phytomer, according to its rank k, is given by $P(k) = b \cdot c^k$. Indeed, in the absence of mortality, the *Bernoulli* process makes all the phytomers of the axis equally probable (with probability b). Concerning survival, the viability factor is multiplied in a chain (with probability c^k).
- By branching: For an axis, if its probability of branching is a, then the probability of existence of a phytomer on this axis, according to its rank k, is given by $P(k) = Pp \cdot a \cdot b \cdot c^k$, where Pp is the probability of existence of the phytomer on the carrier axis. Indeed, the branched axis can only exist if the carrier phytomer exists.

From the base of the tree, by exploring the entire potential structure (in a prefixed order, for example) and applying the rules of phytomer filiation, we can, by propagation, give the probabilities of existence for each phytomer within the potential structure. These probabilities may not be constant: they can be indexed according to various criteria (physiological, chronological, ontogenic ages). Figure 3.27 shows the potential structure generated by identical and synchronized meristem cohorts whose probabilities of existence of phytomers are calculated with the development, branching, and viability parameters a, b, and c, respectively.

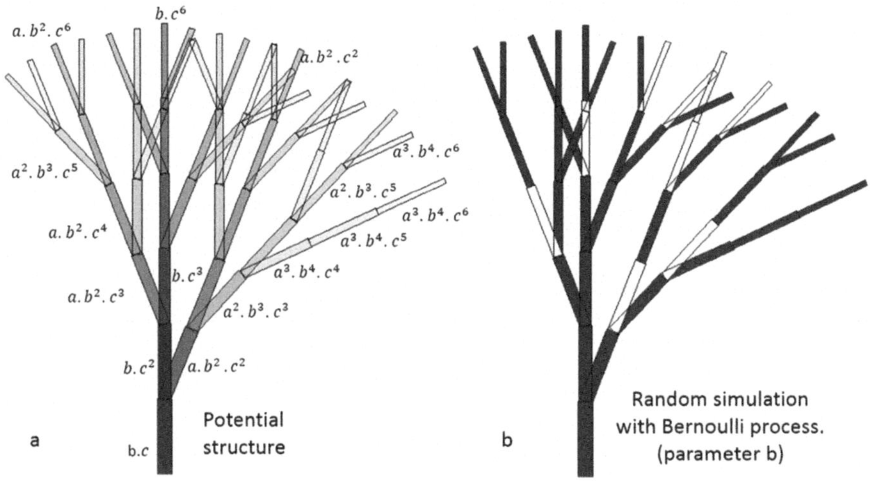

Fig. 3.27 Potential structure (**a**) with six development cycles (DCs) and random simulation (**b**) with a *Bernoulli* process. The existence of phytomers depends on the parameters a (branching), b (*Bernoulli*), and c (viability) of the development. (**a**) In the potential structure, all phytomers are expressed; their coloration here is proportional to the probability of existence. (**b**) A particular random realization (simulation) of the potential structure ($a = c = 1$, $b < 1$)

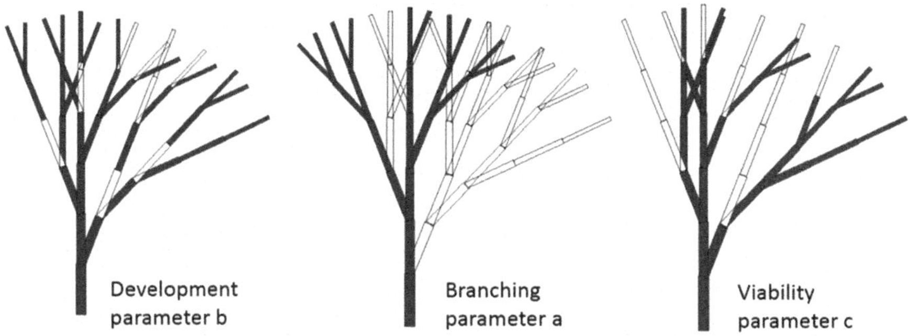

Fig. 3.28 Representation of stochastic simulations in chronological mode with only one parameter taken into account from left to right: pause (b), branching (a), and viability (c)

The effect of each parameter is specific to the structure (Fig. 3.28). The *Bernoulli* process (parameter b) inserts pauses along the axes. The branching probability (parameter a) removes lateral structures. Viability (parameter c) stops the development of the axes.

Figure 3.29 shows two simulated structures, both being realizations of the potential structure in Fig. 3.27a. The sum of the probabilities of existence of the phytomers of the potential structure for the probabilities $a = 0.7$, $b = 0.8$, $c = 0.9$ chosen is $Nt = 16.8$, which is the theoretical average of the number of phytomers produced. The average of 200 random simulations gives $Ns = 17.4$ phytomers,

Fig. 3.29 Two representations of stochastic simulations in chronological and topological mode

which is in good agreement with the value of Nt. The elimination of pauses gives the topological structures performed, the only ones actually observable.

If a large number of stochastic simulations are performed in chronological mode, they completely fill the potential structure. By observing how frequently a location within the potential structure is occupied by a phytomer, we can determine that this frequency will tend on average toward $P(x)$, where $P(x)$ is the probability of existence of the specific phytomer x in the structure. Void entities must be removed to visualize the resulting topological structure (Fig. 3.29). The mathematical expectation of the number of phytomers produced corresponds to the sum of the probabilities of existence of the phytomers within the structure. In the following chapters, we will see that the potential structure is an important tool for preparing data assimilation to be collected in the field in order to estimate developmental and growth parameters.

3.5 Equations and Operators of the Development

When we simulate the development of a plant structure up to a given age, one of the expected results is the enumeration of cohorts of produced phytomers, as well as those of their different types of organs, at all stages of development, and according to their physiological ages. In general, when a topological plant structure is generated by an algebraic grammar, according to simulation rules based on the relationships between the botanical entities, it is described as a sequence of symbols and can be quite extensive. The structure contains all created elements and allows to count them. If the counting is based on the sequential exploration of the entire structure, the required time will be proportional to the number of produced phytomers. The method quickly reaches its limits, because the computational time can increase exponentially with the stage of development (millions of phytomers in large trees!).

The ideal solution would be to bypass this limitation by finding mathematical equations or algorithms that compute the numbers of organs produced at each time step according to the parameters of development, without the need for the simulated topological structure. The calculation time would then no longer depend on the number of organs but on the development time T of the plant structure in DCs.

Finding the equations that calculate the number of organs produced has not drawn particular attention in the field of plant simulation, probably because the virtual image was the main objective, and the plant architectures were of reduced size. The binary trees of Catalan (1838) are an early, ancient example. More recently, it is worthy to mention work of Franquin (1972) who made a theoretical study of combinatorial nature of the production of phytomers in the cotton plant. The case of combinatorial trees has been discussed in depth by Janey (1992). These trees, which can be quite complex, are treated using matrix calculus, and the simulations show that they differ significantly from botanical trees.

In cases of simple structures, it is possible to analytically calculate the mathematical expectations of the numbers of phytomers in chronological and physiological age cohorts, as a function of the number of development cycles (DCs). This analytical formulation allows avoiding simulation and step-by-step exploration of the topological structure. But expressing the production of phytomers by an equation (depending on the number of DCs) seems only possible if the development organization is simple.

In complex cases of development, matrix operators, deduced from the development schedule and functioning by recurrence, make it possible to calculate very efficiently the production of phytomers, i.e., organogenesis, as a function of the chronological age (CA) of the plant. This is called "matrix mode", referring to the use of matrix computation operators. The other mode is called the "list mode," which explicitly enumerates all the phytomers of the structure. Matrix development operators include preformation, neoformation, polycyclism, and reiteration.

3.5.1 The Binary Trees of Catalan: A Precursor Model

Catalan binary trees are stochastic trees whose number of possible structures can be computed for the same number of leaves produced at the end of the axes. A simple equation gives the number $Nf(n)$ as a function of the number of leaves n.

$$Nf(n) = \frac{\binom{2n}{n}}{n+1} \tag{3.52}$$

The numbers obtained are known as the "Catalan numbers". Thus, the following sequence: 1, 1, 1, 2, 2, 5, 14, 42, 132, 429, 1430, 4862, . . ., represents these numbers for n ranging from 1 to 10. For four leaves, there are five possibilities displayed (Fig. 3.30). The number of structures quickly becomes very large.

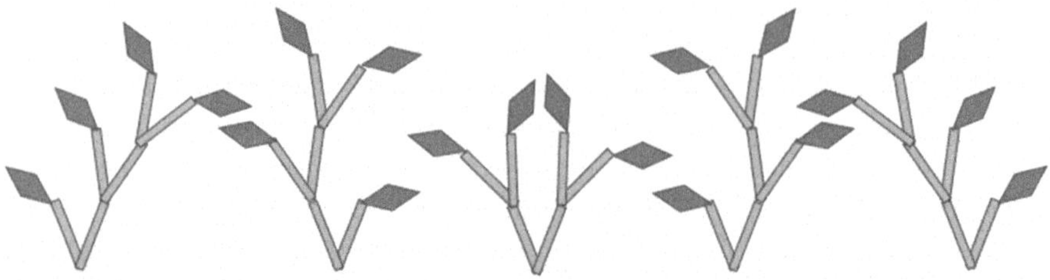

Fig. 3.30 The five binary Catalan trees obtained for four terminal leaves. However, it should be noted that these trees do not come from the same chronological age (CA) (here two and three development cycles (DCs)) and that the leaves belong to different cohorts. So they are not representative of botanical trees

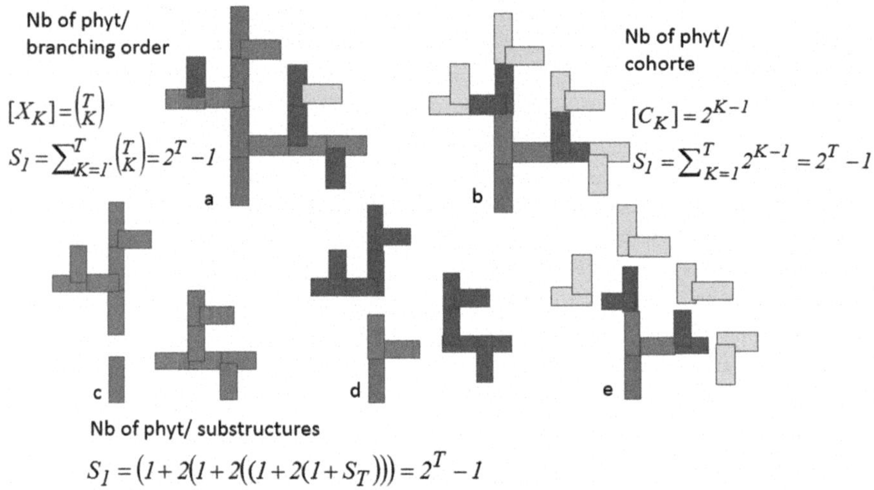

Fig. 3.31 Different possibilities of counting entities produced in a binary structure. Each color represents (**a**) branching order, (**b**) cohort, and (**c**, **d**, **e**) sub-structure

3.5.2 Considerations on Binary Trees

A binary tree at the T cycle of its development produces $2^T - 1$ elements. There are different ways to count these elements—in branching orders, in cohorts, in sub-structures—and can be adapted according to the point of view considered (Fig. 3.31).

By applying the equation from Fig. 3.31, we find for $T = 4$, total entities equal to 15, that is computed from $2^4 - 1$.

In order of branching K, X_K is the number of phytomers in axes:

$$[X_K] = \binom{T}{K}$$
$$S_1 = \sum_{K=1}^{T} \binom{T}{K} = 2^T - 1 \quad (3.53)$$

$X_1 = 4, X_2 = 6, X_3 = 4, X_4 = 1.$

In sorting cohorts at the cycle K, C_K is the number of phytomers in a cohort:

$$[C_K] = 2^{K-1}$$
$$S_1 = \sum_{K=1}^{T} 2^{K-1} = 2^T - 1 \quad (3.54)$$

$C_1 = 1, C_2 = 2, C_3 = 4, C_4 = 8.$

In counting sub-structures by factorization, like Russian dolls, S_K is the number of phytomers in the structure K:

$$S_1 = (1 + 2 \cdot (1 + 2 \cdot (1 + 2 \cdot (1 + S_T)))) = 2^T - 1$$

$S_1 = 15$, $S_2 = 7$, $S_3 = 3$, and $S_4 = 1$ with $S_1 = 1 + 2S_2$, $S_2 = 1 + 2S_3$, and $S_3 = 1 + 2S_4$, so we have: $S_1 = (1 + 2(1 + 2(1 + 2S_4))) = 15$.

The sub-structure method is systematically used in GreenLab model for counting phytomers because of its efficiency.

3.5.3 Example of Computing Simple Botanical Structures

In this case it is possible to find mathematical equations that give the number of phytomers of the structure depending on the age T in DC and the parameters of development.

3.5.3.1 Single-Stemmed Plant Model

This is the simplest case of the unbranched plant (Fig. 3.32a, e.g., palm tree). The mathematical expectation of the total number of phytomers X_T produced from cycle 1 to T depends on the probabilities b (*Bernoulli* process) and c (viability). The computation gives:

$$E[X_T] = b \cdot c(1 - c^T)/(1 - c) \quad (3.55)$$

(See also paragraph 3.1.3)

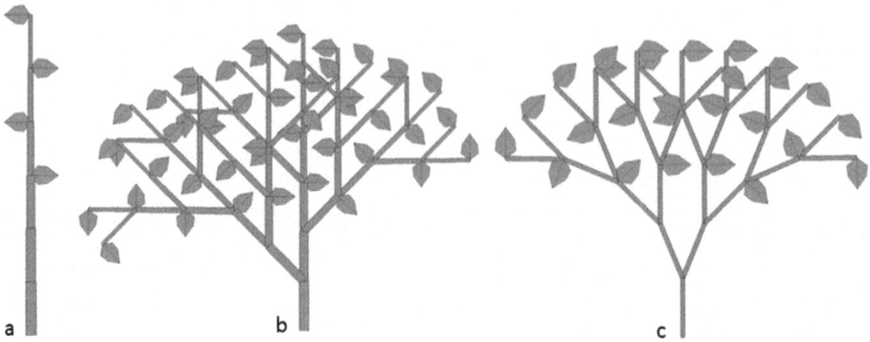

Fig. 3.32 Three simple botanical structures: (**a**) single-stemmed plant, (**b**) monopodial plant, and (**c**) sympodial plant

3.5.3.2 Monopodial Plant Model Limited to a Single Physiological Age

This architectural model corresponds to the plant described in Fig. 3.32b (e.g., coffee tree). K is here referred to the order of branching ($K = 1$ for the trunk). For the purposes of the computation, it is assumed that all axes are reiterations and therefore share the same values of the parameters a, b, and c defined above, regardless of their branching orders. Parameter m is the number of branches for each phytomer. Assuming that the order of branching is unlimited, the following equation, similar to a generating function, gives the expectation of the total number of phytomers produced at the cycle T:

$$E[X_N] = \frac{b \cdot c((c(1 + m \cdot a \cdot b))^T - 1)}{c(1 + m \cdot a \cdot b) - 1} \tag{3.56}$$

The expectation of the number of phytomers produced at the cycle T in the order K, noted $X_T(K)$, is given by the equation:

$$E[X_T(K)] = \frac{b \cdot c(m \cdot a \cdot b \cdot c)^K}{K!} \frac{d^{(K)}}{dc^K} c^K (1 - c^{T-K}) \tag{3.57}$$

The expression of the K-order derivative of the function $c^K \cdot (1 - c^{N-K})$ can develop as:

$$E[X_T(K)] = b \cdot c(m \cdot a \cdot b \cdot c)^{K-1} \left(\sum_{t=0}^{T-K} \binom{K - 1 + t}{K - 1} c^t \right) \tag{3.58}$$

It is interesting to compare the numbers of phytomers calculated by Eq. (3.57) with the averages obtained by computer simulations of the Monte Carlo method using random numbers, as shown in Table 3.5.

Table 3.5 Comparison between the calculated numbers of phytomers on the order K ($K = 1, 2, 3, 4$) (mathematical expectations calculated according to Eq. (3.57)) and the average numbers of phytomers from 200 random simulations at the age 4 DCs with the parameters of m, a, b and c

Simulated trees 200	Order 1	Order 2	Order 3	Order 4	Values m, a, b, c
Theory	1.65	1.42	0.6	0.1	2, 0.5, 0.7, 0.8
Simulations	1.75	1.44	0.67	0.09	

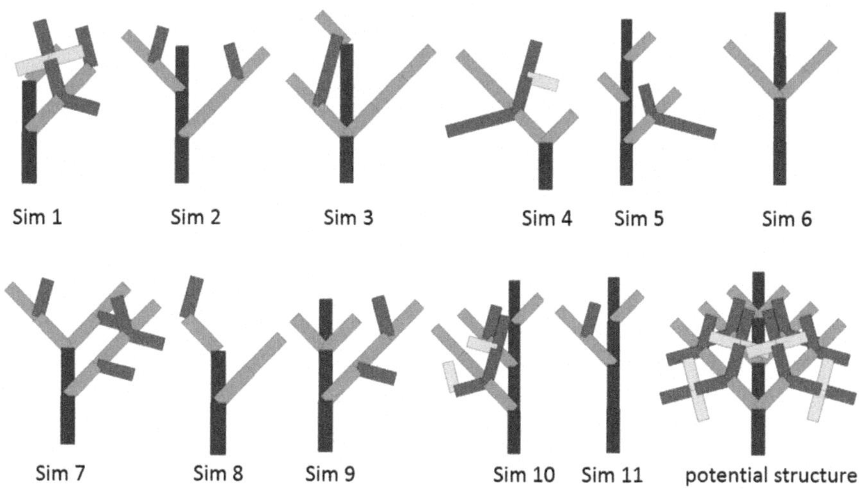

Fig. 3.33 Eleven simulations of a continuous monopodial tree. Parameters for all physiological ages (PAs): $m = 2$, $a = 0.5$, $b = 0.7$, $c = 0.8$. They are all contained in the potential structure

This table shows the average total number of phytomers computed for a sample of 200 random trees in the order 4 of branching and the age 4 DC for numerical values of m, a, b, and c. The results of these random simulations are consistent with the mathematical expectations calculated according to Eq. (3.57).

The high variability of the simulated random trees is illustrated in Fig. 3.33.

3.5.3.3 Monopodial Plant Model with Two Physiological Ages

In this case, there is a supporting structure of PA 1 that sets up branches of PA 2 (e.g., cotton plant). Both systems are reiterated and have their own developmental parameters m_1, a_1, b_1, c_1 for the first system, and m_2, a_2, b_2, c_2 for the second system. The previous enumeration equation, Eq. (3.59), applies to two PAs. At the T cycle of CA, the expectation $E[C_T(1)]$ of the number of phytomers of the PA 1 in the cohort T is written as:

$$E[C_T(1)] = \frac{b_1c_1((c_1(1+m_1a_1b_1))^T - 1)}{c_1(1+m_1a_1b_1) - 1} \tag{3.59}$$

The two sets of parameters contribute to the determination of the number of phytomers produced in the cohort $C_T(2)$ of PA 2.

$$E[C_T(2)] = (m_2a_2b_2c_2)b_1c_1 \cdot \left(\frac{(c_1 + m_1a_1b_1c_1)^T - (c_2 + m_2a_2b_2c_2)^T}{(c_1 + m_1a_1b_1c_1) - (c_2 + m_2a_2b_2c_2)} - \frac{(c_1 + m_1a_1b_1c_1)^T - 1}{(c_1 + m_1a_1b_1c_1) - 1} \right) \tag{3.60}$$

If the numerator and denominator are both null, the value $CT(2)$ is found by applying the Hospital rule.

In a first example, by setting all probabilities to 1 and choosing $m_1 = 2$ and $m_2 = 1$, we count the phytomers produced at the age 4 for both PAs. The result can be checked directly in Fig. 3.34a, b:

$$E[C_4(1)] = \frac{(3^4 - 1)}{3 - 1} = 40$$

$$E[C_4(2)] = \frac{(3^4 - 2^4)}{3 - 2} - 40 = 25$$

In a second example (Fig. 3.35) without reiteration, by setting the probabilities $b_1 = 0.8$ and $c_1 = 1$, and $a_2 = 1$, $b_2 = 0.8$, $c_2 = 0.9$ with $m_2 = 1$, we count the

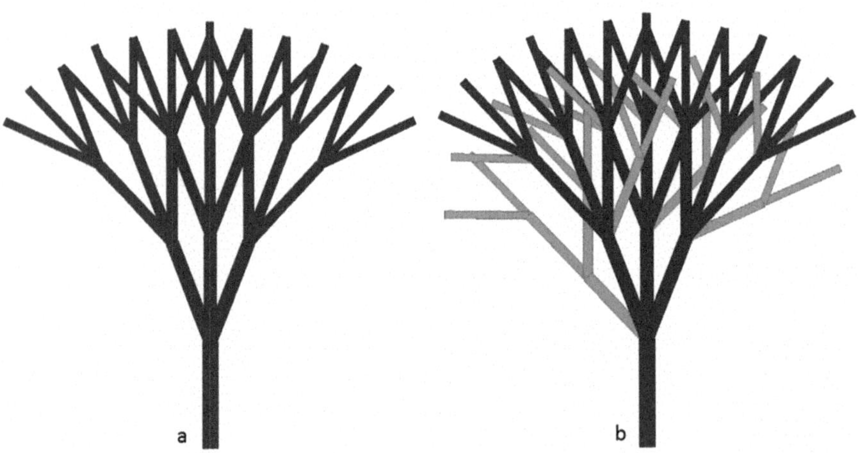

Fig. 3.34 Botanical structures of simple trees calculable by combinatorial equations depending on chronological age (CA). (**a**) With a single physiological age (PA), the number of axillaries per node $m_1 = 2$ and the probabilities associated with the functioning of meristems a_1, b_1, $c_1 = 1$. (**b**) With two PAs, the number of axillaries per node $m_1 = 2$ and $m_2 = 1$ and the probabilities associated with the functioning of meristems a_1, b_1, c_1 and a_2, b_2, $c_2 = 1$

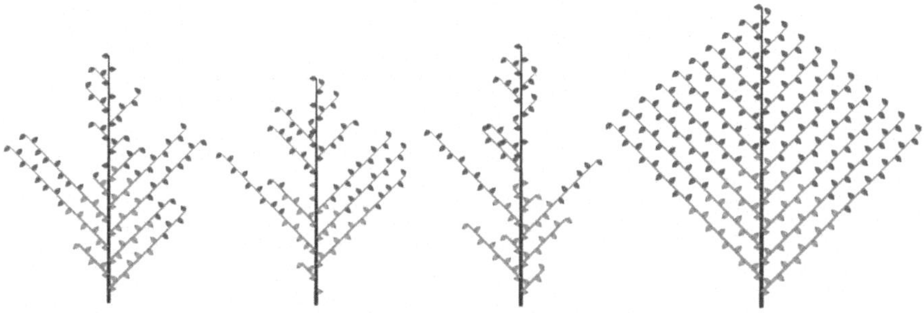

Fig. 3.35 With two PAs, the number of axillaries per node $m_1 = 0$ (no reiterations) and $m_2 = 1$ is determined and the probabilities associated with the functioning of the meristems $a_1 = 0$, $b_1 = 0.8$, $c_1 = 1$ and $a_2 = 1$, $b_2 = 0.8$, $c_2 = 0.9$. Three simulations and the potential structure are displayed

phytomers produced at the age 20 for the two PAs for 30 stochastic simulations, whose mean is compared to the production of the potential structure. Due to mortality, all branches eventually stop their development more or less late.

For stochastic simulations, we have: $E[X_{20}(1)] = 16.3$ and $E[X_{20}(2)] = 64.4$.

For the potential structure, we have: $E[X_{20}(1)] = 16.2$ and $E[X_{20}(2)] = 65.8$.

We check that there is an agreement between the calculation and the simulations.

3.5.3.4 Another Monopodial Plant Model with Mortality Depending of Ontogenic Age

Here, we are interested in the number of living axes that remain on the plant after pruning the axes, when they have reached a fixed maximum size and subsequently die. It is assumed that the life span cycle of the axes in the branching order K, denoted as t_K, is constant. Moreover, let us set: $t_1 > t_2 > t_3 > t_4$. The value t_1 is defined as the maximum age of the tree in DCs. Moreover, in the case of the ontogenic mode, all the meristems of the same branch abort together at the cycle when a second-order branch reaches its maximum size. Viability is no longer stochastic here: the parameter c is replaced by the notion of a fixed lifespan. The parameters m, a, and b are kept.

The following equations are valid for plant age T, where $t_2 < T < t_1$, and for trees that have m axillary per node:

The number of new phytomers that appear in the cycle T for each branching order is:

Table 3.6 Tree production at the age of 25th development cycle (DC) with finite development axes: the number of phytomers is given by Eqs. (3.61) and (3.62)

Branching order	Number of new phytomers at the age of 25th DC	Number of cumulated phytomers
Order 1	1	25
Order 2	12	78
Order 3	45	125
Order 4	58	82
Total	116	310

$$
\begin{aligned}
E[X_T(1)] &= b\binom{T}{0}(=b\cdot T)\\
E[X_T(2)] &= m\cdot a\cdot b^2\binom{t_2}{1}\\
E[X_T(3)] &= m^2\cdot a^2\cdot b^3\left(\binom{t_3+1}{2}+(t_2-t_3-1)\binom{t_3}{1}\right)\\
E[X_T(4)] &= m^3\cdot a^3\cdot b^4\left(\begin{array}{c}\binom{t_4+2}{3}+(t_2-t_4-2)\binom{t_4+1}{2}+\ldots\\ \frac{(t_3-t_4-1)\cdot(t_3-t_4)}{2}\binom{t_4}{1}+(t_2-t_3-1)(t_3-t_4-1)\binom{t_4}{1}\end{array}\right)
\end{aligned}
\tag{3.61}
$$

Total number of living phytomers produced per branching order, at the cycle T, is:

$$
\begin{aligned}
E[SumX_T(1)] &= b\binom{T}{1}\\
E[SumX_T(2)] &= m\cdot a\cdot b^2\binom{t_2+1}{2}\\
E[SumX_T(3)] &= m^2a^2\cdot b^3\left(\binom{t_3+2}{3}+(t_2-t_3-1)\binom{t_3+1}{2}\right)\\
E[SumX_T(4)] &= m^3\cdot a^3\cdot b^4\left(\begin{array}{c}\binom{t_4+3}{4}+(t_2-t_4-2)\binom{t_4+2}{3}+\ldots\\ \frac{(t_3-t_4-1)(t_3-t_4)}{2}\binom{t_4+1}{2}+(t_2-t_3-1)(t_3-t_4-1)\binom{t_4+1}{2}\end{array}\right)
\end{aligned}
\tag{3.62}
$$

Table 3.6 gives the production of the tree at 25 DCs, as shown in Fig. 3.36.

Figure 3.36 shows a simple tree simulated in ontogenic mode with 4 PAs at the age of 25th DC.

3.5.3.5 Sympodial Plant Model

For the sympodial architectural model, the number of phytomers grows exponentially (Fig. 3.32c, *Guayule* plant). The structure consists of a series of branched

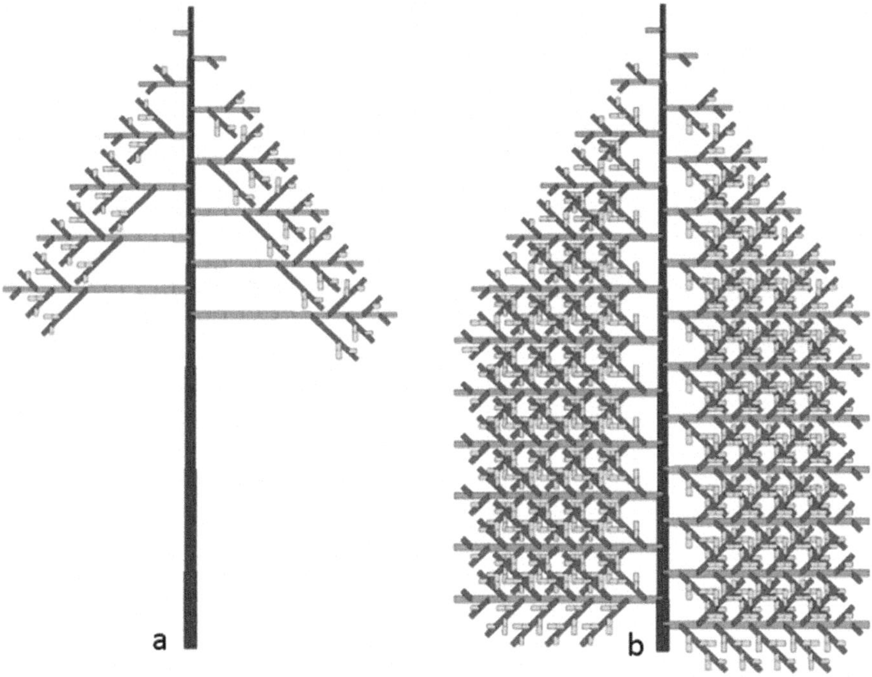

Fig. 3.36 Example of a simple tree simulated in ontogenic mode with four physiological ages (PAs) at the age of 25th DC. (**a**) The pruned tree with living phytomers. (**b**) All the produced axes (including dead and living axes). The probabilities a, b are set to 1 for the potential structure. Age: $N = 25$ DCs. Number of axillaries per node: $m = 1$. Life span of second-order axes: $t_2 = 12$. Life span of third-order axes: $t_3 = 5$. Life span of fourth-order axes: $t_4 = 2$. The structures are pruned at the end of the development. The production of such a tree can be calculated by Eqs. (3.61) and (3.62)

modules. A meristem dies after the installation of a module, and a number of m axillaries take over. Only the parameters m and a are necessary for the construction of the structure. At the cycle T, the expectation of the number of formed phytomers is:

$$E[X_T] = 1 + m \cdot a + (m \cdot a)^2 + \ldots + (m \cdot a)^T = \frac{1 - (m \cdot a)^{T+1}}{1 - m \cdot a} \qquad (3.63)$$

3.5.4 *Example of Computing Double Scale Botanical Structures*

In the case of rhythmic development, it seems impossible to find a mathematical equation that directly gives the number of phytomers in the structure as a function of *T* in DCs. Instead, operators that work through recursive relationships must be used.

3.5.4.1 Simple Case of Preformed Growth Units and Axes without PA Transition

This is a simple example of a double-scale operator that computes rhythmic development and is limited to preformed GUs. In this case, the last formed GUs are not yet branching, and the start of axillary buds functioning is synchronized. It is assumed that there are no transitions in PA due to axis extension or mortality.

The parameters of development allow to write the associated operators. A deterministic case and its stochastic counterpart with three PAs ($mxp = 3$), are presented.

3.5.4.1.1 Deterministic Case

The rhythmic development in Fig. 3.37 presents a double-scale structure of five macrocycles, each composed of five DCs. Depending on the PA, a number of structural pauses are added to the phytomers of the GUs to complete the macrocycles at the age of five DCs. At the base of the delayed ramifications, a number of structural pauses are introduced to synchronize the development of the branches at the beginning.

The matrix $[S]_T$ of dimension $mxp \times mxp$ is defined. This matrix contains, for each row, the number of phytomers per PA contained in the sub-structure corresponding to the row number, at the age *T* expressed in macrocycles. So, for five macrocycles, we have:

$$[S]_5 = \begin{bmatrix} 15 & 20 & 30 \\ 0 & 10 & 10 \\ 0 & 0 & 5 \end{bmatrix} \tag{3.64}$$

On line 1 [15 20 30], the main structure (PA 1) at the age of 5 DCs contains 15 phytomers of PA 1 (stem), 20 phytomers of PA 2 (branches), and 30 phytomers of PA 3 (twigs).

On line 2 [0 10 10], the secondary structure (PA 2) at the age of 5 DCs contains 10 phytomers of PA 2 (branches) and 10 phytomers of PA 3 (branches).

On line 3 [0 0 5], the tertiary structure (PA 3) at the age of 5 DCs contains 5 phytomers of PA 3 (twigs).

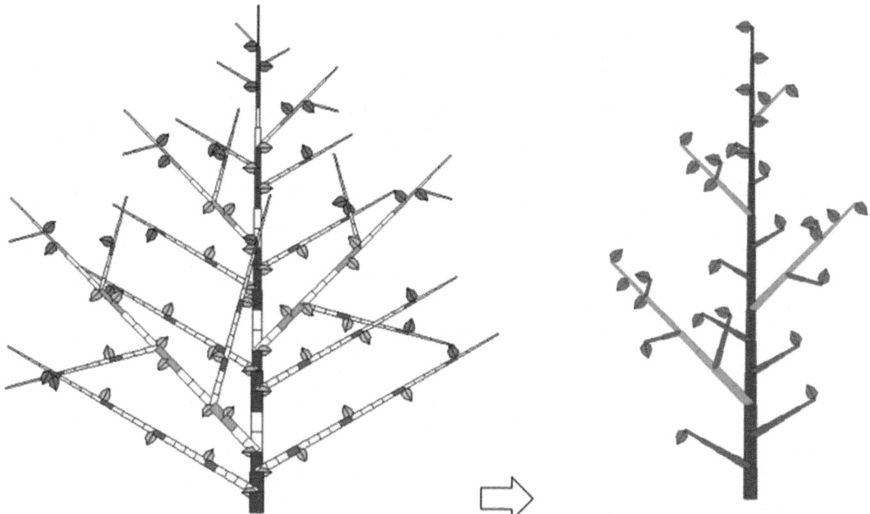

Fig. 3.37 Double-scale botanical structure in chronological and topological modes

A first operator, the diagonal matrix $[U]$, contains the number of phytomers of a GU produced by meristems according to PAs. The variable u_p is the total number of phytomers of the GU, produced each year by a meristem of PA p.

$$[U] = \begin{bmatrix} u_1 & 0 & 0 \\ 0 & u_2 & 0 \\ 0 & 0 & u_3 \end{bmatrix} \tag{3.65}$$

A second operator, the matrix $[M]$, contains the potential branching of the GU for the different types of axes it carries regardless of their positions in the GU:

$$[M] = \begin{bmatrix} 1 & m_{1,2} & m_{1,3} \\ 0 & 1 & m_{2,3} \\ 0 & 0 & 1 \end{bmatrix} \tag{3.66}$$

The variable $m_{i,j}$ gives the number of axillary branches of PA j carried by a GU of PA p.

For the first year:

$$[S]_1 = [U] \tag{3.67}$$

The following relationship is established, which calculates by recurrence in the year T, the matrix $[S]_T$ as a function of $[S]_{T\text{-}1}$:

$$[S]_T = [U] + [M] \cdot [S]_{T-1} \tag{3.68}$$

This development algorithm makes it possible to quickly calculate the annual phytomer production of developing structures from one year to the next.

Numerical Example

The parameters of development that produce the structure in Fig. 3.37 provide the following equations:

$$[U] = \begin{bmatrix} 3 & 0 & 0 \\ 0 & 2 & 0 \\ 0 & 0 & 1 \end{bmatrix}, M = \begin{bmatrix} 1 & 1 & 2 \\ 0 & 1 & 1 \\ 0 & 0 & 1 \end{bmatrix}, \text{ where } [S]_1 = \begin{bmatrix} 3 & 0 & 0 \\ 0 & 2 & 0 \\ 0 & 0 & 1 \end{bmatrix}$$

Applying Eq. (3.68) from year two to year five, we find

$$[S]_2 = \begin{bmatrix} 6 & 2 & 2 \\ 0 & 4 & 1 \\ 0 & 0 & 2 \end{bmatrix}, [S]_3 = \begin{bmatrix} 9 & 6 & 7 \\ 0 & 6 & 3 \\ 0 & 0 & 3 \end{bmatrix}, [S]_4 = \begin{bmatrix} 12 & 12 & 16 \\ 0 & 8 & 6 \\ 0 & 0 & 4 \end{bmatrix}, [S]_5 = \begin{bmatrix} 15 & 20 & 30 \\ 0 & 10 & 10 \\ 0 & 0 & 5 \end{bmatrix}$$

In year 5, we find the production previously counted.

Eq. (3.68) is efficient and allows to count old complex structures. It is not linked to a specific topological structure. The same operator may be common to different plant developments that produce the same number of phytomers. It is the potential for branching of the GUs that must be taken into account, rather than the location of the axes on the GU. This approach makes it possible to obtain simplified development schedule.

3.5.4.1.2 Stochastic Case

The development operator, as defined, also makes it possible to take into account the cases where the functioning of the meristems is stochastic. In the GreenLab tree model, the first phytomer of a GU still exists, and the others are formed with the probability b. The ramifications appear with probability a. The operators $[U]$ and $[M]$ in the stochastic case can be adjusted by calculating the mathematical expectations of the existence of phytomers at given locations:

$$\begin{aligned} [U] &= \begin{bmatrix} 1+(u_1-1)b & 0 & 0 \\ 0 & 1+(u_2-1)b & 0 \\ 0 & 0 & 1+(u_3-1)b \end{bmatrix} \\ [M] &= \begin{bmatrix} 1 & abm_{1,2} & abm_{1,3} \\ 0 & 1 & abm_{2,3} \\ 0 & 0 & 1 \end{bmatrix} \end{aligned} \tag{3.69}$$

In the matrix $[M]$, the potential number of branches should be multiplied by $a \times b$. The term b is included because branching can only occur if the phytomer bearing it exists. In the following example, the numerical values of the automaton are defined:

Fig. 3.38 Potential structure and three random simulations of a structure at three physiological ages (PAs) whose expectation of phytomer production can be calculated by a recursive algorithm using the development operators [*U*] and [*M*]. The theoretical production of phytomers for $T = 5$ with three PAs is [21, 21.7, 54.5]. The average of 20 simulations is [21.6, 21.3, 56.3]

The potential number of phytomers for u_1, u_2, and u_3 is 5, 4, and 3, respectively.

The number of potential branches carried by PA 1 of GUs is 1 for the axes of PA 2 and 2 for the axes of PA 3. The number of potential branches carried by PA 2 of GUs is 2 for the axes of PA 3.

The *Bernoulli* parameters and the branching probability are: $b = 0.8$, $a = 0.8$.

The development operators [*U*] and [*M*] can be written as:

$u_1 = 1 + 4 \times 0.8$; $u_2 = 1 + 3 \times 0.8$; $u_3 = 1 + 2 \times 0.8$; $m_{11} = 1$; $m_{12} = 1 \times 0.8 \times 0.8$; $m_{13} = 2 \times 0.8 \times 0.8$;

$$m_{22} = 1; m_{23} = 2 \times 0.8 \times 0.8; m_{33} = 1.$$

The matrices *U* and *M* then take the following numerical values:

$$[U] = \begin{bmatrix} 4.2 & 0 & 0 \\ 0 & 3.4 & 0 \\ 0 & 0 & 2.6 \end{bmatrix}, [M] = \begin{bmatrix} 1.0 & 0.64 & 1.28 \\ 0 & 1.0 & 1.28 \\ 0 & 0 & 1.0 \end{bmatrix}$$

In the following example, 20 random structures are simulated, and the average production at the age 5 is calculated using the recursive algorithm (Eq. 3.68). Three of the 20 random simulations are shown in Fig. 3.38, next to the potential structure. The number of simulated and calculated average phytomers are comparable (Table 3.7).

3.5.4.2 Complex Case of Axes with PA Transition

Botanical structures are organized into development axes that carry sub-structures. The PA of a structure is given by the age of its development axis at birth. The

Table 3.7 Comparison between the numbers of simulated and computed average phytomers at the age of five years ($T = 5$)

Physiological ages (p)	1	2	3
Average number of phytomers in the 20 simulated structures	21.6	21.3	56.3
Mathematical expectation of the phytomers of production (theoretical number)	21	21.7	54.5

structures are simple or compound according to whether the meristems have a continuous or rhythmic functioning. In the latter case, meristems set up GUs that may or may not function synchronously. The phytomers of GUs can be differentiated and have branches with a PA greater than or equal to that of their developmental axis. Finally, meristems can undergo transitions to higher PAs.

Thus, a primary structure of PA p is composed of a development axis of PA p that carries secondary structures of PA $> p$ and whose CAs increase from the top to the bottom of the developmental axis.

3.5.4.2.1 Computation of Structures in Matrix Mode

We will focus here on simple structures and compound structures that only make synchronized shoots and that are made up of a single growth unit.

We define a structure with mxp PAs. We set here $mxp = 4$.

Operators U, M, N, and J are defined as follows:

Operator U is a diagonal matrix that contains the number of u_i phytomers per GU. In the case of simple structures, the growth unit is the phytomer ($u_i = 1$).

The GUs of PA p have u_p phytomers. These phytomers carry axillary branches of PA $\geq p$. They are noted in number $u_{p,p}, u_{p,p+1}, \ldots, u_{p,mxp}$; these phytomers carry $m_{p,p}, m_{p,p+1}, \ldots, m_{p,mxp}$ branches.

Here the objective is to calculate organogenesis without constructing the plant structure. So only branch productions (numbers of phytomers) are taken into consideration while the details of the GU organization can remain unknown.

Thus, the operator M is a matrix that contains the numbers of branches of PA j carried by a GU of PA i ($j \geq i$). Line i of M contains the detail of the m_{ij} ramifications.

Operator N contains the number of potential repetitions of GU of an axis in the same physiological age before a mutation or abortion.

Operator J contains the transition number corresponding to the PA.

The matrix $[S_T^p] = [0 \cdots s_{p,p}\ s_{p,p+1} \cdots s_{p,mxp}]_T$ contains the number of phytomer types produced according to the CA T and PA p of structure S.

$$[U] = \begin{bmatrix} u_1 & 0 & 0 & 0 \\ 0 & u_2 & 0 & 0 \\ 0 & 0 & u_3 & 0 \\ 0 & 0 & 0 & u_4 \end{bmatrix} \quad [M] = \begin{bmatrix} 0 & m_{12} & m_{13} & m_{14} \\ 0 & 0 & m_{23} & m_{24} \\ 0 & 0 & 0 & m_{34} \\ 0 & 0 & 0 & 0 \end{bmatrix}$$

$$[N] = [N_1 \quad N_2 \quad N_3 \quad N_4] \quad [J] = [J_1 \quad J_2 \quad J_3 \quad J_4]$$

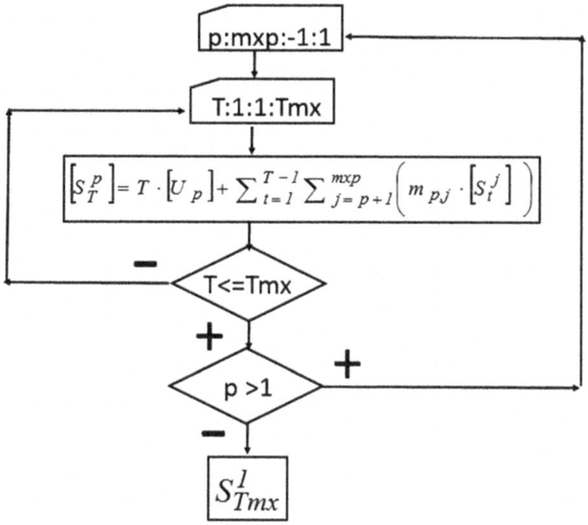

Fig. 3.39 Flowchart of the computation of structures by assembling the sub-structures on the development axes by a double recursive loop over chronological ages (CAs) and physiological ages (PAs)

We start from the maximum PA *mxp* and build the collection of structures that will be assembled top down on the axis of development for all CAs from 1 to *T* macrocycles and repeat recursively until the construction of the main structure 1 for all CAs. An additional loop appears here at the level of the GU, which branches axillary structures that have different PAs (Fig. 3.39). This algorithm (in MatLab form) is quite fast.

At the end of the algorithm, a collection of sub-structures is created for all PAs and CAs.

The main structure $\left[S_T^1\right] = \left[s_{1,1}\ s_{1,2} \cdots s_{1,p} \cdots s_{1,mxp}\right]_T$ contains the numbers of phytomers according to the PAs.

Two cases are possible: either a structure undergo the transition or not.

In case structure is in development before the transition:

$$[S_T^p] = T \cdot [U_p] + \sum_{t=1}^{T-1} \sum_{j=p+1}^{mxp} \left(m_{p,j} \cdot [S_t^j]\right) \tag{3.70}$$

In case structure is in development with the transition $p \rightarrow p + 1$).

$$[S_T^p] = N_p \cdot [U_p] + \sum_{t=T-1}^{T-N_p} \sum_{j=p+1}^{mxp} \left(m_{p,j} \cdot [S_t^j]\right) + [S_{T-N_p}^{p+1}] \tag{3.71}$$

The development of the structure is completed when:

$$T_{max} = \sum_{p=1}^{mxp} N_p \tag{3.72}$$

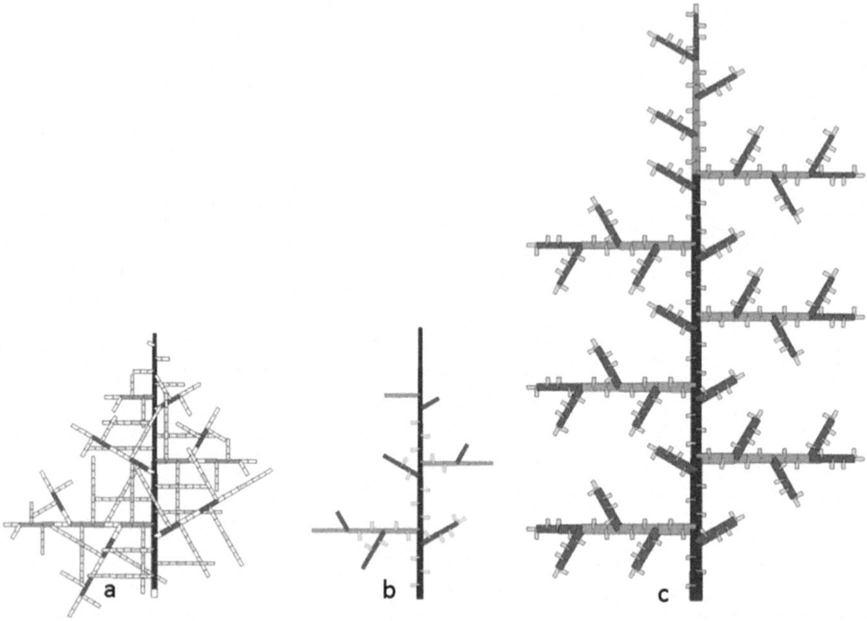

Fig. 3.40 Simulation of the structure under development at four macrocycles: a deterministic case. (**a**) Chronological mode with structural pause, (**b**) topological mode, and (**c**) completed structure at 12 macrocycles

Study of an Example

Figure 3.40 shows a simulated compound structure with macrostates at two stages of development (4 and 12 macrocycles).

The structure is defined by the following operators:

$$[U] = \begin{bmatrix} 5 & 0 & 0 & 0 \\ 0 & 3 & 0 & 0 \\ 0 & 0 & 2 & 0 \\ 0 & 0 & 0 & 1 \end{bmatrix} \quad [M] = \begin{bmatrix} 0 & 1 & 1 & 3 \\ 0 & 0 & 1 & 2 \\ 0 & 0 & 0 & 2 \\ 0 & 0 & 0 & 0 \end{bmatrix}$$

$$[N] = \begin{bmatrix} 6 & 3 & 2 & 1 \end{bmatrix}$$

$$[J] = \begin{bmatrix} 2 & 3 & 4 & 0 \end{bmatrix}$$

From Eq. (3.72), the maximum of macrocycles of the structures is $T_{max} = 12$. The corresponding age in number of DCs including pauses on the axis of development is 60 DCs.

By applying the algorithm in this example, the respective production of phytomers by PA at four macrocycles is [20, 20, 18, 24], and [30, 63, 136, 230] when the structure is completed at 12 macrocycles.

In the stochastic case, a collection of random structures with R repetitions is created for each PA p, and CA T. The larger R is, the closer we get to a true Monte Carlo simulation. This quasi-random mode makes it possible to obtain calculation times that do not depend on the number of phytomers in the structure; instead, they depend on *mxp*, *Tmx*, and *R*.

3.5.4.2.2 Construction of Structures in Matrix or List Mode

Here we aim not only to calculate the number of phytomers produced by the structure, but also to have a detailed construction of the chronological and topological structures. These in turn can be transformed into architecture, if geometrical operators are involved.

In the general case of trees, the annual shoots of the axes have a continuous development during the period favorable to their functioning, with a preformation followed by a neoformation (e.g., Cherry tree). At the end of this period, development stops. It resumes usually at the beginning of the following year. The calculation of the tree must be based on the number of DCs rather than the age in years (macrocycles), which obviously increases the computational cost.

The recursive algorithm must be able to manage this functioning; it must also be able to take into account structural pauses that make it possible to produce delayed and synchronized or immediate ramifications, as well as consider mortality or PA transitions of meristems, etc.

It is then necessary to rely on the details of the development axis to manage the cycle-by-cycle functioning of the meristems. As a result, the development axes and matrices of the sub-structures increase in size with each additional cycle, increasing with the age of the tree.

Three steps are then necessary:

- The construction of growth units.
- The construction of development axes.
- The construction of structures by assembling development axes and axillary structures. As an example, *mxp* is set to 4.

The production of PA p for GUs depends on four operators:

- Operator N, which gives the transition steps in PA:

$$[N] = \begin{bmatrix} N_1 & N_2 & N_3 & N_4 \end{bmatrix}$$

- Operator u, which contains the number of u_p phytomers produced by a GU of PA p:

$$[U] = \begin{bmatrix} u_1 & 0 & 0 & 0 \\ 0 & u_2 & 0 & 0 \\ 0 & 0 & u_3 & 0 \\ 0 & 0 & 0 & u_4 \end{bmatrix}$$

- Operator *Nu*, which contains the numbers of nu_{px} phytomers that bear a PA structure x ($p < x \leq xp$). They define a matrix *Nu*; for example, the term nu_{23} refers to the number of phytomers in a GU of PA 2 that bear PA 3 sub-structures:

$$[Nu] = \begin{bmatrix} nu_{11} & nu_{12} & nu_{13} & nu_{14} \\ 0 & 0 & nu_{23} & nu_{24} \\ 0 & 0 & nu_{33} & nu_{34} \\ 0 & 0 & 0 & nu_{44} \end{bmatrix}$$

The set of nu_{px} phytomers of a line allows to build a GU, which contains potentially u_p phytomers:

$$u_p = \sum_{x=p}^{\mathrm{mxp}} nu_{px} \tag{3.73}$$

The types of phytomers can be distributed in any order within a GU. To simulate acrotony, it is necessary to place at the base of the GU the phytomers that carry the structures of the highest PA, and at the top, those that carry the reiteration of the concerned age.

- Operator *Ma*, the number of axillary branches ma_{px} carried by the type of phytomers mu_{px}, forms an *Ma* matrix:

$$[Ma] = \begin{bmatrix} ma_{11} & ma_{12} & ma_{13} & ma_{14} \\ 0 & 0 & ma_{23} & ma_{24} \\ 0 & 0 & ma_{33} & ma_{34} \\ 0 & 0 & 0 & ma_{44} \end{bmatrix}$$

For each PA, a development axis can be defined that describes the succession of phytomers generated by the functioning of meristems alternating with their pauses. The construction of these developmental axes is carried out using the operators *U*, *Nu*, and *Ma*, to reconstitute step by step the production of the GUs along the axis, i.e., the types of phytomers that alternate with the pauses.

3.5.4.2.2.1 *Initialization of Delays in Development Axes*

Except for the trunk, all development axes come from axillary meristems. In the case of synchronous trees, all axillary meristems distributed along the new annual shoot are in dormancy. They start operating simultaneously the following spring. To achieve such synchronization, each axillary meristem must be assigned a number of pauses indexed to its position in the GU from the base. These are called structural pauses. If the annual shoot is formed by *NA* cycles, then the dormancy of the axillary meristem of the first phytomer at the base of the shoot is $NA - 1$ cycles, that of the next is $NA - 2$ cycles, etc. This ensures the synchronization of the axillary shoots with the activity recovery.

3.5.4.2.2.2 *Attributes of the Development Axes*

After defining the number of initial pauses, an axillary meristem will produce a series of phytomers with their attributes, all of which replicate the development axis. The number of attributes depends on the complexity of the functioning of the meristems that need to be managed.

The development axis is therefore represented as a matrix whose columns correspond to the DCs, and the rows corresponding to the attributes of the phytomers generated by the GU successions for each PA p.

A first attribute $Ar(p, x)$ concerns whether the botanical entity on the development axis in the cycle x is a phytomer or a pause. Consider either in the case of a stochastic simulation or in the calculation of the potential axis. In the first case, a value of 1 characterizes the existence of the phytomer, and value 0 corresponds to a pause from the meristem. In the second case, the probability of the existence of the plant is recorded as attribute.

A second attribute $Ap(p, x)$ concerns the PA of the branch carried by the phytomer at the CA x.

A third attribute $Am(p, x)$ concerns the number of axillary buds carried by the phytomer of the CA x.

In the stochastic case, a fourth attribute $As(p, x)$ gives the identification of the repetitions inside the collection constituted with random sub-structures.

3.5.4.2.2.3 *Assembly of Sub-Structures along the Development Axis*

The recursive method that assembles the sub-structures on the development axes are shown in Fig. 2.7 of Chap. 2.

The main axis of a structure with ramifications, as defined, is the replication of the development axis.

The development axis is therefore a matrix of dimension $nat{\cdot}T{\cdot}\ mxp$, where nat is the number of attributes, T is the CA of the plant, and mxp is the number of PAs. By limiting to the three basic attributes (deterministic model), we can obtain the matrix for the PA p:

$$[AD] = \begin{bmatrix} [Ar] \\ [Ap] \\ [Am] \end{bmatrix} \tag{3.74}$$

The axis of the main structure assembles the axillary structures from the bottom up, following the instructions of the development axis according to their number, PA, and CA from its first phytomer to the terminal phytomer (assuming only one kind of axillary structure per phytomer). The factorization equations Eqs. (3.75) and (3.76) of structures take the following form at PA p and CA T:

$$[S_T^p] = \sum_{t=1}^{T} Ar_t^p (1 + Am_t^p S_x^{Ap_t^p}) \ (T \leq N_p) \tag{3.75}$$

$$[S_T^p] = \sum_{t=1}^{Np} Ar_t^p (1 + Am_t^p \cdot S_x^{Ap_t^p}) + [S_{T-N_p}^{p+1}] \ (T > N_p) \tag{3.76}$$

The main structure $\left[S_T^1\right] = \left[s_{1,1}\ s_{1,2} \cdots s_{1,p} \cdots s_{1,mxp}\right]_T$ contains the numbers of phytomers according to the PAs.

The operation $\left[S_T^1\right] - \left[S_{T-1}^1\right]$ gives the number of phytomers belonging to the cohort of the PA p at the CA T.

When assembling the axillary structure on the main axis, two operations are performed: the calculation of the delay in branching (which ensures the synchronism of meristem functioning at the beginning of the macrocycle [first year]) and the calculation of the CA of the axillary sub-structure. The CA x of the sub-structure depends on its position t counted from the base on the main axis. At the age of T DCs, we will have for continuous development $x = T - t$ DCs. But in the case of a rhythmic development with a macrocycle of NA DCs, for synchronous trees, we will have:

$$x = T - (t - d) - NA$$

where d is the delay in DC for the development of the sub-structure related to its position in the GU, with:

$$d = mod(t, NA) + (1 - sign(mod(t, NA))) \cdot NA$$

where *mod* is the modulo function and *sign* a function that returns 1 or 0 depending on whether the argument is positive or null.

The terminal GU does not yet have any branches. Axillary meristems are dormant. The following year, all branches start at the same time as the main meristem starts to set up a new GU again.

The algorithm proceeds with a double loop are shown in Fig. 3.39:

- From the oldest PA (*mxp*) to PA 1.
- For each PA, the collection of sub-structures is constructed for all CAs from 1 to T.

The ultimate development axis at the PA *mxp* gives a collection of unbranched structures for all CAs from 1 to T.

This algorithm gives two possibilities:

- Matrix mode. This corresponds to the case where in Eqs. (3.75) and (3.76), the addition of the development axes with the sub-structures is cumulated into the form of numbers, and thus the matrix mode is obtained. This mode gives the production of phytomers during the development of the structures, similar to Eqs. (3.70) and (3.71).

Table 3.8 Attributes of axes of development

Existence of the phytomer along the development axis: *Ar(p, t)*															
Chronological age (DC)															
Age phy1	1	1	1	0	0	1	1	1	0	0	1	1	1	0	0
Age phy2	1	1	0	0	0	1	1	0	0	0	1	1	0	0	0
Age phy3	1	0	0	0	0	1	0	0	0	0	1	0	0	0	0
	Year 1					Year 2					Year 3				
Physiological age of the sub-structure carried along the development axis: *Ap(p, t)*															
Chronological age (DC)															
Age phy1	0	3	2	0	0	0	3	2	0	0	0	3	2	0	0
Age phy2	0	3	0	0	0	0	3	0	0	0	0	3	0	0	0
Age phy3	0	0	0	0	0	0	0	0	0	0	0	0	0	0	0
	Year 1					Year 2					Year 3				
Number of axillary buds along the development axis: *Am(p, t)*															
Chronological age (DC)															
Age phy1	0	2	1	0	0	0	2	1	0	0	0	2	1	0	0
Age phy2	0	1	0	0	0	0	1	0	0	0	0	1	0	0	0
Age phy3	0	0	0	0	0	0	0	0	0	0	0	0	0	0	0
	Year 1					Year 2					Year 3				

- List mode: This corresponds to the case where in Eqs. (3.75) and (3.76), the concatenation of the development axes with the sub-structures is cumulated in the form of string details, and thus we obtain the list mode. This mode describes, in the form of a list of items, the details of the chronological structures produced that are stored in prefix order. Exploring the list gives the numbers of phytomers created and possibly two- or three-dimensional (2D or 3D) plant images using geometric operators with computer software.

3.5.4.2.2.4 Application

A synchronous tree simulated with three PAs is represented at age three years. The number of *NA*, DC per year, is 5. The operators are as follows:

$$[U] = \begin{bmatrix} 3 & 0 & 0 \\ 0 & 2 & 0 \\ 0 & 0 & 1 \end{bmatrix};\ [Nu] = \begin{bmatrix} 1 & 1 & 1 \\ 0 & 1 & 1 \\ 0 & 0 & 1 \end{bmatrix};\ [Ma] = \begin{bmatrix} 1 & 2 & 0 \\ 0 & 1 & 0 \\ 0 & 0 & 0 \end{bmatrix}$$

Table 3.8 gives the details of the axis of development.

In Fig. 3.41, *a* and *b* show the assembly of the sub-structures on the development axis, and *c* shows the resulting potential structure with the probabilities of existence for each phytomer. The created structure can be explored, and each phytomer can be listed. Table 3.9 gives the list generated in prefix order for the PA of phytomers within the main structure.

In the stochastic case, the parameters (a_p, b_p, c_p) control the implementation of macrostates during the development by PA of the meristems. With the set $a_1 = 1$,

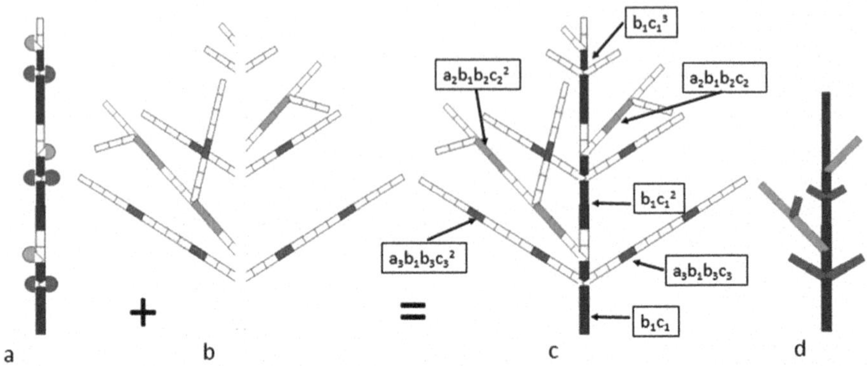

Fig. 3.41 Assembly of the sub-structures with developmental axes. (**a**) Developmental axis: physiological age (PA) 1 with its types of phytomers, its buds, and its annual pauses. (**b**) Structures carried: PAs 2 and 3 with branching delays and annual pauses. (**c**) Potential structure with the probabilities of phytomer existence. (**d**) Topological structure of sub-figure **c**

Table 3.9 List of the physiological ages (PAs) of the phytomer in the chronological structure (prefix order)

1	1	3	3	3	3	3	3	3	3	3	3	3	3	3	3	3	3	3	3	3	3	3
3	3	3	3	3	1	2	2	2	2	3	3	3	3	3	3	3	3	2	2	2	2	2
3	3	3	2	2	2	1	1	1	1	3	3	3	3	3	3	3	3	3	3	3	3	3
3	3	3	1	2	2	2	2	3	3	3	2	2	2	1	1	1	1	3	3	3	3	3
3	1	2	2	1	1																	

$b_1 = 0.8$, $c_1 = 1$, $a_2 = 0.9$, $b_2 = 0.8$, $c_2 = 0.9$, $a_3 = 0.9$, $b_3 = 0.8$, and $c_3 = 0.7$, their values give the probability of existence of each potential phytomer (Table 3.10).

Table 3.10 gives the list generated in prefix order for the probabilities of existence of phytomers.

3.6 Conclusion

Time is discretized into development cycles (DCs) in the GreenLab model. The meristems that build the axes produce phytomers or pauses at each DC in periodic or aperiodic alternation.

The stochastic functioning of meristems can be simulated by a *Bernoulli* process. At each DC, a phytomer is created with a probability b or a pause with a probability $1 - b$.

The chronological age (CA) of an axis is measured in the number of DCs. An axis can be represented in chronological mode with the phytomers and pauses displayed

Table 3.10 List of the phytomer rate of existence in the potential structure (prefix order)

0	0	0	0	0	0.800	0.800	0	0	0	0.403	0	0	0
0	0.282	0	0	0	0	0	0	0	0.403	0	0	0	0.282
0	0	0	0	0.8000	0	0	0.518	0.518	0	0	0	0.261	0
0	0	0	0	0	0	0.466	0.466	0	0	0	0	0	0
0	0.800	0.800	0	0	0	0.403	0	0	0	0	0	0	0
0.403	0	0	0	0	0.800	0	0	0.518	0.518	0	0	0	0
0	0	0	0	0.800	0.800	0	0	0	0	0	0	0.800	0

or in topological mode with the pauses removed, which corresponds to the actual observations.

In addition, viability *c* is defined as the probability of survival of meristem in each cycle. Finally, at its creation, a meristem becomes active with a probability *a*. Parameter triplet *a*, *b*, and *c* fully controls the functioning of a meristem. This meristem generates development axes resulting in simulation from the same statistical distribution. This is classically obtained by the Monte Carlo method. The model makes it possible to define a potential axis that associates each phytomer with its probability of existence. Axis simulations in chronological mode reveal the potential axis. This makes it possible to directly calculate the mean and variance of the number of phytomers resulting from the distribution of the axes.

The branching of the axes generates sub-structures. By means of Monte Carlo simulations, stochastic realizations of structures coming from the same distribution are obtained. In chronological mode, these structures are made up of phytomers and void entities corresponding to pauses or abortions of meristems. A chronological structure is a discretized spatio-temporal representation of the development process of a certain pedagogical aspect. By removing pauses from the chronological structure, we obtain a topological structure that is the actual observable one. The time of appearance of each phytomer is directly visible in the chronological mode but not in the topological mode, and certainly not in the observed natural plant. This is why measurements on plants are automatically biased if the stochastic aspect of the development is not taken into account.

The potential structure is the chronological structure where each element exists with a probability that depends on the parameters of the development. The potential structure contains all the Monte Carlo simulations that represent it. The potential structure makes it possible to calculate the mean and variance of the different types of organs produced without going through simulated structure, which is an advantage for the implementation of inverse methods.

Thus, the method of crown analysis makes it possible to calculate the parameter triplet *a*, *b*, and *c* of the meristems of a plant from samples taken from its architecture. Taking into account these parameter values, it is possible to calculate the detailed production of phytomers for a structure of a given age in cycles, using development operators that operate by recursion. The method makes it possible to restore missing data on plants caused by mortality and pruning of dead branches.

When these operators operate in matrix mode, the production in number of elements is calculated and stored in tables. This fast process makes it possible to compute the plant's demand at each DC (see Chap. 5), to implement the inverse methods, and thus, as we mentioned, to be able to calculate the organogenesis and photosynthesis of the plant without going through the cost of exploring its detailed structure.

When these operators operate in list mode, they generate character strings in prefixed mode that describe the simulated structure in chronological mode. In addition, a computer simulation formalism emerges from the operators and leads to the synthetic representation of the development of structures in the form of a graphical model called the dual botanical automaton (see Chap. 4). Reading this list allows to represent the computed plant by associating 2D or 3D geometric operators. This is the subject of Chap. 14 that explains how to simulate and render 3D computational plants produced by the GreenLab model for Computer Graphics application.

References

Barczi J-F, de Reffye P, Caraglio Y (1997) Essai sur l'identification et la mise en œuvre des paramètres nécessaires à la simulation d'une architecture végétale : le logiciel AmapSim. In : Bouchon J., de Reffye P., Barthélémy D., éds. Modélisation et simulation de l'architecture des végétaux. Paris, Éditions Inra, Sciences Update, p. 255–423

Barczi J-F, Rey H, Caraglio Y, de Reffye P, Barthélémy D, Dong Q, Fourcaud T (2008) AmapSim: an integrative whole-plant architecture simulator based on botanical knowledge. Ann Bot 101(8):1125–1138

Catalan E (1838) Note sur une Équation aux différences finies[J]. Journal de Mathématiques Pures et Appliquées, 1838, 3:508–516

Chavaneau C (1989) Modélisation de la croissance et de l'architecture de l'Abricotier (*prunus armeniaca L.*) D.A.A., spécialité : Agronomie méditerranéenne. ENSA, Montpellier. 77 p

Costes É (1988) Analyse architecturale et modélisation du Litchi (*Litchi chinensis Sonn.*) Contribution à l'étude de son irrégularité de production. L'Ile de la Réunion, Thèse de doctorat. Université de Montpellier. 210 p

Costes É (1990) Modélisation de l'architecture de l'hévéa. Rapport scientifique Cirad-Irca

Dabadie P, de Reffye P, Dinouard P (1991) Modelling bamboo growth and architecture: phyllostachys viridi-glaucescens. J Am Bamboo Soc 8(1–2):65–79

de Reffye P (1979) Modélisation de l'architecture des arbres par des processus stochastiques. Simulation spatiale des modèles tropicaux sous l'effet de la pesanteur. Application au Coffea robusta. Thèse de doctorat, Université Paris Sud Orsay, 195 p

de Reffye P (1981) Modèle mathématique aléatoire et simulation de la croissance et de l'architecture du caféier Robusta. 1ère partie. Étude du fonctionnement des méristèmes et de la croissance des axes végétatifs. Café Cacao Thé 25(2):83–104

de Reffye P, Cognée M, Jaeger M, Traore B (1988) Modélisation stochastique de la croissance et de l'architecture du cotonnier. 1. Tiges principales et branches fructifères primaires. Coton et Fibres Tropicales 43(4):269–291

de Reffye P, Elguero E, Costes E (1991) Growth units construction in trees: a stochastic approach. In: 9th seminar of the theoretical biology school, Solignac, 21-23.09.1989, pp 325–342

de Reffye P, Kang M-Z, Hua J, Auclair D (2012) Stochastic modelling of tree annual shoot dynamics. Ann For Sci 69:153–165

Diao J, de Reffye P, Lei X-D, Guo H, Letort V (2012) Simulation of the topological development of young eucalyptus using a stochastic model and sampling measurement strategy. Comput Electron Agric 80(1):105–114

Franquin P (1972) Modeles mathematiques de structures chez les vegetaux. II Relations de structures. Cahiers Orstom, sér biol 17:3–21

Janey N (1992) Modélisation et synthèse d'images d'arbres et de bassins fluviaux associant méthodes combinatoires et plongement automatique d'arbres et cartes planaires (Doctoral dissertation, Besançon)

Fournier D (1989) Modélisation de la croissance et de l'architecture du Merisier (*prunus avium L.*) DAA spécialité : Agronomie méditerranéenne. ENSA, Montpellier. 77 p

Guédon Y, Heuret P, Costes E (2003) Comparison methods for branching and axillary flowering sequences. J Theor Biol 225(3)301–325

Mialet I (1990) Le cacaoyer, *Theobroma cacao* La modélisation de son architecture et de sa croissance : exemple de l'hybride UPA402* C410 (Doctoral dissertation)
Nicolini E (1998) Architecture et gradients morphogénétiques chez de jeunes hêtres (Fagus sylvatica L. Fagaceae) en milieu forestier. Can J Bot 76(7):1232–1244
Sabatier S, Barthélémy D, Ducousso I, Germain É (1998) Modalités d'allongement et morphologie des pousses annuelles chez le noyer commun, Juglans regia L.'Lara'(Juglandaceae). Can J Bot 76(7):1253–1264

Chapter 4
Plant Structures and Architectures: Modeling and Simulation

Baogang Hu, Philippe De Reffye, Marc Jaeger, Daniel Barthélémy, and Mengzhen Kang

Abstract This chapter introduces a simulation-based approach to create two- and three-dimensional (2D and 3D) computational plant models through simulations that mimic the growth patterns of real plants. When only plant structure is needed, the models are based on geometric rules and do not account for the functional aspects of plant organs, such as photosynthesis. The models provide a topological view in 2D and a realistic representation in 3D, applicable in multimedia and agronomy. The development of the plant structure is simulated through the activity of meristems, employing a botanical automaton, which simplifies the programming of plant development. The GreenLab model utilizes concepts like phytomers, transitions, and states to create a user-friendly interface for defining plant development parameters, including stochastic elements for meristem function. The chapter also discusses the experimental validation of plant development models across various

Electronic supplementary material The online version of this chapter (https://doi.org/10.1007/978-981-97-7541-5_4) contains supplementary material, which is available to authorized users.

B. Hu
National Laboratory of Pattern Recognition, Institute of Automation, Chinese Academy of Sciences (CASIA), Beijing, China
e-mail: hubg@nlpr.ia.ac.cn

P. De Reffye · M. Jaeger
CIRAD, UMR AMAP, Montpellier, France

AMAP, Univ Montpellier, CIRAD, CNRS, INRAE, IRD, Montpellier, France
e-mail: marc.jaeger@cirad.fr

D. Barthélémy
CIRAD, Montpellier, France

M. Kang (✉)
State Key Laboratory for Multimodal Artificial Intelligence Systems, Institute of Automation, Chinese Academy of Sciences (CASIA), Beijing, China

School of Artificial Intelligence, University of Chinese Academy of Sciences, Beijing, China
e-mail: mengzhen.kang@ia.ac.cn

P. De Reffye et al. (eds.), *Plant Model Greenlab for Botany and Agronomy*,
https://doi.org/10.1007/978-981-97-7541-5_4

species using the "crown analysis" method to estimate meristem function parameters from actual plant architectures. The results from their simulations are presented as computer-generated images, demonstrating the model's applicability and accuracy in studying plant architecture and growth.

Keywords Botanical automaton · Physiological states repetition and transition · Architectural models · Reiteration · 2D and 3D plant simulations

4.1 Introduction

Through the simulation of plant structures and architectures, the aim of this chapter is to obtain computational plants in two or three dimensions. They result from a computer program that simulates the development and positions of the organs created by meristems according to geometric rules. Here organs play no functional role and their sizes are given empirically, i.e., according to observations; the photosynthesis that ensures the growth of the organs is not taken into account. A two-dimensional (2D) structure is a simplified planar representation of a plant that gives the topological description of the organization, i.e., the relationships between the different botanical entities (Fig. 4.1a). Using geometric operators, the 2D

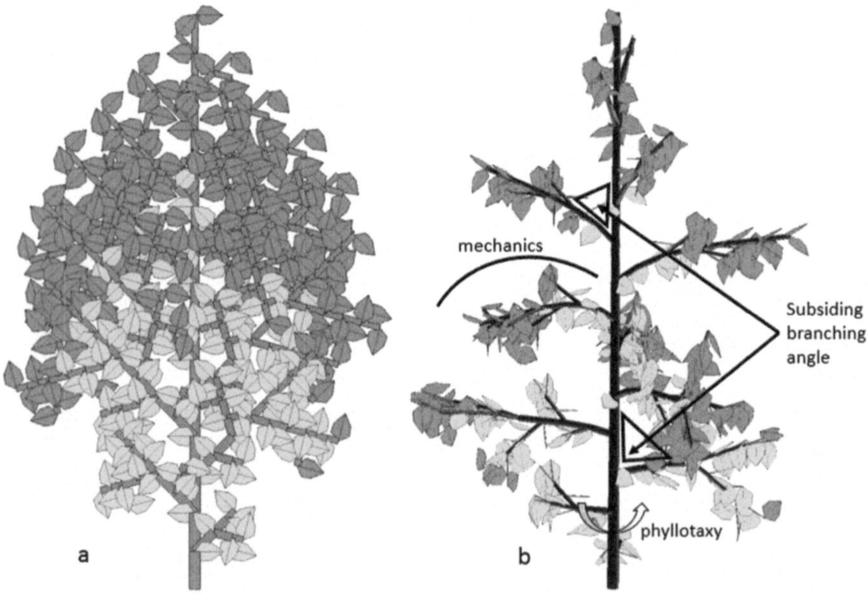

Fig. 4.1 Plant structure and plant architecture: Simulated virtual plant can be visualized as a 2D planar structure (**a**) or as a 3D architecture (**b**) using geometrical operators to integrate traits like branching angles, phyllotaxy, and mechanics. Green leaves are functional; yellow leaves are no longer functional due to overtaking the functioning time (Color figure online)

structure moves to three-dimensional (3D) plant model that conforms to the architecture of the original plant, and whose visualization is intended to be faithful to reality (Fig. 4.1b).

The 3D models of computational plants offer multiple applications via computer graphics, such as multimedia, the representation of architectural-landscape projects where plants coexist with buildings, etc. They are also useful in agronomy, for example, for calculating the light interception of a canopy or plant transpiration.

The Simulation program simulates the functioning of meristems and gives the number of organs in cohorts. Each cohort contains organs of different types produced by the plant at a given age. Cohorts are results of crop production related to the functioning of meristems.

In the modeling presented in this book (GreenLab model), the development process is based on a botanical automaton derived from the reference axis technique (de Reffye et al. 1995). The reference axis keeps the trace of the running of the botanical automaton. They are equivalent from a methodological point of view.

The automaton builds the plant structure according to the chronological age (CA) and physiological age (PA). The numbers of organs in plant structure are either counted by exploring the topological structure (list mode) or calculated directly by operators through its organ cohorts (matrix mode). Finally, plant development and growth are coupled through the calculation of demand (Chap. 5).

In Chap. 2, the notion of a development cycle (DC) to control the step-by-step functioning of meristems was discussed. At each DC, a meristem produces a new phytomer if the functioning mode is deterministic, or it produces a phytomer with a probability if the mode is stochastic. In the latter case, the simulated plants of the same age have different architectures but belong to the same clone. The stochastic aspect does not disturb the progress of morphogenesis in general but diversifies it in detail.

4.2 Modeling and Simulation of the Development of Plant Structure

4.2.1 The Use of Graphical Model

The botanical knowledge for describing plant development is translated into mathematical terms. GreenLab model uses the visual form of the graphical models widely adopted in Bayesian and Markovian schemes.

"A graphical model is a probabilistic model for which a graph expresses the conditional dependence structure between random variables". This type of model is commonly used in probability theory and statistics.

The GreenLab model proposes a dual-scale automaton to describe the stochastic process of plant architectures. The approach uses the graph-based notations so that it is rather compatible with botanical knowledge about topological information. Compared to the language-based approach, such as L-system, the dual-scale automaton provides an intuitive means for either designing or understanding the architecture model of plants. Simulations of almost all kinds of plant architectures defined by botanists confirm the advantages of using the graph-based notations in the GreenLab model (Zhao et al. 2003).

This botanical automaton expresses visually the repetition at the phytomer level of physiological ages (PAs), and their transitions either by branching or by continuing the axis. In the case of rhythmic development (Chap. 2), the automaton is on a double scale. The introduction of mathematical development operators makes it possible to count the numbers of the different types of organs produced directly without requiring explicit simulation of the topological structure, which significantly reduces calculation costs.

4.2.2 From the Reference Axis to the Botanical Automaton

The reference axis (Chap. 2) is a linear representation of all the successive transitions in PA that a plant's meristems undergo, from seed to terminal flowering (Fig. 2.7). All the axes of the plant are inscribed within the reference axis. They are born more or less old according to their initial physiological ages, and all die at the same stage of ultimate differentiation.

This concept, close to botanical analysis, has the advantage of visualization. The AMAPSim software (Barczi et al. 1997; Rey et al. 1998) developed at International Center for Agricultural Research for Development (CIRAD) uses this concept. It is a generic tool that has been used to simulate the architectural development of different tree species, such as cherries, apricots, and elms trees. However, the reference axis describes only the development of plants and does not consider the chronological aspect in the simulation. When meristem pauses occur and photosynthesis is still going on, the reference axis is unable to address this situation.

Based on the same botanical assumptions but considering the dynamic development process, the botanical automaton (Zhao et al. 2001, 2003) operates in a chronological mode and can express explicitly the positioning of void entities (pauses) in the simulated structures according to phytomers. The chronological mode makes it possible to address the organ growth, i.e., the production of biomass and its distribution among various organs to ensure their expansion. With the "reference axis," a leafy axis is made up of a series of phytomers, whereas with the botanical automaton, it is described by a symbol with a repetition number of phytomers, which constitutes a factorization of the development process.

4.2.3 Definition and Organization of the Botanical Automaton

The basic unit of the leafy axes is the phytomer (Chap. 2). In herbaceous plants and shrubs, meristems continuously produce phytomers over finite or indefinite periods of time, and the scale of the axis is single (phytomer level). A phytomer is identified by its rank from the base of the axis. In trees with rhythmic functioning of meristems, the axes are made up of growth units (GUs), which in turn are made up of phytomers. The scale is then described double: the GU level and the phytomer level. A phytomer is identified by the number of the GU from the base of the axis, and by its rank within the GU.

The automaton creates cyclically phytomers corresponding to different states, generating "development axes". In continuously growing plants (herbaceous plants and shrubs), a micro-state of the botanical automaton corresponds to a type of phytomer with its associated PA. In rhythmically growing plants (trees), two scales are needed to distinguish the states of phytomers in the automaton: a "micro-state" that describes each type of phytomer, and a "macro-state" that describes a growth unit (GU) that consists of a set of phytomer(s) with the same PA. A macro-state is generated by a macro-DC and a phytomer is generated by a (normal) DC. A fixed number of DC is needed for a macro-state. Therefore, the total number of DCs of an axis equals to the number of macro-cycles of the axis multiplied by the number of DCs per macro-state. In temperate trees, the macro-state corresponds to the annual growth that can generate one or more GUs.

4.2.4 Different Types of Phytomers

The phytomers of a leafy axis can be morphologically different, although they belong to the same PA. They may or may not bear leaves, fruits, or branches. A phytomer that bears a branch is represented by a micro-state consisting of an internode, a node with one or more axillary buds, which have either the same PA (in the case of reiteration) or a greater PA (in the case of branches) on the node. The combinations between the PAs of the internode and the bearing branch define the type of phytomer (Fig. 4.2). Cotton plants and coffee trees have phytomers that can bear both reiterations and branches.

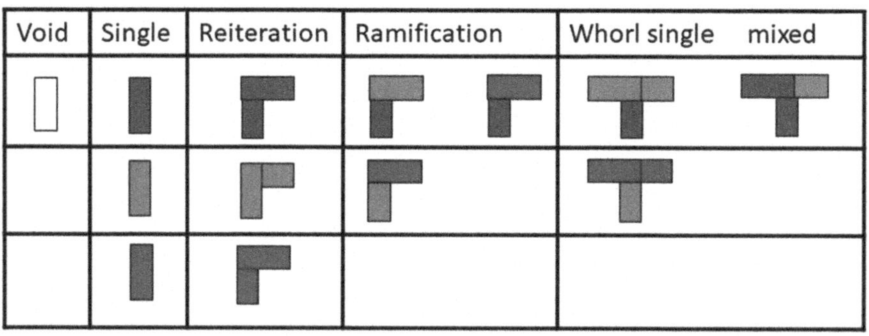

Fig. 4.2 Combinations of possible micro-states for phytomers with structures at three physiological ages (PAs). A void entity corresponds to a virtual phytomer during a pause and displays in chronological mode. The phytomer can remain unbranched or bear one or several branches of different PAs

4.2.5 Graphical Models of Micro-States

The functioning of a meristem can be represented by a botanical automaton that produces a series of micro-states through repetition that represent the different types of phytomers (Fig. 4.2). The result is a "development axis". In the stochastic case, these phytomers are done with a probability b and pauses with a probability $1 - b$. Phytomers are created alternately with void entities. A *Bernoulli* process is generated (see Chap. 3, and Fig. 4.3a).

If we consider the viability c, which is the survival rate of a meristem at each cycle, the process becomes compound (see Chap. 3, and Fig. 4.3b).

The meristem can also change by a transition of PA after a certain number of repetitions of its functioning (Fig. 4.3c).

Finally, a meristem can give birth to axillary meristems that will produce branches of different PAs with a branching probability a (Fig. 4.3d).

If the functioning is continuous, the automaton is simple and produces micro-states. If the functioning is rhythmic, the automaton is dual-scale and produces macro-states that represent the GUs. The development of axes in macro-states is also stochastic and follows a *Bernoulli* process, to which a survival rate can be added. The development axis is therefore made up of macro-states that are composed

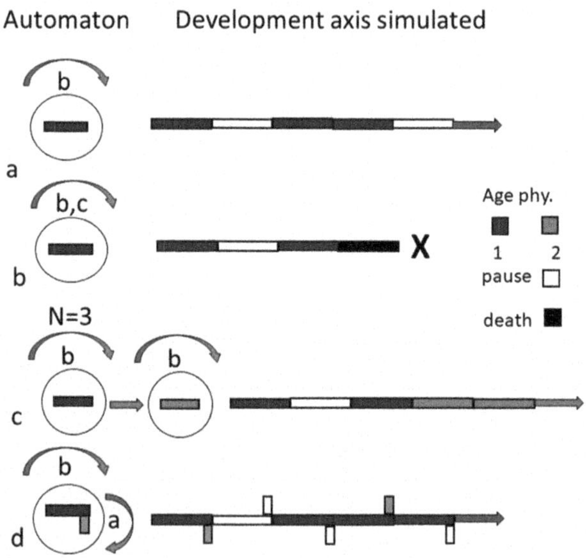

Fig. 4.3 Graphical model of the functioning of micro-states. A colored rectangle means a normal activity of meristem, a blank one means a pause, and a black one means the death of meristem. (**a**) The repetition of the micro-states with a probability b generates a *Bernoulli* process. (**b**) The consideration of viability c eventually leads to the death of the meristem. (**c**) The transition in PA occurs after a number of N repetitions (here $N = 3$). (**d**) The probability of branching a produces branched micro-states in proportion a and unbranched in proportion $1 - a$. There cannot be any ramification on void micro-states

of micro-states. In the following graphical models of micro- and macro-states, the functioning probabilities, which are implicit, are not displayed.

4.2.6 Production of Micro-States in Continuous Development

In continuous development, the same micro-state repeats itself to produce an axis of development, according to a law of occurrence, or changes types according to an alternation law as in Markov chains (Fig. 4.4).

4.2.7 Production of Macro-States in Rhythmic Development

In a rhythmic development axis, the GU defines a macro-state, which is considered as a meta-phytomer. We consider here the sequence of successive macro-states along the axis.

The same meristem sets up GUs (macro-states) whose chronological and physiological ages evolve separately, the latter being able to repeat or undergo a transition from one state to another. The PA of the axillary buds determines the possible types of micro-states for branching (Fig. 4.5). Leaves and fruits are attributes that characterize the type of phytomer; these attributes are necessary to be explicit if we are interested in plant production. The number of organs produced is determined by the types of phytomers.

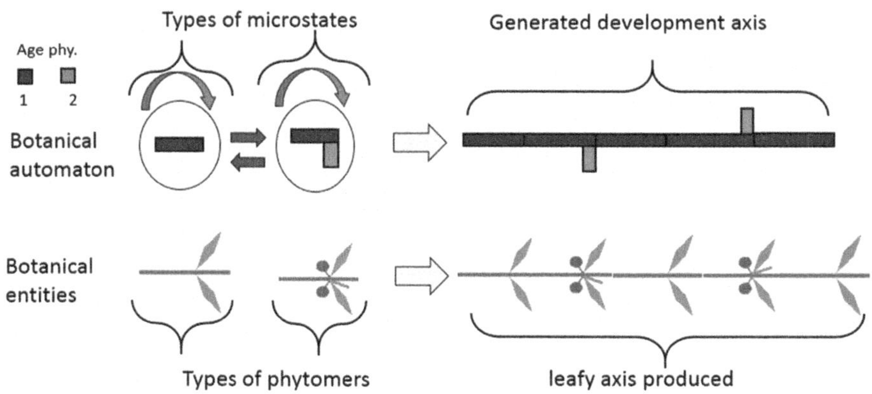

Fig. 4.4 Construction of a leafy axis with continuous development with two types of phytomers. The two types of phytomers, with the same physiological age (PA), differ in axillary production. Each type corresponds to a micro-state of the botanical automaton that simulates the functioning of the terminal meristem. Its operation generates series, repetitions, and alternations of micro-states, according to predefined rules, which produces a development axis, corresponding to the leafy axis composed of phytomers

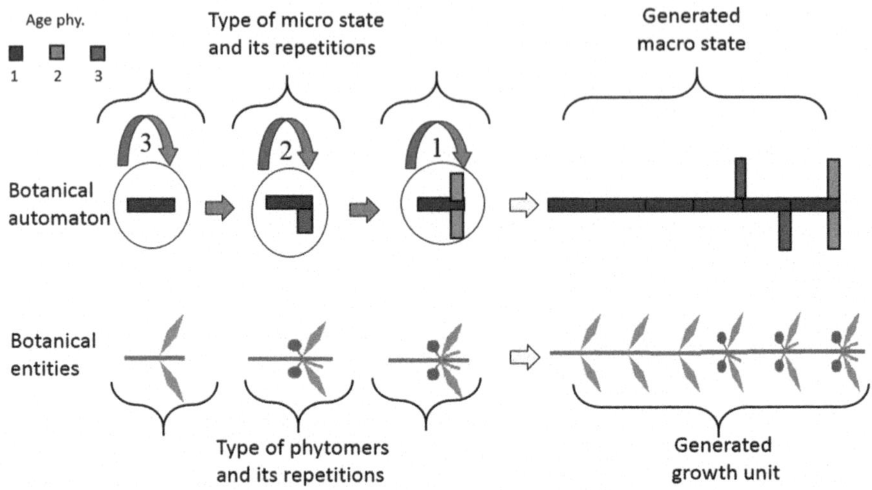

Fig. 4.5 Construction of a growth unit (GU) with different types of phytomers by repetition and alternation. Each type of phytomer is matched by a micro-state of the botanical automaton that simulates the functioning of the terminal meristem. Its operation generates alternating series of micro-states, according to predefined rules for producing a macro-state, homologous to the GU

The automaton is complicated in the case where there are several types of branched axes per node (for example, coffee, cotton). The generation of alternating series of micro-states is carried out according to the laws of occurrence (state occupation) and transition between these states. The nature of these laws (deterministic or stochastic) is indifferent; they must be identified in reality. The AMAP model often uses Markov chains. We will see, however, that it is the average number of different micro-states produced that is important for plant production, but not the patterns of their alternation.

Acrotony is a widespread phenomenon (Chap. 2). To model the gradient due to the acrotony along a GU, in the GreenLab model, the succession of micro-states in the macro-state is ordered from unbranched micro-states (older PA) to micro-states with increasingly vigorous branches (younger PA).

Finally, micro-states are assembled into macro-states (Fig. 4.6) to simulate GUs. Their succession by repetition constitutes the axis of development; the latter simulates the leafy axis. Along the same axis, the macro-states produced are similar according to a law of occurrence of the states as long as a transition does not modify their PA.

4.2.8 *Production of Structures*

Plant structures are made up of a development axis that branches into axillary sub-structures (Yan et al. 2003).

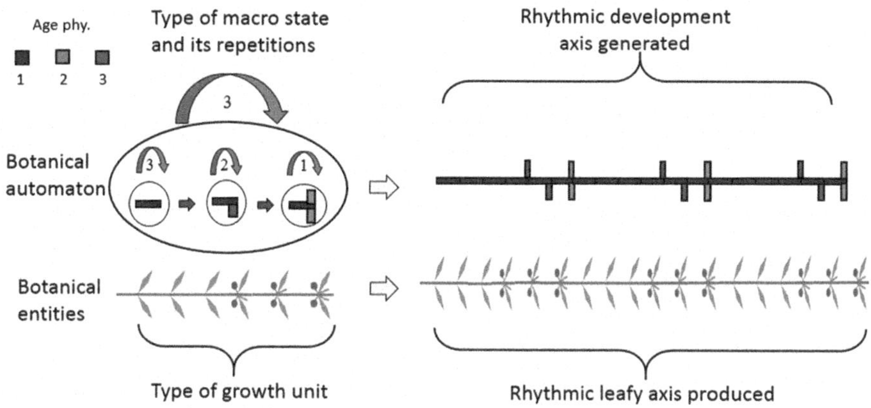

Fig. 4.6 Construction of a rhythmic leafy axis formed by growth units (GUs) with different types of phytomers. The botanical automaton builds macro-states that are made up of micro-states. Their repetitions generate a rhythmic development axis that represents the leafy axis

If the meristem of the leafy axis remains the same PA, it creates a series of identical phytomers. However, when its functioning changes, a transition in PA is observed. In this case, the development axis is considered composite. The increase in PA can be by one or several units. The structure can be simple if it is made up only of micro-states, or compound if it is made up of macro-states.

4.2.8.1 Simple Structure

Let us take the example of a simple botanical automaton with three micro-states and three PAs (Fig. 4.7). The first state corresponds to the PA 1 that gives birth to a stem with continuous development for eight cycles. The second state corresponds to the PA of the branches (PA = 2) that functions over a limited period of six cycles. The third state corresponds to short fruiting shoots that operate for a short period of two cycles. An incremental transition of a unit of PA takes place when the repetitions of micro-states reach their limit values, which corresponds to the completion of the development axis.

4.2.8.2 Compound Structure

A compound structure is made up of GUs, because the development is rhythmic. The botanical automaton is on a double scale. A GU corresponds to a macro-state; it is formed during a macro-cycle (which is made of DCs). A phytomer corresponds to a micro-state and is formed during a DC. Macro-states that simulate GUs are made up of several different micro-states. The duration of a macro-cycle is the same regardless of the PA of the meristem and the number of associated DCs for each meristem.

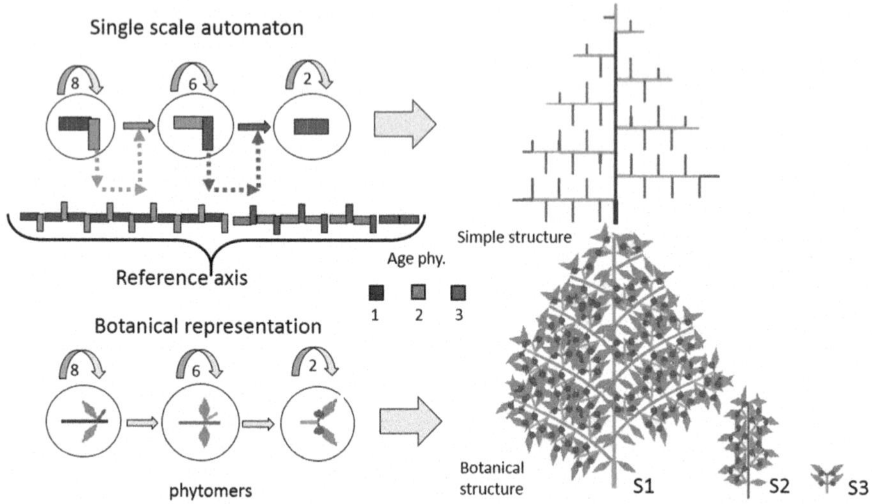

Fig. 4.7 The botanical automaton (left) and its corresponding simple structure (right). Simulation of a simple structure with three physiological ages (PAs) by a simple botanical automaton with three micro-states. The track of its functioning is a reference axis. The micro-states correspond to the phytomers. Structure *S*1 (the whole plant) is in development and is simulated at the tenth cycle. It is composed of a development axis with branched structures (*S*2) on it whose chronological age increases from the top to the bottom. Sub-structures *S*2 (branches) and *S*3 (short shoots) are also shown completed within six and two cycles, respectively. The *S*1 structure extends into *S*2 by transition after eight cycles. The structure *S*2 extends into *S*3 by transition after six cycles. (VisualPlant software, CASIA, Zhao et al. 2003)

Therefore, the development of the structure's axes is synchronized. If the architecture is the only target, pauses of development can be omitted as shown in the example of the reference axis. However, if photosynthesis is considered, the number of DCs should include pauses because it should be the same for all macro-cycles of different PAs.

The repetitions of macro-states generate the axes of development at different PAs. Depending on its type, each phytomer can have one or more axillary structures with an equal or higher PA. The alternation between the types of axillary structures on a GU is simulated by the botanical automaton. The laws that control the repetition of micro-states and macro-states may be arbitrary. Usually, transitions correspond to an increase in the PA of a unit. To be clear, consider a structure with three PAs that gives rise to four types of phytomers, three types of GUs, and three types of axillary sub-structures (Fig. 4.8). It is assumed that the first two structures have no fruiting and the last one is fruitful.

The steps in the construction of this structure composed of nine growth macro-cycles are explained in Fig. 4.9, from the point of view of the automaton for botanical representation with the displayed organs.

The age of a compound structure is counted in number of macro-cycles.

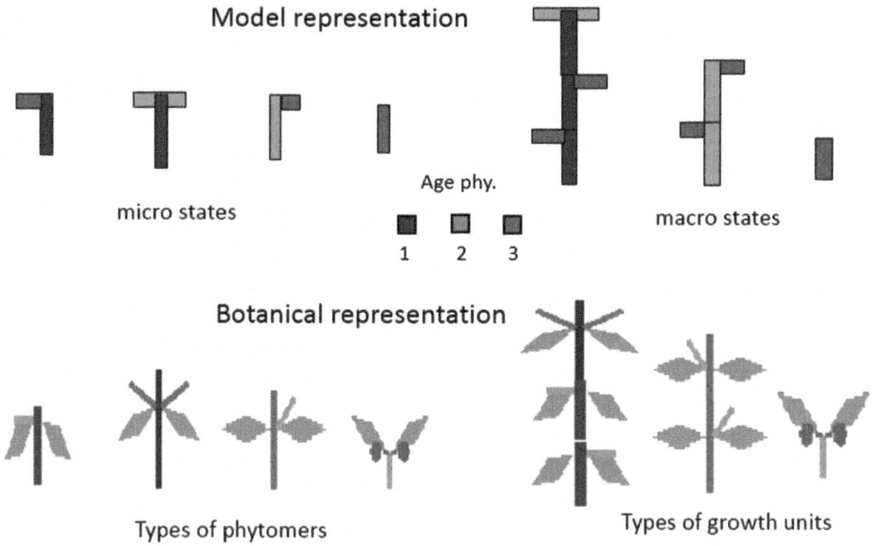

Fig. 4.8 Elements of a compound structure: correspondences between the micro-state and the phytomer, and between the macro-state and the GU. In micro-states, carrier-supported nodes are represented in color according to their PA; organs are omitted

4.2.9 Reiteration

Reiteration, in the botanical sense, is the duplication of a structure. The main structure before reiteration is called S_1. If S_1 is totally duplicated, it is defined as total reiteration (Fig. 4.10); otherwise, if the structure of PA k (k being greater than 1), S_k, is duplicated, it is defined as partial reiteration (Fig. 4.11). There are two extreme possibilities:

- The main bud transmits its ontogenic and physiological ages to the axillary bud. Stem and reiterations are synchronized in their development and the generated structure is monopodial (Fig. 4.10).
- The main bud only transmits its PA to the axillary bud. The ontogenic age is reset. Stem and reiterations are desynchronized in their development and the generated structure is sympodial (Fig. 4.12).

These two possibilities can easily be simulated by the botanical automaton described so far.

4.2.9.1 Total or Partial Monopodial Reiteration

The order of reiteration is introduced, similar to the order of branching. Order 0 of the reiteration corresponds to the non-reiterated structure (corresponding to the

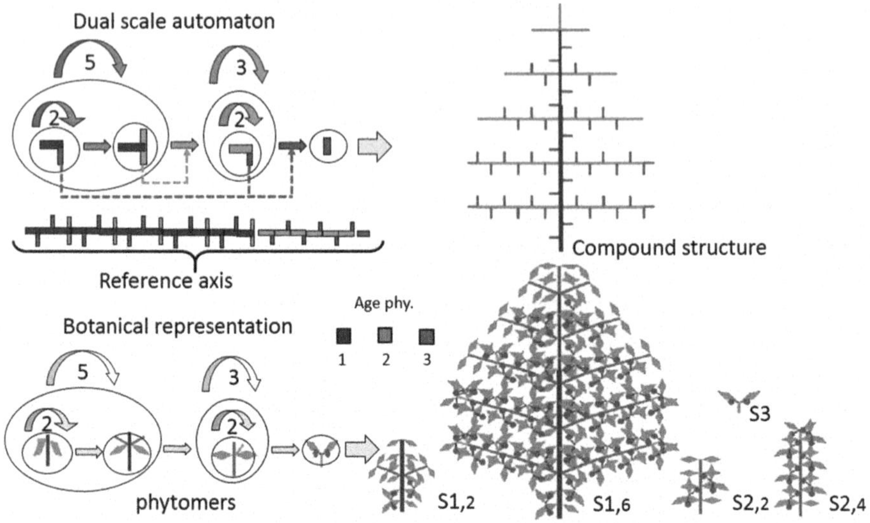

Fig. 4.9 Correspondence between a compound structure and its botanical automaton. Simulation of a structure composed at three physiological ages (PAs) by a double botanical automaton with four micro-states. The automaton is on a double scale with macro-states being composed of repeated micro-states. The track of its functioning is a reference axis. Thus, the main meristem of the stem (in blue) is programmed to make five macro-states in PA 1. It has a transition from PA 2 (in green) to the sixth macro-state. At this stage, the stem is transformed into a branch of PA 2. The meristems of the branches make three macro-states (in green) and have a transition into the ultimate PA (in red) forming a short axis composed of a single micro-state. Note that the short axes are found on both the stem and the branches. $S(1, 2)$ and $S(1, 6)$ are botanical representations of the plant (structure $S1$) with two and six development macro-cycles. $S(2, 2)$ and $S(2, 4)$ are representations of the branches (structures $S2$) with two and four macro-cycles (completed after transition). $S(3)$ is a short axis (structure $S3$ completed). (VisualPlant software) (Color figure online)

"architectural unit" in botany); order 1 corresponds to the first level of reiteration, etc. The order of reiteration is considered to be limited. The maximum order of possible successive repetitions of PA φ is denoted as R_φ. If N_φ is the number of macro-states (GUs) of the development axis of PA φ, we have by construction $R_\varphi < N_\varphi$.

The automaton makes it possible to simulate another notion of architectural botany: intercalation. During intercalation, structures of a higher PA appear within the reiteration, thus gaining another order of branching. For example, structures of PA 2 can be branching order 2, 3, or 4, because of reiteration (Fig. 4.10). Thus, PA is no longer directly related to the order of branching.

The complexity of the structure increases with the increase in the order of reiterations. The tree becomes a "colony" of structures (Fig. 4.10).

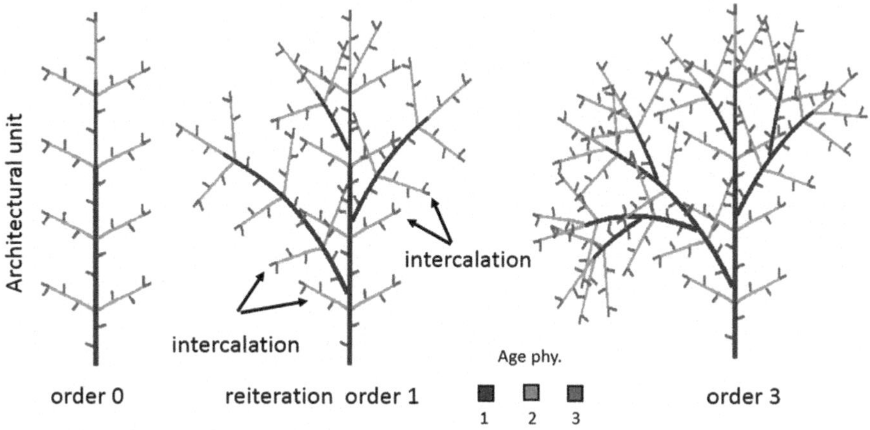

Fig. 4.10 Monopodial reiteration of *S*1 with orders 0, 1, 3, involving the ontogenic age and intercalation of *S*2 and *S*3. The compound structure *S*1 (not reiterated) is the "architectural unit" in the botanical sense. In reiteration order 1, we observe the first level of reiteration of *S*1. At reiteration order 3, there is a complete repetition of *S*1. It should be noted that the trunk and the reiterations have simultaneous transitions in physiological age (PA), which characterizes the effect of ontogenic age. (VisualPlant software)

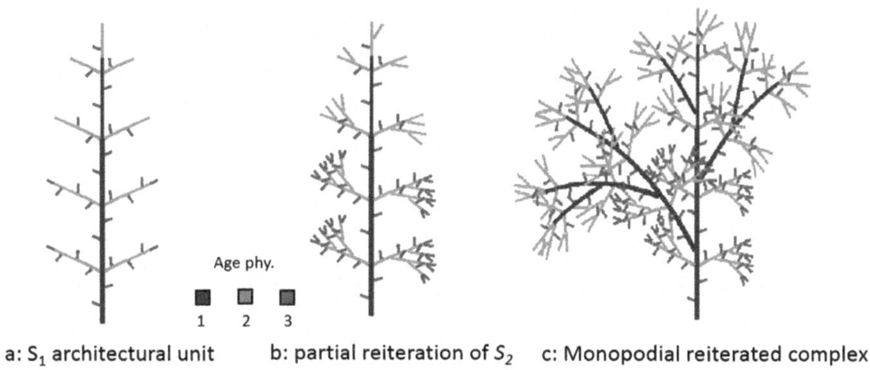

Fig. 4.11 Structure *S*1 (**a**) forms a finished architectural unit (AU). The simulated plant here duplicates its S_2 structures (partial reiteration) (**b**). If all PAs reiterate, we obtain a reiterated complex structure (**c**). (VisualPlant software)

4.2.9.2 Reiterated Partial and Reiterated Complex

Partial repetition is the duplication of a structure with a PA greater than 1. For total repetition, a structure that reiterates over all PAs is defined botanically as a reiterated complex (Fig. 4.11).

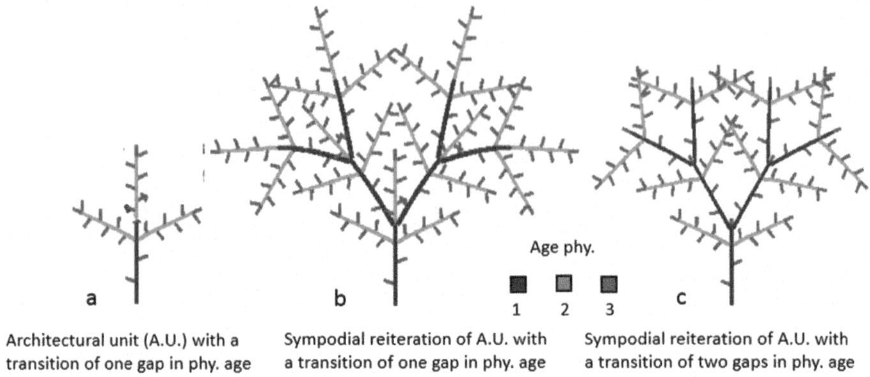

Fig. 4.12 Sympodial reiterations of a structure. (**a**) Architectural unit (AU). (**b**) Sympodial reiteration of the AU with one gap of transition in the axis of development. (**c**) Sympodial reiteration of the AU with two gaps of transition. The development axes are limited here to a single macro-state; they could have several. (VisualPlant software)

4.2.9.3 Sympodial Reiteration

If the ontogenic age (CA) of the main bud is not transmitted to the reiteration, it is reset and development will have a sympodial character. The resulting structure will be composed of a succession of forks. The development of the structure is limited only by the maximum order of reiterations allowed. In addition, each PA may have its own monopodial or sympodial branching pattern. According to the laws of transition between PAs in the botanical automaton, reiterated structures may have a more or less discontinuous appearance (Fig. 4.12).

4.2.10 Reference Axis and Architectural Metamorphosis

The reference axis (Chap. 2) is the linear representation of all PA transitions of meristems from the seed to the terminal flowering (Fig. 2.7). All plant axes are contained within the reference axis. They emerge with varying degrees of maturity according to their initial PAs, and all die at the same stage of ultimate differentiation (except for accidental mortality). As previously mentioned, the reference axis is the trace of the functioning of the botanical automaton.

Each development stage is associated with a PA that controls the values of the parameters of meristem function and organ dimensions. These parameters may be stable or may gradually change from one stage to the next. In this system, transitions in PA are therefore linked to development cycles. The disadvantage of the multiplication of PAs is compensated by the possibility of interpolating between series of measurements carried out at different stages of development.

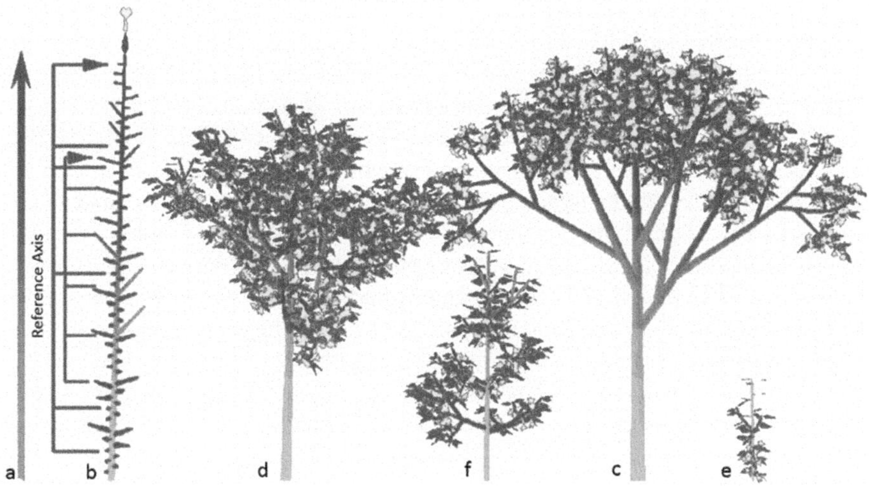

Fig. 4.13 Metamorphosis of the architecture of a tree during its development under the effect of aging meristems and controlled by a reference axis. The green color represents initial vegetative stages, and the red color represents terminal floral stages. The differentiation of the meristems into leaves and flowers is placed at the end of the reference axis. We continuously move from an initial differentiation stage to the final stage along an axis (**e**, **f**, **d**, **c**). The branching generates a discontinuity in PA between carried axes and carrier axes. Axillary branches are born older than the bearing stem by skipping steps, which also shortens their life span. (**a**) PA gradient. (**b**) Reference axis. The jumps of the axillaries are shown. Four stages of development, i.e., (**e**), (**f**), (**d**), (**c**), are represented. Stage **e** is the young plant. Stage **f** is the young tree with a strongly hierarchical structure around the trunk and where the architectural model is expressed. Stage **d** is characterized by the appearance of reiterations that allow the tree to explore space. Stage **c** is the maturity stage. The peripheral axes tend toward the same morphology (short axes) characterizing the terminal floral stage (Color figure online)

The age of the tree corresponds to the number of GU of the trunk. The number of phytomers in a GU is directly derived from the statistical distribution observed at the development stage considered. The main axis (trunk) is represented by a series of GUs.

The GU is also characterized by a series of axillary branches, generally organized according to the acrotony rule (Chap. 2). The axillary buds are born physiologically older than the main bud. Axillary buds get "jumps" forward along the reference axis (Fig. 4.13). These buds will therefore function from the stage in the reference axis that reproduces their functioning. Depending on the jump, an axillary branch will have a varying life span. If the jump is local, it will mark a repetition of the functioning of the carrying branch and will allow to simulate a reiteration. If the jump is terminal, i.e., at the end of the reference axis, it will be a short branch with a short life span, because reaching the end of the axis means the end of the functioning for a meristem. Organs (leaves, fruits) also have their place in this system. They are derived from the meristems positioned at the end of the reference axis and are emitted over a single DC.

With the reference axis, the PA of apical meristems through the interplay of transitions catches up with the PA of axillary meristems, which characterizes the aging of the plant structure. At the end of the process, there is no longer any difference between a carrier and a carried branch. When all the meristems have completed their evolution, and have therefore reached the end of the reference axis, the structure is complete and can no longer evolve. For any age smaller than the number of steps of the reference axis, we obtain a structure in development. The stochastic simulation of the functioning of meristems makes it possible to simulate the observed probability laws and to restore the polymorphism of the GUs. However, the numerical expression of these laws is only known for the measured ages of plant development. Thanks to the linear interpolation between the values obtained over the different observed development stages, it is possible to follow and simulate the evolution of the functioning of meristems in a continuous way during the implementation of plant architecture. This makes it possible to follow the architectural metamorphosis over long periods of development (Fig. 4.13).

4.2.11 Automaton and Architectural Models

The botanical automaton of the GreenLab model makes it easy to program the development of a plant and obtain its architecture. Thanks to the notions of PA, transition, micro-, and macro-state, practical user interfaces make it possible to define all types of phytomers, with their transition and branching laws, including their stochastic components that control the functioning of meristems. This was done for the first time with the VisualPlant software developed by Xing Zhao at LIAMA (Zhao et al. 2001, 2003).

4.2.11.1 Simulated Architectural Models

The architectural models of botany contain information on continuous or rhythmic development, monopodial or sympodial branching, axial or apical flowering, tropisms of the axes (orthotropic or plagiotropic), etc. (Hallé et al. 1978; Hallé and Keller 2019). With the definition of the attributes of organs and tropisms, the flexibility of the botanical automaton allows to simulate all architectural models (Fig. 4.14). The 9 models represented illustrate a selection from the collection of 26 architectural models.

4.2.11.2 Reiteration of Architectural Models

Architectural models allow reiterations. The architecture becomes more complex without increasing the number of PA (types of axes) (Fig. 4.15).

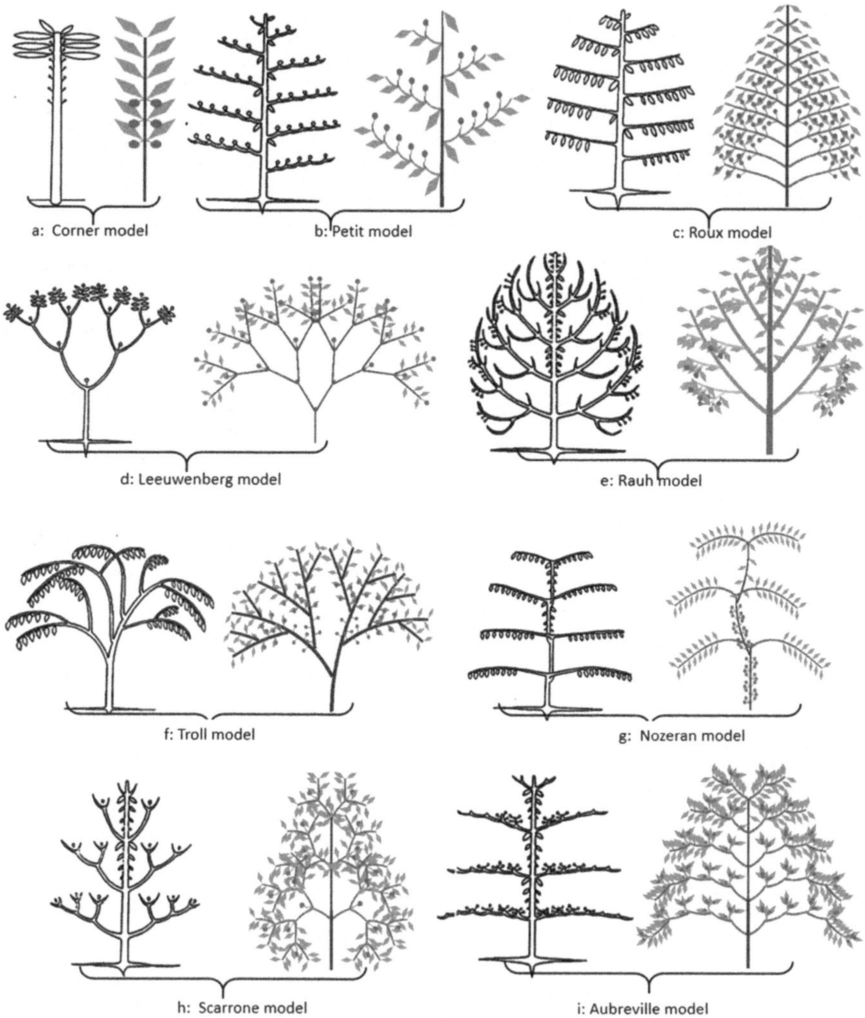

Fig. 4.14 Simulations of architectural models (Zhao et al. 2001). Comparisons between botanical drawing (from AMAP) and 2D simulations (from VisualPlant software). (**a**) Corner model (palm tree), (**b**) Petit model (cotton plant) (**c**) Roux model (coffee tree), (**d**) Leeuwenberg model (frangipani tree), (**e**) Rauh model (pine tree), (**f**) Troll model (beech tree), (**g**) Nozeran model (cacao tree), (**h**) Scarrone model (mango tree), and (**i**) Aubreville model (*Terminalia* tree)

4.3 Applications to the Simulation of 3D Plant Architecture

The simulation of meristem functioning is done by means of a computer program; it builds a topological structure, sets up the phytomers, and links them together. Exploring this structure makes it possible to count the number of botanical entities produced. Starting from the topological structure in a planar mode, the addition of

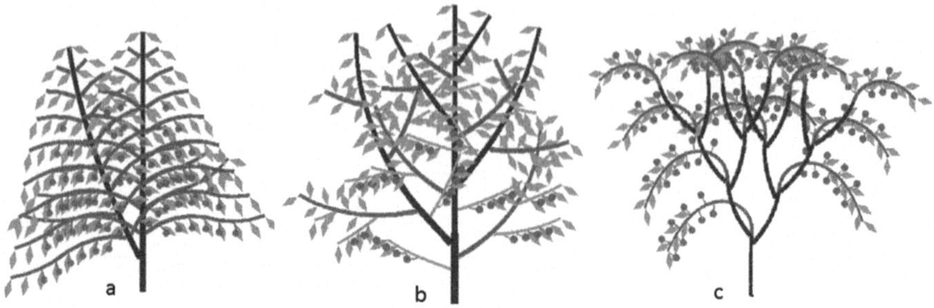

Fig. 4.15 Reiteration of three architectural models: (**a**) Roux model, (**b**) Rauh model, and (**c**) Mangenot model. (VisualPlant software)

geometric operators that integrate translations and rotations (phyllotaxis), as well as the shapes and sizes of the organs, makes it possible to move to the three-dimensional architecture of the plant (see Chap. 14). Finally, the plant can be visualized in the form of computer graphics images (Fig. 4.1). To remind, the simulation here is carried out by the development of the plant under the action of meristems, but not by the growth in biomass under the action of photosynthesis (the organs play no functional role here).

The plant development model has been experimentally validated on many species where the parameters of meristem functioning have been estimated from measurements made on architectures with the "crown analysis" method (Chap. 3). The results are presented below in the form of computer-generated images from the AMAP software (Barczi et al. 1997, 2008; Rey et al. 1998) and QingYuan software (Hua and Kang 2011). These two software programs simulate the stochastic functioning of meristems and PAs with their transitions. The first is based on the notion of the reference axis (Chap. 2) and the second is based on the notion of the botanical automaton, which is the mathematical form.

4.3.1 Simulation of Cotton Plant Architecture

Cotton (*Gossypium hirsutum*) is a annually grown shrub with two PAs. The stem (PA 1) is with continuous growth, and each node has branches. Its basal vegetative branches are reiterations, followed by fruiting twigs (PA 2) with sympodial branching. Each module blooms individually. The parameters of cotton development were published by de Reffye et al. (1988). The parameter identification of meristem functioning was carried out on a sample of 50 plants for the same variety. Chapter 3 details the calculation algorithms applied to cotton (crown analysis). The stems and branches have a rhythm ratio and distinct *Bernoulli* processes. The vegetative branches have a delayed development, whose law is identified from the sample. Additional measurements of plant geometry can be used to simulate stochastic plants in accordance with natural variability (Fig. 4.16).

Fig. 4.16 Three cotton plants showing variability from random simulations (AMAP software). The bearing structure is pyramidal. A distinction is made at the base between vegetative branches, whose delayed branching law gives very variable developments

*4.3.2 Architecture Simulation of Coffee Tree (*Coffea canephora*)*

The coffee tree is a shrub whose cultivation is sustainable. It has two PAs. The orthotropic stem (PA 1) has a continuous and indefinite development. Each phytomer has two leaves and potentially two plagiotropic branches (PA 2). These branches have an immediate branching and continuous development that is stopped by an increasing mortality rate associated with the age of the branch. Supernumerary meristems can give rise to reiterations of the stem. The plant studied is a cloned *Coffea canephora*. The meristems of the stems work according to a classical *Bernoulli* process. The meristems of the branches have a more complex functioning. On one hand, a rhythm ratio is highlighted, and the *Bernoulli* and viability parameters are variable (de Reffye 1981a, b). The identification of meristem function parameters was carried out on a sample of 250 plants from the same clone. Chapter 3 details the calculation algorithms applied to coffee (crown analysis). The architecture simulation considers all the parameters of the functioning of the meristems identified on the plant, to which the organs are added with their shapes and dimensions as measured (Fig. 4.17).

4.3.3 Simulation of Development and Architecture of the Japanese Elm Tree

This tree has a development that gives at the end a spherical shape, due to the fact that all the peripheral annual shoots of the crown come from the same distribution

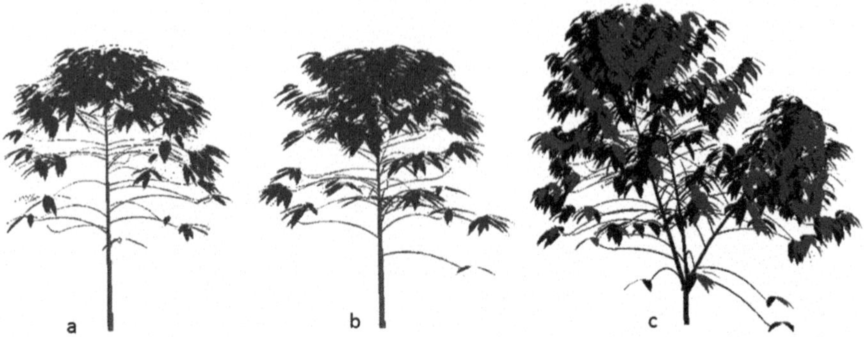

Fig. 4.17 (**a** and **b**) Two random stem simulations of a *Coffea canephora* clone (AMAP software). Note the effect of mortality in branch meristem produces natural pruning. The reduction of the *Bernoulli* parameter for the branches gives a columnar growth form. (**c**) Simulation of the same coffee tree with basal reiterations, as observed in its growth pattern

that evolves during the growth. Sampled shoots from the crown allow it to be characterized at each stage of development. The evolution of the parameters that characterize the functioning of the shoot, such as neoformation and polycyclism (several GUs per annual shoot), can be monitored by interpolation between stages (see Chap. 3).

Six stages of development were studied, ranging from 1 to 100 years, on which the parameters of GU development were estimated. Intermediate values are estimated by interpolation, which makes it possible to continuously monitor the construction of the architecture. Thus, before the age of 25 years, the plant's GUs are complex and make polycyclism and neoformation. After 25 years, polycyclism disappears. The neoformation gradually disappears and is totally absent at the age of 100 years. In Fig. 4.18, we display the simulated architecture at three development stages (3, 25, 100 years) as well as randomly selected simulated annual shoots from the respective crowns. At the age of 100 years, the tree has generated more than one million phytomers.

4.3.4 Genericity of the Developmental Algorithm

The method was applied to various plants (herbs, bushes, trees). The architectural analysis of a plant, presented in Chap. 3, gives the parameters of the functioning of meristems. In addition to the plants studied in the previous sections, diverse plants have been modeled, such as *eucalyptus* (Diao et al. 2012), Aleppo pine (Caraglio et al. 2007), oil palm with its root system (Jourdan and Rey 1997), bamboo (Dabadie et al. 1991), *Arabidopsis* (Christophe et al. 2008), and sunflower (Rey 2003). The sampling technique, based on the notion of PA and used in the crown

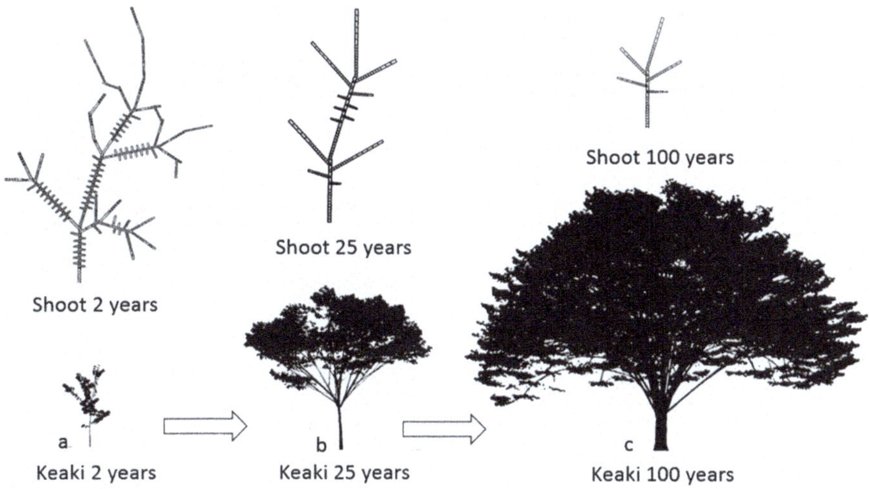

Fig. 4.18 Three stages of development of Japanese elm: (**a**) 2 years. (**b**) 15 years, and (**c**) 100 years. All the growth units (GUs) at the periphery of the tree come from the same distribution. This changes gradually with the age. An individual stochastic simulation of a shoot at the periphery of the tree is displayed above the plant (de Reffye et al. 1991). The development of GUs gradually decreases with the age of the tree (AMAP software)

analysis of the trees, has proven to be very generic and adaptable to all studied plants.

In addition, a survey of the dimensions of botanical entities and branching angles allows geometric operators implemented in a computer program to reconstruct the plant's 2D structure or 3D architecture (Chap. 14). The development of the plant is restored at all stages of the development in its stochastic components.

It is then possible to precisely evaluate the effects of genetic variability or environmental factors (stress, planting density, etc.) on the functioning of meristems, mainly concerning mortality and the ramification of branches. This allows the plant's architecture to be predicted under various environmental constraints.

AMAP software has been the pioneer in simulating plant architecture. Figure 4.19 shows various plants that have been studied and simulated for agronomic applications.

4.4 Conclusion

Chapter 3 explained how to simulate the stochastic development of the axes with development cycles. This chapter completes it by modeling the differentiation of the phytomers created by the meristems, either by extension or by branching of the axes according to the concept of PA. At each DC, a meristem can make a pause, produce a

Fig. 4.19 Architecture of various plants simulated by AMAP or GreenLab software: (**a**) root system of the palm tree (Jourdan and Rey 1997), (**b**) pea (Rey 2003), (**c**) *Arabidopsis* (Christophe et al. 2008), (**d**) sunflower (Rey 2003), (**e**) tobacco (Poisson et al. 1997), (**f**) bamboo (Dabadie et al. 1991), (**g**) *Cecropia* (de Reffye et al. 1988), (**h**) *Eucalyptus* (Coudurier et al. 1993), (**i**) pine (Caraglio et al. 2007), (**j**) apricot tree (Chavaneau 1989), (**k**) cherry (Fournier 1989), (**l**) palm (Atallaoui et al. 2016), and (**m**) *Cryptomeria* (de Reffye et al. 2008)

new phytomer according to a *Bernoulli* process, or abort. The PA of the main meristem can be repeated or changed through transitions, as well as that of the axillary meristems that have been put in place. The functioning can be continuous or rhythmic. The dual botanical automaton allows to generate all known architectural models with their reiterations.

The reference axis is the track of the functioning of the botanical automaton, which is the mathematical form. The AMAPsim software is based on the use of the reference axis, which make it possible to simulate the stochastic development of plants. But if the simulation is limited to the development of plant architecture, it does not integrate the notion of chronological age that manages the appearance of pauses in the functioning of meristems. Based on the same botanical concepts, the Gloups software based on the botanical automaton have taken over by improving calculation performance through factorization and considering the source–sink relationship involved in photosynthesis (Chap. 5).

If the aim is to couple the architecture and development, computer simulation makes it possible to obtain computer graphic images of 3D plants that are realistic from the point of view of the organogenesis with pre-defined organ dimensions, without assigning any functional roles to organs as sources and sinks.

References

Atallaoui K, BenMehaia R, Simozrag A, Lecoustre R, Rey H (2016) Modeling vegetative and reproductive architecture of date palm (Phoenix dactylifera L.) The case of two Algerian cultivars 'Deglet Nour'and 'Litima'

Barczi J-F, de Reffye P, Caraglio Y (1997) Essai sur l'identification et la mise en œuvre des paramètres nécessaires à la simulation d'une architecture végétale: le logiciel AmapSim. In: Bouchon J, de Reffye P, Barthélémy D, éds. Modélisation et simulation de l'architecture des végétaux. Paris, Éditions Inra, Sciences Update, pp 255–423

Barczi J-F, Rey H, Caraglio Y, de Reffye P, Barthélémy D, Dong Q, Fourcaud T (2008) AmapSim: an integrative whole-plant architecture simulator based on botanical knowledge. Ann Bot 101(8):1125–1138

Caraglio Y, Pimont F, Rigolot E (2007) *Pinus halepensis* Mill. Architectural analysis for fuel modelling. In: Proceedings of the international workshop MEDPINE. CIHEAM-IAMB, Bari, pp 43–59

Chavaneau C (1989) Modélisation de la croissance et de l'architecture de l'Abricotier (*prunus armeniaca L.*) D.A.A., spécialité: Agronomie méditerranéenne. Montpellier, ENSA. 77 p

Christophe A, Letort V, Hummel I, Cournède P-H, de Reffye P, Lecœur J (2008) A model-based analysis of the dynamics of carbon balance at the whole-plant level in Arabidopsis thaliana. Funct Plant Biol 35(11):1147–1162

Coudurier T, Barczi J-F, Barthélémy D, de Reffye P (1993) Modélisation et simulation de l'architecture de l'Eucalyptus hybride Saligna en plantation au Congo. (Rapport ATP n° 37/93). Cirad-Gerdat, Montpellier, p 28

Dabadie P, de Reffye P, Dinouard P (1991) Modelling bamboo growth and architecture: phyllostachys viridi-glaucescens. J Am Bamboo Soc 8(1-2):65–79

de Reffye P (1981a) Modèle mathématique aléatoire et simulation de la croissance et de l'architecture du caféier Robusta. 1ère partie. Étude du fonctionnement des méristèmes et de la croissance des axes végétatifs. Café Cacao Thé 25(2):83–104 de Reffye P (1981b) Modèle mathématique aléatoire et simulation de la croissance et de l'architecture du caféier Robusta. 2ème partie. Étude de la mortalité des méristèmes plagiotropes. Café Cacao Thé 25(2):219–230

de Reffye P, Cognée M, Jaeger M, Traore B (1988) Modélisation stochastique de la croissance et de l'architecture du cotonnier. 1. Tiges principales et branches fructifères primaires. Coton et Fibres Tropicales 43(4):269–291

de Reffye P, Dinouard P, Barthélémy D (1991) Modélisation et simulation de l'architecture de l'Orme du Japon *Zelkova serrata* (Thunb.) *Makino* (Ulmaceae): la notion d'axe de référence. Comptes Rendus du 2e Colloque international sur l'arbre, Montpellier, 10–15 Septembre 1990. Naturalia Monspeliensia, Hors-Série, pp 251–266

de Reffye P, Houllier F, Blaise F, Barthélémy D, Dauzat J, Auclair D (1995) A model simulating above- and below-ground tree architecture with agroforestry applications. IUFRO World Congress. 20, 1995-08-06/1995-08-12, Tampere, Finland. Agrofor Syst 30(1-2):175–197

de Reffye P, Jaeger M, Cournède P-H (2008) Le monde végétal. In: Comité International des Jeux Mathématiques (ed) Maths nature express. CILM Paris, pp 26–35

Diao J, de Reffye P, Lei XD, Guo H, Letort V (2012) Simulation of the topological development of young eucalyptus using a stochastic model and sampling measurement strategy. Comput Electron Agric 80(1): 105–114

Fournier D (1989) Modélisation de la croissance et de l'architecture du Merisier (*prunus avium L.*) DAA spécialité: Agronomie méditerranéenne. ENSA, Montpellier, 77 p

Hallé F, Keller R (2019) Mais d'ou viennent les plantes ? Actes Sud

Hallé F, Oldeman R-A-A, Tomlinson P-B (1978) Tropical trees and forests. Springer Verlag, Berlin/Heidelberg/New York, 441p

Hua J, Kang M-Z (2011) Functional tree models reacting to the environment. In: ACM SIGGRAPH 2011 posters. ACM, Vancouver

Jourdan C, Rey H (1997) Modelling and simulation of the architecture and development of the oil-palm (*Elaeis guineensis* Jacq.) root system. Plant Soil 190(2): 235–246

Poisson C, Rey H, Caraglio Y, Rabaste S, Barczi J-F, Guedon Y (1997) Modélisation de l'architecture et de la croissance de 5 espèces du genre *Nicotiana. Annales du Tabac*, France. Section 2(29): 37–54

Rey H (2003) Utilisation de la modélisation 3D pour l'analyse et la simulation du développement et de la croissance végétative d'une plante de tournesol en conditions environnementales fluctuantes (température et rayonnement). Doctoral dissertation, ENSA, Montpellier

Rey H, Barczi J-F, Caraglio Y, de Reffye P (1998) AmapSim: un outil de construction et simulation de plantes numériques tridimensionnelles: principes et applications. In: *Architecture et modélisation en arboriculture fruitière*, 11ème colloque sur les recherches fruitières, Montpellier, France 5-6/03/1998. Inra-Ctifl, pp 296–306

Yan H-P, de Reffye P, Pan C-H, Hu B-G (2003) Fast construction of plant architectural models based on substructure decomposition. J Comput Sci Technol 18(6):780–787

Zhao X, de Reffye P, Xiong F-L, Zhan Z-G (2001) Dual-scale automaton model for virtual plant development. Chin J Comput 24(6):608–617

Zhao X, de Reffye P, Barthélémy D, Hu B-G (2003) Interactive simulation of plant architecture based on a dual-scale automaton model. In: de Reffye P, Zhao X (eds) Plant growth modeling and applications. Proceedings PMA03: 2003' International symposium on plant growth modeling, simulation, visualization and their applications, Beijing, China, October 13–16, 2003. Tsinghua University Press/Springer, Beijing, pp 144–153

Chapter 5
Modeling Plant Growth: Deterministic Case

Philippe De Reffye, Ep Heuvelink, Véronique Letort, and Mengzhen Kang

Abstract This chapter introduces a detailed mathematical model for simulating plant growth, with a focus on integrating physiological crop models to accurately represent biomass production and distribution. The GreenLab model allows for the analysis of individual plant growth at the phytomer level, reconstructing the plant's three-dimensional (3D) architecture from a two-dimensional (2D) planar structure formed by meristem function. Based on a source-sink functioning process, organs within the model are assigned specific roles in biomass production and allocation. The model calculates the biomass demand by evaluating the sink strength of organs produced per cohort, considering their physiological and chronological ages. The model uses a recurrence mechanism starting from the seed to track the growth and development of plants over time. The integration of allometric relationships ensures that the organ morphology is consistent with their functional roles. The use of physiological age allows to decompose the plant structure in separated pieces of phytomer axes, known as organic series. Organic series contain the plant growth memory. The organic analysis method infers growth parameters from plant

P. De Reffye
CIRAD, UMR AMAP, Montpellier, France

AMAP, Univ Montpellier, CIRAD, CNRS, INRAE, IRD, Montpellier, France

E. Heuvelink
Horticulture and Product Physiology, Wageningen University and Research, PB, Wageningen, The Netherlands
e-mail: Ep.Heuvelink@wur.nl

V. Letort
Mathématiques et Informatique pour la Complexité et les Systèmes, CentraleSupélec-Université Paris-Saclay, Gif-sur-Yvette, France
e-mail: veronique.lechevalier@centralesupelec.fr

M. Kang (✉)
State Key Laboratory for Multimodal Artificial Intelligence Systems, Institute of Automation, Chinese Academy of Sciences (CASIA), Beijing, China

School of Artificial Intelligence, University of Chinese Academy of Sciences, Beijing, China
e-mail: mengzhen.kang@ia.ac.cn

P. De Reffye et al. (eds.), *Plant Model Greenlab for Botany and Agronomy*,
https://doi.org/10.1007/978-981-97-7541-5_5

architecture, aiding in data restoration. This mathematical model not only bridges the gap between functional–structural and physiological crop models but also provides a practical approach to crop production and agronomic applications.

Keywords Crop models · Photosynthesis · Source-sink functioning · Growth equations · Cohorts · Organic series · Free and limited growth

5.1 Introduction

This chapter aims to put plant growth into mathematical equations. In this approach, particular attention is paid to staying as close as possible with physiological crop models, to keep strong compatibility. These models are usually photosynthesis driven and calculate the biomass produced by a plant stand and its distribution in organ compartments (leaves, internodes, fruits, roots, etc.). Mathematical modeling in GreenLab makes it possible to study plants at the level of the architecture of the individual plant, and more precisely at the level of the phytomer and the organs. The stand becomes a set of phytomers composed of organs distributed over a set of plants.

The architecture of a plant is reconstructed in three-dimensional (3D) using geometric operators from a two-dimensional (2D) planar structure, which is generated by the functioning of meristems. This structure is created using scaled objects representing the organs (Chaps. 4 and 14). These organs are now assigned functional roles in terms of biomass production and distribution to calculate plant growth. Biomass is synthesized by leaves (source organs) and then distributed to all the functioning organs belonging to the plant architecture (sink organs). There are new organs created in the structure, as well as old ones that continue to expand in volume (including leaves). By feedback, the updated leaf area increases the plant's growth. The plant development model resulting from the functioning of meristems (Chap. 3) makes it possible to calculate the number of organs produced per cohort (being defined as a layer in plant crown composed of organs of the same physiological and chronological ages), to obtain the "biomass demand" of the plant by considering their sink strength. The system works by recurrence from the seed. The morphology of organs depends on their biomass, density, and shape coefficients (allometries). The architecture of the plant produced by simulating the growth is not visually different from that provided by simulating the development with empirically pre-scaled organs. However, the organs now play their functional roles. This allows a new approach to crop production and new agronomic applications.

5.2 Basic Biological Processes of Plant Growth and Production

The growth of a plant is dependent on two biological processes that interact:

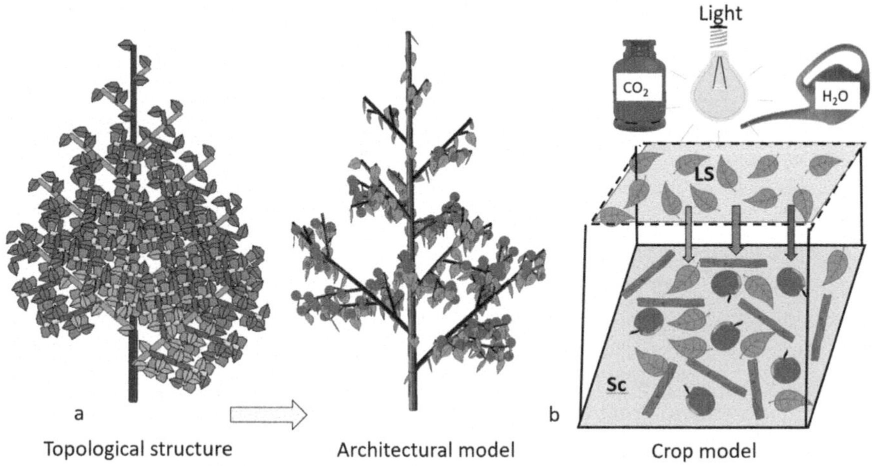

Fig. 5.1 Architectural plant model and physiological crop model. (**a**) Architectural model for development, a topological representation of the planar structure of a plant with different types of axes corresponding to several physiological ages (PAs) (4 here, represented by the blue, green, red, and yellow color, respectively). The architecture of the plant is obtained by giving the botanical entities their geometric positions, shapes, and dimensions. (**b**) Physiological crop model for growth and interception of light by a plant canopy. Plants are not considered as individuals: what matters is the ratio between the active leaf surface (*LS*) and the ground area (*Sc*). The biomass produced is distributed in proportion in different plant compartments according to their sinks (leaves, internodes, and fruits) without knowing the position of the organs in the plant structure

- Development, which builds the leafy axes to form a topological structure (Fig. 5.1a) and depends on the functioning of meristems. Botany studies plant architecture as a composition of different types of axes, whose combination generates architectural models (Hallé et al. 1978). The plant is considered as an individual, and its architecture is seen as the result of a morphogenetic program, which does not interact much with the climatic environment in the absence of stress. Photosynthesis is ignored in this process.
- Growth, which depends on photosynthesis through the interaction of source–sink relationships among organs and causes organ expansion. Agronomy studies the production of a stand per unit area, using physiological crop models involving environmental factors and the interception of light by foliage. The plant is not considered as an individual in this context (Fig. 5.1b). The different types of organs are grouped into compartments (e.g., leaves, stems, fruits) at the plant stand level, expressed per unit ground area (often 1 m^2). Stand growth is, therefore, the result of both biomass produced per unit area and time, and its distribution into the various plant compartments.

The modeling approach described in this book (GreenLab model) synthesizes the two approaches by combining botanical architectural models with agronomic crop models to produce a computational plant model adapted to the needs of agriculture.

5.2.1 Axis Elongation and Thermal Time

Plant growth fluctuates as a result of variations in environmental factors (light, temperature, etc.). The elongation of the axes is irregular when measured according to the "calendar time" (days). It depends not only on the functioning of the meristems that create new phytomers but also on their expansion. If the temperature drops, the elongation slows down. Agronomists observe that the cumulative number of phytomers created by a meristem on a stem is a linear function of the sum of the daily average temperatures (called "thermal time") received by the crop. Each created phytomer corresponds to a development cycle (DC). The study of physiological crop models is then greatly simplified using thermal time, as the development rate becomes more stable.

While it seems relevant to discretize the developmental stages of a plant whose meristems produce integer numbers of phytomers in cycles, on the other hand, the expansion in organ volume remains continuous and has no explicit marker of time. The growth time of the organs, however, is related to the thermal time as for the development of the axes. The expansion of organs that depends on photosynthesis slows down when the temperature drops. The expansion duration of the organs also stabilizes when the thermal time is used as the unit of measurement instead of the calendar time. To synchronize growth and development in the plant's functioning, the GreenLab model uses the development cycle (DC) as a modeling time unit. The growth cycle (GC) is then equal to the development cycle (DC). In each cycle, meristems initiate new phytomers, and leaves produce biomass that is distributed in the plant according to the competition among different sink organs. This method makes it possible to monitor the growth of a plant, even when its development stops with the terminal flowering of the stem, as in the case of wheat, sunflower, and maize (Guo et al. 2006) (Fig. 5.2). The linear relationship between the number of cycles and thermal time is established during the development stage when phytomers accumulate, and then extrapolated beyond.

The plant's functioning is thus synchronized for development and growth.

5.2.2 Photosynthesis and Biomass Production

The fresh biomass (fresh matter) of the organs consists mainly of water (80–95%) absorbed from the soil, and dry matter (5–20%) mainly derived from photosynthesis, A small part of the dry matter consists of minerals necessary for physiological functioning (nitrogen, phosphorus, potassium, iron, etc.) but negligible in weight.

The dry biomass (dry matter) of the organs consists of the part of the fresh biomass, which remains when all its water content is removed by desiccation.

Biomass production depends mainly on environmental factors: light, temperature, water, and CO_2 concentration of the air.

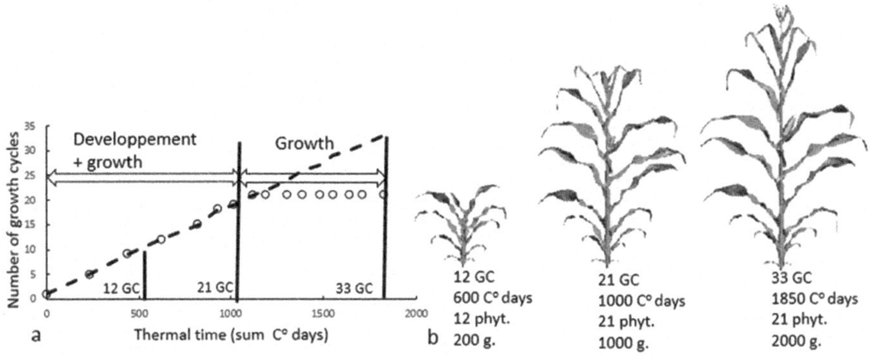

Fig. 5.2 Synchronization of growth and development cycles. Observations on the development and growth stages of a maize plant grown in China. "o" represents the number of observed phytomers at different stages of growth, "—" represents number of theoretical growth cycles (GC). Source: Ma Yun-Tao, Chinese Agricultural University (CAU). (**a**) The duration of the first stage is between germination and the formation of the terminal tassel, which stops the production of new phytomers but not the growth in cycles. During this period, a linear relationship between thermal time and the number of cycles is established, which is equal to the number of phytomers. In the second period, the relationship is extrapolated to the duration of the growth in the number of cycles. Note that the biomass produced doubles after the development has stopped. (**b**) Three growth stages of 12, 21, and 33 development cycles (DCs) are shown, with the corresponding number of phytomers, temperature sum, and weight of plants

The raw sap composed of water and mineral salts is extracted from the soil by the roots. By a physical phenomenon of suction, water is sucked through the xylem vessels to the surface of the leaves, where 90% of its volume evaporates. The remaining 10% contributes to the plant's fresh biomass. Sugars (assimilates) from photosynthesis are distributed to the organs to ensure their growth through specialized channels throughout the plant (phloem).

The activity of photosynthesis depends on the energy carried by the light and the surface of the foliage that intercepts it. de Wit (1978) firstly put forward the relationship between the biomass produced per m^2 of plants per day and the incident light energy:

$$Q(t) = C \cdot PAR(t) \cdot (1 - \exp(-k \cdot LAI(t))) \cdot \Delta t \qquad (5.1)$$

where *Q(t)* is the dry biomass produced by the stand on day *t* during the period Δt; PAR *(t)* (photosynthetically active radiation) is the amount of the sun's radiation that can be used for photosynthesis, measured in megajoules (MJ)/time unit/m^2; *C* is the light use efficiency (a conversion coefficient); *k* is the light extinction coefficient, a coefficient depending on the horizontality of the leaves; LAI, called leaf area index, is the leaf area (m^2) of the stand of plants per m^2 of ground area. The function $(1 - \exp(-k \cdot LAI))$ is derived from Beer-Lambert's law, which calculates the fraction of light intercepted by the foliage. It allows the overlapping rate of the leaves within the canopy to be considered in the calculation of photosynthetic production. This limits the interception of light by the leaf surface. The function gives the percentage of ground area shadow by the foliage per m^2. For LAI > 3, the plant intercepts almost all incident light. In the absence of stress (thermal, water), light use efficiency is a constant for a species.

For tomato cultivation in the Netherlands, Heuvelink (2006 personal communication) states that the overall radiation received by the stand in summer averages 17 MJ/m^2 per day; but only 50% of the radiation (PAR) is usable for photosynthesis and this rate decreases by a further factor of 0.7 if the plant is grown in a greenhouse, as light must pass through the greenhouse cover. The radiation received by the stand is therefore calculated as: $17 \times 0.5 \times 0.7 = 6.0$ MJ/m^2 per day. In winter, the global radiation drops to 2 MJ/m^2 per day, which is why supplemental lighting is necessay.

Since the temperature is controlled (≈ 20 °C) in modern greenhouses, it is usually not a limiting factor.

Beer Lambert's law of light extinction is generally well verified in greenhouse crops (Fig. 5.4).

The LAI is high during the production period: LAI > 3, so $(1 - \exp(-3 * 0.7)) \cong 0.9$, which means that about 90% of the light is intercepted by the foliage (with $k = 0.7$, typical for a horticultural crop).

A typical value for the light use efficiency of a tomato stand is 3 g/MJ PAR. It is stable and does not depend on the planting density. But it is a value that can change depending on the climatic factors in the greenhouse (e.g., about 30% higher at 1000 ppm CO_2 compared to 400 ppm CO_2).

We deduce from this that the crop production of dry matter per m^2 for a tomato stand in summer in a Dutch greenhouse is: $Q = 3 \times 6 \times 0.90 \approx 16$ $g/m^2/day$.

This value also depends on the plant species, but an order of magnitude of $Q \approx 20$ g/m^2 per day of dry matter is reasonable for the dry mass increase of a canopy in general in summer, which gives at a dry matter content of 6%: $Q \approx 270$ $g/m^2/day$ for the synthetized fresh matter.

If we consider the planting density d and a $k{\cdot}\text{LAI} > 3$ that produces a quantity Q of biomass per m^2 independent of the planting density, each plant will grow on average each day by an amount q inversely proportional to d: $q = Q/d$. The more plants there are, the more they are reduced, the limiting factor is the maximum production Q of the stand per unit area that does not depend on LAI and thus of planting density.

Figure 5.3 gives the effects of temperature and light on plant development and growth.

A genetic program controls step by step the functioning of the meristems to generate architectural development. In Fig. 5.3, after five repetitions of the same phytomer, the meristem transforms into a flower. The higher the ambient temperature, the faster the plant develops. The more incident light the plant receives, the bigger it grows.

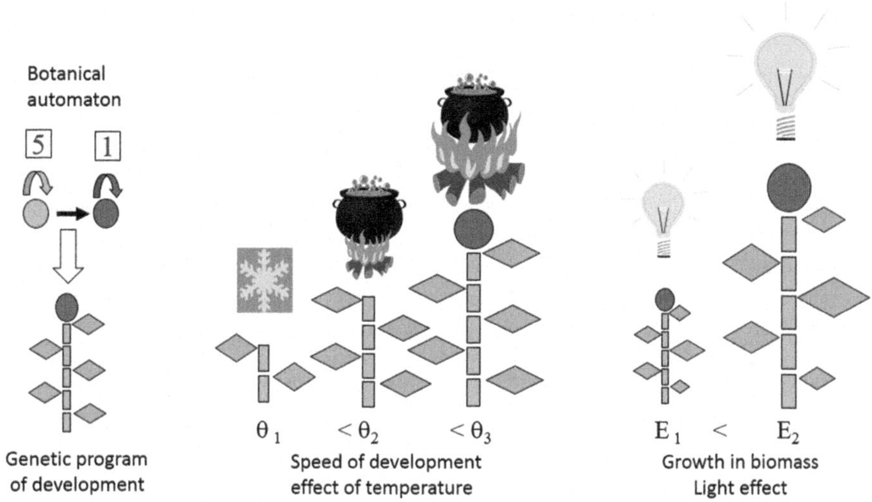

Fig. 5.3 Effect of light and temperature on plant development and growth

5.2.3 Common Pool

5.2.3.1 Use of Sugars by the Plant

Photosynthesis produces sugars (assimilates) that serve both as components and as combustible materials for the plant system. Before being used and distributed to the plant's organs, the synthesized sugars are initially stored in a virtual tank called a *common pool.*

Once transformed into cellulose, they constitute the cellular walls of the plant that build its structure. They form the largest proportion of the plant's dry matter. This material, once produced, no longer participates in metabolic processes; it is referred to as fixed carbon. When converted into starch, these sugars are stored as reserves within the vacuoles of the cells, appearing as grains. They are used for physiological functions of the plant system (transport of sugars in the structure, respiration), supplying the energy necessary for growth.

In a favorable environment, the proportion between the component and the combustible is stable. Depending on the type of component, the production cost is different. For instance, synthesizing oil requires twice as much sugar as producing cellulose, on a weight basis.

In this book, we focus only on "net photosynthesis," which refers to the portion of cellulose-derived sugars that contribute to produce the plant's architecture within a stable climatic environment.

5.2.3.2 Storage of Sugars by the Plant

The expansion of the organs in volume is ensured by their access to the sugars produced by photosynthesis. To simplify, everything happens as if a common biomass pool that is directly accessible by the organs. The notion of a pool has the advantage of not having to simulate the transport of the material produced by the leaves throughout the structure to feed the organs via the network of branches. The computer simulation of this transportation is costly in terms of calculation time. The assumption of a common pool has been validated as a good approximation and is frequently used in physiological crop models.

The plant's organs are divided into two categories:

- *Source organs:* They produce biomass through photosynthesis by intercepting light at their surface, using intracellular organelles called chloroplasts. These organs are also sinks because they are made of fresh biomass. The leaf is the main source organ. Its flattened shape optimizes light interception. Other organs may also be involved in photosynthesis, such as the sheaths of grass leaves, pods of legumes, siliques of cruciferous plants, and even young internodes. At the origin of growth, the seed is the initial source organ. As they become senescent, the organs can return part of their biomass to the common pool and thus become sources again: this is the process of remobilization. Organs like tulip bulbs or potato tubers are source organs at the start of cultivation.
- *Sink organs*. All organs (even leaves) are sinks that attract the produced biomass. The sink strength of an organ corresponds to the extent of its needs to sustain its expansion. It varies according to its type and developmental stage. In general, the sink strength increases from the creation of the organ to a maximum, then decreases and finally becomes null when the organ reaches maturity. Some sinks are particular: the woody rings of the leafy axes at the origin of secondary growth, and the roots. In GreenLab, the latter is simulated as a mere compartment because they cannot be sufficiently detailed. Another hidden sink is the reserve compartment consisting essentially of starch grains stored in the cells.

The geometrical shape of an organ is dependent on its volume (which depends on its weight and density) and on shape coefficients called allometries:

- The leaf is considered as a flattened volume whose thickness ε is constant. The surface is in these conditions proportional to the volume;
- The internode is considered as a cylinder defined by the surface of its section s and its height h. Its volume v ($v = h{\cdot}s$) depends on the quantity of biomass q and the density d: $v = q/d$; the shape coefficient β characterizes the elongation of the cylinder, and the coefficient γ is the allometry between s and h according to the relationship: $h/s = \beta{\cdot}v^{\gamma}$. The dimensions of the cylinder are then linked to the volume and allometries:

$$h = \sqrt{\beta} \cdot (v)^{\frac{1+\gamma}{2}}$$

$$s = \sqrt{\frac{1}{\beta}} \cdot (v)^{\frac{1-\gamma}{2}}$$

$$v = h \cdot s \tag{5.2}$$

Finally, to simplify, we can assume a fruit to be an ellipsoid, the half axes being the shape coefficients.

5.3 An Example of a Physiological Crop Model: TomSim

Numerous agronomic experiments on tomato have been carried out to identify the rules that underly the functioning of the TomSim model (Heuvelink 1995, 1996a, b). The production modeling is based on Eq. (5.1). The leaf area index (LAI) and compartment weights are recorded during crop growth.

5.3.1 *Light Interception by the Canopy in the Greenhouse*

The stability of light use efficiency and the validity of the law of light interception (Beer-Lambert) are well verified in the tomato experiments underlying TomSim. The LAI determines production per m^2 (Fig. 5.4).

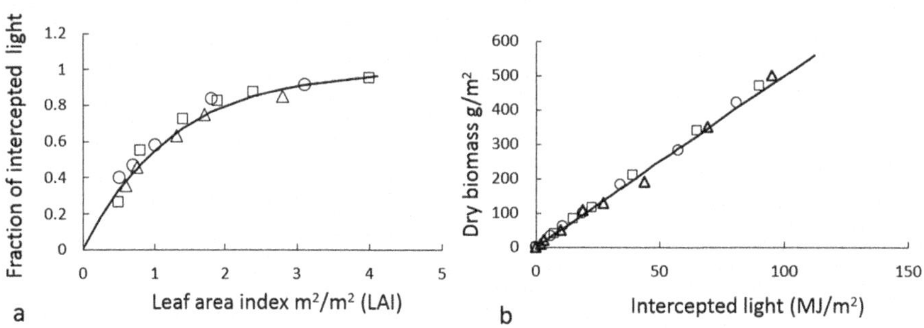

Fig. 5.4 (**a**) The computation of the light interception with Beer-Lambert's law for a chrysanthemum population of three planting densities. (**b**) Under the same conditions, the biomass production by the stand is proportional to the cumulative intercepted light. The light use efficiency (LUE) was determined to be five g·DM/MJ/PAR. The markers correspond to the measurements (three different planting densities) and the curves represent the theoretical values. (Source: Ep Heuvelink, Wageningen University)

5.3.2 *Effect of Planting Density on the Individual Plant and Population Growth*

The reference cultivated area for production is a square meter (m^2). At low-density ranges, increasing the planting density obviously produces more biomass per m^2, because the canopy of plants has not yet closed. We see that production is growing exponentially. As soon as the LAI becomes large (greater than 3), the biomass production per m^2 is hardly dependent on planting density (Fig. 5.5a). Individual plants tend to have a weight that is inversely proportional to the planting density (Fig. 5.5b).

Another representation of the effect of planting density on the single plant and on the stand is illustrated below (Figs. 5.33 and 5.34).

5.3.3 *Effect of Biomass Partitioning on the Growth*

During tomato growth, Heuvelink carried out controlled fruit removal in the bunch to observe the effect on total biomass production for the population (Fig. 5.6). In all these experiments, an LAI of more than 3 was obtained. If the harvest index (HI) depends on the number of fruits per bunch, the total biomass production per m^2 is invariant because it is the LAI that controls the total biomass production (Eq. 5.1) and not the partitioning of the biomass.

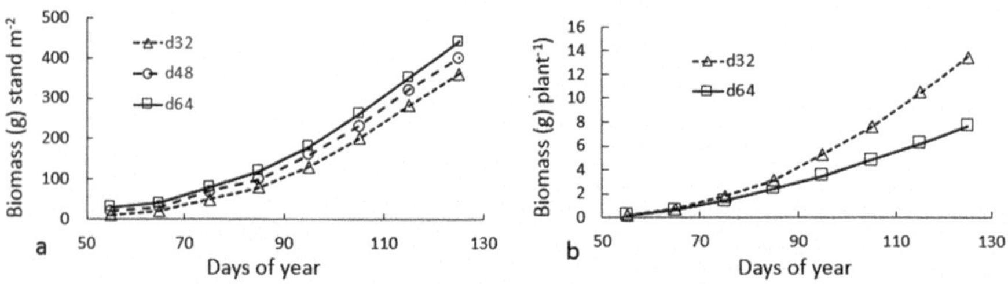

Fig. 5.5 Effect of planting density on individual and population growth. (**a**) Growth of a chrysanthemum population at three planting densities (32, 48, 64 plants/m^2): at the beginning, it is the total leaf area that counts and the growth is exponential. Biomass production per m^2 obviously increases with the density. After about 90 days, growth becomes nearly linear and close for the three densities (parallel lines) because the plant canopies are all closed. The LAI is high, and the ground surface is completely covered with the foliage. It is then the m^2 that produces and is no longer the sum of the surface area of the leaves. (**b**) Growth of an individual plant for two densities (32, 64 plants/m^2). A plant with a density of 64 plants/m^2 (□) weighs as much at the beginning of growth as a plant with a density of 32 plants/m^2 (Δ), because the plants are isolated from each other. At the end of the cultivation, plant weight is almost twice as high for the widest spacing (roughly the same total biomass per m^2 (panel a) but only half the number of plants). (Source: Ep Heuvelink, Wageningen University)

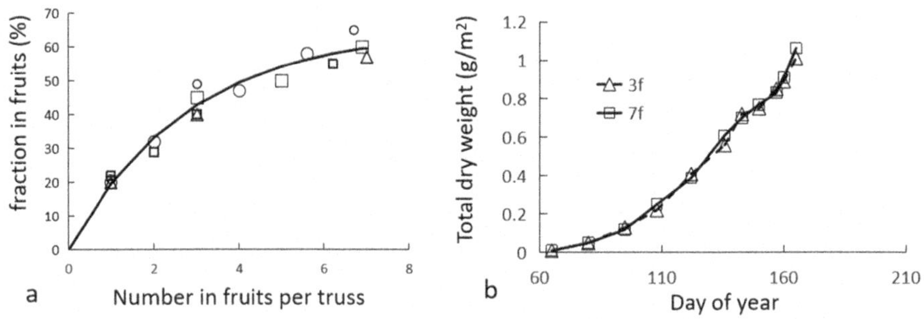

Fig. 5.6 (**a**) Harvest Index (HI) (% fruit weight/plant weight) for greenhouse tomato crops grown for a period of about 100 days starting at the anthesis of the first truss, as a function of the number of fruits per truss for three planting densities. At low numbers of fruits per truss, HI is proportional to the number of fruits, and it saturates at a maximum of 60% of the plant's weight. (**b**) Total weight of the plants at two different fruit loads. The number of fruits per truss has no influence on the total weight of the plants. This is because at both fruit loads LAI was high (LAI >3), so total biomass production was the same, and only partitioning is influenced by the fruit load. (Source Ep Heuvelink, Wageningen University)

This phenomenon is found in the production of cucumbers and peppers. Many fruit abortions cause the HI to vary greatly over time. But since the LAI is high, biomass production is constant as light interception is saturated (i.e., at its maximum). The biomass of the compartments varies in a complementary way, under the constraint that their sum (the total biomass of the plant) is constant.

It should be noted that if the fruit is removed in some plants, there may be physiological reactions. Thus, if the ear of maize is removed, the plant dies because the short expansion time of its other organs does not allow the excess sugar to be stored elsewhere. The plant dies of being poisoned (personal communication of Guo Yan, CAU, 2006). On the other hand, in sunflowers, where the duration of organ expansion is long, if the terminal cap is removed, the last (normally small) leaves become very large, as well as the stem that bursts under the pressure of excess supply.

5.3.4 Effect of Planting Density on Biomass Partitioning

In tomatoes, as in other cultivated plants (e.g., maize in Ma et al. 2008, see Chap. 11), planting density influences the total weight of the individual plant, but does not influence the partitioning of biomass among the different organ compartments, and therefore the HI remains unaffected (Fig. 5.7).

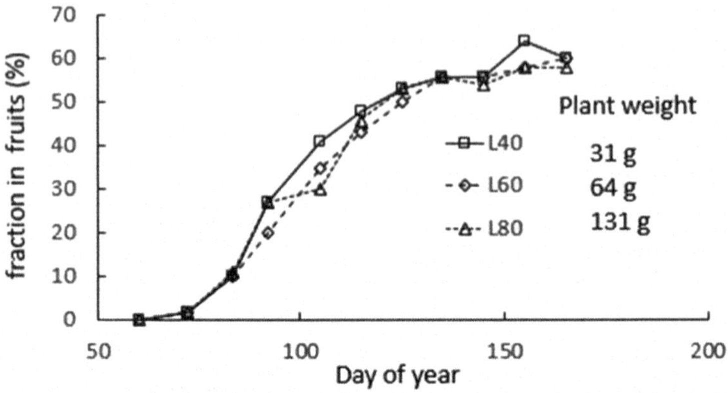

Fig. 5.7 Patterns of the HI during tomato growth at three planting densities: (spacing between plants on a row: $L = 40$ (squares), 60 (diamonds), 80 (triangles) cm). The distance between rows was 80 cm for all three densities. The three profiles corresponding to the three densities are similar, indicating that partitioning is independent of planting density. It should be realized that all trusses were pruned to 7 fruits per truss. On the other hand, the weight of the individual plants is obviously dependent on the density (131, 64, 31 g, respectively). (Ep Heuvelink, Wageningen University)

5.3.5 *Effect of the Position of Organs in the Growth*

To prove the efficiency of the common pool concept, one experiment was conducted on locating trusses on tomato plants grown with two similar stems for one plant. In the first experiment, every second truss was left for both stems. In the second experiment, all the trusses were left on the first stem, while all the trusses are removed from the second stem. Both experiments have the same number of plants and organs. If the common pool is verified, they must have the same growth, and therefore the same weight per compartment, which is what was observed (Fig. 5.8).

5.3.6 *How the Model Works*

The TomSim physiological crop model (Heuvelink 1999) is calibrated and validated on many greenhouse experiments. It functions on a daily time step (photosynthesis is calculated at an hourly step).

Once these rules have been defined, and the system parameters have been calibrated, the TomSim model grows the plant population according to the flowchart in Fig. 5.9. The respiration is considered here, but more frequently, only net photosynthesis is considered.

- The plant leaf area is computed from the leaf biomass divided by the thickness (specific leaf weight: SLW); from plant leaf area and planting density, the LAI (leaf area per m^2) can be calculated;

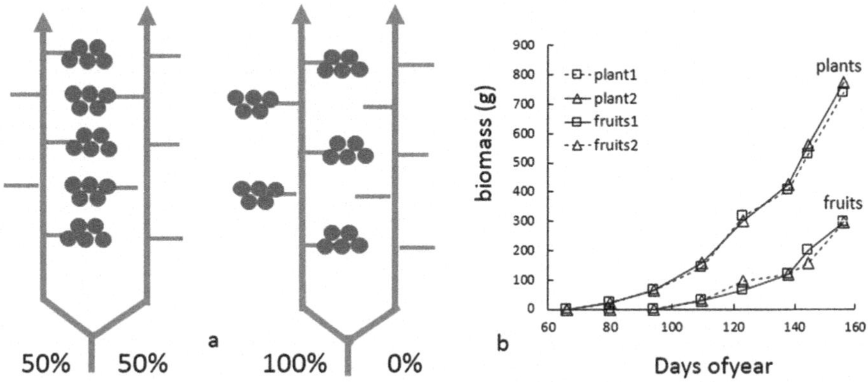

Fig. 5.8 Tomato plants grown with two stems originating from cotyledon axillary buds. (**a**) In 50–50 treatment, on each stem half of the trusses were kept, while in 100–0 treatment, all trusses were kept on one stem and all trusses were removed from the other stem, resulting in plants with the same number of trusses but located differently. (**b**) The two types of plants have the same total plant growth and the same fruit harvest index (HI) (Δ is 50–50 and □ is 100–0), which validates the common pool hypothesis. (Source: Heuvelink 1995)

Fig. 5.9 Flowchart of the growth of plant population per m^2 functioning on a daily time step. The biomass that goes to the leaf compartment results in increased leaf area, and hence increased light interception. (Source: Ep Heuvelink, Wageningen University)

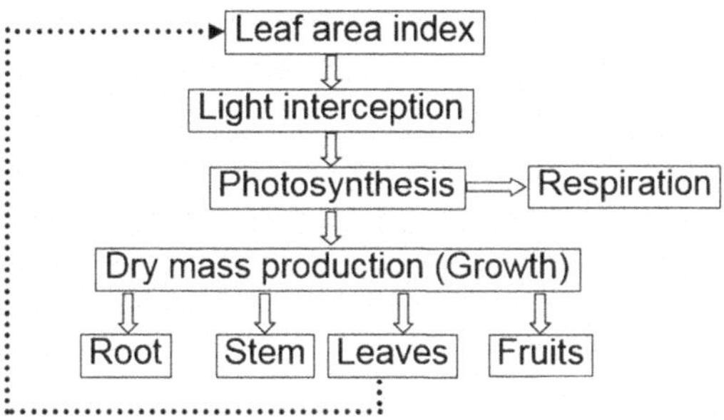

- Biomass production depends on LAI, light intensity (PAR), CO_2, and temperature, taking effect in Eq. (5.1);
- The biomass produced can be considered as forming a common pool from which the organs draw for their growth according to their sink strength relative to the total sink strength of all organs together.

Biomass distribution among organ compartments is independent of biomass production. In particular, it does not depend on planting density. It is often stably fractionated according to sink strengths measured directly from the ratio of daily weight increases in vegetative compartments (leaves, stems, and roots) and fruit organs. The leaf compartment provides feedback on the growth.

The TomSim model, which uses the rules validated by the above experiments, gives a good prediction of crop production according to the time of year for all organ compartments (fruits, leaves, stems, and roots) (Fig. 5.10). In greenhouse conditions, the limiting factor is light: the biomass production in the Netherlands is three times less in the winter period than that in the summer period.

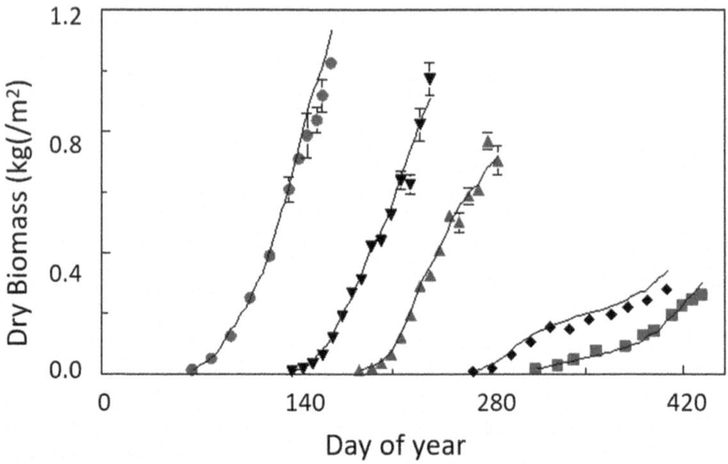

Fig. 5.10 Measured (symbols) and simulated (lines) dry biomass produced per m^2 for a tomato crop planted at different times of the year using the TomSim model. The growth in late autumn (black diamond) and winter (marker ☐ pink) is strongly slowed down by a lack of light (Heuvelink 1995) (Color figure online)

5.3.7 Review of Physiological Crop Models

Based on simple concepts of biomass production (using eco-physiological factors), and its partitioning in plant compartments, physiological crop models (Marcelis and Gijzen 1998; Marcelis et al. 1998) provide good harvest predictions.

In greenhouses, normally water and temperature are not limiting factors because they can be controlled. The main limiting factor is photosynthetically active light (PAR), which varies greatly over the seasons and is, at best, only 70% transmitted into the greenhouse. The proportionality parameter between the produced biomass and the intercepted light is called *light use efficiency* (LUE). The LUE depends mainly on the CO_2 concentration in the greenhouse and has little dependence on the PAR. For better growth prediction, changes in environmental factors must be incorporated into each simulation step.

In field conditions, the main limiting factors are temperature and water, with sufficient light during the growing season. The use of thermal time compensates for the variation in temperature and makes it possible to linearize the development of the crop according to the sum of daily temperatures. Empirically, the biomass produced per m^2 is proportional to the water transpired by the canopy, which depends on the evapotranspiration value (PET). The proportionality coefficient between the biomass produced and the transpired water is called *water use efficiency* (WUE). The WUE is generally stable. Eq. (5.1) remains valid for computing biomass production per m^2 because the canopy transpiration also depends on the LAI, according to the same formulation as Beer-Lambert's law. When the transpiration is computed, parameter C is no longer the LUE, but the WUE. For example, as an order of magnitude for plants, 500 L of transpired water per m^2 give about 10 kg of fresh matter or 1 kg of dry matter. In the event of water stress, a water module must be developed that calculates the transpired water as a function of soil moisture and is integrated into Eq. (5.1). The Pilot software (Mailhol et al. 1996) thus predicts the production of different crops according to climatic conditions with rainfall data.

Due to their functioning at the stand and organ compartment level, physiological crop models have several limitations:

- Difficulties in modeling the effect of stresses (water, light, etc.) that lead to organ abortions (meristems, leaves, fruits, branches);
- Difficulties in synchronizing the different modules of the physiological crop model, such as soil moisture model or organogenesis model;
- Lack of information on the position of organs within the crop;
- No consideration of stochastic aspects in the development;
- No consideration of secondary growth;
- Lack of representations (visualizations) of the simulated plants, which is especially relevant for ornamental plants.

5.3.8 An Example of Functional–Structural Plant Modeling Close to Physiological Crop Models: The GreenLab Model

Figure 5.11 positions the approach of the GreenLab model in relation to physiological crop models and functional–structural plant (FSPM) models.

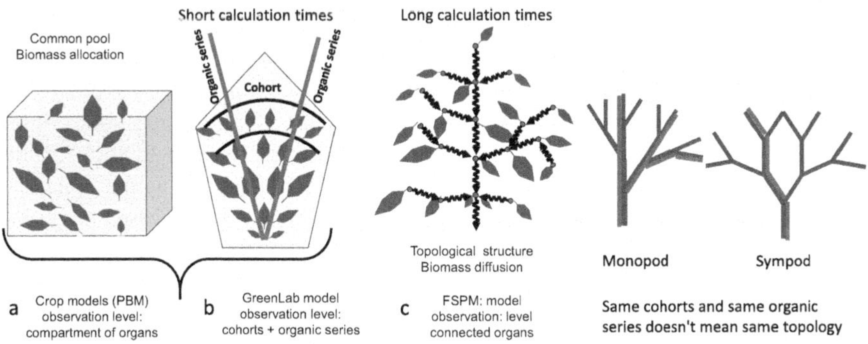

Fig. 5.11 Different approaches to modeling plant growth. (**a**) Physiological crop models gather organs into compartments and use a common pool. (**b**) The GreenLab model uses cohorts and organic series of organs that are orthogonal to each other and also use a common pool. (**c**) Functional–structural plant models (FSPM) use individual organs and source–sink relationships through a topological structure

In most physiological crop models (CM), the organs are grouped into compartments. The number of organs is not of main interest. At each time step of the growth, an allocation of the biomass taken from the common pool goes to the compartment according to a proportion, so the history of the growth is lost if there is no concept of the cohort. The process is not precise, but it is immediate, and simple equations make it work.

In functional–structural plant models (FSPM), each organ is connected to the topological structure. Propagation of the biomass from the source organs takes place in the structure towards the sink organs according to a short time step. The "Common pool" hypothesis can be used or not. Each organ works independently as in a multi-agent system. There is no implicit concept of cohorts. The process is accurate but time-consuming because it requires software to simulate the propagation of biomass within the structure.

GreenLab is a functional–structural plant model, and aims to model plant growth at the phytomer level while remaining compatible with physiological crop models. Plant development is built on the functioning of meristems. A hierarchical and anatomical structure is created. The notion of potential structure also makes it possible to compute the number of organ types per cohort, even in the stochastic case (Chap. 3).

The number of organs produced at each time step is computed by recursive algorithms using a botanical automaton (Chap. 4). These organs are organized into cohorts according to organ type and chronological and physiological ages, with the number of cohorts being a product of them. All organs in a cohort are identical. The development history is preserved, and a strong factorization is carried out, since only one representative organ of each cohort needs to be simulated. At each step of the development, the biomass collected from the common pool goes into the cohorts according to the relative sink strength and the number of organs. The process is based on equations and in the case of Matrix Mode does not require browsing the whole plant structure (as in List Mode), to simulate the biomass partitioning, so it is fast. In GreenLab, the plant is decomposed into spare parts of organs (see below organic series). Thus, organ cohorts provide a functional structure, different and simpler than the anatomical and hierarchical structure, for modeling source-sink interactions. This is characteristic of integrative biology. The construction of the topological structure is optional, and it does not directly participate in the functioning. GreenLab makes it possible to switch from organ-compartment-based functioning (e.g., physiological crop models) to individual-organ-based functioning (e.g., functional–structural models) while maintaining the knowledge acquired from the former.

In the following sections, it is assumed that the climatic conditions are stable and non-stressful to be able to study the model's behaviors under so-called potential growth conditions.

5.4 Plant Demand: The Function of the Development

The plant's demand at a chronological age (CA) t corresponds to the sum of the demands of the active sinks of the growing structure. It is the sum of the products of the numbers of organs (computed by the botanical automaton in the cohorts defined according to their types, physiological and chronological ages) and the values of their sink functions.

5.4.1 *Scheme of the Functioning of the Organs*

The organs are created by meristems and then expand due to their sink strengths, which allows them to capture biomass from the common pool. The duration of this expansion is measured in growth (or development) cycles (GC or DC). At maturity, the organ sink ceases to compete with others and is no longer part of the plant's demand.

5.4.2 *Definition and Form of Sink Functions*

The organs can grow over several expansion cycles. The sink strength of an organ of type o with PA φ and expansion duration T_o is modeled, according to its expansion cycle x by the equation:

$$P_o^{\varphi}(x) = p_o^{\varphi} \cdot F_o\left(\frac{x}{T_o}\right) \tag{5.3}$$

where p_o^{φ} is the sink strength of the organ; x is its chronological age (CA); φ is its physiological age (PA); o represents the organ type (a: leaf, i: internode, f: fruit). $F_o\left(\frac{x}{T_o}\right)$ is the variation function of the sink related to its expansion duration. This function is normalized: $\max(F_o) = 1$.

An empirical function is defined in advance, the form of which, controlled by the parameters to be estimated, is flexible enough to adapt to numerical changes in sink values during organ expansion. In GreenLab, this function is a beta law (Fig. 5.12): it shows great plasticity and gives a good prediction of organ expansion. It is defined by two parameters a and b on the interval [0, 1]:

$$\beta(x) = \frac{(a+b-1)! \cdot x^{a-1} \cdot (1-x)^{b-1}}{(a-1)! \cdot (b-1)!} \quad 0 < x < 1 \tag{5.4}$$

This law has the following mean (m) and variance (v) in x [0 1]:

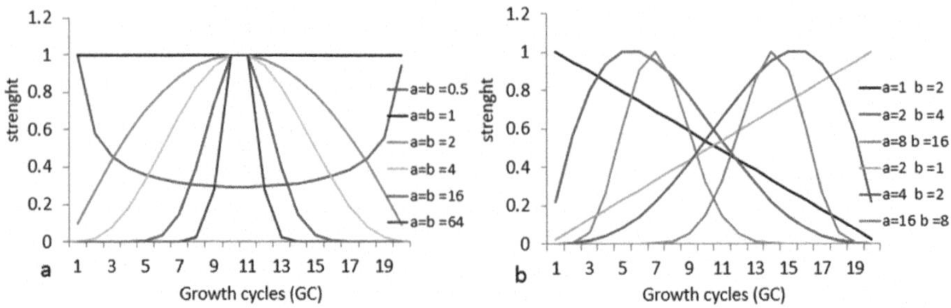

Fig. 5.12 Shapes of sink functions of discretized beta law, normalized over an interval T_o of 20 cycles. (**a**) Symmetrical shape with $a = b$. The values given for a and b are 0.5, 1, 2, 4, 16, and 64. For $a < 1$, the function has a U-shape. For $a = 1$, the function is constant. For $a > 1$, the function has a symmetrical bell shape centered in $m = 10$ ($m = 20a/(a + b)$). The more a and b increase, the less the curve is spread. (**b**) Asymmetrical shape. Examples are given that the pairs of parameters a and b of the beta law are chosen as an example: (1, 2), (2, 4), and (8, 16). The larger $a + b$ is, the less the curve is spread

$$m = \frac{a}{a+b}$$
$$v = \frac{a \cdot b}{(a+b)^2 \cdot (1+a+b)} \tag{5.5}$$

The notations are modified to make the formulation of this equation more appropriate by making it discrete. Note T_o is the domain of x for the CA of the organ expressed in cycles: ($T_o \geq 1$ and $0 \leq x \leq T_o$). The coefficient C_o is calculated so that the maximum of the function is normalized to 1.

We define:

$$F_o\left(\frac{x}{T_o}\right) = C_o \cdot \left(\frac{x+0.5}{T_o}\right)^{a-1} \cdot \left(1 - \frac{x+0.5}{T_o}\right)^{b-1} \cdot \left(\frac{1}{T_o}\right) \tag{5.6}$$

So in the interval [1 T_o], we have the mean:

$$m = \frac{a}{a+b} T_o \tag{5.7}$$

This unimodal function can have various shapes, which may be more or less asymmetrical and spread out. The sink variation function is assumed to be independent of PA. It depends only on its type o. The influence of PA appears only in the multiplicative factor, the sink strength itself. Since sink strengths are relative (see next paragraph), a reference must be set. It is assumed that the sink strength of leaves with PA 1 is equal to the unit: $p_a^1 = 1$.

It is possible to empirically determine the sink function of an organ by monitoring its growth under non-limiting assimilate supply, the so-called potential growth

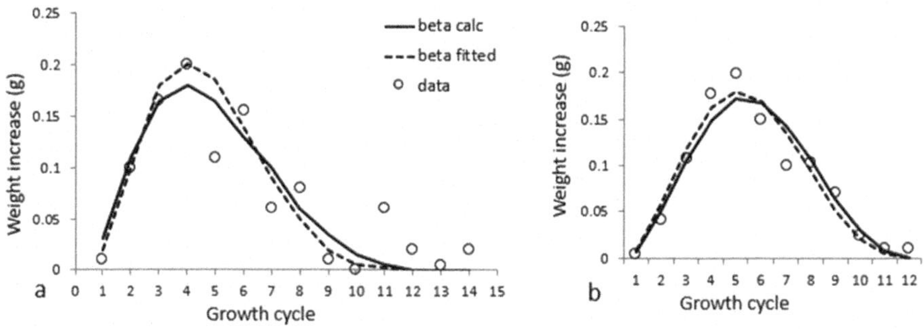

Fig. 5.13 Observed (symbols) increase in weight of a tomato fruit (**a**) and sweet pepper (**b**) as a function of its age measured in cycles (o). The curves represent beta laws. Fitted curves (dotted lines) are close to those calculated from the organs on the stem by inverse method (solid lines) without growth monitoring. (Sources: China Agricultural University (CAU) and Wageningen University)

(Heuvelink 1996b). Establishing the relationship between fruit diameter and fruit weight makes it possible to determine the sink function of fruit without sacrificing it during its growth.

However, the method remains often impractical and applies only to fruits (pepper, tomato). In general, the parameters of the sink functions are called "hidden" parameters. In GreenLab, these parameters are estimated using an inverse method, from measurements on the plant architecture without growth monitoring (Chap. 9). In practice, the expansion time T_o is empirically observed or assessed, and the parameters a and b are estimated from the measurements taken on the architecture.

Figure 5.13 shows an example of the increase in weight of tomato fruit and sweet pepper fruit, measured periodically during their expansion using the allometry between their diameter and weight. The pattern can be described by a beta law. This law is also directly identifiable from the weights of organs scattered on the stem of a plant (without growth monitoring) and whose parameters are identified by the inverse method at the plant level (Chap. 9). The two laws obtained are very similar. This shows the coherence of the model with two different computation methods.

5.4.3 *Plant Demand as the Sum of the Sinks of the Organ Cohorts*

A botanical study defines the number of organ types per phytomer. From the number of phytomers produced by the botanical automaton, we deduce the numbers of leaves, internodes, and fruits produced in each cycle. The number of phytomers created at each cycle for each PA is computed, either by enumeration during simulating organogenesis in list mode or by calculation using the corresponding development equation in matrix mode. With the latter, we can deduce the number of organs $N_o^\varphi(t)$ produced for a type o at CA t for PA φ. By definition, it constitutes a cohort or set of homologous elements that appeared at the same time. Hence, the expression of the plant's demand at CA t is:

$$D(t) = \sum_{o,\varphi} \left(\sum_{i=1}^{t} N_o^\varphi(t-i+1) \cdot P_o^\varphi(i) \right) \tag{5.8}$$

where organ type o, of PA φ and aged i cycles have a sink strength $P_o^\varphi(i)$ which depends on the maturity i in DC of the organ. They appeared at cycle $t - i + 1$ and are in the number of $N_o^\varphi(t-i+1)$. The expression of $D(t)$ is obtained by a convolution that is quick to compute.

5.5 Growth Equations

Same as the development, the growth can be formulated by mathematical equations.

5.5.1 *Biomass Partitioning Equations*

The expansion by weight of an organ of type o, of PA φ and aged i cycles of development in a plant of CA t (therefore at cycle t) is written:

$$\Delta q_o^\varphi(i,t) = P_o^\varphi(i) \cdot \frac{Q(t-1)}{D(t)} \cdot \Delta t \tag{5.9}$$

Here $\Delta t = 1$ because the growth cycle is the same with the DC.

Organ expansion depends on the sink strength $P_o^\varphi(i)$, the biomass $Q(t - 1)$ produced in cycle $t - 1$, which is distributed within the plant structure whose total demand is $D(t)$. Note that the sinks in this system are relative because they appear in the numerator and denominator of the expression (Eq. 5.8). It is, therefore, necessary to choose a reference organ. Only the leaves always exist, while in herbaceous plants, the internodes are not measurable at the rosette stage. For this reason, the sink strength of a leaf blade of PA 1 is normalized to 1.

The biomass accumulated in an organ created at i DCs in a plant aged t DC is the sum of expansions:

$$q_o^\varphi(i,t) = \sum_{j=i}^{t} P_o^\varphi(j-i+1) \cdot \frac{Q(j-1)}{D(j)} \tag{5.10}$$

The weight of organ $q_o^\varphi(i,t)$ that appeared at the age $t - i + 1$ is represented $N_o^\varphi(t-i+1)$ times in its cohort.

Equation (5.10) can be written in matrix form. A growth operator is defined as one that connects organ biomass to produced biomass and plant demand through the sink functions.

There are as many operators as there are types of organs o, multiplied by the maximum number of PAs ($mx\varphi$). For a tree, there are most often 12 operators (3 types of organs × 4 PAs). The operator's matrix dimension increases as the number of DCs t increases.

$$\begin{bmatrix} q_o^{\varphi}(1,t) \\ q_o^{\varphi}(2,t) \\ \vdots \\ q_o^{\varphi}(t,t) \end{bmatrix} = \begin{bmatrix} P_o^{\varphi}(1) & P_o^{\varphi}(2) & \cdots & P_o^{\varphi}(t) \\ 0 & P_o^{\varphi}(1) & \cdots & P_o^{\varphi}(t-1) \\ 0 & 0 & \ddots & \vdots \\ 0 & 0 & \cdots & P_o^{\varphi}(1) \end{bmatrix} \begin{bmatrix} \frac{Q_0}{D_1} \\ \frac{Q_1}{D_2} \\ \vdots \\ \frac{Q_{t-1}}{D_t} \end{bmatrix} \tag{5.11}$$

This system describes the individual expansion of organs according to their CA of appearance according to their type o, their PA φ, and the evolution of supply on demand: $Q(i-1)/D(i)$. This quotient is equivalent to an internal pressure of the synthesized biomass that must be allocated to the various organs of the botanical structure. It is referred to here as "trophic pressure".

Each organ is a representative of the cohort in which it was born. The number of organs corresponding to rank k is variable: it depends on the number of phytomers produced in the cohort at age k by the botanical automaton. The chronological sequence of organ biomass (left part of the operator in Eq. (5.11)) defines the description of a series of homologous organs along a vegetative axis for their weights.

This series $q_o^{\varphi}(1,t), \ldots, q_o^{\varphi}(t,t)$ is called the "organic series" (Buis and Barthou 1984). It describes the profile of the vegetative axis for the biomass of an organ o positioned according to the rank k of its phytomer.

Since we know the number of organs $N_o(k)$ associated with a rank k, we can also know the total accumulated biomass $SQ_o^{\varphi}(t)$ in the compartment constituted by these physiologically aged organs φ:

$$SQ_o^{\varphi}(t) = \sum\nolimits_{k=1}^{t} N_o^{\varphi}(k) \cdot q_o^{\varphi}(k,t) \tag{5.12}$$

The sum of all the results $SQ_o^{\varphi}(t)$ for all organs o and PA φ gives at the end the total biomass of the plant.

The plant can be broken down into a small number of categories of axes, which in turn are divided into separate organic series. The description of these series gathers all the necessary information contained from development and growth. This system makes it possible to define very efficient targets for the calibration of the source-sink model from the experimental data (Chaps. 9 and 10). The analysis of the plant decomposed in separate pieces, such as cohorts or organic series (see below), bypasses the complicated use of the topological structure.

5.5.2 *Considerations on Organic Series*

Buis and Barthou (1984) defined and studied the evolution of an organic series mathematically:

> "All organs of the same morphological nature generated by the same primary meristem during the development of a leafy axis and on which the same morphogenetic character is considered".

For example, this could refer to the weight or surface area of a leaf, the length of the internode, etc. The size of each organ depends on its rank in the series, the CA of the axis and its creation date. The system generates a family of curves that gives the profile of the organic series according to the age of the leafy axis and the dates of creation of the phytomers. Thus, the same types of organs in the same series have different dimensions depending on their rank.

However, the study of Buis and Barthou (1984) remains descriptive because it is based on allometric relationships between the dimensions of the organs that increase according to kinetics based on thermal time and not on source–sink relationships. Nevertheless, the thermal time is often sufficient to follow step by step the development of the axis and the expansion of the organic series generated, as in maize (leaf length, sheaths, internodes) (Fournier and Andrieu 1998).

GreenLab generalizes the notion of organic series, which is no longer necessarily linked to the physical notion of one single axis. The axial notion of organic series, depending on the position of the phytomers on an axis, can be extended to the chronological notion that describes the evolution of the phytomers in successive cohorts. The organic series is "orthogonal" to the notion of cohort in relation to time. The organs can be sorted according to a time step (developmental cycle, DC), which then replaces the notion of rank in the axis. The rank of the series corresponds to the age of the cohort, and the number of representatives of the same organ varies from one cohort to another.

Figure 5.14 shows the same single-stemmed plant at four different growth stages and the evolution and expansion of the resulting organic series.

Time organic series are also defined in monopodial or sympodial structures. In Fig. 5.15, starting from the top of the axes, the same organic series is found identically throughout a plant structure organized according to the principle of the PA. Indeed, the same rank of phytomer from the apex of the living branches and the same PA is characteristic of the same cohort. In the case of monopodial branching (Fig. 5.15a), only the first branch has a complete series, other series are top-down truncated depending on the age at which the branches appear in the structure.

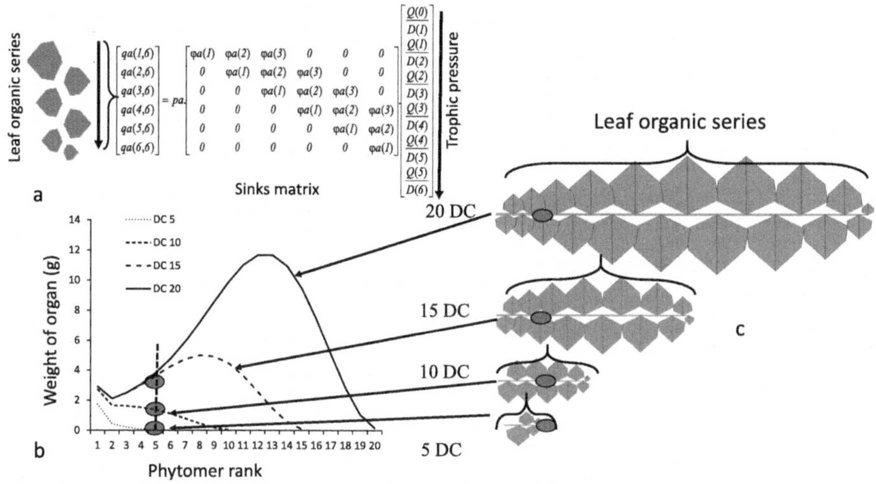

Fig. 5.14 Evolution of the organic series of leaves during a single-stemmed plant growth. This evolution is twofold. On the one hand, we have the development of the series, which increases by adding one phytomer at each cycle, and on the other hand, the individual growth of the organs according to their rank. (**a**) Relationship between organic series and evolution of trophic pressure through the matrix of sinks (with an expansion duration $t_x = 3$ DCs). (**b**) Curves of development and growth of the leaves organic series for periods of 5, 10, 15, 20 DCs. The red dot refers to the rank 5 of the phytomer on the axis whose behavior is particularly examined. On this rank, the expansion of leaf is completed at the age $t_x = 15$ DCs. (**c**) Simulation of the plant for the age of 5, 10, 15, 20 DCs corresponds to the series displayed with the mark of phytomer 5

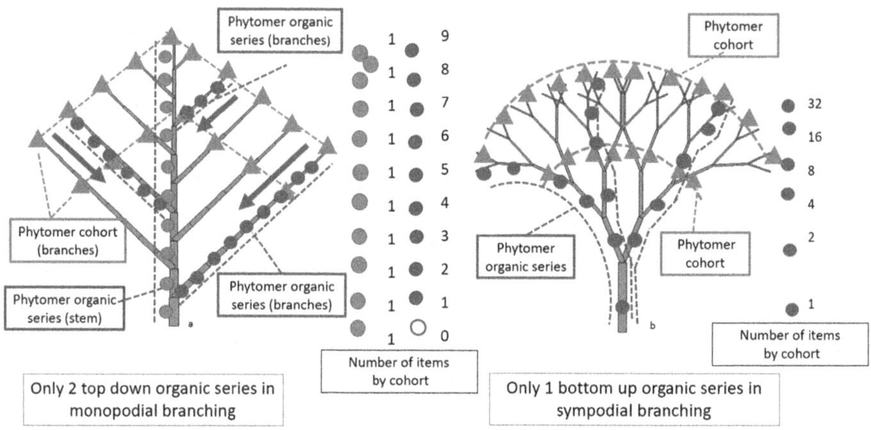

Fig. 5.15 Visualization of organic cohorts and series in plant architecture according to the GreenLab model. (**a**) Case of monopodial branching, (**b**) Case of sympodial branching. The number of organic series depend only on the physiological ages. The phytomers of the same nature and created at the same time form a cohort. The organic series are orthogonal to the cohorts. They can be found in both monopodial and sympodial structures. The same truncated series can be replicated many times in a structure. In this example, there are only two organic series for the monopodial structure: stem and branches, and only one for the sympodial structure

The organic series contains the entire history of source–sink relationships in plant growth. They are effective for data assimilation because they can be found in any architectural model.

5.5.3 Dynamics of Biomass Production of an Individual Plant

5.5.3.1 Production Equation of an Individual Plant

Consider Eq. (5.1) of the production of a stand.

$$Q = C \cdot \mathrm{PAR}(t) \cdot (1 - \exp(-k \cdot \mathrm{LAI}(t))) \cdot \Delta t$$

The reference cultivated area S_c of a crop stand is 1 m^2, which is implicitly included in Eq. (5.1). It is made explicit in Eq. (5.13); the LAI is written as Sf/S_c; S_f is the leaf surface of the canopy and S_c is the surface area of the stand on the ground:

$$Q = C \cdot \mathrm{PAR} \cdot S_c \cdot \left(1 - \exp\left(-k \cdot \frac{Sf}{S_c}\right)\right) \cdot \Delta t \qquad (5.13)$$

To link the growth to the development for an individual plant, the expression must be expressed in DC, measured in thermal time. During the DC t, the light energy received by the canopy is denoted as PAR(t) and the leaf area of the canopy is assumed to stay constant during one DC and is denoted as $Sf(t)$.

This equation can be adapted to an individual plant.

$$Q(t) = C \cdot E(t) \cdot Sp(t) \cdot \left(1 - \exp\left(-k \cdot \frac{Sf(t)}{Sp(t)}\right)\right) \qquad (5.14)$$

where $E(t)$ is the growth factor (light (PAR) or transpired water (PET) as appropriate) and C is the resource use efficiency (water use efficiency or light use efficiency, respectively). They are supposed to be known.

The leaf area and weight of a plant can be measured for each DC by sampling from the population. In cycle t, we can, therefore, know the increase in biomass $Q(t)$ of the individual plant produced by its leaf area $Sf(t)$. On the other hand, S_c, which is the surface area of the stand, is not defined a priori for an individual plant. Let $Sp(t)$ be this unknown value assigned to the individual plant in the cycle t. It can be estimated by solving Eq. (5.14) at each DC, i.e., by assuming the changes in known $Q(t)$ and $Sf(t)$ and looking for the values of $Sp(t)$ that give equality to each cycle.

$Sp(t)$ becomes the production surface of the plant (that is the computed surface area occupied by the plant on the ground in the stand), defined as the "solution" of Eq. (5.14) in the cycle t. It is the effective production surface that intercepts light, considering the overlapping of the leaves and its spreading on the ground. The ratio

$\frac{Sf(t)}{Sp(t)}$ is an optimized individual *LAI*. However, this requires some comments.

The expression $S\varphi(t) = Sp(t) \cdot \left(1 - exp\left(-k \cdot \frac{Sf(t)}{Sp(t)}\right)\right)$ measures the surface of the shadow on the ground projected by the foliage, which quantifies the light intercepted with the sun at its zenith; it is call the photosynthetic surface.

Sp(*t*) is first a calibration parameter. While Beer-Lambert's law may not be applicable if the foliage is heterogeneous, the production area is always a solution of Eq. (5.14). If the area of interception of light by the foliage is the total leaf area (leaves do not overlap), *Sp*(*t*) takes on an infinite value and loses its topographical significance. This situation concerns the juvenile stage of the plant, where all the leaves are illuminated. It corresponds to a limited Taylor series development of Eq. (5.14) at order 1. Biomass increases become proportional to the leaf area. Eq. (5.14) is then approximated as:

$$Q(t) \approx C \cdot k \cdot E(t) \cdot Sf(t) \tag{5.15}$$

A priori, the production area of the plant *Sp*(*t*) varies with each cycle in Eq. (5.14). Due to a high planting density *d* per unit area of stand S_c, when competition between plants makes them touch each other and limits the expansion of the projection of foliage on the ground to an average individual area *Sd* ($Sd = S_c/d$), then *Sp*(*t*) should stabilize around $Sp(t) \approx Sd$.

Indeed, if the expression $k \cdot \frac{Sf(t)}{Sp(t)}$ is greater than 3, the term $\exp\left(-k \cdot \frac{Sf(t)}{Sp(t)}\right)$ is negligible and Eq. (5.14) is written as:

$$Q(t) \approx C \cdot E(t) \cdot Sp(t); Q(t) \approx C \cdot E(t) \cdot Sd \tag{5.16}$$

At the start of plant growth, leaves do not overlap, *Sp(t)* is very large and is not involved in biomass production (Eq. 5.14). When the plants come into contact, *Sp(t)* takes the fixed value *Sd*. *Sd* becomes an approximate solution of Eq. (5.14). The individual production area is the area of ground occupation by the plant, which is equivalent to the projected surface of the foliage and can be written as $Sp \approx Sd$. In the case of high planting density, neighboring plants touch each other soon, as if *Sp* was constant.

The equivalence $Sp \approx Sd$ in GreenLab has been verified and validated for various field crops (tomato, maize, beet, Sect. 5.8.2.1) and for various planting densities (Chaps. 9 and 10). The role of *Sp* is discussed in more detail below.

In order to study the behavior of GreenLab during the growth, it is assumed that the parameters *E*(*t*) and the production area *Sp*(*t*) per cycle are constant. They are represented by *E* and *Sp*, respectively. We choose $E = 1$ and *Sp* as the solution of Eq. (5.14). $r = 1/C$ is defined as a calibration coefficient.

5.5.3.2 Leaf Surface Expression

The functioning time of the leaves (in cycles) is indicated by the parameter t_a. Let us consider it here equal to the duration of their expansion, denoted as t_x. In order to calculate the leaf area of the plant, it is necessary to sum the masses of the functional leaves and divide by their thickness ε, which is assumed to be constant. The functional leaf area $Sf(t)$ can be computed by Eq. (5.17), where the index φ refers to the PA (ranging from 1 to $mx\varphi$). Within a cohort of leaves that appeared at CA i, the number of leaves $N_a^\varphi(i)$ and their individual biomass are recorded. Two cases must be distinguished for the computation of the area according to whether the age of the plant t on DC exceeds the lifespan of the leaves t_a or not. Indeed, one case $t < t_a$ corresponds to the beginning of the plant's growth, as long as no leaves have died yet. On the other hand, in the other case $t \geq t_a$, only leaves whose CA is less than $t - t_a + 1$ are still functional. A unique equation can describe these two cases, the distinction being represented by the limits of the index of the second sum in the expression below. The leaf area is computed as:

$$Sf(t) = \frac{1}{\varepsilon} \cdot \sum_{\varphi=1}^{mx\varphi} \sum_{i=\max(1,\, t-t_a+1)}^{t} N_a^\varphi(i) \cdot q_a^\varphi(i,t) \tag{5.17}$$

5.5.3.3 Dynamic Growth Equation

By expressing the leaf surfaces as a function of the previous Q/D ratios (Eq. 5.17), we obtain the generic recurrence Eq. (5.18) that characterizes the increase in biomass $Q(t)$ at cycle t of a computational plant according to GreenLab:

$$Q(t) = \frac{E \cdot Sp}{r} \cdot \left(1 - \exp\left(-\frac{k}{\varepsilon \cdot Sp} \cdot \sum_{i=\max(1,t-t_a+1)}^{t} \sum_{\varphi=1}^{mx\varphi} N_a^\varphi(i) \cdot \sum_{j=i}^{t} \frac{P_a^\varphi(j-i+1) \cdot Q(j-1)}{D(j)}\right)\right) \tag{5.18}$$

In real plants, the weights of the plant and its compartments are measurable; their simulated values can be computed from the sum of the increase in biomass per cycle.

5.5.4 *How Recurrence Works During the Growth*

Consider a simplified plant with only one PA, without secondary growth or root system and without fruit. The number of leaves per phytomer is 1. The functioning time of the leaves is t_a and the expansion time of the organs (leaves and internodes) t_x

equals to t_a. The sink functions of the leaves and internodes are, respectively, $P_a(t)$ and $P_e(t)$. The seed biomass is Q_0.

5.5.4.1 Production and Allocation of Biomass during the First Cycle

- Computation of the demand:

There is only one phytomer created from the seed. The number of leaves produced is therefore $N_a(1) = 1$ (case of monocotyledonous plant) and the number of internodes $N_e(1) = 1$. The demand for the first cycle is, therefore:

$$D(1) = N_a(1) \cdot P_a(1) + N_e(1) \cdot P_e \tag{5.19}$$

- Biomass and leaf construction:

The amount of biomass taken by a leaf $q_a(1, 1)$ in the first cycle can be computed, the demand of the plant being $D(1)$ and the volume of biomass given up by the seed being Q_0. The notation $q_o(i, j)$ refers to the mass at the cycle j of the organ o (a for leaves, e for internodes) that appeared at the cycle i.

$$q_a(1,1) = \frac{P_a(1)Q_0}{D_1} \tag{5.20}$$

The surface of the leaf is written as (Sect. 5.3.2):

$$Sf(1) = \frac{q_a(1,1)}{\varepsilon} \tag{5.21}$$

- Internode construction:

The amount of biomass from the seed that goes to the internode equals to:

$$q_e(1,1) = \frac{P_e(1)Q_0}{D_1} \tag{5.22}$$

The allometric relationships (Sect. 2.3.2) are used to obtain the shape:
The length of the internode:

$$h_{1,1} = \sqrt{b_e}(q_e(1,1))^{\frac{1+\gamma}{2}} \tag{5.23}$$

and the surface of its section:

$$s_{1,1} = \sqrt{\frac{1}{b_e}(q_e(1,1))^{\frac{1-\gamma}{2}}} \quad (5.24)$$

- Computing the biomass production:

Once the surface of the leaf has been established, the biomass production can be computed by Eq. (5.14):

$$Q(1) = \frac{E \cdot Sp}{r} \cdot \left(1 - \exp\left(-k\frac{Sf(1)}{Sp}\right)\right) \quad (5.25)$$

5.5.4.2 Generalization by Recurrence to Cycle *T*

It is assumed that the recurrence equation is validated until the cycle $t - 1$.

- Computation of the Demand:

The numbers of leaves and internodes produced in the cycle t by the botanical automaton of the architectural model are, respectively, $N_a(t)$ and $N_e(t)$. The CA of the organs can potentially be up to t cycles for organs that appeared in the first cycle: the sum of the cohort sinks must, therefore, be between 1 and t. Eq. (5.8) gives:

$$D(t) = \left(\sum\nolimits_{i=1}^{t} (N_a(t-i+1) \cdot P_a(i) + N_e(t-i+1) \cdot P_e(i))\right) \quad (5.26)$$

The terms of this sum correspond to the demand of the organs of age i in DC (sink functions $P_o(i)$), which are in number $N_o(t - i + 1)$ as they appeared in the cycle $t - i + 1$.

- Leaf construction.

We move from cycle t—1 to cycle t by increasing the size of Eq. (5.11) of one unit, which gives the volume of a single leaf per cohort using leaf sink function $P_a(i)$:

$$\begin{bmatrix} q_a(1,t) \\ q_a(2,t) \\ \vdots \\ q_a(t,t) \end{bmatrix} = \begin{bmatrix} P_a(1) & P_a(2) & \cdots & P_a(t) \\ 0 & P_a(1) & \cdots & P_a(t-1) \\ 0 & 0 & \ddots & \vdots \\ 0 & 0 & \cdots & P_a(1) \end{bmatrix} \begin{bmatrix} \frac{Q_0}{D_1} \\ \frac{Q_1}{D_2} \\ \vdots \\ \frac{Q_{t-1}}{D_t} \end{bmatrix} \quad (5.27)$$

The left vector represents the organic time series of leaf weights.

The surface *Sfa* of a leaf of the cohort that appeared in the cycle i is written as:

$$Sfa(i,t) = \frac{1}{\varepsilon} q_a(i,t) \tag{5.28}$$

The leaf surface of the plant is written as:

$$Sf(t) = \frac{1}{\varepsilon} \sum\nolimits_{i=\max(1,\, t-t_a+1)}^{t} N_a(i) \cdot q_a(i,t) \tag{5.29}$$

- Building of the internodes of the tth cycle.

  We have the volume of the internodes by Eq. (5.10) using leaf sink function $P_e(i)$:

$$\begin{bmatrix} q_e(1,t) \\ q_e(2,t) \\ \vdots \\ q_e(t,t) \end{bmatrix} = \begin{bmatrix} P_e(1) & P_e(2) & \cdots & P_e(t) \\ 0 & P_e(1) & \cdots & P_e(t-1) \\ 0 & 0 & \ddots & \vdots \\ 0 & 0 & \cdots & P_e(1) \end{bmatrix} \begin{bmatrix} \frac{Q_0}{D_1} \\ \frac{Q_1}{D_2} \\ \vdots \\ \frac{Q_{t-1}}{D_t} \end{bmatrix} \tag{5.30}$$

  The left vector represents the organic time series of internodes weights. Allometries give the lengths and surfaces of the sections of the internode. Length from the internode i to the cycle t:

$$h_{i,t} = \sqrt{b_e}(q_e(i,t))^{\frac{1+\gamma}{2}} \tag{5.31}$$

  Surface of the section from the internode i to the cycle t:

$$s_{i,t} = \sqrt{\frac{1}{b_e}}(q_e(i,t))^{\frac{1-\gamma}{2}} \tag{5.32}$$

- Biomass production at the cycle t.

  Biomass production at the cycle t is given by Eq. (5.14):

$$Q(t) = \frac{E \cdot Sp}{r} \cdot \left(1 - \exp\left(-k\frac{Sf(t)}{Sp}\right)\right)$$

Recurrence works and the plant, therefore, grows from cycle to cycle by increasing the size of the growth equation by one unit at each CA and by alternating the feedback between plant development (organ creation) and the production and distribution of biomass that ensures their expansion.

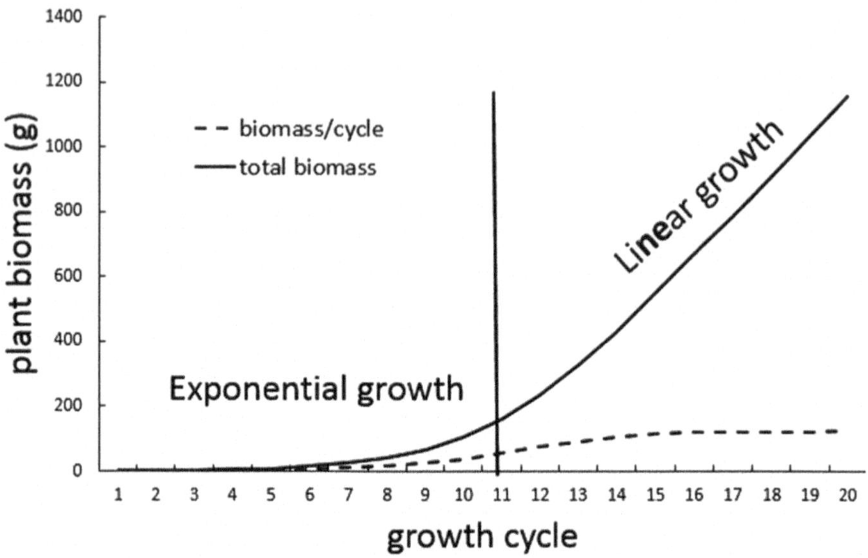

Fig. 5.16 Biomass production per cycle (dashed line) and cumulative weight (solid line) during the growth of a computational plant according to GreenLab (Eq. 5.14) in the case of high planting density (*Sp*). Initially, the growth is exponential (Eq. 5.15) and smoothly becomes linear (Eq. 5.16), because of the effect of the density

5.5.5 How the GreenLab Model Works

GreenLab at the individual plant level works similarly to the TomSim model at the stand level. According to Eq. (5.14), under normal conditions, the biomass produced at each cycle at the beginning of growth grows exponentially. Then, when the limiting effect of the production area *Sp* is reached, the biomass produced at each cycle becomes constant. Correlatively, the weight of the plant, i.e., the sum of $Q(t)$, increases exponentially at the beginning of the growth and then linearly with time under the effect of the limitation due to *Sp* (Fig. 5.16), as long as the senescence of the leaves does not decrease the LAI. This is well verified in cultivated plants, including beetroot (Lemaire et al. 2008; Lemaire 2010) (Fig. 5.30).

GreenLab is available at the organ level. A physiological crop model at the compartment level can be considered as a "projection" of GreenLab at the compartment level, where the sum of the organs reconstructs the compartments. Once these are reconstituted, GreenLab gives the same behavior as physiological crop models (Tomsim, Pilote) for the compartments (leaf indices, harvest indices, etc.). But GreenLab runs at a finer spatial scale, because it simulates, from the seed, the development and growth of the plant structure at the level of the phytomer, which makes it possible to rebuild the plant architecture. Figure 5.17 is thus an extension of Fig. 5.9.

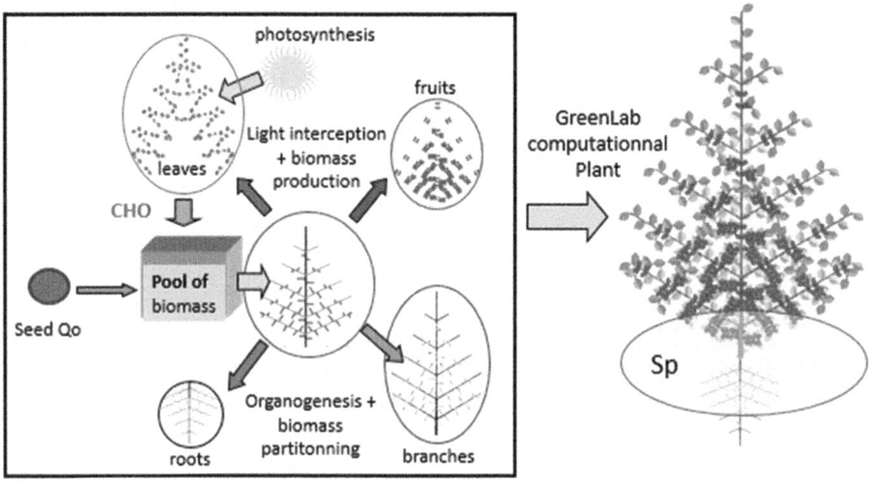

Fig. 5.17 Flowchart of the functioning of GreenLab

5.6 Behavior of the Model: Case of Free Growth

The mathematical study of the model's behavior is an important step in its construction. It makes it possible to establish the influence of the parameters on the growth of the plant. The model equations make it possible to validate the performance of the computer program that numerically simulates the growth and architecture of the plant without explicitly incorporating these equations.

Free growth refers to the case where biomass production is proportional to leaf area, which occurs at the beginning of a plant's growth when the leaves do not yet overlap (Eq. 5.15). The production surface *Sp* is not involved.

Equation (5.14) can be studied in simplified cases. At first, we consider plants with phytomers consisting of an internode with one or more leaves, possibly several fruits and axillary meristems. There is no root system and no secondary growth. Only one PA is also considered. The sinks of the leaf organs and internodes are constant and have the same expansion time t_x equal to the functioning time t_a of the leaves.

By Taylor development limited to the first order, the simplified Eq. (5.18) with a single PA becomes:

$$Q(t) = \frac{E}{\varepsilon \cdot r} \cdot \sum\nolimits_{i=\max(1,\, t-t_a+1)}^{t} N_a(i) \cdot \sum\nolimits_{j=i}^{t} \frac{P_a \cdot Q(j-1)}{D(j)} \tag{5.33}$$

The demand of the plant is the sum of the sinks of the organs that compose it. By designating the number of leaves (n_a), internodes ($n_i = 1$) and fruits (n_f), and their sink strengths P_a ($P_a = 1$), P_e and P_f, respectively, the sink strength of the phytomer P_p is written as:

$$P_p = n_a \cdot P_a + P_e + n_f \cdot P_f \tag{5.34}$$

where n_a is the number of leaves per plant ($n_a = 1, 2, \ldots$) and $N_p(i)$ is the number of phytomers with active leaves in the cycle i. Hence, the number of leaves:

$$N_a(i) = n_a \cdot N_p(i) \tag{5.35}$$

It is assumed here to simplify that all organs have the same duration of expansion t_x.

The demand equation is written as:

$$D(i) = \sum\nolimits_{j=1}^{i} N_P(j) \cdot P_p \tag{5.36}$$

Let us assume hereafter:

$$A = r \cdot \varepsilon \cdot P_p \tag{5.37}$$

The coefficient A includes the action of three parameters on the growth ($1/r$ is resource use efficiency, and ε is the thickness of the leaf). Parameter E represents the energy supplied to the plant by the environment during each DC, which here is assumed to be constant and normalized with $E = 1$.

Equation (5.33) is simplified with this notation:

$$Q(t) = \frac{n_a \cdot E}{A} \cdot \left(\sum\nolimits_{i=\max(1,\, t-t_a+1)}^{t} Q(i-1) \right) \tag{5.38}$$

This recurrence equation is effective in studying the behavior of the model. To make it simple, we fix $n_a = 1$ (one leaf per phytomer). Two cases can be distinguished.

5.6.1 *Conditions for Exponential Growth for* $\mathbf{T < t_a}$

When $t < t_a$, all leaves are functional since the age of the plant is less than the duration of their functioning.

For $t_a = 1$ (leaf function and expansion only last one cycle), Eq. (5.38) becomes:

$$Q(t) = \left(\frac{E}{A}\right)^t \cdot Q_0 \tag{5.39}$$

And for $t_a > 1$

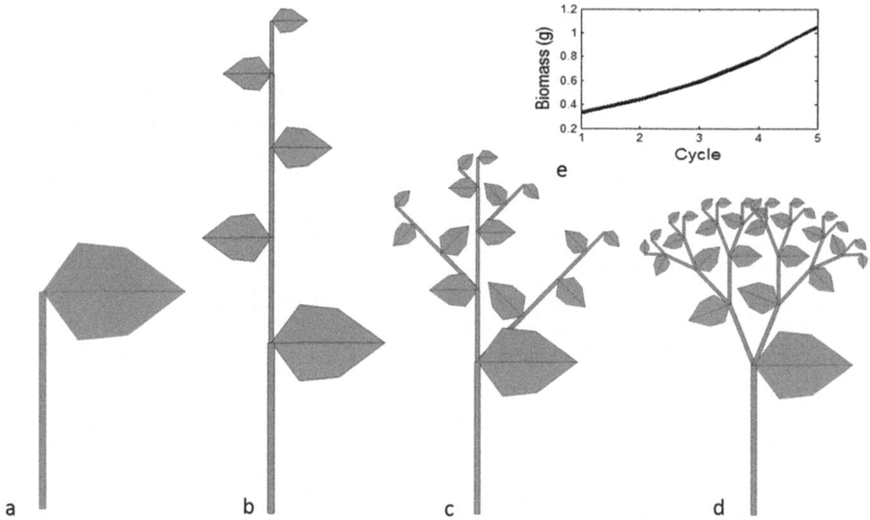

Fig. 5.18 Growth of architectural models with all organs expanding. Plant (**a**) has only one growing phytomer and therefore no development. Plants (**b**, **c**, **d**) follow the developments of a single-stemmed, branched monopodial, and branched sympodial plant, respectively, (**e**), the four plants produce the same total biomass and leaf area with exponential growth

$$Q(t) = \frac{E}{A} \cdot \left(1 + \frac{E}{A}\right)^{t-1} \cdot Q_0 \tag{5.40}$$

These equations explain the exponential growth pattern generally observed in the juvenile growth stage of a plant until the age limit for leaf function is reached. Moreover, the number of phytomers is not involved in the final equation, which means that the development has no influence on the growth. It means that whatever their architectural model, plants with the same functional parameters will have the same trajectory of biomass production. The sum of the weights of the phytomers is independent of the architectural model. Therefore, the more phytomers there are, the smaller their weights are (Fig. 5.18).

For $t_a = 1$, there can be a meta-stable equilibrium if $E/A = 1$.

But if $t_a > 1$, the growth is necessarily exponential according to Eq. (5.40).

In Fig. 5.18, the organs have constant sinks equal to 1, t_a and t_x are 5 DCs. The demand of the plant **a** is $D(5) = 2$, that of the plant **b** is $D(5) = 2 \times 5$, that of the plant **c** is $D(5) = 2 \times (5 + 10)$, and that of the plant **d** is $D(5) = 2^{5-1}$.

These four plants produce the same biomass, and they have the same growth in terms of plant compartments, but different development and, therefore, different architectures. However, in a physiological crop model, these differences are not distinguished.

5.6.1.1 Seed Effect on the Growth

5.6.1.1.1 Influence of the Seed on Plant Growth

We notice in Eqs. (5.39 and 5.40) the importance of Q_0, which is the amount of biomass provided by the seed at the origin of the growth. The plant biomass is proportional to the seed biomass at the beginning of the growth as long as $t \leq t_a$. This is well verified in reality.

In 1920, Professor Achille Urbain, Director of the French Museum (1942–1949), in his thesis on botany (Urbain 1920), carried out experiments to remove albumen of cotyledons to a greater or lesser extent depending on the nature of the seed on various species. He compared the development and growth of the resulting plants with a control plot. The removal of seed reserves significantly reduces both development and growth. The plants obtained were dwarf plants because they produced fewer phytomers, and the organs were atrophied (Fig. 5.19).

More recently, we have observed the germination of cocoa seedlings after weighing the seeds at the beginning. Figure 5.20a shows the effect of the seed on the beginning of growth on a young cocoa tree.

It may not be sufficient to assign a single cycle of seed biomass distribution from seed to seedling, especially in the case of big seeds, such as avocado or coconut. In this case, the seed gradually empties itself over several cycles according to a specific pattern. If we assume an exponential decrease, the quantity released to cycle t is $Q_{\text{seed}}(t) = Q_0 \cdot c^t$. The parameter c is the biomass rate released by the seed at each cycle. In the general case, the photosynthesis will gradually take over (Fig. 5.20b and solid line in Fig. 5.20d). In the absence of photosynthesis, the biomass available per cycle tends to zero (Fig. 5.20c and dotted line in Fig. 5.20d).

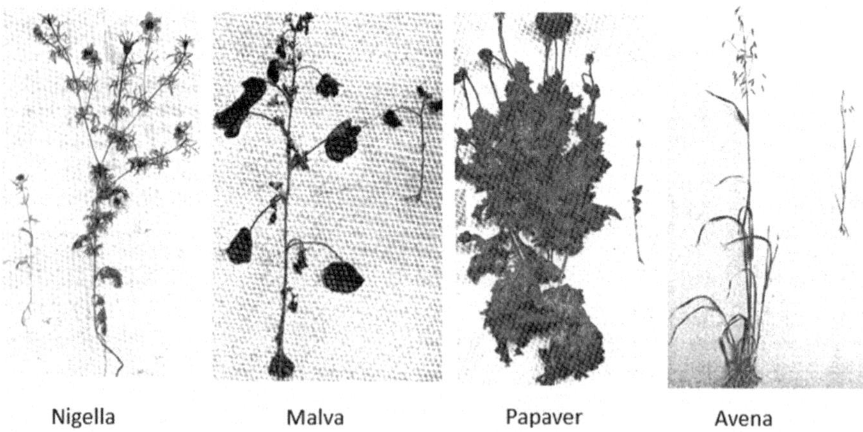

Fig. 5.19 Effect of removing a large part of the albumen from seeds on plant development and growth by Achille Urbain (1920). This means drastically reducing the reserves for germination. We get dwarf plants. On the photographs of four plant species (Nigella, Malva, Papaver, Avena), we can compare the controls with the ex-albuminated plants

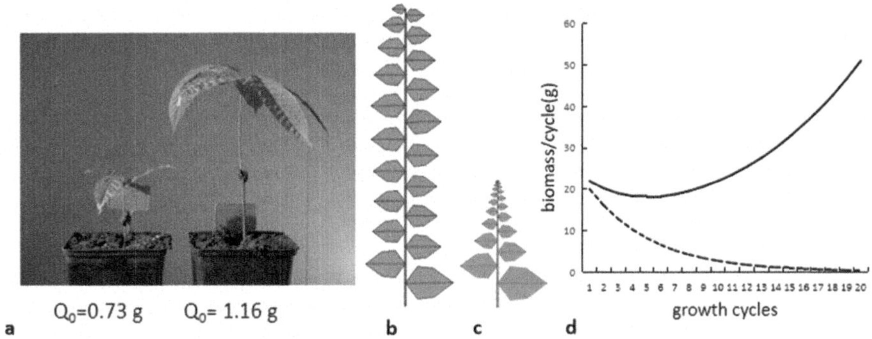

Fig. 5.20 (**a**) Effect of seed weight on seedling size. (**b**, **c**, **d**) Simulation of the effect of seed emptying in several cycles, with photosynthesis relay (**b** and solid line in **d**) or without relay (**c** and dotted line in **d**)

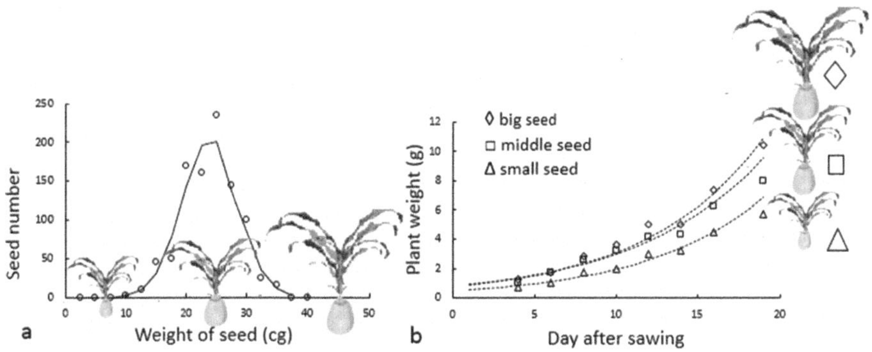

Fig. 5.21 Effect of seed size on the growth of seedlings on maize. (**a**) The weight of seeds has a normal distribution. (**b**) Seedling growth is verified as exponential (free growth, Source Ma Yun-Tao, China Agricultural University (CAU))

5.6.1.1.2 Influence of Seed Weight on Maize Growth at Early Stage

The quantitative effect of the seed on maize growth was specified by Ma Yun-Tao (personal communication, CAU, 2006). From a batch of 950 seeds of the same variety, seed weights varied between 0.15 and 0.35 g (Fig. 5.21a). The seed weights were distributed according to a normal distribution ($\mu = 0.24$ g, and $\sigma = 0.045$ g).

Three weight batches of seeds are made up: big seeds 0.32–0.34 g, medium seeds 0.24–0.26 g, and small ones 0.1–0.18 g, the germination of which is monitored over a period of 20 days.

At this stage, the seedlings have developed five leaves. The increase in biomass of seedlings is measured every two days (Fig. 5.21b). The seedling weight grows exponentially, as predicted by the model for free growth (Eq. 5.15). The weights of the small seeds $Q_s(j)$ and the big seeds $Q_b(j)$ observed are adjusted according to the number of days d after germination using the equation $Q(i) = a \cdot \exp(b \cdot i)$. The coefficients a and b are calculated by regression.

$$Q_s(i) = 0.48 \cdot \exp(0.14 \cdot i); \quad Q_b(i) = 0.76 \cdot \exp(0.14 \cdot i).$$

It can be seen that the coefficients b are the same, so the ratio of seedling weights during the growth is constant. In addition, the weight of a seedling is an increasing function of the measured seed weight.

The model is well verified here in this simple experimental case.

5.6.1.1.3 Influence of Seed Weight on Germination of Pterocarpus Tree

In Ivory Coast, Adji Beda studied the effect of seed size on germination of Pterocarpus tree. Figure 5.22 shows the appearance of the seeds and their weight distribution. This distribution tends to a normal distribution as for maize.

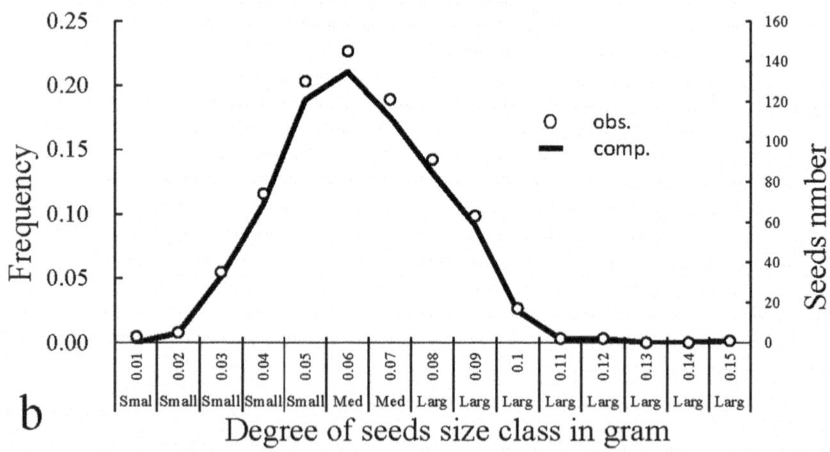

Fig. 5.22 Photos and distribution of seed size of Pterocarpus: (**a**) Photos of seeds. (**b**) Sorting seeds into large, medium, and small categories from left to right and distribution of seeds number and frequency according to their size

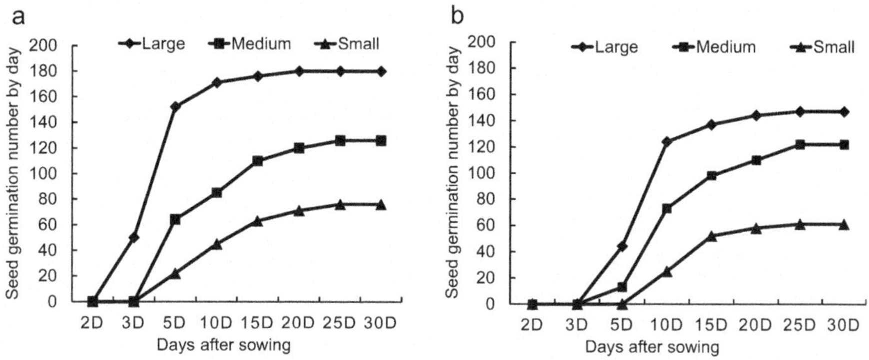

Fig. 5.23 Germination dynamics of three sizes of seeds of *Pterocarpus* tree at Daloa (**a**) and Korhogo (**b**)

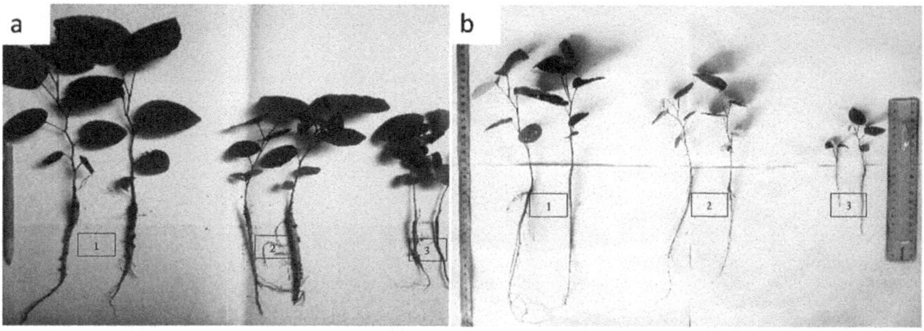

Fig. 5.24 Influence of seed reserves (large, medium, small) on seedling size and influence of the station environment (stress) on seedling size at the stations of Daloa (**a**) and Korhogo (**b**)

The following of the germination rates as a function of time shows that they increase with the volume of the seed (Fig. 5.23). This generates a heterogeneity in the seed dormancy and thus in the development of the seedlings at the stand level. Here two stations are represented: Daloa and Korhogo. At the Daloa station, seeds with large size get 100% of germination, while small seeds have less than 50% germination. At the Korhogo station, only 88% of the seeds germinate due to water stress, which is reflected in the germination rates.

The size of the young seedling is directly proportional to the seed reserve, as observed in cocoa tree (Fig. 5.24). Moreover, the same seed batch gives seedlings of different sizes depending on the environmental stress (climate, soil) of a station. Thus, in Korhogo the seedlings are smaller than those in Daloa.

In conclusion, from a modeling point of view, it is essential to integrate the effect of seed reserves (Q_0) to initiate the growth correctly. This aspect is absent in physiological crop models. The seed effect generates heterogeneity in the stand at two levels. The first is the variation of the seed size which modifies plant growth within the stand, the second is the variation of the dormancy ending which modifies plant development within the stand (see maize section in Chap. 11). However, the seed effect disappears progressively with the effect of planting density.

5.6.1.2 Influence of the Removal of Axillary Meristems

At the juvenile stage, the model predicts that growth does not depend on the architecture. Therefore, if axillary meristems are removed from a branched plant at birth, the resulting single-stemmed plant should weigh the same as the branched control plant. This fact is verified experimentally on plants such as coffee trees in the young stage. Removal of axillary meristems during the growth results in an increase in the volume of organs of the stem to an equivalent biomass of the whole unpruned plant (Fig. 5.25). Removing meristems reduces the demand and therefore increases the volume of the remaining organs. Similar experiments on other species, such as Cotton, give the same type of behavior.

Fig. 5.25 Coffee trees with branches (control) and pruned in free growth [S. Sabatier, AMAP]. All leaves are functional for these plants at early stage. The production is proportional to the leaf area. Axillary meristems have been removed for pruned coffee trees. In accordance with the simulated prediction of GreenLab, pruned coffee trees have larger phytomers, but keep the same total weight as that of the control coffee trees

5.6.2 *Condition for Steady Growth, where* $T > t_a$ *and* $t_a \geq t_x$

In this case, the growth time of the plant exceeds the functioning time of the leaves, which wither and no longer participate in the functioning at the end of t_a cycles after their creation. It is assumed here that the functioning time of the leaves t_a is equal to or greater than the expansion time of the organs t_x: $t_a \geq t_x$. Moreover, to simplify, it is assumed that the sinks are constant and equal to 1, thus the phytomer has only one leaf and no fruit.

Let us consider the conditions required for the growth to stabilize. It is sufficient that demand D is limited and that the Q/D ratio (the trophic pressure) tends towards a limit. For the demand to be limited, the life of the axes must be limited. After a while, the newly created phytomers exactly replace the old ones belonging to dead axes. Starting from the growth Eq. (5.33), in order to achieve an equilibrium, the following results are established in different cases:

- If $t_a = 1$ and $t_x = 1$.

We have the production from Eq. (5.39):

$$Q(t) = \left(\frac{E}{A}\right)^t \cdot Q_0.$$

Growth is steady only for $E/A = 1$. In this case, the production $Q(t)$ is constant and is equal to Q_0. Otherwise, the growth becomes exponential (increasing or decreasing).

- If $t_a > 1$ and $t_x = 1$.

The leaves work several cycles, but the expansion of the organs only lasts one cycle. As long as $t \leq t_a$, the production is written from Eq. (5.40):

$$Q(t) = \frac{E}{A} \cdot \left(1 + \frac{E}{A}\right)^{t-1} \cdot Q_0$$

When $t > t_a$, the leaves stop working after t_a cycles. The growth will stabilize for each cycle if: $\frac{E}{A} = \frac{1}{t_a}$. After the number of t_a DCs, the production will then tend towards the limit value:

$$Q_l = \frac{2 \cdot Q_0}{1 + t_a} \tag{5.41}$$

- If $t_a > 1$ and $t_a = t_x$,

As long as $t \leq t_a$, the production is exponential and is given by Eq. (5.40). The growth will stabilize beyond $t > t_a$ if the following equality is verified:

$$\frac{t_a + 1}{2} \cdot \frac{E}{A} = 1 \tag{5.42}$$

The production limit Q_l is found by iteration on Eq. (5.38).

- If $t_a > 1$ and $t_a \geq t_x$, this is the general case.

As long as $t \leq t_a$, the production is exponential and is given by Eq. (5.40). The growth stabilizes for $t > t_a$ if the following equality is verified:

$$E \cdot \frac{1}{A} \cdot \left(\frac{t_x + 1}{2} + t_a - t_x\right) = 1 \tag{5.43}$$

This production limit Q_l is also found by iterations on Eq. (5.38).

In summary, the equilibrium equations of the production depend on the functioning time t_a and the expansion time t_x of the organs, a structural coefficient A and the climatic factor E. The growth is, therefore, stabilized in free growth whatever the architectural model if the demand is limited and if the condition of equality given by Eq. (5.41) is satisfied. Otherwise, the production increases or decreases exponentially.

5.6.3 *Study of the Growth of Two Architectural Models*

Here we study in particular the growing behavior of two widespread branched architectural models, with monopodial development (e.g., coffee tree representing the Roux model) and sympodial development (e.g., manioc representing the Leeuwenberg model).

5.6.3.1 Model with Monopodial Branching

This model has two categories of axes corresponding to the two PAs: main stem and branches. Each phytomer on the main stem has a leaf and a branch. Sinks are same for the two PAs.

A duration $t_a = t_x$ is set for the functioning and expansion times of the organs and a lifetime t_2 for axes of order 2 (branches). The plant demand and the biomass production stabilize for $t > t_2$ if the condition (Eq. 5.41) is verified. Indeed, since the lifetime of the leaves is t_a, the number of active leaves becomes constant and is: $N_a(i) = t_a \cdot (t_2 + 1)$ for all $i > t_2$. With the values of the parameters $t_\mathrm{a} = t_\mathrm{x} = 5$, $\varepsilon = 0.05$, $r = 30$, $p_a = 1$ and $p_e = 1$, the equality (Eq. 5.28) is checked. We have:

$A = 0.05 \times 2 \times 30$, we have: $\frac{1}{3} \times \left(\frac{5+1}{2} + 5 - 5\right) = 1$

With these parameters, production stabilizes and so does the Q/D ratio, which represents the trophic pressure. We show by simulation that the production depends increasingly on the lifetime t_2 of the branches. For $t > t_a$ and $t > t_2$, the larger the t_2,

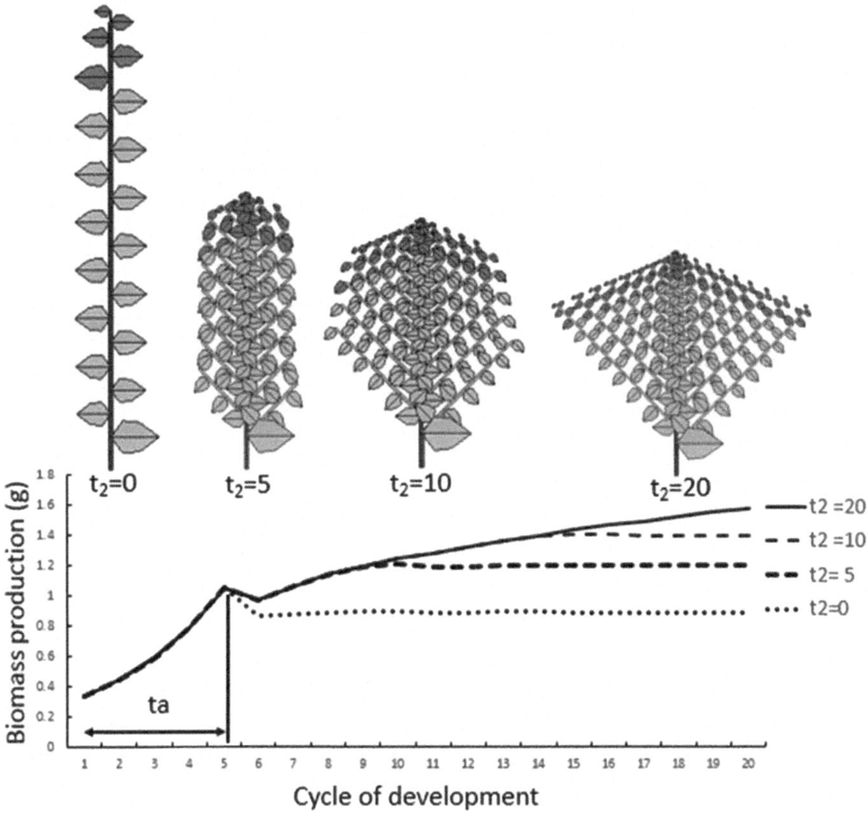

Fig. 5.26 Stabilization of the growth of the Roux model when the condition (Eq. 5.41) is verified. Yellow leaves are no longer functional. Beyond $t > t_a$, they are different lifetime values t_2 for the branches: $t_2 = [0, 5, 10, 15, 20]$. The value $t_2 = 0$ corresponds to the single-stemmed plant. As long as $t < t_a$, the plants have the same exponential growth. For $t > t_2$, the larger the t_2, the stronger the production at equilibrium, which is reached later, and the smaller organs are. Branched architectures correspond to the lifetime of the axes $t_2 = [0, 5, 10, 15]$ for the chronological age (CA) of 20 DCs with $t_a = 5$ and $t_x = 5$. Stabilization of the organic series is observed when the age of the plant exceeds the lifetime t_2 of the branches. Simulated plants are sorted in ascending order of weight (Color figure online)

the higher the production $Q(t)$. However, an equilibrium is reached when t is sufficiently large (Fig. 5.26). If t_2 is unlimited, the growth increases indefinitely.

Simulations in the model (Fig. 5.26) show that the volume of phytomers decreases when t_2 increases, while the biomass production increases. For $t_2 = 0$, the plant is single-stemmed, and the phytomers reach their maximum volumes. It should be noted that biomass production for all t_2 values is the same as long as $t \leq t_a$.

This fact is verified experimentally, for example in cotton (Fig. 5.27): pruned cotton is taller than branched cotton but weighs less. There is a change in the functioning when $t = t_2$. Below this ($t < t_2$), branched or pruned plants have the

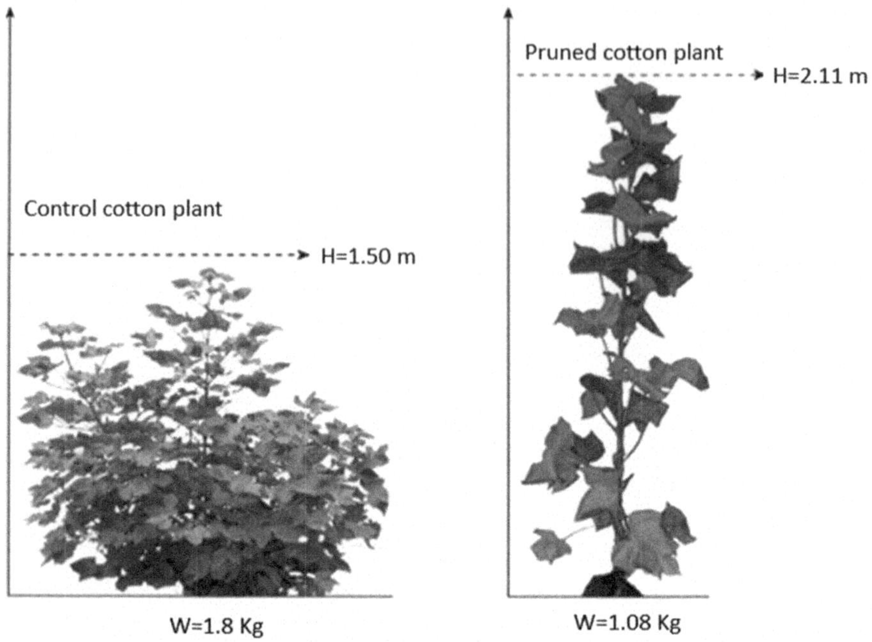

Fig. 5.27 Pruning experiments on cotton trees (personal communication, experiment by Li et al. 2010). The removal of axillary meristems from cotton stems results in an increase in the number and volume of phytomers on the main stem for the same development period. The weight of the control cotton plant (1.80 kg) is about twice as heavy as that of the pruned cotton plant (1.08 kg), whose height reaches 2.11 m, compared to 1.50 m for the control plant, with larger organs. According to GreenLab, if the duration of axis growth is longer than the duration of leaf functioning, the control plant weighs more than the pruned plant

same production (coffee tree, Fig. 5.26). Beyond that ($t > t_2$), plants with longer branches have larger biomass.

5.6.3.2 Model with Sympodial Branching

When condition (41) is verified, a limited demand is a sufficient but not necessary condition for stabilizing production. Thus, in the case of a sympodial model whose development is sympodial and exponential, it is possible to establish an equilibrium condition that can be verified by simulation. For $t_a = t_x$, if M (with $M > 1$) represents the number of branches per phytomer, this condition is written as:

$$\frac{E}{A} \cdot \left(\frac{M^{t_a}}{M^{t_a} - 1} \cdot t_a - \frac{1}{(M-1)} \right) = 1 \tag{5.44}$$

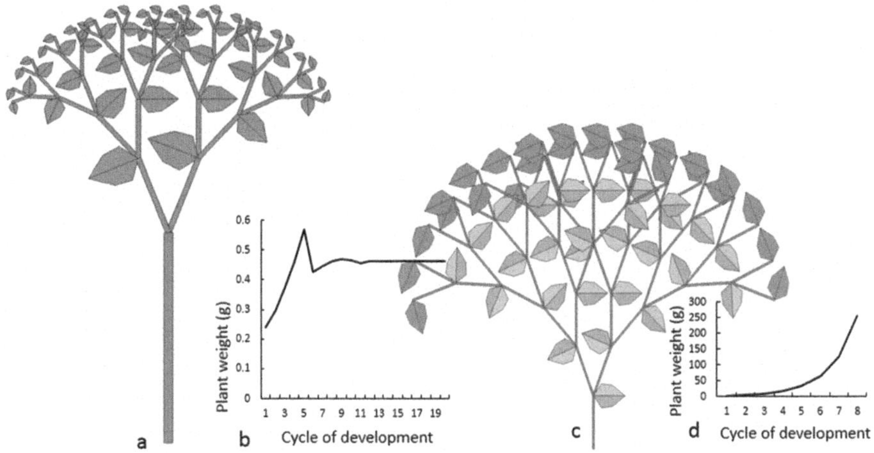

Fig. 5.28 Study of the behavior of a sympodial model in free growth. The duration of leaf function and organ expansion is 5 development cycles (DCs) here. The condition of Eq. (5.29) is checked for the values $M = 2$, $t_a = t_x = 5$, $r = 41.613$, $\varepsilon = 0.05$, and $P_p = 2$. Growth stabilizes, although the volume of phytomers tends towards zero. (**a**) Simulated plant at $t = 6$ DC. (**b**) Evolution of the produced biomass that stabilizes after 10 DC of the development. **c** here we set $t_a = t_x = 1$. The growth equation is written for the values $M = 2$, r = 5, $\varepsilon = 0.05$ and $P_p = 2$. $Q(t) = 2^{t-1} \cdot Q_0$, and the demand is $D(t) = 2^{t-1}$, so Q/D is constant, organs keep the same size, and growth is exponential

The equilibrium here depends on the number of branches per phytomer M (if $M = 1$, it can be simplified to a model with a single sympodial axis). Although the demand is growing exponentially, the foliar biomass produced per cycle is stabilizing, while organ volume is decreasing exponentially (Fig. 5.28a, b). In general, in this model, the development and growth are exponential. The organs can even be of constant biomass if Q and D have the same evolution, which naturally gives an exponential growth of the biomass (Fig. 5.28c, d).

5.6.3.3 Sink Source Organs and Stabilized Production Relationships

In the general case, a phytomer is a set of organs consisting of an internode, one or more leaves, and possibly several fruits. It is assumed here that all organs have constant sink values.

Equation (5.41) gives the condition of free growth at equilibrium. Each of the variables that constitute the coefficient A can be adjusted to achieve this equilibrium.

Figure 5.29a shows a case of equilibrium based on a complex phytomer consisting of one internode, two leaves, and six fruits. The architectural model here is a monopodial branched plant with a branch life span limited to $t_2 = 10$ DCs.

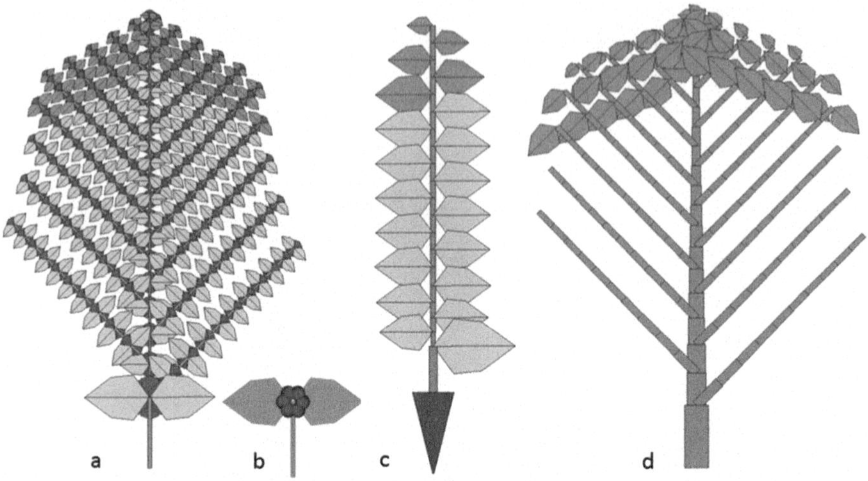

Fig. 5.29 Plants with complex phytomers, with secondary growth or root system. (**a**) Plant reaching a steady production with complex phytomers consisting of one internode, two leaves, and six fruits. (**b**) Detail of a phytomer. In this example, all the sinks of the organs are identical, only their numbers change. The expansion times are $t_x = 3$ DCs and the functioning lifespan of the leaves $t_f = 4$ DCs, and the branches $t_2 = 10$ DCs. Eq. (5.28) is checked for $r = 13.33$, with $\varepsilon = 0.05$. The demand for the expanding phytomer is $P_\varphi = 2 + 1 + 6 = 9$. $A = r \cdot \varepsilon \cdot P_\varphi$. Eq. (5.28) gives: $E \cdot \frac{na}{A} \cdot \left(\frac{t_x+1}{2} + t_a - t_x\right) = 1 \Rightarrow \frac{2}{13.33 \times 0.05 \times 9} \times \left(\frac{3+1}{2} + 4 - 3\right) = 1$. (**c**) We introduce a constant root sink $p_r = 4$. The sinks of the internodes and leaves are 1. The functioning and expansion times of the leaves are $t_a = 5$ and $t_{xa} = 5$. The expansion time of the internodes is set to $t_{xe} = 1$. The demand per cycle is therefore $D = 5 + 1 + 4 = 10$ and the growth is steady for the values of the chosen parameters. (**d**) The secondary growth on a branched model is visible. The parameters are $p_a = 1$ for the leaf sink, $p_e = 0.5$ for the internode sink, and $p_c = 0.5$ for the ring elements. We set: $t_a = t_x = 5$, $\varepsilon = 0.05$, $r = 30$, $t_2 = 10$, which gives a growth to the equilibrium

5.6.3.4 Special Sinks: Rings and Roots

In plants, there are particular sinks that do not correspond to well-located organs. They are the rings that ensure the secondary growth of the woody axes, and the roots whose architecture is complex and difficult to access.

Modeling of root system architecture has been a subject of research at CIRAD (Jourdan and Rey 1997), INRA (Pagès et al. 2004) in France, and CAU in China (Zhang and Li 2009). There are root architectural models (Atger and Edelin 1994) in their organizations. However, the notion of root phytomer does not exist in the literal sense. From an experimental point of view, the study of root systems is hampered by the difficulty of access. GreenLab considers the root system as a compartment with a variable sink, whose expansion can be indefinite. For some studied species (Chap. 9, beetroot, tomato, cucumber, etc.), the weight of the root system was available and was, therefore, integrated into the source–sink relationships. Figure 5.29c shows a single-stemmed plant whose growth is stabilized with source–sink parameters, including a root system treated as a constant sink.

Chapter 8 discusses the rings and provides the principles of secondary growth modeling in GreenLab. Here we consider the demand for the rings related to the architecture in the case where each functional leaf is associated with an element of ring. This element is also a sink. It has a shape of a cylindrical tube of biomass that travels along the edge of the axis from the leaf to the base of the tree. These sets of elements (pipes), grouped in bundles, extend from the ends of the tree axes to the base of the trunk, forming the rings. This model is known as the "pipe model" (Chap. 8). The demand for rings is the product of the number of active leaves multiplied by the sink strength of a ring element. Its sink strenght is constant and lasts for only one cycle. In this case, it is easy to define a production equilibrium as in the previous cases (Fig. 5.29d) to test the model's behavior.

5.6.4 *Case of Rhythmic Growth*

In the case of trees, the rhythmic development that forms GUs complicates the functioning. However, it is possible to study its behavior in two contrasting cases.

5.6.4.1 Synchronous Trees with Preformed Growth Units

The preformed part of the GUs is first formed during the year as a series of embryonic phytomers in apical meristems, in several DCs with negligible demand. A bud is thus formed. During its bud burst in spring, the sinks of the organs become effective, and there is a simultaneous expansion of all preformed phytomers (which are identical in our example here): a new GU is formed from the reserves of the previous year. The leaves and pith of the internodes no longer evolve, and the shape of the GU is stabilized. Its photosynthetic functioning produces the biomass in reserve for the following year. The expansion period is considered negligible compared to the functioning period. Without loss of accuracy, all functioning cycles can be concentrated in a single cycle that aggregates the expansion of the GU and cumulative photosynthesis. Its development scheme is:

- a development period of several DCs to form embryonic phytomers of the GU;
- a DC to ensure the expansion of the GU and its subsequent production;
- a DC to mark the winter pause (by convention).

Trees such as chestnut trees or maple trees fit well into this framework. Finally, this type of functioning is like that of the continuously growing shrub, but on a double scale. All the phytomers of the GU behave as if there were only one meta phytomer composed of an internode and a leaf, and the growing conditions are those of a shrub with $t_a = 1$ and $t_x = 1$ (Fig. 5.30a).

The simplified growth equation is written as Eq. (5.39):

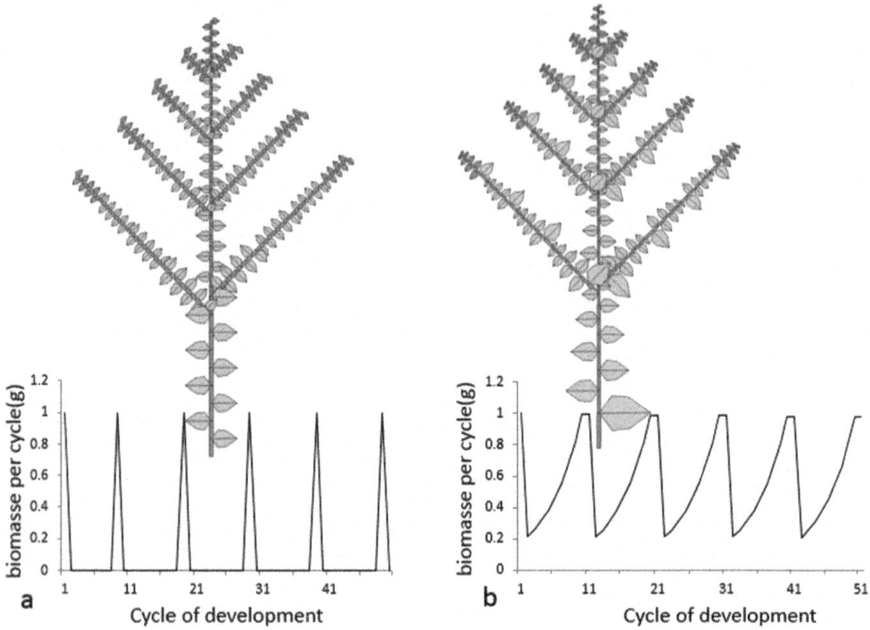

Fig. 5.30 Two theoretical cases of possible stabilization in the rhythmic growth of trees with an annual growth divided into 10 development cycles (DCs). In both cases, a cyclic stabilization of production can be achieved for a set of parameters. The stabilized cycles of production $Q(t)$ are displayed below the corresponding trees. (**a**) Growth from a preformed GU. Development and growth are not synchronized (Chap. 6). Growth is delayed. Embryonic phytomers preformed in 9 DCs have simultaneous expansion at DC 9, and therefore have the same volume. DC 9 concentrates the photosynthesis into one cycle. A pause (DC 10) separates two years (Parameter $r = 10$). (**b**) The phytomers have been formed one by one. Development and growth are simultaneous. Phytomers have different volumes depending on the trophic pressure Q/D at the time of their creation. At the end of the GU, all the leaves stop working together, and a pause ends the year (Parameter $r = 46.9$)

$$Q(t) = \left(\frac{E}{A}\right)^{t-1} \cdot Q_0$$

It gives the growing behavior of the model and the equilibrium condition $E/A = 1$.

5.6.4.2 Synchronous Trees with Neoformed GUs

The phytomers produced by meristems immediately expand after their formation. It is assumed that the leaves remain functional until $t = t_a$ whatever their time of appearance. The functioning of the first leaf of the GU of rank 1 lasts t_a DCs, the second of rank 2 lasts $t_a - 1$ DCs, and the functioning of the last leaf of rank t_a lasts only 1 DC. In addition, the expansion time of the phytomers $t_x = 1$ is fixed. Under these conditions, the growth is driven by Eq. (5.40) established previously.

To reach stabilized cyclical growth, A must be the solution of Eq. (5.40):

$$Q(t) = \frac{E}{A} \cdot \left(1 + \frac{E}{A}\right)^{t-1} \cdot Q_0$$

Which gives the following equation to solve (setting $Q_0 = 1$):

$$A + E - \left(\frac{A^{t_a}}{E}\right)^{\frac{1}{t_a - 1}} = 0 \tag{5.45}$$

We have $A = r \cdot \varepsilon \cdot P_p$; we can choose r as unknown and set: $E = 1$, $\varepsilon = 0.05$, $P_p = 2$, $t_a = t_x = 9$ as an example.

The value $r \approx 47$ gives a stabilized cyclical growth rate each year (Fig. 5.30b).

These behavioral studies are useful for understanding plant growth and monitoring the proper functioning of simulation software.

5.7 Behavior of the Model: Case of Limited Growth

The production area *Sp* limits the growth. When plants are grown at a high planting density *d* on a surface S_c, the ground is completely covered by the foliage, and the available ground area per plant tends towards the value $Sd = S_c/d$, which also becomes the production area Sp. In this case, biomass production per plant is no longer proportional to its leaf area *Sf*, as in the case of free growth, but to the available ground area per plant *Sd*. The leaves are in fact in excess compared to the produced photosynthesis.

According to Eq. (5.18), there is necessarily an upper limit value for biomass production per plant in each cycle:

$$Q(t) < \frac{E \cdot Sp}{r} \tag{5.46}$$

There are, therefore, two possibilities (Fig. 5.31) for the plant to reach an equilibrium during the growth. One is in free growth, and the other is in limited growth. The first equilibrium is metastable and gives dwarf plants; the second is limited by an upper limit which corresponds to normal growth conditions, constrained by the available area *Sp* per plant in the stand. This upper limit does not depend on the plant architectural model, but the true limit Q_l depends on it and is found by iteration on Eq. (5.18). This limit is related to the solution of Lambert equation.

The evolution of the Q/D ratio (trophic pressure) controls the size of the phytomers. Figure 5.32 shows the architecture of three plants single-stemmed, monopodial with $t_2 = 10$ and sympodial, respectively. The influence of *Sp* on the

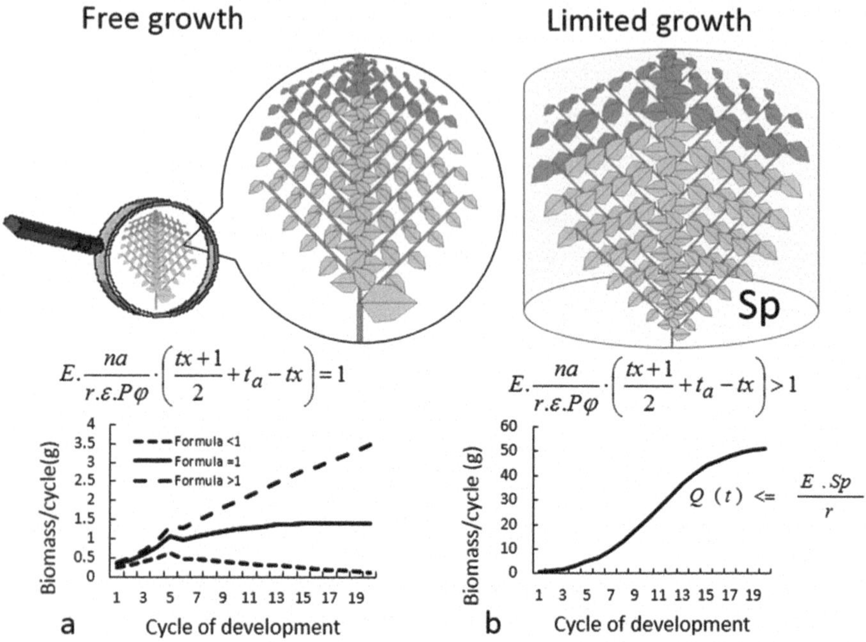

Fig. 5.31 Two cases of stabilized growth and development. The branches are limited to a development of 10 phytomers. The functioning of the leaves is limited to five cycles. Plants (**a**) and (**b**) are on the same scale. Note that the first case gives dwarf plants in metastable growth equilibrium. (**a**) *Sp* is very large and the relationship (Eq. 5.27) $\frac{(t_a+1)}{2}\cdot\frac{E}{A}=1$ is verified: it is the free growth proportional to the leaf area. It should be noted that the volume of the organs decreases from the seed to stabilize with the organic series. If the equation value is smaller or larger than 1, the growth decreases or increases exponentially. (**b**) *Sp* is small and $\frac{(t_a+1)}{2}\cdot\frac{E}{A}>1$. This is the growth limited by the light intercepted with $Q(t)\le\frac{E}{r}\cdot Sp$. It should be noted that the volume of organs grows from the seed to stabilize

growth limitation means that the biomass produced per cycle reaches an equilibrium, whatever the architectural model. But the demand also stabilizes, leading to the organic series stabilizing from the top. For the single-stemmed plant (Fig. 5.32a), it stabilizes when $t > t_a$. The crown of the stem becomes invariant. For the monopodial plant (Fig. 5.32b), whose branches stop their development after t_2 cycles, the demand stabilizes after $t_2 + t_a$ cycles. The crown of the plant also becomes invariant. For the sympodial plant (Fig. 5.32a), whose development is exponential (2^t), the architecture becomes fractal, because the biomass produced per cycle is stabilized.

The architecture of the plant certainly influences its production. In the model, this is traduced by the fact that *Sp* depends on the bearing of the tree (columnar, tabular).

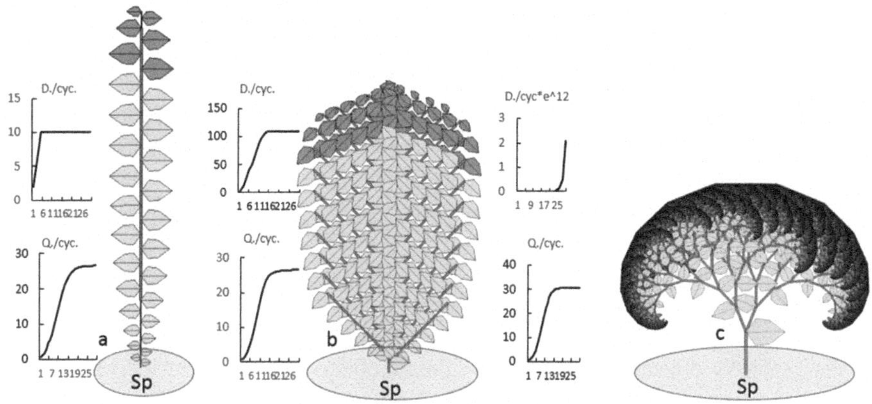

Fig. 5.32 Architectures of three plants. (**a**) Single-stemmed, (**b**) Monopodial with $t_2 = 10$, and (**c**) Sympodial dichotomous with $t_a = t_x = 5$ DCs, $p_a = p_e = 1$, $E = 1$ and $r = 15$. The plants (**a**) and (**b**) have a production per cycle that stabilizes because of $Sp = 500\ cm^2$. It is bounded by ($E \cdot Sp/r = 33$). The plants (**a**) and (**b**) have a stabilizing demand. The crowns of these two plants become invariant. Plant (**c**) has an exponentially growing demand. The architecture becomes fractal

5.8 Influence of Environmental Parameters

5.8.1 Influence of the Factor E

In ecophysiology, the factor E links the growth to the resource related to the environmental parameters. It is associated with the parameter r, which characterizes the efficiency of converting this resource into biomass. The biomass $Q(t)$ synthesized at cycle t is proportional to parameter E (Eq. 5.18). Relationship Eq. (5.41) shows that growth can only occur if E exceeds a certain threshold that depends on source-sink parameters.

In real conditions, the parameter E can change with each growth cycle and is then defined as a variable $E(i)$. In the absence of data about the environmental variables and in the absence of strong stress, the factor E can then be normalized to 1; the parameter r is used for calibration, and the study of source–sink relationships remains unchanged. This is of great interest, as the growth and architecture become independent of environmental parameters. It becomes possible to study the source–sink relationships of a plant collected in an environment (light and temperature) whose history is unknown.

5.8.2 Influence of the Parameter Sp

The parameter Sp makes it possible to solve the growth at each time step, by balancing the two members of Eq. (5.13) if the weight of the plant and its leaf area is known or estimated.

$$Q(t) = \frac{E \cdot Sp(t)}{r} \cdot \left(1 - \exp\left(-k \cdot \frac{Sf(t)}{Sp(t)}\right)\right)$$

In practice, it is not possible to calculate the value of *Sp*(*t*) at each time step. To solve this problem, an empirical function is chosen that follows the evolution of *Sp*(*t*) as a function of CA, with a small number of parameters that can be identified by an inverse method. Two cases are considered.

5.8.2.1 Cases of High Planting Densities

This is generally the case for field crops; the number of plants per m^2 depending on the type of plant varies from a few units to a few dozen.

At germination, the ratio *Sf*(*t*)/*Sp*(*t*) is small and Eq. (5.12) becomes:

$$Q(t) = \frac{E \cdot k \cdot Sf(t)}{r}$$

The value of *Sp* does not play roles. When this ratio increases, it suddenly tends towards the limit of the available surface area per single plant $Sd = S_c/d$ given by the cultivated area *Sc* of the stand and the planting density *d* and we find:

$$Sp \approx Sd.$$

The solution of Eq. (5.13) is, therefore, *Sp*(*t*) = *Sd*.

When *d* increases, the area *Sd* decreases as $Sd = S_c/d$. Assuming unchanged source–sink relationships, the biomass *q* produced by the individual plant decreases and the biomass *Q* produced by the stand increases. This value tends towards a maximum $Q_l = E{\cdot}Sd/r$.

Figure 5.33 shows a simulation of the relationship between the biomass of the individual plant and those of the stand according to the planting density.

This behavior has been well verified for field crops. A verification of the relationship between the biomass produced by a single plant and that produced by its stand for a planting density is given with beetroot (Fig. 5.34, Lemaire et al. 2008).

The weight of a single plant depends on the planting density, but the harvest of the stand per m^2 is the same.

Repeated experiments over several years and on various crops—tomato (Dong et al. 2007, 2008), maize (Ma et al. 2008), beetroot (Lemaire et al. 2008), and wheat (Kang et al. 2008)—confirm the suitability of the model for agronomy. The leaf area and weight of the plant are available on several different dates. We can, therefore, find the numerical values of the parameters *r* and *Sp* by an inverse method which best fits Eq. (5.12).

For example, Fig. 5.34 shows that the biomass production of a beet (Lemaire et al. 2008) grown at a density *d* = 5 plants/m^2 (giving *Sd* = 2000 cm^2) is adjusted for the

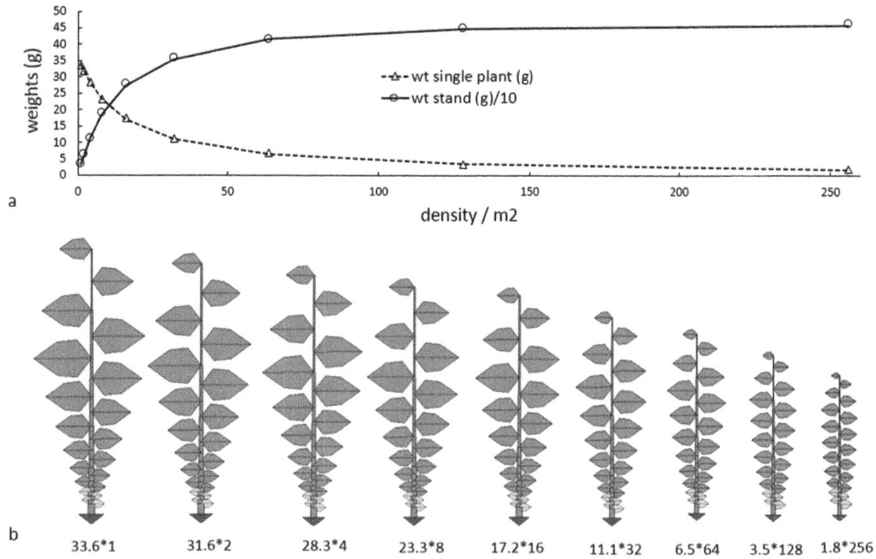

Fig. 5.33 The parameters of the simulated plant at 20 development cycle (DC) are: $t_a = 15$ DCs, $t_{xa} = t_{xi} = 5$ DCs, $t_{xt} > 20$ DCs. $p_a = 1$, $p_e = 0.5$; p_c (ring) $=0.5$; p_t (root tape) $= 4$. Starting from a planting area $S_c = 10{,}000$ cm^2 (1 m^2), the planting density is doubled each time from $d = 1$ to $d = 256$. The volume of the individual plant gradually decreases. (**a**) The biomass produced per individual plant (triangle marker) and the biomass produced/10 plants by the stand (round marker) are displayed. This tends towards a maximum: $Q_1 = E \cdot S_c / r = 500$ g. (**b**) The architecture of the plant is displayed according to the planting density

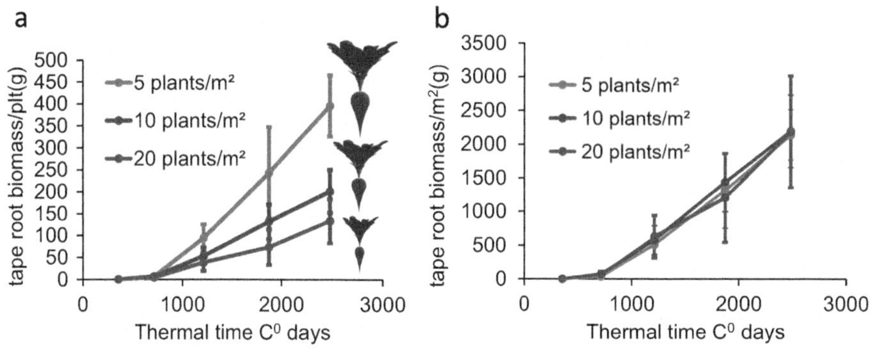

Fig. 5.34 (**a**) Weight of the taperoot of an isolated plant for three densities (5 plants/m^2, 10 plants/m^2, 20 plants/m^2) and for three growth dates. (**b**) weight of the harvest per m^2 in tape roots for the three densities (5 plants/m^2, 10 plants/m^2, 20 plants/m^2) and for the three growth dates

values: $r = 177$ and $Sp = 2055$ cm^2, by the inverse method. The Sp and Sd values are, therefore, very close. The model can retrieve the planting density only from the leaf area and the biomass produced. Figure 5.35 clearly shows the behavior of the two-stage growth as shown in Fig. 5.16. In the first stage (DC $<$ 40), which is

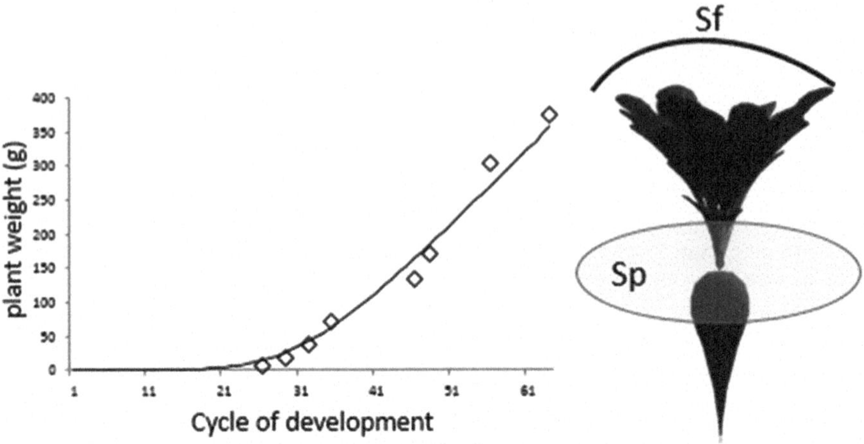

Fig. 5.35 Adjustment of the beet weight according to the leaf area during the growth, grown at a density of 5 plants/m^2 by Eq. (5.12). The solid line and the symbols (◊) represent the calculated data and the observed data, respectively. The average available surface area per plant is, therefore, $Sd = 2000$ cm^2. There are eight sampling dates (in DC). The leaf area is measured as well as the weight of the plant with the tape-root (average of 18 plants per date). The assessed parameters r and Sp of the growth equation are $r = 177$ and $Sp = 2055$ cm^2. The value of Sp corresponds well to the value of Sd (5 plants/m^2). (Source: Data from Sébastien Lemaire, ITB, 2010)

exponential growth, it is the leaf area Sf that drives the process; in the second stage, which is linear growth, it is the available area Sd that limits the production with $Sp = Sd$.

The numerical value of Sp is a hidden parameter in the production Eq. (5.17) of GreenLab.

It is easier to consider the weight of a plant (easily measurable) for computations rather than the production of biomass per cycle. The latter has a sigmoidal shape (Fig. 5.32) and its integration gives precisely the weight of the plant.

In Chap. 9, experiments confirming the above results are shown on tomatoes, maize, beet, and wheat.

They are an experimental validation of GreenLab. In the case of high planting density, Sp is always close to Sd (Fig. 5.36). When the plants are isolated, the value of Sp becomes lower than the value of Sd. It corresponds to the value of the projection of the leaf crown on the ground. At the young stage, the assessed value of Sp is very high (all leaves see light) and is not actually computable by the inverse method.

5.8.2.2 Case of Low Densities: Variation of *Sp*

In cases of low densities, the problem is complex. Between the initial free growth and the limited growth, there should be a progressive evolution of Sp which starts from a high initial value, decreases to a minimum under the effect of the self-shading

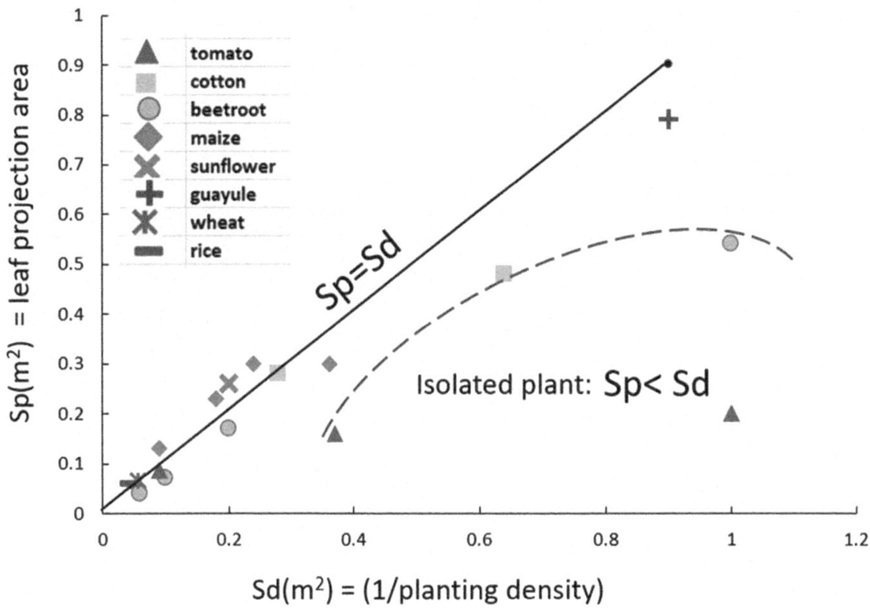

Fig. 5.36 Comparison between the available areas *Sd* by the planting density *d* ($Sd = 1\ m^2$/plant) and the calculated production areas *Sp* for four types of plants: tomato (○), maize (Δ), beet (◊), wheat (*). All dots are around the bisector for high planting densities ($Sp = Sd < 0.3\ m^2$/plant which corresponds to a planting density > 5 plants/m^2), or for crowns in contact. Isolated plants move away from the bisector ($Sd > Sp$). *Sp* no longer corresponds to an available surface, but to a surface that balances production with the leaf surface considering self-shading. (Sources: data from Sébastien Lemaire (2010) in ITB, Ma Yun-Tao (Ma et al. 2008) and Dong Qiao-Xue (Dong et al. 2007, 2008) in China Agricultural University (CAU) and Feng Lu (Feng et al. 2013) in CIRAD)

of the leaves, then increases under the effect of the extension of the crown projection. If the thickness of the latter is sufficient to obscure the light, then the surface of projection on the ground *Sh* of the crown is an estimate of *Sp*. If the plant is isolated and the growth is indefinite, *Sp* grows indefinitely. If it meets its neighbors, then the production surface stabilizes due to the stabilization of the projection surface on the ground of the crown.

Olivier Taugourdeau (Taugourdeau et al. 2013) (CEF, Canada) was able to estimate the value of the projection surfaces on the ground *Sh* of sugar maple tree canopy during its growth. They are approximated as ellipses whose two axes are measured over a period of 10–80 years (see Chap. 10). The adjustment of the organic series can be achieved only by assuming that the evolution of *Sp* follows a bi-parametric U-shaped curve, the parameters of which are to be assessed at the same time as the other source-sink parameters of the maple. The comparison of the optimized *Sp* values with the estimated ground projection surfaces *Sh* is satisfactory, and experimentally validates GreenLab for trees over a long growth time. After 20 years, the LAI is strong and, as expected, the measured crown projection surfaces

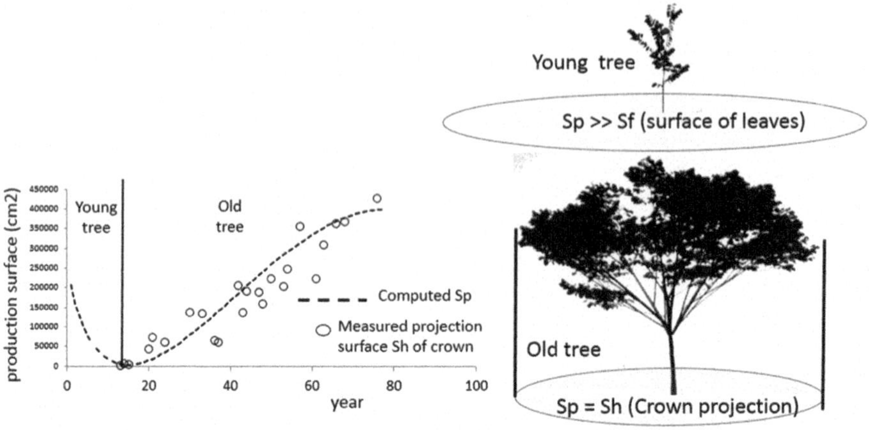

Fig. 5.37 Evolution of *Sp* along the chronological age (CA). Comparisons between *Sh* values estimated from measurements of maple crown projection surfaces (circles) and *Sp* values calculated by the inverse method (dashes) by adjusting the organic series of GU. There is a satisfactory agreement between the prediction and the observations. For young stages, *Sp* is not measurable, but its value is calculable. (Source: Data from O. Taugourdeau (2013), CEF, Canada)

are acceptable *Sp* solutions (Fig. 5.37). But at the young stage, no estimation of *Sp* by direct measurement is possible. Only the inverse method can optimize its value.

The study of Teak growth (Chap. 10) has shown the behavior for the evolution of the parameter *Sp* which follows a U-shaped curve as a function of time. The formulation of this U-curve is explained in this Chap. 10.

5.8.2.3 Pool of Reserve Implementation

In the current version of the Gloups software, the GreenLab model has a simple common pool with no storage. At each growth cycle, a quantity of biomass is produced by the photosynthetic organs (leaves, sheaths, etc.) and is placed into a common pool which is redistributed in its entirety to the expanding organs. There are, therefore, no remobilizable reserves over several cycles except in the following two cases:

- At the beginning of growth, the seed gives up the quantity Q_0 to create the young seedling. This can be given up to the plant either in one cycle or over several cycles (see Sect. 5.6.1.1, Fig. 5.20d). In this case, the quantity given up in each cycle is supposed to be formulated as: $\Delta q_0(i) = Q_0 \cdot (1 - k_o)^{i-1} \cdot k_o$ which decreases exponentially until all the seed's reserve is exhausted. k_o is a damping coefficient. If it is zero (usual case), all the biomass is given up at the time of the first growth cycle.

- The phenomenon of remobilization (see Chap. 9, Sect. 6.10 on rice) allows the organs to return part of their biomass to the common pool when their individual operating time is exceeded.

The rule is that all biomass in the common pool is taken up by the expanding organs during the growth cycle. If photosynthesis is suppressed, growth stops. In the case of continuous growth, a single cycle without photosynthesis is enough to cause the death of the plant. In the case of rhythmic growth (trees) in the autumn, the biomass of the common pool is put on hold, through structural pauses for the renewal of growth in the following spring (see Sect. 6.4.1).

It is possible to implement a more flexible system that incorporates a reserve pool, denoted as Q_s, which accumulates biomass at each growth cycle in proportional to the photosynthetic production. In the case of photosynthesis stops, the system can continue to support growth by drawing from the reserve pool.

Unfortunately, it seems very difficult to validate such a system by agronomic experiments. Anatomically, the reserves are stored as starch grains within cell vacuoles, taking the place of water. The volume and density of the cells change little and we cannot directly measure the reserves by deforming the volumes of the organs in the organic series. Since the profiles of the organic series being unaffected, we cannot estimate the functioning of the reserves by an inverse method. Indeed, given that the fits of the series are of good quality with a common pool system without reserves, there is not enough information left to refine the model with a reserve compartment.

Let Q_s be the common pool, we can conjecture the need to define two parameters a and b that are used, respectively, for filling and emptying the common pool Q_s. We pose:

$$\Delta Q_s(i) = a_{Qr} \cdot Q(i-1)/D(i)$$

$Q(i)$ is the biomass synthesized at DC i and $\Delta Q_s(i)$ is the amount added to the reserve pool Qs. a_{Qr} is the sink of the reserve compartment. The amount $Q(i) - \Delta Q_s(i)$ is used for organ expansion in cycle i. If photosynthesis is removed, the reserve pool returns the quantity $b_{Qr} \cdot Q_s(i)$ to the common pool at each cycle. b_{Qr} is the emptying coefficient of the reserve pool. There remains the quantity $(1 - b_{Qr}) \cdot Q_s(i)$ in the reserve pool.

Such a system with parameters a_{Qr} and b_{Qr} cannot be estimated using the organic series by an inverse method. But the values can be fixed a priori for simulation purposes. Probably experiments with cuttings or pruning would be good candidates to study the effect of reserve behavior. For example, during pruning, the biomass comes mainly from the reserves stored in the root system.

For the purpose of this study, a subroutine in Gloups was developed to simulate the behavior of the plant interacting with a reserve compartment (Fig. 5.38). The light radiation $E(i)$ at cycle i varies throughout the growth period, with values ranging from 1 to 2. A threshold $Eo = 1$ is set below which photosynthesis stops and therefore the reserve pool feeds the common pool.

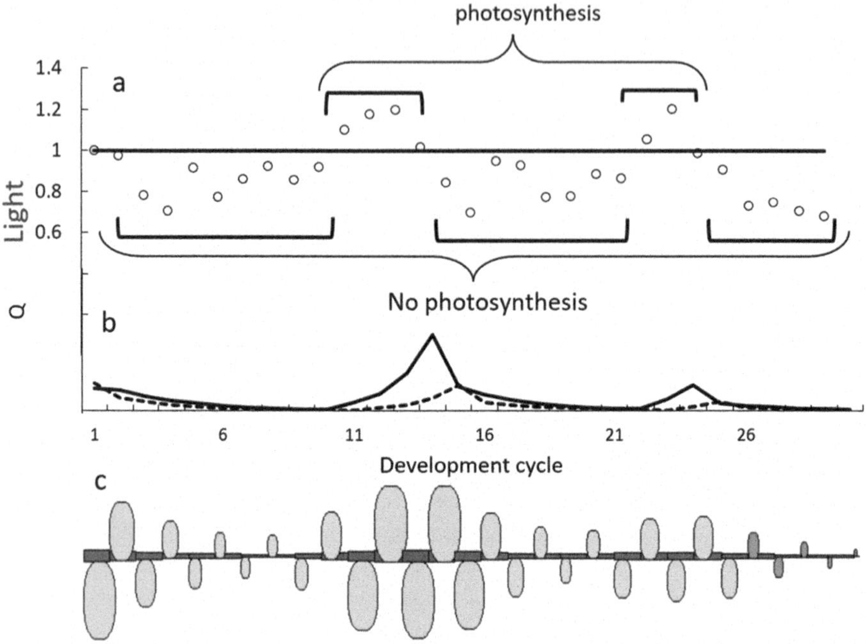

Fig. 5.38 Effect of the reserve pool: (**a**) Occurrence of photosynthesis during growth. (**b**) Parallelized evolution of the biomasses of the common pool and the reserve pool. (**c**) Plant growth resulting from the action of the reserve pool (biomass production solid line, reserve pool dotted line)

The parameters of the system are $t_a = t_x = 5$; $Q_0 = 1$; $r = 5$; $Sp = 2000$; $p_a = p_e = 1$ (sinks of the leaf organs and internodes); $a_{Qr} = 5$ (sink of the reserve compartment); $b_{Qr} = 0.5$ (emptying coefficient of the reserve compartment).

Between DC 1 and DC 10, there is no photosynthesis; the growth is ensured by the seed, which feeds the reserve pool that is being empty. Between DC 11 and DC 15, DC 22 and DC 24 photosynthesis is active and the reserve pool fills up. Between the DC 16 and DC 21, DC 27, and DC 30, the reserve pool is getting empty. Without the reserve pool, growth would have stopped after the first cycle.

5.9 Application of Organic Series Analysis on Cultivated Plants

The adequacy of the GreenLab model to the crops is shown below by the quality of the fittings obtained when compared to the observed organic series within the model. These adjustments are more significant on the performance of the GreenLab model than 3D plant simulations that only provide a geometric appearance. Fitting the organic series with the GreenLab model using the inverse method is explained in Chap. 9.

5.9.1 Plasticity of Organic Series

Among the collection of plant species studied, a high variability was observed in the profile of organic series, i.e., the distribution of organ volumes by rank along the stem. This is fully explained by the source–sink interactions during the growth. The evolution of supply and demand (trophic pressure) literally sculpts plant architecture. It is omnipresent in the form of organic series.

The volume of an organ (see Sect. 4.2) depends on:

- The expansion time t_x;
- The sink function $P_o(x) = p_o \cdot F_o(x)$, where p_o is the sink strength of the organ and $F_o(x)$ is the variation of the sink strength normalized to its mode. This function is empirical. Its formulation is chosen to best capture numerical variations in observed sink strength.
- The evolution of the supply-to-demand Q/D during the expansion of the organ.

In the plants studied, the duration of organ expansion varies significantly. Often constant (cotton, tomato), it may depend on its position from the base in the organic series (sunflower, beetroot). Another example is that of leaves, which is almost immediate (2–3 DCs) for grasses such as wheat or rice, and about a hundred DCs for sunflower.

For internodes, the primary growth (the pith) is quite short (a few DCs). In herbaceous plants, it can be delayed. In this case, a rosette is formed before the elongation of the internodes of the stem. In woody plants, secondary growth is effective if the axis is not dead, it can occur on the trunk for hundreds of years.

The flowers can be in a lateral or terminal position on the axes. Fruits often have delayed expansions (cucumber).

In GreenLab, only the weight of the roots is considered if they exist. They are modeled as a compartment with a sink function that is active throughout the growth. If the weight of the roots is not measured, it is considered not to disrupt the internal source–sink relationships inside the shoots, which allows the aerial part to be studied independently (see Chap. 9).

Sink functions are expressed during the expansion time of the organs. They are hidden in the architecture and can only be estimated by the inverse method by fitting the organic series. Most often, their shape is an asymmetrical bell curve sink adjusted by a beta law. Exceptionally, in plants such as peppers or tomatoes, fruit expansion can be monitored and measured directly using diameter–weight allometries. The observed sink variations, when compared to those calculated by the inverse method, are found to be close, which validates the sink model. This approach avoids laborious growth monitoring.

5.9.2 Organic Series and History of Plant Growth

Organic series taken from the architecture of a plant implicitly contain the history of the evolution of its functioning. Their analysis allows to calculate the source-sink parameters and thus to reconstitute the supply and demand evolution of the plant during its growth.

Figure 5.38 shows the fittings by the GreenLab model of the organic series of leaves on the stems of five cultivated single-stemmed plants, whose profiles are different and characteristic of their specific functioning.

The first three plants in Fig. 5.39 have indefinite growth and the last two have finite growth, which ends with terminal flowering. All these plants begin with exponential growth, which is well simulated by the model. Projection areas are limited by planting density, which limits biomass supply. The senescence of the organs eventually limits the demand. Growth then quickly becomes stationary, resulting in the following results:

- Monopodial cotton (Li et al. 2010) ends up replicating the same phytomer. Leaves have a constant functioning duration. The profile of the top of the organic series of leaves becomes invariant. The newly expanding leaves at the beginning are small (Fig. 5.38a);

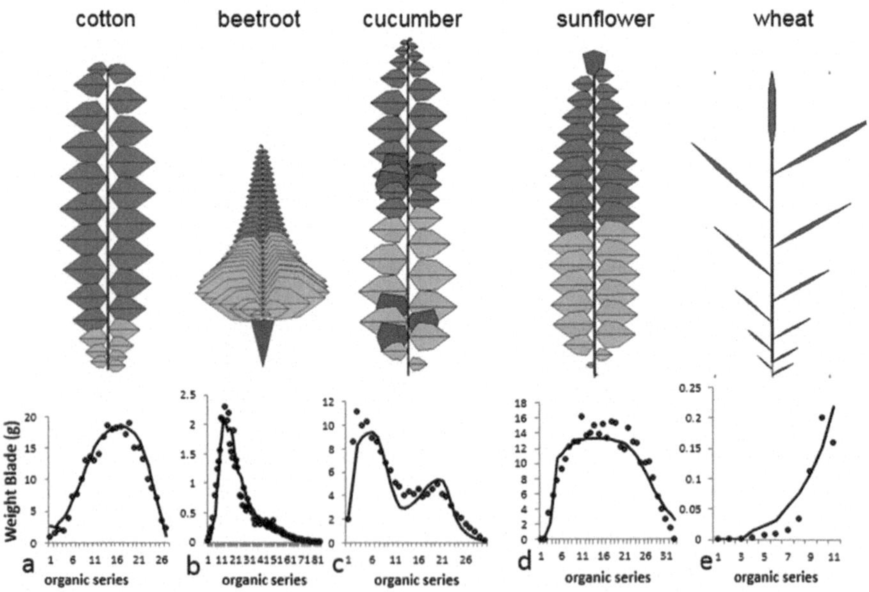

Fig. 5.39 Observations (dots) and fittings (lines) by the GreenLab model, organic series of leaves on the stem in five cultivated plants. There is a high degree of plasticity in the profiles of the leaf organic series according to the species. They depend on source–sink interactions between organs. (**a**) Cotton stem. (**b**) Beet rosette. (**c**) Cucumber stem. (**d**) Sunflower stem. (**e**) Wheat stem

- The root sink of the beetroot (Lemaire et al. 2008), initially negligible, is becoming increasingly important and concentrates all the demand. It eventually prevents the expansion of new created leaves that remain embryonic (Fig. 5.38b);
- Cucumber fruits (Mathieu et al. 2007, 2012) are an important sink. They lead to the abortion of the new fruits during their activity, until the end of their expansion. Fruiting can then resume. The demand for the fruits inhibits the growth of new phytomers and shrink the leaves. The position of the fruits depends on supply/demand ratio (Fig. 5.38c);
- The expansion of sunflower organs (Rey 2003) is done over a very long period (90 DCs). Leaves have variable functioning duration from the base of the stem. The appearance of the capitulum creates a new, strong sink. This slows down the growth of the last leaves, which remain small. If the capitulum is cut off at birth, these last leaves become larger (Fig. 5.38d);
- In wheat (Feng et al. 2013), the leaf expansion time is very short (2 DCs). The last leaves are therefore the largest, because the sink of the terminal ear has no influence, since they have already completed their growth (Fig. 5.38e).

These five examples will be presented in detail in Chap. 9.

5.10 Conclusion

The architecture of a plant depends not only on the activity of the meristems that generate its development but also on the biomass produced by the leaves and its distribution in the structure. The volumes and sizes of organs are dependent on the simultaneous evolutions of the demand D and the biomass production per cycle Q, and more specifically, on the trophic pressure (Q/D ratio).

The botanical automaton produces the cohorts of organs (organs of the same physiological and chronological age) which, together with their sinks, allow the computation of the plant demand. The architecture of any plant that meets the criteria of PA and the common pool can be decomposed into organic series, which contain the history of growth.

The study of the mathematical model makes it possible to identify two modes of growth: free growth (no competition with neighboring plants) and limited growth.

In free growth, biomass production is proportional to the leaf area. This situation is realized in the early stages of growth and results in exponential growth. All the leaves of the young seedling intercept the light.

In the case of high planting densities, the available area per plant is limited and a double limitation in the search for the light by the leaves occurs. The leaves of one plant overlap with each other and meet those of neighboring plants. The limited growth occurs and depends on the light intercepted by the foliage, which corresponds to its shadow on the ground, the sun at its zenith. This can be calculated using the notion of production area Sp, which is experimentally validated. In the case of high planting density d on a cultivated area S_c, the growth gradually becomes linear and proportional to the surface area available per plant *Sd* ($Sd = S_c/\mathrm{d}$). GreenLab computes a production area $Sp \approx Sd$ by an inverse method, which is a validation of the model (see Chap. 9).

The *Sp* production area method also ensures the passage from the individual plant to the stand level, and thus bridging the junction between a functional–structural model and a physiological crop model (Chap. 11). In addition, *Sp* is the solution of the production equation, which avoids calculating the interception of light by computer methods that use ray-tracing or radiosity on 3D models of simulated plants, a process that is difficult to implement in practice despite its academic interest.

The deterministic model of a functional plant, the subject of this chapter, is only found on single-stemmed plants such as maize and sunflower. But the more general stochastic model has a similar behavior (Chap. 6).

Validation of the model on several crops and inverse methods for computing the source-sink parameters of GreenLab from measurements on organic series are presented in Chaps. 9, 10 and 11).

References

Atger C, Edelin C (1994) Premières données sur l'architecture comparée des systèmes racinaires et caulinaires. Can J Bot 72(7):963–975

Buis R, Barthou H (1984) Relations dimensionnelles dans une série organique en croissance chez une plante supérieure. Rev Biomath 85:1–19

Li D, Letort V, de Reffye P, Zhan Z-G (2010) Modeling branching effects on source-sink relationships of the cotton plant. In: Li B.-G., Jaeger M., Guo Y (ed) The third international symposium on plant growth modeling, simulation, visualization and applications (PMA 09), 9-13 november 2009, Beijing, China. Los Alamitos, IEEE Computer Society, pp 293–302

Dong Q-X, Louarn G, Wang Y-M, Barczi J-F, de Reffye P (2008) Does the structure function model GreenLab deal with crop phenotypic plasticity induced by plant spacing? A case study on tomato. Ann Bot 101(8):1195–1206

Dong Q-X, Wang Y-M, Yang L, Barczi J-F, de Reffye P (2007) GreenLab-tomato: a 3D architectural model for tomato development. N Z J Agric Res 50:1229–1233

Feng L, Rey H, Mailhol J-C, Kang M-Z, de Reffye P (2013) Concept and calibration of virtual wheat including stochastic tillering. In: Risto Sievänen, Eero Nikinmaa, Christophe Godin, Anna Lintunen, Pekka Nygren (ed) Proceedings of the 7th international conference on functional-structural plant models (FSPM2013), Saariselkä, Finland, 9–14 June 2013. Vantaa: Finnish Society of Forest Science, pp 279–281

Fournier C, Andrieu B (1998) A 3D architectural and process-based model of maize development. Ann Bot 81(2):233–250

Guo Y, Ma Y-T, Zhan Z-G, Li B-G, Dingkuhn M, Luquet D, De Reffye P (2006) Parameter optimization and field validation of the functional–structural model GreenLab for maize. Ann Bot 97(2):217–230

Hallé F, Oldeman R-A-A, Tomlinson P-B (1978) Tropical trees and forests. Springer Verlag, Berlin, Heidelberg, New-York. 441p

Heuvelink E (1995) Dry matter production in a tomato crop: measurements and simulation. Ann Bot 75(4):369–379

Heuvelink E (1996a) Dry matter partitioning in tomato: validation of a dynamic simulation model. Ann Bot 77(1):71–80

Heuvelink E (1996b) Re-interpretation of an experiment on the role of assimilated transport resistance in partitioning in tomato. Ann Bot 78(4):467–470

Heuvelink E (1999) Evaluation of a dynamic simulation model for tomato crop growth and development. Ann Bot 83(4):413–422

Jourdan C, Rey H (1997) Modelling and simulation of the architecture and development of the oil-palm (*Elaeis guineensis* Jacq.) root system. Plant Soil 190(2):235–246

Kang M, Evers J-B, Vos J, de Reffye P (2008) The derivation of sink functions of wheat organs using the GREENLAB model. Ann Bot 101(8):1099–1108

Lemaire S (2010) Système dynamique de la croissance et du développement de la betterave sucrière (Beta vulgaris L.). AgroParisTech (Ph. D. Thesis)

Lemaire S, Maupas F, Cournède P-H, de Reffye P (2008) A morphogenetic crop model for sugar-beet (beta vulgaris l.). In: International symposium on crop modeling and decision support: ISCMDS 2008, April 19-22, 2008, Nanjing, China, 2008

Ma Y-T, Wen M-P, Guo Y, Li B-G, Cournède P-H, de Reffye P (2008) Parameter optimization and field validation of the functional–structural model GREENLAB for maize at different population densities. Ann Bot 101(8):1185–1194

Mailhol J-C, Revol P, Ruelle P (1996) Pilote: un modèle opérationnel pour déceler l'apparition de stress hydrique. In: ICID 16th international congress on irrigation and drainage: workhop on crop-waterenvironment models, pp 167–182

Marcelis and Gijzen (1998) Evaluation under commercial conditions of a model of prediction of the yield and quality of cucumber fruits. Sci Hortic 76:171–181

Marcelis L-F-M, Heuvelink E, Goudriaanc J (1998) Modelling biomass production and yield of horticultural crops: a review. Sci Hortic 74:83–111

Mathieu A, Cournède P-H, Barthélémy D, de Reffye P (2007) Condition for the generation of rhythms in a discrete dynamic system. Case of a functional structural plant growth model. In: T. Fourcaud and X. P. Zhang. Second international symposium on plant growth modeling, simulation, visualization and applications, PMA06, 13-17 november 2006, Beijing, P.R. China. Los Alamos, IEEE Computer Society, 26–33

Mathieu A, Letort V, Cournède P-H, Zhang B-G, Heuret P, de Reffye P (2012) Oscillations in functional structural plant growth models math. Model Nat Phenom 7(6):47–66

Pagès L, Vercambre G, Drouet J-L, Lecompte F, Collet C, Le Bot J (2004) Root Typ: a generic model to depict and analyse the root system architecture. Plant Soil 258(1):103–119

Rey H (2003) Utilisation de la modélisation 3D pour l'analyse et la simulation du développement et de la croissance végétative d'une plante de tournesol en conditions environnementales fluctuantes (température et rayonnement). Doctoral dissertation, Montpellier, ENSA

Taugourdeau O, Delagrange S, de Reffye P, Messier C (2013) Modelling sugar maple development along its whole ontogeny: modelling hypotheses and calibration methodology. In: Eds. Risto Sievänen, Eero Nikinmaa, Christophe Godin, Anna Lintunen, Pekka Nygren. Proceedings of the 7th international conference on functional-structural plant models (FSPM2013), Saariselkä, Finland, 9–14 June 2013

Urbain A (1920) Influence des matières de réserve de la graine sur le développement des plantes phanérogames. Librairie générale de l'Enseignement, Impr. Nemourienne, 1920

Zhang W, Li B-G (2009) General structural model of crop root system based on the dual-scale automaton. In: Li B.-G., jaeger M. And Guo Y. (Ed) 2010. Proceedings of plant growth modeling, and their applications (PMA09). Beijing, China, November 9–13, IEEE CPS, pp 161–164

Chapter 6
Plant Growth Modeling: Stochastic Case

Philippe De Reffye and Mengzhen Kang

Abstract This chapter presents the stochastic simulation of plant growth and development with GreenLab, addressing the challenges of controlling variability in stochastic plant simulations and verifying the validity of these simulations. It includes the growth of temperate and tropical trees with synchronous and asynchronous development. It introduces a method for simplifying complex plant topologies into crowns and organic series based on physiological age and common pool concepts, enabling more accurate parameter estimation through data assimilation and the inverse method. This chapter emphasizes the potential structure as a crucial tool for validating simulation results and for establishing effective inverse methods to compute sink-source parameters from organic series. It is applicable to a wide range of architectural models in botany and agronomy. The Gloups software is used to simulate and visualize plant structures, demonstrating the breakdown of plant structures into axis categories and the creation of organic series. This validation is crucial for the development of inverse methods and for assessing the GreenLab model's accuracy with natural plants.

Keywords Stochastic modeling · Continuous and rhythmic structures · Synchronous and asynchronous development

---

Electronic supplementary material The online version of this chapter (https://doi.org/10.1007/978-981-97-7541-5_6) contains supplementary material, which is available to authorized users.

P. De Reffye
CIRAD, UMR AMAP, Montpellier, France

AMAP, Univ Montpellier, CIRAD, CNRS, INRAE, IRD, Montpellier, France

M. Kang (✉)
State Key Laboratory for Multimodal Artificial Intelligence Systems, Institute of Automation, Chinese Academy of Sciences (CASIA), Beijing, China

School of Artificial Intelligence, University of Chinese Academy of Sciences, Beijing, China
e-mail: mengzhen.kang@ia.ac.cn

P. De Reffye et al. (eds.), *Plant Model Greenlab for Botany and Agronomy*,
https://doi.org/10.1007/978-981-97-7541-5_6

6.1 Introduction

The method of stochastic simulation of tree development and tree growth with the GreenLab model has been published in de Reffye et al. (2012), Kang et al. (2008, 2016, 2018).

But stochastic approach to plant growth (biomass production and partitioning) raises several issues:

- how to control the variability of random plants simulated by a Monte Carlo method?
- how to verify the validity of these simulations?

The topological structures of the natural plants are often voluminous and complex. It cannot be recorded easily, because of missing parts due to mortality, not to mention its number of branches. Chapters 3 and 5 have shown that, if the plant is organized according to the notions of physiological age and common pool, then it can be broken down into crowns and organic series which contain the whole history of development and growth. These notions make it possible to define a sampling strategy for the plant's architecture. Data assimilation is necessary to calculate the parameters of the development and growth of the plant using the inverse method.

The stochastic simulation of random plants (using the Monte Carlo method) (Cieslak et al. 2011) allows quickly building a virtual database (crowns, organic series) with its statistical characteristics; the corresponding database is difficult to obtain by measurements on real plants. The simulated samples in the database are virtual realizations of the model, but different from data extracted from a set of natural plants (Cieslak et al. 2022).

Simulating with the Monte Carlo method does not give the binomial law parameters that control the variability. Similarly, stochastic simulation of a sample of random plants is not sufficient, because it only implicitly contains the values of the model parameters, even though the average realization can be computed from simulation. Instead, the computation with the aid of the potential structure explicitly contains model parameters by its mathematical formulation. The mathematical law makes it possible to study the model's behaviors and to implement the inverse method to assess the functional parameters efficiently.

As shown in Chap. 3, the potential structure contains all possible random simulations of the plants in chronological mode, and each phytomer of this potential structure is associated with its probability of existence. The potential plant is not the average plant; the notion of the average plant is not so precise. Although the average plant can be approached in monopodial structures, it remains problematic in sympodial structures (see Sect. 6.4.4.3).

On the other hand, the numerical analysis of the potential structure makes it possible to extract mathematically the statistical characteristics of the crowns and organic series. The computed results must match well with those resulting from the simulations, to verify that the mathematical model is operational. This verification can make the model cope with data coming from natural plants.

The software computing plant growth and architecture are usually complex, and may not be reliable due to programming errors. Therefore, matching simulated results with their computed counterparts is essential to prove that simulations and computations work properly together.

The purpose of this chapter is to show that the potential structure is a useful tool for (1) validating the simulation results and (2) proposing the conditions of an effective inverse method for computing sink–source parameters from organic series. The method is generic, because it is applicable for most of the architectural models encountered in Botany and Agronomy (Chaps. 2 and 4), even in their stochastic forms. Numerous examples are available to demonstrate it. Simulations of the computational plants are implemented using the Gloups software, and whose planar structures are displayed for better understanding.

6.2 Stochastic Development and Growth of the Leafy Axes

Plants that grow with a single stem, whether naturally in the field (such as palm trees, maize, sunflowers) or artificially in greenhouses (such as pruned tomatoes, cucumbers) have a deterministic development. The terminal meristem functions according to thermal time (Chaps. 2 and 5) and the stem produces a fixed number of phytomers for a given development period. On the other hand, branched plants most often have a development with random behavior: leafy axes of the same chronological and physiological ages produce a variable number of phytomers (Chap. 3). The presence of branches in the leafy axes is often random; the axes may abort at any stage of their development unpredictably.

The eco-physiological functioning of plant growth of the GreenLab model, as defined in Chap. 5, is not stochastic. Source–sink relationships are written as a deterministic dynamic system with finite difference equations whose dimension size increases with the development cycle (DC).

However, as we have seen in Chap. 3, there can be variability in the time interval between the appearance of two successive phytomers, which are set up by the same terminal meristem (Chap. 3, Fig. 3.1). The *Bernoulli* process is well adapted to account for the variability observed in the number of phytomers created during a given functioning period of the meristem, as illustrated in Fig. 6.1.

The stochastic development that disrupts demand brings irregularities in biomass production and the volume of organs produced along leafy axes (organic series). Thus, the observed volume of phytomers does not continuously change from the apex of the stems. The random pauses in the functioning of meristems mean that organs of the same rank can have variable volumes because the age of the phytomer does not necessarily correspond to its insertion rank under the apex. The stochastic development introduces fuzziness in determining the chronological age of phytomers in the plant architecture.

When a stochastic plant is simulated, the chronological structure is characterized by a random alternation of real phytomers and pauses, which is a subset of the

Fig. 6.1 Morphological stages of the apex of a coffee stem. (**a**) Beginning of the installation of a new phytomer from a preformed part in the meristem. (**b**) The new phytomer is expanding. (**c**) The growth of the new phytomer is complete. Meristem produces a new invisible preformed phytomer, which will expand during the next development cycle (DC). This phenomenon affects all the apexes of the branched structure. In general, in the case of continuous growth, the apexes are in various possible morphological states (**a**, **b** or **c**). (Photos © S. Sabatier, CIRAD)

potential structure. This chronological structure is not directly observable as its representation is internal; only the topological structure is observable, which is the result of eliminating meristem pauses. In the topological representation of a real plants, the information related to the development time of the plant is lost, which makes the strategy of plant measurements complicate.

When a large number of simulations are performed, and the chronological structures obtained in the potential structure are superimposed, their stacking eventually occupies all the sites reserved for phytomers in that structure. Each site will be filled in proportion to the number of realized phytomers divided by that of all potential phytomers, which corresponds to the probability of obtaining a phytomer at that site (Chap. 3, Fig. 3.27). Meristem pauses have two possible origins: structural or random (Chap. 2). They affect the development of the axes and the growth of the organs: the functioning leaves ensure photosynthesis, and the biomass continues allocating to existing organs according to whether the meristems are active or dormant.

In this chapter, we study the influence of the distribution of meristem pauses on plant growth in two cases: the deterministic case (the periodic development with structural pauses) and the stochastic case (the *Bernoulli* process with random pauses).

6.3 Growth with Continuous, Periodic, and Deterministic Development of the Axes

The periodic functioning of meristems can be translated by a periodic series of 1s and 0s, which determines the creation of phytomers (Chap. 3). The rhythm ratio w is defined as the proportion of phytomers produced over the number of cycles including pauses. Here we study the influence of rhythm on free growth (the projection area Sp is very big) and look for the equilibrium conditions in biomass production.

We have seen that the following relationship gives the equilibrium conditions for free growth (Chap. 4).

$$E \cdot \frac{n_a}{A} \cdot \left(\frac{t_x + 1}{2} + t_a - t_x\right) = 1 \quad (6.1)$$

E is the climatic factor; n_a is the number of leaves per phytomer; t_a is the functioning time of the leaves; t_x is the expansion time of the organs. $A = r \cdot \varepsilon \cdot P_p$, where $1/r$ is water or light use efficiency, ε is leaf thickness, and P_p is the demand of the phytomer (sum of the sinks of the organs constituting the phytomer). The sinks of the organ are constant in that example and the sinks of the organs are equal to 1.

In the case of a single-stemmed plant, growth cannot stabilize because the number of functional leaves (and therefore the demand) fluctuates according to the development rhythm, due to the placement or suppression of phytomers. Nevertheless, if Eq. (6.1) is verified, these oscillations are generally stable and periodic (Fig. 6.2).

In the case of a branched model whose branches have limited development, the oscillations in biomass production disappear because the demand becomes constant when the first installed branch has reached its maximum length (Fig. 6.2b). Equation (6.1) always holds giving the conditions for the free growth equilibrium in this case (independent of w). The level of biomass production for each organ decreases as w increases, while the volume of organs increases at equilibrium (Fig. 6.2c).

6.4 Growth with Continuous and Stochastic Development of the Axes

The stochastic functioning of meristems affects the number of phytomers produced by a plant. The phytomer number is variable from one plant to another for the same time period during its growth. Demand cannot reach a stable equilibrium and production will fluctuate. The probabilities that affect the functioning of meristems (*Bernoulli* process, mortality, branching) contribute to the disruption of biomass production. Each simulated plant, therefore, has a unique developmental trajectory. However, what can be said about the average behavior of a population? The aim here is to establish a theoretical model that computes the values towards which the mean of the stochastic simulations converges: this model is provided by the potential structure.

6.4.1 Growth of the Potential Structure

The potential structure was defined in Chap. 3 as the structure that contains all the possible realizations of stochastic simulations in chronological mode. The structure is, therefore, deterministic. Each phytomer of this structure has a probability of

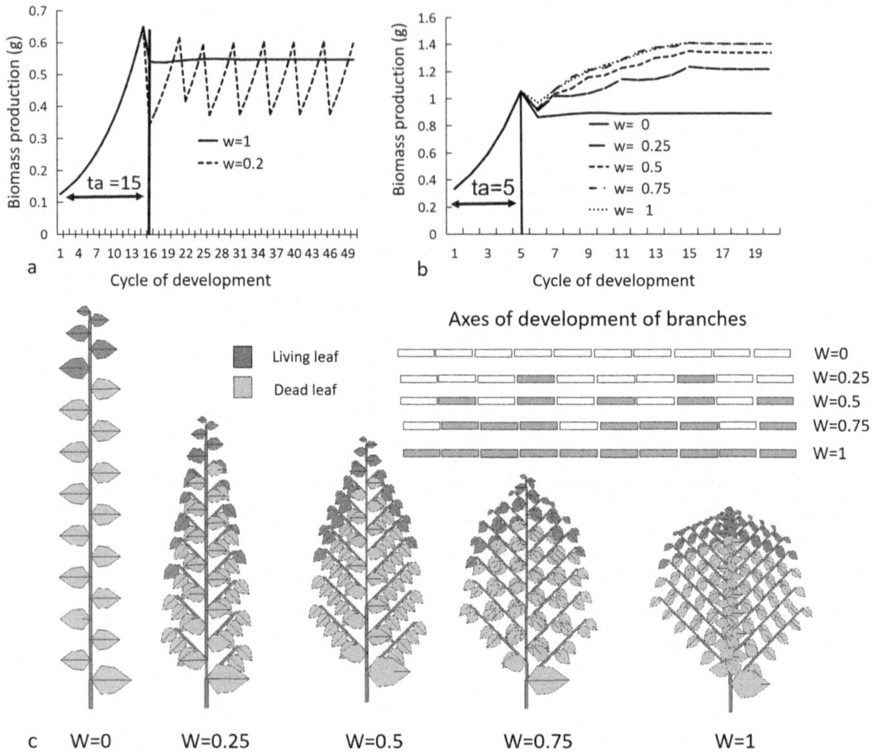

Fig. 6.2 Effect of rhythm ratio w on plant growth (case of free growth, Gloups software). (**a**) Single-stemmed case: the parameter values are $r = 80$, $t_a = t_x = 15$ DCs, the organ sinks are constant and equal to 1, $w = 0.2$. Stable oscillations occur for $t > t_a$, due to leaf mortality. (**b**) Monopodial development: the parameter values are $r = 30$, $t_a = t_x = 5$ DCs. The organ sinks are constant and equal to 1. Branch development is limited to $t_2 = 10$ DCs. The demand stabilizes after $t_2 + t_a$ DCs. The equilibrium condition $\frac{(t_a+1)}{2} \cdot \frac{E}{A} = 1$ is verified for each case, so it does not depend on the variations in rhythm ratio ($w = 1, 0.75, 0.5, 0.25, 0$). Biomass production is an increasing function of w. The lowest and highest curves correspond to $w = 0$ (single stem) and $w = 1$, respectively. The *Q*/*D* ratio (trophic pressure) is a decreasing function of w (data not shown). (**c**) Plant size with monopodial development under different rhythm ratio ($w = 1, 0.75, 0.5, 0.25, 0$). As a consequence, the smaller the w is, the higher the plant is, and the less biomass it produces

occurrence that is computed from the functioning probabilities of the meristems given by the botanical automaton. The plant demand is the sum of the sinks of the phytomers, each multiplied by their respective probabilities of existence. Growing a potential structure is, therefore, equivalent to growing a deterministic structure, but the demand now includes the probabilities of how meristems work.

6.4.2 Study of the Growth of a Single-Stemmed Plant According to a Bernoulli Process, in Free Growth

Here we simulate the growth of a single stem whose leaves work on t_a cycles and whose phytomers have an expansion on $t_x = t_a$ cycles with constant sinks over the expansion period (the value of the sinks is set to $p_o = 1$ for the leaf and the internode). Meristem sets up the phytomers with a probability b at each DC.

N simulations over a development time t generate N single-stemmed structures. For each rank on the chronological axial structures, where the chronological age of the phytomers is the same, there is a certain number of realized phytomers. Their average biomass is obtained from the sum of the biomass divided by their number. For the chronological age t in cycles, there are $t + 1$ classes of topological structures that have x phytomers ($0 \leq x \leq t$), whose frequency $n(x)$ are given by the binomial law:

$$n(x) = N \cdot \binom{t}{x} \cdot b^x \cdot (1 - b)^{t-x} \tag{6.2}$$

Even if they contain the same number of phytomers, all stems grow differently due to the varying locations of pauses. Once the pauses in the chronological structures are removed, there are 2^N morphological structures (stems) that are different from the point of view of organic series.

6.4.2.1 Growth of the Potential Structure, Potential Demand for a Single-Stemmed Plant

In the single-stemmed plant, the potential structure is simplified into a simple axis of potential development. All the phytomers of this axis have a probability of existence b. The demand for the potential axis is expressed by weighting the sink of the organs with the probability of existence b. The expected theoretical values are indicated with the symbol θ.

Thus, potential demand can be written at the DC t as the following equations:

$$\begin{aligned} D^{\theta}(t) &= b \cdot t \cdot P_p(t < t_a) \\ D^{\theta}(t) &= b \cdot t_a \cdot P_p(t \geq t_a) \end{aligned} \tag{6.3}$$

where P_p is the demand of the phytomer, the sum of organ sinks for the phytomer. It can be seen that the potential demand of the stochastic case is smaller than the demand of the deterministic case ($b = 1$). The equation for potential growth in free growth mode is, therefore, written:

$$Q^{\theta}(t) = \frac{E}{A} \cdot \left(\sum_{i=t-t_a+1}^{t} \sum_{j=i}^{n} b \cdot \frac{Q_{j-1}^{\theta}}{\sum\limits_{k=j}^{t} b \cdot P_p} \right) \quad (t > t_a)$$

Parameter b, as one can see, can be simplified, which gives:

$$Q^{\theta}(t) = \frac{E}{A} \cdot \left(\sum_{i=t-t_a+1}^{t} \sum_{j=i}^{n} \frac{Q^{\theta}(j-1)}{\sum_{k=j}^{t} P_p} \right) (t > t_a) \tag{6.4}$$

Potential growth is, therefore, not affected by stochastic development. It is equal to the production by deterministic development with $b = 1$. Note that while the production is not affected, the volume of potential organs is larger when the demand is smaller.

6.4.2.2 Comparison for Growth Between Potential Output and Average Stochastic Simulations

Stochastic simulations are carried out on a single-stemmed plant, using a *Bernoulli* parameter value of $b = 0.7$ as an example. At each cycle t, a phytomer can form or a pause can occur. For each simulation s of a sample of N plants, at the DC t, the biomass $Q(t, s)$ and the demand $D(t, s)$ will take random values depending on the distribution of pauses in the *Bernoulli* process. For each cycle t, based on the simulated N sample plants, the average of the productions $QS(t) = \frac{\sum_s Q(t,s)}{N}$ and demands $DS(t) = \frac{\sum_s D(t,s)}{N}$ can be obtained. We are interested in the evolution of the simulated $QS(t)$ and $DS(t)$ averages from random plants compared to the respective theoretical values $Q^{\theta}(t)$ and $D^{\theta}(t)$ provided by the potential plant.

Figure 6.3a, b illustrates the oscillations in demand and biomass production of a single-stemmed plant resulting from 5 among 100 individual stochastic simulations with a *Bernoulli* parameter b ($b = 0.7$). Potential values (lines) are compared to simulated averages (dotted lines) during plant growth. As long as the age of the plant t has not exceeded the functioning age of the leaves t_a, despite fluctuations in individual demands, production per plant is identical. Beyond t_a, individual productions (thin dotted line) vary, but it can be seen that the simulated average values (thick dotted line) are stabilizing and are close to the corresponding potential values. The values of five random individual simulations for the demands and biomasses produced per cycle are displayed with thin dotted lines (Fig. 6.3a, b), whose leafy axes structures are represented in chronological mode (Fig. 6.3c), showing strong oscillations. Figure 6.3d shows the corresponding potential structure.

Potential growth and potential demand (solid lines) match well with the simulated average of growth and demand (thick dotted lines), respectively (Fig. 6.3a, b).

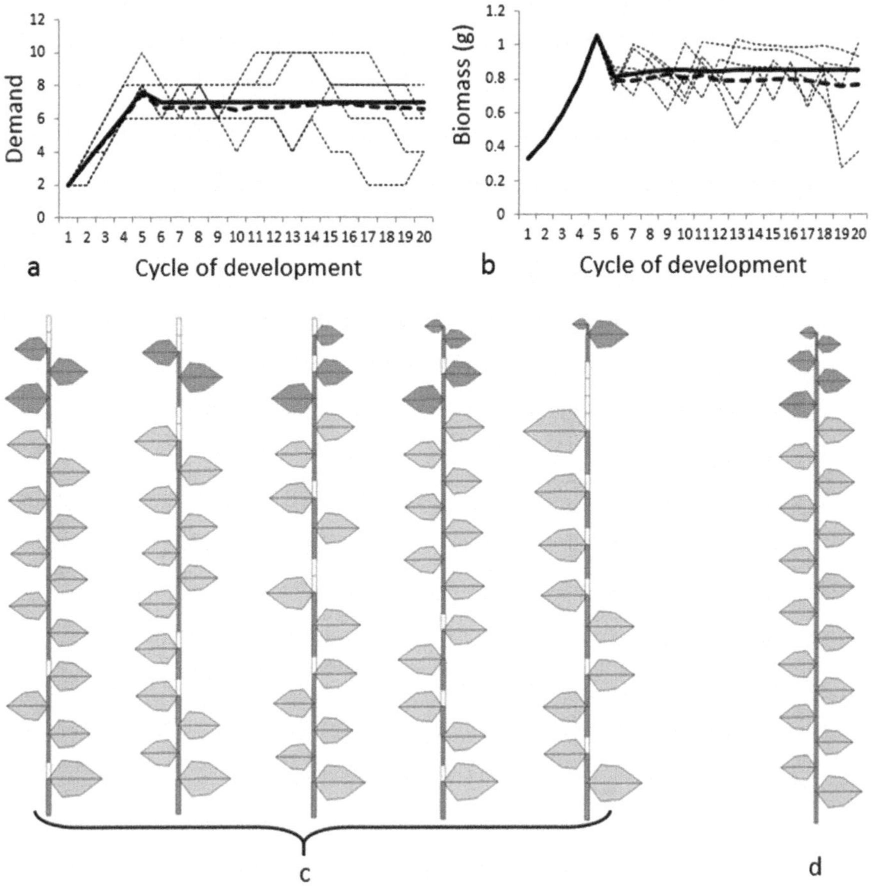

Fig. 6.3 Chronological and free growth (very large *Sp*) simulations of stochastic growth at the equilibrium of a single-stemmed plant

In general, the averages of the simulated supply/demand ratio tend towards the supply/demand ratio of the potential structure with a good approximation. To do this, it is sufficient that the distributions of biomass Q and demand D are close to normal distributions with lower average variances compared to their average means.

So we have:

$$\frac{\overline{Q(t-1)}}{\overline{D(t)}} \approx \frac{Q^{\theta}(t-1)}{D^{\theta}(t)} \tag{6.5}$$

As a consequence, it can be seen that the average of the chronological organic series obtained by stochastic simulation provides a good approximation of the theoretical organic series obtained from the potential axis (Fig. 6.5a):

$$\begin{bmatrix} \overline{q_o(1,t)} \\ \overline{q_o(2,t)} \\ \vdots \\ \overline{q_o(t,t)} \end{bmatrix}_{\text{chrono}} \approx \begin{bmatrix} q_o^{\theta}(1,t) \\ q_o^{\theta}(1,t) \\ \vdots \\ q_o^{\theta}(t,t) \end{bmatrix}_{\text{pot.}} \tag{6.6}$$

6.4.2.3 Organization of Topological Organic Series on Leafy Axes

Topological organic series from simulations are the result obtained from the suppression of pauses in the chronological axis, which corresponds to the observed leafy axes. Phytomers with the same rank in relation to the top of the leafy axis do not necessarily have the same rank in the chronological series, so the age of the phytomer can be different. At rank k from the top, the age of the phytomer is distributed according to a negative binomial law $BN(k, b)$, which is truncated by the age of the axis. The age of the phytomer becomes $k + x$, where x is the number of pauses before reaching rank k (x varies from zero to the maximum age of the axis). The topological organic series is the result of the distribution of pauses on the leafy axis. Thus, an organ at the top of an axis may be small if its carrier phytomer is just formed, or big if the carrier phytomer is old because of successive pauses in the terminal meristem (Fig. 6.4).

For organs located at the phytomer with rank k from the top of the stochastic axes, to compare the theoretical means of the weights of the organs $q_o^{\theta}(k,t)$ with the averages obtained by simulation or measurement, it is necessary to consider the values $q_o(k + x, t)$, taken from the potential axis at positions $k + x$ from the top, multiplied by the probability of obtaining x pauses (given by negative binomial law $BN(k, b)$ truncated as in Eq. (6.7)).

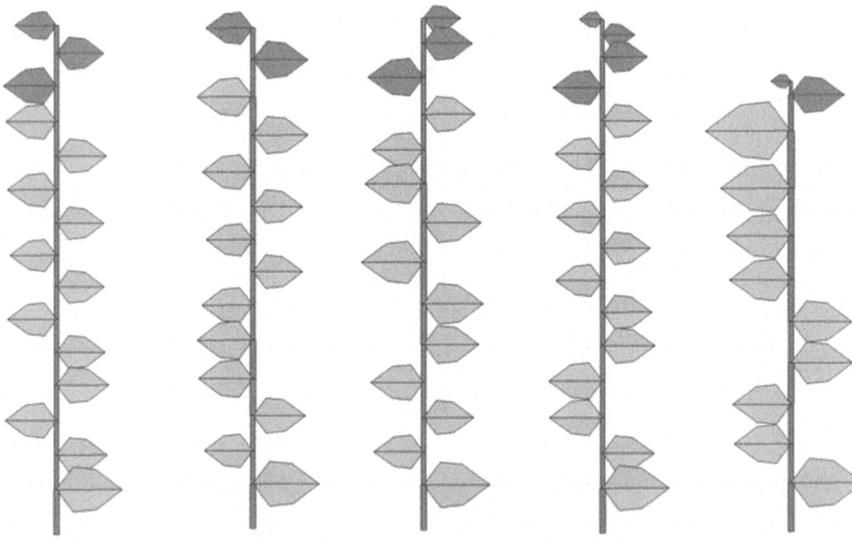

Fig. 6.4 Topological leafy axes corresponding to the chronological leafy axes in Fig. 6.3c. The effect of meristem pauses is seen on the size of the terminal leaves

The sum of the probabilities of pauses, multiplied by the weight of the organ according to the rank from the top in the potential series, is closed to the weight of the average organ according to the rank from the top in the topological series observed in the simulations. Hence, the following expression considers truncation related to the age of the axis:

$$\overline{q_o(k,t)_{\text{topo}}} \rightarrow \frac{\sum_{x=0}^{t-k} \binom{k+x}{x} \cdot b^k \cdot (1-b)^x \cdot q_o^\theta(k+x,t)}{1-\sum_{x=t-k+1}^{\infty} \binom{k+x}{x} \cdot b^k \cdot (1-b)^x} \tag{6.7}$$

Equation (6.7) shows that the theoretical mean weight of the organ $\overline{q_o(k,t)}$ at the rank k from the top can be computed by all the weights of the organs $q_o^\theta(k+x,t)$ at with organ CA $k+x$ on the potential organic series, multiplied by the probability of existence

$$P(k+x) = \binom{k+x}{x} \cdot b^k \cdot (1-b)^x$$

It is important to note that the sizes of the organs in the potential structure are larger because the demand is smaller due to the probability b of the existence of the phytomer. In Monte Carlo simulations, the biomass produced does not depend on the probability b. Therefore, the size of the organs is necessarily increased because of the decrease in their number. One important consequence is that when calculating the theoretical leaf surface area with the potential structure, it must be multiplied by b to adjust for the reality of photosynthesis. This is because at each rank, there is a probability b of obtaining a leaf. When constructing the theoretical organ series based on Eq. (6.7), it matches the average organ series resulting from the simulations where the pauses are suppressed.

In Fig. 6.5a, 100 Monte Carlo simulations are performed over 30 DCs. The dotted line displays the weight of the organs of the potential (chronological) axis according to the rank from top to bottom. The symbols (o) are the average values of simulations that are in good concordance with the dotted line. The solid line is the theoretical topological organic series derived from the series of the potential axis (dotted line) by the Eq. (6.7). The symbols (◊) display the average weights of the topological organic series according to the rank after pauses have been removed. The matching with the dotted line is excellent and results from the correct approximation given by Eqs. (6.5) and (6.6). There is some dispersion of points at the base of the simulated organic series. This is due to the fact that the number of phytomers produced in simulation per rank from the top decreases with the increase in the rank. With 30 cycles, the probability of obtaining all 30 phytomers with a success probability $b = 0.7$ is $p(x = 30) = 0.7^{30}$ ($=2.2 \cdot 10^{-5}$). A large number of simulations is required to obtain it.

Figure 6.5a shows the good matching between stochastic simulations and the theoretical model for both chronological and topological organic series. Thus, the transition from the potential chronological axis to the observed topological axis leads to a transformation of the data on organic series.

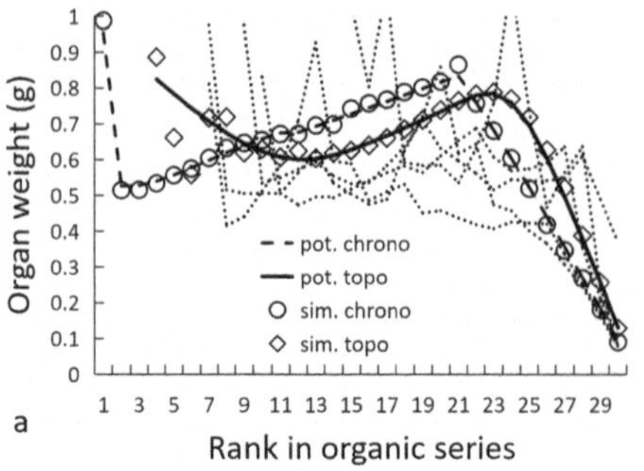

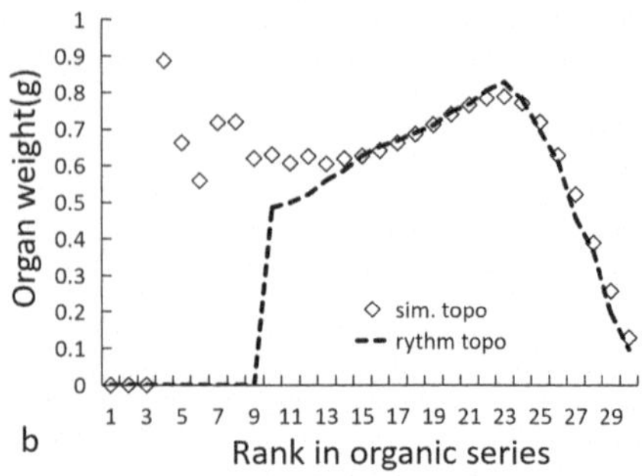

Fig. 6.5 Computed and simulated chronological and topological organic series in the stochastic case. (**a**) Comparison between the average of 100 stochastic simulations (○) in the chronological mode of a *Bernoulli* process ($b = 0.7$) on 30 DCs, and the theoretical distribution (dotted line) according to the rank in the potential structure. The agreement is the result of the correct approximation provided by Eq. (6.6). Comparison between the mean by the rank of the 100 simulated topological organic series (◊) sorted from the top (after removing pauses in the respective chronological series) and the theoretical distribution (solid line) which uses the transformation given by Eq. (6.7). The simulated and computed series are consistent. Thin dotted lines from five Monte Carlo individual simulations show the evident variation in organs weights along the ranks of the organic series. (**b**) Comparison between the mean of the organic topological series (◊) (same to Figure **a**) from the stochastic simulation and the deterministic topological series (dotted line) from a rhythm ratio $w = 0.7$. The two series correspond well near the apex of the axis

This result influences the assimilation of measurement data to estimate the model parameters. When computing the source–sink relationships of a plant from observations made on the organic series measured on the architecture, all observations should be treated as part of the stochastic model. This stochastic method eliminates the previously unresolved problem of the methodology of measuring vegetative axes (clones, varieties) of the same age but with varying numbers of phytomers. To address the question "how to organize the data?" when the stochastic functioning of meristems is known (Chap. 3), it is sufficient to group the axes according to their physiological age and establishes the organic series of the average organ weight according to their rank from the top.

We can see in Fig. 6.5b that the organic series obtained from a rhythm or a *Bernoulli* process with the parameter $w = b$ are close near the top of the axis. This allows switching from the stochastic model to the deterministic model with good approximation, which is simpler to process numerically. Such a procedure has been used for cotton plants (see Chap. 10).

6.4.3 *Study of the Growth of a Single-Stemmed Plant According to a* Bernoulli *Process, in Limited Growth*

This case is the normality in plant growth with the relationship (Chap. 5):

$$E \cdot \frac{n_a}{A} \cdot \left(\frac{t_x + 1}{2} + t_a - t_x\right) > 1$$

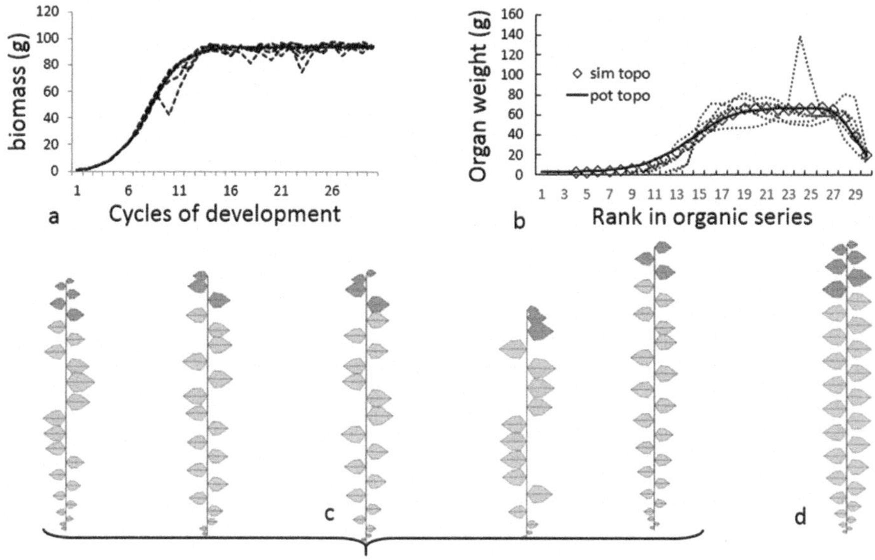

Fig. 6.6 Five simulations of the stochastic limited growth of a single-stemmed plant. The parameters are those of Fig. 6.5a with $b = 0.7$, except for $r = 10$ and $Sp = 1000$. (**a**) Biomass production per cycle, an exponential production is observed at the beginning of the growth, which increases until being constant when the ratio Sf/Sp exceeds a threshold. Five individual stochastic growth curves (fine dotted lines) are displayed. The mean growth (from 50 simulations) overlaps with potential growth (solid line). (**b**) Organic series of the leafy axis: The mean weights of organs in the simulated topological organic series (◇) are compared with the values provided by the potential structure (solid line). There is a good match. Thin dotted lines come from five Monte Carlo individual simulations, showing the significant variation in organ weights along with the ranks of the organic series. (**c**) Five plants from stochastic simulations. Although the simulated plants are highly variable, the average weight of the phytomers sorted by the rank tends to be the same theoretical value provided by the potential axis. (**d**) The potential structure

Growth is limited by planting density. Here, the value of Sp (the production area) is small and the growth stabilizes after several cycles as the ratio of leaf area to production area (Sf/Sp) saturates biomass production. We have:

$$Q = E \cdot \frac{Sp}{r} \cdot \left(1 - \exp\left(-k\frac{Sf}{Sp}\right)\right) \to E \cdot \frac{Sp}{r} \text{(the upper bound)}.$$

In this case, the growth curve has a sigmoid shape. The fluctuations in demand are the same as in the case of the free growth mentioned above (Fig. 6.3a), but the consequences on the variability of biomass production per plant are less, as can be seen in Fig. 6.6a. It is because the saturation of biomass production is reached due to the limited production surface Sp. Therefore, the growth of individual plants is less affected by stochastic development in the case of higher planting density.

The growth of the potential plant match well with the average of the simulated stochastic growth (Fig. 6.6a). The means of the simulated organic series sorted by the rank from the top along the axis are also well adjusted by the potential organic series (Fig. 6.6b), despite a high variability expressed in the five simulations, as displayed in Fig. 6.6c. Figure 6.6d shows the potential single-stemmed plant.

6.4.4 Study of the Stochastic Model of a Monopodial Plant

This type of plant is characterized by a stem–branch dimorphism (two physiological ages) and offers the simplest expression of the stochastic growth of branched plants. The coffee tree is an emblematic example. The values b_1 and b_2 can be assigned to the stems and branches, respectively, to the parameters of the *Bernoulli* process. The sinks of the phytomers are denoted as p_1 and p_2, respectively. To simplify, without losing the value of the results, we assume $b_1 = b_2 = b$ and $p_1 = p_2 = p$, and assign the same function values to the different organs for the two physiological ages (duration of functioning and expansion).

As with the single-stemmed plant model, the aim is to compare the means of stochastic simulations for development and biomass production with those resulting from the growth of the potential structure. It is also necessary to compare the averages of the organic series of organ biomass, simulated and sorted according to their rank, with the respective values provided by the potential structure. Suppose there are N repetitions of simulated plants, the possible chronological ages decline along the same stem, which provides many repetitions of their growth for the branches. Indeed, from the apex to the base of the stem, we observe the same branch at increasing stages of development. At the rank k from the top of the stem, the age of the branch is distributed according to a negative binomial law $BN(k, b)$. Similarly, for any branch on the stem, all the phytomers of the branches located at the rank k_2 under the branch apexes can be grouped, because they follow the same distribution law of chronological age, which is the negative binomial law $BN(k_2, b)$. However, if a vegetative axis has completed its growth due to terminal meristem mortality, its organic series should not be considered because the negative binomial law is no longer valid for the ranks.

Thus, in order to study its functioning of a plant at a given age, we can simplify the complex architectural development of the plant by describing the organic series of phytomers into two types of leafy axes: stems and branches. This simplifies the assimilation of plant data, since it is not necessary to detail the descriptions of their architectures. Instead, one need only to record the weights of the organs of the phytomers of physiological ages according to their respective ranks under the apexes. The number and location of measurements are not important if each type of data (phytomer of a given physiological age, located at a certain rank) is sufficiently repeated. The stochastic botanical automaton (Chap. 3) makes it possible to determine the number of types of organs produced $N_o^{\varphi}(i)$ according to physiological age φ to the DC i. When the number of organs stabilizes, due to the limited life span of the branches (demand becomes constant), production also stabilizes in free growth, if the condition $\frac{E}{A} \cdot \frac{t_a+1}{2} = 1$ is verified.

As discussed below, the *Bernoulli* process assigned to stem and branch meristems produces highly variable stochastic structures.

6.4.4.1 Effect of the *Bernoulli* Process in Free Growth on Biomass Production

On a branched plant, the probability of obtaining a phytomer is b for each phytomer on the stem and b^2 for each phytomer on the branch. In this example, b takes the value 0.9.

A branch can only exist if the carrying phytomer of the stem is not a pause.

The demand and biomass produced from the averages of the simulations evolve towards equilibrium and they are in good agreement with the values computed from the potential structure (Fig. 6.7a, b), despite the significant individual variability (Fig. 6.7c),

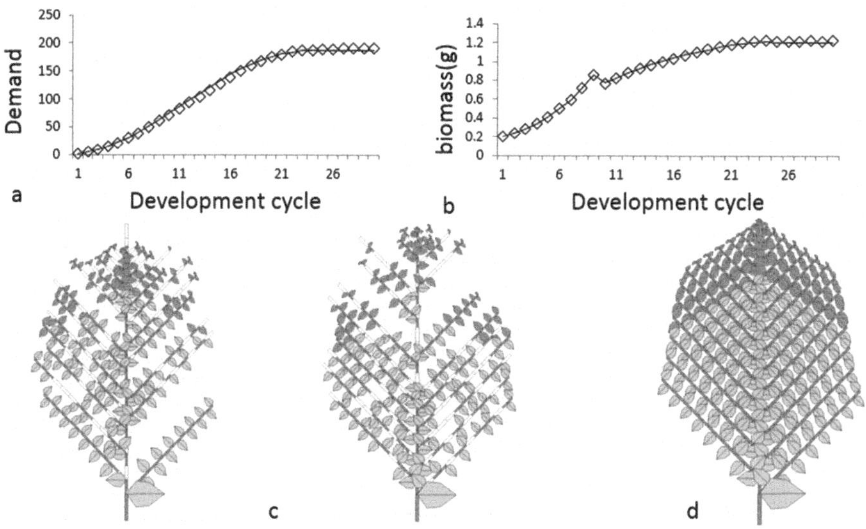

Fig. 6.7 Simulation of the stochastic growth of a monopodial plant with 30 DCs in free growth and equilibrium mode. The branches have a development time limited to 15 DCs. The equilibrium is reached when the condition $\frac{E}{A} \cdot \frac{t_a+1}{2} = 1$ is verified, whatever b value. Fifty stochastic simulations were performed. (**a**) Demand per cycle and (**b**) biomass per cycle, compared the averages of demand and biomass production from stochastic simulations (◊) with potential demand and production (solid lines), respectively. The matching is excellent, especially when compared to the single-stemmed plant, where the demand fluctuations are more marked because there are fewer phytomers involved. (**c**) Two of the stochastic simulations in chronological mode, and (**d**) the corresponding potential structure. Each plant of the 50 simulated structures can be positioned in the representation of the potential structure (**d**)

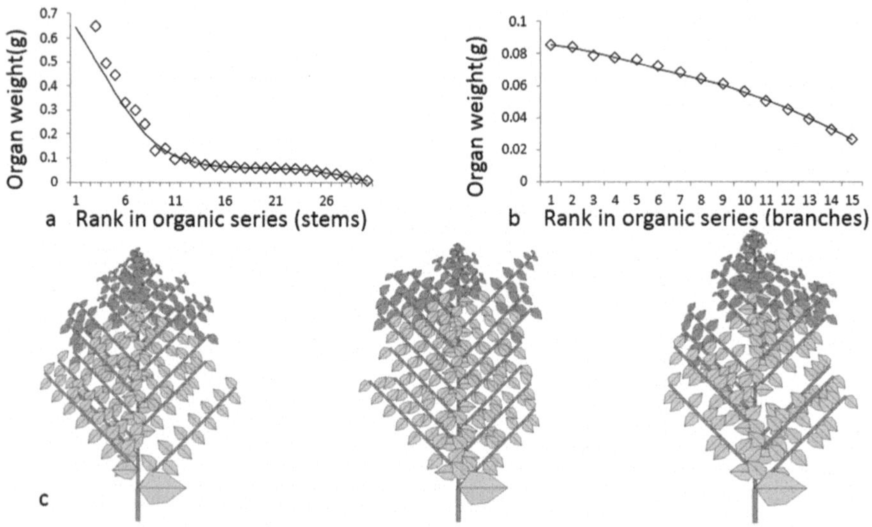

Fig. 6.8 Organ weights in the example of stochastic growth simulation in Fig. 6.7. (**a**) Organ weights of the organic series on the stems. (**b**) Organ weights of the organic series on the branches (sorted from the top). These two figures show the organic series on the vegetative axes (topological mode) for the weights of the organs of the stem and branch phytomers (theory solid line, (◇) simulations) relative to the data in Fig. 6.7. These series correspond to those predicted by the potential structure using the negative binomial law. (**c**) A visualization of three topological structures from 50 simulations of 30 DCs

With regard to organ weights, the topological average organic series of stems and branches of simulated plants of both physiological ages match well with those resulting from computations on the potential structure according to the negative binomial law (Fig. 6.8a, b). As the organs have the same sink function in this example, the organs of leaf and internode have the same organic series. It should be noted that the profiles of the stem and branch series are different, because the branches only develop over 15 cycles. The organic series for branches are sorted from the apexes of the living branches only. The dead branches produced by the simulation are excluded from the organic series as mentioned above. The method, thus, makes it possible to create coherent sets of measurements on shrubs whose architectures are apparently heterogeneous.

6.4.4.2 Study of the Monopodial Plant in Limited Growth with Stochastic Functioning for Development, Mortality, and Branching

With regard to mortality, there is an important difference between the computed potential structure and the simulated chronological structures. In computation, there is no cessation of meristem functioning in potential structure, but a probability of survival (viability) is considered. In the simulation, mortality leads to the end of the axis's functioning, which no longer grows. Only the simulated branches that remain alive are taken into account to define the organic series in comparison with those from the potential structure.

An example is shown in Fig. 6.9. In Fig. 6.9, all stochastic aspects of meristem functioning are considered: *Bernoulli* process for development (parameter b), branching, and mortality (parameters a and c). In addition, a rhythm ratio (parameter w) is assigned to the branches. Eventually, the development of the structure depends on the quadruplet set of parameters w, a, b and c. Branch mortality rapidly limits organ production. The growth is exponential at first and then tends towards a limit value imposed by *Sp*, forming a sigmoid curve. Branch mortality leads to long-term stabilization of the demand. The *Q*/*D* ratio (trophic pressure) tends towards a limit (not shown) and the organic series tend towards a constant profile. A crown is formed finally. It should be noted that demand has not yet fully stabilized at 30 DCs (Fig. 6.9a). On the other hand, biomass supply reaches a stable level (Fig. 6.9b) because of the production area *Sp*. The convergence between the simulation means and the corresponding values derived from the potential structure is satisfactory for demand, biomass (Fig. 6.9a, b), and organic series (Fig. 6.9c, d). Organ volumes increase at the beginning of plant growth, go through a maximum and decrease at the end, where they are still in the expansion phase if the branch remains alive. Limited growth mitigates the effects of random oscillations on organ weights. This mode is normal if we are interested in the growth of plants under high density.

6.4.4.3 The Issue of Average Tree

In case of the absence of mortality of the meristems ($c = 1$), and an immediately forced branching ($a = 1$), it is possible to replace the *Bernoulli* parameters b by the rhythm ratio w, for the analysis of the organic series. The averages of the simulated and random organic series are then compared with those coming from the average deterministic tree using w instead of b parameters with a good approximation (see Fig. 6.5b). Data processing is then simple, as it does not require any more sophisticated statistical data processing. This is useful if the method based on stochastic series analysis is not yet available. The notion of the average tree can then replace the use of the potential structure. However, these applications are limited to herbaceous and shrubby plants in the young stages and in the absence of meristem mortality. This method has been successfully applied to cotton plants (Chap. 10). Figure 6.10 shows three random plants, the average plant, the potential structure, and the average organic series from simulations compared to those from the average tree and the potential structure.

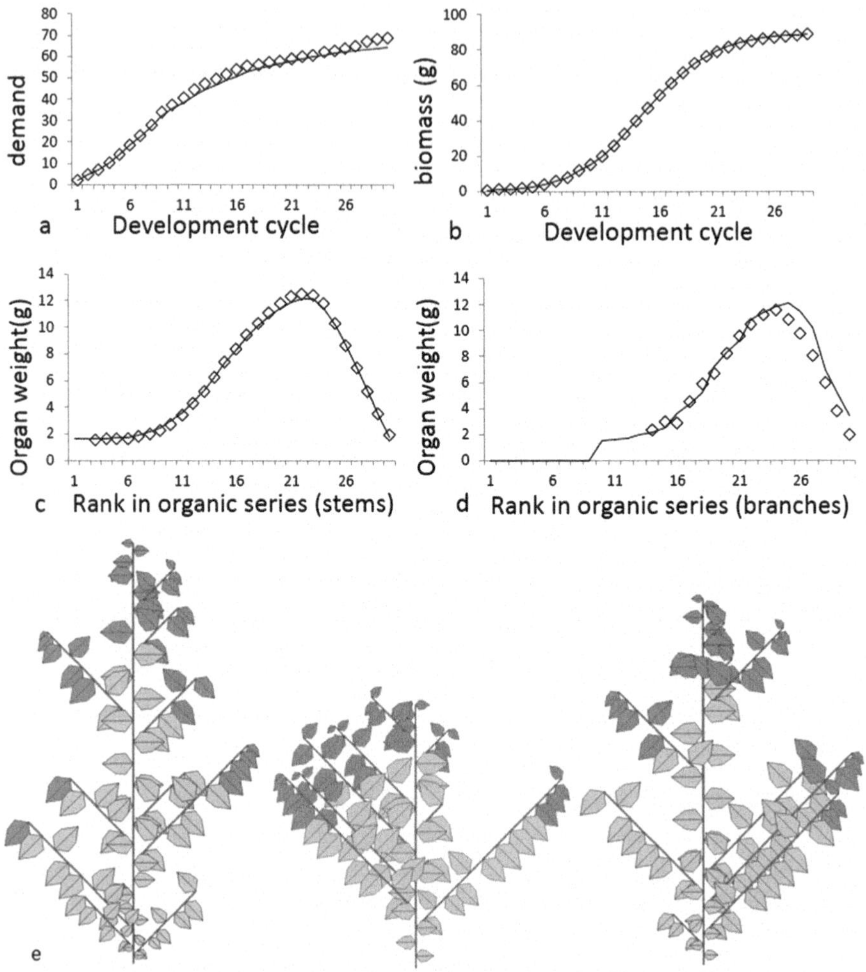

Fig. 6.9 Stochastic growth in the limited mode of the monopodial plant. Meristems have a triple stochastic function (growth, branching and mortality). Functioning time: $t_a = t_x = 10$ DCs. The rhythm ratio $w = 0.75$ for the branches in relation to the stem. The parameters $r = 20$, $Sp = 2000$, $a = 0.7$ for immediate branching, $b = 0.8$ for stem and branch development, $c = 0.9$ for branch viability, the main stems have no mortality. The sinks of the leaves and internodes are equal. The averages of the 50 random simulations (◊) of demand (**a**) and biomass production (**b**) are in good agreement with the values derived from the potential structure (curves). Demand gradually approaches a limit because of the diffuse mortality. The growth is sigmoid. The weights of the organic series of the simulated (◊) and computed vegetative axes fit correctly for stems (**c**) and branches (**d**). Part (**e**) shows three plants from the 50 simulations, illustrating the high variability of structures

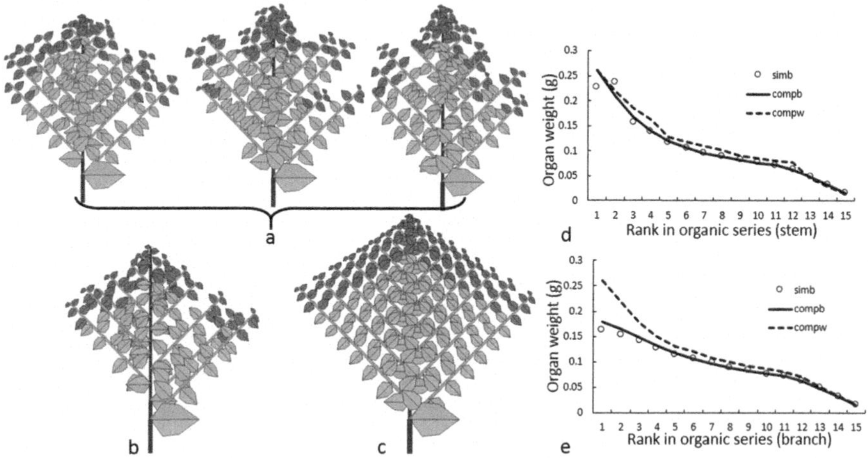

Fig. 6.10 The use of the average plant with $w = 0.8$ $a = 1$, $c = 1$. (**a**) Three random plants with $w = 1$, $b = 0.8$ (from a sample of 50) simulated at 20 DCs. (**b**) The average tree simulated with $b = 1$ and $w = 0.8$. (**c**) The potential structure. (**d**, **e**) The average of the stochastic organic series (o) is compared to the organic series of the average tree with rhythm ratio $w = 0.8$ (- - dotted line); the matching is good near the apexes. Naturally, the theoretical organic series (— solid line) computed from the potential structure, match better with the ones coming from simulations

6.4.4.4 Case of Delayed Branching

Delayed branching can be observed for example on the basal branching of the vegetative branches of cotton plants or those of soybeans, where branches do not develop immediately. Let parameter a be a constant probability of breaking the dormancy of the meristem at each cycle, the probability of the existence of thc phytomer created at cycle n on a delayed axis is written as $p(n) = 1 - (1 - a)^n$. It can be easily extended to cases with a variation in a.

By integrating this equation into the potential structure, we obtain the probability of the existence of all phytomers, and accordingly the potential demand of the structure by associating sinks with it. All the growing axis tips are homologous and can be grouped to obtain the biomass of the organic series from the apex. The method allows analyzing stochastic structures with complex branching modes. Figure 6.11 shows the relationship between the simulations and the computation of organic series for phytomer weights.

6.4.4.5 Effect of Reiteration

Reiteration, i.e., the case that branches are featured by the same physiological age with the bearing axis, increases the complexity of simulated topological structures. However, this complexity does not increase the number of organic series. The

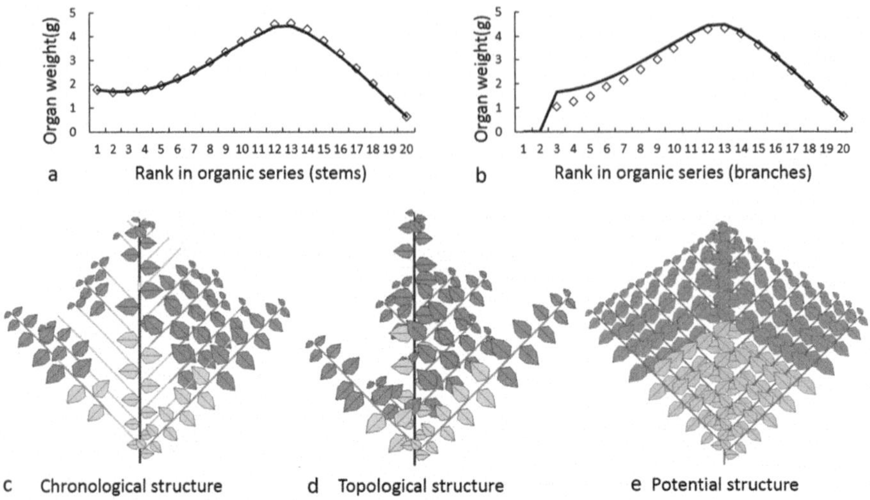

Fig. 6.11 Effect of the delayed branching on the constitution of organic series. The branched plant grows according to a *Bernoulli* process with a parameter $b = 0.9$ and a delayed probability of branching $a = 0.1$. The functioning of the leaves and the expansion of the organs with constant sinks is done throughout $t_a = t_x = 10$ DCs. The parameters $r = 20$, $Sp = 2000$. 100 simulations of plants are run. The averages of the simulated (solid curve) and computed (◇) organic series match well for stems (**a**) and branches (**b**). (**c**) A stochastic simulation in chronological mode. (**d**) Topological mode. (**e**) Associated potential structure. On the chronological structure (**c**), we can visualize the pauses that characterize the delay in branching

organic series of the trunk and its reiterations can be collected together. Reiteration can, therefore, be taken into account in the same way.

We choose here a monopodial plant with two physiological ages. The phytomers of the trunk have two axillary buds: one for the branch (always present), one for the stem reiteration. Buds function with a development probability $b = 0.8$. The stem can reiterate until order 5 with a probability $a = 0.25$. Thirty plants are simulated at the age of 15 DCs (Fig. 6.12).

The architectures obtained are very variable. Simulated organic series is constructed by averaging the weights of the organs according to their rank from the top of the axis. There are four types of organic series because there are two physiological ages and two types of organs (leaves, internodes). The averages of the simulated organic series and those computed from the potential structure match well. The series of internodes on trunks and branches are displayed.

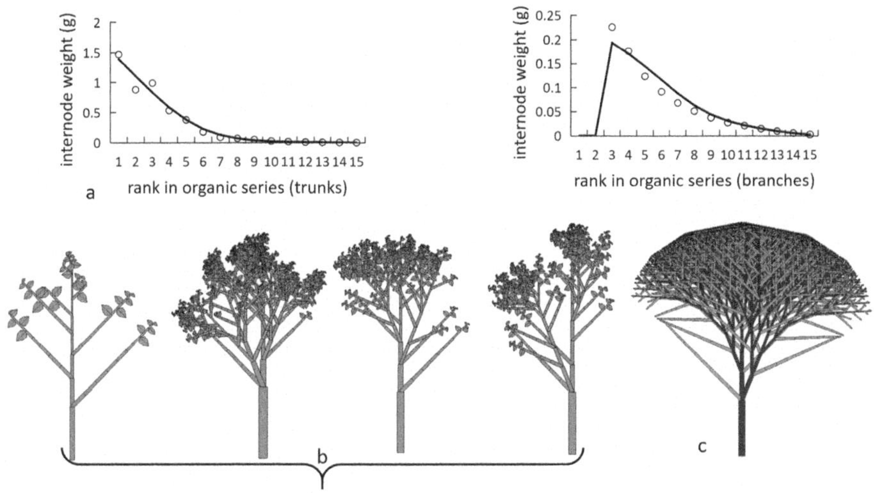

Fig. 6.12 Simulation of a monopodial plant with its reiterations. (**a**) Averages of the simulated organic series (o) and computed series (—) for the internodes of stems and branches. They match well. (**b**) Four random simulations. The first one did not reiterate. (**c**) Potential structure. Cultivated plants like coffee or cotton plants follow this pattern (see Chap. 10)

6.4.5 Study of the Stochastic Model of a Sympodial Plant with One Physiological Age

If we go through the phytomers from the base to the end of the tree, whatever the path, we describe the same organic series for the leaves or internodes (see Chap. 5, Fig. 5.15b). The tree-to-tree averages of the organ weights performed on the ranks, which correspond to the cohorts, tend towards the value of organ weight for the same rank in the potential structure. In this example, there is only one physiological age and two types of organic series (leaves and internodes). Cultivated plants like cassava or guayule follow this pattern (see Chap. 10).

We take a computational sympodial plant as an example (Fig. 6.13). Randomness is focused on the probability of branching ($a = 0.9$). There are two possible branches per phytomer.

6.5 Growth with Rhythmic and Stochastic Development of the Axes

6.5.1 Presentation

Some trees like cypresses or eucalyptus have continuous growth. There are no visible markers of growth stops along the axes. But most often the growth of trees

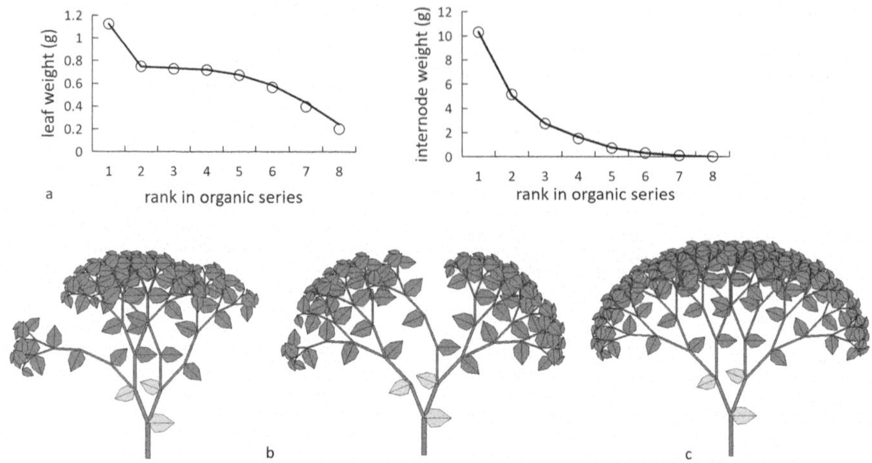

Fig. 6.13 Organic series in the case of a sympodial plant. All paths that join the base at the tips of the plant describe the same organic series. (**a**) The average of the simulated series (o) matches well with the computed series (—). (**b**) Two random simulations (with free growth: $t_a = t_x = 5$ DCs, $r = 10, a = 0.9$) at the age of 8 DCs. sink blade $P_a = 1$, sink internode $P_i = 1$, sink ring $P_c = 0.5$. (**c**) Potential plant

is rhythmic. The meristems periodically produce growth units (GU) that are delimited on the axes by markers (scars) that show the growth stops.

Two cases should be considered, which often correspond to temperate and tropical trees.

In temperate trees (e.g., maple tree), the growth has a winter stop and a recovery of vegetation in spring (see Chap. 2, Sects. 2.2.6 and 2.2.7). The meristems of the trees have limited duration of functioning (called here period and set to be constant) during which they set up a preformed or neoformed GU. Meristems normally produce a new GU in spring, except in the case of abortion. This produces a synchronism of functioning among the buds of the tree which gives birth to cohorts of leafy axes, components of the tree crown. The period of meristem activity may be constant or follow a distribution of DCs that can be adjusted well by a positive or negative binomial law (Chap. 3). Monocyclism corresponds to the case where the meristem sets up only one GU each year (e.g., maple, beech tree). If several GUs are made to form the annual shoot, it is called polycyclism (walnut tree). Following the formation of GUs, a meristem pause occurs to complete the annual growth period, with the duration of pause being complementary to that of the activity. The annual growth period has a fixed number of DCs. To simplify, the age of the tree is measured at the end of these annual periods, without going into the details of intra-annual cycles. The functioning of the organs definitively stops at the end of the year.

In tropical trees (e.g., hevea tree), buds produce several GUs during the year, but not all of them function at the same time. A proportion remains dormant and

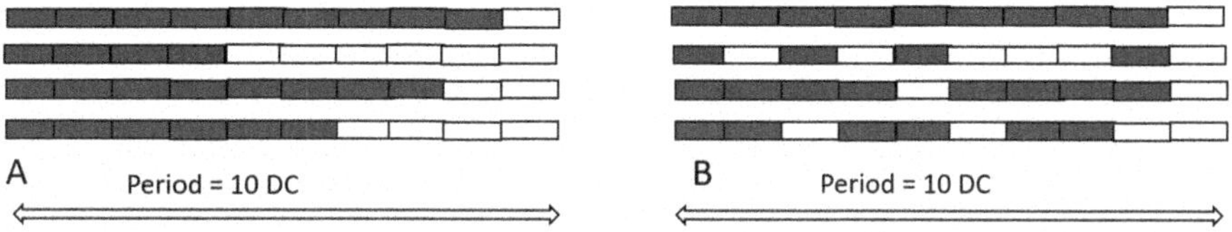

Fig. 6.14 Two types of functioning of the meristems which give identical growth unit (GU) for phytomer distribution. Four stochastic repetitions are displayed. (**a**) Variation in the functioning time and deterministic development. (**b**) Constant functioning time and development according to a *Bernoulli* process. The two types of functioning can be combined

functions again after a delay of one or more periods, and bud functioning becomes asynchronous. The setting up of GUs follows a *Bernoulli* process. A meristem produces at each period a new GU or a void GU in proportion b or 1-b.

But the phytomers within a GU also follow a distribution. The variation in the number of phytomers per GU can depend on two processes (Fig. 6.14) that lead to a compound distribution. Two cases are analyzed:

1. During the growth period, the meristem produces a variable number Nx of DCs distributed according to a binomial distribution (N, b) that corresponds to the number of phytomers formed, because meristem functioning is continuous and without pauses during the development.
2. During the growth period, the meristem produces a fixed number N of DCs, but a *Bernoulli* process with a parameter of b_u introduces pauses in proportion $1 - b_u$. This makes that the numbers of phytomers follow the binomial law (N, b_u).

The two processes thus give identical results of the distribution of the phytomers, if $b = b_u$ (Fig. 6.14).

Moreover, these two processes can be combined: the functioning time is variable and during the functioning time, the phytomer setting up follows a *Bernoulli* process. This gives a compound law, which is a Binomial law (N, $b*b_u$) as a known statistical result.

It is important to note that in the case of B (Fig. 6.14), a single potential axis accounts for all the stochastic simulations, while in case A, there is a distribution of potential axes, each of which can be subject to a *Bernoulli* process. The favorable case B is that of the positive binomial distribution; it simplifies the computations of the parameters with the inverse method. The case A is that of the negative binomial law and needs to take account of the distribution of the potential axes (see Sect. 6.5.8).

In this chapter, for simplicity, only positive binomial distributions will be considered. A single potential structure corresponds to the stochastic simulations of a given tree.

The duration of the organs' functioning can overtake the period of growth, and can cover one or more successive GUs at the tip of an axis.

The development axis as defined in Chap. 3 integrates the entire functioning of meristems characterized by their physiological age and their stochastic behavior, i.e., the result of phytomer creations or pauses cycle by cycle (it generates a chronological structure).

The leafy axis is the visible part of the functioning. Once the pauses are removed, it is a topological axis. The beginnings of GUs are periodically positioned on the development axis because the activity and pause periods are complementary and define the renewal period T_{GU} of the GU. It should be recalled that to ensure synchronization of vegetation recovery, the development axes of the branches insert on the main axis have structural pauses at the beginning, indexed on the position of their insertion on the GU (Chap. 2). Experimentally, it is possible to date the GUs of the vegetative axis using morphological markers associated with development stops (scars, etc.).

We consider here the synchronous and asynchronous functioning of the meristems and the matching of the results which come either from a Monte Carlo simulation, or from the computations which use the potential structure. It is verified that this concordance applies to the architectural models described in Chap. 2.

6.5.2 Study of the Rhythmic Development of Synchronous and Asynchronous Axes

We consider here a single stem formed by a series of GU. The annual functioning time of the meristems is T_{GU} DCs, a fixed renewal time of a GU. This period is divided into two parts. The first of T_d DCs duration characterizes the period of meristem activity (GU development) during which it produces phytomers. The second of $(T_{GU} - T_d)$ DCs duration characterizes the dormancy of the meristem following the period of activity. This dormancy consists of successive pauses corresponding to void phytomers entities in the development axis.

6.5.2.1 Counting the Number of Phytomers in a GU

Within the T_d activity period, random pauses may occur according to a *Bernoulli* process with parameter b_u (Chap. 3). The number of GU phytomers produced during the activity period T_d is the result of the composition of the distribution in DCs of the T_d activity period with the *Bernoulli* process. A condition for all GUs to function synchronously at the beginning of the year is to set $b_u = 1$ for the creation of the first phytomer; it takes its nominal value b_u for the following DCs.

On the leafy axis, all the morphological markers (scars) associated with the formation of GUs and the phytomers that compose them can be observed. Once the leafy axis has been constructed, pauses, on the other hand, cannot be observed afterward. The rank of the GU from the beginning of the leafy axis characterizes its age. We can count the number of GUs per axis as well as the number of phytomers per GU formed on the development axis during a fixed development period. In the case of temperate trees, the synchronous functioning of the meristems means that the number of GUs on an axis corresponds to its age in developmental periods. But in tropical trees, asynchronism due to meristems pauses introduces a *Bernoulli* process in the emission of successive GUs. The double scale botanical automaton, therefore, has two interlocking *Bernoulli* processes. One uses the parameter b ($b < 1$) at the axis level for the emission of a GU, and the other uses the parameter b_u at the GU level for the emission of a phytomer. According to observation, $b = 1$ for temperate trees.

6.5.2.2 Variability Inside Growth Units

Growth units bear branches with variable physiological ages. The types of phytomers $u_{i,j}$ are thus characterized by two indices. The first index i refers to the physiological age of the bearing axis, and the second index j refers to the physiological age of the branch (Chap. 4). It is assumed that there is a potential GU that has n_{uc} phytomers and in which all types of $u_{i,j}$ phytomers are present. All observable GUs derive from this potential unit expressing only a part of the potential GU.

In the case of acrotony (Chap. 2), the physiological age of the branches increases from the top of the GU to the bottom; the vigor and life span of the carried branches decrease as they descend along with the GU. The types of phytomers $u_{i,j}$ are grouped by zones with a number $n_{i,j}$, which is botanical data for the considered species. With the parameter b_u, the number of phytomer types $u_{i,j}$ is distributed according to the binomial law ($n_{i,j}$, b_u). The proportion of $u_{i,j}$ to the amount of all emission phytomers in this GU is maintained as $p_{i,j} = n_{i,j}/n_{uc}$. Note that if we are only interested in the axillary production of GUs, knowing the proportions of the different types of branches (and therefore the numbers) is sufficient, and it is not necessary to know exactly where they are placed (except in the rare cases of anticipated shoots). There is, therefore, no need in this case to know the rules of alternation between the series of axillary structures of different categories (according to Markov Chains for example), which constitute an important simplification for the production rules.

During the growth period of T_{GU}, the period T_d varies according to a distribution law. We start from the potential GU with all its phytomers (in number T_d), and for each value from the distribution, we redefine the dimension of the GU. Due to the notion of thermal time, the number of phytomers of a potential GU is necessarily limited, and the variation in the number of phytomers per GU can only be achieved by truncation (which can result from stress). This must be modeled according to the observations.

Here we show a case among various possibilities. With the action of stress with index s ($0 \leq s \leq 1$), the dimension of the GU in DCs takes the rounded value of $n_{uc}' = s \cdot n_{uc}$. In the chosen model, the size of the areas $n_{i,j}'$ does not vary upwards from the base of the GU, but it undergoes a distal truncation. This model appears to be well represented in nature (maple, cedar) that highly stressed GUs have a small number of phytomers with only short axes, unlike unstressed GUs with strong branches in the distal position, which may even be reiterations (Fig. 6.15).

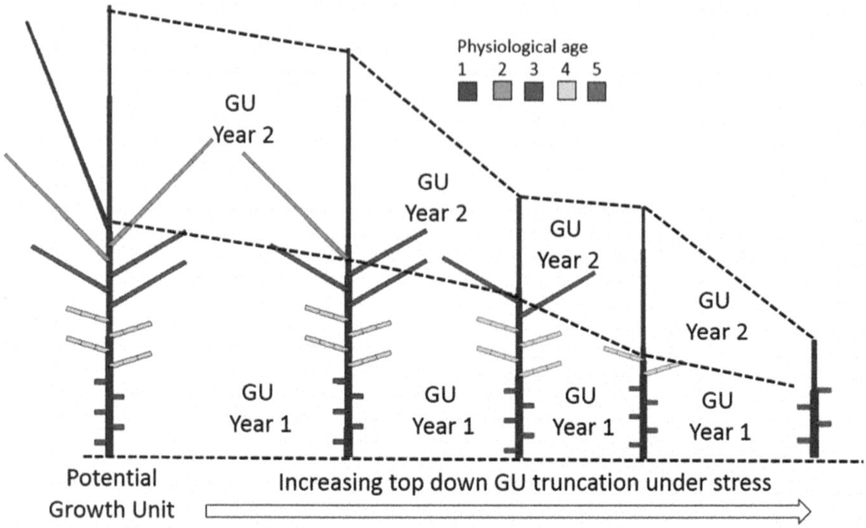

Fig. 6.15 Variability of the growth unit (GU) for the number of phytomers and proportions of branched axis types over two successive years. The shoots of the year are not branched. The branching is delayed until the following year. The potential GU and four simulations are represented using the distal truncation method of the potential GU. The GU is truncated and the types of phytomers are more or less expressed. Long GU wears reiterations and short GU only has short axes

The value of stress index s can be locally random, which leads to distribution in the number of DCs. Its evolution during growth may also be the result of a variation in the Q/D ratio.

6.5.2.3 Forming the Simulated Development Axes and the Computed Development Axis

The simulation of an annual GU composed of T_{GU} DCs is done over the two periods T_d and $T_{GU} - T_d$. The first concerns the functioning of the meristem and the second is the complementary pause, which closes the annual development. Two cases are possible (see also Fig. 6.14).

6.5.2.3.1 Constant Duration of GU Development

For example, $T_d = 10$ DCs and $T_{GU} = 15$ DCs are selected. Three random simulations of development axes are displayed to illustrate the effect of the parameter b_u. Here, we set $b_u = 0.6$ which inserts functional pauses during the T_d period, while the following ones are structural. Two consecutive years of function for three simulations give:

$Ads_1 = 1011101010\ 00000,\ 1100101111\ 00000, \ldots$
$Ads_2 = 1101011100\ 00000,\ 1011101010\ 00000, \ldots$
$Ads_3 = 1010110111\ 00000,\ 1111000111\ 00000, \ldots$

The potential development axis is unique and is written as:

$Adp = z_1z_2z_3z_4z_5z_6z_7z_8z_9z_{10},\ 00000,\ z_1z_2z_3z_4z_5z_6z_7z_8z_9z_{10},\ 00000, \ldots$

where z_j is the probability of occurrence of a phytomer during the period T_d. We have:

$z_1 = b_u = 1$ for the first phytomer of the GU and $z_j = b_u = 0.6$ for the others of the T_d period.

6.5.2.3.2 Variable Duration of GU Development

In this example, the development time (T_d) changes according to a binomial distribution $B(12,\ 0.8)$, so the maximum duration is $T_d = 12$ DCs. For the annual functioning time of the meristems, the same parameter is selected as above ($T_{GU} = 15$ DCs). We set $b_u = 1$ to visualize the variation in the development time. Three random simulations of the development axes give:

$Ads_1 = 111111000000\ 000,\ 111111111111\ 000, \ldots$
$Ads_2 = 111111111111\ 000,\ 111111111000\ 000, \ldots$
$Ads_3 = 111111100000\ 000,\ 111000000000\ 000, \ldots$

The axis of potential development is unique and written as:

$Adp = z_1z_2z_3z_4z_5z_6z_7z_8z_9z_{10}z_{11}z_{12}\ 000,\ z_1z_2z_3z_4z_5z_6z_7z_8z_9z_{10}\ z_{11}z_{12}\ 000, \ldots$

where z_j is the probability of occurrence of a phytomer at the rank j during the period T_d. In this example, $z_j = 1$.

6.5.3 Photosynthetic Functioning of Stochastic GU

In the case of synchronous trees, the development of GUs is synchronized. The development model of GUs resulting from the creation of phytomers during a T_d period was studied in Chap. 3. This development is followed by their simultaneous or delayed expansion, which constitutes growth. Three cases are possible:

- Delayed expansion: The expansion of the phytomers is delayed and starts simultaneously after the constitution of an embryonic part. In this case, the GU is preformed (e.g., maple, chestnut tree);

- Immediate expansion: The expansion of the phytomers starts immediately and sequentially follows their formation by meristems. In this case, the GU is neoformed (e.g., michelia, cacao tree);
- Mixed expansion: There is a preformed part that may be extended by a neoformed part. In this case, the GU is mixed (e.g., poplar, cherry tree).

For a preformed GU, the expansion of phytomers in GU is deferred to the next growing season, and is carried out simultaneously for all phytomers. All phytomers should be identical because they share the same Q/D (trophic pressure) history, but empirically they can also be differentiated according to their rank in GU.

For a neoformed GU, the expansion of the phytomers is sequential. They are all different because their expansion does not share the same Q/D (trophic pressure) history.

All preformed parts of the GUs of the same tree stop their expansion simultaneously. To simplify, this expansion is then considered to occur in a single growth cycle when the development cycles reach the T_d value, and the production of photosynthesis is accumulated in this cycle. This means that for this type of tree, the functioning time of the leaves is $t_a = 1$ and the expansion time of the organs is $t_x = 1$.

This simplification is reasonable considering that the photosynthetic functioning time of the GU from spring to autumn is much longer than that of its expansion in spring (a few weeks). During photosynthetic functioning, the surface of the leaves does not change and the biomass production results from the accumulation of radiation. To limit the number of DCs in computation, we choose $T_{GU} = T_d + 1$. If the hidden development of the preformed parts in the buds occurs during T_d cycles, only the last T_d cycle ensures growth and functioning. Finally, an additional cycle is added to mark the annual pause. Then all organs of GU stop expansion and functioning in synchronous trees (but not for asynchronous trees).

Concentrating the expansion of the GU and its functioning on a single cycle is like simulating a compartment system with a metaphytomer (the GU). It is the PhytoFacto mode of the GreenLab model. This method allows simulating the growth of large trees (e.g., maple, Chap. 10).

6.5.4 Potential Demand for Stochastic Growth Units

The potential demand for a GU at DC t is the scalar product between the existence probabilities of the phytomers constituting it and their respective sinks.

The probability of the existence of phytomers on the GU from the base depends on the statistical distribution of the number of DCs per GU. It is a compound law, resulting from the distribution of the development time (meristem activity) of the GU and the *Bernoulli* process of the parameter b_u that creates the phytomers.

In the case of a preformed GU, as mentioned above, a phytomer is a sink not at its creation but at its expansion, i.e., when the number of DCs of GU development

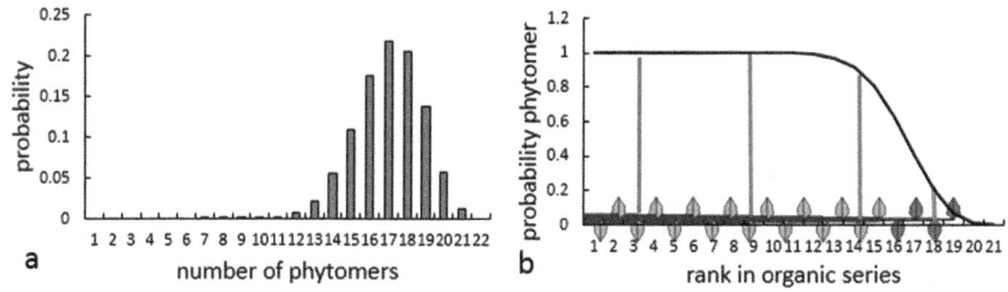

Fig. 6.16 Example of a G_{GU} function is the binomial law (20, 0.8). (**a**) Probability to get i DCs. (**b**) Probability z_i to observe phytomer i

reaches the value T_d. Then all the phytomers of the GU expand simultaneously, creating the demand. The GU's demand is, therefore, void until its development is completed. Moreover, all new GUs expand simultaneously.

In the case of a neoformed GU, each phytomer contributes as a sink to the demand of the GU at its creation. Let G_{GU} be the distribution function of the number of DCs of the GU development. Recall that z_i is the probability of phytomer existence as a function of its rank i from the base. Let x be the age of the GU in DC.

For preformed GUs, we have $z_i = 0$ if $x < T_d$

And

$$z_i = b_u \cdot \left(1 - \sum_{j=1}^{i} G_{UC}(j)\right) \quad \text{if } x \geq T_d \tag{6.8}$$

For neoformed GUs, for all x, we have again:

$$z_i = b_u \cdot \left(1 - \sum_{j=1}^{i} G_{UC}(j)\right)$$

If the number of DCs in the distribution is constant, we obviously have $z_i = b_u$.

As an example of a G_{GU} function, the binomial law (20, 0.8) is chosen. Figure 6.16a gives the probability of obtaining i phytomers per GU and Fig. 6.16b gives the probability z_i of observing phytomer i.

A potential development axis is built from the periodic sequence of GUs in which phytomers appear with the probabilities z_i. Similarly, the potential structure is built by assembling the axes. It is possible to compute the potential output of these structures with the GUs function, through the propagation of development, branching, and mortality effects.

The sink P_p (i) of the phytomer depends on its rank i $(i \leq x)$ and its probability of existence in GU z_i. The demand of the GU at the age $x = T_d$ on DC is written as:

$$D_{GU}(T_d) = \sum_{i=1}^{Td} P_p(i) \cdot z_i \tag{6.9}$$

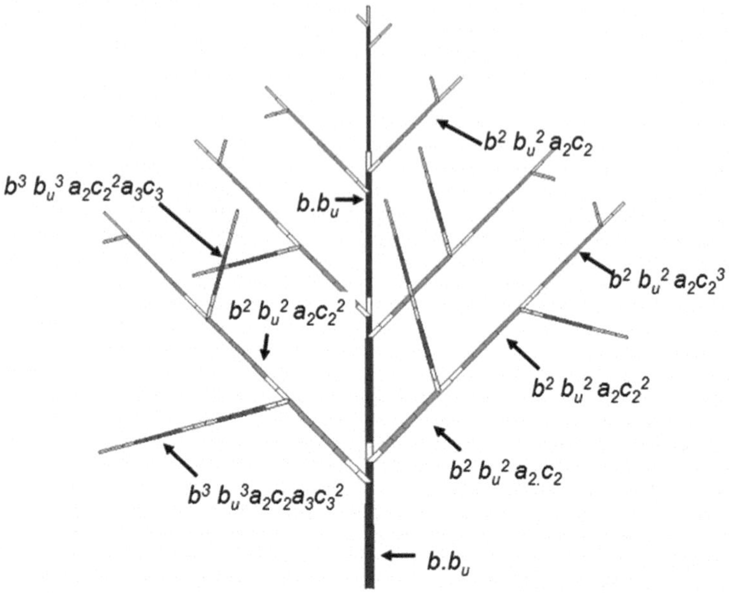

Fig. 6.17 Probabilities of the existence of phytomers in potential rhythmic structure. In this example, the duration of the annual growth period is 7 DCs for the entire tree, which is represented over four periods. The age of the tree is, therefore, 28 DCs. The development time *Td* of the GUs is 6 DCs for physiological age (PA) 1 (blue), 5 DCs for PA 2 (green) and 4 DCs for PA 3. Structural pauses at the beginning of the branched axis and those at the end of the periods allow for the resynchronization of all meristems at the beginning of the growth period. These pauses have a probability b of creating a GU and a probability b_u of creating a phytomer during their activity periods (for the first phytomer of a GU: $b_u = 1$). For temperate trees, $b = 1$, and for tropical trees, $b < 1$. Axes of physiological age 2 and physiological age 3 have branching probabilities a_2 and a_3, respectively, and the viability of meristems at the unit level in these axes is c_2 and c_3, respectively. The main axis does not die ($c_1 = 1$, not noted in the figure). The probability of the existence of a phytomer depends on the path from the base of the tree to the phytomer itself, which can be easily calculated. The sum of the probabilities gives the expectation of the number of phytomers created, and when multiplied by the associated sinks, it gives the potential demand of the structure (Color figure online)

6.5.5 *Definition of the Potential Structure of Synchronous and Asynchronous Trees*

With the defined rules, it is possible to make the assembly of the development axes, to build the potential structure of synchronous trees where all the phytomers are characterized by their respective probabilities of existence. Equation (6.8) gives the probability of the existence of the phytomer according to the rank in the GU. We assume here $z_i = b_u$ to simplify. In Fig. 6.17, we specify the expression of the probability of the existence of a given phytomer as a function of its position in the structure. The probability is zero at the location of structural pauses. At branches, the probability depends also on the sequence of branches of probabilities a_p according to physiological ages. To simplify, there is no mortality during the construction of the GU. This occurs only at the end of the development period of duration T_d. The succession of GUs along the axes depends on the viabilities c_p and the *Bernoulli* process with parameter b_u. The probability of the existence of the phytomer of an axis of physiological age p, located outside the domain of structural pauses and on a GU of the rank i, is therefore $a_p{\cdot}b_p.b_u.c_p^{\,i}$. This probability, linked to the formation of the axis, is to be multiplied by the probability of the existence of the phytomer that carries it. For the first phytomer of a GU, $b_u = 1$ is fixed.

6.5.6 Development and Growth of Synchronous Stochastic Structures

In synchronous growth, the time factor can be identified on one axis (counting the number of GUs), while simultaneous events on the other axes of the structure can be grouped into cohorts. The basis for the analysis of synchronous rhythmic trees is the period of annual growth T_{GU}. In the case of monocyclism, the age of a development axis in annual periods is characterized by its number of GUs. All GUs are synchronized with the resumption of meristem activity (in spring in temperate trees). The GUs of a tree that share the same physiological age and appear at the same chronological age are similar. From a stochastic point of view, they come from the same distribution. These units can be sorted according to the rank of their phytomers from the base. Compound (GU) and single (phytomer) organic series are bottom-up analyzed because the setting up of the GUs by the meristems is deterministic as the development is synchronous. The meristem development inside the GUs is based on a *Bernoulli* process with the parameter b_u.

In synchronous trees with monocyclic annual shoots, the function of the development axes is completed with the following rules:

- As all GUs start expansion simultaneously, the structural pauses at the origin of the axillary buds depend on their position in the GU, which allows all axillary meristems to be synchronous at the beginning of the activity period. The first axillary buds formed by a GU remain dormant longer before the bud pause. There is also a branching rate, denoted as a, which depends on physiological age.
- There is viability, denoted as c, which depends on physiological age, and may limit axis development as a result of the abortion of terminal meristems. Here, for simplicity, the probabilities are assumed constant for a given physiological age. They can be variable, which is considered in the GreenLab model.
- As mentioned earlier, there is no *Bernoulli* process at the level of the GUs ($b = 1$) and no mortality at the phytomer level inside a GU.

6.5.6.1 Calculation of the Growth of Rhythmic and Synchronous Structures

According to the double-scale botanical automaton of the GreenLab model (Chap. 4), a phytomer is identified firstly by the rank of its GU on the axis, secondly by its rank in the GU (but not by its rank on the axis). A GU can be described in organic series from the base, organ by organ, according to the rank of the phytomer (simple organic series). An axis can be described, GU by GU, according to the rank of the GU from the base, each data being the weight of the organ compartments inside the GUs (compound organic series). The GUs are sorted according to their chronological age and physiological ages, and each GU can be further decomposed into (simple) organic series that are sorted according to organ ranks from the base. All the first phytomers of GUs that appear at the same time are synchronous. This system makes it possible to construct simple and compound organic series extracted from complex plant structures.

6.5.6.2 Growth of Synchronous Monocyclic Structures

The structures have a period of T_{GU} growth cycles. Figure 6.17 is used as an example, and the probabilities intrinsic to the functioning of meristem are displayed (branching, mortality rates, etc.). Stochastic simulations are performed and the average simple and compound organic series are compared with the respective values computed by the model from the potential structure.

6.5.6.2.1 Structures with the Expansion of Preformed GU

To develop the GU, T_d DCs are counted to form the embryonic phytomers (sinks are negligible). The last cycle creates the demand and ensures simultaneously the expansion and functioning of all phytomers of GU, so they are all identical in this simple example. A pause closes the growth period. Maple and beech trees (see Chap. 3, Fig. 3.24a) follow this pattern.

Consider the following example. Fifty stochastic plant structures with five growth periods (1 growth period = 1 year) and seven DCs per period ($T_{GU} = 7$) are simulated (Table 6.1). There are three physiological ages. The potential output of the botanical automaton is shown in Fig. 6.17 (that one has only four growth

Table 6.1 Distribution of growth units (GUs) into axillary phytomer types by area

Physiological age	Type of axillary phytomers at the distal part of GU	Type of axillary phytomers at the basal part of GU
1	$U_{12} = 2$	$U_{13} = 4$
2	$U_{22} = 1$	$U_{23} = 4$
3		$U_{33} = 4$

Table 6.2 Average number of simulated phytomers computed at period 5

Physiological ages	Computed	Simulated
1	25	25.4
2	46.1	42.1
3	18.3	21.4

periods). The development time of the GU of physiological age 1 is $T_{d1} = 6$ DC, that of PA 2 is $T_{d2} = 5$ DC, and that of physiological age 3 is $T_{d3} = 4$ DC. Structural pauses at the end of GU development, completed at $T_{GU} = 7$ DCs.

We choose the distribution of the types of phytomers inside the GU in two areas defined by the botanical automaton.

These areas define the types of branches carried by the GU and here only the distal areas have branches (to generate acrotony). The number of phytomers per GU follows a binomial distribution shifted by 1 (the first phytomer of the GU is forced). The expansion of the GUs' phytomers is simultaneous. Meristem parameters are $b_u = 0.8$ for the *Bernoulli* process, $a_2 = a_3 = 0.8$ and $c_2 = c_3 = 0.9$, respectively, for the branching and viability rates for the axes of physiological age 2 and physiological age 3 in the authorized areas. Table 6.2 shows the good agreement between the averages of the numbers of phytomers produced by each physiological age over the 50 simulations and the values predicted by the potential structure.

Knowing that the parameter b_u at the first phytomer is forced to 1, to synchronize the expansion of the GUs at the beginning of the period, the GU of the trunk has an average of $1 + 0.8 \times 5 = 5$ phytomers and, at the age of 5, 25 phytomers are produced by the trunk. The functioning parameters are: $r = 10$, $\varepsilon = 0.05$, $E = 1$, Sp is very large (free growth) and the organ sinks are equal to 1. The functioning time of the leaves is $t_a = 6$ DCs and the expansion time is $t_x = 6$ DCs (but as it occurs at the last cycle of T_d, there is in fact only the GC T_d that functions). When the GU of physiological age 1 has exceeded 6 DCs, the source-sink functioning stops for all phytomers and the GUs pause until $T_{GU} = 7$ DCs, to complete the period. In this example (Fig. 6.18), considering the parameters, according to Eq. (5.27) in Chap. 5, we have: $\frac{t_a+1}{2} \cdot \frac{E}{A} = 1$. Even with a stochastic development, we get free growth at equilibrium and the growth produces 1 g per period. This case corresponds to Sect. 5.5.4.1 of Chap. 5 but with an additional physiological age 3. It should be noted that if the development occupies 6 DCs, the growth occupies only 1 DC. The same reasoning applies to physiological ages 2 and 3 whose development periods have, respectively, 5 and 4 DCs and are completed terminally by pauses of 2 and 3 DCs.

Let us consider the compound organic series composed of the weights of the organ compartments in the GUs (In this example, the weights of organs (leaves and internodes) inside a GU are identical because their expansion is synchronized). Figure 6.19 shows the good agreement on the means of the simulated and computed compound series. Recall that the computed compound series is based on the potential structure, and it is a necessary condition for computing the system parameters by the inverse method.

Figure 6.20 shows four stochastic simulations among the 50 simulated plants (the associated potential structure is displayed in Fig. 6.19). There is high variability

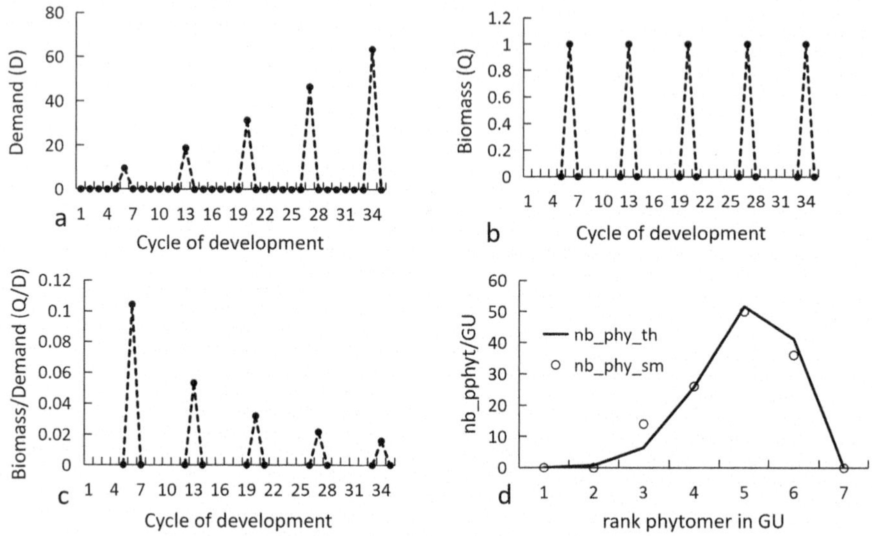

Fig. 6.18 Main results of a synchronous structure in free growth at equilibrium. The structure produces 1 g per period. (**a**) Demand is concentrated on the expansion cycle of preformation; it increases during the development due to the branching and lengthening of the axes. (**b**) Free growth at equilibrium makes constant biomass production. (**c**) The *Q*/*D* ratio decreases and therefore the volume of phytomers (all identical inside a growth unit (GU)). (**d**) The distribution of the number of phytomers per GU/axis for trunk follows well the binomial law *B*(5, 0.8) shifted by 1 (mean of simulations: circle, theoretical value: solid line)

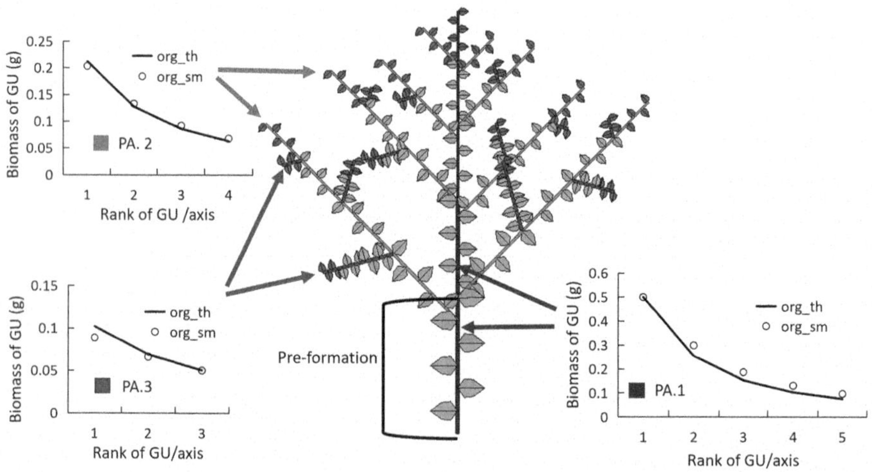

Fig. 6.19 Comparisons between the computed weights of the compound series of growth uiits (GUs) (—) which are the compartments of organs in the GU, and the averages of the respective simulated series (○) obtained with 50 simulations, for the 3 physiological ages over five growth periods. The matching is correct. The organs of the same GU are identical because the GU is preformed and their expansions are grouped. The potential structure is displayed

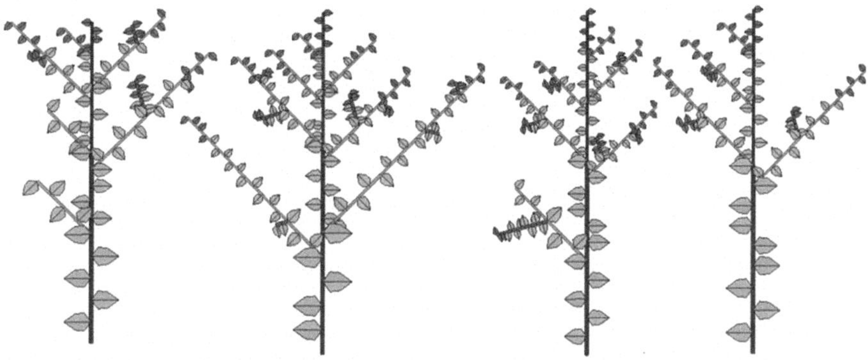

Fig. 6.20 Four stochastic simulations of a five-period structure with preformed growth unit (GUs). The number of phytomers per GU varies according to the *Bernoulli* process. The effects of branching and mortality rates are visible and result in high structural variability. The potential structure (Fig. 6.19) makes it possible to compute the theoretical organ weights, towards which the simulated averages organ weights tend. Phytomers inside a GU are identical

Table 6.3 Distribution of growth units (GUs) into axillary phytomer types

Physiological age	Type of axillary phytomers at the distal part of GU	Type of axillary phytomers at the basal part of GU
1	$U_{12} = 2$	$U_{13} = 7$
2	$U_{22} = 2$	$U_{23} = 5$
3		$U_{33} = 5$

among the simulated plants. The stochastic processes that control meristems offer here a complete variability in the development of the axes, their branching, and their mortality.

6.5.6.2.2 Structures with the Expansion of Neoformed GUs

In this case, the development of the growth unit requires T_d DCs to form the phytomers. Their expansion and functioning are sequentially at each DC and stop simultaneously at the T_d cycle. A pause closes the growth period. Cocoa (Chap. 3, Fig. 3.25d) and Michelia trees follow this pattern.

All phytomers have different biomass depending on their rank in the GU, since the expansion of phytomers is sequential and depends on the supply on demand *Q*/*D* in each DC. The stop of the development of GUs is simultaneous at T_d DCs, and the distribution of phytomers is a binomial law.

Consider the following example (Table 6.3). The period is composed of $T_{GU} = 10$ DCs. Three physiological ages are retained. The potential GU is composed of 9, 7, 5 phytomers for physiological ages 1, 2, and 3, respectively. The neoformation laws are, respectively, the binomial laws *B*(9, 0.8), *B*(7, 0.8) and *B*(5, 0.8). The *Bernoulli*

Table 6.4 Average number of simulated phytomers computed at period 5

Physiological age	Computed	Simulated
1	29.2	29.8
2	73.9	74.8
3	226.8	228.4

process is thus set at the value $b_u = 0.8$. The structure of the GUs is divided into two parts: meristem activity and pause.

These fields define the types of branches that are carried, and generate acrotony. Phytomers U_{33} do not carry branches. It is assumed here that the rate of branching is $a = 1$ if the phytomer exists, and the mortality rate is zero ($c = 1$). The defined botanical automaton makes it possible to compute the number of phytomers produced at each period (year) of development. Random simulations of the computational plant over five periods are performed. Table 6.4 shows the good matching between the averages of the numbers of phytomers produced by physiological age in the 50 stochastic simulations, and the values computed by the potential structure.

The functioning parameters of the organs are defined as follows: plant grows under conditions of free growth (very large Sp), $r = 47$, and the functioning and expansion times of the organs are $t_a = t_x = 9$ DCs. When the GU exceeds 9 DCs, the source-sink functioning stops simultaneously for all phytomers, and the GUs pause until $T_{GU} = 10$ DCs. Because of the neoformation and the simultaneous stop of the growth of all phytomers in the GU, the functioning times of the organs inside a GU differ. If x is the rank of the phytomer from the base of the GU and t_a is the functioning time of the leaf, the actual functioning time will be $T_d - x + 1$ if $t_a > T_d - x + 1$ (*with* $1 \leq x \leq T_d$). The same reasoning applies to the expansion time t_x for all organs.

All sinks here are constant and equal to 1. According to Eq. (5.30) in Chap. 5, considering the parameters, to get a constant production during free growth, we must have:

$$A + E - \left(\frac{A^{t_a}}{E}\right)^{\frac{1}{t_a - 1}} = 0$$

We have: $A = r \cdot \varepsilon \cdot P_p$ ($r = 0.47$; $\varepsilon = 0.05$; $P_p = 2$).

The coefficient A includes the action of three parameters on the growth ($1/r$ is resource use efficiency, and ε is the thickness of the leaf). Parameter E represents the energy supplied to the plant by the environment during each DC, which here is assumed to be constant and normalized to $E = 1$. The parameter r is the inverse of the resource use efficiency that is used to calibrate the model. P is the sink strength of the phytomer.

Even with stochastic development, we get periodic free growth. This case corresponds to Sect. 5.5.4.2 of Chap. 5 with an additional physiological age of 3. The evolution of supply and demand is given in Fig. 6.21. There is a dispersion

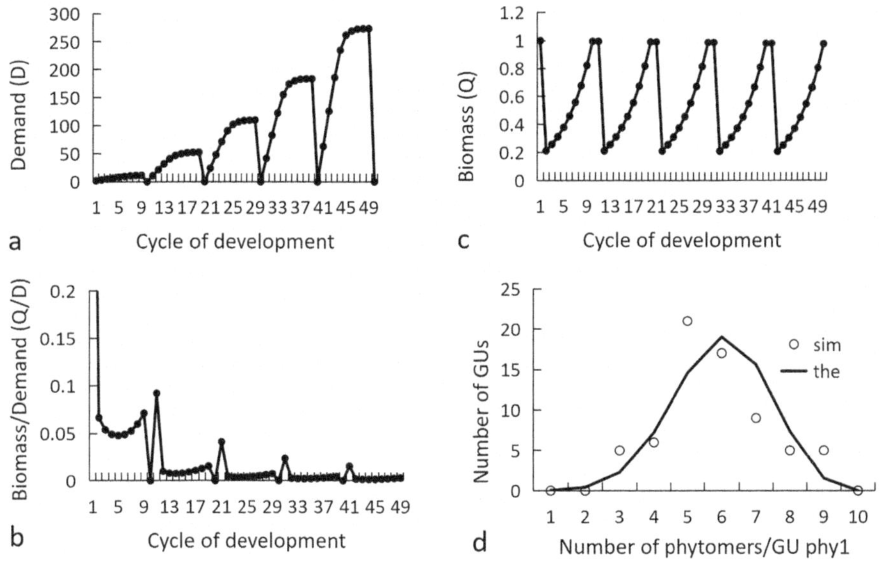

Fig. 6.21 Changes in demand (**a**), supply (**b**) and *Q*/*D* ratio (**c**) during the growth of a synchronous and rhythmic structure with neoformed GUs. The free growth (with $r = 47$) generates a periodic evolution of biomass production. (**d**) Comparison between the distribution of the number of phytomers per GU on the trunk and the theoretical distribution (observed data = ○, computed frequency = curve)

between the simulated and theoretical distribution of the number of phytomers per GU in the trunk, due to the small sample size.

The cyclical aspect is obvious. Demand returns to zero at the end of the growth period, but not the biomass production, which remains latent in the form of reserves for the following period. The *Q*/*D* ratio varies at each DC, so all the organs have different biomass. The results are then compared between the averages of the simulations (50 random plants) and the computations (with the potential structure) over five growth periods for the simple (plant organs) and compound (GU compartments) organic series that are grouped according to their chronological and physiological ages.

The simple organic series simulated according to the rank of the phytomer from the base of the GU, are sorted bottom up, and their average weights are compared to the computed values. The latter requires transforming the chronological series of the potential structure into topological series, using the negative binomial law (Fig. 6.22). The matching between the simulations and computations shows a good fitting according to the potential structure.

There is also satisfactory agreement on the matching of the cumulative weights of the organs (compartments) in the GUs that form compound organic series along the axes (Fig. 6.23), both for average simulation and computation.

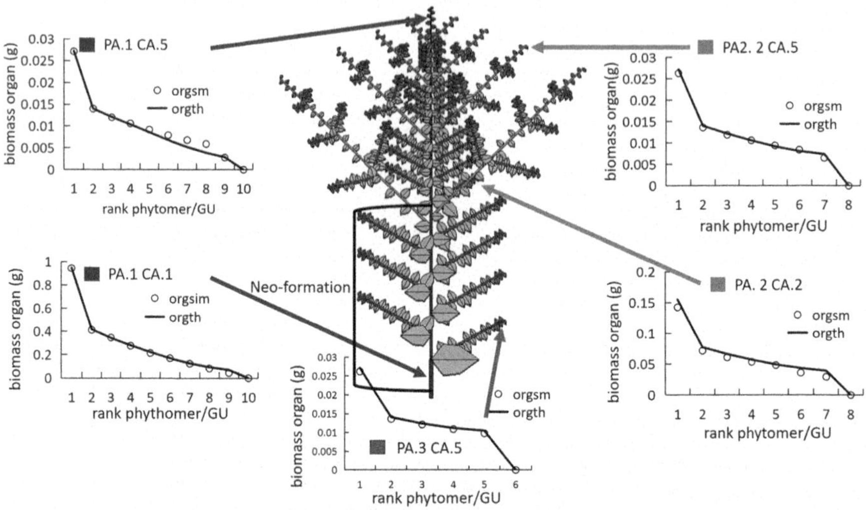

Fig. 6.22 Simulated (○) and computed simple organic series (—) for physiological ages (PAs) (1, 2, 3), and for chronological ages (CAs) (1, 2, 5) expressed in five periods in a GU with neoformation. The potential structure is displayed. The averages of the simulations match well with the values computed from the potential structure

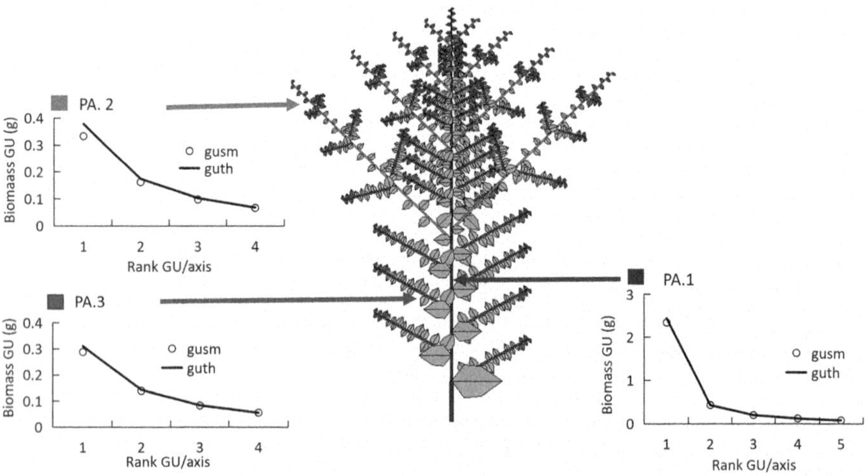

Fig. 6.23 Total organ weights per growth unit (GU) (compound organic series) in simulated (○) and computed (—) GUs, by chronological ages (CAs) and physiological ages (PAs), per periods of the growth. Adequacy is a consequence observed in Fig. 6.22. The potential structure is displayed

Figure 6.24 shows four stochastic simulations of the computational plant (an associate potential structure is displayed in Figs. 6.22 and 6.23). Since the branching is systematic and there is no mortality, all the variability comes from

Fig. 6.24 Four stochastic simulations of a five-period structure with neoformed growth units (GUs). The number of phytomers per GU follows a binomial law according to the *Bernoulli* process. There is no mortality in this example and all the phytomers (PA1 and PA2) can produce branches

the distributions of neoformed GUs whose development pause follows a binomial law specific to each PA. This law is consistent with the *Bernoulli* process. Type 3 axes are found branched on both the trunk and branches.

6.5.6.2.3 Structures with Expansion of Preformed GU That Can Be Extended with a Neoformed Part

To develop the GU, T_d DCs are counted to form the phytomers. T_p DCs are assigned to the preformed part and T_n DCs to the neoform part ($T_d = T_n + T_p$). A pause closes the growth period T_{GU} ($T_{GU} = T_d + 1$). Normally, in this case the preformed part does not have branches. Trees such as cherry trees, poplars (Chap. 3, Fig. 3.25a, c), follow this pattern.

Table 6.5 Distribution of growth units (GUs) into axillary phytomer types by area

Physiological age	Neoformed and distal part of GU	Neoformed and basal part of GU	Preformed and basal part of GU
1	$U_{12} = 3$	$U_{13} = 3$	$U_{10} = 4$
2	$U_{23} = 3$	$U_{24} = 3$	$U_{20} = 4$
3			$U_{30} = 4$

Table 6.6 Average number of simulated phytomers at period five

Physiological ages	Computed	Simulated
1	32	32.9
2	80.6	79.6
3	82.9	86.9

Considering the following example, a stochastic tree with four growth periods (i.e., 1 period = 1 year) and 11 DCs per period ($T_{GU} = 11$) are simulated. There are three physiological ages (Table 6.5). The development time of the GU of physiological age 1 is $T_{d1} = 10$ DCs, with a preformed part $T_{p1} = 4$ DCs and always a neoformed part $T_{n1} = 6$ DC, that of physiological age 2 is $T_{p2} = 4$ DCs and $T_{n2} = 6$ DCs with a neoformation rate $a_{nf} = 0.5$, and that of physiological age 3 is $T_{p3} = 4$ DCs without neoformation. Structural pauses at the end of GU development, complete at $T_{GU} = 11$ DC.

We choose the distribution of the types of phytomers inside the GUs in three areas defined by the botanical automaton.

These areas define the types of branches that are carried and the preformed part has no branches. Phytomers U_{12} bear one PA_2 branch, phytomers U_{13} bear one PA_3 branch, Phytomers U_{23} bear one PA_3 branch, Phytomers U_{24}, U_{10}, U_{20}, U_{30} have no branches.

The number of phytomers per GU follows a binomial distribution that is shifted by 1 (the first phytomer of the GU is forced). The expansion of the phytomers of the preformed GU is simultaneous, while the expansion of the following phytomers of the neoformed part is sequential. Meristem parameters are $b_u = 0.8$ for the *Bernoulli* process. There is no mortality ($c = 1$) and all branches of corresponding phytomers are set in place ($a = 1$). Table 6.6 shows the good agreement between the averages of the numbers of phytomers produced by each physiological age over the 30 simulations and the values predicted by the potential structure at period 4. As shown in Fig. 6.25, there is a dispersion between the simulated data and theoretical distribution due to the small sample size.

The functioning parameters of the organs are defined as follows, under limited growth ($Sp = 1000$), $r = 30$, the functioning and expansion times of the organs are $t_a = t_x = 10$ DCs. When the GU exceeds 10 DCs, the source-sink functioning stops simultaneously for all phytomers and the GUs pause until $T_{GU} = 10$ DCs. As a result of the neoformation and the stop of the growth of the GU, the functioning times of the organs differ in the GU. If x is the rank of the phytomer from the base of the GU and t_a is the functioning time of the leaf, the actual functioning time for rank x: $T_d -$

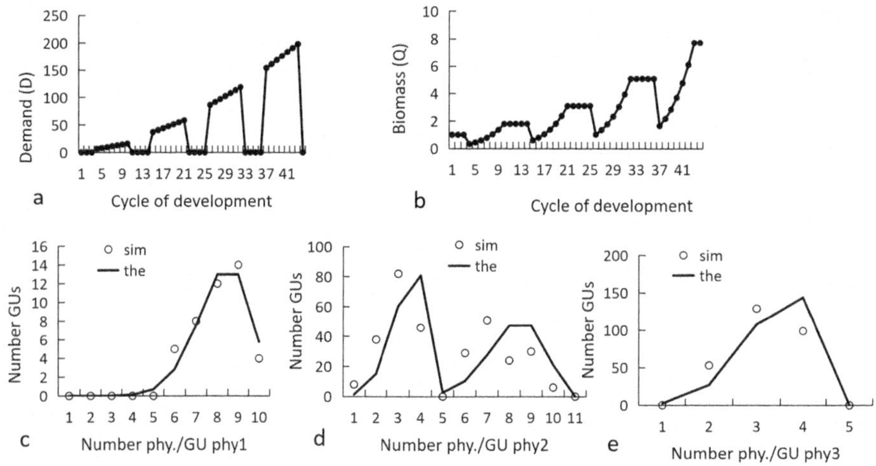

Fig. 6.25 Results come from 30 random simulations of a synchronous and rhythmic structure with pre-neoformed growth units (GUs). (**a**) Demand evolution. (**b**) Biomass evolution during the growth. The limited growth (with $Sp = 1000$) generates a periodic and increasing evolution of the biomass production, until a future stabilization of the cycles. (**c**) Comparison between the GUs of physiological age (PA) 1 (the trunk) follows a unimodal distribution because all GUs get neoformation. (**d**) The GUs of PA 2 (branches) follow a bimodal distribution because only 50% ($a_{pn} = 0.5$) of GUs get neoformation. Others remain in a preformation state. (**e**) The GUs of PA 3 (the short axis) follow a unimodal distribution because all GUs stay in preformation (observed data = ○, computed frequency = curve)

$x + 1$ if $t_a > T_d - x + 1$ (with $1 \leq x \leq T_d$). The same reasoning applies to the expansion time t_x for all organs.

The simple organic series, simulated according to the rank of the phytomer from the base of the GU, are sorted bottom up (void entities are removed) and their average weights are compared to the computed values using the potential structure. This comparison requires transforming the chronological series of the potential structure into topological series using the negative binomial law (Fig. 6.26).

This example is a mixture of the two previous ones. The phytomers of the preformed part are all the same size because their expansion is simultaneous. In contrast, the phytomers in the neoformed parts are all different sizes because their expansion is sequential. There is also satisfactory agreement on the matching of the cumulative weights of the organs (compartments) in the GUs that form the compound organic series along the axes (Fig. 6.27).

Figure 6.28 shows four stochastic simulations of the computational plant (an associate potential structure is displayed in Figs. 6.25 and 6.26). Since the branching is systematic and there is no mortality, all the variability comes from the distributions of pre- and neoformed GUs, whose development stop follows a bimodal law specific to each physiological age. This law is consistent with the *Bernoulli* process. The axes of type 3 are branched on both the trunk and branches.

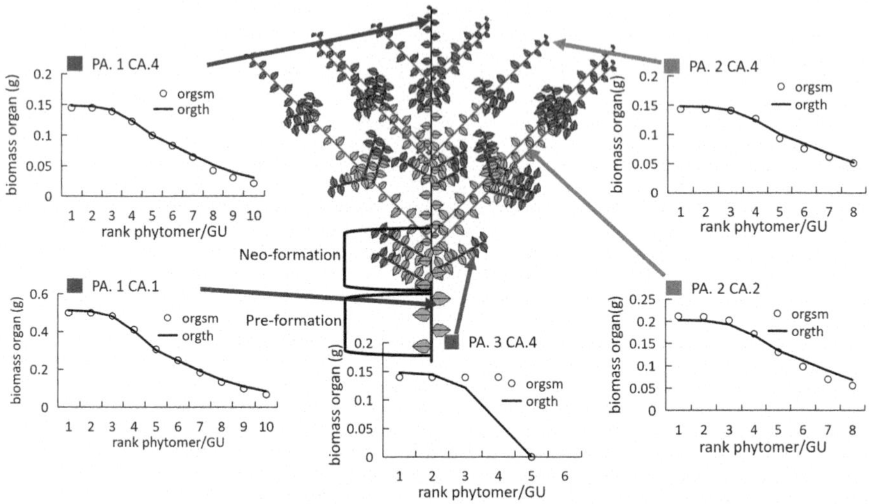

Fig. 6.26 Simulated (○) and computed (—) simple organic series for physiological ages (PAs) (1, 2, 3) and chronological ages (CAs) (1, 4) expressed in four periods in a growth unit (GU) with pre- and neoformation. The averages of the simulations match well with the values computed from the potential structure. The potential structure is displayed

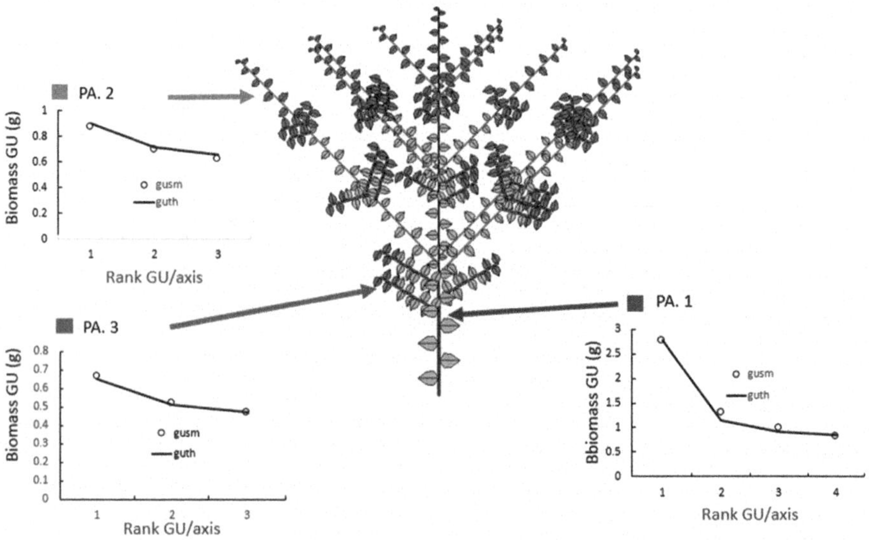

Fig. 6.27 Total organ weights per growth unit (GU) (compound organic series) in simulated (○) and computed (—) GUs, by chronological ages (CAs) and physiological ages (PAs) for four growth periods. Adequacy is a consequence observed in Fig. 6.26 (the sum of phytomers per GU gives the compound organic series). The potential structure is displayed

Fig. 6.28 Four stochastic simulations of a four-period structure with preformed and neoformed growth units (GUs). The number of phytomers per GU varies according to the *Bernoulli* process inside the bimodal (pre- and neoformed) distribution. The effect of bimodal GUs is visible and results in high structural variability. The potential structure is displayed in Fig. 6.26

6.5.7 Development and Growth of Asynchronous Stochastic Structures

The stochastic parameters (rhythm ratio, *Bernoulli*, viability, branching) that played on the phytomer level, are, therefore, also found at the level of the GU (see Chap. 3,

Sect. 3.3.2). The creation of GUs now depends on a *Bernoulli* process with parameter b.

6.5.7.1 Traits of the Potential Structure of Asynchronous Trees

At the GU level, there is a *Bernoulli* parameter that sets up a new GU with a probability b or a void GU with a probability $1 - b$, for the extension of the axis. This parameter does not play in synchronous structures because it is implicitly set to 1. This introduces in simulations a series of random pauses (void GU) of length T_{GU} in DCs which alternate with the T_{GU} phytomers of the created GUs.

It is also necessary to add a rhythm ratio w which differentiates the emission rates of GU between PAs at the same level. This parameter creates periodic alternations of structural pauses and GUs of length T_{GU} DCs along the leafy axes (see Chap. 3, Sect. 3.3.1.8).

Finally, this structure is equivalent to the monopodial structure with continuous growth in Sect. 6.4.4.2., but on an upper scale. The GU becomes a meta-phytomer and its creation also depends on a quadruplet of parameters w, a, b, c. Within the GU, the parameter b_u determines the number of phytomers (see Chap. 8, Sect. 8.6).

6.5.7.2 Computation of the Growth of Rhythmic and Asynchronous Structures

Previously we have seen that a phytomer is identified first from the rank of its GU on the axis, and second from its rank inside the GU from the bottom (but not from its rank on the axis). But here we cannot count the GUs from the base of the axis because the number of GUs per chronological age varies according to a *Bernoulli* process. They must, therefore, be numbered from the top of the axis and proceeded at the GU level using the negative binomial law, as for the phytomers level of continuous structures (see Sect. 6.4.2.3). Thus, the axis is described top down, GU by GU, according to its rank from the top by the organs sorted as compartments of the GUs (compound organic series). The GUs are sorted according to their chronological ages and physiological ages and decomposed into (simple) organic series according to the phytomer ranks bottom up from the GU base. All the first phytomers of the GUs that appeared at the same time are considered synchronous.

6.5.7.3 Growth of Asynchronous Structures

These structures have a period of T_{GU} DCs. Figure 6.30 is used as an example.

Compared to synchronous monocyclic structures, note that there are two additional parameters:

Table 6.7 Distribution of growth units (GUs) into axillary phytomer types

Physiological age	Neoformed and distal part of GU	Neoformed and basal part of GU
1	$U_{12} = 2$	$U_{13} = 3$
2		$U_{23} = 5$

Table 6.8 Average number of simulated phytomers computed at growth period nine

Physiological ages	Computed	Simulated
1	28.8	30.6
2	57	59

- the *Bernoulli* parameter b which randomly alternates void entities of duration T_{GU} with GUs of duration T_{GU} along an axis.
- the rhythm ratio w that controls the emissions of GUs by the terminal meristem and periodically and structurally introduces empty entities of duration T_{GU} along the axis. The rubber tree (see Chap. 3, Sect. 3.3.1.8) follows this pattern.

Stochastic simulations are performed and the average organic series are compared with the respective values computed by the model from the potential structure. As mentioned in Sect. 6.5.1 the uniqueness of the potential structure is available only with the case of positive binomial distribution of phytomers per GU.

Development of GU may be neoformed or preformed. We chose here the neoformed functioning. In the previous synchronous case, the functioning duration of organs stops at DC $= T_{GU}$, because the meristem is pausing. Here as the development is asynchronous, the duration of functioning of organs can overtake T_{GU}. So, we can have $t_a > T_{GU}$ and $t_x > T_{GU}$. The organ functioning may cover the period of several successive GUs.

Considering the following example in the case of neoformation, 100 stochastic structures with nine growth periods and five DCs per period ($T_{GU} = 5$) are simulated (Table 6.7). There are two physiological ages. The development time of the GU of physiological ages 1 and 2 is $T_d = 5$ DCs, thus, $T_d = T_{GU}$. This means that there is no structural pause at the end of GU development.

We choose the distribution of the types of phytomers inside the GU in two areas defined by the botanical automaton.

These areas define the types of branches carried by GUs and here only the distal areas have branches (to generate acrotony). The number of phytomers per unit follows a binomial distribution shifted by 1 (the first phytomer of the GU is forced). The expansion of the GUs' phytomers is simultaneous. Meristem parameters are $b = 0.8$ for *Bernoulli* process at GU level, $b_u = 0.8$ for the *Bernoulli* process at phytomer level, $a_2 = 0.8$ for branching probability, $w = 0.67$ for rhythm ratio of branches at GU level, and $c_2 = 0.9$, respectively, for the viability rate for the axes of physiological age 2. Table 6.8 shows the good agreement between the averages of the numbers of phytomers produced by each physiological age over the 100 simulations and the values computed by the potential structure for nine growth periods.

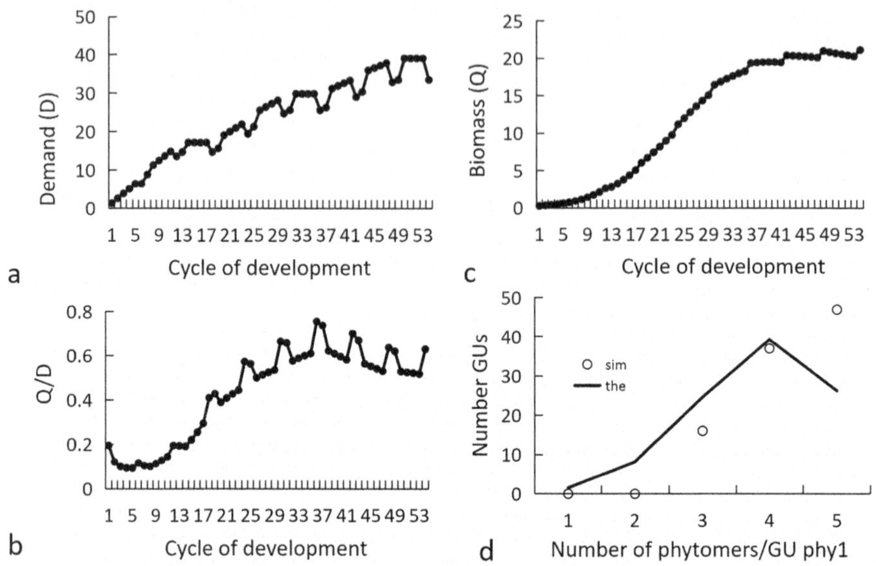

Fig. 6.29 Changes in demand (**a**), biomass production (**b**) and Q/D ratio (**c**) during the growth of an asynchronous and rhythmic structure with neoformed GUs in the case of limited growth of asynchronous trees. (**d**) Comparison between the distribution of the number of phytomers per growth unit (GU) in the trunk and the theoretical distribution (observed data = ○, computed frequency = curve)

The functioning parameters are $r = 40$, $\varepsilon = 0.05$, $E = 1$, $Sp = 1000$ (limited growth) and the organ sinks are equal to 1. The functioning time of the leaves is $t_a = 12$ DCs and the expansion time is $t_x = 8$ DCs (t_a and $t_x > T_{GU}$).

Figure 6.29 shows the evolution of demand and biomass production over the 12 growth periods corresponding to 12*5 = 60 DCs, as well as the distribution of the number of phytomers by GU. The effect of mortality and limited growth is observable through the damping of demand and biomass production per DC.

The pauses inserted in the simulations of chronological structures are removed, to obtain (1) the averages of the compound organic series (organ compartments in the GU) at the axis level, and (2) the simple organic series (the organs inside the GUs). The averages of the simulated organic series are compared with those extracted by computing the potential structure. Figure 6.30 shows the matching between the simulated and the computed simple organic series. Matching simulation and computation of GU is less good here because the quintuplet w, b, b_u, a, c play a role, the simulation sample should increase a lot. If only b and b_u are concerned, the matching becomes as good as in Fig. 6.22.

There is a better agreement on the matching of the cumulative weights of the organs (compartments) in the GUs that form compound organic series along the axes (Fig. 6.31), both for average simulations and computations.

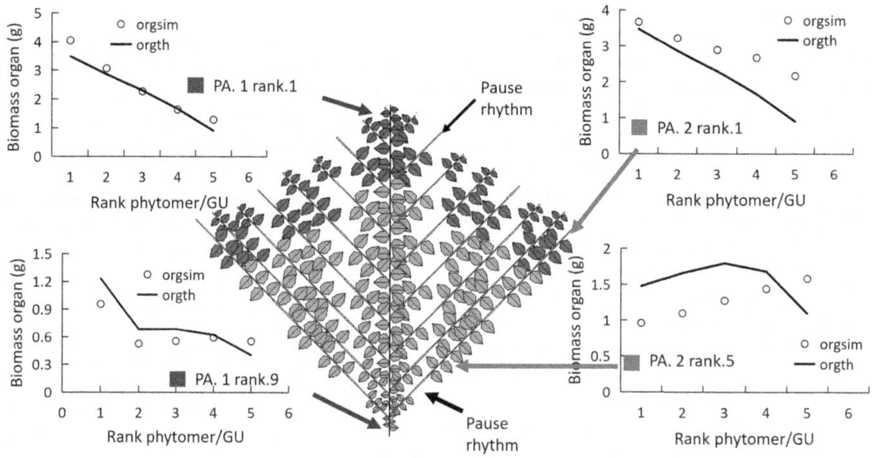

Fig. 6.30 Simulated (○) and computed (—) simple organic series for physiological ages (PAs) (1, 2), and growth unit (GU) ranks from the top (ranks 1, 9 for the stem, ranks 1, 5 for branches) expressed in nine periods in a GU with neoformation. The averages of the simulations match rather well with the values computed from the potential structure despite the high variability due to the quintuplet w, b, b_u, a, c and the small size of the sample. The potential structure is displayed. Structural pauses due to w are visible

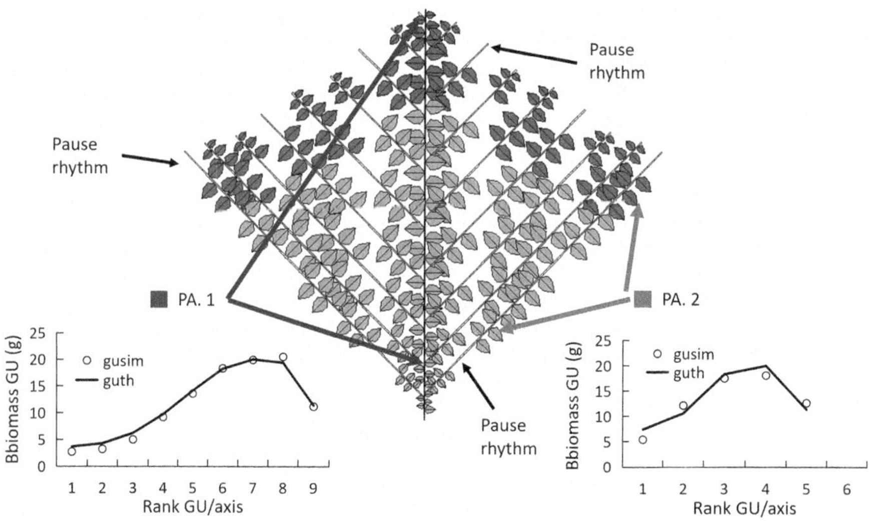

Fig. 6.31 Total organ weights per growth unit (GU) (compound organic series) in simulated (○) and computed (—) GUs, by rank from the top and physiological ages (PAs) for growth periods. Matching simulated and computed weight of organ compartments of GU is good. The potential structure is displayed

Fig. 6.32 Four stochastic simulations of a nine-period rhythmic and asynchronous structure with neoformed growth units (GUs). The number of phytomers per GU varies according to the *Bernoulli* process (binomial distribution). One can see missing and dead branches. The corresponding potential structure is displayed (Figs. 6.29 and 6.30)

Figure 6.32 shows four stochastic simulations of the computational plant (an associate potential structure is displayed in Figs. 6.30 and 6.31). There is a high variability due to branching and mortality processes.

6.5.7.4 Growth of Synchronous–Polycyclic Structures

The case of polycyclism is the most complex and has both synchronous and asynchronous traits. Walnut and elm trees follow this pattern (see Chap. 2, Sect. 2. 2.6 and Chap. 4, Sect. 4.2.3).

Thus, in the case of polycyclism, the botanical automaton is triple scale: the axis is made up of shoots, the shoots are made up of GUs, and the GUs are made up of phytomers. To simplify, here we suppose that GUs belonging to a shoot have identical development.

The growth period is the year and lasts T_{GU} DCs. During this period, a shoot of one or several GUs is set in place by the meristem. All axes stop their development and the organs stop their functioning at the end of the period. Growth starts again in the following period (spring) after a rest. So, there is a synchronization of meristems at the start of the period.

The period is divided into N_{GU} sub-periods, each of t_{GU} duration in DCs ($t_{GU} < T_{GU}$), which can set up N_{GU} GUs. The first GU is always created (unless there is an abortion of the bud). The others are set up alternately with void entities of t_{GU} duration in DCs according to a probability b and $1 - b$. The distribution of sub-periods is, therefore, a shifted binomial law $(N_{GU} - 1, b) + 1$. The development of GUs inside the annual period is, therefore, asynchronous.

Polycyclism in GUs by physiological age is variable. In general, PA 1 has the highest number (up to 5 for Japanese elm tree) but short axes are always monocyclic.

The annual shoots (duration T_{GU}) and the GUs (duration t_{GU}) are clearly identifiable along the axes of elm or walnut trees. Since the growth of shoot is synchronous, the numbering of the shoots can be done either top-down or bottom-up along the axis. But inside a shoot, the GUs are numbered bottom up. It is necessary to use the negative binomial law with the potential axis to match computations and simulations using the organic series.

Considering the case of polycyclism, 50 stochastic structures with four growth periods and 22 DCs per period ($T_{GU} = 22$) are simulated. The age of the structure in DCs is thus $22*4 = 88$ DCs. There are three physiological ages. Polycyclism is 3 GUs for physiological age 1, 2 GUs for PA 2, 1 GU for PA 3. The development time of the GUs of PAs 1, 2 and 3 is $T_d = 6$ DCs. After the T_d DCs, we add 1 DC to end the GU with a pause. So $t_{GU} = 7$ DCs. This means that there is a structural pause at the end of each shoot development.

For the physiological age 1, the number of DCs used for the growth of shoot is $3*t_{GU} = 21$ DCs. So, we have $T_{GU} = 3*t_{GU} + 1 = 22$ DCs. One structural pause ends the shoot.

For the physiological age 2, the number of DCs used for the growth of shoot is $2*t_{GU} = 14$ DCs. So, we have $T_{GU} = 2*t_{GU} + 8 = 22$ DCs. Eight structural pauses end the shoot.

For the physiological age 3, the number of DCs used for the growth of shoot is $1*t_{GU} = 7$ DCs. So, we have $T_{GU} = 1*t_{GU} + 15 = 22$ DCs. Fifteen structural pauses end the shoot.

Table 6.9 Distribution of growth units (GUs) into axillary phytomer types

Physiological age	Neoformed and distal part of GU	Neoformed and basal part of GU	Polycyclism
1	$U_{12} = 2$	$U_{13} = 4$	3
2	$U_{23} = 1$	$U_{24} = 5$	2
3		$U_{34} = 6$	1

Table 6.10 Average number of simulated and computed phytomers at period four

Physiological ages	Computed	Simulated
1	49.9	51.7
2	215.6	219.8
3	115	126.4

As branching is delayed by one year on the shoot, all axillary branches have structural pauses at the beginning to ensure the synchronism of branching. Figure 2.13 of Chap. 2 displays the partitioning of structural pauses among the potential structure over three growth periods of a computational plant with the polycyclism.

The development of GU is neoformed in this example (Table 6.9). The functioning duration of organs stops at DC $= T_{GU}$, because the meristem is pausing. We have $t_a = 21$ DCs and $t_x = 21$ DCs. The organ functioning duration may cover several successive GUs but is less than T_{GU}.

We choose the distribution of the types of phytomers inside the GU in two areas defined by the botanical automaton.

Phytomers U_{12} and U_{23} have branches and there is no mortality.

Note that there are only three physiological ages, but phytomers U_{24} and U_{34} exist. They are necessarily unbranched because physiological age 4 does not exist.

These areas define the types of branches carried by GUs and here only the distal areas have branches (to generate acrotony). The number of phytomers per unit follows a binomial distribution shifted by 1 (the first phytomer of the GU is forced). The expansion of the GUs' phytomers is sequential (case of neoformation). Meristem parameters are $b = 0.8$ for the *Bernoulli* process at GU level, $b_u = 0.8$ for the *Bernoulli* process at phytomer level. Table 6.10 shows the good agreement between the averages of the numbers of phytomers produced by each physiological age over the 50 simulations of plants and the values computed from the potential structure.

The functioning parameters are: $Q_o = 1, r = 60, \varepsilon = 0.05, E = 1$, Sp very big (free growth), $t_a = 21$, $t_x = 21$ DCs, and the organ sinks are equal to 1.

Figure 6.33 shows the evolution of demand and biomass production over the four growth periods, corresponding to $4*22 = 88$ DCs, as well as the distribution of the number of phytomers by the shoot of physiological age 1.

The pauses inserted in the simulations of chronological structures are removed, to compute (1) the averages of the compound organic series (organ compartments in the GU) at the axis level, and (2) the simple organic series (the organs inside the GUs). The averages of the simulated organic series are compared with those

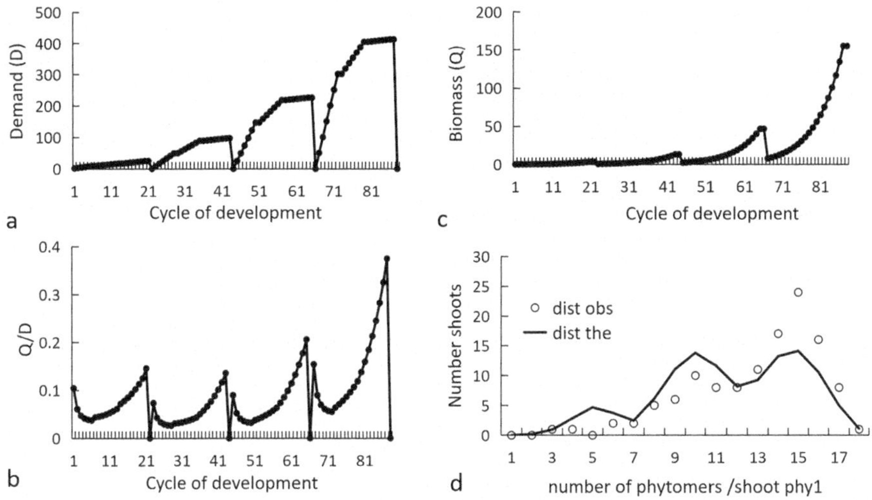

Fig. 6.33 (**a**) Changes in demand, (**b**) biomass production and (**c**) *Q*/*D* ratio during the growth of polycyclic structure with neoformed growth units (GUs) in the case of free growth of polycyclic trees. (**d**) comparison between the distribution of the number of phytomers per GU in the trunk and the theoretical distribution (observed data = ○, computed frequency = curve). Three waves of polyclism are visible on trunks (there are only two waves for branches and one for twigs, not shown)

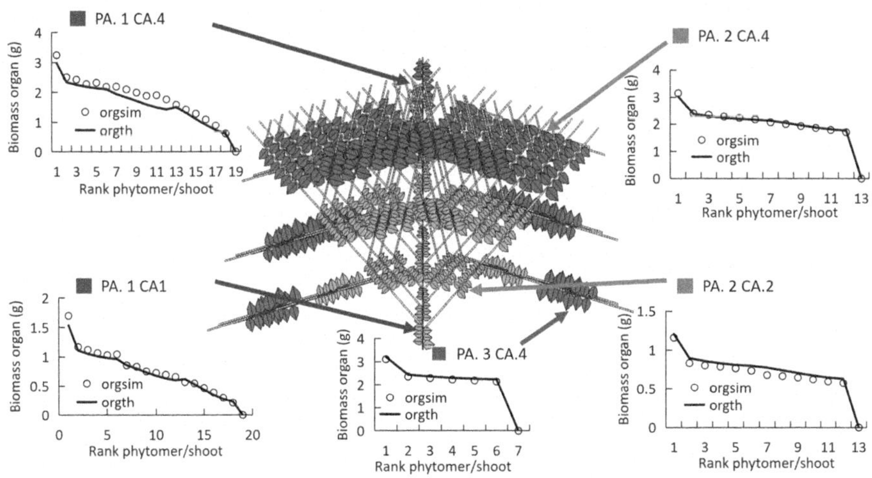

Fig. 6.34 Display of simulated (○) and computed (—) simple organic series for physiological ages (PAs) (1, 2, 3), and shoot ranks (ranks 1, 4 for the stem, ranks 2, 4 for the branches, rank 2 for the twigs) expressed in four periods in an axis. Stems may have 1, 2, 3 growth units (GUs) per shoot, branches may have 1, 2 GUs per shoot, twigs have 1 GU per shoot. The averages of the simulations match well with the values computed from the potential structure. The potential structure is displayed. Structural pauses due to delays for branching are visible

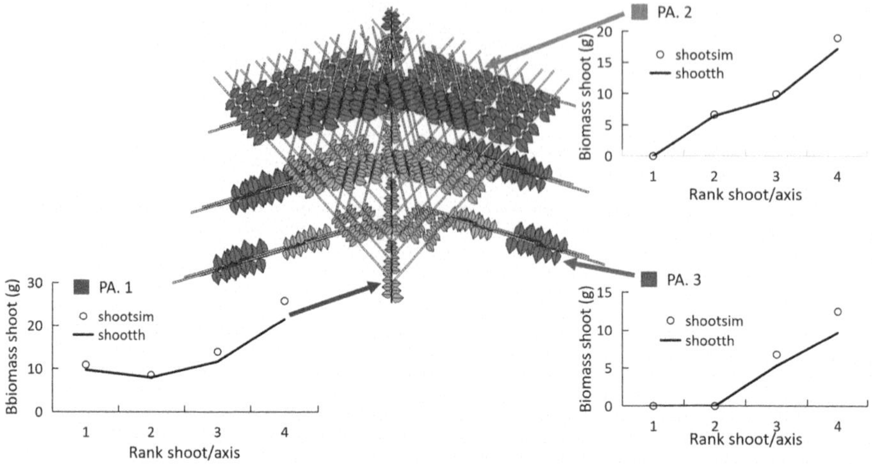

Fig. 6.35 Total organ weights per shoot (compound organic series) in simulated (○) and computed (—) shoots, by rank from the top and physiological ages for growth periods. (Adequacy is a consequence observed in Fig. 6.33 because a good fitting on simple organic series involves a good fitting on compound organic series). The potential structure is displayed

extracted by computing the potential structure. Figure 6.34 shows the good matching between the simulated and the computed simple organic series.

Consequently, there is a good matching for the cumulative weights of the organs (compartments) in the shoots that form compound organic series along the axes (Fig. 6.35), both for average simulations and computations.

Figure 6.36 shows four stochastic simulations of the computational plant (an associate potential structure is displayed in Fig. 6.34). The variability is due to a variable number of GUs per shoot, and a variable number of phytomers per GU.

6.5.8 Case of Growth Units with Negative Binomial Distributions of the Number of Phytomers

If the distributions of the number of phytomers in the GUs are positive binomial laws, then it is relevant to assume a *Bernoulli* process for the development. All GUs have the same development time *N*, and the variability comes from the *Bernoulli* b_u parameter. This is the case, for example, for maple, walnut, horse-chestnut trees. This case is easily detected because the mean *M* of the distribution of phytomers per GU is lower than the variance *V*. In the opposite case ($V > M$), we find ourselves with a negative binomial law. The consequence is that the parameter *N* is no longer fixed and follows a distribution. Let us suppose that this distribution is a positive or negative binomial distribution (*N*, *b*). To each $N \subset (N, b)$, a *Bernoulli* process with

Fig. 6.36 Four stochastic simulations of a four-period polycyclic structure with neoformed growth units (GUs). The number of GUs per shoot varies according to a *Bernoulli* process (parameter b). The number of phytomers per GU varies according to a *Bernoulli* process (parameter b_u)

parameter b_u can be associated, which gives a particular binomial distribution (N, b_u). Probability theory teaches us that the distribution of the number of phytomers per GU is a compound binomial distribution of the same nature and parameter (N, $b*b_u$).

It is practically impossible to distinguish the separate actions of parameters b and bu.

If $V < M$ then it makes sense to choose fixed N; $b = 1$; $b_u < 1$ and the (N, b) distribution is binomial positive.

If $V > M$ then it makes sense to choose N variable $b < 1$; $b_u = 1$; and the (N, b) distribution is binomial negative.

In this last case, the notion of a potential structure no longer exists. It must be replaced by a collection of stochastic structures, from which we extract the average organic series. The organic series obtained by simulation are fitted to the target

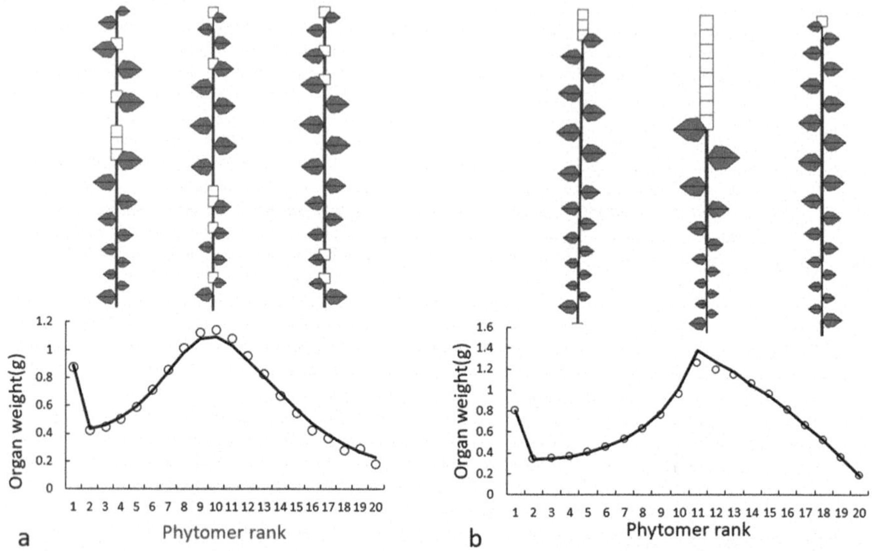

Fig. 6.37 Comparison between two development modes that give the same distribution of phytomers per growth unit (GU) but different organic series. (**a**) The breaks are randomly distributed. (**b**) The pauses are at the end of the GU. Three stochastic simulations are displayed per mode. The organic series obtained are different

organic series extracted from the observations. The inverse method of computing the parameters remains possible, but it is more computationally heavy. This is illustrated by two examples which have the same distribution of the number of phytomers per GU that follows a positive binomial distribution (20, 0.8). But the development is done in two different ways.

Case where $V < M$

In this case, the GUs have the same development time, the pauses are randomly distributed (Fig. 6.37a). The number of phytomers/GU follows the binomial distribution (20, 0.8). The fitting (bottom up) of the simulated organic series is done using a potential axis.

Case where $V > M$

In this case, the GUs have different development times, the pauses are positioned at the top of the GU to complete its development (Fig. 6.37b). The number of phytomers per GU follows the same binomial distribution (20, 0.8).

Although the phytomer distributions are the same in both cases, the growth and development are different as shown in the organic time series (Fig. 6.37). The fitting of the simulated organic series is done using a collection of potential axes (50 repetitions).

These two cases are discussed in Chap. 10 with Maple and Mahogany trees.

6.6 Conclusion

Considering physiological ages, any plant structure can be broken down successively into a small number of axis categories, made as Russian dolls, consisting of shoots, GUs, and phytomers.

For plants with continuous growth (e.g., coffee trees), there is a single development scale: the phytomer. For plants with rhythmic growth and monocyclic (e.g., maple trees), two development scales appear: the phytomer and the GU. The GU can be used as a compartment for observation and measurement. For plants with rhythmic growth and polycyclism (e.g., walnut trees), three development scales exist: phytomer, GU and annual shoot; the shoot can be used as a compartment for measurements.

Finally, a small number of simple or compound organic series (equal to the number of physiological ages multiplied by the number of organ types) are built that contain all the information related to the history of source–sink relationships during plant growth. So, the values of Q, D, and Q/D for each DC are computed from the static description of plant architecture.

The interest of building organic series is that they are independent of the complexity of the architectural model with its reiterations and of the stochastic features. Organic series are identifiable on the plant by the botanists or agronomists. The sampled data are sufficient to estimate the weight of the plant without measuring all organs' weights.

The missing biomass due to mortality becomes computable because of estimated probabilities coming from bud functioning. This method, therefore, does not require a full description of the tree architecture. So, for each experiment, a sample of reasonable size for measurements can be established. Recalled that in establishing organic series, only plants of the same chronological age should be considered, and dead axes should be excluded. Although stochastic development increases the complexity of the model, the types of measurements to be performed on stochastic plants are the same as on deterministic plants, which is an advantage for the users.

This chapter shows that the use of potential structures is suitable for many kinds of plant structures that cover most of the architectural stochastic models of plants found in nature.

In all studied cases, the average of the organic series obtained from Monte Carlo simulations matches well with the theoretical computed organic series extracted from the potential structure. This is a necessary condition to set up an inverse method that enables to compute the hidden functional parameters of the plant system. In the case of GUs with negative binomial phytomer distributions, the potential structure is no longer unique, and the distribution of chronological stochastic structures must be considered, which implements the inverse methods more cumbersome. This case will be presented in Chap. 10.

The stochastic simulation of the model is of great interest. It makes it possible to obtain computational plants that are pure instantiations produced by the GreenLab model. The efficiency of data assimilation and inverse methods can thus be tested under the best conditions. Once the fitting of Monte Carlo simulations works properly, it is possible to cope the method with natural plants to characterize their functional parameters. This enables us to assess the good fitting between the plants and the GreenLab model. This is the subject of Chaps. 9, 10, and 11.

References

Cieslak M, Seleznyova A-N, Hanan J (2011) A functional–structural kiwifruit vine model integrating architecture, carbon dynamics and effects of the environment. Ann Botany 107(5):747–764

Cieslak M, Khan N, Ferraro P, Soolanayakanahally R, Robinson S-J, Parkin I, McQuillan I, Prusinkiewicz P (2022) L-system models for image-based phenomics: case studies of maize and canola. In Silico Plants 4(1):diab039

de Reffye P, Kang M-Z, Hua J, Auclair D (2012) Stochastic modelling of tree annual shoot dynamics. Ann Forest Sci 69:153–165

Kang M-Z, Cournède P-H, de Reffye P, Auclair D, Hu B-G (2008) Analytical study of a stochastic plant growth model: application to the GreenLab model. Math Comput Simul 78(1):57–75

Kang M-Z, Hua J, de Reffye P, Jaeger M (2016) Parameter identification of plant growth models with stochastic development. IEEE International Conference on Functional-Structural Plant Growth Modeling, Simulation, Visualization and Applications (FSPMA2016), Qingdao, China, pp 98–105. https://doi.org/10.1109/FSPMA.2016.7818294

Kang M-Z, Hua J, Wang X-J, de Reffye P, Jaeger M, Akaffou S (2018) Estimating sink parameters of stochastic functional-structural plant models using organic series-continuous and rhythmicdevelopment. Front Plant Sci 9:1688. https://doi.org/10.3389/fpls.2018.01688

Chapter 7
Modeling of Acropetal and Basipetal Development in Herbaceous Plants

Mengzhen Kang and Philippe De Reffye

Abstract This chapter presents the case of herbaceous plants whose growth and development show great architectural variability. The GreenLab model is adapted to account for these differences, with herbaceous plants forming new phytomers much faster than woody plants. The chapter also discusses the phenological stages of rosette-forming herbaceous plants, which have a distinct vegetative phase followed by a reproductive phase. The expansion of phytomers does not immediately follow their creation; instead, branches remain preformed until flowering is triggered. In contrast to woody plants, the growth and development of organ cohorts in herbaceous plants are not synchronized. The expansion of preformed organs is delayed and depends on their creation time, their type, and their position within the plant's topological structure. The organic series need to be adapted using simple or compound organic series. This complexity necessitates calculations to be performed in list mode rather than matrix mode, as previously described. Despite these challenges, the computational cost remains manageable due to the reduced development cycle of herbaceous plants compared to trees.

Keywords Herbaceous · Acropetally and basipetally of inflorescences · Simple and compound organic series

---

Electronic supplementary material The online version of this chapter (https://doi.org/10.1007/978-981-97-7541-5_7) contains supplementary material, which is available to authorized users.

M. Kang (✉)
State Key Laboratory for Multimodal Artificial Intelligence Systems, Institute of Automation, Chinese Academy of Sciences (CASIA), Beijing, China

School of Artificial Intelligence, University of Chinese Academy of Sciences, Beijing, China
e-mail: mengzhen.kang@ia.ac.cn

P. De Reffye
CIRAD, UMR AMAP, Montpellier, France

AMAP, Univ Montpellier, CIRAD, CNRS, INRAE, IRD, Montpellier, France

P. De Reffye et al. (eds.), *Plant Model Greenlab for Botany and Agronomy*,
https://doi.org/10.1007/978-981-97-7541-5_7

This chapter presents the case of herbaceous plants whose growth and development require specific adaptations of the GreenLab model (Kang et al. 2010). It should be recalled that the term "development" refers to the creation of phytomers due to the functioning of meristems, while the term "growth" refers to the expansion of these organs as a result of the acquisition of biomass.

Herbaceous plants show great architectural variability. They are characterized by the absence of woody secondary growth. Various greenhouse crops such as tomatoes and cucumbers, which are single-stemmed plants subjected to systematic pruning of axillary meristems, have axial flowering, indefinite development, and form stems of several meters long that can be rolled up. Many field crops, such as rapeseed and chrysanthemums, form leaf rosettes at the base, topped by a well-developed inflorescence whose axes have a terminal flowering.

The development of herbaceous stems, such as those of maize and sunflowers, is fast (about 1–2 new phytomers per day). In woody plants, such as coffee or cotton, instead, at a rate of about 1–2 new phytomers every 15 days. The effects of the development speed are significant on its growth. For the same duration of functioning in calendar time, the number of DCs will be much higher in an herbaceous plant than in a woody plant. The annual growth of a sunflower over a few months requires 120 DCs, whereas a coffee tree only requires 20 DCs per year. To simplify, we suppose that a given plant has a growth cycle associated with a DC. If this principle of the GreenLab model remains unchanged from one species to another, the duration of the respective cycles in thermal or calendar time can be subject to significant variations. But we can define a different rhythm of development. For example, a rate $w = 0.25$ defines four growth cycles for one development cycle. This makes it possible to refine the consideration of climate data.

Phenologically, rosette-forming herbaceous plants undergo two main developmental phases: a vegetative stage followed by a reproductive stage. In the vegetative phase, all phytomers are set up according to the rules of the botanical automaton, but their expansion does not immediately follow their creation. Branches remain preformed, pending expansion until flowering is triggered.

If the expansion of growing organs is immediate and coincides with their creation by development, or is delayed by a fixed number of DCs, the growth-related organ cohorts are the same as those generated by the development. But in the case of herbaceous plants, the expansion of the preformed organs of the branches is triggered with a delay that depends not only on the time of their creation but also on their type and position within the topological structure generated by the development schedule. In other words, the development of organ cohorts no longer coincides with the growth of organ cohorts. This prevents the decomposition into substructures based on the notion of PA, as described in Chap. 2, so the calculations of the growth can no longer be performed in matrix mode; they must be performed in list mode. On the other hand, since herbaceous plants have a reduced development compared to trees, the cost of computation remains reasonable. If the preformed branches expand top-down from the top of the stem, a basipetal pattern (*chrysanthemum*) is presented;

instead, if the expansion is bottom-up from the basis, an acropetal pattern (*Spilanthes*) is presented.

7.1 Study of Herbaceous Development

Most often, in herbaceous plants, the axes have an apical flowering that stops their development. If apical flowering depends only on the developmental stage of the leafy axis, a structure shown in Fig. 7.1a is obtained. But more frequently, under the influence of an ontogenic age, see Chap. 2 (Sect. 2.2.8), flowering is synchronized throughout the structure by the simultaneous transition of apical vegetative meristems into floral meristems, which stops the plant's development (Fig. 7.1b).

The development of herbaceous architecture is visible only after the branches have ended their expansion. In the basipetal pattern, the most developed axillary branches, i.e., those with the most phytomers, are located at the lowest position on the stem, because they appear first. These branches often remain embryonic in the preformed state, awaiting expansion. However, their phytomers can be counted with a binocular microscope, and their structures are already constructed. There is therefore an apparent development that corresponds to the visible plant structure, and also a complete development that takes into account the preformed parts that remain hidden and awaiting expansion.

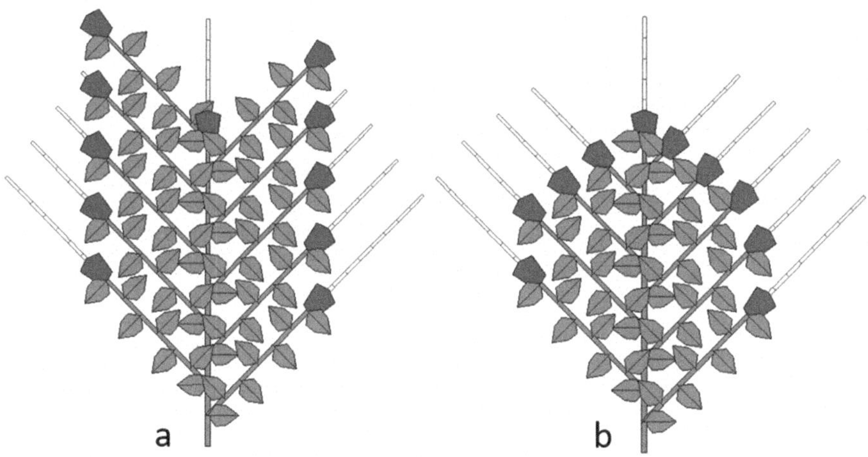

Fig. 7.1 Two patterns of development of herbaceous plants. (**a**) Complete development of a structure with chronologically limited axes by terminal flowering. (**b**) Complete development of a structure with axes limited by terminal flowering by an ontogenic age. When the main axis flowers, the other axes flower together even if they have not completed their developments. Void entities show the cycles of meristem functioning

7.1.1 An Example: The Apparent Development of Rapeseed Architecture

In herbaceous plants, the classic development pattern often follows a marked basipetal developmental pattern. After a leafy stage in a rosette form, the stem lengthens by creating new phytomers and eventually by lengthening the internodes, pushing back the leaves of the rosette, then sets up a terminal inflorescence. We consider here the example of rapeseed that follows this pattern with a raceme inflorescence. The expansion of preformed branches occurs starting at the apex and spreads to the base, inversely to the development pattern. The most basal branches remain embryonic (Fig. 7.2), because at the end of the growth, most leaves are senescent, starting with the rosette, which stops photosynthesis.

The number of vegetative phytomers before the flowering of the main axis can be observed on the second-order branches, which helps to highlight the structure of the floral crown. Flowering is synchronized on all axes although their apparent development depends on their rank of the stem.

Fig. 7.2 Rapeseed architecture. Three plants are of the same growth stage with different apparent developments. Each photograph is associated with a corresponding apparent development schedule. The stressed plant (**a**) has made only a leafy stem with a terminal raceme. The plant grown at normal density (**b**, equivalent to 30 plants per m^2) have branches to order 2. The isolated plant (**c**) branches to order 3. The series of branches of plant (**b**) is spread out in subfigure (**d**); it shows on the left the terminal raceme, then the series of basipetal branches, ended by a raceme. The number of leaflets on the branch depends on the distance from the branch to the apex of the stem. The last visible branches are at the beginning of the expansion. Only the first two leaves are deployed; the entire branch is only visible in the terminal meristem using a binocular microscope. The lower internodes of the stem have lost their leaves, and the branches remain embryonic in the axillary meristems. (Photos © P. de Reffye, CIRAD)

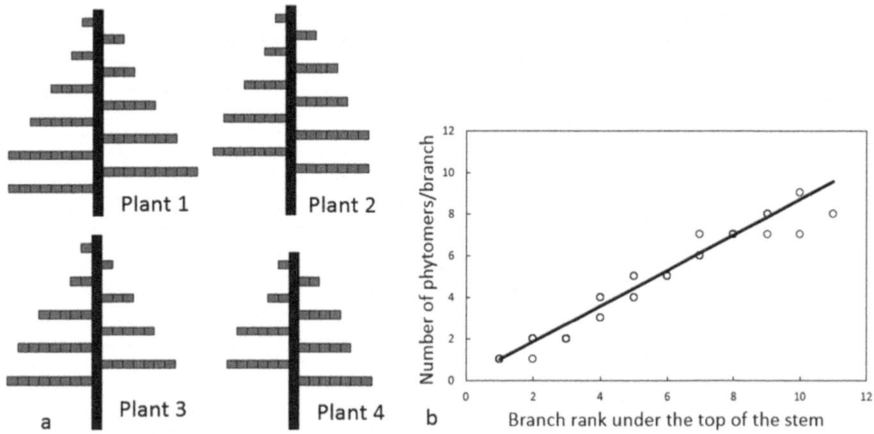

Fig. 7.3 Development and growth of rapeseed. (**a**) Topological structures of four crowns of rapeseed. Only the phytomers that carry leaves are displayed. All axes have terminal racemes that have formed at the same time and stop the vegetative development. The branches at the base can be detailed only under a binocular microscope. (**b**) Evolution of the number of phytomers of the order-2 branches from the top of the stem in rapeseed crowns in Fig. 7.3a. The linear increase observed comes from the uniform development of the crown during its embryonic stage and the synchronization of the flowering of the axes

Local growing conditions determine the extent to which the potential development is complete in embryonic form. Only a portion of the preformed structure will expand in a marked basipetal pattern, resulting in apparent development. Thus, under stressful conditions, the plant may not branch (Fig. 7.2a). On the other hand, an isolated plant can express its entire development in a favorable environment (Fig. 7.2c). Under normal growing conditions, the density per m^2 limits branching to order 2 (Fig. 7.2b).

It is difficult to count the number of phytomers branching beyond the 12th rank under the apex of the stem because they have not yet begun their expansion. It is noted that the number of phytomers per lateral axis increases linearly with rank, reaching a maximum of around eight phytomers. Figure 7.3a shows the topological structures of the measured plants that correspond to Table 7.1. And Fig. 7.3b shows the number of phytomers in branches along the stems.

The stochastic aspect of development is visible although not very marked. The analysis of crowns (Chap. 3), from Table 7.1, gives the following *Bernoulli* parameters for the meristems of second-order stems and branches: $b_1 = 0.92$, $b_2 = 0.92$ with a rhythm ratio $w = 0.89$.

This slight stochastic behavior disturbs the regularity of the structure of the plant. The rhythm ratio shows an effect of the PA related to the order of branching.

The same analysis can be carried out on the floral subcrowns formed by the order 3 branching of order 2 on the isolated plant, as shown in Fig. 7.2c, which carries three orders of branching. The basipetal pattern is found recursively on order 2. Only the very first branches of order 3 under the apex have their expansion, while the

Table 7.1 Number of phytomers on second-order branches according to their rank from the top of a rapeseed stem (four observations in the field)

Rank under apex	Stem 1	Stem 2	Stem 3	Stem 4
1	1	1	1	1
2	2	1	2	2
3	2	2	2	2
4	3	3	4	4
5	4	5	4	5
6	5	5	5	5
7	6	7	6	7
8	7	7	7	7
9	8	8	7	7
10	9	?	?	?
11	8	?	?	?
12	?	?	?	?

Below rank 8, phytomers can only be counted with a binocular microscope. Below around rank 10, the branches are too embryonic to be counted (noted with?)

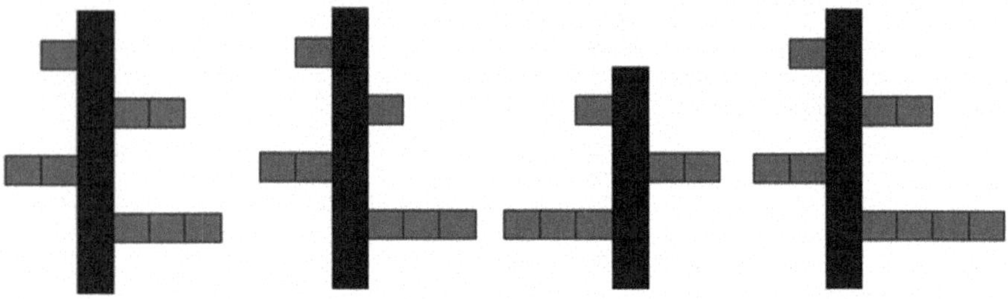

Fig. 7.4 Examples of subcrowns of order 2 branches and measured on rapeseed plants

others remain embryonic. Note that the number of phytomers tends to reach a maximum of 4 (Fig. 7.4).

The analysis of subcrowns for order 2 gives similar values of *Bernoulli* parameters to those of order 1: $b_1 = 0.92$, $b_2 = 0.96$, and $w = 0.89$.

Note: in the analyzed sample of rapeseed, there were no visible order 4 branches, as they remained embryonic. But they may appear under optimal cultural conditions (isolated and well-irrigated plants).

7.1.2 *Generic Schedule of Herbaceous Growth and Development*

Other herbaceous plants, such as chrysanthemum or *Spilanthes*, show the same types of development and growth with variations. We can deduce a generic development schedule for rosette-form herbaceous plants:

- Meristems have a continuous and slightly stochastic functioning. They set up a preformed structure with well-identified types of axes, in particular by a rhythm ratio and by a maximum number of phytomers;
- Organ expansion of the same phytomer is not generally synchronized. The leaves appear first, followed by the elongation of internodes that space the successive leaves, and finally, the terminal inflorescence forms at the end of the axis;
- Depending on their location in the basipetal structure, the organs will trigger their expansion after a delay that depends on their position in the topological structure rather than the chronological age of their creation.

Figure 7.5a shows the sequential development of a branched plant with three physiological ages (PAs). Physiological ages 1, 2, and 3 flowers after 10 DCs, 6 DCs, and 3 DCs, respectively. At age 9, the crown of the structure is still developing and only the axes that have reached the limit of their development have a terminal inflorescence. At age 10, all remaining vegetative meristems flower simultaneously under the influence of ontogenetic age, and vegetative development is complete. If the meristems are transformed into simple flowers, like in chrysanthemum, the structure does not evolve anymore. If they turn into a raceme, like in rapeseed, an entirely floral development will continue, fed by the underlying vegetative structure.

Figure 7.5b shows a branched plant with three PAs whose development and growth are desynchronized. The plant has the same development as that of Fig. 7.5a; the type and number of organs are the same. Their expansion time is 20 DCs and the functioning time of the leaves is 25 DCs. At DC 10, the development of the structure is complete, identical to that of Fig. 7.5a, but the preformed branches remain hidden in the embryonic state at the axils of the leaves, waiting for the expansion. At DC 35, all the leaves are dead and therefore the growth is complete. The plant is observed in terms of apparent development because the expansion of phytomers does not simply follow the order of their creation.

The order of organ expansion of the same phytomer is in this example following the leaf-internode-flower sequence, although they were created at the same time. Figure 7.5b shows that: at DC 6, only the leaves are visible on the main stem, and the plant is in the rosette stage; at DC 10, the stem elongates, the leaves are repelled and the flowering of the last phytomer is apparent; at DC 18, the leaves of the order 2 branches are visible and functional at the axils of the leaves on the stem; at DC 24, these leaves are pushed out of the stem by the expansion of the internodes of the second-order axes, which are set up in a basipetal way. The apical flowering is apparent, and the axes of order 3 are put in place in a basipetal pattern on the order 2. Leaves die from the bottom because their functioning time (25 DCs) is exceeded; at DC 35, all leaves are dead, and the growth is stopped. The entire topological structure on DC 10 (Fig. 7.5a) is eventually visible, but the plant shape is different because of the de-synchronization between development and growth.

In a basipetal developmental pattern, after a rosette stage, the stem elongates and pushes the leaves from the bottom to the top. The lengthening of order 2 axes appears in the inverse order of their creation by the development process, that is to

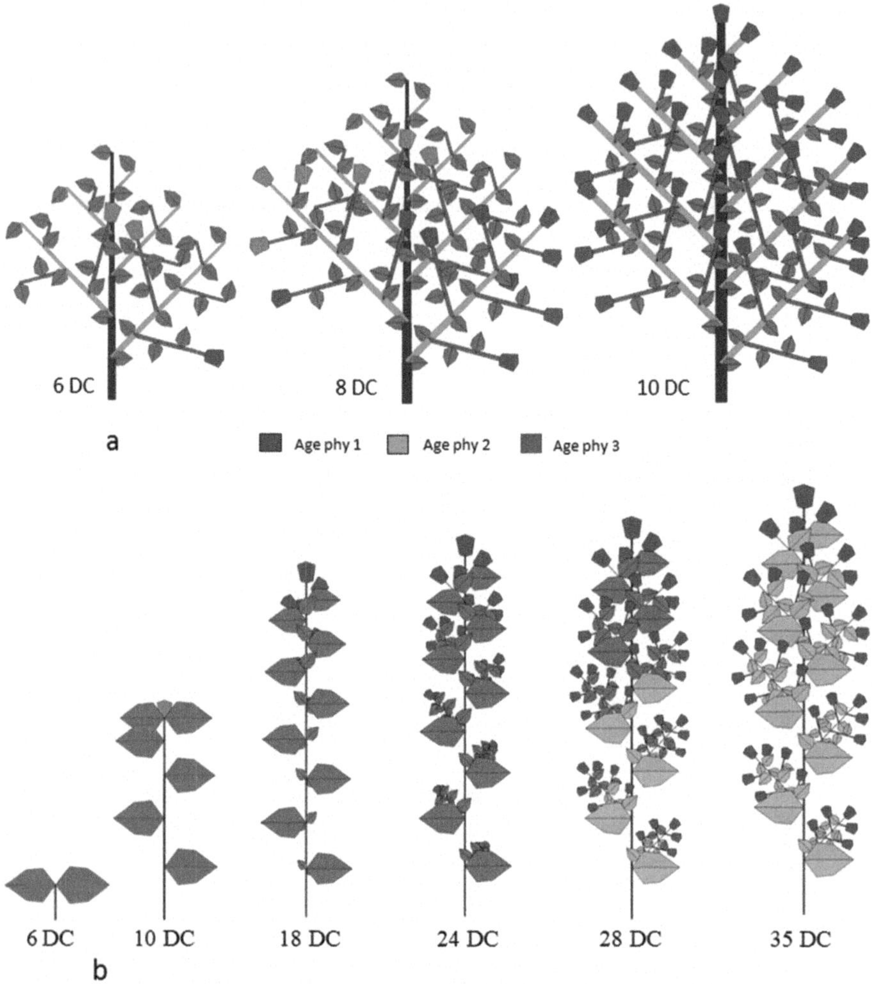

Fig. 7.5 Growth of herbaceous plants with the basipetal pattern. (**a**) Sequential development of the herbaceous topological structure according to the order of organ creation in cohorts organized according to chronological ages (CAs) and physiological ages (PAs). There are three PAs and three developmental stages (6, 8, 10 DCs). (**b**) Growth of an herbaceous plant with apparent development from the same topological structure as shown in Fig. 7.5a, but organ expansion follows a basipetal pattern, leading to the apparent disappearance of cohorts

say, from the apex to the base according to the basipetal schedule. The same phenomenon is reproduced on the order 3 axes recursively concerning order 2 (Fig. 7.6b). Although organs of the same type have the same sink function regardless of their PA, because of the expansion delay laws related to the basipetal pattern, organs of the same cohort for the development have different dimensions. Thus, the leaves of the order 2 axes can be very small, although they are the same leaves from a functional point of view as those of the main stem. As a consequence,

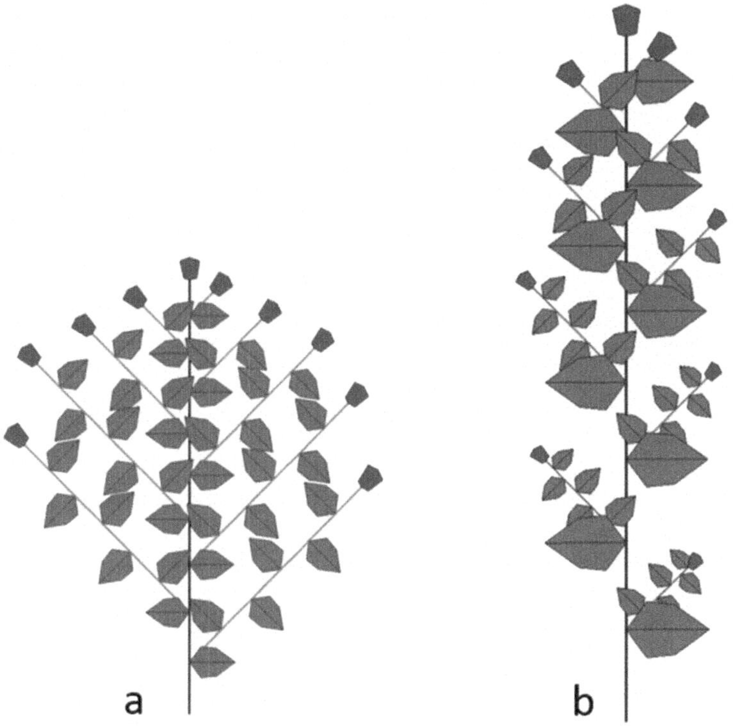

Fig. 7.6 Two plants on the same scale and sharing the same topological structures and source-sink parameters. Figure 7.6a and b differ only for the expansion delay laws of organs. Figure 7.6a corresponds to a plant whose organ expansion immediately follows its creation by meristems (neoformation). Organ cohorts of the same kind at each DC are set in place by the development schedule. Figure 7.6b shows a plant whose potential development is identical to that of Fig. 7.6a, but its apparent development is different because of the delay laws in organ expansion that depend on the organs' positions in the plant structure. There are no more organ cohorts during the growth process. The plant in Fig. 7.6b has a stronger growth than the plant in Fig. 7.6a, because the source organs have an expansion that precedes that of the sink organs

the organic series, as observed in shrubs, are no longer operative, since the cohorts of organs are no longer homogeneous. On the other hand, along the stem, the organs of the phytomers constitute simple organic series, and the branch organ compartments constitute compound organic series, which are measurable.

In herbaceous plants, the order of organ expansion (leaf-internode-fruit) in a phytomer favors the appearance of the source organs before that of the sink organs, and consequently, the basipetal pattern of the branching favors the source production. From the point of view of the sink, if the organs expand as soon as they are created, the demand for the plant will increase rapidly (Fig. 7.6a), which slows down growth. In the basipetal pattern, the appearance of the source-sink organs (the leaves) is ahead of the sink organs (internodes and fruits), which keeps the demand at a lower level. As illustrated in Fig. 7.6b, it is the law of delay, or the source-sink ratio

Q/*D*, that makes the difference between the dimensions of the organs of the same branching order, rather than the sink strength. This strategy gives priority to the sinks of the source organs and is well adapted to herbaceous plants, which have a short life span and need to grow rapidly in height. This is observed in herbaceous plants such as rapeseeds and chrysanthemums. This is equivalent to the single-stemmed stem size experiments of the branched plants studied in Chap. 5. As a consequence, the growth of the plant in Fig. 7.6a is weaker than that of the plant in Fig. 7.6b.

7.2 Generic Equations Controlling Delays in Organ Expansion

It is assumed that the main stem has a development limited to T_{1mx} DCs at the onset of flowering. At this point, the terminal meristem turns into a flower. If the developmental system is ontogenetic, all the apical meristems of the branches also turn into flowers simultaneously, and all the axes reach their definitive development. These depend on the position of the branching. Thus, the number of phytomers of an order 2 axis appearing at rank i on the stem is $T_2(i) = T_{1mx} - i$. That of an axis of order 3 appearing at rank j on an axis of order 2 appearing at rank i on the stem is: $T_3(i, j) = T_2(i) - j$. That of an axis of order 4 appearing at rank k on an axis of order 3 which appeared at rank j on the branch of order 2 is: $T_4(i, j, k) = T_3(i, j) - k$, etc. Here we limit the branching to order 3. However, the observations show that the axes of the architecture of the herbaceous plants do not have indefinite development and are limited in chronological age. The limits of development of the axes according to their order of ramifications are denoted as T_{1mx}, T_{2mx}, T_{3mx}, T_{4mx}, etc.

It is assumed that the delay in the expansion of an organ carried by a phytomer depends on its position in the structure, identified by the successive positions of the branches and the respective ranks of the phytomers carrying them (Fig. 7.7).

In herbaceous plants, the order of branching can be identified as the PA in simple cases. Branching can be identified by its position on the bearing axis. In the case of basipetal development, the further from the apex, the more delayed its expansion will be. The phenomenon is recursive and propagates across branching orders. The following concepts are represented in Fig. 7.7.

The term "shortest distance to seed" (SDS) of a phytomer is called the shortest path from seed to the phytomer, which corresponds to its date of creation. This path is equivalent to a path prefixed in the theory of graphs (mathematical trees). Let i, j, k, x denote the lengths of the DC paths belonging respectively to the branch orders 1, 2, 3, 4 (Fig. 7.7).

Similarly, we define the "shortest distance to crowns" (SDC) path obtained from the seed but passes through all the apices of the axes bearing the phytomer until this is defined (Fig. 7.7).

Thus, the expansion time of an organ, defined by its SDS, depends on the SDC. The SDC is derived from the exploration of the potential structure starting from the

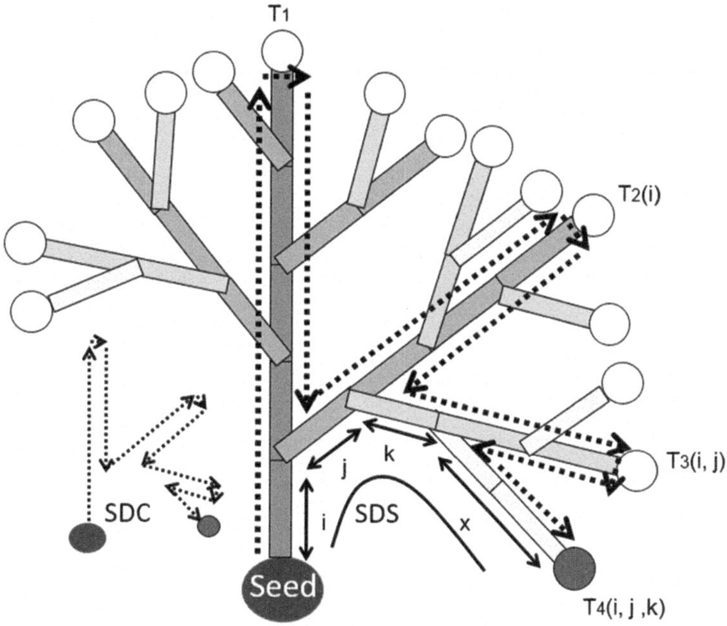

Fig. 7.7 Representation of the position of an organ within a structure. An organ can be spotted by the path of the smallest distance to the seed (SDS), starting from the base of the plant and extending to the organ itself. At each order of branching, we record the number of phytomers of the prefixed segments. Thus, the organ O (red) occupies the position $O(i, j, k, x)$ in a structure with four orders of branching. The length of the stem in the phytomer is T_1. That of an axis of order 2 depends on its position on the stem, it is denoted as $T_2(i)$. Similarly, the length of an axis of order 3 is denoted as $T_3(i, j)$, and so on. The creation of the organ O is done at the age $t = i + j + k + x$. The moment of initiation of an organ expansion according to its position in the plant structure depends on the path of the Smallest Distance to the crown (SDC). Starting from the stem base, this path is counted by systematically reaching, in priority, the apices of the axes before turning back to the position of ramification that leads to the organ (Color figure online)

base, in such an order that requires first to reach the apices of the axes of development, before turning back to the ramification that leads to the organ (Fig. 7.7). We can associate this exploration with a speed of path. The expansion delay in the GreenLab model is computed from a function whose parameters are: T_{1mx}, $T_2(i)$, T_{2mx}, $T_3(i, j)$, T_{3mx}, i, j, k, x. Note that in a whorl, the organs have the same prefixed coordinates i, j, k, x. $T_2(i)$ and $T_3(i, j)$ are, respectively, the lengths of the axes (in the number of phytomers) of order 2 carried by the phytomer i of the stem, and order 3 carried by the phytomer j of the 2th-order ramification.

This method is an alternative to the one that comes from ecophysiology and is already used in the L-systems (Prusinkiewicz et al. 1988). The L-system consists in spreading step by step a signal coming from the apex in the architecture which gains the ramifications to trigger flowering. But to propagate a signal in a topological structure by simulation is more computationally intensive than to calculate the delay of organ expansion by a system of equations. Moreover, the signal simulation does not make it easy to apply the methods of parameter estimation (Chap. 8).

Thus, consider the organ located at the prefixed coordinates (*i*, *j*, *k*, *x*) in Fig. 7.7, according to the definition, SDS and SDC are:

$$SDS = i + j + k + x.$$

$$SDC = T_1 + [T_1 - i] + T_2(i) + [T_2(i) - j] + T_3(i,j) + [T_3(i,j) - k] + x \quad (7.1)$$

The organ expansion time, T_d (*i*, *j*, *k*, *x*), is to be associated with the SDC.

7.2.1 Simple Case of Organ Expansion Delay

In a simple case, we consider the basipetal flowering sequence where flowers start growth one by one.

Most often the branches have a limited number (bound) of phytomers beyond a certain rank under the apex of the bearing axis, which is denoted as T_{mx}.

Length of order 2 developmental axes, $T_2(i)$, bounded by T_{2mx}, is dependent on its rank *i* from the stem base:

$$T_2(i) = \begin{cases} T_1 - i & i > T_1 - T_{2mx} \\ T_{2mx} & i \leq T_1 - T_{2mx} \end{cases} \quad (7.2)$$

Length of order 3 development axes, $T_3(i, j)$, bounded by T_{3mx}, is dependent on the rank *j* from the base of axis 2, which located at rank *i* from the base of the stem:

$$T_3(i,j) = \begin{cases} T_2(i) - j & j > T_2(i) - T_{3mx} \\ T_{3mx} & j \leq T_2(i) - T_{3mx} \end{cases} \quad (7.3)$$

The system is recursive and we get $T_4(i, j, k)$, similarly.

Considering Fig. 7.7, it is determined that the expansion of an organ located at the position (*i*, *j*, *k*, *x*) will be triggered, according to the formula, when its age is greater than T_d:

$$\begin{aligned} T_d(i,j,k,x) &= SDC = T_1 + T_1 - i + T_2(i) + T_2(i) - j + T_3(i,j) + T_3(i,j) - k + x \\ &= (2T_1 - i) + (2T_2(i) - j) + (2T_3(i,j) - k) + x \end{aligned} \quad (7.4)$$

Furthermore, the organ was created at time T_a, which is the length of the SDS path that leads to the organ whose position is: T_a (*i*, *j*, *k*, *x*) = *i* + *j* + *k* + *x*.

The local delay $T_g(i, j, k)$ in the growth of this organ since its creation is:

$$T_g(i,j,k,x) = SDC - SDS = T_d(i,j,k,x) - T_a(i,j,k,x)$$

It is a linear function that relates the distances of the branching points to the apices of the orders considered. The number of indices used depends on the order of branching:

For order 2, we have: $T_g(i) = T_d(i,j) - T_a(i,j) = 2(T_1 - i)$;

For order 3, we have:

$$T_g(i,j) = T_d(i,j,k) - T_a(i,j,k) = 2[(T_1 - i) + (T_2(i) - j)] \tag{7.5}$$

For order 4, we have:

$$T_g(i,j,k) = T_d(i,j,k,x) - T_a(i,j,k,x) = 2[(T_1 - i) + (T_2(i) - j) + (T_3(i,j) - k)]$$

Therefore, all the organs of the same type on the same axis have the same delay in expansion regardless of their rank x; the delay only depends on the position of this axis in the topological structure.

7.2.2 General Case

Since the organs can behave differently according to their type, it seems advisable to assign go parameters associated with the expansion of a type of organ "o" whose coordinates are i, j, k, x according to the branching order so that you can modify the SDC path. Moreover, the flowering sequence is not necessarily sequential.

The following notations are used: $o = l$ for the leaves, $o = e$ for the internodes and $o = f$ for the fruits. The expansion time To_g of the organ o after its creation depends on its position in the SDS coordinates in the topological structure according to the order of branching and a constant To_c, we have:

$$\begin{aligned}
\text{order } 1 &: To_g = To_c \\
\text{order } 2 &: To_g(i) = go_2(T_1 - i) + To_c \\
\text{order } 3 &: To_g(i,j) = go_2(T_1 - i) + go_3(T2(i) - j) + To_c \\
\text{order } 4 &: To_g(i,j,k) = go_2(T_1 - i) + go_3(T_2(i) - j) + go_4(T_3(i,j) - k) + To_c
\end{aligned} \tag{7.6}$$

The parameters go_i are real numbers that generalize the equations from the previous paragraph and modulate the expansion delay for a type of organ denoted as o, according to the order of branching 2, 3, or 4. The order of the stem being noted as 1. The axial organic series, whose homogeneity according to the PA is no longer assured, may have very different profiles, depending on whether they are leaves, internodes, or fruits by changing the parameters go_i.

This formulation of the expansion delay law provides great flexibility in the apparent development of plant architecture from the same pattern of potential development.

7.2.2.1 The go_i Parameters of the Organs are Identical in the Same Phytomer

In this case, the organs on the same axis have similar growth curves.

Figure 7.8 shows the different reactions of plant growth according to the parameter values *go* (with $To_c = 0$), with all the other botanical or functional parameters being identical.

For $go = 0$, the plant has no delay in elongation for its phytomers, characterized by development in neoformation. This is the normal functioning of a shrub plant. The demand increases too fast compared to the supply of photosynthesis, so the plant is stunted. The whole topological structure is given in Fig. 7.8a.

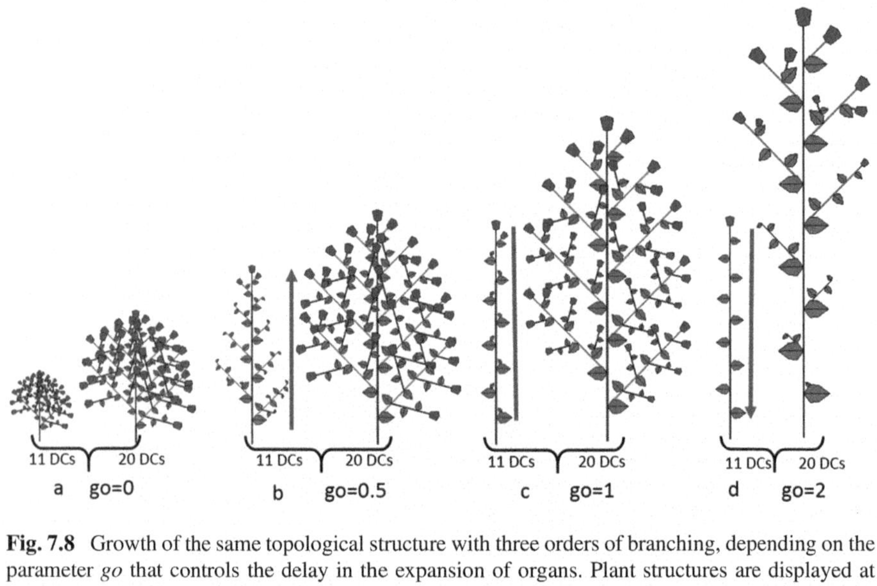

Fig. 7.8 Growth of the same topological structure with three orders of branching, depending on the parameter *go* that controls the delay in the expansion of organs. Plant structures are displayed at 11 DCs and 20 DCs of growth. At 10 DCs, all the development is completed. All organs are created, but some remain invisible, waiting for their expansion to be triggered according to the delay law. After 10 DCs, only the growth of phytomers occurs. The plants (**a**, **b**, **c**, **d**), and (**e**) are at the same chronological age and on the same scale. The functioning time of the leaves t_a is 25 DCs, and the duration of organ expansion is $t_x = 20$ DCs. The sink of the leaves and internodes is constant at the value 1, and that of the fruits is constant at the value 5. (**a**) $go = 0$. Phytomers have immediate expansion following their creation. The plant is stunted. (**b**) $go = 0.5$. Branches start their expansion one by one from the base. (**c**) $go = 1$. All branches start simultaneously their expansion at T_l DCs. (**d**) $go = 2$. Branches start their expansion one by one from the top. It will be noted that the law of delay modifies the organ sizes of the ramifications whereas they are the same as those of the stem

For $0 < go < 1$, the plant exhibits acropetal development. The branches appear late on the stem starting from the base. This delay allows the stem to trigger its growth more effectively (Fig. 7.8b).

For $go = 1$, all branches start simultaneously their expansion after T_1 DCs (Fig. 7.8c).

For $go \geq 1$, the plant exhibits basipetal development. The branches appear late on the stem, starting from the top after T_1 DCs. Basipetal development becomes more pronounced as go increases; the apparent development is truncated concerning the complete development. The growth of the stem is favored (Fig. 7.8c and d).

The case $go = 0$ (the expansion of the organs is immediately after their creation) is unfavorable to favor the growth of the stem, because plant demand grows too quickly due to the ramification. The bigger the go, the more the expansion of the branching organs is delayed, and the more the stem increases rapidly. At the limit (very large go), we obtain a single flowering stem at its end and no branching reaches its expansion time.

7.2.2.2 Case of Divergent Branching

The expansion of lateral structures can start from a point of the main stem both bottom-up and top-down after T_1 DCs of the growth. The parameter γo $(0 < \gamma o < 1)$ controls the location of the starting point of the expansion on the stem after T_1 DCs of the growth. Eq. 7.6 becomes Eq. (7.7).

$$\begin{aligned} \text{order 1}: & \; To_g(i) = To_c \\ \text{order 2}: & \; To_g(i) = T_1 + \gamma o_2 \mid T_1 - i \mid + To_c \\ \text{order 3}: & \; To_g(\mathrm{i,j}) = T_1 + \gamma o_2 \mid T_1 - \mathrm{i} \mid + T_2 + \gamma o_3 \mid T_2(\mathrm{i}) - \mathrm{j} \mid + To_c \\ \text{order 4}: & \; To_g(i,j,k) = T_1 + \gamma o_2 \mid T1 - i \mid + T_2 + \gamma o_3 \mid T_2(i) - j \mid + T_3 \\ & + \gamma o_4 \mid T_3(i,j) - k \mid + To_c \end{aligned} \tag{7.7}$$

This case is shown in Fig. 7.9 with parameters $\gamma o(i)$. Figure 7.9 illustrates these cases.

7.2.2.3 The Parameters go_i of the Organs are Different in the Same Phytomer

In herbaceous plants, as we have seen previously, during growth, the leaves of the branch often appear first, followed by the internodes, and finally the fruits. The parameters go of the expansion delay, allow for precise control over the phenomenon, varying according to organ type.

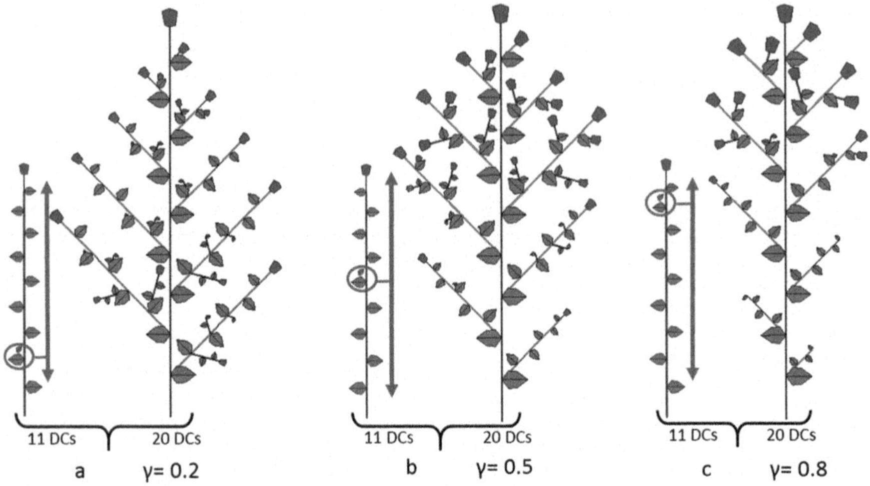

Fig. 7.9 Divergent cases of branching. After a delay of T_1 DCs, the expansion starts at a time either bottom-up or top-down from a starting point situated on the stem at $T_I{\cdot}\gamma o$ ($0 < \gamma o < 1$). If $\gamma o < 0.5$, the branching appears acropetal, and if $\gamma o > 0.5$, the branching appears basipetal. The values of the source-sink parameters are those of Fig. 7.8

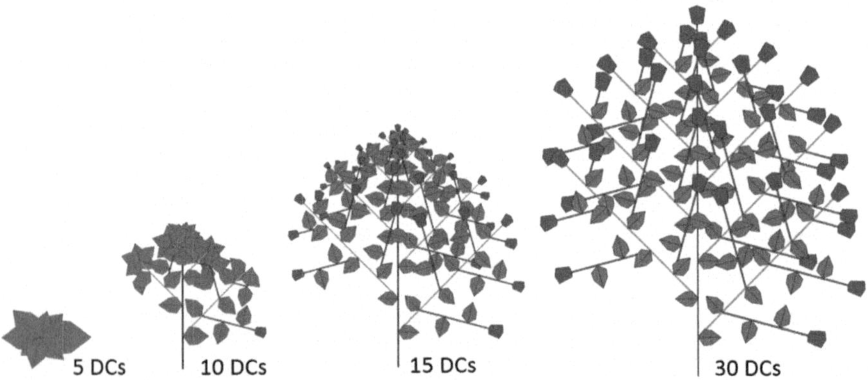

Fig. 7.10 The leaves have an immediate expansion and the internodes and fruits have a simple delay in the expansion (5 DCs here). At first, the plant forms a leaf rosette. The apices also form leaf rosettes because of the time required for internode expansion. The values of the source-sink parameters are those of Fig. 7.8

7.2.2.3.1 Simple Constant Delay in Organ Expansion

If the expansion of the internodes is delayed by a constant number of DCs relative to the leaves of the phytomer, it is observed that this functioning automatically induces the initial rosette formation on the apices. The leaves are then pushed out of the rosette by the expansion of the internodes (Fig. 7.10). Often, there is also a delay in the expansion of the petioles (or sheaths in grasses) compared to the limb of the leaf (beet, corn, sunflower), which makes it possible to favor the expansion of leaves and therefore to promote biomass production at the beginning of plant growth.

7.2.2.3.2 Basipetal Pattern Differentiated According to Organ Type

In this case, the parameter *go* that controls basipetal development takes different values depending on the organ type. Figure 7.11 shows a plant with $gl_i = 1$ for the leaves, $ge_i = 1.5$ for the internodes and $gf_i = 2$ for the fruits. These settings cause the organs of the phytomer to appear at different times within the structure, according to their positions. The leaves of the order 2 axes first appear at the axils of the order 1 axis (Fig. 7.11a), before being pushed away from the stem by the expansion of the internodes that carry them (Fig. 7.11b). The same process then occurs recursively on the order 3 axes concerning the order 2 axes (Fig. 7.11c). This phenomenon occurs for example during the growth of the chrysanthemum as an example in Fig. 7.12, which is a similar simulation of Fig. 7.11.

For chrysanthemum in Fig. 7.12, we also note that the components of axis 2 are smaller than those of axis 1. This is automatically induced by basipetal development and the *Q*/*D* ratio, even when the sink functions are the same, which is also shown in simulations b, c, d of Fig. 7.8. Thus, in herbaceous plants, the size of the organs not

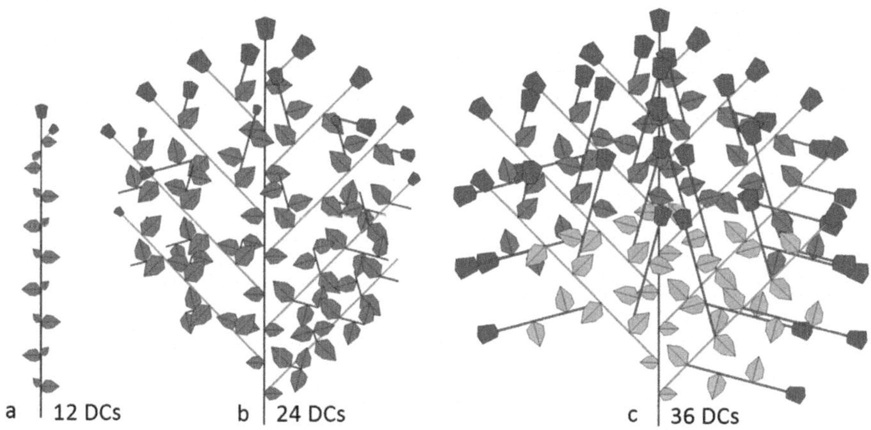

Fig. 7.11 Example of a growth simulation where the organs of a phytomer follow different delay laws for the expansion: $gl_i = 1$ for the leaves, $ge_i = 1.5$ for the internodes, and $gf_i = 2$ for the fruits in orders 2 and 3. The basipetal development is marked. After 10 DCs, the stem flowers and the leaves of axis 2 appear in the axils of the leaves of the stem. Then, the expansion of the internodes repels the leaves. Finally, the flowers appear at the end of the axis and bloom. The values of the source-sink parameters are those of Fig. 7.8. Leaf senescence begins from the base after 30 DCs

Fig. 7.12 Four stages of chrysanthemum growth (photos © Wageningen University). (**a**) End of rosette stage. The leaves of the stem are expanding, and the internodes begin to expand. (**b**) Stage of stem construction. The internodes have expanded, pushing back the leaves they are wearing by removing the rosette. The axils of the stem appear in the leaves of axes 2 whose internodes have not yet begun their expansion. (**c**) Basipetal stage. Axis 1 has initiated its flower bud and the internodes of axes 2 extend starting from the top of the stem and push the small leaves of the axes 2 away from the stem. The flower buds appear in an apical position. (**d**) Terminal floral stage. Basipetal development manifests itself fully. Flowering propagates up and down in the plant in a recursive way according to the branching orders

only depends on sink strength like the shrubs, but also strongly depends on the delay laws to the expansion.

7.2.3 *Case Studies*

Delays in organ expansion may take a variety of expressions.

7.2.3.1 Simultaneous Expansions of Same Organs under the Effect of an External Event

This phenomenon is not specific to herbaceous plants. Some axillary meristems, created at different dates, are dormant and begin to function simultaneously under

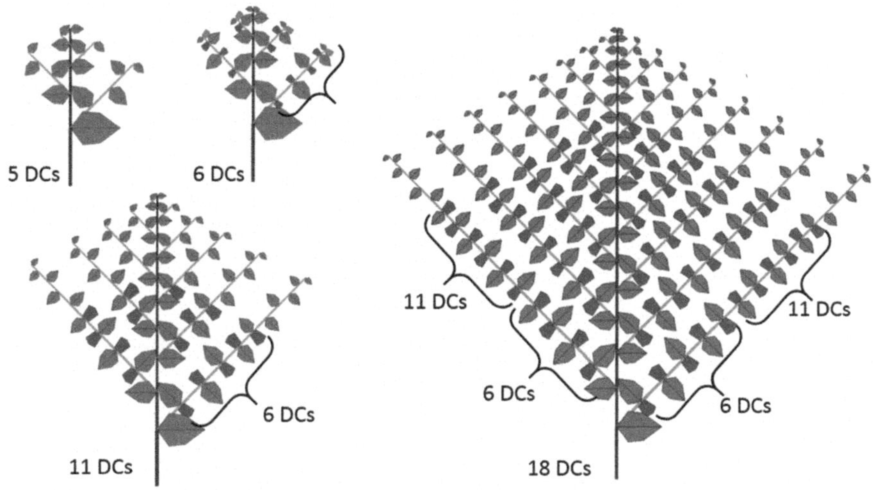

Fig. 7.13 Two waves of synchronous fructifications in a branched structure with continuous development. Only branches are flowering. A first flowering is programmed at DC 6 and a second is at DC 12. All the phytomers that have flower buds on these dates are blooming. Beyond DC 12, the flower buds are waiting for the start of a new flowering period

the effect of an external event. One example is synchronous temperate trees whose preformed GUs lengthen in the spring (e.g., horse chestnut tree). Another example is the case of flowering shrubs such as the coffee tree, which occurs after a rain that follows a dry season. All the flower buds of the leaf axils initiated in the previous months and staggered in their creation are in bloom simultaneously. The organic series of leaves and internodes are different from those of fruits. All fruits appear at the same time and have the same biomass, but their flower buds were initiated at different dates during the development.

Under these conditions, the moment of organ expansion is no longer related to their situation in the topological structure but at a fixed date To_d. The expansion time of an organ at the position *i, j, k, x* belonging to the SDS path (Fig. 7.7) is therefore:

$$To_g(i,j,k,x) = To_d - To_a(i,j,k,x) \text{ with } To_d > To_a(i,j,k,x)$$

Figure 7.13 shows flowering occurring at two different dates (6 DCs and 12 DCs) during the growth of a branched structure. The emblematic example is coffee. Only branches bear fruits. The organic series of fruits corresponding to the same flowering period are homogeneous because they belong to the same cohort, while those of leaves and internodes are heterogeneous because they belong to different cohorts.

Fig. 7.14 Architecture of rapeseed with the terminal racemes. (Photo Wang Xiu-Juan.)

7.2.3.2 Simulating Herbaceous with Basipetal Flowering

If the apical flowering does not give a simple flower, but an inflorescence, this one can develop in the form of a raceme on an indeterminate time, nourished by the underlying vegetative structure. The secondary axes have a delayed expansion according to the basipetal development pattern, and they are prolonged by racemes which are increasingly reduced as their development is further delayed. The stop of raceme growth corresponds to the end of photosynthesis due to the senescence of the leaves. This type of functioning is typical of racemes that extend vegetative axes (rapeseed) (Fig. 7.14). Rapeseed is an emblematic example of this kind of growth (see also Fig. 7.2).

The development of rapeseed follows an ontogenic pattern (see Chap. 2). After a rosette stage, the stem extends. The secondary axes first remain preformed and then elongate in a basipetal order. The terminal flowering of the stem triggers the flowering of all the apices of the branches that give birth to racemes with indefinite development (Fig. 7.15).

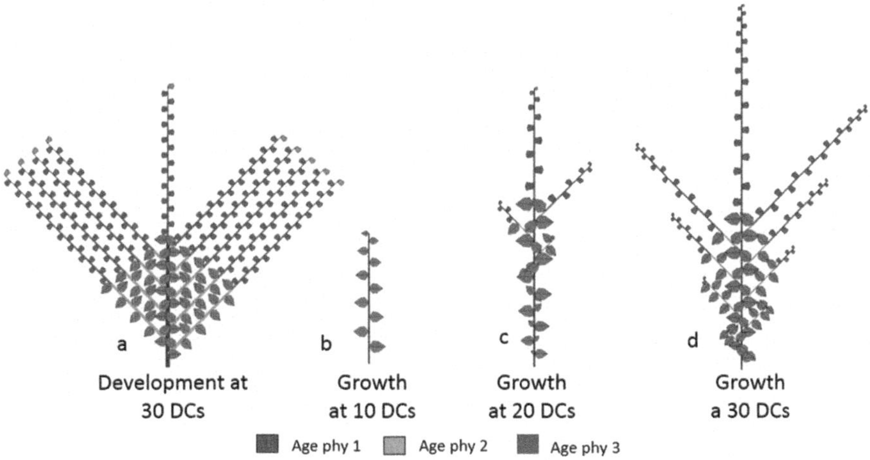

Fig. 7.15 Growth of an herbaceous plant with a basipetal growth and putting in place racemes that extend the vegetative axes. After a rosette stage, the stem elongates and sets up a terminal inflorescence at the phytomer 10, which is a raceme with indeterminate development. The potential and ontogenic development (**a**); The apparent development of the expansion delay caused by basipetal growth at step 10 DCs (**b**), 20 DCs (**c**) and 30 DCs (**d**)

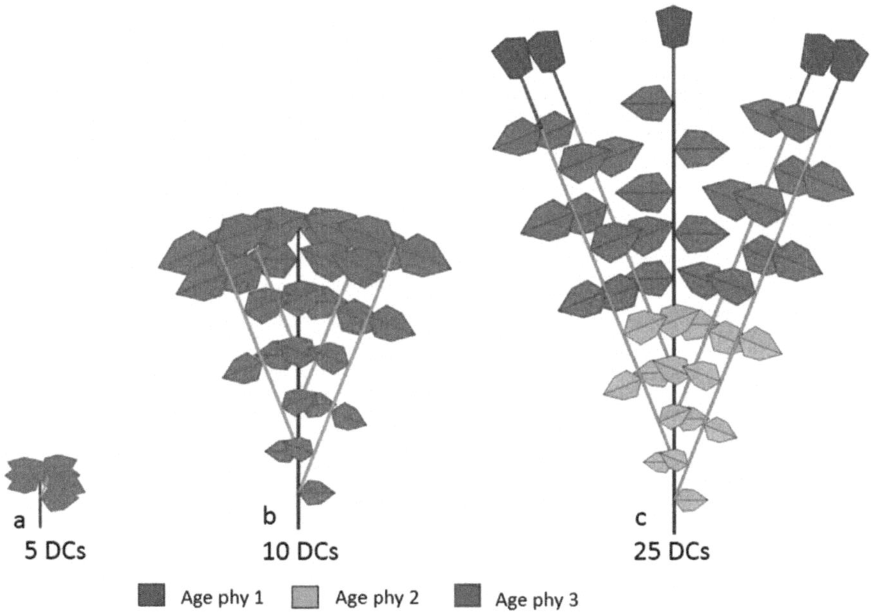

Fig. 7.16 Development and growth of an herbaceous plant. After a rosette stage, the stems lengthen and apical synchronous blooms form spikes that complete the development of the vegetative structure, which consists of a stem with tillers that are basal reiterations. From this point on all the biomass is absorbed by the ears. The leaves reach their life span and die starting from the base. Growth stops when all the leaves are dead

7.2.3.3 Simulating Grass Plants

Grass plants are generally characterized by tillers that are reiterations of the main stem at the base of the plant (e.g., wheat, rice). These tillers are usually synchronous with the main stem and bloom simultaneously. The expansion duration of the vegetative organs (leaves, sheaths, internodes) is short ($\approx$ 2 DCs), and the volume of these organs along the stem roughly reproduces the value of supply on demand at their creation. After an initial stage in a rosette, the phytomers lengthen, and the ears of the stem and tillers are put in place (Fig. 7.16), which stops the vegetative development. From this point, all the biomass is allocated to the ears. Leaf mortality, starting with the rosette, spreads progressively from the base to the apex, which ends plant growth.

7.3 Simple and Compound Stochastic Organic Series

The basipetal pattern disrupts the organic series in plant. On the one hand, the organs of the same cohort are no longer homogeneous and, on the other hand, the ramifications in the expansion phase have embryonic organs. The only measurable organs are those of the main stem because they are sufficiently large and can provide simple organic series. The ramifications whose expansion is often incomplete have organs that are too small to be individually detailed, although they can be counted, and the crown analysis performed (Chap. 3). On the other hand, it is relatively easy to dissect an expanding branch and to group all the organs of the same type in one compartment (Fig. 7.17). This group of organs can then be weighed without much difficulty. We can, therefore, on the stem, decompose an herbaceous plant into simple organic series (organs of the stem) and compound organic series (composed of organ compartments of branches). These series of compartments are identified by the rank of the branch inserted in the axil of a leaf on the stem from the basis or the

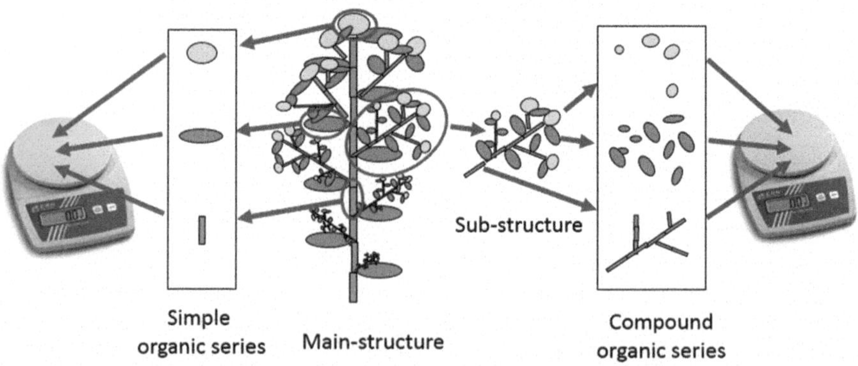

Fig. 7.17 Simple and compound organic series from an herbaceous plant with basipetal pattern

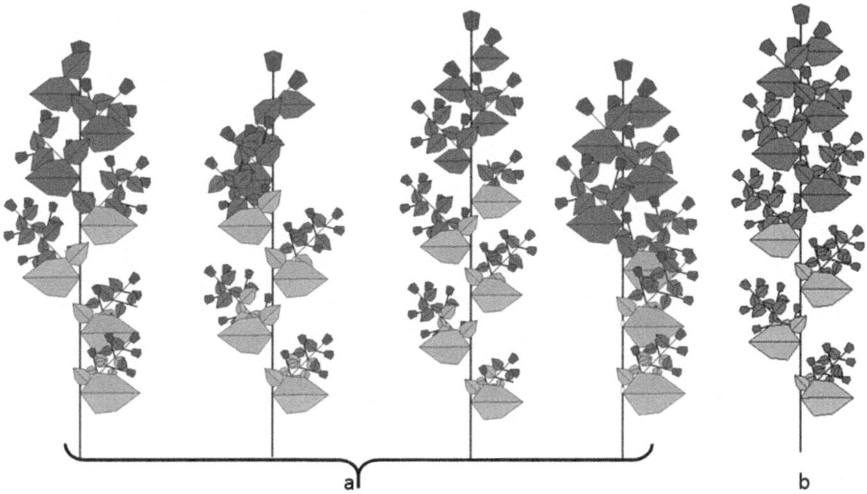

Fig. 7.18 (**a**) 4 stochastic simulations at 30 DCs of the plant extracted from a sample of 30 simulated stochastic plants. (**b**) Potential structure (Chap. 3). The meristems are given a *Bernoulli* parameter $b = 0.9$ for the 3 PAs. All plants are different, 4 random simulations among 30 simulated plants are displayed

top, according to the chosen measurement strategy. Most of the time the herbaceous plant structures have a more or less irregular architecture due to the stochastic functioning of the edifying meristems.

In the following example of a simulated plant, the functioning of meristems is characterized by a *Bernoulli* process ($b = 0.9$). The development time of the stem, second and third-order branches are limited to 10 DCs, 6 DCs, and 3 DCs, respectively. Thus, the simulated plants can reproduce the observed variability in the field.

The source-sinks parameters are: leaf functioning time $t_a = 25$, organ expansion duration $t_x = 20$ for all organs; $r = 20$; $Sp = 1000$ (limited growth). Sinks of leaves and internodes $p_a = p_i = 1$; sink of fruits $p_f = 5$.

Parameter of delays: $g_{o2} = 1.5$ for all organs of order 2; $g_{o3} = 1$ for all organs of order 3.

Moreover, there is an additional delay of 5 DCs for fruit expansion added to the one computed by Eq. 7.6.

Four simulations of the same plant ($b = 0.9$) are given in Fig. 7.18. The probabilities attached to the functioning of meristems can be computed using the crown analysis method (Chap. 3) as for rapeseed. Potential structure (see Chap. 3) is shown for comparison.

The average organic series, simple and compound organic series by type of organ are represented in Fig. 7.19

The good matching (Fig. 7.19) in the comparison of simple and compound organic series, simulated and computed from potential structure, makes the method an effective tool for the calibration of source-sink relationships in the case of

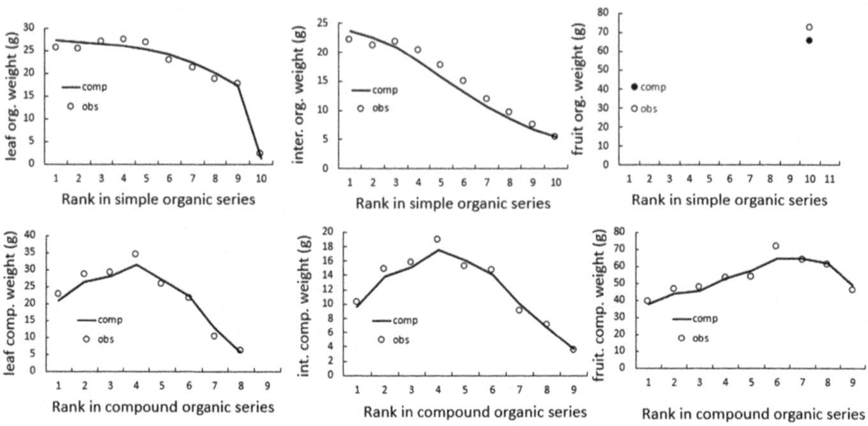

Fig. 7.19 Comparison of the simulated averages (○) of simple and compound organic series for the three types of organs (leaves, internodes, and fruits) and the corresponding values derived from the potential structure. There is a good matching between the simulated and computed values

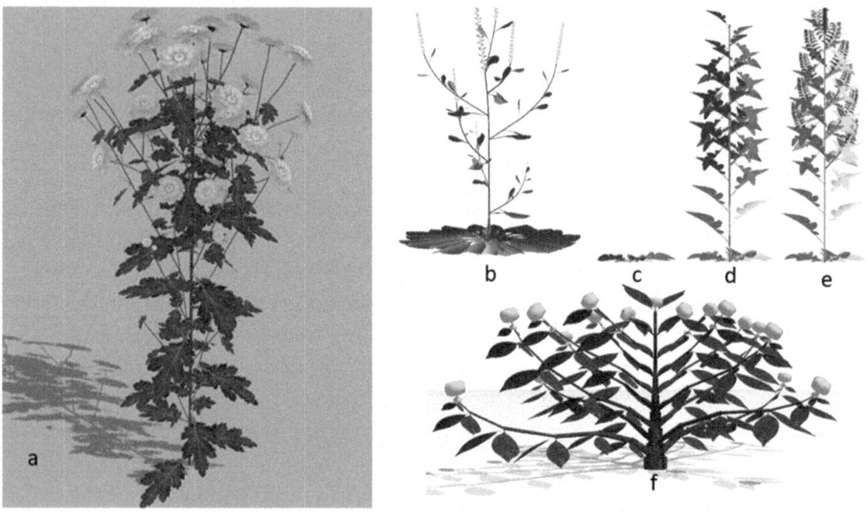

Fig. 7.20 Simulation of growth and architecture with basipetal development in three plants modeled with the GreenLab model. (**a**) Chrysanthemum (QingYuan software). (**b**) *Arabidopsis* (Digiplante software). Three developmental stages of rapeseed (**c**) rosette, (**d**) vegetative growth, (**e**) raceme expansion) (Digiplante software). (**f**) *Spilanthes* (QingYuan software)

basipetal development. Herbaceous plants of this type modeled by the GreenLab model are the *arabidopsis* (Christophe et al. 2008), *rapeseed* (Jullien et al. 2011; Wanget al. 2019), *chrysanthemum* (Kang et al. 2006, 2012), and *Spilanthes* (Vavitsara et al. 2017), as shown in Fig. 7.20.

7.4 Simulation of Agronomic Plants with the GreenLab Model

The modeling of basipetal plants such as chrysanthemum, rapeseed, and Spilanthes has allowed to develop Eqs. (7.6 and 7.7) that control the expansion of the organs according to their position in the plant architecture. The 2D or 3D growth of these plants is simulated by the software, Digiplante, QingYuan, or Gloups (Fig. 7.20).

7.5 Conclusion

The architectures of herbaceous plants are rather regular with few stochastic traits. But a big part of the development can remain hidden because the expansion of branches is more or less delayed according to their positions in the plant structure. In this case, the visible part is the apparent development, and to have the complete development, it is necessary to add the preformed parts of the ramifications awaiting expansion in the axillary meristems. The preformed parts are observable after dissection using a binocular microscope.

Delay laws of organ expansion are controlled by a small number of parameters associated with organ type and their branching order. They allow accounting for many plasticities observed in the apparent development of rosette-form herbaceous plants. Their parameters depend on the coordinates of the organs in the structure and can be calibrated empirically or by heuristic methods. According to the parameter values, great plasticity of the apparent development is possible, going from the basipetal to the acropetal development.

Stress due to competition for space (density), or resources (water), strongly modifies basipetal development, as shown in rapeseed (Fig. 7.2), or the effect of drought on *Spilanthes* (Chap. 9).

References

Christophe A, Letort V, Hummel I, Cournède P-H, de Reffye P, Lecœur J (2008) A model-based analysis of the dynamics of carbon balance at the whole-plant level in Arabidopsis thaliana. Funct Plant Biol 35(11):1147–1162

Jullien A, Mathieu A, Allirand J-M, Pinet A, de Reffye P, Cournède P-H, Ney B (2011) Characterization of the interactions between architecture and source-sink relationships in winter oilseed rape (Brassica napus) using the GreenLab model. Ann Bot 107(5):765–779

Kang M-Z, de Reffye P, Heuvelink E (2010) Modelling growth of inflorescence. In: Li BG, Jaeger M, Guo Y (eds) Proceedings of plant growth modeling, and their applications (PMA09). Beijing, China, November 9–13, IEEE CPS, p 311–310

Kang M-Z, Heuvelink E, Carvalho S-M, de Reffye P (2012) A virtual plant that responds to the environment like a real one: the case for chrysanthemum. New Phytol 195(2):384–395

Kang M-Z, Heuvelink E, de Reffye P (2006) Building virtual chrysanthemum based on sink-source relationships: preliminary results. III international symposium on models for plant growth, environmental control and farm management in protected cultivation, 718, p 129–136

Prusinkiewicz P, Lindenmayer A, Hanan J (1988) Developmental models of herbaceous plants for computer imagery purposes. ACM SIGGRAPH Comp Graphics 22(4):141–150

Vavitsara M-E, Sabatier S-A, Kang M-Z, Ranarijaona H-L, de Reffye P (2017) Yield analysis as a function of stochastic plant architecture: case of Spilanthes acmella in the wet and dry season. Comput Electron Agric 138(1):105–116

Wang X-J, Li D, Lin B-G, Hua J, Kang M-Z, Zhang D-Q, de Reffye P, Wang F-Y (2019) Stochastic simulation of branch morphological structure in oilseed rape. Scientia Sinica Vitae 49(1):67–76

Chapter 8
Secondary Growth (for Woody Plants)

Véronique Letort, François Houllier, Philippe De Reffye, Frédéric Blaise, and Mengzhen Kang

Abstract The chapter presents the dual specificity of tree development and growth, marked by axis mortality and secondary growth. Terminal meristem death leads to the loss of leafy axes, and presents sampling challenges due to the disappearance of tree parts over time. Secondary growth, characterized by ring formation, is primarily associated with woody plants and represents the increase in stem diameter. In contrast to herbaceous plants, tree internodes exhibit ongoing radial growth as long as their axis remains alive, with annual rings indicating the tree's age. The chapter introduces the secondary growth module of the GreenLab approach, which refines traditional forest models by incorporating plant architecture and biomass. The GreenLab model simulates ring formation based on biomass distribution from leaves

---

Electronic supplementary material The online version of this chapter (https://doi.org/10.1007/978-981-97-7541-5_8) contains supplementary material, which is available to authorized users.

V. Letort
Mathématiques et Informatique pour la Complexité et les Systèmes, CentraleSupélec-Université Paris-Saclay, Gif-sur-Yvette, France
e-mail: veronique.lechevalier@centralesupelec.fr

F. Houllier
IFREMER, Plouzané, France
e-mail: daniel.barthelemy@cirad.fr

P. De Reffye
CIRAD, UMR AMAP, Montpellier, France

AMAP, Univ Montpellier, CIRAD, CNRS, INRAE, IRD, Montpellier, France

F. Blaise
CIRAD, Montpellier, France
e-mail: frederic.blaise@cirad.fr

M. Kang (✉)
State Key Laboratory for Multimodal Artificial Intelligence Systems, Institute of Automation, Chinese Academy of Sciences (CASIA), Beijing, China

School of Artificial Intelligence, University of Chinese Academy of Sciences, Beijing, China
e-mail: mengzhen.kang@ia.ac.cn

P. De Reffye et al. (eds.), *Plant Model Greenlab for Botany and Agronomy*,
https://doi.org/10.1007/978-981-97-7541-5_8

using fast algorithms, bypassing the need to simulate the propagation process. These advancements have been integrated into various software applications, including GreenScilab, Digiplante, QingYuan, and Gloups, aiming to improve the accuracy and efficiency of forest growth modeling.

Keywords Dendrometry · Biomass allocation · Secondary growth simulation · Fast algorithm

Tree development and growth are characterized by a double specificity: axis mortality and secondary growth. Terminal meristem mortality leads to the death and pruning of the leafy axes, which results in more and more missing data, and poses a sampling problem since parts of the tree structure have disappeared at the time of measurement.

Although herbaceous plants may have girth growth (e.g., Chap. 9, Spilanthes), secondary growth with rings mainly concerns woody plants and characterizes the increase in diameter of the leafy axes. Unlike herbaceous plants, in which the radial and longitudinal expansion of the internodes are simultaneous and generally brief concerning the duration of the plant's development, in trees, the radial growth of the internodes continues as long as their axis is alive. In temperate zones, a radial section taken at the base of a trunk often shows a stack of concentric rings, characterizing its annual increase in diameter. Secondary growth rapidly envelops the internodes initiated by primary growth. The pith itself can often be distinguished from the surrounding rings in a cross-section of the stem (Fig. 8.1). The process of ring formation is associated with the alternation of periods of cessation of growth during colder seasons. So, the number of rings of a section at the base of a trunk is directly related to the age of the tree. In monocyclic trees, after the expansion of the annual shoots during spring, development quickly stops, with all the preformed phytomers having reached their final lengths. The biomass subsequently produced during the summer period will mainly ensure the continuous formation of rings, fruit growth, root system, and storage of reserves until autumn.

The purpose of this chapter is to present the secondary growth module of the GreenLab approach (at the phytomer level). It was initially developed to refine the classical forest models that linked diameter growth to crown size without taking into account plant architecture and biomass. The pioneer model, AmapHydro (de Reffye et al. 1997), was the first one to simulate the formation of rings in vegetative axes as a function of the propagation of biomass produced by leaves in plant architecture. The method has been incorporated into the GreenLab model and optimized by fast algorithms that directly compute a biomass allocation instead of simulating the propagation process itself. These algorithms have been implemented in the software GreenScilab (Kang et al. 2006, 2009), Digiplante (Cournède et al. 2008, 2013), QingYuan (Hua and Kang 2011), and Gloups (de Reffye and Jaeger 2013).

Fig. 8.1 Secondary growth of woody axes. (**a**) Cross-section of a trunk. We distinguish successively, starting from the periphery: the bark, the light rings of the sapwood, the dark rings that constitute the heartwood, and the pith in the center, remains of the initial internode. The difference in color may not be indicative of the conductive capacity of the associated vessels. The change in color results from changes in the chemical composition of the cells. (**b**) Sculpture by the artist Giuseppe Penone that longitudinally clears the pith from the rings of a trunk. This coat not only the pith, but also the branches at birth that are attached to it. The whorls are clearly visible (Color figure online)

8.1 About Dendrometry

Dendrometry is a forestry science that deals with the measurement of the quantifiable physical characteristics of trees (or stands) to produce wood production models. It concerns:

- the shape and volume of the trunk (diameter at the base and height of the trunk);
- the study of the relationships between the crown and the trunk;
- the study of ring stacks to date trees and go back as far as possible in time to the climatic data that determined growth;
- the ground area, which gives the total area of the sections of trunks at the base of a forest stand per hectare. Surprisingly, this area does not exceed a few dozen square meters per hectare.

This science, which dates back to the nineteenth century, focused primarily on describing monospecific stands, subject to standard forest treatments and establishing production tables around some reference forestry stations.

In the twentieth century, models incorporated a finer description of the tree, including a description of the crown (usually the geometric shape of its envelope) and its influence on the distribution of growth rings in the stem. These two factors are important in determining wood quality (Mitchell 1975). While maintaining the same level of simplified crown cover description, some models have integrated physiological processes (light interception, biomass production, and allocation) or some integrated growth and competition indices (e.g., Pretzsch 2002). In these models, individual wood production is closely linked to spatial competition induced by neighboring trees in a forest stand. The AmapHydro model (and its successors) share some characteristics of these models, although adapted to take into account the architecture of the trees.

8.1.1 Characterization of the Crown

The description of a tree crown is based on the following characteristics:

- its length from the top of the tree to the last living branch;
- its projection surface on the ground, bounded by a polygon centered on the base of the trunk with a variable large number of rays.
- the surface area of the leafy envelope, or the volume of the last layer of leaves produced in a year, modeled by parametric equations that characterize the shape of the crown.

8.1.2 Stem-Crown Relationship

Based on experimental samplings, Shinozaki et al. (Shinozaki et al. 1964) investigated vertical distributions of leaves and wood in stands of different types. They proposed a principle called the *pipe model* whereby each unit quantity of leaves is associated with a tube (pipe) of constant cross-section running to the base of the trunk, and having both conductive and mechanical support roles. The assembly of these tubes automatically produces the external ring in the axes of the structure during each growing season (Fig. 8.2).

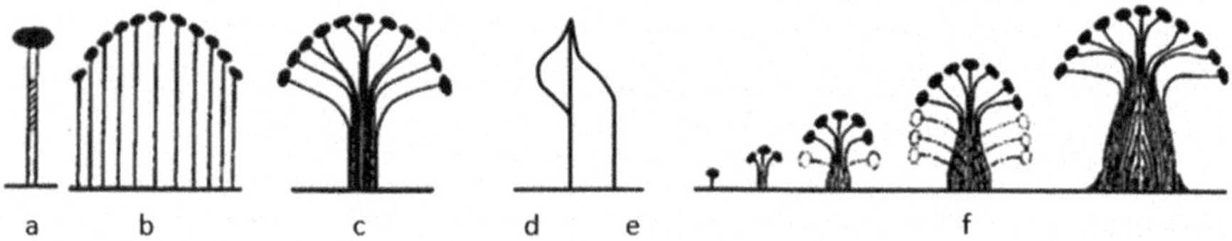

Fig. 8.2 The *pipe model* (Shinozaki et al. 1964). (**a**) Each unit quantity of leaves is associated with a tube of constant cross-section running to the base of the trunk having both conductive and mechanical support roles. (**b**) Representation of woody tubes. (**c**) Assembly of tubes to visualize the shape of the trunk. (**d**) Vertical distributions of leaf densities (left). (**e**) Nonphotosynthetic organs (right). (**f**) Growth of the trunk with the elimination of tubes whose leaves at the ends are dead

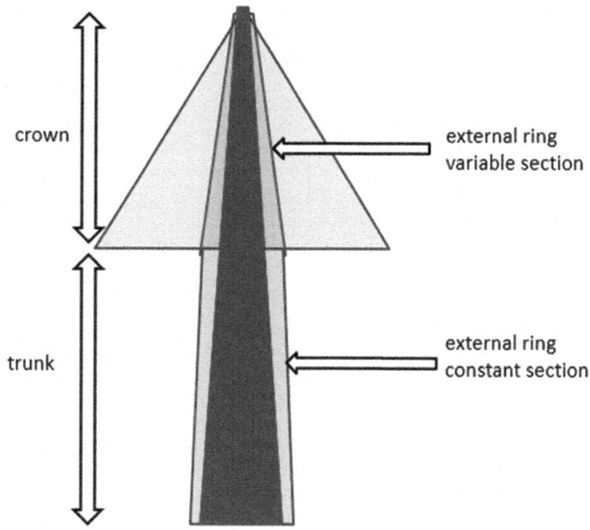

Fig. 8.3 Variation of the surface of the external ring along an axis according to Pressler's law. The cross-section of the external ring increases linearly in the crown from top to bottom (in proportion to the leaf area distribution) and becomes constant below

Similarly, Pressler's Law (1865) reports that the surface area of the last ring formed at a given level of the stem is proportional to the leaf mass above that level. The surface of the ring thus increases from the top to the base of the crown, beyond which it becomes constant (Fig. 8.3).

8.1.3 Competition Between Trees

Competition among trees affects the dendrometric characteristics of the trunk-crown system.

The height of trees is little affected by competition, as long as the trees have access to light. This observation is a corollary of Eichorn's law (1904), which states that the height of dominant trees is little affected by the planting density.

Competition from neighboring trees slows down the radial extension of the tree crown when branches begin to touch or overlap each other. This competition for space results in the death of some branches or, at least, a reduction of their extension. The planting density therefore limits the volume of the crown by a truncation, which influences the volume of the rings.

A Silvicultural Production Model: The CEP Model

The above principles are sufficient to build simple dendrometric production models that can integrate various forestry management practices. In these models, tree crowns are represented by simple shapes, polyhedral, conical, warhead, etc., i.e., simple enough architectures.

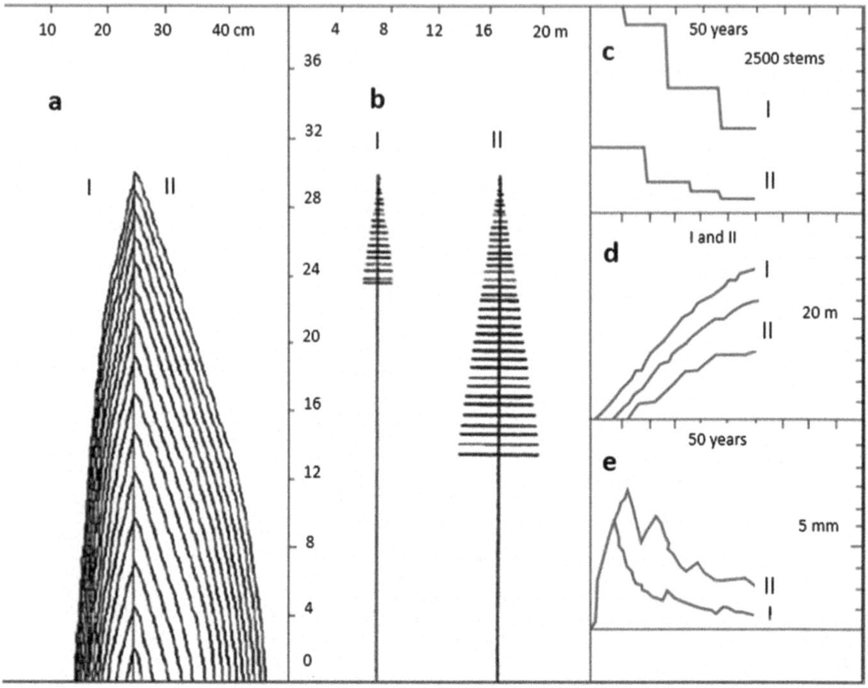

Fig. 8.4 CEP model was developed by Leban and Houllier in 1991. We represent the two average trees from two different silvicultural scenarios. In scenario I, the planting density is high, resulting in reduced crown tops and narrow trunks; in scenario II, the density is low, resulting in more developed crown tops and large diameter trunks. (**a**) Longitudinal half-cut of two medium trees. (**b**) Representation of crown top shape at the end of the simulation. (**c**) Silvicultural scenarios followed by the number of stems per hectare. (**d**) Height growth and rise of the crown base (only the latter changes according to the scenario). (**e**) Evolution of the width of the ring in the stem sections at the base of the tree

These models estimate the diameter growth of trunks by simulating the stacking of annual rings. To this end, the convex envelope of the crown allows an estimation of the leaf mass that produces the biomass of the rings by photosynthesis. On these principles, Mitchell (1975) simulates the wood production of Douglas fir. It experimentally verifies that the surface area of the annual external ring grows linearly from the top of the tree to the base of the crown, and then remains constant below (Fig. 8.3). This observation is in accordance with Pressler's law. Contacts between neighboring trees limit the development of the tree crown, thus the surface area of the rings, and ultimately the shape of the trunk. Thus, these simulations provide predictions for future wood production. Adapted to spruce by Leban and Houllier (Leban et al. 1991; Houllier et al. 1995), the CEP model allows for characterizing the shape and volume of trunks according to different silvicultural scenarios (Fig. 8.4).

8.2 Integration of Dendrometry Models with Structural-Functional Models

Many tree models aim to provide an accurate 3D visual rendering of form and structure for the needs of computer graphics. Since the diameter of the tree axes is an important component for the realism of the representation, these models include secondary growth modules, often in the form of an empirical adjustment. Thus, by fairly simple means from a botanical point of view but based on classical parametric functions in computer graphics, computer programs such as Onyx-Tree (Bosanac and Zanchi 2005), Xfrog, (Linterman and Deussen 1997, 1999) and the works of (Palubicki et al. 2009) produce images of virtual trees with compelling realism. In contrast, the visual rendering is not the primary objective for structural-functional models: they focus on a realistic computation of secondary growth as an ingredient of the biomass balance of the tree.

Thus, functional-structural models can be expected to meet the needs of silviculture by integrating dendrometry knowledge. The aim here is to take into account both the architecture of the tree and the empirical laws highlighted in dendrometry (which ignores the notion of growth unit!) to model secondary growth. But the question is what level of detail is suitable for forestry applications? Indeed, only the trunk is often the main product of interest; its description depends on the evolution of the crown during its growth. The progress made in computing power and the appearance of botanical descriptors such as the notion of physiological age (PA) allows, on the one hand, an economic and optimal sampling measurement strategy and, on the other hand, the implementation of factorization methods, particularly in the GreenLab model. It is therefore possible to compute the thickness of the rings formed at any point in the tree as soon as the model can simulate the dynamics of biomass production and allocation. Thus, if the architecture of the tree is simulated mathematically and computerized, it is possible to access structures describing the spatial arrangement of the leaf mass and the phytomers of the vegetative axes. The allocation of biomass in rings can then be explicitly addressed under various laws, in particular Pressler's law. In doing so, a formal link is established between the internal structure of the leafy axes and the external architecture of the growing tree.

One of the first structural-functional models to take into account the legacy of dendrometric models is the Lignum model (Perttunen et al. 1996, 1998). The model outputs are the volume of the trunk and foliage, the height of the tree, its diameter at the base, and its 3D representation. Lignum incorporates a few botanical concepts regarding the development of tree architecture (not taking into account the detailed architectural unit (AU), stochastic characteristics, axis mortality, pre-formation, and neoformation, etc.). Its purely simulation-based approach leads to significant computation times, as the topological structure must be explicitly represented, and the functioning is detailed as closely as possible to the biological context to be able to implement development and growth through the source–sink relationships that ensure the circulation of biomass in the structure.

8.3 AmapHydro Model and Secondary Growth

The AmapHydro model (de Reffye et al. 1997) simulates ring stacking in leafy axes, related to tree development and growth. The development module in AmapHydro is a simplified one: the tree topology is described only at the GU scale, considered as a meta-phytomer. The top node of the GU carries all the branches of the GU. The model is therefore limited to simple architectures (conifers, maples, etc.).

In the Amaphydro model, organ biomass is not considered and all the internodes are assumed to have the same length. But the skeleton of the tree architecture is described, and all the leaves are positioned in this architecture. Thus, it provides the necessary framework to simulate the distribution of secondary growth according to the principles introduced in the previous paragraph. The model assumes that each leaf produces a ring element of constant volume (similar to a tube) that spreads to the ground through the direct path inscribed in the tree architecture. The surface area of the element section is therefore inversely proportional to the length of the path from the leaf to the ground. The ring elements are bundled together to form the rings as in the case of the pipe model (Fig. 8.2). Dead axes no longer produce ring elements.

The rings of an internode therefore consist of a number of ring elements equal to the number of functional leaves seen upstream. The cross-sectional area of a given ring is equal to this number multiplied by the surface area of the section of the ring element. To simulate the internal growth in the diameter of the axes, it is therefore sufficient to spread the production of a leaf from its position, with a tube over all the underlying internodes along the path to the base of the trunk at each DC, and repeat the operation for all the leaves of the crown.

The internal structure of the trunk is ultimately obtained (Fig. 8.5). The center is the pith, i.e., the internode created from the apical bud during the primary growth phase. The other rings pile up concentrically at a rate of one new ring per year.

The pith of the internode that appears at a cycle i is considered a cylinder with a volume denoted as $q_m(i, t)$ (Chap. 5). For a plant age t, the vector $q_m(*, t)$ gives the profile of the organic series of the stem pith. In addition, at age t, the internode that appears at cycle i is coated with a new volume of ring denoted as $q_c(i, t)$. Note that the tree is assumed to be observed at the end of the growing season, so the first internode (i.e., the first component of the table) consists of the pith itself plus the first ring.

The stacking of rings on a given internode i is denoted as $q_c(i,*)$ with $i \leq t$. The volume of the internode $q_e(i, t)$ is, by definition, the sum of the volume of the pith and the volume of the rings that are deposited on it. We have:

$$q_e(i,t) = q_m(i,t) + \sum\nolimits_{j=i}^{t} q_c(i,j) \tag{8.1}$$

The organic series $q_e(*, t)$ describes the volume of the stem at the level of the plant. In dendrometry, we are interested in the height and diameter of the stem. It is, therefore, necessary to switch from volume to internode dimensions by

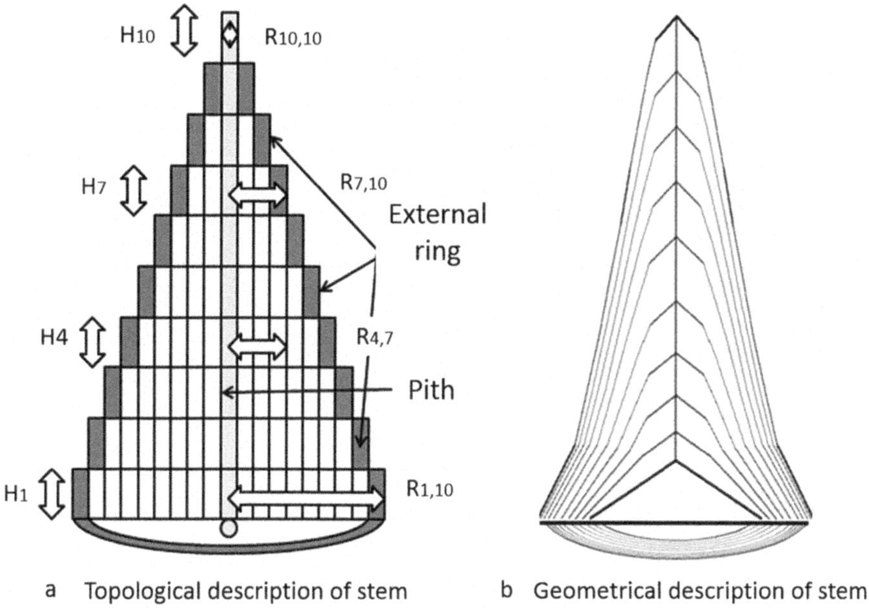

Fig. 8.5 Representation of the internal structure of a woody stem in AmapHydro. (**a**) Topological representation. The internodes are numbered at the axis level (and not at the growth unit (GU) level) from the bottom. For rings, two indices are required: the index of the internode and the index of the ring (greater than or equal to that of the internode). (**b**) The geometric representation of the stem profile depends on the volumes of the internodes and the allometric relationships. Smoothing avoids a profile that is too "telescopic"

allometries (Chap. 5). The length $h(i, t)$ of the internode is equal by definition to that of the pith, i.c.:

$$h(i,t) = \sqrt{\beta} \cdot (q_m(i,t))^{\frac{1+\gamma}{2}} \tag{8.2}$$

where β and γ are allometric coefficients.

This gives the surface area of the internode section:

$$s(i,t) = \frac{q_e(i,t)}{h(i,t)} \tag{8.3}$$

The diameter of the internode is finally obtained from this section by

$$d(i,t) = 2 \cdot \sqrt{\frac{s(i,t)}{\pi}} \tag{8.4}$$

At age t, the geometric profile of the stem is thus given by the series $h(*, t)$ and $d(*, t)$. These computations can be applied to any axis of the tree architecture in

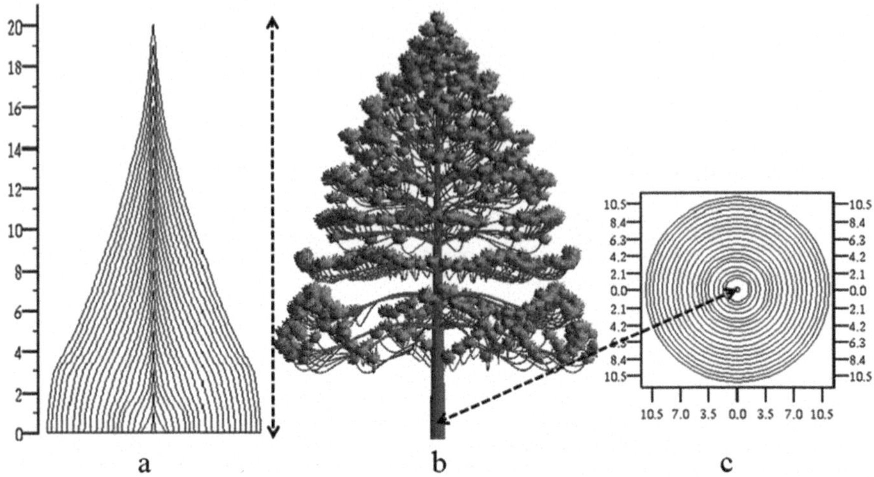

Fig. 8.6 Architecture and secondary growth simulated by AmapHydro. Each leaf of the crown produces a ring element similar to a tube that runs along the path to the base of the tree. The length of this tube is that of the path it takes. Given that the volume of each tube is fixed, we can deduce its section. The assembly of these tubes into bundles gives the surface area of the rings on all the internodes, in particular, on the trunk (the diameters are dilated for better visibility). (**a**) Pine tree at 20 years old. (**b**) and (**c**) Sections of the trunk, longitudinal and radial at the base, showing the rings resulting from secondary growth

relation to its CAs and PAs. When the axis is dead, it no longer has any functional leaves, and its secondary growth is stopped at the same time as its development. The AmapHydro model provides the shape of a trunk in longitudinal or radial sections, as simulated from the distribution of leaves in the tree, which builds up the bundles of the ring elements in the plant architecture (Fig. 8.6).

The stacking of rings in the stem can be considered as a memory of functioning over time, since the ring thicknesses are associated with the condition of the tree at the time they appeared. Dendrometry is precisely based on this property. The effect of climate can be simulated and recovered (Fig. 8.7) from these rings. Similarly, by inversion, the ring stack profile allows reconstructing the number of functional leaves upstream of each stem GU, and thus reconstituting the volume of the leafy crown without having access to its precise structure.

It should be noted that the direct fitting of a stem profile by a parametric function has a limited interest because this function is not directly related to the functioning of the ring construction.

With AmapHydro, the secondary growth of the internodes is related to the architecture of the tree. The classical empirical laws of the longitudinal and radial evolution of the surfaces of rings appear as emergent properties of the model. The surface area of the longitudinal ring grows almost linearly from the top of the tree to the base of the leafy crown, beyond which it remains constant. Linearity is ensured by the senescence of the leaves and the pruning of the branches, whose foliage then becomes stable in volume. The surface area of the concentric rings at the base of the trunk increases from the pith, reaches a maximum, and then decreases slowly. This

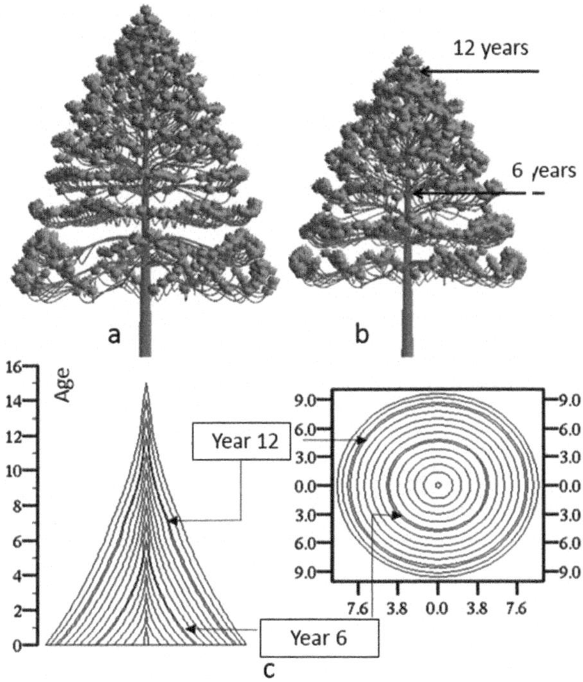

Fig. 8.7 Annual stress and consequence on the growth simulated by AmapHydro. The 16-year-old tree is stressed in years 6 and 12. This leads to a decrease in the growth of its length and thickness in these years. (**a**) Control tree. (**b**) Stressed tree. (**c**) Visualization of stress in the trunk (longitudinal and radial sections)

reflects the initial exponential growth; then when demand stabilizes as a result of crown formation, it reflects the effect of elongating the bare part of the stem combined with an invariant crown.

Each ring element is a sink like any other organ. Its lifespan is limited to a single growth cycle. Each leaf is associated with a ring element that participates in the tree demand. The entire bundle of ring elements constitutes the secondary growth compartment, and its demand is equal to the demand for an element multiplied by its number. Two applicative modules have been introduced in AmapHydro: a mechanics module allows to study the balance of the standing trees, and a simulated trunk sawing module allows to compute and visualize virtual boards with their ring drawings (Chap. 12).

8.4 GreenLab Model and Secondary Growth

Cumbersome computations are a major limitation in tree simulations based on the creation and exploration of topological structures for the propagation and distribution of biomass in tree rings. The development of the GreenLab model addresses this issue by integrating botanical and agronomic knowledge from AmapHydro. But it replaces the simulated propagation of biomass in the structure with a more efficient approach: using equations that directly compute the allocation of biomass to the phytomers.

8.4.1 Adaptation of the Notion of a Common Pool of Biomass for Trees

In herbaceous plants, the notion of a common pool is commonly accepted (Chap. 5); however, in a large woody structure such as a tree, the length of the paths of sugars synthesized by the leaves poses a challenge to this hypothesis. The plant biomass in trees is produced in the peripheral, leafy crown and then migrates down through the architecture like a hydrographic network along the axes towards the base. The synthesized biomass is first distributed in the leafy phytomers of the crown and then in those of the underlying axes below, where it forms the rings.

What feeds an individual leafy phytomer of the crown is proportional to the total produced biomass, divided by the number of phytomers, as if it was self-feeding with its production. Therefore, it can be considered that in the leafy crown, expanding organs have access to the same common reserve pool. On the other hand, the internodes of the underlying axes, which are subjected only to secondary growth, receive nutrients from only see part of the above leaves that feed them. This part can be computed using the substructure method (Chap. 2) and constitutes a "window" over the common pool. Each internode is thus assigned a countable number of leaves that feed its secondary growth (Fig. 8.8a and b).

Keeping the same biological assumptions, GreenLab has been transformed into a computational form that was entirely processed by simulation in the AmapHydro model. First, the multi-scale factorization of substructures (Cournède et al. 2006) based on PA drastically reduces the time required to compute the development of a tree. Second, the secondary growth of internodes is no longer computed by spreading the biomass allocated to the ring from the leaves along with the topological structure, but by directly allocating a fraction of the biomass of the ring compartment. Indeed, we know by computing the number of living leaves above each phytomer on the main axis for a given CA and PA sub-structure. By equalization, the biomass allocation of rings is done throughout the tree structure without having to go through it as many times as the number of leaves, which considerably reduces the computation time.

8.4.2 Secondary Growth and Plants with Continuous Development of the Architecture

Herbaceous plants and shrubs have continuous development, but only the latter, which are woody, have secondary growth. Some trees, such as cypresses or eucalyptus, also have continuous development, but this is an exception; in general, trees show rhythmic development (Chap. 2). Regardless of the volume of secondary growth to be distributed, several algorithms can be proposed to simulate different patterns of biomass circulation in the rings. To improve the performance, these

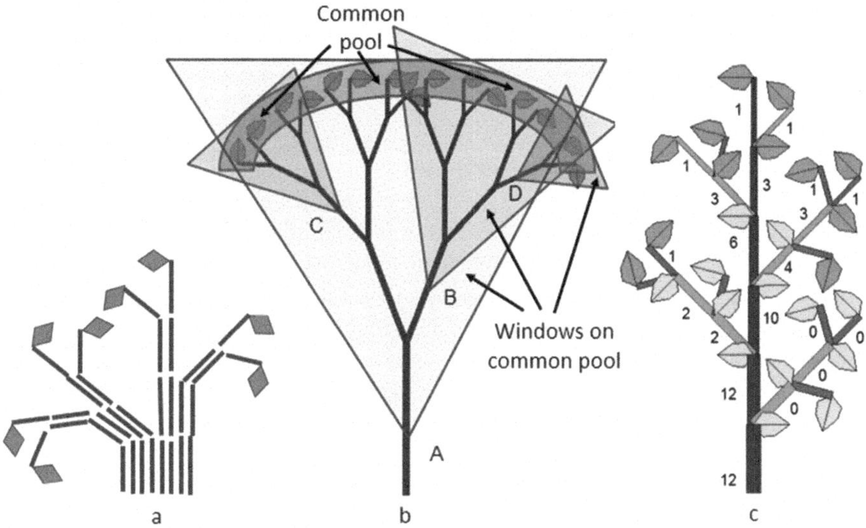

Fig. 8.8 Adaptation of the common pool for tree rings. (**a**) Arrangement of the ring elements associated with the functional leaves. These elements are a sap flow that spreads uniformly from the leaves to the base of the tree (pipe model). (**b**) The pool is made up of the leafy crown (green half ring). The leafy phytomers are in the pool. Outside, point A at the base of the tree "sees" the whole pool, but the other points B, C, D "see" only a part of the pool, which is a window on the pool. The thickness of the ring of a particular internode is proportional to the size of the associated window, which simulates Pressler's law or its equivalent pipe model. The notion of a common pool can thus be preserved in large plant structures. (**c**) A detailed study of the number of functional leaves that a phytomer "sees" upstream according to its location in a structure. The surface area of its external ring of a phytomer is proportional to the number of leaves it "sees". The number of leaves (displayed under each internode) can be computed by an operator, which does not require exploring the entire structure. The life span of the leaves is 2 cycles. The branches have a limited lifetime (3 DCs for physiological age (PA) 2, 1 DC for PA 3). The dead axes no longer have functional leaves and no more rings (Color figure online)

algorithms can also be implemented by exploiting the substructure factorization. Indeed, it is known that the sugars synthesized by the leaves are not only distributed downwards. Part of them goes up in the tree architecture. Two main modes of ring growth can be distinguished: Pressler and uniform modes. These two modes can be linearly combined, which generates a continuum of different model behaviors. What matters, in the end, is that it is possible to compute by an inverse method the coefficients of the propagation laws that best fit the profile of the organic series of the internode biomass.

A *pc*-sink is assigned to the ring element. For fine-tuning of the biomass distribution for secondary growth, a secondary sink pc_2 can be assigned depending on the PA of the phytomer. The secondary growth for a given type of phytomer can also be canceled by assigning a null sink to it. This is particularly useful for the short axes which have the highest PA ($p = mx\varphi$). This option is necessary because if the numerous internodes of the short axes had secondary growth, they would capture the majority of the biomass $Qc(i)$ allocated to the rings, thus preventing the trunk from growing. Their secondary growth must be negligible, as observed in real trees.

8.4.2.1 Implementation of Pressler's Law

The computation of the surface area of a ring is done in three steps.

The first step is to compute the plant demand, including the demand for ring elements. Each functional leaf corresponds to a ring element with an associated constant sink that lasts one growth cycle. This requires counting (Chap. 3) the types of entities that make up the topological structure and adding all the sinks of the organs associated with the phytomer (Chap. 4). Let D(t) be the total demand of all organs and $D_c(t)$ represent the demand of the ring compartment. $D(t)$ is expressed by Eq. 8.8 in Chap. 5, which is recalled here.

$$D(t) = \sum\nolimits_{o,\varphi} \left(\sum\nolimits_{i=1}^{t} N_o^{\varphi}(t-i+1) \cdot p_o^{\varphi}(i) \right) \tag{8.5}$$

The index o refers to the type of organ (a, leaf, e, internode, f, fruit and c, ring element). The index φ takes all the PA values. We set: $j = (t-i+1)$.

Then $N_o^{\varphi}(j)$ is the number of organs of type o and PA cohort φ that appear at age j. $p_o^{\varphi}(i)$ is the sink associated with age i. The demand for rings $D_c(t)$, included in $D(t)$, is written as:

$$\mathrm{D}_c(t) = N_a^{\varphi}(t) \cdot p_c \tag{8.6}$$

It is the sum of the sinks of the ring elements, whose number is the same as the number of alive leaves (one alive leaf gives one ring element), but whose expansion time is fixed to a single cycle.

The second step is to compute the biomass of the ring compartment. We know the biomass $Q(t-1)$ of the reserve that will supply the sinks for cycle t. Therefore, the reserve allocated to the ring compartment is computed as:

$$Q_c(t) = \frac{D_c(t)}{D(t)} \cdot Q(t-1) \tag{8.7}$$

For simplification, a density equal to 1 is assigned to the volumes here.

The third step is to compute the biomass allocated to the internodes. The quantity $Q_c(t)$ is shared among all internodes that "see" at least one living leaf located above them in the plant architecture (Fig. 8.8b). The demand for the ring of an x-numbered internode in the topological structure is proportional to its length $h(x)$ and the number of leaves $n_a(x)$ above it, according to Pressler's law. For any position x of the phytomer in the topological structure, the relative demand $D_c^r(t)$ of all the internodes concerning the biomass pool reserved for rings can be computed as:

$$D_c^r(t) = \sum_x n_a(x) \cdot h(x) \tag{8.8}$$

The volume of the ring of the internode x, $q_c(x)$ is therefore written as:

$$q_c(x) = \frac{n_a(x) \cdot h(x)}{D_c^r(t)} \cdot Q_c(t) \tag{8.9}$$

The section surface of the ring along an internode is invariant and is computed as:

$$s(x) = \frac{n_a(x)}{D_c^r(t)} \cdot Q_c(t) \tag{8.10}$$

The length $h(x)$ depends on the pith allometry coefficients according to formula (8.2). However, empirically it can be seen that their values have little influence on the weight of the internode x and thus on the p_c value.

Each position x within the topological structure corresponds to a pair (i, t) that gives the rank of the internode in its organic series.

Let n be the number of times that a couple (i, t) is associated with x in the topological structure. We can then obtain:

$$q_c(i,t) = \frac{1}{n} \sum q_c(x) \tag{8.11}$$

The variable n varies in the stochastic case but the average profile of the volume of the rings of the internodes in the organic series can still be obtained. The volume of the internode is the sum of the volume of the pith and the volumes of the concentric rings (Eq. 8.1). Its diameter is then deduced.

The total volume of the internode $q_e(i, t)$ is equal to its length $h_e(i, t)$ multiplied by its section $s_e(i, t)$, which is the sum of the volume of the pith $q_m(i, t)$ and the volume of the rings $\sum_{j=i}^{t} q_c(i,t)$.

So we have:

$$q_e(i,t) = q_m(i,t) + \sum_{j=i}^{t} q_c(i,t) \tag{8.12}$$

And knowing that $h_e(i, t)$ is known thanks to the allometries of the pith of the internode, we get.

$$q_e(i,t) = h_e(i,t) \cdot s_e(i,t) \tag{8.13}$$

We deduce the diameter of the internode in the organic series:

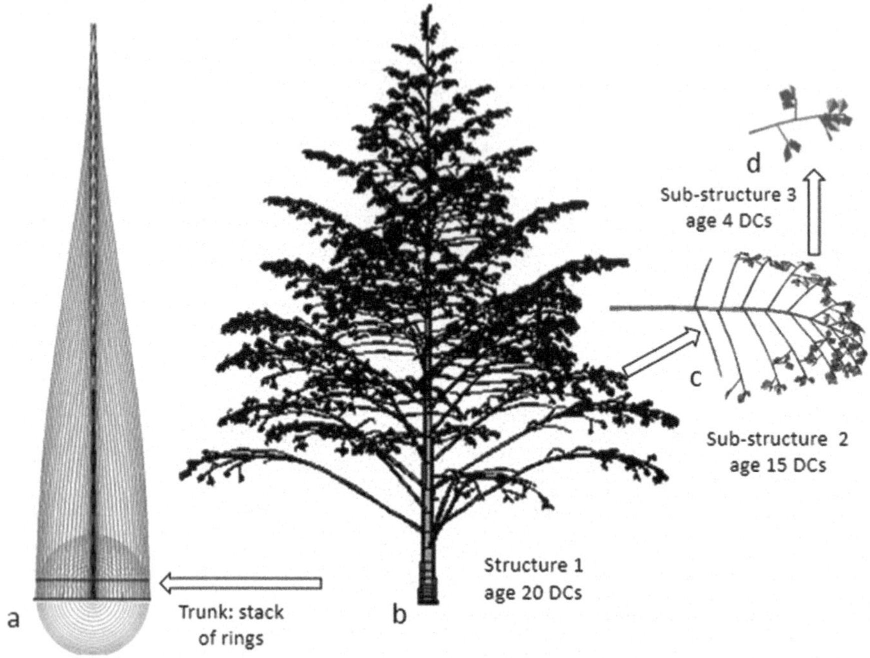

Fig. 8.9 Construction of the architecture of a tree and the internal structure of the rings of woody axes by the substructure method (Yan et al. 2003), using the GreenScilab software (Kang et al. 2006, 2009). With this software, 3D structures can be displayed separately at different scales (rings are dilated for more visibility). Here, the tree is 20 years old and has four PAs. (**a**) Simulation of the internal structure of the trunk with its stacks of rings. (**b**) Simulation of the tree architecture. (**c**) and (**d**) Visualization of the primary and secondary branch substructures

$$d_e(i,t) = 2\sqrt{\frac{q_e(i,t)}{\pi \cdot h_e(i,t)}} \tag{8.14}$$

Figure 8.9 shows the result of secondary growth in a 3D tree architecture according to Pressler mode (see also Fig. 8.10a).

In this secondary growth mode, the tips of the axes are thin.

Based on this substructure factorization method, the first software called GreenScilab (Kang et al. 2006, 2009) was developed at LIAMA. It can compute secondary growth instead of simulating it, as was done with AmapHydro (Fig. 8.6). The results obtained are equivalent but with a significant reduction in computation time. Moreover, here the complete growth of the tree is simulated, and the expansion of the organs related to the weight is ensured by the source-sink relationships.

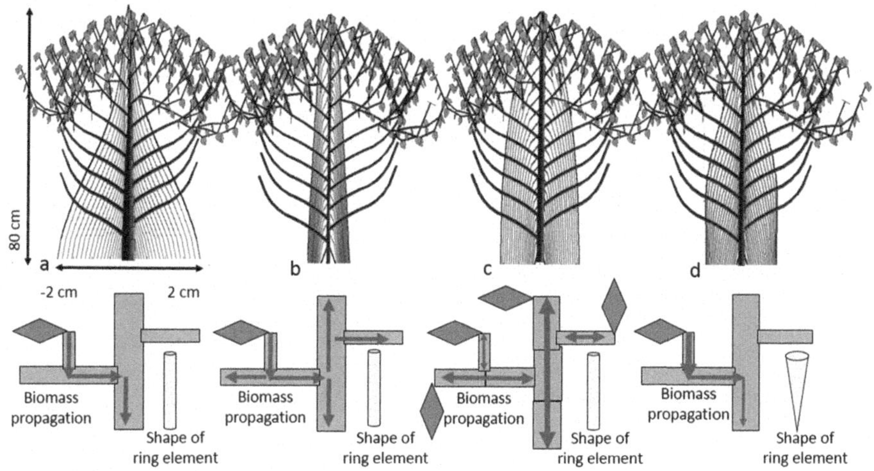

Fig. 8.10 Different possible modes of biomass propagation for ring growth (Gloups software). Plant architectures are simulated. A diagram below shows the mode of biomass distribution in the rings for the same quantity Q_c to be spread out. The diameters are exaggerated and superimposed on the architecture, to better see the profile of the trunks. (**a**) The ring element runs from the leaf to the bottom of the tree (Pressler mode), trunk shape is rather conical. (**b**) Uniform propagation of the synthesized biomass from the leaf in the whole plant structure: trunk shape is rather cylindrical. (**c**) Implementation of the storage of the biomass in the axes of the same PA. Trunk has three internodes and four leaves ($N_b = 4$, $N_i = 3$). Branch Order 2 has three internodes and three leaves ($N_b = 3$, $N_i = 3$). Branch order 3 has one internode and one leaf ($N_b = 1$, $N_i = 1$). (**d**) Pressler mode but the section of the ring element is no longer constant, it decreases or increases along the axes

8.4.2.2 Implementation of Uniform Propagation of Biomass (Common Pool Mode)

In this approach, it is assumed that the biomass produced by a leaf is allocated to all internodes of the total plant structure that "see" at least one living leaf upstream (Fig. 8.10b) and in proportion to their lengths. This can be considered as the "common pool" mode.

The demand stipulates that each leaf distributes its production to all nodes of the structure. So:

$$D_c^r(t) = \sum_x \text{sign}(n_a(x)) \cdot h(x) \tag{8.15}$$

where x denotes a position of the internode in the structure. The expression sign ($n_a(x)$) returns the value 1 or 0 depending on whether there is at least one living leaf or not upstream of internode x. In this case, the thickness of the last ring is the same on all the internodes of the tree.

The volume of the ring on the internode x is then obtained:

$$q_c(x) = h(x) \cdot \frac{\text{sign}(n_a(x))}{D_c^r(t)} Q_c(t) \tag{8.16}$$

The diameter profile of the axes is significantly modified compared to the Pressler mode that generates more conic shapes (Fig. 8.10b). Each phytomer in the tree receives biomass from all leaves (no window on the common pool). The axes then have a quasi-cylindrical shape and are thin.

Observations on young pines with this algorithm show that uniform propagation fits rather well the shape of the trunk, while Pressler mode is more adequate for older trees.

8.4.2.3 Implementation of the Storage of the Biomass in the Axes of the Same PA

In this method of ring construction, the biomass Q_c is allocated to compartments at the scale of axes defined by PA p and level of reiteration r.

The biomass allocated for the rings of a given axis, defined by the indices p and r (level of reiteration) (i.e., $Q_c(p, r)$) at CA i, is proportional to the number of leaves seen by this axis. The trunk internodes ($p = 1$) see all the leaves of the tree, but those of the axes $p > 1$ and $r > 1$ only see part of them. A single internode ring receives a quantity of biomass proportional to its length and $Q_c(p, r)$ divided by the total number of internodes in the compartment. This mode tends to reduce the conical shape of the organic series of the axes internodes (Fig. 8.10c).

An internode numbered x in the structure list has $N_b(x)$ leaves. The number of similar internodes of the compartment with indices p and r is $N_i(x)$. The length of the internode is $h(x)$, and the sink for the ring element of the compartment to which it belongs is $p_{c2}(x)$. So the relative demand $D_c^r(t)$ can be written as:

$$D_c^r(t) = \sum\nolimits_x h(x) \cdot p_{c2}(x) \cdot \frac{N_b(x)}{N_i(x)} \tag{8.17}$$

The volume of the ring on the internode x is then obtained:

$$q_c(x) = h(x) \cdot p_{c2}(x) \cdot \frac{N_b(x)}{N_i(x)} \cdot \frac{Q_c(t)}{D_c^r(t)} \tag{8.18}$$

In this mode for secondary growth, the tips of the axes can be thick.

8.4.2.4 Variation Around Pressler Law

In the standard case, the cross-section of the ring element is constant, as in the "pipe model". However, Mitchel (1975) noted a progressive decrease in the section surface of the external ring along the trunk in the case of trees growing in poor environmental conditions. This is due to a decay of the ring sink P_c which changes with the age of the tree, the ring element keeping a constant cross-section along the path from a leaf to the collar. A damping coefficient ($k_C < 1$) is introduced to decrease the biomass allocated to the secondary growth, called modified Pressler law.

The sink P_c of the ring element becomes $P_c(\mathrm{N}) = P_c \cdot (k_c)^N$. The study of the maple tree reveals the necessity of this mode to fit the trunk profile (Chap. 10). Another case occurs if the cross-section area of the ring element is not constant. It could decrease or increase along the path that leads from the leaves to the collar (Fig. 8.10d). This is the case of the Mahogany tree (Chap. 10). Both Maple and Khaya trees need to use the modified Pressler law.

8.4.2.5 Mixture of Biomass Propagations

Regardless of the volume of secondary growth to be distributed, a mixture of these different propagation modes can be implemented. Secondary growth is thus a combination of Pressler's (Sect. 8.4.2.1) and uniform propagation modes described above (Sect. 8.4.2.1), to better adjust the profiles of the observed trunks.

In this approach, it is assumed that a fraction λ of the biomass allocated to rings follows Pressler's mode (Fig. 8.10a) and that the remaining fraction $(1 - \lambda)$ is allocated uniformly to all internodes of the total plant structure that have at least one living leaf upstream (Fig. 8.10b), in proportion to their lengths. It is, therefore, necessary to compute two demands.

The first demand, corresponding to the case $\lambda = 1$, is the same as the so-called Pressler's mode, which is already explained in Sect. 5.2.1:

$$D_{1c}^{\,r}(t) = \sum\nolimits_x n_a(x) \cdot h(x) \tag{8.19}$$

The second demand, corresponding to the case $\lambda = 0$, stipulates that each leaf distributes biomass to all nodes of the structure, as in Sect. 5.2.2:

$$D_{2c}^{\,r}(t) = \sum\nolimits_x \mathrm{sign}(n_a(x)) \cdot h(x) \tag{8.20}$$

where x denotes a position of the internode in the structure.

The volume of the ring on the internode x is then obtained by linearly interpolating the two modes:

$$q_c(x) = h(x) \cdot \left(\lambda \cdot \frac{n_a(x)}{D_{1c}^{\,r}(t)} + (1 - \lambda) \cdot \frac{\mathrm{sign}(n_a(x))}{D_{2c}^{\,r}(t)} \right) \cdot Q_c(t) \tag{8.21}$$

According to the values of the coefficient λ, for the same biomass allocated to rings, the diameter profile of the axes is significantly modified (Fig. 8.11). The biomass produced by the entire crown is allocated to each internode in proportion to

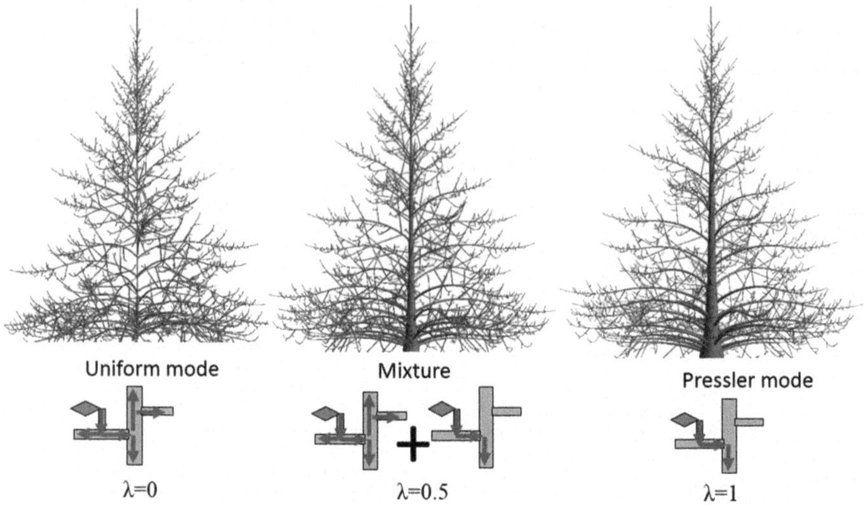

Fig. 8.11 Influence of the coefficient λ on the diameter of vegetative axes (Digiplante software). The linear combination of two laws of biomass propagation in the rings allows a better fitting of the trunk profile

the leaf area above its position within the plant architecture with a ratio λ and in proportion to the total number of leaves with a ratio $1-\lambda$. If $\lambda = 0$, each phytomer in the tree receives biomass from all leaves. The axes then have a cylindrical shape and are very thin. If $\lambda = 1$, we find Pressler's law, and the axes have a conical shape. All intermediate situations are possible for the same amount of biomass distributed in the rings (Fig. 8.11).

Observations on real trees have allowed fitting the parameter λ. For young pines, it was found that λ is close to 0, while it was found closer to 1 for older trees (Guo et al. 2009). This result seems logical. When the tree structure is small, the biomass can circulate everywhere, while on older trees, the biomass allocation of a leaf for rings should not "go up" easily to the axes tips.

More precisely, we can see that the effect of λ is mainly reflected on the trunk (PA 1) and the most peripheral branches. Intermediate branches receive biomass upstream about as much as they do downstream (Fig. 8.12). This algorithm was implemented in Digiplante software (Cournède et al. 2008), QingYuan (Hua and Kang 2011), and Gloups (de Reffye and Jaeger 2013).

8.4.2.6 Effect of Branch Mortality and Reiteration on Rings Formation

The lifespan of the branches affects the growth in weight and height of trees (Chap. 5, Fig. 5.27). This effect also affects secondary growth. Increased branch

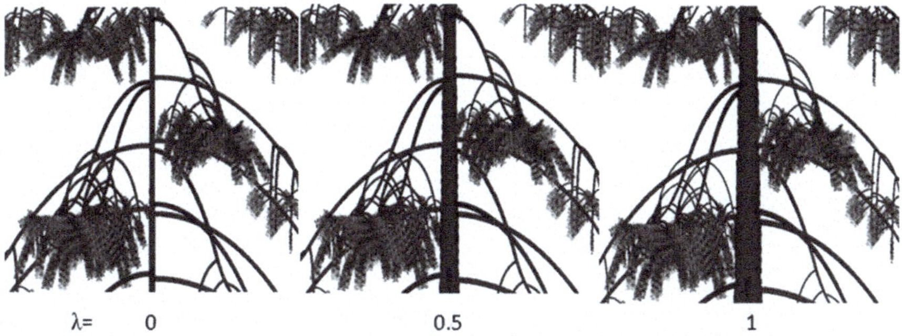

Fig. 8.12 Local influence of the coefficient λ. Detail of whorl for simulated conifers. For $\lambda = 0$, the axes are of comparable thickness. For $\lambda = 1$, it is mainly the trunk that benefits from the distribution of biomass in the rings. The peripheral branches are empty and the intermediate branches receive biomass about as much as they produce (QingYuan software, Hua and Kang 2011)

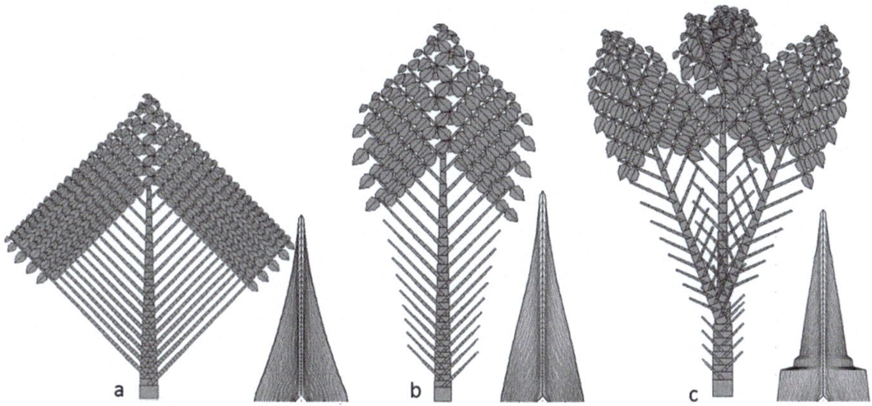

Fig. 8.13 Effects of branch mortality and reiteration on the secondary growth in the case of Pressler's mode (Gloups software). Plant architectures are simulated at age 40, with corresponding ring profiles are shown. (**a**) Control tree without branch mortality. (**b**) Branch development stops at 12 DCs. The trunk is taller, but its diameter and the biomass are smaller. (**c**) The tree trunk reiterates. The diameter of the trunk shows discontinuities at the location of the reiterations, which results from Pressler's law

mortality decreases trunk diameter and increases tree height. This is verified experimentally. In general, increasing planting density increases branch mortality and significantly increases tree height compared to open field growth (Fig. 8.13 a and b). Reiteration causes a discontinuity in the diameter of the trunk, which is a consequence of Pressler's law (Fig. 8.13c).

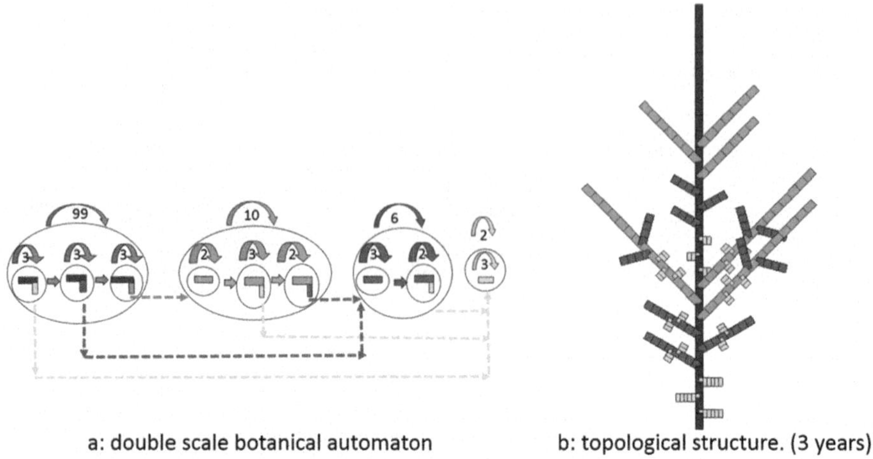

Fig. 8.14 The selected double-scale botanical automaton. (**a**) Produces a topological structure with four physiological ages (PAs). (**b**) The growth units (GUs) of the trunk (blue) have nine phytomers, and the development is unlimited. The GUs of the primary branches (green) have seven phytomers, and the development of the branches is limited to 10 GUs. The GUs of the secondary branches (red) have five phytomers, and the development of branches is limited to six GUs. The GUs of the short axes (yellow) have three phytomers, and the development of the axes is limited to two GUs. Acrotony characterizes the branching pattern (Color figure online)

8.4.3 Secondary Growth and Plants with Rhythmic Development of the Architecture

The secondary growth model remains the same, but the functioning of the double-scale botanical automaton complicates the simulation of the ring profile where the GUs are well identified.

8.4.3.1 Case Study with a Double-Scale Botanical Automaton

Trees can grow with either preformed or neoformed GUs, with the same development pattern. Organic series can be simple or compound. As an example, a double-scale botanical automaton is considered here (Fig. 8.14). These trees conform to the development and growth of those in Chap. 6 (Sect. 6.5.6.2.1 and 6.5.6.2.2) but with the secondary growth included.

8.4.3.2 Synchronous Trees with Preformed Growth Units

In the case of preformed GUs, the expansion of the GU is considered to take place in the last cycle of its development. Therefore, all phytomers are the same size. Annual

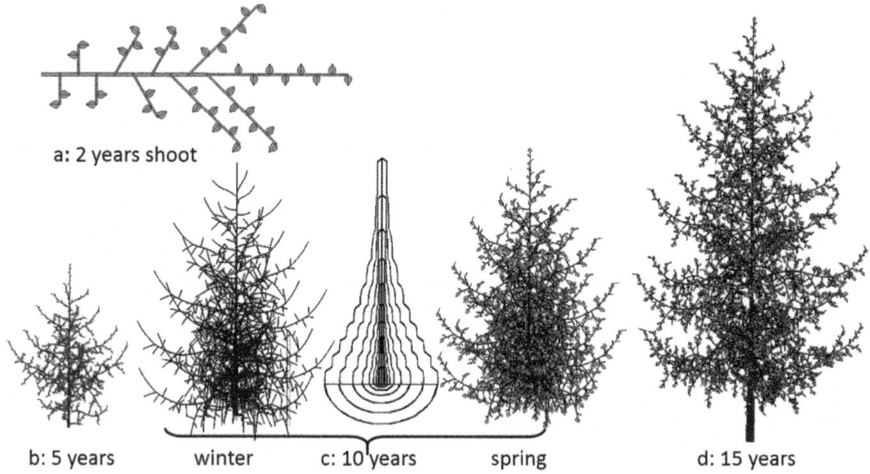

Fig. 8.15 Simulation of a tree with four physiological ages (PAs), with acrotony branching and preformed growth units (GUs). (**a**) GU of the trunk at two years old. The phytomers have the same size because of the pre-formation. (**b**) 3D visualization of the tree at five years old. (**c**) Tree at 10 years old (winter and spring), visualization of the stack of longitudinal and radial rings (strongly magnified) at 10 years. Effect of rhythmic growth is visible. (**d**) Tree at 15 years old. The dead axes are pruned (Gloups software)

photosynthesis is concentrated on this last cycle. A pause cycle is added to separate the annual growth periods. In this case, the GU can be considered as a meta-phytomer consisting of an internode and a leaf whose biomasses are the sum of the respective biomasses of the organs for the phytomers that make up the GU. These form the compound series of the GU, which are sufficient to characterize the growth.

Figure 8.15 shows the growth of a synchronous tree with pre-formed GUs based on the automaton in Fig. 8.14.

8.4.3.3 Synchronous Trees with Neoformed GUs

This case is the most complex because it is necessary to take into account in detail the functioning of meristems during annual DCs, which generate structural and random pauses. At each DC (which is also a growth cycle here), new ring elements are formed to ensure the expansion of secondary growth. The development is periodic. Each year is characterized by several cycles that end with the winter pause. The pith of the GUs is no longer cylindrical because all the phytomers of a shoot have a different expansion depending on the variation of *Q*/*D* ratio.

Figure 8.16 shows the growth of a synchronous tree with neo-formed GUs based on the automaton in Fig. 8.14.

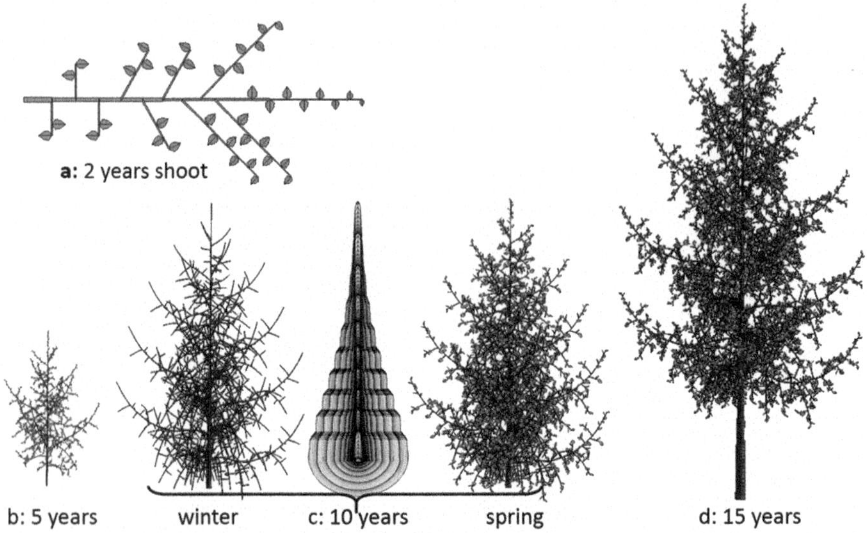

Fig. 8.16 Simulation of a tree with four physiological ages (PAs), with acrotony branching and neoformed growth units (GUs) (Gloups software). (**a**) GU of the trunk, two years old. The phytomers have different sizes because of the neo-formation. (**b**) 3D visualization of the five years old tree. (**c**) 3D visualization of the 10 years old tree (winter and spring), visualization of the stack of longitudinal and radial rings (strongly magnified) at 10 years. Effect of rhythmic growth is visible. (**d**) 3D visualization of the 15 years old tree. The dead axes are pruned

8.4.3.4 Comparison Between Ring Stacks in the Cases of Pre- and Neoformation

The preformation hypothesis implies that the expansion of the GU is carried out in a single cycle, and the GU pith has a cylindrical shape. In the case of neoformation, the internodes of the GU of the year expand in each cycle, which gives a conical shape to the pith of the GU. The following year, the ring covering the pith is thicker at the bottom than at the top of the axis, as it is strengthened by the biomass input of the successive branches encountered (Fig. 8.17).

8.5 Study of Organic Series in the Case of Secondary Growth and Branch Mortality

While the stochastic development model based on the botanical automaton covers most of the observed variety of tree architectures (pre- and neoformation, polycyclism, etc.), the formalism for the growth model is confronted with two difficulties linked to the complexity of these architectures, especially when secondary growth and branch mortality are included in the model. The first one concerns the sampling of the measures to be carried out. The second one concerns the costs in computing

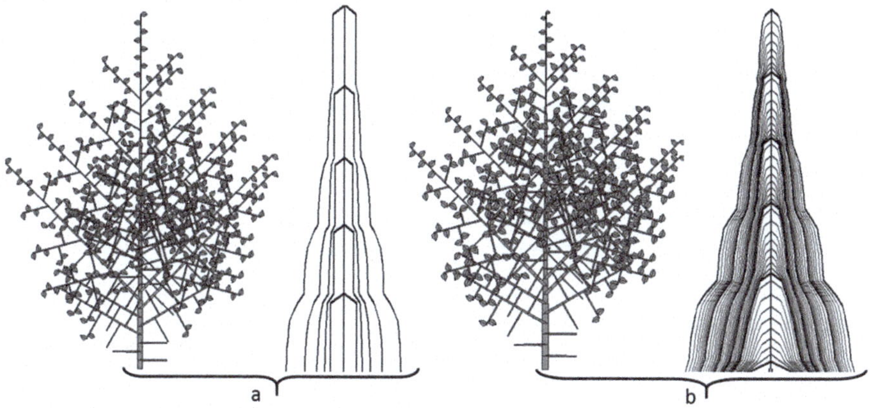

Fig. 8.17 2D representations of pre- (**a**) and neo-formed (**b**) plant structures generated by the botanical automaton of Fig. 8.14 at five years. Stacks of rings are magnified. The pith of the growth unit (GU) is cylindrical in the case of pre-formation and conical in the case of neo-formation (Gloups software)

time and volume of data related to the large tree structures which largely exceed the capacities of ordinary PCs.

In the simulation mode, the rings are formed only on the realized phytomers, while the void entities of the chronological structure do not have any. The organic series are constituted according to their rank after the deletion of the empty entities.

In the computation mode, based on the potential structure, there are no empty entities except for the structural pauses that result from rhythm ratios and branching delays (see Chaps. 2, 3, and 6).

8.5.1 Miscellaneous Aspects About Meristem Mortality

The leafy axes stop developing after the abortion of the terminal meristem. The terminal meristem may abort either randomly depending on its viability (see Chap. 3, Fig. 3.8) or undergo a state transition in a deterministic way after a fixed number of DCs (see Chap. 2, Fig. 2.5). In the latter case, two cases can be simulated regarding the cessation of axis development: either the axes of a structure reach their maximum development, independently of each other, or the whole structure stops its development when the main axis stops (see Chap. 7, Fig. 7.1).

Branch mortality induces some loss of data since the branches that have fallen cannot be measured. Therefore, the sampling cannot be done at the level of organs for the whole architecture since they are incomplete. Instead, only simple or compound organic series (in the case of GUs) can be used.

In Chap. 6, it is shown that measurements on the organic series should only be carried out on living branches. As the plant demand is computed, only the living branches can be grouped into homogeneous sets and analyzed by the crown analysis method. They contain all the information needed to compute the source-sink parameters of the tree by an inverse method. The GUs are therefore grouped according to their PA and CA and are detailed in simple or compound organic series.

The data in Chap. 6 are recalled here and the mortality is considered to affect only the branches but not the stems.

Case of Continuous Growth (Simple Scale Automaton)
The simple organic series (phytomers) of the living axes are sorted by their rank from the top.

Case of Rhythmic Growth (Double Scale Automaton)
The compound organic series (GUs) of the living axes are sorted by their rank from the top. Inside a GU, the phytomers are simple organic series.

A phytomer is indexed according to the rank of the GU on the branch from the top of the axis and according to its rank from the base of its GU. In the case of neoformation, a simple organic series contains information related to organ expansion.

Within the living axes, the GUs are compound organic series (made of the total weights of their phytomers) and are sorted according to their rank from the top of the axis.

In the compound organic series, the total weight of the organs in the GU is considered as a compartment. The information related to the expansion of the organs is lost, but in the case of preformation it is no longer necessary, since it is considered for simplicity that the expansion of all the phytomers of the GU is synchronous and takes place in a single cycle (Chap. 6). All phytomers are considered identical.

It appears that the study of trees with preformation is greatly simplified. Indeed, a GU can be considered as a metaphytomer.

A major sampling problem arises with the combination of the *Bernoulli* process and mortality. Meristem mortality is effective in the case of the simulation, but it does not exist in the potential structure: in that latter case, it is replaced by the notion of meristem viability. Strictly speaking, the distribution law of the GU of a branch located at k GUs from the top of the stem is a compound law. If p is *Bernoulli*'s parameter of the stem, at k GUs from the top, the number of GUs formed N follows a negative binomial law k, p.

N being fixed, the number of GUs formed on the branch is a compound law N, b, c, where b is the *Bernoulli* parameter of the branches, and c is the viability of the meristems (see Chap. 3). If the computation of this distribution is theoretically solved, it is too tedious, in practice, to make the necessary measurements, and there is a loss of information resulting from mortality. However, it is not necessary to measure everything on the branches: a limited number of data are sufficient to be able to compute the source–sink parameters of the structure by the inverse method. By eliminating dead branches from the measurement, we notice that the simulated living branches express all the homologous branches of the potential structure. It is, therefore, relevant to assume that the ends of the living branches match those of the potential structure. More or less important deviations should appear beyond that.

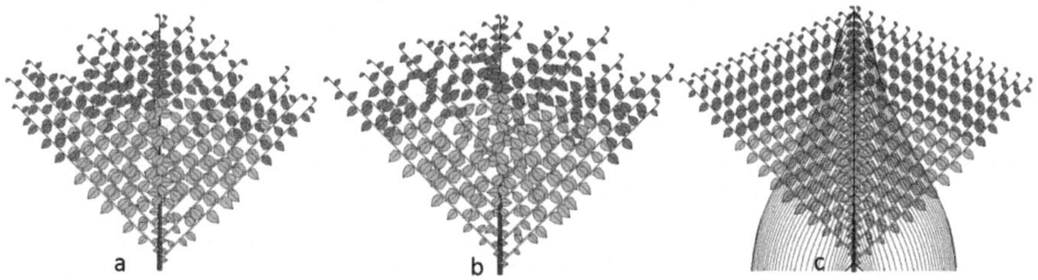

Fig. 8.18 (**a**) and (**b**) Two random simulations of a monopodial plant with a continuous growing monopod (30 DCs). (**c**) Potential structure and profile of rings with oversized diameters (Gloups software)

This can be verified by simulation and therefore provides a good solution to the problems of tree sampling.

8.5.2 Simulated Case Studies with Continuous Growth

8.5.2.1 Case with *Bernoulli* Process Without Mortality

The first step is to study the model properties including secondary growth but without branch mortality ($c = 1$).

A monopodial woody plant (Fig. 8.18) is simulated with the following characteristics:

Number of PAs: 2; *Bernoulli* parameters of meristems: $b = 0.8$ for stem and branches; number of simulated plants at 30 DCs: 50; duration of leaf functioning $t_a = 10$ DCs; duration of leaf and internode organ expansions: $t_x = 5$ DCs; sink strength of leaves $p_a = 1$; sink strength of internodes $p_i = 1$; sink strength of rings $p_c = 0.75$. In addition, $r = 10$ and $Sp = 1000$ cm^2. The growth is limited. The average organic series of the axes are built.

Figure 8.19 compares the simulated organic series with their computed counterparts from the potential structure. The good agreement is consistent with the results of Chap. 6. The novelty is the inclusion of secondary growth in the model, particularly visible on the internodes of the stem.

8.5.2.2 Case with *Bernoulli* Process and Mortality

In the second step, we study the secondary growth with the mortality of branch meristem ($c = 0.95$). Figure 8.20 shows two simulated random topological structures and the potential structure. It is observed that branches with dead meristems have fewer and well-developed living leaves (green), whereas the tips of living branches (red circle) have small developing new leaves.

Figure 8.21 compares the simulated organic series with their computed counterparts from the potential structure. But only living simulated branches are taken into

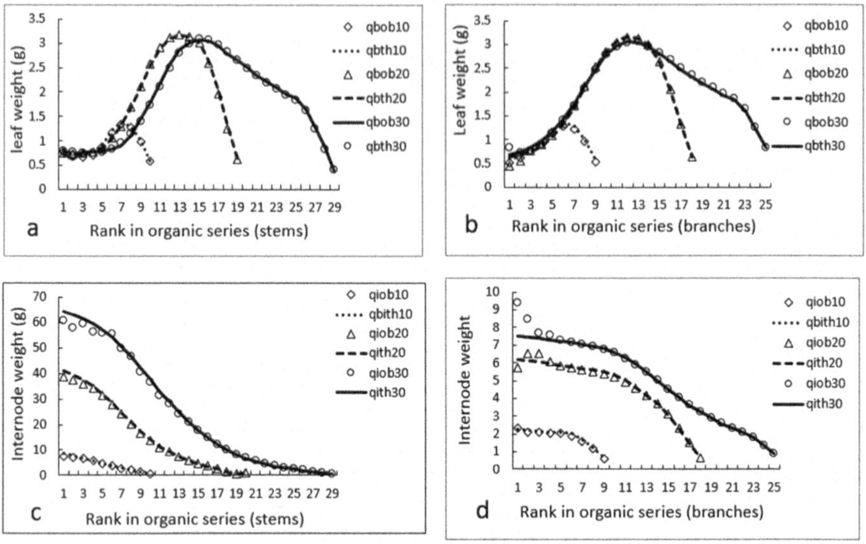

Fig. 8.19 Three growth stages are overlaid at 10, 20, 30 DCs. (**a**) and (**b**) Organic series of leaves on the stems and branches. (**c**) and (**d**) Organic series of internodes on the stems and branches. The secondary growth is very apparent. The simulations are in good agreement with their theoretical counterparts extracted from the potential structure (Gloups software)

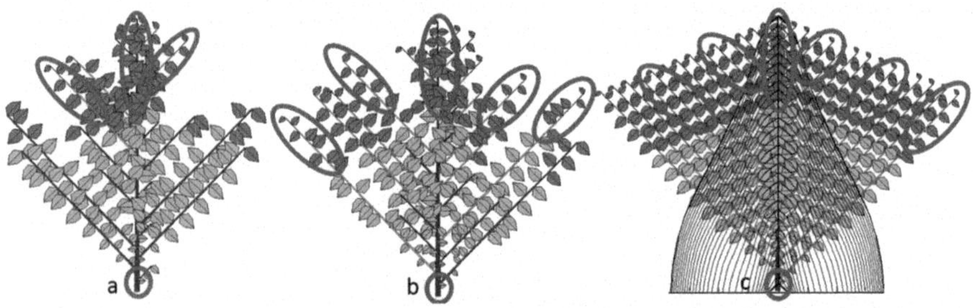

Fig. 8.20 (**a**) and (**b**) Two random simulations of a monopodial plant with continuous growth (30 DCs) and branch meristem mortality. (**c**) Potential structure and profile of rings with exaggerated diameter. The potential structure has no mortality. It can be seen that the tips of alive branches have small leaves that have just appeared, whereas the tips of dead branches have well-developed leaves because they continued to grow after the terminal meristem died. Only the ends of the organic series of living branches should be taken into account, as well as the basal part of the trunk, to compute the source–sink parameters. Measurements on the plant are taken in the areas circled in red (Gloups software)

account. It can be seen that, according to the assumption, only the branch ends are correctly fitted.

In the sampling strategy, therefore, only the last phytomers at the end of the axes will be measured. The phytomer at the base of the stems must always be preserved as an indispensable reference for Pressler's law. However, the computation slightly overestimates the weight of its internodes. The number of data with this sampling

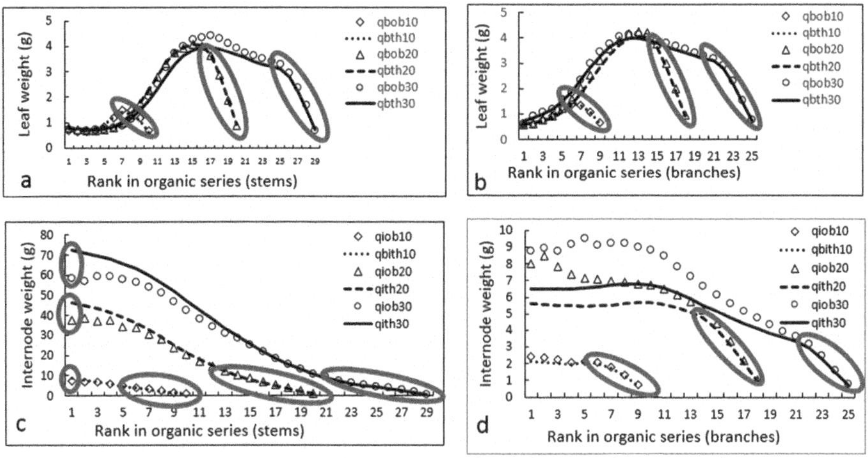

Fig. 8.21 Three growth stages are overlaid at 10, 20, 30 DCs. (**a**) and (**b**) Organic series of leaves on the stems and branches. (**c**) and (**d**) Organic series of internodes on the stems and branches. The secondary growth is very apparent. The organic series of leaves are little disturbed, but those of internodes are, especially for those of branches. Only the ends of branches circled in red are taken into account for the computation of the parameters (Gloups software) (Color figure online)

Table 8.1 Source-sink parameters of the simulated plant

Parameters	r	Sp	p_c	p_b	p_I
Original values	10.0	1000	0.75	1	1
Fitted	10.5	1024	0.57	1	1.05

Table 8.2 Source–sink parameters of the simulated plant

Parameters	r	Sp	p_c	p_b	p_I
Original values	5	1000	1	1	1
Fitted	5.1	906	0.94	1	1.00

strategy is sufficient to compute the source–sink parameters of the simulated plant with a good approximation.

Using the inverse method presented in Chap. 9, which computes the source–sink parameters from the simulated organic series, it can be seen that the parameters are correctly estimated (Table 8.1).

8.5.2.3 Study of a Particular Case

In the case of temperate trees with preformation, the meristems do not undergo a *Bernoulli* process ($b = 1$). Each year a meristem makes a new GU or dies ($c < 1$). Crown analysis provides the stochastic parameters (branching and mortality) of meristem functioning. The GU can be assimilated to a meta-phytomer. A year corresponds to a cycle. The duration of leaf functioning is therefore $t_a = 1$ DC and

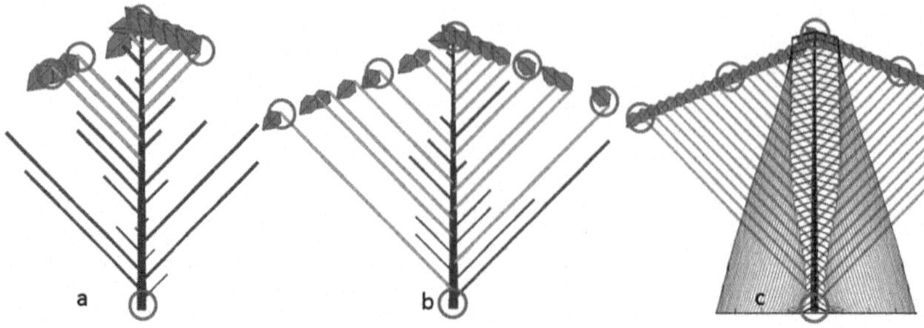

Fig. 8.22 (**a**) and (**b**) Two random simulations of a monopodial plant with continuous growth (40 DCs) and mortality of branch meristems ($c = 0.95$). (**c**) Potential structure and profile of stem rings with an exaggerated diameter. The functioning time of the phytomers is 1 DC. Dead leaves have been removed, living branches are green and dead branches are gray. Measurements on the plant are taken at the last phytomer of the living branches and at the basal growth unit (GU) of the stem, in the areas circled in red (Gloups software) (Color figure online)

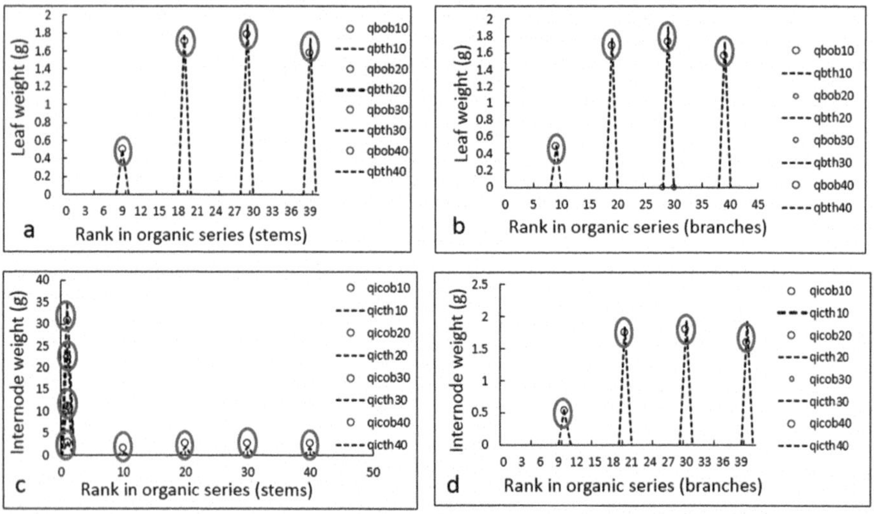

Fig. 8.23 Four growth stages are overlaid at 10, 20, 30, 40 DCs. (**a**) and (**b**) Organic series of leaves on the stems and branches. (**c**) and (**d**) Organic series of internodes on the stems and branches. The secondary growth is very apparent. Only phytomers at the end of living branches and at the beginning of the stem are taken into account (Gloups software)

the duration of organ expansion is also $t_x = 1$ DC. The parameter values of the simulation are shown in Table 8.2.

The measurements on this type of tree concern only the weight of the leaves and internodes of the terminal GUs, which form a minimum organic series with a single item for the branches and with two items for the trunks for which only the weights of the leaves and internodes of the basal and terminal GUs are measured (Figs. 8.22 and 8.23).

The same tree at different developmental stages should be examined to carry out the study. The minimum measurements made on the different types of axes are sufficient to recompute the source–sink parameters. The parameters are correctly estimated (Table 8.2). This case is a simplification in the Phyto_Facto mode, as described in Sect. 6.3.

The method is, therefore, quite effective, despite the limitations of the sampling (limited number of measurement points). This strategy has been successfully applied to the Maple tree (Chap. 10).

8.5.3 Simulated Case Studies with Rhythmic Growth (Preformed GUs)

Most temperate trees exhibit rhythmic growth with preformed GUs. There is no *Bernoulli* process in the creation of new GUs. Each year, a meristem always creates a new GU (no pause, $b = 1$) or it aborts ($c < 1$). For simplification, the number of phytomers per GU is constant if the phytomer distribution per GU is a positive binomial law.

A monopodial rhythmic woody plant with preformed GUs is simulated with the following characteristics:

Number of PAs: 2; the number of DCs per year: 6; the number of phytomers per GU: 5; the number of branches per GU on the stem: 2; *Bernoulli* parameters of meristems: $b = 1$ for stem and branches; viability of branch meristems: $c = 0.95$; the number of simulated plants at 25 DCs: 30. Sinks are constant. Sink strength of leaves

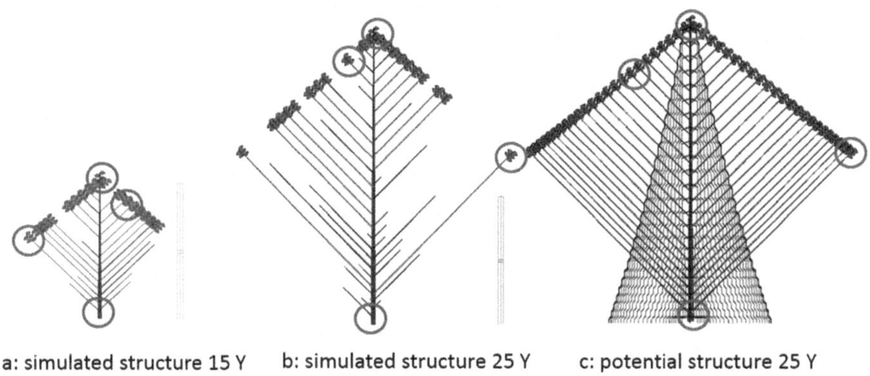

Fig. 8.24 (**a**) and (**b**) Two random simulations of a monopodial plant with rhythmic growth (15 and 25 DCs) and mortality of branch meristems ($c = 0.95$). (**c**) Potential structure and profile of stem rings with exaggerated diameter. The functioning time of the GUs is 1 year. Dead leaves have been removed, living branches are green and dead branches are gray. Measurements on the plant are taken at the basal GU of the stem, and the terminal GU of the living branches in the areas circled in red (Gloups software) (Color figure online)

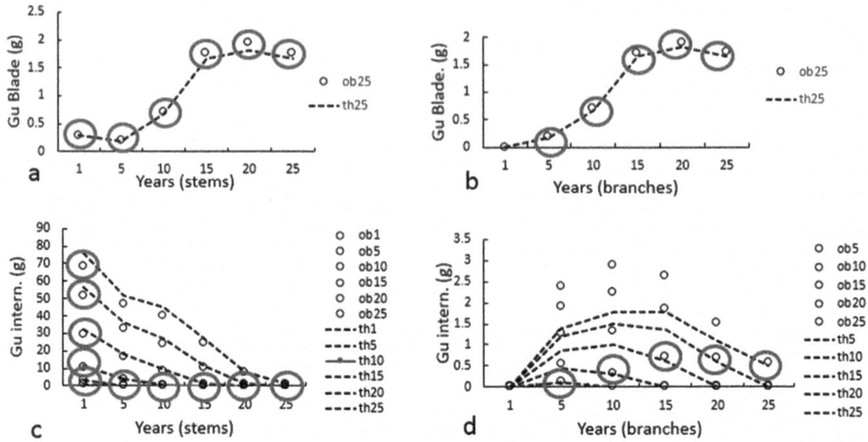

Fig. 8.25 Six stages of growth are overlaid at 1, 5, 10, 15, 20, 25 years of growth. (**a**) and (**b**) Compound organic series of the sum of leaves per growth unit (GU) of stems and branches. (**c**) and (**d**) Compound organic series of the sum of internodes per GU of stems and branches, secondary growth is very apparent. Only phytomers at the end of living branches are taken into account (Gloups software)

$p_a = 1$; sink strength of internodes $p_e = 0.5$; sink strength of rings $p_c = 2$ (Pressler mode). In addition, $r = 4$ and $Sp = 1000\ \text{cm}^2$. The growth is limited.

Figure 8.24 shows two simulated random topological structures at 15 and 25 years and the potential structure. Virtual measurements on simulations were carried out on GUs in years 1, 5, 10, 15, 20, and 25.

As already mentioned, the preformed GUs have an expansion that lasts only one DC, which is the last cycle of the GU development. The phytomers of a GU are all similar. In this case, it is not necessary to detail them. Therefore, we are interested only in the compounded organic series along the axes, i.e., on the sum of organ weights (leaves and internodes) of the GU. This computational plant, with rhythmic growth, is similar to the plant of Sect. 6.5.2.3 with continuous growth, and the GU is like a meta-phytomer.

Under these conditions, the weight of leaves does not change once the GU is set up, and they disappear in the following year. In contrast, the weight of internodes of the GUs changes every year due to secondary growth.

Measurements on simulated plants are carried out only on living branches, which necessarily affects the comparison with measurements extracted from the potential plant structure.

Figure 8.25 compares the simulated compound organic series with their computed counterparts from the potential structure.

Note that the simulated weight of the leaves per GU along the axes is in good agreement with the computed weight from the potential structure. For the sums of internodes per GU on the stem, the average weight of simulations agrees with that of the computation. For those of the branches, only the terminal GUs are well fitted, as

Table 8.3 Source–sink parameters of the simulated plant

Parameters	r	Sp	p_c	p_b	P_e
Original values	4	1000	2	1	0.5
Fitted	4.55	1102	1.6	1	0.31

expected. But it is quite sufficient to estimate the values of the source–sink parameters of the system.

Estimating the model parameters from the compound and simulated organic series using the method of Chap. 9 gives the following results (Table 8.3).

Given the small sample size (30 plants), the parameter values are quite well recomputed.

The method of construction of the organic series in the case of preformation, which uses only living branches, can therefore be preserved in the case of trees with meristem mortality. This method was successfully used to fit the maple tree with GreenLab (Chap. 10).

8.5.4 Simulated Case Studies with Rhythmic Growth (Neoformed GUs)

Unfortunately, in the case of neoformation, it is not possible to concentrate photosynthesis on a single expansion cycle. In each cycle, the development and the growth processes must be coupled. Moreover, in preformation, if the development lasts about 10 DCs (Maple tree), in the case of neoformation (poplar), the development is counted in ten times of DCs. The computing and memory capacities of the basic PCs are then quickly exceeded after about 10 years of growth.

The graphic representation of the topological structure of the simulated plant and its architecture becomes a challenge. Nevertheless, for small trees, it can be verified that the method used for the trees with preformation is also suitable for the case of neoformation.

However, it is also necessary to use the simple organic series generated by the GU, because the compound organic series are insufficient to calibrate the model, due to the visible manifestation of sink functions at the level of the phytomers, which are all different.

A monopodial rhythmic woody plant with neoformed GUs is simulated with the following characteristics:

Number of PAs: 2; the number of DCs per year: 10; the number of phytomers per GU: 9; the number of branches per GU on the stem: 2; *Bernoulli* parameters of meristems: $b = 1$ for stem and branches; viability of branch meristems: $c = 0.95$; the number of simulated plants at 16 DCs: 30; functioning duration of leaf: $t_a = 9$; expansion duration of leaf and internode: $t_x = 5$. When the GUs reach the end of development, all organ expansions stop. Sinks are constant. Sink strength of leaves

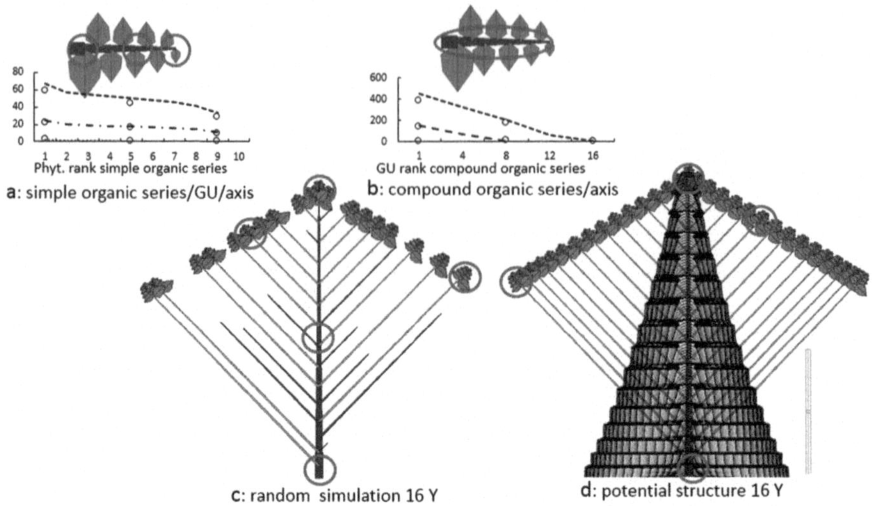

Fig. 8.26 Measurements made on the tree. (**a**) Detail of the weights of the internodes at the ranks of the growth units (GUs) 1, 5 and 9 of the displayed basal stem at the tree ages 1, 8 and 16 years. All the phytomers have different sizes. The cessation of growth at the end of the development, in this particular case, makes the size of the phytomers decrease. (**b**) Evolution of the total weight of the internodes along the stem at the ranks of the GUs 1, 8 and 16 at the tree ages 1, 8 and 16. (**c**) A random simulation of the tree at 16 years. A distinction is made between dead and living branches and the locations of the measurements on the architecture. (**d**) Potential structure of the tree and exaggerated display of the ring stacks (Gloups software)

$p_a = 1$; sink strength of internodes $p_e = 0.5$; sink strength of rings $p_c = 2$ (Pressler mode). In addition, $r = 20$ and $Sp = 5000\ \text{cm}^2$. The growth is limited.

The tree is simulated at the age of 16 years. Sample measurements from the simulations are virtually taken for ages 1, 8 and 16 years. For the branches, the terminal GU (1 year) and, on the stems, the median (8 years) and basal (16 years) GUs were additionally measured. Simple organic series of the GU are measured on the first, fifth, and ninth internodes of the GU. It is not necessary to measure all the phytomers. Compound organic series measured at the same ranks of the GUs on the axes contain the sum of organ weights. In this case, there is redundancy, because a simple organic series is sufficient. Nevertheless, it is useful to note the consistency of the computations (Fig. 8.26).

Figure 8.27 shows the weights of the leaves according to their positions in the GU at the ranks 1, 5, and 9, at tree ages 1, 8 and 16 years. Note that these measurements do not change for a given GU rank, and are definitive as the leaves die each year. The average of the 30 random simulations matches well with the homologous values derived from the potential structure.

Figure 8.28 shows the weights of the internodes according to their positions in the GU at ranks 1, 5, and 9, at three ages 1, 8, and 16 years. Note that these measurements change for a given GU rank because of the effect of the secondary growth. The average of the 30 random simulations matches well with most of the homologous

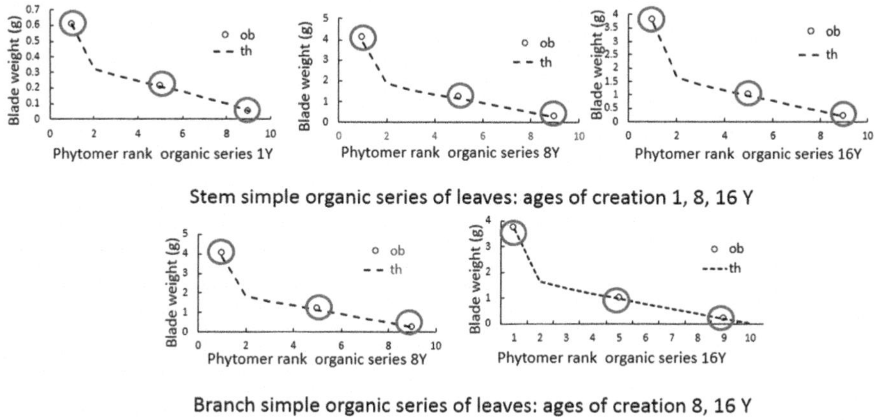

Fig. 8.27 Simple organic series of leaves inside a growth unit (GU) for stems and branches at different ranks on the axis and different ages of the trees. Observed (o) and computed values (—) are in red circles (Gloups software) (Color figure online)

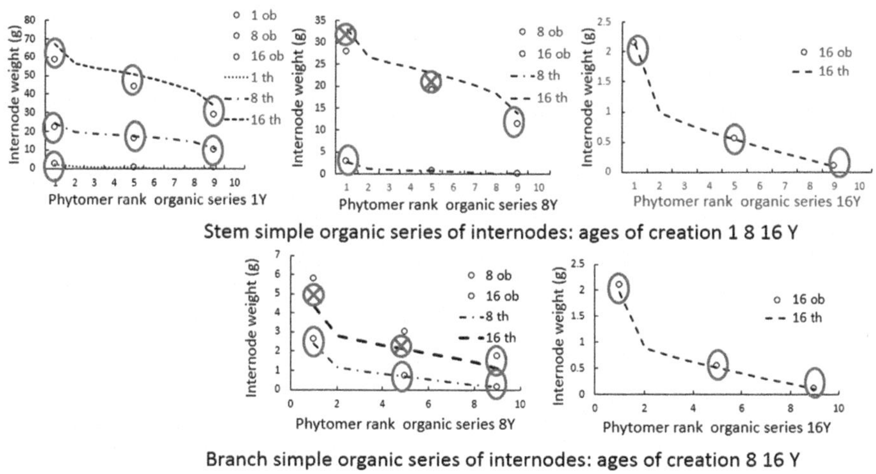

Fig. 8.28 Simple organic series of internodes inside a growth unit (GU) for stems and branches at different ranks on the axis and different ages of the trees. Observed (o) and computed values (—) are in red circles (Gloups software)

values derived from the potential structure. Matchings are good for basal and terminal GUs of stems and terminal GUs of branches.

The results at the level of the organic series composed of GUs are redundant with those of the simple series. It can be seen from Fig. 8.30 that the averages of the

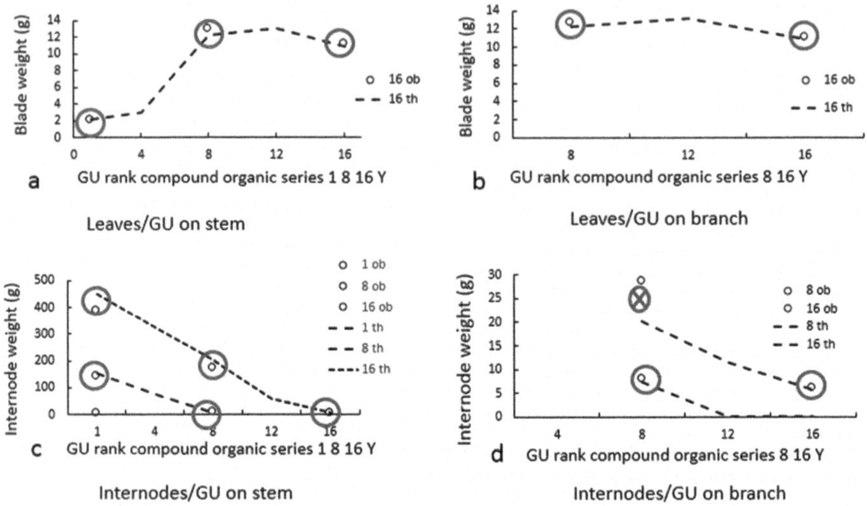

Fig. 8.29 Three growth stages are overlaid at 1, 8 and 16 years. (**a**) and (**b**) Compound organic series of the sum of leaves per growth unit (GU) on the stems and branches. (**c**) and (**d**) Compound organic series of the sum of internodes per GU on the stems and branches. The secondary growth is very apparent. Only phytomers at the end of living branches are taken into account (Gloups software)

Table 8.4 Source–sink parameter of the simulated plant

Parameters	r	Sp	p_c	p_b	p_l
Original values	20	5000	2	1	0.5
Fitted	21.3	5039	1.7	1	0.55

simulated organic series are in good agreement well with those of the computation. For the preformation of the branches, only the terminal GUs are well fitted. Here we put fewer checkpoints, to show that it is useless to multiply them (Fig. 8.29).

Fitting the model parameters from the compound and simulated organic series gives the following results (Table 8.4).

Given the small sample size (30 plants), the parameter values are quite well recomputed with a high degree of accuracy.

The method of construction of the organic series in the case of neoformation, which uses only living branches, can therefore be preserved in the case of trees with meristem mortality. This method was successfully used to fit the young Teck tree with GreenLab (Chap. 10), although no branch mortality was observed at this growth stage.

8.6 Fast Algorithms for Tree Architecture Simulation

The simulation of tree growth and architecture raises computational challenges.

The first problem is the sampling of plants. Mortality is a common occurrence in the plant structures, and the amount of missing data may sometimes exceed the data available for the measurement.

A second problem is that data on fruiting is often lacking, which biases the calculation of source–sink relationships.

The crown analysis is a generic method, which proposes a good approach for the parameter estimation of meristem functioning, even in cases where there is a significant amount of missing data.

This chapter shows that by selecting only the living branches, it is possible to estimate the source–sink parameters from the sampled organic series using the inverse method. However, it is important to take measurements on the terminal parts of the organic series of branches and stems, and on the basal parts of the stems.

Even after establishing the methodology of data acquisition and parameter estimation, simulating the development and growth of trees in their maturity raises further problems. The tree parameters are assumed to be stable or have a measurable drift modeling the senescence of trees implies simulating a radical change in functioning with the appearance of many delayed reiterations. Indeed, adult trees produce hundreds of thousands (or even millions) of phytomers, and their age in numbers of DCs can exceed a thousand. Computationally, it requires storing huge hypermatrices and gigantic topological structures, which is beyond the capacity of the standard PCs.

Face to this complexity, several solutions have been developed. One approach is to represent these trees at the scale of the GU only, without going deeper to the phytomer scale. This makes the measurements simplified since only the total weight of leaves and internodes of the GUs needs to be measured. Trees that make only preformed GUs, like the maples, fall into this category.

There are four possible computation modes.

8.6.1 List Mode

The list mode is the standard mode in plant modeling. It is ideal for herbaceous plants because it is the only one that can handle the absence of cohorts in organ expansion (see Chap. 7). In the case of trees, the notion of cohorts is generally valid. The list mode is operational on trees less than 10 years old (e.g., Teak tree in Chap. 10). Additionally, the 3D plant architecture can be displayed in list mode.

8.6.2 Matrix Mode

The matrix mode can be used to compute trees that can be several tens of years old and up to 100 years. It is ideal for estimating functioning parameters. While it retains the notion of organic series, it cannot generate the topological structure, and thus, cannot visualize its architecture. The tree is decomposed into various categories of axes, which constitute many separate parts. The method makes it possible to compute the number of produced phytomers, the simple and the compound organic series, the biomass of the different compartments (leaves, trunks, and branches), and to visualize the structure of the trunk (which is a compound organic series) with its stacks of rings. This is sufficient and goes even beyond the possibilities of most classical forest models.

8.6.3 Phyto-Facto Mode

The Phyto-Facto mode makes full use of the notion of metaphytomer present in list mode. The metaphytomer assembles all the phytomers of the GU to form a single meta-internode, conserving the same dimension in length and diameter, and a single leaf with the same surface area as the total number of leaves in the GU. All the axes carried by the GU, according to the principle of acrotony, are concentrated at the top of the metaphytomer. The size of the system in DCs is drastically reduced since only one year corresponds to one DC.

The produced biomass in Phyto-Facto mode is the same as in list mode, and the number of metaphytomers multiplied by the number of phytomers per GU gives the total number of produced phytomers in list mode.

The Phyto-Facto mode generates a topological structure whose dimensions are comparable to those produced in the list mode. It allows efficient factorization of the 3D architecture. The compound organic series of the trunk and branches, and the trunk structure with the stack of rings are preserved (Fig. 8.30).

Whenever the topological structure is still used, it is not practically possible to analyze trees exceed 50 years with this mode on a standard PC.

8.6.4 Phyto-Facto*Matrix Mode

The matrix mode can be combined with the Phyto-Facto mode. A very powerful algorithm is thus obtained, but the topological structure is lost. Despite this, we keep the compound organic series of the axes at the GU level and the trunk structure for the stacks of rings. This approach allows the estimation of source–sink parameters to be applied to trees that are older than 100 years.

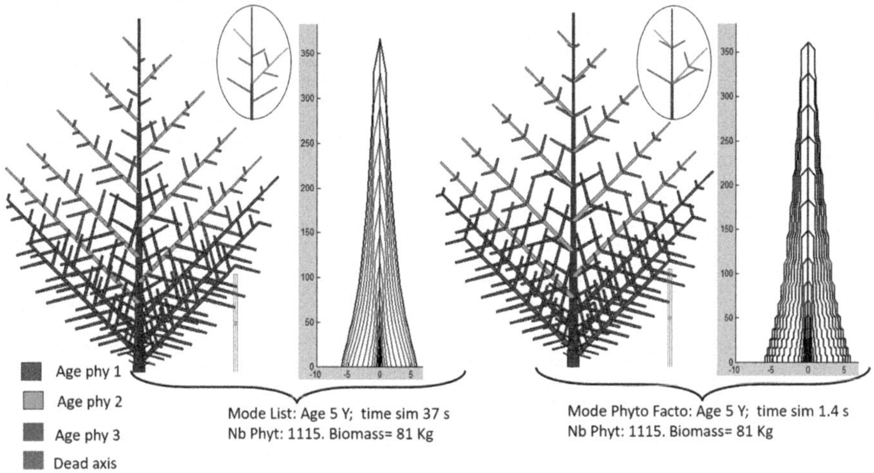

Fig. 8.30 Computation of a 15-year-old tree structure using both list mode and Phyto-Facto mode. Both modes give identical results for the number of produced phytomers and the produced biomass. The dimensions of the two structures obtained are also identical. The Phyto-Facto method dramatically reduces computation time (Gloups software)

Table 8.5 Performances of the different simulation modes

		List	Matrix	Phyto-Facto	Phyto-Facto*Matrix
Years	Number of phytomers	Time (s)	Time (s)	Time (s)	Time (s)
2	60	2.1	1.6	1.1	1.22
4	591	8.9	2.2	1.2	1.23
6	2334	41.3	3.1	1.5	1.28
8	6045	137.3	4.2	2.3	1.32
10	12,352	355.7	5.7	4.1	1.35
12	21,042	856.3	7.2	8.1	1.41
20	71,886	?	15.2	69.8	1.64
30	140,136	?	30.2	377.2	2.11
40	208,386	?	50.1	1239.5	2.79
60	344,886	?	104.1	?	4.55
100	607,888	?	286.2	?	11.66

Note: It was pointed out that trees with neo-formation generally require a lot of DCs to be simulated and that if the matrix mode remains operational, the Phyto-Facto mode cannot be applied, which excludes the study of source–sink relationships for old trees with neoformation. Nevertheless, if we limit ourselves to the compound organic series related to the GUs, we can still analyze them as if they came from preformed GUs and analyze the result. So, the fitting of the target file for the tree of Fig. 8.26 gives (Table 8.6):

8.6.5 Comparison of the Performance of the Different Modes

Table 8.5 shows the production in numbers of phytomers of the tree in Fig. 8.15 generated by the botanical automaton from Fig. 8.14. At the age of 100 years, more

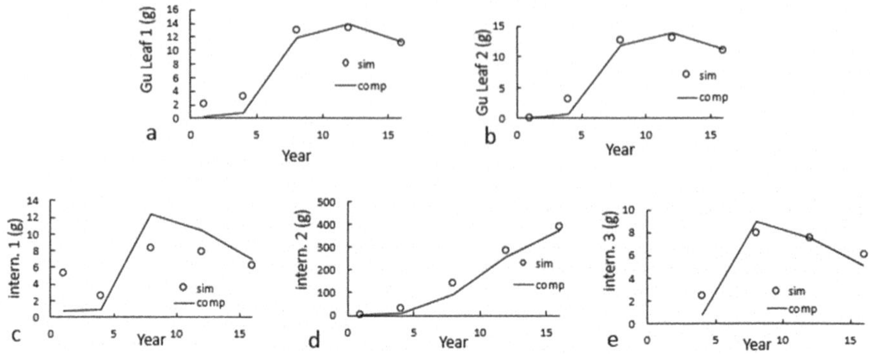

Fig. 8.31 Fitting compound organic series from neoformed GUs, as if they were preformed growth units (GUs). (**a**) and (**b**) Compound organic series of leaves at the tip of stems and branches for the 1, 4, 8 and 12 years. (**c**) and (**e**) Compound organic series of internodes at the tip of stems and branches for the 1, 4, 8 and 12 years. (**d**) Compound organic series of internodes at the basis of stems for the 1, 4, 8 and 12 years. Secondary growth is visible

Table 8.6 Computing source–sink parameter of the neoformed plant as if it was preformed

Parameters	r	Sp	p_c	p_b	p_l
Original values	20	5000	2	1	0.5
Fitted	2.0	1900	2.3	1	0.4

than 600,000 phytomers are produced. On a standard PC, the list mode cannot reasonably simulate exceed 20 years (computing time > 5 h in 2020). The Phyto-Facto mode reachs its limits at around 50 years. These computational limits are due to the management of the generated topological structure inside the limited computer memory.

The matrix mode, on the other hand, remains very efficient. It is particularly useful for the computation of functioning parameters by the inverse method. As previously said, it breaks the structure into separate parts, and the axes of all PAs are restored. In particular, the trunk and its ring stacks are represented, which is sufficient for most of the forestry applications. Thus, at 100 years in matrix mode, the computation time of the simulated tree is about 5 minutes. By applying the Phyto-Facto mode together, this duration decreases to about 10 seconds. The method thus allows for studying the functioning of adult trees, as shown in the study of the maple tree (Chap. 10).

Fitting computed compound organic series with the simulated ones is quite acceptable (Fig. 8.31). But the environmental parameters r and Sp are far from the original ones. They are no more functional parameters but calibration parameters. Nevertheless, the optimized sink strength values are close to the original ones. Moreover, the method reconstructs rather well the compound organic series. So, the method could be worth being applied.

Fig. 8.32 Influence of the ring sink and the parameter λ on the secondary growth. (**a**), (**b**) and (**c**) Effects of the sink of the ring elements on the growth of the tree: (**a**) If the sink is too weak, the tree grows too high, which jeopardizes its mechanical balance and decreases wood production. (**b**) The sink of the rings ensures the best wood production and the stability of the tree. (**c**) The sink is too strong, which prevents the growth (QinYuan software). (**d**) and (**e**) Effects of the parameter λ on the mechanical stability of the tree. The two trees (**d**) and (**e**) have the same functional parameters and make the same amount of wood for the rings. The only difference is the distribution that depends on λ in the architecture. For $\lambda = 0$, the stability is no longer guaranteed because the trunk is too thin. It is the buckling. For $\lambda = 1$, the biomass is accumulated on the trunk and allows the mechanical balance of the tree (Digiplante software)

8.7 Interaction Between Plant Growth and Secondary Growth

The sinks of the ring elements contribute to the plant demand. So, their increase induces a decrease in the *Q*/*D* ratio, and thus a decrease of the biomass that returns to the leaves and, ultimately, a decrease of the overall tree growth (Fig. 8.32a, b, c). A too strong ring sink reduces the growth and, therefore, wood production. A sink of ring element that is too small also reduces wood production since not enough biomass is allocated to the woody parts; the tree is also too slender, which jeopardizes its stability. The parameter λ, which modifies the distribution of biomass in the rings, changes the profile of the trunk and therefore the mechanical balance of the standing tree (Fig. 8.32d and e). There is a trade-off, which can be determined by optimization, between mechanical stability and wood production. This trade-off is studied in (Qi 2009).

Fig. 8.33 Images of secondary growth with the Pressler law. (**a**) Pine tree of 20 years (Digiplant software). (**b**) A poplar with reiterations of 30 years (QinYuan software)

8.7.1 Images of Secondary Growth

Digiplante (ECP) and Qinyuan (LIAMA) software, both written in C++, integrates models for the secondary growth. However, the Gloups software (MatLab) in its list mode, reaches its limit when simulating growth around 20 years old. Figure 8.33 shows this by depicting a pine tree and a poplar tree of 30 years old.

8.8 Conclusion

The secondary growth modeling in the functional-structural model GreenLab proposes its approaches for two issues: wood production allocated to the secondary growth and biomass distribution in the rings. It is recommended on trees, to work with the compound organic series of the axes at GU level to simplify the sampling.

Wood Production (Global), Allocated to the Secondary Growth
Each living leaf provides its contribution in the form of a ring element. This entity can be considered as an organ associated with the phytomer like the others (leaves, internodes, fruits). Each ring element has a dedicated sink. The number of ring elements corresponds to the number of functional leaves with the number N_a. A

single expansion DC is associated with each ring element. The sum of the sinks of the ring features is the demand of the secondary growth D_c which is added to the sum of the other sinks to form the overall demand D of the tree. Knowing the biomass Q produced by the photosynthesis, we can deduce the volume of biomass Q_c to be distributed in the rings: $Q_c = Q \cdot D_c / D$.

Biomass Distribution in Rings

The ring element can have different shapes, as proposed in the following:

- In Pressler mode, each leaf diffuses biomass, under the form of a ring element, which is a biomass entity with a constant cross-section that extends along the path through the tree that leads from the leaf to its base (collar). The volume of such an entity q_c is: $q_c = Q_c / N_a$. All the ring elements converge into a bundle, and the surface area of the ring on a given phytomer is proportional to the cross-section of the number of ring elements passing through it. The total biomass of these entities is the secondary growth compartment. The model mechanistically reproduces the notions of "pipe model" or "Pressler's law", which states that the surface area of a ring is proportional to the active leaf mass carried upstream. Note that the cross-section of a ring element may vary.
- In the uniform propagation mode, the ring element receives biomass from the ring compartment in proportion to the length of the internode and the number of leaves in the plant. In this case, all the ring elements have the same cross-section at a given age.
- A linear combination of the above two modes can be chosen to better model the biomass uplift in the crown: they can be combined to better approximate the diameter profile of the trunks and other vegetative axes.
- Other modes are possible. But note that the methods mentioned above have the advantage of being computable by fast algorithms (Matrix mode).

The parameters that control the secondary growth can be estimated from observations on trees. To this end, the organic series of internodes that geometrically describe the axis profile are fitted by an inverse method. The method distinguishes the contributions of the pith and the rings to the diameter of the internodes. The complexity of tree architectures raises several difficulties when applying the estimation procedure to real experimental datasets. Trees are more difficult to measure than shrubs and herbaceous plants. It is necessary to have regular, mono-specific and even-aged plantations. Data acquisition is cumbersome and requires specific equipment (ladder, nacelle, etc.). The usual dendrometric data from forest models are not detailed enough for the GreenLab model. For example, the height and the diameter of the trunk at the base are insufficient to describe the profile of a trunk.

Instead, measurements on trees are made according to sampling methods based on the notion of axis type, which makes them independent of the architectural model and of the size of the tree (see Maple Tree, Chap. 10). The crown analysis makes it possible to program the botanical automaton and to compute the development and demand of the tree. A sampling of the organic series should only be carried out at the last GUs taken at the tips of the living axes. Not only is this sufficient to assess the parameters of the tree, but it is also more economical from a practical point of view.

The growth can then be estimated by fitting organ cohorts and the measured organic series, by the respective cohorts and organic series computed by the model using the inverse method. In particular, the fitting of the organic series of lengths and diameters extracted from the measurements on the axes makes it possible to estimate the parameters of secondary growth and to reconstruct the stacks of rings in the vegetative axes, especially in the trunk, an important element in forestry.

The first woody plants studied with GreenLab (cotton, coffee, pine, beech, teak, maple, mahogany) made it possible to experimentally validate the ring model and the dedicated estimation procedure that was proposed (Chap. 10).

References

Bosanac B, Zanchi P (2005) Onyx TREE professional. http://www.onyxtree.com/

Cournède P-H, Kang M-Z, Mathieu A, Barczi J-F, Yan Y-P, de Reffye P (2006) Structural factorization of plants to compute their functional and architectural growth. Simul Trans Soc Model Simul Int 82(7):427–438

Cournède P-H, Mathieu A, Houllier F, Barthélémy D, de Reffye P (2008) Computing competition for light in the GreenLab model of plant growth: a contribution to the study of the effects of density on resource acquisition and architectural development. Ann Bot 101(8):1207–1219

Cournède P-H, Chen Y, Wu Q, Baey C, Bayol B (2013) Development and evaluation of plant growth models: methodology and implementation in the Pygmalion platform. Math Model Nat Phenom 8(4):112–130

de Reffye P, Jaeger M (2013) Modèles mathématiques du développement et de la croissance de l'architecture des plantes. Le cas du modèle GreenLab. In: Varenne F, Silberstein M (eds) Modéliser & simuler: épistémologies et pratiques de la modélisation et de la simulation. (Tome 1, volume 2). Éditions Matériologiques, Paris, pp 625–658. (Sciences &, Philosophie)

de Reffye P, Fourcaud T, Blaise F, Barthélémy D, Houllier F (1997) A functional model of tree growth and tree architecture. Silva Fennica 31(3):297–311

Eichorn F (1904) Relationships between dominant height and stand volume. All Forst Jagd 80:45–49

Guo H, Letort V, Lei X-D, Lu Y-C, de Reffye P (2009) Applying GreenLab model to adult Chinese pine trees with topology simplification. 2009 Third international symposium on plant growth modeling, simulation, visualization and applications, pp. 58–61. https://doi.org/10.1109/PMA.2009.24

Houllier F, Leban J-M, Colin F (1995) Linking growth modelling to timber quality assessment for Norway spruce. For Ecol Manag 74:91–102

Hua J, Kang M-Z (2011) Functional tree models reacting to the environment. ACM SIGGRAPH 2011 Posters. ACM, Vancouver

Kang M-Z, Heuvelink E, de Reffye P (2006) Building virtual chrysanthemum based on sink-source relationships: preliminary results. In: III international symposium on models for plant growth, environmental control and farm management in protected cultivation, 718, p. 129–136

Kang M-Z, Qi R, de Reffye P, Hu B-G (2009) GreenScilab: a toolbox simulating virtual plants in the Scilab environment. 2009 IEEE international workshop on open-source software for scientific computation (OSSC 2009)

Leban J-M, Houllier F, Goy B, Colin F (1991) La qualité du bois d'épicéa commun en liaison avec les conditions de croissance. Forêt Entreprise 80:13–26

Lintermann B, Deussen O (1997) A modelling method and user Interface for creating plants. Graph Interface 1997:189–198
Lintermann B, Deussen O (1999) Interactive structural and geometrical modeling of plants. IEEE Comput Graph Appl 19(1):56–65
Mitchell K-J (1975) Dynamics and simulated yield of Douglas fir. For Sc Monogr 17. 39 p
Palubicki W, Horel K, Longay S, Runions A, Lane B, Měch R, Prusinkiewicz P (2009) Self-organizing tree models for image synthesis. ACM Trans Graph 28(3):58
Perttunen J, Sievänen R, Nikinmaa E, Salminen H, Saarenmaa H, Väkevä J (1996) LIGNUM: a tree model based on simple structural units. Ann Bot 77:87–98
Perttunen J, Sievänen R, Nikinmaa E (1998) LIGNUM: a model combining the structure and the functioning of trees. Ecol Model 108(1–3):189–198
Pressler R (1865) Das Gesetz der Stammbildung. Arnoldische Buchhandlung (Leipzig), 153 p
Pretzsch H (2002) The single tree-based stand simulator SILVA: construction, application and validation. For Ecol Manag 162:3–21
Qi R (2009) Optimization and optimal control of plant growth: application of the GreenLab model for the decision help in agriculture. Doctorat École centrale de Paris
Shinozaki K, Yoda K, Hozumi K, Kira T (1964) A quantitative analysis of plant form—the pipe model theory. I: basic analyses. Jpn J Ecol 14:97–105

Chapter 9
Validation of the GreenLab Model on Crops (Case of Herbaceous Plants)

Baogui Zhang, Yuntao Ma, Mengzhen Kang, Véronique Letort, and Philippe De Reffye

Abstract This chapter focuses on the adequacy of the model to the reality of crop plants (herbaceous). The GreenLab model aims to compute biomass production and distribution among plant organs using source–sink parameters. This can be assessed in the form of graphs in which measurements on plants (plant crowns and organic series) are correctly fitted by the corresponding mathematical expressions of the model, using the parameters optimized by the inverse method. This chapter describes the use of the inverse method to fit model outputs to measured plant data. This process involves both directly measurable and hidden parameters, with the latter estimated from organic series data. Finally, this chapter discusses the application of inverse methods on cultivated plants and the integration of stochastic development for phenotypic variability. The analysis of organic series is essential for linking plant measurements to the model's mathematical expressions, allowing for accurate growth modeling and parameter estimation across different developmental stages. Validation of the GreenLab model is established on various cultivated plants (maize, sunflower, rice, etc.).

B. Zhang · Y. Ma
College of Land Science and Technology, China Agricultural University, Beijing, China
e-mail: zhangbg@cau.edu.cn; yuntao.ma@cau.edu.cn

M. Kang (✉)
State Key Laboratory for Multimodal Artificial Intelligence Systems, Institute of Automation, Chinese Academy of Sciences (CASIA), Beijing, China

School of Artificial Intelligence, University of Chinese Academy of Sciences, Beijing, China
e-mail: mengzhen.kang@ia.ac.cn

V. Letort
Mathématiques et Informatique pour la Complexité et les Systèmes, CentraleSupélec-Université Paris-Saclay, Gif-sur-Yvette, France
e-mail: veronique.lechevalier@centralesupelec.fr

P. De Reffye
CIRAD, UMR AMAP, Montpellier, France

AMAP, Univ Montpellier, CIRAD, CNRS, INRAE, IRD, Montpellier, France

P. De Reffye et al. (eds.), *Plant Model Greenlab for Botany and Agronomy*,
https://doi.org/10.1007/978-981-97-7541-5_9

Keywords Herbaceous · Inverse method · Application on cultivated plants · Model validation

9.1 Introduction

We place ourselves here in a simple eco-physiological framework and consider only the carbon fixed in the plant. Therefore, the net photosynthesis without taking into account respiration is considered in the current study, which is the case of many basic crop models. In addition, the plants studied were not subjected to stress and were grown in a homogeneous physical environment with uniform planting density. These conditions are necessary to develop the model under stable parameters. The impact of stress can be assessed later by analyzing the effects on the organic series and therefore the functional parameters.

The first purpose of the GreenLab model is to extract the demand and the production of biomass of a plant from the architecture at each step of the growth and its distribution among the competing organs.

This is done by computing the source–sink parameters that drive plant growth. The Gloups software is not only a simple growth simulator but also a source–sink solver. The method is obviously limited to plants for which the model is well adapted. The GreenLab assumptions (growth cycle, common pool, physiological age, etc.) are required. The ability to identify plant crowns, cohorts, and organic series is necessary. The even-aged plants must be clones or selected cultivars that produce stands inside a homogeneous environment for soil and a regular planting density, so each plant is a repetition that can be considered as a realization of the same growth process. These conditions are generally realized in agronomy for many crops.

To validate the model, agronomic experiments were carried out in France, China, Africa, and the Netherlands, in both fields and greenhouses, on various cultivated crops. The accumulated measurements obtained on these plants have formed an important database that has been long and difficult to obtain, along with various projects. This database is a valuable reference because if the models can evolve or change, they will always have to face the reality of the data resulting from agronomic experiments.

This chapter focuses on the adequacy of the model to the reality of crop plants (herbaceous). This can be assessed in the form of graphs in which measurements on plants (plant crowns and organic series) are correctly fitted by the corresponding mathematical outputs of the model, using the parameters optimized by the inverse method.

The age of the plant does not correspond necessarily to a calendar duration, but rather to a cumulative number of growth cycles which are the growth stages for its architecture construction.

It takes one Development Cycle (DC) for a meristem to create a phytomer. Its duration in thermal time is given by the phyllochron. As the growth of organs continues, even if its development stops, a Growth Cycle (GC) is then the time step that controls organ growth. For simplifying the computations, the durations of DC and GC are set to be equal in our model. Thus, in the rest of the chapter, we will use DC to control the elongation of the axes and GC to control the growth of the organs, knowing that these cycles are equivalent in duration.

Under non-limiting conditions, the resource use efficiency $1/r$ and $E(t)$, which represents the radiation received per GC, vary a little around an average from one cycle to another. Variations within a GC are invisible to the model, but those from one growth cycle to another can also be invisible because they are buffered over the expansion time of the organs.

Source–sink relationships in the GreenLab model are governed by thermal time, which is expressed in the number of DCs. We can, therefore, normalize the parameter E to be 1 to represent these relationships. In this case, the parameter r is just a simple calibration coefficient, without any eco-physiological significance. The biomass production and partitioning during the development are not modified. The study of plant growth and architecture thus gains a degree of independence from eco-physiology. The link between the model and the environment (light, water) will be done in Chap. 11.

Even if the environmental conditions were not limiting, this did not prevent the stochastic aspects in development. On one hand, these aspects appear in the functioning of the buds (e.g., mortality), and on the other hand, in the local environmental variations (e.g., irregularity in the density of plantation). At first glance, the complexity of a plant structure with its stochastic variability seems unaffordable. However, solutions emerge through the concepts of the common pool, physiological age (PA), organic series, and potential structure that are suitable for all architectural models and plant sizes (Chaps. 5, 6, 7, 8).

The notions of cohorts and organic series break the plant architecture down into separate components that contain all the necessary information, and define a target to fit by the model. It is not necessary to record its whole architecture. Unlike compartments of organs that need to be complete (without missing data due to abortion), organic series can be measured by sampling in the plant architecture.

The information contained in the compartments when they are available is obviously redundant with that of the organic series, but they have the advantage of giving a more global vision of the growth. Moreover, they do not require the tedious description of the topological structure of the axes. A good point of view is to choose a stage of development in which the organic series with detailed information for each organ is described, and to use compartments data for each type of organ for other stages. This strategy allows controlling the invariance of the parameters during the growth.

9.2 Parameter Estimation

Chapters 5, 6, 7 and 8 define the functional parameters of the organs that allow the GreenLab model to simulate the growth of plants. The aim of this chapter is to estimate the functional parameters for crop plants using an inverse method applied to the measurements.

Simulation of plants according to the model plays an important role in the development of inverse methods. Computer plants are pure realizations of plant growth by models, and simulate natural plants in the field. Fitting the model to the data resulting from the simulation must give back the values of the set of parameters used in the simulation. Simulation provides as much data as needed to develop methods for sampling plants and to test the accuracy and sensitivity of parameters. The development of the inverse method on simulated plants allows the transition to the study of cultivated plants under non-limiting conditions.

Some parameters, such as organ functioning and expansion durations and allometries, are directly accessible via measurements on the plant. Others, such as resource use efficiency $1/r$, production surface Sp, sink functions, are hidden parameters and cannot be obtained from measurements directly. They should be estimated from the data of measured organic series. However, in a few cases, the model shows that they can still be measured directly. Thus, the matching between the values of the measured and computed parameters constitutes a validation of the method. This is particularly the case for the area S_p (see Chap. 5 Figs. 5.13, 5.33) or the sink functions relating to fruits, where the values measured directly or computed by the inverse method are close. The coefficient k of the Beer-Lambert's law, which measures the horizontality of the leaves, is empirical and often fixed. In practice, the GreenLab model is not very sensitive to this value and is fixed by default at $k = 1$, except in certain cases as for rice and wheat which have erect leaves.

An important question is whether to consider only the above-ground part of the plants, ignoring the growth of the below-ground part. Most of the time, the root data is missing because of the tedious work for the root measurement. This does not prevent most of the classical crop models to make good predictions of the harvest and the various compartments of the above-ground part (seeds, fruits, leaves, etc.).

Similarly, the analysis of the organic series of the above-ground part of the plant can often give good results in the fitting of the model. This can be explained if the percentage of root biomass is small compared to that of the plant or if the shoot–root ratio of the synthesized biomass is relatively stable during the growth process.

Simulation is of great help in testing these hypotheses about the shoot–root interactions. Moreover, successful modelling of the growth of the above-ground part of plant architecture is an interesting result, which can lead to many potential applications in agronomy.

So, in this chapter, with few exceptions (beetroot, cucumber, etc.), the roots are not considered due to the lack of root data.

9.3 Implementation of Inverse Methods on Cultivated Plants

In deterministic development (case of herbaceous plants), a single plant may be sufficient to assess the sink–source parameters. The simulated structure is also the potential structure. Because of the possible variability of environmental parameters (*E* and *Sp*), it is more appropriate to have a small number of repetitions (≈ 5). In stochastic development (case of woody plants), it is necessary to work on a collection of homogeneous plants (clones, varieties). The simulated random structures are all contained in the potential structure (Chaps. 3 and 6). Only the latter allows the inverse method to be used.

The first fittings were made on single-stemmed plants. In branched plants, in the absence of a *Bernoulli* process (parameter $b = 1$) and mortality (parameter $c = 1$) in the functioning of meristems, only the branch remains in random location on the stem. By defining a "mask" for the simulation that only allows the localized development of the branches observed in the structure, the plant is constrained to a deterministic functioning. The "mask" file is a 0–1 series that indicates which phytomer in organic series can bear a branch or a fruit, so that the simulated plant architecture is the same as the observed one. This is useful when identifying sink–source parameters from a small sample of a simple plant structure as for crops. This made it possible to validate the GreenLab model on sweet peppers, rice, etc.

The integration of stochastic development into the formulation of source–sink relationships was developed and implemented with the Gloups software (Kang et al. 2018), to calibrate all deterministic or stochastic plant architectures in their phenotypic variability.

To study the growth of a plant, its development pattern must first have been defined (the architectural model with its PAs) and the parameters of the functioning of meristems must have been estimated by the method of crown analysis (Chap. 3). They are used to compute the potential structure and plant demand (Chaps. 5 and 6), which are essential for studying growth modelling based on source–sink relationships.

9.4 Analysis of Organic Series

The aim here is to model mathematically the relationships between the different organic series or compartments and the parameters of the GreenLab model.

9.4.1 More Details about the Concept of Organic Series

The averages of the simulated organic series generally match with a good approximation of the organic series computed from the potential structure (Chap. 6).

In the GreenLab model, measurement on plant phenotypic data can be linked to a mathematical expression from the model equations (Fig. 9.1), for simple measurements such as the weight of an organ or its dimensions, or for compound measurements such as the weight of organ compartments, the plant height, the leaf area, and the weight of organs in Growth Units (GUs) or in branches that are compound organic series (case of trees). Each measurement in an organic series corresponds to an average performed in the corresponding cohort whose number of elements is known. The fitting of the dynamic growth model to a given plant therefore consists in estimating the values of the parameter set that best fits together with the computed and simulated organic series. It is possible to operate single-fitting on a single development stage, which does not guarantee the stability of the parameters during the growth, or multi-fitting on several stages, which makes it possible to check parameter stability and increases accuracy. In the computation of growth parameters, "data assimilation" is how organic series measured on the architecture at different development stages are provided to the model.

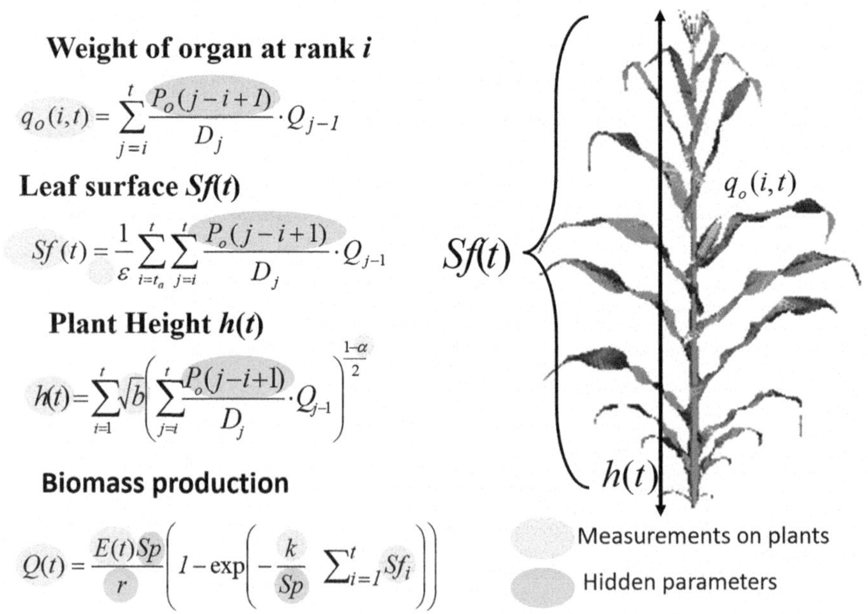

Fig. 9.1 Measurable data (in green) in organic series and their mathematical expressions according to the GreenLab model. They range from the simple weight of an organ q_o to the leaf surface *Sf*, the height *h* of the plant, or the synthesized biomass *Q*. The parameters in red color, which are hidden, must be estimated by the inverse method

Let a plant be of age t in DCs, we consider the PA φ and the organ of type o that constitutes the organic series. The weight of an organ $q_o^{\varphi}(i,t)$ located at rank i in an organic series depends on the variation in its sink and the evolution of the Q/D ratio during its growth. The dimensions of the system increase with each DC of a unit. The link between the organic series and the growth operator is recalled below (Chap. 5).

$$\begin{bmatrix} q_o^{\varphi}(1,t) \\ q_o^{\varphi}(2,t) \\ \vdots \\ q_o^{\varphi}(t,t) \end{bmatrix} = \begin{bmatrix} P_o^{\varphi}(1) & P_o^{\varphi}(2) & \cdots & P_o^{\varphi}(t) \\ 0 & P_o^{\varphi}(1) & \cdots & P_o^{\varphi}(t-1) \\ 0 & 0 & \ddots & \vdots \\ 0 & 0 & \cdots & P_o^{\varphi}(1) \end{bmatrix} \begin{bmatrix} \frac{Q_0}{D_1} \\ \frac{Q_1}{D_2} \\ \vdots \\ \frac{Q_{t-1}}{D_t} \end{bmatrix} \tag{9.1}$$

The vector $[q_o^{\varphi}(1,t), q_o^{\varphi}(2,t), \ldots q_o^{\varphi}(t,t)]$ constitutes the organic series of the organ type o, at the PA φ and the chronological age (CA) t of the plant, and the elements $P_o^{\varphi}(j)$ correspond to the sink functions of the organ type o, at the PA φ to be estimated.

The organic series is orthogonal to the notion of the cohort. The rank of the phytomer is in correspondence with the CA of the cohort. The measurement corresponds to the average value in the cohort. The number of organs per cohort, known by the functioning of the botanical automaton, varies during the development. The notion of organic series in GreenLab is not necessarily linked to the notion of leafy axis or architectural model. It is linked to the growth of an organ type along with successive cohorts according to CA, in a defined context of supply/demand evolution. This allows many adaptations for measurements.

9.4.2 Measurements on Plants, Organ Compartments, Simple and Compound Organic Series

A source–sink model requires that measurements should be made at the organ level and concern their weights and dimensions. Biomass data can be fresh weight or dry weight. Therefore, individual organs or organ compartments must be weighed.

Crop models such as Stics (Brisson et al. 1998), TomSim (Heuvelink et al. 1999), Pilot (Mailhol et al. 1996) use the notion of organ compartments. With compartment data, time-related production information in the architecture is lost. The organ cohorts are mixed in the same compartment, and the harvest index (economic part of the biomass) incorporates too much information on development and growth while it always remains stable (see Chap. 5 Fig. 5.11). In reality, the presence of mortality causes missing data, which no longer allows the compartments to be reconstructed.

An organic series implicitly contains the CAs of the plant and the memory of the biomass production in the volume of successive organs. The size of organs along

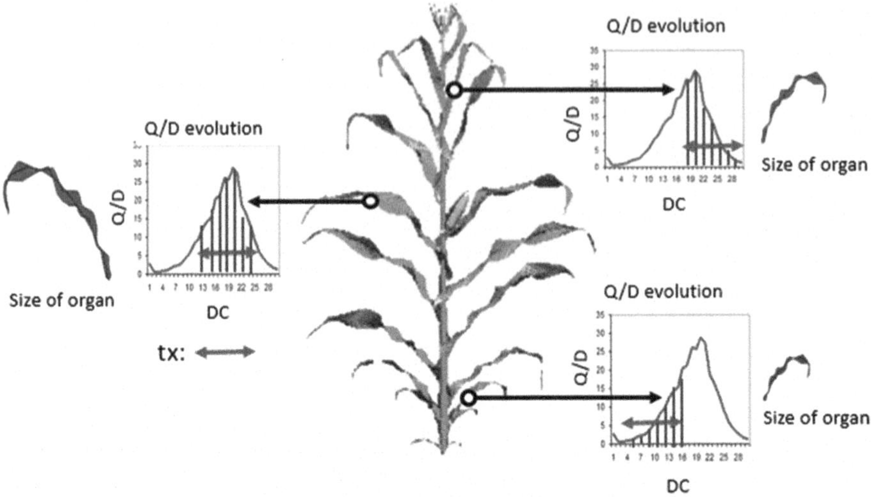

Fig. 9.2 Size of organs and *Q*/*D* evolution (case of maize leaves by Ma Yun-Tao): the ratio *Q*/*D* varies during the plant growth (red curve). The leaf expansion duration t_x is smaller than the plant development duration. The size of the organ depends on the sum of the t_x *Q*/*D* (blue sticks) inside the development time

organic series depends on the evolution of the source–sink ratio *Q*/*D* along with the CA (Fig. 9.2).

The organic series of the same PA is homogeneous if the ranks are sorted from the top of the living leafy axes (Chap. 5 Fig. 5.15). The architectural analysis makes it possible to take a relevant sample of the categories of axes required for the analysis of the data by the model. Organic series may be incomplete, but there must obviously be more data than parameters to assess, and all parameters must be implicitly presented in the data. The measured set of organic series is a target to which a homologous output of the model must be fitted by optimizing the parameters.

For a given PA and CA, there are generally three organic series to be identified, including leaves, internodes, and fruits (Fig. 9.3). Thus, for a monopodial model of continuous growth with two PAs (stems and branches, e.g., the coffee tree belonging to the Roux model) and analysis on one of its growth stages, it will be necessary to take the following data by sampling: 3*2*1 = 6 organic series. Sometimes it is necessary to consider that a leaf is a compound of a blade and a petiole (eg., maize).

The calculated organ biomasses q_i are sorted according to their order of creation t_j and their insertion ranks *j* in the organic series. The functions f_o related to the types of organs *o* compute their biomasses. *E* is the driving factor. The vector *A* contains the estimated parameters of the development and those known empirically (i.e., ε, the specific leaf weight). The vector *P* contains the unknown source–sink parameters *r*, *Sp*, Q_0 and the triplets of the organ function sinks (P_o, B_{o1}, B_{o2}), which are estimated by the Weighted Least Squares Method (WLSM). The difference between the computed and the measured biomasses should be minimal.

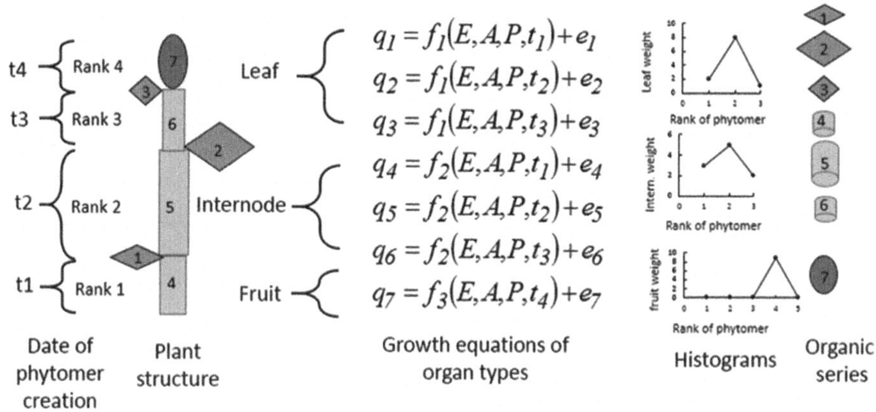

Fig. 9.3 From plant structure to organic series

9.4.3 The Weighted Least Squares Method (WLSM)

Once the data collection is completed, they are organized according to a target. Each element of which corresponds to a respective output of the model (Fig. 9.1). The goal is to minimize the error, which is calculated as the sum of the squares of the differences between the measured and computed data, by optimizing the system parameters. The Gauss-Newton method for non-linear least squares (Arbenz and Wohlauser 1986) is effective for estimating model parameters from organic series. The principle of its implementation in the GreenLab model was discussed in Zhan et al. (2003) and Guo et al. (2006). Least squares can be generalized by introducing a weighting matrix corresponding to the inverse of the covariance matrix of observational errors (Cournède et al. 2011).

Suppose that we have n observations, y_i corresponds to n model outputs at DC t_i. There are n_f functions f_o corresponding to the different organ types o whose parameters are the source and sink parameters P to be estimated. Note that t_i keeps the same value for organs belonging to the same rank (or cohort).

Moreover, suppose that the additive error is e_i, the functions f_o $(P_1, P_2, \ldots, P_m, t_i) + e_i$ in the GreenLab model most often represent the weight of an organ associated with a single organic series or the weight of a compartment belonging to a compound organic series. Deviations from the model are referred to as e_i. All y_i are the different outputs of the model (Fig. 9.1) that are fitted together. The grouping of y_i into simple or compound organic series decreases the number of functions f compared to the number of observations n. Each y_i corresponds to a function f_o assigned to an organ.

Thus, the histograms of all compartments and organic series belong in fact to the same set from which they are extracted. These histograms are not fitted one by one separately by the model, but all together as a unique one at the same time because they are the various outputs of the same functioning which depend on the same set of parameters.

The values (P_1, P_2, …, P_m) estimated using the least squares method minimize the expression:

$$err = \sum_{i=1}^{n} (y_i - f_o(P_1, \ldots, P_k, \ldots, P_m, t_i))^2 \tag{9.2}$$

We empirically give an approximate starting solution: (P_{10}, P_{20}, …, P_{m0}) and we obtain a first better estimation of the parameters (P_1, P_2,…, P_m) denoted as (P_{11}, P_{21},…, P_{m1}). The system is linearized in the vicinity of the solution by approximation:

$$\begin{aligned} f_o(P_{11}, P_{21}, \cdots, P_{m1}, t_i) &= f_o(P_{10}, P_{20}, \cdots, P_{m0}, t_i) \\ &+ \sum_{i=1}^{n} \frac{\delta f_o}{\delta P_k}(P_{10}, P_{20}, \ldots, P_{m0}, t_i)\Delta P_k + e_i \end{aligned} \tag{9.3}$$

with:

$$P_{k1} = P_{k0} + \Delta P_k (k = 1, 2, \ldots, m) \tag{9.4}$$

We put:

$$[X] = \begin{bmatrix} \frac{\delta f_1}{\delta P_1} & \frac{\delta f_1}{\delta P_j} & \cdots & \frac{\delta f_1}{\delta P_m} \\ \frac{\delta f_2}{\delta P_1} & \frac{\delta f_2}{\delta P_2} & \cdots & \frac{\delta f_2}{\delta P_m} \\ \cdots & \cdots & \cdots & \cdots \\ \frac{\delta f_n}{\delta P_1} & \frac{\delta f_n}{\delta P_2} & \cdots & \frac{\delta f_n}{\delta P_m} \end{bmatrix}, [P_0] = \begin{bmatrix} P_{10} \\ P_{20} \\ \cdots \\ P_{m0} \end{bmatrix}, \tag{9.5}$$

$$[P_1] = \begin{bmatrix} P_{11} \\ P_{21} \\ \cdots \\ P_{m1} \end{bmatrix}, [\Delta P] = [P_1] - [P_0]$$

In the general case of a complex model, the formulas of the derivatives cannot be easily obtained. But their numerical values can be approximated by the following numerical approximation with a sufficiently small value dp_k compared to the parameters p_k:

$$\frac{\delta f_i}{\delta P_k} = \frac{f_i(P_1, \ldots, P_k + dP_k, \ldots, P_m) - f_i(P_1, \ldots, P_k, \ldots, P_m)}{dP_k} \tag{9.6}$$

We now consider the linear system to be solved: $[\Delta Y] = [X] \cdot \Delta P + [E]$ with

$$[\Delta Y] = \begin{bmatrix} y_1 - f_i(P_{10}, P_{20}, \ldots, P_{m0}, t_1) \\ y_2 - f_i(P_{10}, P_{20}, \ldots, P_{m0}, t_2) \\ \cdots \\ y_n - f_n(P_{10}, P_{20}, \ldots, P_{m0}, t_n) \end{bmatrix} \; and \; [E] = \begin{bmatrix} e_1 \\ e_2 \\ \cdots \\ e_n \end{bmatrix} \tag{9.7}$$

The classic solution of this system is given by:

$$\Delta P = (X^T \cdot X)^{-1} \cdot X^T \cdot \Delta Y \tag{9.8}$$

We then obtain a first iteration that gives a better estimate of the system solution:

$$P_1 = P_0 + \Delta P \tag{9.9}$$

The process is repeated until the components of ΔP are small enough ($|\Delta P| < \varepsilon$). In most cases, less than 10 iterations are sufficient.

The problem can then be posed of the precision of the parameters to be estimated and the sensitivity of the model to these parameters. This is an issue that no longer concerns the field of modelling, but also the field of statistics. In the case where the parameters are not correlated with each other, ordinary least squares can provide both the parameter values and the associated standard deviations.

But in case the data have different high values, the method does not work properly. In this case the values observed correspond to the different organic series of the organs (leaves, internodes, fruits). The series with high values will be better fitted than the series with low values. This is often the case, for example, for the organic series of the leaves compared to those of the fruits. The WLSM assigns weights to the different series to balance the data.

Each organic series has a mean and a variance. The weight of a series is defined as the inverse of the variance. (If the series has only one element, the weight is the inverse of the mean). The diagonal matrix $\sum$ contains the variances of the organic series.

We define the diagonal matrix $\sum^{-1}$ formed by the inverse of $\sum$.

Formula 9.8 becomes:

$$\Delta P = (X^T \cdot \sum^{-1} \cdot X)^{-1} \cdot \sum \cdot X^T \cdot \Delta Y \tag{9.10}$$

With $P_1 = P_0 + \Delta P$ until $|\Delta P| < \varepsilon$.

We define as $\hat{\sigma}^2$ the variance of all the data from the set of organic series

$$\hat{\sigma}^2 = \frac{\sum_{i=1}^{n} (y_i - \hat{y}_i)^2}{n - p} \tag{9.11}$$

The Matrix $[VP]$ contains the variances of the parameters of the P matrix.

$$[VP] = \sigma^2 (X^T \cdot X)^{-1} \tag{9.12}$$

$[VP]$ is diagonal if the covariances of the parameters are null.

The standard deviations of the parameters of $[P]$ deduced from $[VP]$ give an idea of the accuracy obtained on these parameters.

We get the coefficient of determination by:

$$R^2 = 1 - \frac{n-1}{n-p} \frac{\sum_{i=1}^{n} (y_i - \widehat{y}_i)^2}{\sum_{i=1}^{n} (y_i - \overline{y}_i)^2} \tag{9.13}$$

Other methods can be used to fit the model, such as heuristics algorithm: Simulated Annealing, Particles Swarm Optimization (PSO). These methods are less efficient, but do not need to compute the derivatives.

9.4.4 *Adaptation of Least Squares Method to Plant Growth and Plant Architecture*

Figure 9.3 shows how to build organic series coming from the plant structure. The parameter estimation method was first validated on simulated plant samples, before being used on real plants. In the deterministic case, the WLSM obviously finds the exact values of the parameters.

In the stochastic case, the parameters of meristem functioning must first be calculated using the crown analysis. The method estimates the parameters by optimizing them to best fit the model to the observations on organic series. It is obvious that the more developmental stages selected, and the more plants collected, the more accurate the estimation of the parameters will be.

The computed organic series must therefore be adjusted as best as possible by the inverse method to their simulated counterparts, which serve as the target. The y_i of Eqs. (9.1) are the weights of the measured organs, sorted according to their types, the ranks and PA, and the parameters P_j are those of the source functions or sinks of the organs. Each organ represents the average of a cohort that corresponds to a rank in the organic series. The profile (i.e., histogram) of a series does not correspond to a given parametric function; it is the result of a numerical computation performed by the equations of the model on each element.

9.4.4.1 Parameters Estimated Directly from the Observations

The parameters that control the development must be known beforehand to be able to compute the source–sink relations, they are:

- The functioning time of leaves t_a and the expansion times of the organs (leaves, internodes, fruits, roots) t_{xo} are only empirically accessible by observing the plant. Indeed, t_a and t_{xo} are not computable parameters, and must be determined empirically. The parameters r and t_a are closely correlated. Thus, the demand must be fixed before assessing the values of the source–sink parameters.
- The DC at which the organ finishes its expansion is deduced either by daily measurements of the increase in size of an organ or by observing the phenological stage of wilting. Sometimes, the expansion times of the organs vary from the base of the stem until they stabilize. These measurements require careful monitoring of the development; most often these variations can be adjusted empirically so that the histograms of the observed and computed organic series fit correctly at the beginning of the stem.
- The allometries that control the shape of the organs must be known (e.g., the thickness, or specific leaf weight (SLW) of the leaf, ε).
- The evolution of the environmental parameter E, the driving factor (light, PET), must be known. If daily values fluctuate around a constant value, considering the average gives a good result. In the absence of data, we can assume $E = 1$. Only the resource use efficiency ($1/r$) is changed; the other parameters will not be modified.

9.4.4.2 Parameters Computed with the WLSM

In all studied plants, there are less than twenty parameters to be estimated at the same time.

These parameters are not directly observable and cannot be extracted from the data (simple and compound organic series, compartments) by computation. They are:

- The seed biomass Q_0, which initiates the growth by creating the first phytomer. The action of the seed can cover several cycles in the early stages of growth.
- The environmental parameters: r, Sp.
- The parameters of the sink functions of o-type organs: P_o, B_{o1}, B_{o2}. The expansion time t_{xo} is given as well as its possible variation. The coefficients B_{o1}, B_{o2} that control the sink variation are assumed invariant with CA and PA, for an organ type o. Because of the overlapping action of the sinks on each other in the expression of the demand during the development, sometimes these two coefficients cannot be assessed together. Only one parameter (B_{o1}) can be computed and the second (B_{o2}) must be fixed empirically, because the WLSM cannot estimate them at the same time. Indeed, the ratio $B_{o1}/(B_{o1} + B_{o2})$ is sufficient to produce generally a good fitting in the large domain of variation in B_{o2}.

The sink strength P_o generally depends on the PA φ. Thus, often the organs of the branches are smaller than those of the trunk. To avoid having too many parameters to assess, it is convenient to use the observed ratios of biomass between branch organs and stem organs of the same age, as these ratios do not depend on the variation in the expansion times. Thus, we can set $P_o(\varphi) = k_o(\varphi) \cdot P_o(1)$. Since the parameters $k_o(\varphi)$ are known empirically, there are only the sinks of the stem organs to be computed; those of the branches are known by the established allometries $k_o(\varphi)$.

9.4.4.3 Sampling Strategy

In the case of monocaulous plants, like maize or sunflower, it is possible to measure all the necessary information for each organ. However, it becomes practically impossible to do so on small shrubs even like cotton plants with more than 500 phytomers. There are too many phytomers, and the mortality of the axes and organs results in a lot of missing data. The use of the compartments becomes impossible (because they are incomplete), but the organic series show their interest, because they remain measurable and that they don't need to be complete. It is, therefore, possible to study the growth of trees in plantations, provided that the sampling method of the architectural model has been developed. Once the parameters of the development are known (by the crown analysis in Chap. 3), the sampling of the organic series is up to the willing of the experimenter. Note that only living branches should be measured, as the series analysis begins from the top to the bottom along the stem. Their statistical features on weight and number of organs do not depend on the position of the branch in the architecture, if the GreenLab hypotheses are valid. If the sampled series implicitly contains all the source–sink relations, and the number of data is greater than the parameters to be estimated, then theoretically the determination of the hidden source–sink parameters that drive the plant growth is possible.

9.4.4.4 Practical Use of the WLSM

Even if the WLSM is undoubtedly the most efficient inverse method with a strong convergence, its practice remains nevertheless delicate and it can be acquired by training, especially with the targets generated by the simulation whose source–sink parameters are known in advance. One problem is the possibility of several solutions that gives local minima. This should not be the case of the GreenLab model, which on many examples has always given a unique solution.

Another problem is to find an initial set of parameter values to launch the computation. If we are too far from the solution, the convergence fails. A good method is to start computing with only a small number of parameters (1–3) always including the resistance r and the sink strengths to find a solution, which can be set as a new starting point by adding another set of parameters and so on, at the end all the parameters are most often found.

The GLOUPS (GreenLab Operators Used for Plant Simulation) software is used in this book to fit the crop plants with the GreenLab model. It is written in Matlab, and was developed by P. de Reffye and M. Jaeger.

9.4.4.4.1 The Numerical Results of the Computation

(a) The coefficient of determination R^2 and the number of coupled data give the significant threshold of the fitting and the number of parameters involved.
(b) The values of the hidden parameters, their standard deviations, and the coefficients of variation (CV) are presented by an array. The standard deviations are provided by the WLSM, but they are correct only if the parameters are not statistically correlated. In any case, they give a good indication of the accuracy. It is better that the CV should be less than 20%.
(c) The global parameters are Q_0, r, Sp. Note that Sp cannot be calculated at the start of growth as the leaves do not overlap. Using a large value (e.g., $Sp = 100{,}000\ \mathrm{cm}^2$) is sufficient for initial calculations.

Additionally, there are sink coefficients for organs of type o: (blade ($o = b$), petiole ($o = p$), internode ($o = i$), ring ($o = c$), female fruit ($o = f$), male fruit ($o = m$), root ($o = r$)). For each type of organ, the sink function $P_o(\varphi)$ depends on the type of organ o and the PA φ (See Chap. 5).

9.4.4.4.2 The Graphical Results of the Computation

- Global overview of the simulation:
  - Number of phytomers produced by the PA along with the DCs.
  - Comparison between the observed and computed weight data.
  - Evolution of the driving factor E along with the DCs.
  - Evolution of the ratio (leaf surface/surface of production, Sf/Sp) along with the DC. This ratio is approximately equal to the individual optimized LAI.
  - Evolution of the demand D along with the DCs.
  - Evolution of the biomass production Q along with the DCs.
  - Evolution of the ratio Q/D along with the DCs that gives the internal trophic pressure of the plant.
  - Evolution of the reserve Q_{pool} if a compartment of the common pool is created. For normal conditions, Q_{pool} is null. All the biomass produced in each cycle are used for growing organs.
  - Evolution of the duration in DC for the functioning time of the leaves, for the organs in activity.
  - Evolution of the sink variation in DC for each type of organ in activity.
  - Evolution of the biomass of compartments in DC for each type of organ in activity, for the observed and computed data.
  - Organic series for each type of organ and for each PA in use for the observed and computed data (single or multi-fitting).
  - Stack of rings in the main stem (radial and longitudinal in the case of woody plants).
  - Two-dimensional (2D) or three-dimensional (3D) display of the simulated plant architecture.

9.5 Studies of Plant Growth and Plant Architecture of Crop Plants

The results obtained on the modelling of the growth of various crop plants are presented here. Each plant as presented in this chapter makes a new contribution to the validation of the GreenLab model and allows researchers to increase the experience of the model.

The information on the experiments can be found from the publications for each type of plant to keep clear and avoid tedious details here. In most cases, the environmental conditions are not limiting and the climate driving factor can be normalized ($E = 1$) in this chapter. Normalizing E does not change the biomass partitioning and the sink parameters. Only the parameter r is affected. Moreover, the notion of cycle based on thermal time makes the development independent of the calendar time. Thus, as we said previously, the architectural methods gain autonomy concerning eco-physiology. It is not necessary to know the environmental parameters to study step by step the development and growth of a plant. Taking account of the environmental factor E (PAR or PET) is done in Chap. 11. Depending on the case, either fresh or dry matter is chosen for analysis.

What counts here is the ability of the model to correctly fit the observed separate parts of the plant architecture, by optimizing the hidden parameters. The ultimate goal is to restore by simulation the 2D structure or 3D architecture of the plant, with the optimized parameters, in order to visualize the good concordance between the observed and simulated plants. The 3D visualization in agronomy does not constitute a validation of the model as it is the case for computer graphics. Most of the time it is a legitimate satisfaction for the researcher, to observe the result of the modelling undertaken. However, obtaining mock-ups is still necessary for some applications (radiative transfer, landscaping, etc.).

9.5.1 Case of Lettuce Study (Kang Meng-Zhen, LIAMA, CASIA)

9.5.1.1 Plant Description

In herbaceous plants such as lettuce, the first phytomers often have short internodes whose weight is not measurable. A rosette of leaves is formed, it implicitly contains the memory of the plant's growth. No root data was available in this experiment.

9.5.1.2 Cultivation Condition

A single rosette of lettuce is studied here on a single stage of development ($T = 36$ DCs) that is deterministic. Fresh weight is measured. The environmental factor,

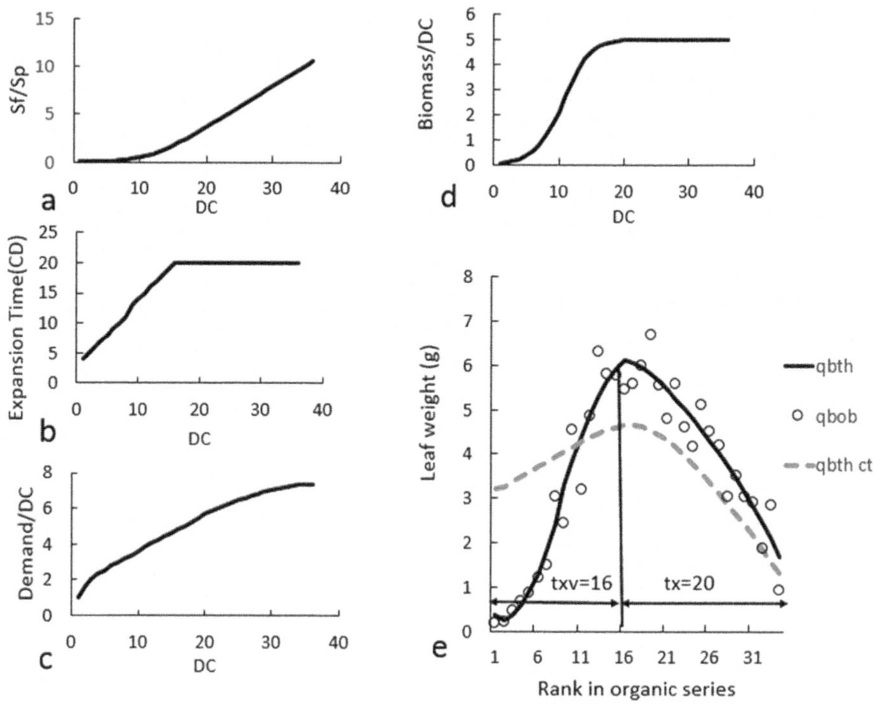

Fig. 9.4 Fitting of lettuce. (**a**) Evolution of individual LAI (*Sf/Sp*). (**b**) Bottom-up variation in leaf expansion duration. (**c**) Demand per DC (*D*). (**d**) Biomass per DC (*Q*) and (**e**) the observed (q_{bob}) and fitted organic series (q_{bth}) of the leaf. The circles are the observed values of the weights of leaves $q_{bob}(k)$ in the rosette at rank *k*. The black solid line is the fitting of the computed weights of leaves $q_{bth}(k)$ with the WLSM. The gray dotted line is the suppression of the variation in the expansion duration from rank 1 to 16 (q_{bth} *ct)*. No more fitting with observations is possible that shows the importance of the t_x variation along the axis

being unknown, is normalized to $E = 1$. The leaf rosette produces a single organic series. The thickness of the leaf (SLW) ε is 0.024 g/cm^2. The sink strength of the leaf is fixed and normalized to $P_b = 1$. The five hidden parameters involved are Q_0, r, Sp, B_{b1}, B_{b2}. No available data for the roots.

We consider that all the leaves are functional, which fixes the value of the photosynthetic functioning time $t_a = T$. Because the first leaves put in place by germination are the oldest, they should be the largest if their expansion time is equal to the age of the plant *T*, and the sink function of the leaf was stable. The bell shape of the organic series of leaves (Fig. 9.4e) clearly shows that the first leaves that remain small have a shorter expansion time. The first leaves of the rosette remain very small, and therefore their expansion time is reduced compared to the stabilized value t_x, which has an impact on the demand.

It is tedious to control the expansion time of each leaf according to its rank in the organic series. On the other hand, it can be assumed that it evolves linearly from an initial value at the base of the stem to a stable value from a certain rank k_x onwards. This occurs necessarily near the top of the bell curve of the leaf organic series. Its distance from the base x gives the interval of variation in DCs of the expansion time. As for the stabilized expansion time, it is at least k_x. The peak of the organic series of leaves is about the rank 16 of the rosette from the base (Fig. 9.4e). This indicates a duration of leaf expansion $t_x = 36\text{–}16 = 20$ DCs.

The model needs to well control the variation in the demand by controlling the variation in the expansion times that modify the values of the sink functions. Let's choose a minimum value $t_m = 4$ for the expansion duration (DC) of the leaf in the first rank of the rosette. Let's choose the stabilization of the t_x duration in the rank k_x. With these parameters, we can obtain a linear evolution of the expansion time duration $t_{xd(i)}$ depending on rank i by the equation:

$$\begin{aligned} t_{xd}(i) &= t_m + (t_x - t_m) \cdot \left(\frac{i-1}{k_x - 1}\right) \; for \; 1 \leq i \leq k_x \\ t_{xd}(i) &= t_x \qquad for \; i > k_x \end{aligned} \tag{9.14}$$

So, if in the rank i, the expansion time is $t_{xd}(i)$. The value of the sink variation function after x DCs of expansion is given by the expression (see Chap. 5):

$$F_o\left(\frac{x}{t_{xd}(i)}\right) = C_o \cdot \left(\frac{x+0.5}{t_{xd}(i))}\right)^{a-1} \cdot \left(1 - \frac{x+0.5}{t_{xd}(i)}\right)^{b-1} \cdot \left(\frac{1}{t_{xd}(i)}\right) \tag{9.15}$$

9.5.1.3 Results on Plant Functioning

Table 9.1 gives the parameter values computed by the WLSM.

The fitting is good: $R^2 = 0.97$, number of data = 36, number of parameters = 5.

The five hidden parameters are computed in Table 9.1, except for the sink strength of the leaf P_b which is fixed to 1. Note that the model computes an estimate of the seed reserves (Q_O) to initiate the growth process. From the analysis of the rosette, the GreenLab model computes the evolutions of the demand and the produced biomass, as well as their distribution along with the DCs to reproduce at best the organic series observed at the chosen stage.

Figure 9.4a, d, c gives the evolution of individual LAI (*Sf/Sp*), Biomass per DC (*Q*), Demand per DC (*D*). We can observe that even if the fitting is good (Fig. 9.4e), the accuracy of the parameters given is low because the CV values are high (CV > 20%). The computed seed biomass is also imprecise.

Table 9.1 Computed source–sink parameters of lettuce

Param	r	Sp	Q_O	P_b	B_{b1}	B_{b2}
Mean	99.9	513	0.26	1	0.76	1.64
SD	24.5	140	0.23	0	0.44	1.09
CV (%)	24	27	89	0	57	66

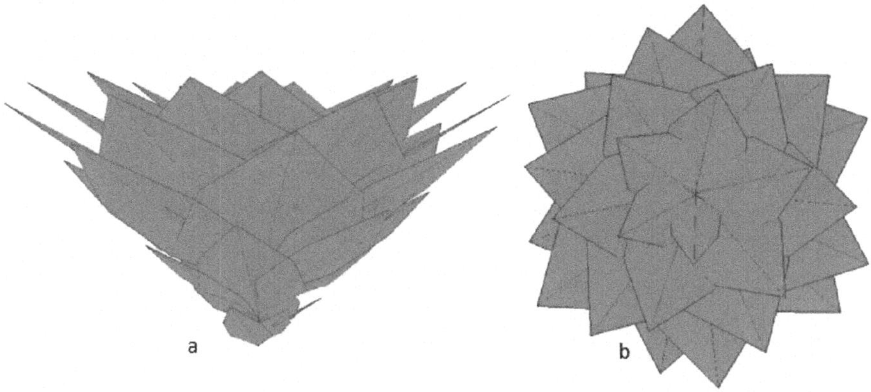

Fig. 9.5 Simulation of the lettuce at 36 DCs of growth. (**a**) Side view. (**b**) Top view (Gloups software)

9.5.1.4 Simulation of the Lettuce

We can simulate the 3D plant architecture of Lettuce at the age of 36 DCs. There is only one organic series (the leaves). The leaves are arranged along an axis with a phyllotaxy angle of 144° and an insertion angle of 60°. Figure 9.5 shows a side view and a top view of the simulated architecture.

9.5.1.5 Discussion

Demand increases almost linearly with the DCs (Fig. 9.4c). The value of the production surface $Sp = 513$ cm^2 indicates that the rosette is like an unclosed 23 cm square that is realistic. The *Sf/Sp* ratio (individual LAI Fig. 9.4a) becomes high because the leaves pile up overlapping in the rosette. One consequence is that biomass production Q saturates after 20 DCs, because of the ratio $Sf/Sp > 3$ (Fig. 9.4d and see Chap. 5).

Without the variation in the expansion time $t_{xd}(i)$ of leaf, it is impossible to fit the model. If $t_{xd}(i)$ is constant and set to 20 DCs, the gray dotted curve in Fig. 9.4e is obtained, which shows how impossible to properly fit the organic series with GreenLab. Despite the good fitting of the model to the lettuce rosette, there is no certainty that the parameters are invariant during the growth, especially since only one stage of development is analyzed, and the coefficients of variation in the parameters are high.

9.5.2 Case of Arabidopsis *Study of B. Wenden (University of Edinburgh)*

9.5.2.1 Plant Description

The beginning of the growth for the *arabidopsis* (*Arabidopsis thaliana*) is in a rosette form. The plant is reduced to its simplest expression: an organic series of leaves (no roots available in this experiment).

9.5.2.2 Observations on the Development

Modelling a clone of *arabidopsis* with the GreenLab model was first carried out by Christophe et al. (2008) (INRA) and subsequently by Chew et al. (2014) (University of Edinburgh). The environmental factor is normalized to $E = 1$. Leaf weights are in dry matter (g). The specific leaf weight (SLW) is $\varepsilon = 0.002$ g/cm^2. The sink strength of the leaf is normalized to $P_b = 1$. Like for lettuce, the five hidden parameters involved are Q_0, r, Sp, B_{b1}, B_{b2}.

We have the averages of five plants per repetition with synchronous growth and four developmental stages. The number of leaves produced at a given developmental stage is characterized by a mean and a variance, which defines a *Bernoulli* process with $b = 0.86$ (Fig. 9.6a). The four stages correspond to 9, 14, 21, 37 DCs, respectively. The measured series are fitted with the computed series from the

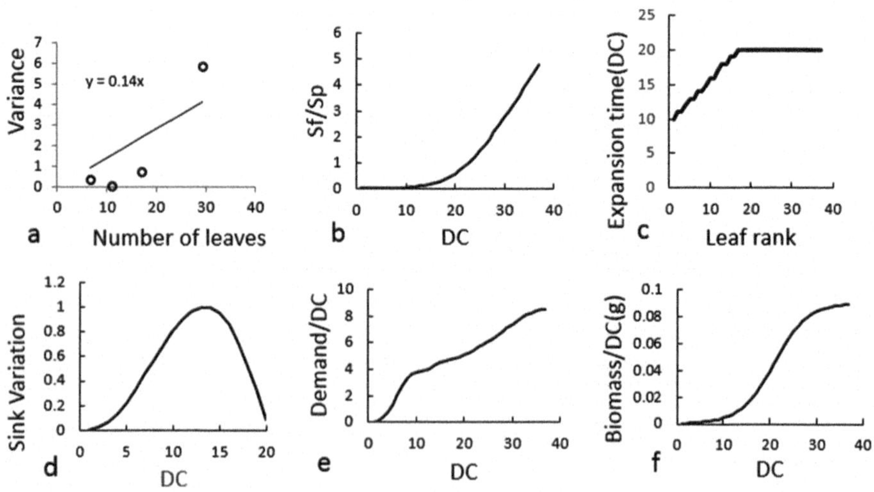

Fig. 9.6 The functioning of *arabidopsis*. (**a**) Link between mean and variance of leaves during plant development. This gives the value of the *Bernoulli* parameter $b = 0.86$. (**b**) Evolution of individual LAI (*Sf/Sp*). (**c**) Bottom-up variation in leaf expansion duration. (**d**) Sink variation in the leaf (Beta law shape). (**e**) Demand per DC (D). (**f**) Biomass production per DC (Q). The evolution is sigmoid shape because the growth is limited by *Sp* (According to B. Wenden)

Table 9.2 Computed source–sink parameters of *Arabidopsis* (multi-fitting)

Param	r	Sp	Q_0	P_b	B_{b1}	B_{b2}
Mean	1644	147	0.001	1	3.3	2.3
SD	150	20	0.001	0	0.4	0.35
CV (%)	9	13	41	0	13	15

potential structure that gives the average of the weights as a function of rank k using the negative binomial law $BN(k, b)$ (Chap. 6).

We consider that all the leaves are functional, which sets the value of the photosynthetic functioning time at $t_b = T = 37$ DCs. The expansion time is variable for the first leaves. Empirically, by varying the expansion time $txd(i)$ linearly from the value $t_m = 10$ DCs to the stabilized value $t_x = 20$ DCs in $k_x = 17$ ranks of phytomers, we obtain a good fitting (Fig. 9.6c) with the optimized parameters of Table 9.2. The target to be fitted by the multi-fitting consists of the weights of the compartments and the organic series of leaves at the four stages of development.

9.5.2.3 Results on Plant Functioning

The four stages of growth are analyzed together, which implies a multi-fitting process of the WLSM and assumes that the hidden and functioning parameters are invariant during the growth.

These measurements produce 73 weights of organs in the organic series and compartments to be fitted simultaneously during the growth process.

The fitting is good: $R^2 = 0.99$. Number of data = 73. Number of parameters = 5.

The five hidden parameters are computed with the WLSM in Table 9.2 except for the sink strength of the leaf P_b that is normalized to 1. The coefficients of variation are satisfying (CV < 20%) except for seed values. From the analysis of the rosette, the GreenLab model calculates the evolutions of the demand and of the produced biomass, as well as its distribution according to the DCs, in order to reproduce at best the organic series observed at the selected stage.

Figure 9.6 gives the evolution of individual LAI, the evolution of leaf expansion duration along the stem, the sink function, the demand per DC, and biomass production per DC.

Figure 9.7 shows the multi-fitting between the leaf compartment weights (a) and the leaf weights in the organic series (b) for the four selected growth stages 9, 14, 21, 37 DCs, respectively.

With the optimized parameters, the model completely restores the evolution of the organic series of the leaves of the rosette and that of the leaf compartment (the latter being redundant since it is the sum of the weights of the leaves) at the same time.

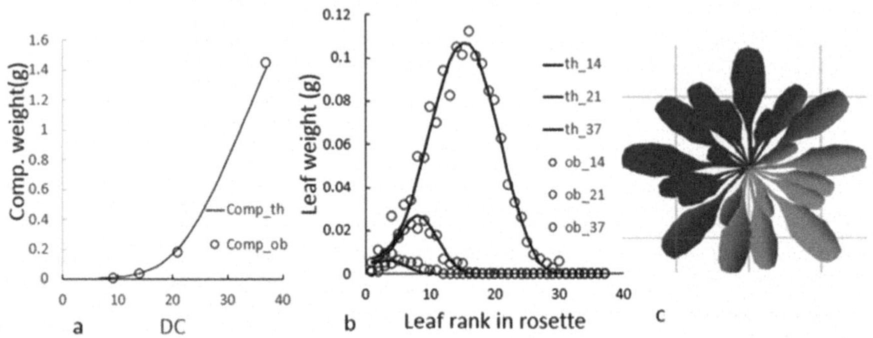

Fig. 9.7 Multi-fitting of leaf weights for the compartment and organic series for the four selected growth stages 9, 14, 21, 37 DCs (Gloups software). (**a**) Fitting evolution of the weight of the leaf compartments during the growth. (**b**) Fittings of the organic series of the leaves by the model. "○" represents the observations, "—" represents the theoretical values. (**c**) Simulation of the rosette (Digiplante software)

9.5.2.4 Simulation of the Rosette of Arabidopsis

Using the parameters computed with the WMLS, the rosette is simulated at 37 DCs (Fig. 9.7c).

9.5.2.5 Discussion

Surprisingly, the lack of data on the root system does not prevent excellent modelling of rosette growth. From this it can be deduced that the balance of shoot–root partitioning of the biomass during the growth is either stable or the weight of root system is negligible.

The demand increases with the DCs (Fig. 9.6e). The value of the production surface $Sp = 147\ \text{cm}^2$ indicates that the rosette is an unclosed 12 cm square that is realistic. The Sf/Sp ratio (individual LAI in Fig. 9.6b) becomes high because the leaves pile up overlapping in the rosette. One consequence is that biomass production Q saturates after 30 DCs, because the ratio $Sf/Sp > 3$ (Fig. 9.6f and see Chap. 5).

The perfect fit of the observations by the model for the four development stages shows that the parameters in Table 9.2 are invariant during the growth process, which is an important result for the robustness of the model. In principle, it is useless to consider the compartments if we provide the complete organic series, as these are redundant. Nevertheless, if available, they allow a more synthetic vision of the growth.

Multi-fitting not only verifies that the parameters of the growth model are stable during the growth, but also significantly reduces the CV of the parameters. So, if only the last development stage (37 DC) is taken into account, the single fitting gives Table 9.3.

Table 9.3 Computed source–sink parameters of *Arabidopsis* (single-fitting)

Param	r	Sp	Q_0	P_b	B_{b1}	B_{b2}
Mean	1921.5	169.4	0.004	1	4.52	3.27
SD	307	42.8	0.002	0	1.25	1.10
CV (%)	16	25	58	0	27	33

It can be seen that the CVs of the parameters for single fitting have doubled compared to the value of Table 9.2 for multi-fitting and therefore the accuracy has decreased.

The case of the *arabidopsis* is the simplest form of the GreenLab model with a meristem whose development is stochastic and a single rosette of leaves.

9.5.3 *Case of Beetroot Study of S. Lemaire (ITB)*

9.5.3.1 Plant Description

Beet is a biennial plant. In the first year, it is considered to have only two compartments: the rosette of leaves and the tape root.

9.5.3.2 Cultivation Condition

Experiments on beet were carried out at the ITB (Institut Technique de la Betterave) in 2006 by S. Lemaire and F. Maupas (Lemaire 2009a, b; de Reffye et al. 2009). They have focused on the effect of planting density in the absence of stress. The climate factor is normalized to $E = 1$. Organ weights are in dry matter. The tape root weights are more than three times the weight of the leaves when mature. The total weight of the roots is unknown, and it is assumed here that the weight of the tape root is a suitable default value, which remains to be checked. ITB has carried out an experiment with a planting density of 10 plants/m^2. Although beetroot has a simple morphological appearance, its development requires more details for successful modelling.

9.5.3.3 Development Scheme

Cultivated beets are a mix of genotypes. There is, therefore, genetic variability to be considered. This is manifested by differences in the speed of individual development (Fig. 9.8), and by potential variability in the source–sink parameters. If it is possible to consider a mean value for the source–sink parameters, it is not the same for the development because it involves the method of construction of the potential structure.

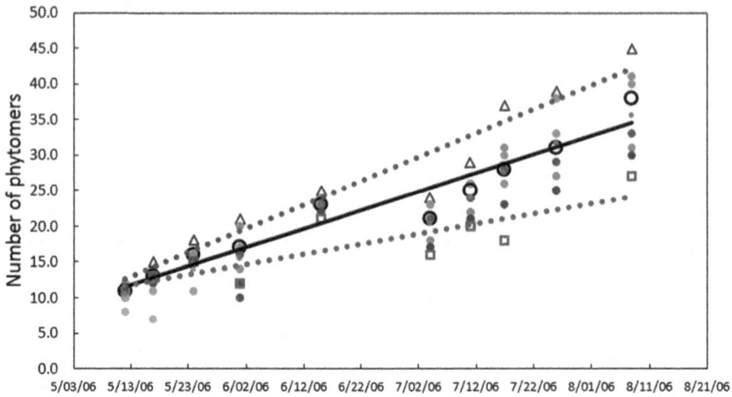

Fig. 9.8 Development of the beet rosette for isolated plants of different genotypes (same treatments). Number of leaves produced as a linear function of calendar time. The plants have different growth rates. The black circles (○) give the average rhythm linear trend; the blue triangles (Δ) and the red squares (□) give the extreme individuals rhythms

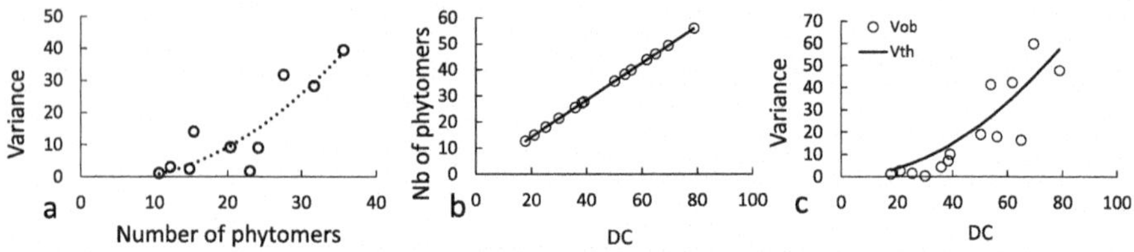

Fig. 9.9 The relationship between the mean and the variance of the number of produced phytomers in beet. (**a**) Quadratic relationship between the mean and the variance of the number of phytomers as a function of the measurement date. (**b**) Modelling of the mean number of the rosette leaves as a function of the number of DCs. (**c**) Modelling of the variance as a function of the number of DCs

Beet phyllochron shows two distinct stages when the planting density is high; it slows down slightly when the plant rosettes come into contact. As it is reasoned in the development cycle, this phenomenon can be neglected. Indeed, when the rhythm ratio decreases, the plant produces more biomass per cycle, but in the case of beet, this biomass mainly goes to the roots because the leaf area per plant no longer evolves.

Each plant has a linear development as a function of thermal time, which varies from one plant to another due to the intrinsic genetic variation (Fig. 9.8). In other words, the rhythm w is variable. This has an influence on the top-down construction of the average organic series. At row k, there is a mixture of dates in DC.

For 14 observed development dates, the mean and variance of the number of leaves for 10 randomly selected plants were computed (Table 9.4).

The relationship between the mean and the variance of the number of phytomers produced has a quadratic appearance (Fig. 9.9a).

Table 9.4 Observations of the mean-variance pairs of the number of phytomers for 14 development stages of beet

Mean X.	Variance V_X	Computed DC
12.600	1.187	17
14.933	2.440	21
18.000	1.359	25
21.375	0.238	29
25.500	4.286	35
27.750	10.000	38
27.25	6.98	44
35.625	18.905	49
39.857	17.867	55
38.250	41.286	53
43.875	42.286	60
46.125	16.333	64
49.375	59.810	68
56.000	47.700	77

The fastest growing plant will have $w = 1$. The number of leaves of this plant will, therefore, correspond to the number of DCs that determine the age of each plant in the population. The average plant gets $w < 1$. The effect of w is to create a periodic series of 0s and 1s (see Chap. 3). The number of phytomers produced during N DCs for the plant i with a rhythm ratio w_i is described as follows:

$$X_i = w_i \cdot N \tag{9.16}$$

It is now assumed that w_i are distributed according to a normal distribution with mean and standard deviation σ_w. It is demonstrated that the mean $\overline{X}$and variance V_X of the number of phytomers are described as a function of the number of cycles N, the rhythm $\overline{w}$, and its standard deviation σ_w according to the equations:

$$\bar{X} = \bar{w} \cdot \bar{N}; V_X = \bar{w}^2 \cdot V_N + \bar{N}^2 \cdot \sigma_w^2; \text{but} V_N = 0; \text{so} V_X = \bar{N}^2 \cdot \sigma_w^2 \tag{9.17}$$

We deduce:

$$V_X = \frac{\sigma_w^2}{\bar{w}^2} \bar{X}^2 \tag{9.18}$$

The variance is thus well quadratically linked to the mean. We pose as a model:

$$w = \frac{1 + \sigma_w U}{1 + 3\sigma_w} \tag{9.19}$$

U is a normal variable centered [0, 1]. We have a set $\bar{w} + 3\sigma_w = 1$ which is the maximum allowed value at w (at 3 times the standard deviation, we cover more than

99% of the distribution). So, we have: $\overline{w} - 3\sigma_w < w < \overline{w} + 3\sigma_w < 1$. This gives $\overline{w}$ depending on σ_w and we express X and V depending of σ_w

$$w = \frac{1}{1 + 3\sigma_w}; X = \frac{1}{1 + 3\sigma_w} N; \quad V_X = \left(\frac{\sigma_w}{1 + 3\sigma_w}\right)^2 N^2 \tag{9.20}$$

Pairs of T (the growth stages) and (Xi, V_{Xi}) (the measured values) are arranged, each of which gives an estimate of σ_w. It is, therefore, possible to calculate the mean value from Eq. (17).

$$T \cdot \sigma_w{}^2 = \sum\nolimits_{i=1,T} \frac{V_{Xi}}{(X_i)^2}; \sigma_w = \sqrt{\frac{\sum_{i=1,N} \frac{V_{Xi}}{(X_i)^2}}{T}} \tag{9.21}$$

The relation between the average number of phytomers X_i and the number of DC N_i is written as:

$$N_i = (1 + 3\sigma_w) \cdot X_i \tag{9.22}$$

This gives: $\sigma_w = 0.135$

Hence, the numerical relationships between mean, variance, and number of cycles are shown in Fig. 9.9b and c.

$$\bar{X} = \frac{1}{1 + 3 * 0.135} N \to 0.71 * N; V_X = \left(\frac{0.135}{1 + 3 * 0.135}\right)^2 N^2 \to 0.0092 * N^2$$

9.5.3.4 Results on Plant Functioning

For beets, the organic series of leaves, leaf compartments, and root compartments (taproot) are used. It is assumed that the taproot accounts for most of the root weight. Samples were taken from the 17, 21, 25, 29, 35, 38, 44, 49, 53, 55, 60, 64 DCs, representing 12 growth stages used previously for studying the development.

The average rhythm ratio is $w = 0.71$, the leaf thickness ε is 0.012 g/cm^2. The functioning time of the leaves (before wilting) is estimated as 60 DCs. The functioning time of a leaf is referred to here as the time that manages both the duration of photosynthesis and the expansion time. The expansion time duration of the tape root is considered until the end of growth at harvest, with $T_r = 114$ DCs.

To reduce the sample size of the organic series, the weights of the last 15 leaves of the rosette were measured for each of the 12 growth stages. The weights of the leaf and root compartments were also measured.

9.5.3.4.1 Multi-Fitting of Rosette and Taproot Together

The weight of the root system is collected as a compartment and weighed at each observed development stage, so that the organic series of leaf weights of the rosette, and the weights of the leaf and root compartments are the available targets in multi-fitting. The variation law of the sink of the leaves needs to follow the evolution of the bottom-up expansion duration according to the rank in the rosette. It also needs two parameters B_{b1} and B_{b2} to be controlled (the sink strength of the leaf is defined to $P_b = 1$).

For the root system, it is necessary to estimate the sink strength P_r and only one parameter of variation B_{r1} is to be computed. Indeed, the sink of the roots is an increasing function in time, which makes it possible to set $B_{r2} = 1$.

These measurements produce 264 weights of organs in the organic series and compartments to be adjusted simultaneously during the growth process. There are seven parameters (Table 9.5) to be estimated.

The fitting is good: $R^2 = 0.97$, number of data = 264, number of parameters = 7).

The accuracy of the parameters is good, with the CVs around 10%.

The computed production area *Sp* (1111 cm^2) is close to the available area *Sd* (1000 cm^2), which corresponds to a planting density of 10 plants/m^2. This is a validation of the GreenLab model. Figure 9.10a shows the evolution of the individual LAI (*Sf*/*Sp*) that reaches the value of 3, which corresponds to a saturation of biomass production per DC. This can be seen in Fig. 9.10c. The taproot demand eventually captures all the biomass, and the surface expansion of the rosette is blocked (Fig. 9.10b and c).

Table 9.5 Computed source–sink parameters of shoot and root on beet (multi-fitting)

Param	r	Sp	P_b	P_r	B_{b1}	B_{r1}	B_{b2}	B_{r2}
Mean	154	1116	1	551	1.5	5.2	2.3	1
SD	5.2	89	0	115	0.1	0.23	0.3	0
CV (%)	3	8	0	20	6	4	13	0

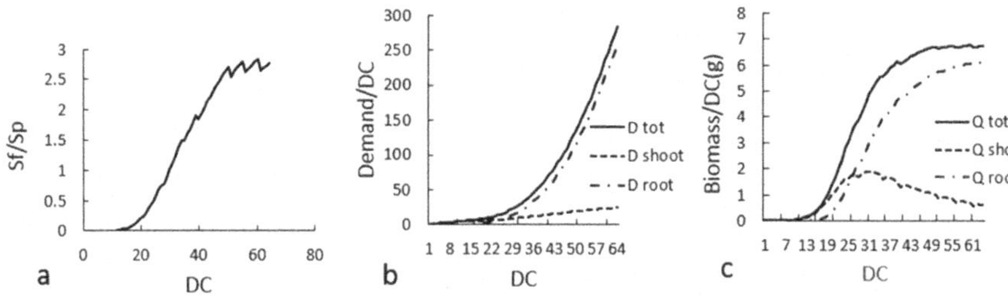

Fig. 9.10 Computed LAI, demand and biomass of beetroot. (**a**) Evolution of the individual LAI (*Sf*/*Sp*). The LAI reach *Sf*/*Sp* = 3. (**b**) Evolution of the total Demand/DC, shoot demand per DC (rosette), and root demand per DC. (**c**) Total biomass reaches saturation when *Sf*/*Sp* - > 3. Eventually the root system captures all the biomass

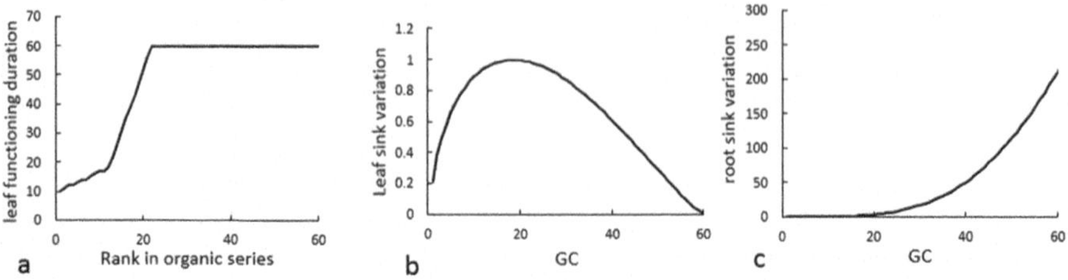

Fig. 9.11 Fitting of beet. (**a**) Evolution of the expansion time of the leaves along the stem. (**b**) Sink variation in the leaf. (**c**) Sink variation in the taproot

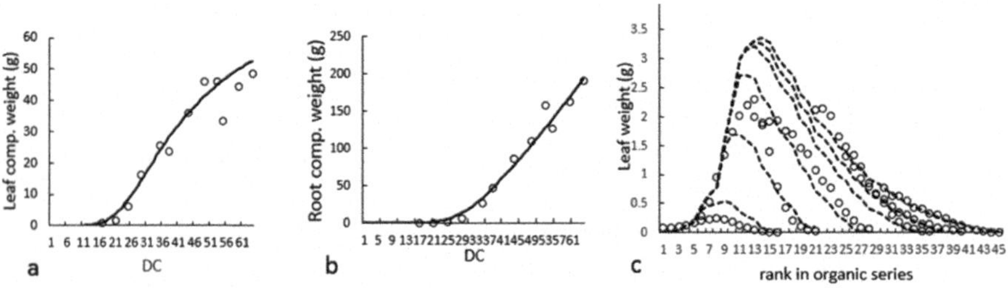

Fig. 9.12 Evolution of the weights for the compartment and organic series of leaf and taproot. (**a**) Evolution of leaf compartment weight. (**b**) Evolution of taproot compartment weight. (**c**) Organic series of leaves for six selected growth stages (for more clarity only half of stages has been represented). "○" represents the observations and "—" represents the computations (Gloups software)

There is high variability in the duration of leaf expansion at the base of the rosette (Fig. 9.11a). The expansion is solved empirically so that the base of the organic series is correctly adjusted. For leaves ranked from 1 to 20, with two control points, the expansion duration is changed from 10 DCs to 60 DCs. The WLSM gives the sink variations in the leaf and the tape root (Fig. 9.11b and c).

The parameters of Table 9.5 allow fitting the observed leaves (Fig. 9.12a) and taproot (Fig. 9.12b) compartments correctly as well as the organic series of leaves (Fig. 9.12c) with the homologous computed values (only the last 15 leaves of the organic series are fitted).

9.5.3.4.2 Rosette Fitting Alone

One question is whether multi-fitting is possible by considering only the rosette of leaves as for *Arabidopsis*.

Three parameters are retained: r, Sp, B_{b1} (B_{b2} is set empirically to 2.32). The WLSM converges and gives Table 9.6.

The fitting is bad: $R^2 = 0.48$, number of data = 240, number of parameters = 3.

Table 9.6 Computed source–sink parameters of rosette alone on beet (multi-fitting)

Param	r	Sp	P_b	B_{b1}
Mean	298.866	1022.177	1	2.61
SD	7.918	170.365	0	0.314
CV (%)	2.6	16.7	0	12

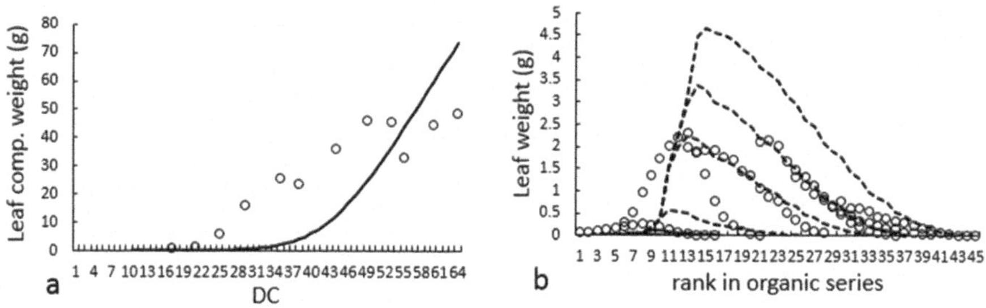

Fig. 9.13 Fitting with the rosette alone (leaves and compartments, no tape root) using the model. There is no fit between the observed and the computed data. (**a**) Leaf weight per DC in the compartment. (**b**) Leaf weight per rank in organic series (Gloups software)

Note that the parameter r is increased because there is less biomass (no roots) to produce. However, the surface Sp is well estimated.

Despite the WLSM converges, it is not possible to adjust the observed and computed data of the rosette alone in multi-fitting, compared with the case of *Arabidopsis* (Fig. 9.13)

The reason is that the shoot–root balance in biomass is strongly modified during the growth. The demand of the root gradually replaces that of the shoot, to the point where the leaves of the rosette can no longer grow and remain embryonic (see Fig. 9.10). Thus, the rosette expansion is stopped. Compared with the case of the *Arabidopsis*, the difference is that its shoot–root balance was not disturbed enough to prevent a good multi-fitting on leaf organic series of the rosette alone.

9.5.3.4.3 The Search for the Root Ghost Sink

In the demand of the plant, all the sinks of the organs are involved. Therefore, the compartment and the rosette of the leaves implicitly reflect the action of the demand of the roots. It is, therefore, possible to compute the root sink function, which can be considered a ghost sink, without using root system data. In the case of the beet, the procedure works well. Table 9.7 gives the parameters of the beetroot without using the taproot data.

Tables 9.5 and 9.7 give very similar results. In the case where the root compartment strongly disturbs the shoot–root balance, it is possible to estimate it only from the rosette without using the taproot data.

Table 9.7 Computed source–sink parameters of shoot and root without using taproot data on beet (multi-fitting)

Param	r	Sp	P_b	B_{b1}	B_{b2}	P_r	B_{r1}	B_{r2}
Mean	156.8	1123.7	1	1.56	2.284	558.9	5.32	1
SD	3.5	39.6	0	0.097	0.28	79.4	0.15	0
CV (%)	2.3	3.5	0	6.2	12.3	14.2	2.9	0

Table 9.8 Comparison between the observed values of root weights and those estimated by a ghost sink

DC	17	21	25	29	35	38	44	49	53	55	60	64
Observed root weight (g)	0.2	0.2	1.5	5.8	25.6	46.2	85.9	109.2	157.3	126.1	162.5	189.9
Calculated root weight (g)	0.2	1.3	5.1	13.1	32.8	45.4	74.4	101.4	124.3	136.0	165.9	190.1

Table 9.8 compares the observed values of root weights and those computed solely from leaf compartments and organic series for the 12 selected growth stages.

The estimation which is imprecise at the beginning of the growth is gradually becoming satisfactory.

9.5.3.5 Simulation of Beetroot

Simulations of 2D and 3D beet using the parameters computed by the WLSM are shown in Fig. 9.14. The 2D simulation allows better visualization of the organic series of leaves.

9.5.3.6 Discussion

The beet study is the first time to consider the shoot–root interaction in a detailed way during the growth. The WLSM estimates the functional parameters that control the development and growth. The fitting of the compartments and organic series at all measured stages is of good quality. These parameters are, therefore, invariants that characterize the functioning of the beet. The balance between the shoot and root is strongly altered during the growth, which disturbs the organic series of leaves. This deformation makes it possible to find the parameters and the evolution of the root compartment, even in the absence of direct root data. ITB has focused on the effect of planting density on beet growth in the absence of stress. The results are shown in Chap. 11.

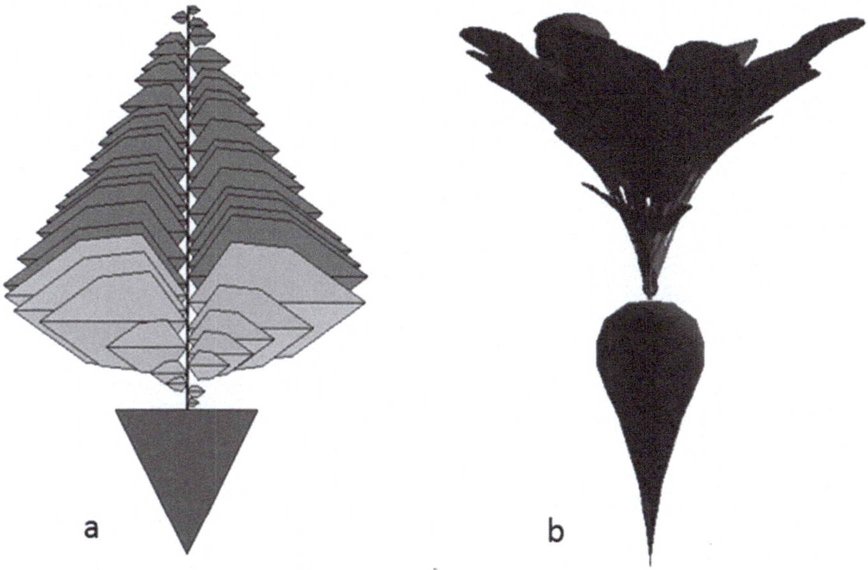

Fig. 9.14 Beet simulation. (**a**) 2D simulation, the leaf organic series is visible (Gloups software). (**b**) 3D simulation (Digiplante software)

9.5.4 Case of Maize Study of Ma Yun-Tao (CAU)

Maize is one of the first plants calibrated by the GreenLab model and has contributed well to its validation in agronomy. It has also been the most studied plant. The relationships between the development, growth, architecture, and environmental factors (climate, density) have been the subject of repeated experiments over six years at China Agricultural University (CAU) (Song et al. 2003; Guo et al. 2006; Ma et al. 2007, 2008) and at CIRAD and IRSTEA (Feng et al. 2014).

9.5.4.1 Plant Description

Maize is a single-stemmed plant for cultivated varieties. After an initial rosette stage, the stem elongates and the development stops after about 20 phytomers (a fixed number depending on the variety) following terminal flowering and the formation of the male flower called the tassel. Growth continues and mainly concerns the maturation of the female ears, or cobs, of which there are typically one or two, appeared in the middle of the stem. The basic phytomer consists of an internode, a leaf made of a blade and a sheath. The sheath must be considered because its biomass is not negligible. Some median phytomers carry a cob. The stem ends with the tassel at its tip.

9.5.4.2 Cultivation Condition

The plants have grown in the field with a planting density of 5 plants/m^2. They were collected at different growth stages, the measurements were on blades, sheaths, internodes, cobs, and tassels. Data on the root system was not available. Five observation dates were selected between germination and maturation. Six plants were harvested and averaged at each stage and, therefore, formed independent samples. First, the analysis is done in a normalized climate ($E = 1$). In Chap. 11, the climatic effect is studied by recording the values of $E(i)$ in each cycle, taking observed transpiration (PET) and water use efficiency (WUE) as the growth factor, as well as the effect of planting density on biomass production and partitioning.

9.5.4.3 Development Scheme

Maize development is deterministic and in a good linear relationship with thermal time (see Chap. 5, Fig. 5.2). The creation of a phytomer corresponds to the development cycle. There are four types of phytomers. After a leaf stage of six phytomers (which can emit roots), the stem elongates. The subsequent nine phytomers have the potential to carry a cob, but only the last two can develop. The next five phytomers do not have cobs. The terminal meristem is finally transformed into a tassel at cycle 21, which stops the development, but the growth continues beyond that and concerns mainly the cob. Growth cycles are no longer materialized by phytomers. They are obtained by extrapolating the linear relationship between thermal time and the number of phytomers obtained during the development phase. Normally it is equivalent to reasoning in development cycle (DC) or growth cycle (GC). But in the case of maize, the development stops at 21 DC, while the growth continues. It is, therefore, relevant to use the GC.

9.5.4.4 Growth Scheme

To correctly adjust the base of the organic series, it is necessary to vary the organ expansion durations, which is shown in Fig. 9.16. For computing the demand, it is essential to consider variations in the observed expansion time of organs. They are estimated empirically:

- The blades and sheaths pass from 5 to 11 expansion GCs in rank 9 of phytomers from the base. The same functioning times are assigned based on these expansions. A delay of two GCs is attributed to the emergence of the sheaths that appear after the blades.
- The internodes pass from 2 to 20 expansion GCs in rank 9 of phytomers from the base. The first very short internodes have indeed almost no expansion;
- There can be several cobs from rank 6, as illustrated by the botanical automaton. But up to the 13th rank, they are either aborted or very small. The plant, therefore,

has only one or two cobs in equal numbers. They concentrate almost all the fruit biomass. When there are two cobs, since they are adjacent, to simplify, their biomass is concentrated on the 15th rank to avoid the use of stochastic analysis without significant loss of accuracy. We, therefore, consider that there is only one cob. This cob shows an expansion delay of four GCs and an expansion time of 18 GCs.

An interesting experiment is to suppress the cob at its creation. It is observed that the plant quickly withers and dies. The reason is that the expansion time of the leaves of grasses is reduced and the surplus biomass produced cannot be absorbed by the plant, which is poisoned by the excess of sugars (Guo Yan, personal communication).

Simple organic series are easily established by weighing, according to their rank counted from the base, the blades, sheaths, internodes, and ears. The weights of redundant compartments are also considered in parameter optimization.

9.5.4.5 Results on Plant Functioning

9.5.4.5.1 Empirically Measured Parameters

The specific leaf weight (SLW) ε, which corresponds to a thickness, is deduced from the ratio of fresh weight and its surface, which is 0.024 g/cm^2.

9.5.4.5.2 Computed Parameters

Measurements are available at stages 8, 14, 18, 24, 30 GCs. These measurements produce 445 data of organs in the organic series and compartments to be adjusted simultaneously during the growth process. Fourteen parameters (Table 9.9) are estimated according to a multi-stage calibration (multi-fitting).

The fitting is good: $R^2 = 0.98$, number of data = 445, number of computed parameters = 14.

The values of the 14 parameters are invariants that control well the fittings of the organic series during the growth period (the parameter P_b is fixed to 1).

In this experiment, all the parameters of the source and sink functions of organs (P_o, B_{o1}, B_{o2}) are computed (*b*: blade; *p*: sheath; *i*: internode; *f*: cob; *m*: tassel). The computed production area *Sp* (2181 cm^2) is close to the available area corresponding to the planting density *Sd* (2000 cm^2), which is a validation of the model.

The sink strength of the fruits (P_f) is less clear. This is because the fruit is present only once per stem, while the vegetative organs of the phytomer (leaves, sheaths and internodes) are presented 21 times. For the seed, its effect dissipates during the growth; for the fruit, it captures all the demand at the end of the growth, and makes it difficult to be accurate because of its big value; and for the tassel, with a short expansion period (two DCs), only one constant sink is assigned.

Table 9.9 Computed source–sink parameters of maize using the complete set (multi-fitting)

Param	Mean	SD	CV (%)
r	23.4	1.4	6
Sp	2181	177	8
Q_0	1	0.3	25
P_{b1}	1	0	0
P_{p1}	1	0.1	14
P_{i1}	1.5	0.2	11
P_{f1}	367.7	2076.3	564
P_m	10.4	1.8	17
B_{b1}	2.6	0.5	19
B_{p1}	3.6	0.7	18
B_{i1}	3.6	0.4	10
B_{f1}	2.9	0.9	29
B_{b2}	7.7	1.8	22
B_{p2}	14.1	3.1	21
B_{i2}	3.8	0.5	13
B_{f2}	1.2	2.9	247

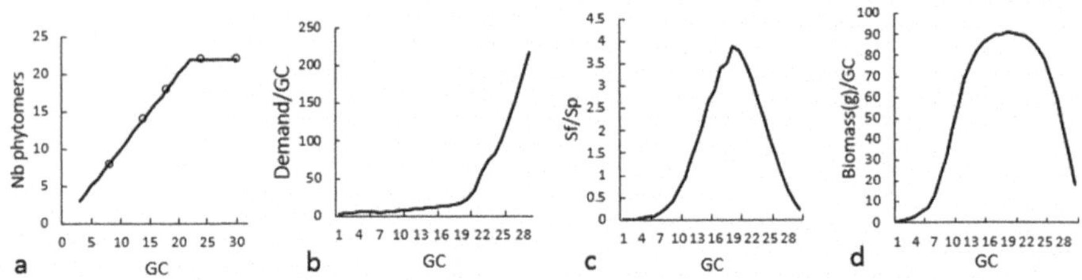

Fig. 9.15 History of the development and growth of maize. (**a**) Number of phytomers per GC. Development stops at 21 DCs (equals to 21 GCs). (**b**) Evolution of demand per GC. The appearance of the cob of maize in cycle 15 gradually captures all the demand. (**c**) Evolution of individual LAI. Production saturates when $Sf/Sp > 3$. (**d**) Evolution of biomass production per GC. After exponential growth, the plants reach saturation, and then decrease due to leaf mortality

Figure 9.15 shows the evolution of the number of phytomers, plant demand, individual LAI (Sf/Sp), and produced biomass per cycle as a function of GC.

As mentioned above, to correctly fit the base of the organic series, it is necessary to empirically vary the expansion duration of organs, which is shown in Fig. 9.16. This makes it possible to refine the demand.

Three parameters (P_o, B_{o1}, B_{o2}) are used to calculate the sink variation of the organs with the cycles, and to account for the shift to expansion if necessary (Fig. 9.17).

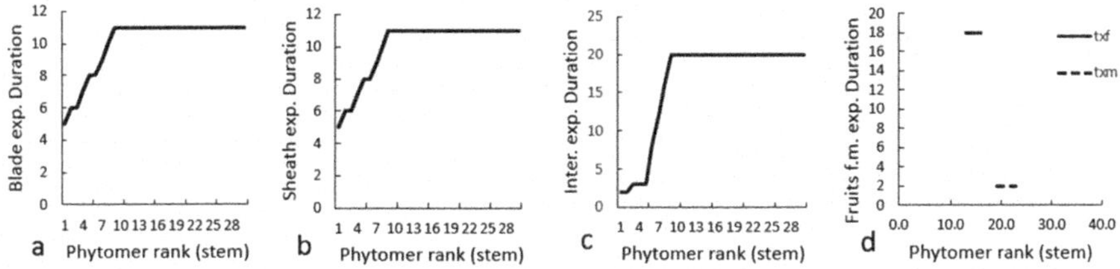

Fig. 9.16 Evolution of organ expansion time from the base of organ series of maize. (**a**) Blade. (**b**) Sheath. (**c**) Internode. (**d**) Male and female fruits

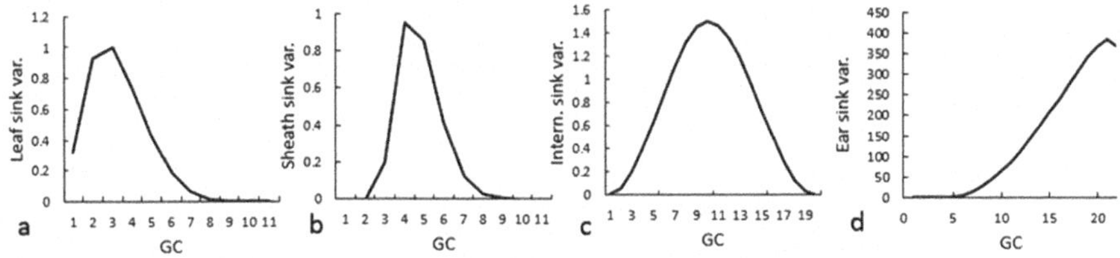

Fig. 9.17 Sink variation during organ expansion. (**a**) Sink variation in blade. (**b**) Sink variation in sheath (there is a delay of two GCs before the expansion). (**c**) Sink variation in internode. (**d**) Sink variation in cob (there is a delay of four GCs before the expansion)

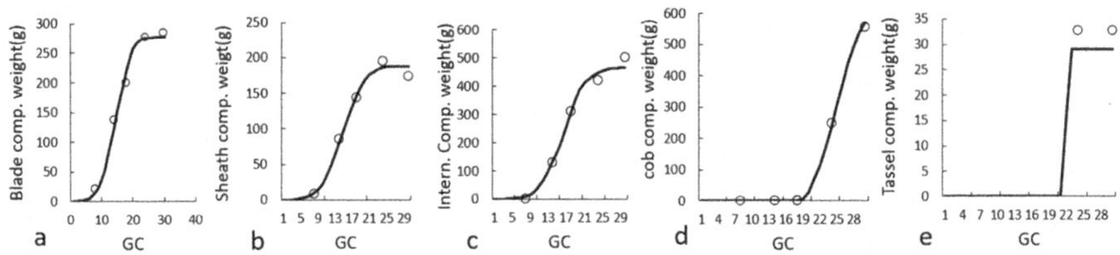

Fig. 9.18 Multi-fitting of the observed organ compartments (○) and computed organ compartments by the model (—) for the five measurement dates (8, 14, 18, 24 and 30 GCs). (**a**) Blades. (**b**) Sheaths. (**c**) Internodes. (**d**) Cobs. (**e**) Tassel

9.5.4.6 Comparison Between the Observed and Computed Data

The organ compartments and the observed organic series are used as the target data for the WLSM, which can optimize the source–sink parameters as well as possible by the model. The fitting results are good for both compartments (Fig. 9.18) and organic series (Fig. 9.19) with a high regression coefficient ($R^2 = 0.98$).

For the fruits cob and tassel, organic series and compartments are identical because there is only one organ on the stem.

Remark: If one tries to fit young maize plants (<10 DCs) with the model, it is impossible to compute with the WLSM, the value of *Sp* as it grows indefinitely. This shows that the interception of light occurs over the whole leaf surface.

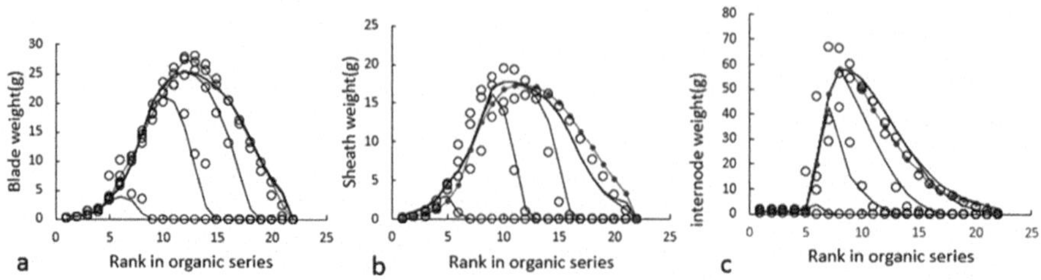

Fig. 9.19 Maize: multi-fitting of the observed (○) organic series and computed organic series by the model (—) for the five measurement dates (8, 14, 18, 24 and 30 GCs). (**a**) Blades. (**b**) Sheaths. (**c**) Internodes

Competition between plants is not yet apparent at this stage. This is applicable to young plants which validates the model.

9.5.4.7 Sampling Optimization

There were 445 measurements made on maize with five stages of growth considering five types of organs, along with their corresponding compartments and organ series. Too many measurements to be carried out may seem incompatible with the means available to the experimenter. This experimental sample allows for an exploration of the extent to which the measurements can be simplified without compromising the accurate estimation of the source–sink parameters involved.

9.5.4.7.1 Subsets of Sampling: (C = Compartment; S = Organic Series)

- Set 1, 5C*5S, the set is complete (number of measurements: 445, number of estimated parameters: 14): five growth stages (8, 14, 18, 24, 30 GCs) for the compartments and the organic series, all parameters are computed.
- Subset 2, 0C*5S (number of measurements: 420, number of estimated parameters: 13): five growth stages for the organic series without the compartments.
- Subset 3, 3C*1S_R (number of measurements: 71, number of estimated parameters: 10): three growth stages for the compartments and the last stage for the organic series (one phytomer measurement out of 5).
- Subset 4, 0C*1S (number of measurements: 110, number of estimated parameters: 11): the last stage for the organic series without the compartment.
- Subset 5, 5C*0S (number of measurements: 25, number of estimated parameters: 6): five growth stages for the compartments without the organic series. The parameters of the sink variation law are known because they have already been estimated by a previous experiment of type 5C*5S.

Table 9.10 Source–sink parameters of maize (Subset 3)

Param	r	Sp	Q_0	P_b	P_p	P_i	P_f	P_m	B_{b1}	B_{p1}	B_{i1}	B_{f1}	B_{m1}
Mean	28.2	3353.0	4.04	1	0.91	2.7	565.2	11.4	3.7	2.2	5.4	3.2	1
SD	1.08	217.7	0.7	0	0.02	0.10	209.6	0.6	0.2	0.06	0.12	0.2	0
CV (%)	4	7	20	0	2	8	3	10	5	2	2	8	0

- Subset 6, 5C*0S_E (number of measurements: 25, number of estimated parameters: 6): five stages for the compartments without the organic series. Sink parameters are considered constant.

In any case, it should be remembered that the sink strength of the leaves (P_b) is normalized to 1, while that of the tassel does not vary with $B_{m1} = 1$ and $B_{m2} = 1$.

Depending on the type of sampling, not all the parameters are computable. In this case, some of them must be estimated empirically. This is the case of the seed (Q_0) and the second parameter of the sink variation laws (B_{o2}). We can see empirically that the ratio $B_{o1}/(B_{o1} + B_{o2})$ is stable. Thus, in a wide range of variations (B_{o2}), the model can be fitted well. Empirically, we will look for the smallest possible value of B_{o2}.

The values and the coefficients of variation for the parameters of seven measurement sets are summarized in Table 9.10. We observed that all CVs are <20% which means a quite good fitting.

9.5.4.7.2 Results on Subsets

All the fittings are good, but the accuracy of parameters varies according to the sampling size. The parameter values of a kind of organ are relatively stable among the different subsets.

The organs of seed and fruits are special because they are single. However, the other organs, called vegetative organs (vo), are repeated for the 21 phytomers. Thus, the CVs are higher, and the accuracy is lower for these parameters. The sink of the cob is about 300 times higher than the sink of the blade, so practically it takes all the biomass, when the variation is at its maximum.

- Set 1, 5C*5S. The set is complete (number of measurements: 445, number of estimated parameters: 14). All the parameters can be computed (Table 9.9). Parameters r and Sp have a low CV (< 10). The CVs of the vegetative organs (vo) are <=20. Figures 9.18 and 9.19 show good multi-fitting.

Then we fix B_{b2}, B_{p2}, B_{i2}, B_{f2} to the values obtained from set 1, and compute the other parameters. We found that the computed parameter values are the same as that of set 1, but the CVs of the parameters decrease two times (CV < 10). This is because the more parameters are computed, the better is the fitting, but the greater the imprecision on them. In fact, the two parameters B_{o1} and B_{o2} characterize the sink variation in the o-type organ; with a sink strength P_o, the most important is the ratio

Table 9.11 Stability of the ratio $B_{o1}/(B_{o1} + B_{o2})$

Organs		B_{o2}		2^*B_{o2}
Blade	B_{b1}	2.6	B_{b1}	4.5
	B_{b2}	7.6	B_{b2}	14
	$B_{b1}/(B_{b1} + B_{b2})$	0.25	$B_{b1}/(B_{b1} + B_{b2})$	0.25
Sheath	B_{p1}	3.6	B_{p1}	7.3
	B_{p2}	14.05	B_{p2}	28
	$B_{p1}/(B_{p1}+ B_{p2})$	0.25	$B_{p1}/(B_{p1}+ B_{p2})$	0.24
Internode	B_{i1}	3.6	B_{i1}	5.7
	B_{i2}	3.8	B_{i2}	7
	$B_{i1}/(B_{i1}+ B_{i2})$	0.49	$B_{i1}/(B_{i1}+ B_{i2})$	0.44
Cob	B_{f1}	2.9	B_{f1}	3.4
	B_{f2}	1.2	B_{f2}	2
	$B_{f1}/(B_{f1}+ B_{f2})$	0.72	$B_{f1}/(B_{f1}+ B_{f2})$	0.61

$B_{o1}/(B_{o1} + B_{o2})$, which is stable. So, by doubling the initial parameter values of B_{o2} found in the experiment, and then fixing them, the WLSM recalculates the parameter values of B_{o1}, and provides equivalent quality of fittings. The errors with the WLSM are 18.7 for B_{o2} and 19.7 for 2^*B_{o2}.

Table 9.11 shows the stability of the ratio $B_{o1}/(B_{o1} + B_{o2})$, when B_{o1} is recomputed by the WLSM when the value of B_{o2} increases twofold.

- Subset 2, 0C*5S (number of measurements: 420, number of estimated parameters: 14). In this subset, the compartments of organs are skipped. The results on parameters are the same with the complete set 5C*5S (Table 9.9) because the compartments are redundant. The compartments give a more global view, while the organic series offer a more local view of the fitting.
- Subset 3, 3C*1S_R (number of measurements: 71, number of estimated parameters: 11). Here the sample size is reduced a lot by selecting only three growth stages (8, 18, 30 GCs) for the compartments and measuring the organ weights only one out of five ranks in the last organic series. The last organic series allows the estimation of parameters B_o and the compartments allow to verify the stability of source–sink parameters during the growth. In this experiment, there are six times fewer data to adjust than in the complete set, but the WLSM cannot calculate the parameters B_{o1} and B_{o2} at the same time. Therefore, it is necessary to fix empirically the parameters B_{o2}. It is wise to choose the smallest values for B_{o2} to obtain a good fit. Increasing them does not significantly improve the quality of the fit. We fix the B_{o2} to the fitting values obtained in the complete set 5C*5S (Set 1).

The fitting is good, and the error is low ($R^2 = 0.99$, Error = 1.34). Compared to Table 9.10, the values r, Sp, and Q_0 are slightly increased. These values are statistically correlated.

Figures 9.20 and 9.21 show the good quality of the fitting. The computed parameter values remain close to those of the complete set. Compared to

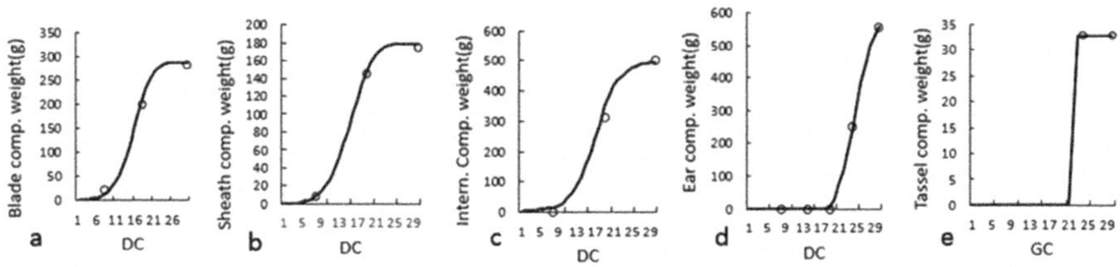

Fig. 9.20 Weights of observed (○) and computed by the model (−) of the organ compartments for five growth stages. (**a**) Blades. (**b**) Sheaths. (**c**) Internode. (**d**) Ear. (**e**) Tassel

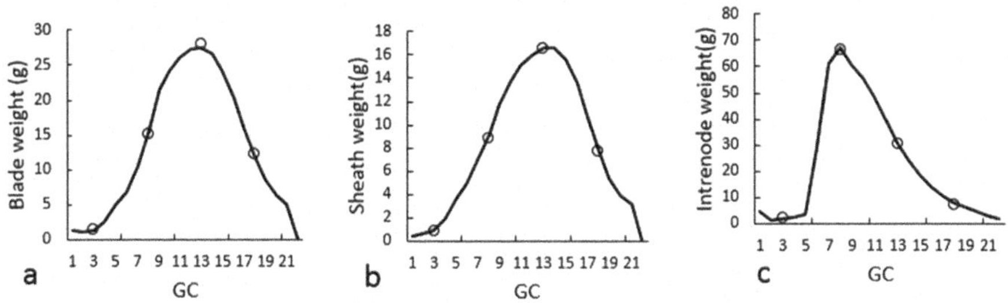

Fig. 9.21 Weights of observed (○) and computed by the model (−) of the organic series on the last stage of growth. (**a**) Leaves. (**b**) Sheaths. (**c**) Internodes

Table 9.10, the parameters *r*, *Sp,* and Q_0 slightly increase. They are indeed statistically linked. This reduced subset 3C*1S_R allows us to compute the sinks and their variations using the organic series of the last stage of growth quite well and to verify using the compartments that the source–sink parameters are stable during the growth.

- Subset 4, 0C*1S (number of measurements: 110, number of estimated parameters: 11): This subset includes the last growth stage of the organic series without the compartments. The fitting on the organic series is very good ($R^2 = 0.99$). The values of the computed parameters are close to those of the previous sample Subset 3, but we cannot verify the stability of the parameters because we do not have several stages of growth. This type of sample is, however, interesting because it corresponds to an isolated observation of a plant in the natural environment. We can get a first impression of its growth mode.
- Subset 5, 5C*0S (number of measurements: 25, number of estimated parameters: 7): This subset consists of five growth stages of the compartments without the organic series. They need to be complete (without missing data due to the abortion). The lifespan and appearance time of the organs are known. We suppose that the parameters B_{o1} and B_{o2} have been estimated before. Seven parameters (Table 9.12) can be computed, including the source parameters *r*, *Sp*, *Qo* and the four sink parameters P_p, P_i, P_f, P_m. The fitting on the compartments is very good ($R^2 = 0.99$).

Table 9.12 Source–sink parameters of maize (Subset 5)

Param	r	Sp	Q_0	P_b	P_p	P_i	P_f	P_m
Mean	24.3	2211.6	1.3	1	0.95	1.5	327.7	0.8
SD	5.67	704.3	1.3	0	0.03	0.062	67.2	13.3
CV (%)	23	31	99	0	3	4	2	1600

Table 9.13 Source–sink parameters of maize (Subset 6)

Param	r	Sp	Q_0	P_b	P_p	P_i	P_f	P_m
Mean	13.63	1185.6	0.17	1.00	0.86	1.69	55.72	9.99
SD	5.86	619.9	0.39	0.00	0.04	0.09	8.58	43.44
CV (%)	0.43	0.5	2.27	0.00	0.05	0.05	0.15	4.35

It is not possible to find the variations in the sinks during the expansion of the organs without the organic series. This sampling allows to reduces many measurements. This is the case, for example, for density experiments where the values of the parameters B_{o1} and B_{o2} are already computed for a reference density. One can then observe the effect of the density on the seven computable parameters only from the compartment data.

Subset 5 obviously gives values close to those of the complete Set 1 for the seven corresponding parameters, which shows the interest of this type of sampling. Exception is for P_m (Tassel) that is completely missed. But the weight compartment value is 10 times less than the others.

- Subset 6, 5C*0S_E (number of measurements: 25, number of estimated parameters: 7): This subset includes only five growth stages of the compartments without the organic series, but the parameters Bo_1 and Bo_2 have not been estimated previously. The lifespan and appearance time of the organs are known. Thus, the sink parameters are supposed to be constant (Table 9.13).

The fitting on compartments is quite good ($R^2 = 0.99$; Fig. 9.22a), as for Fig. 9.18. It is surprising to obtain a good fitting with constant sink parameters. In addition, the organ sinks are not proportional to the weight of the compartments, where the weight of the blades is taken as a reference value of 1 (Table 9.14).

A comparison with Table 9.9 (complete Set 1) shows that the values of the parameters no longer correspond to the functioning of the plant. They are only calibration parameters for growth computation and are not directly related to eco-physiology. Figure 9.22 clearly shows that the evolution of maize compartments with variable sinks is close to that of maize with constant sinks. However, the organic series are different, as shown in the simulated architectures. This clearly shows the importance of considering the organic series to build a FSPM.

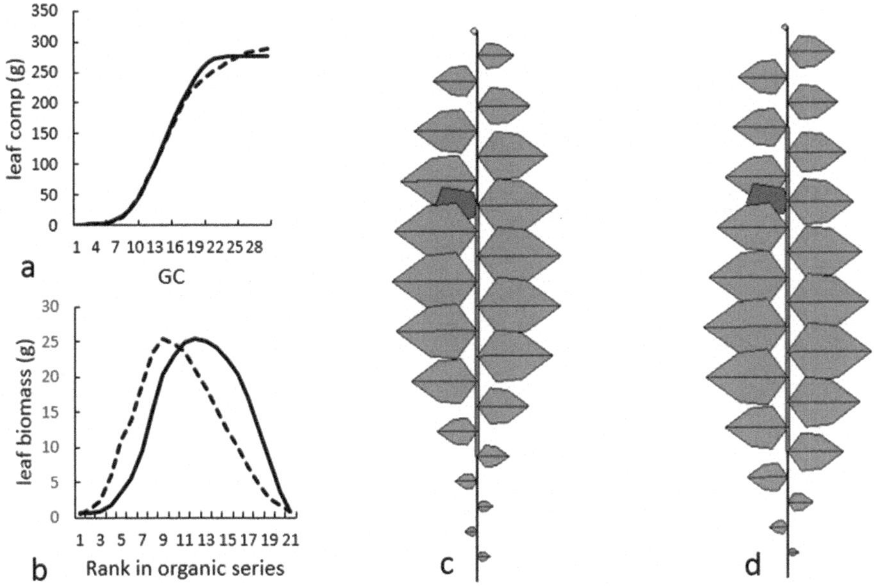

Fig. 9.22 Comparison between simulated maize with variable and constant sinks. (**a**) Evolution of leaf compartment (—) variable sink, (– –) constant sink. (**b**) Leaf organic series (—) variable sink, (– –) constant sink. (**c**) 2D simulation of maize with variable sink. (**d**) 2D simulation of maize with constant sinks. The organic series are clearly visible (Gloups software)

Table 9.14 Weight of organ compartments in relation to leaf weights, the compartment sinks

Param	r	Sp	Q_0	P_b	P_p	P_i	P_f	P_m
Set 1	23.5	2182.2	1.0	1.0	1.0	1.5	385.6	10.4
Subset 6	13.63	1185.6	0.17	1.00	0.86	1.69	55.72	9.99
Weight ratio				1.0	0.6	1.8	2.0	0.1

9.5.4.8 Simulation of Maize

Simulations of 3D maize using the parameters computed by the WLSM are shown in Fig. 9.23 at growth stages of 12, 21, and 30 GCs.

9.5.4.9 Discussion

Obviously, it is desirable to have a consequent number of growth stages and complete descriptions of the organic series to estimate the parameters under the best conditions. However, for common agronomic applications, the investment in measurements is quite heavy. It is important to have the optimal measurement

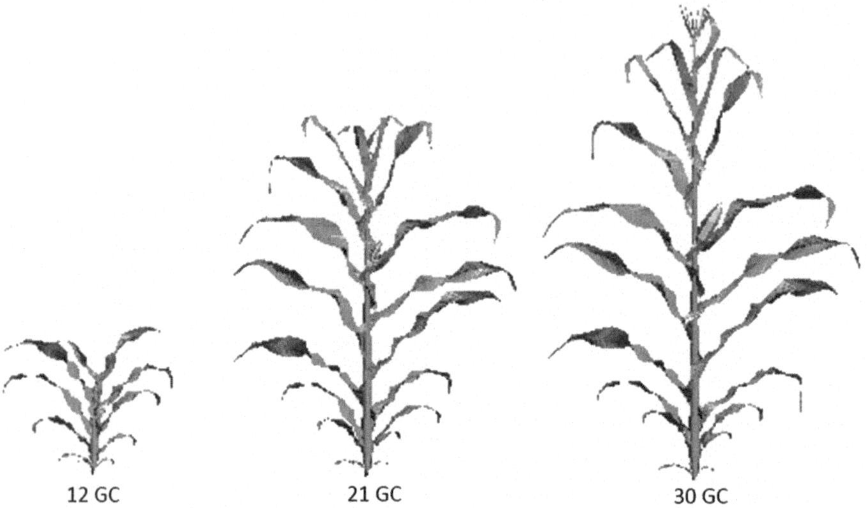

Fig. 9.23 3D growth simulations of maize for three growth stages (Ma Yun-Tao, CAU)

strategies. The large database of maize provides good solutions, and reasonable sample sizes can yield acceptable parameter values.

Weighing the organs is destructive to the plant. It is often possible to replace it with allometries established between the weight of the organs and their dimensions measured on the plant (Okoma et al. 2016).

As a plant progresses through the developmental stage, the first formed organs (especially the leaves), wither and disappear, resulting in incomplete sets. It is advisable to replace the missing data with the organs available from the previous development stages, which have reached their maximum expansion. This completes the organic series and the compartments of the further stages.

In the case of only one development stage available, only the organic series can be used. In most cases, the compartments are not necessary because they are redundant with the organic series, but they are interesting because they give a more global vision of growth.

The computation of the parameters of the sink functions is the most delicate part of the WLSM. In the four examples mentioned above (lettuce, arabidopsis, beetroot, maize), the triplets (P_o, B_{o1}, B_{o2}) are computed correctly, because there are several development stages and many data available. Often, the number of data is insufficient, and B_{o2} must be fixed to obtain P_o and B_{o1}. The value of B_{o2} is not very sensitive in a wide variation range. Thus, it is possible to obtain comparable quality fittings, since the ratio $B_{o1}/(B_{o1} + B_{o2})$ is stable when B_{o2} is modified. The strategy is to choose a minimum value for B_{o2} and then to fix it as long as you study the same plant, because B_{o1} depends on it.

The sample size can be reduced by fixing the value of B_{o2}. Thus, in Subset 4, which keeps only three development stages and measures the organs of the phytomers at every five intervals of organ ranks in the organ series, the measurements are reduced to one-sixth of the complete sample, yet 11 source–sink parameters can still be computed. This makes the GreenLab model usable in practice, and thus make better use of agronomic experiments to compute plant functioning parameters.

9.5.5 *Case of Sunflower Study of Herve Rey (CIRAD)*

9.5.5.1 Plant Description

Sunflower (*Helianthus annuus* L.) is a single-stemmed plant for cultivated varieties. After an initial rosette stage, the stem elongates and the development stops following terminal flowering, which is the formation of the capitulum. The growth continues and mainly concerns the maturation of the capitulum (Rey 2003; Rey et al. 2006).

9.5.5.2 Cultivation Conditions

The data come from experiments carried out at the INRA-Lepse station in Montpellier, France. The sunflower plants grew in the field at a density of five plants per m^2. During the growth period, five observation dates were selected between the germination and the maturation with a single plant per sample to form the simple organic series and organ compartments. No data on the root system were available for these studies.

9.5.5.3 Development Scheme

Sunflower development is deterministic and in a good linear relationship with thermal time (Fig. 9.24a). The development stops at the end of 33 phytomers. Note that the first three phytomers each have two opposite leaves, but the following ones have only one leaf. Here, the opposite leaves are grouped as a single leaf for simplicity. The phyllochron is particularly short. About 1.5 phytomers are formed per day. The extension of the line that links thermal time to the number of phytomers allows by extrapolation to assign dates in growth cycles (DC to GC), which helps control the growth of different samples following the cessation of development. Below 33 GCs, the cycles are both development and growth cycles, beyond that they are only growth cycles. The relationship between the development cycle and thermal time makes it possible to set the five observation dates in CA, expressed in GCs: 20, 40, 63, 84, 120. This plant, therefore, only takes up a quarter of its time for development.

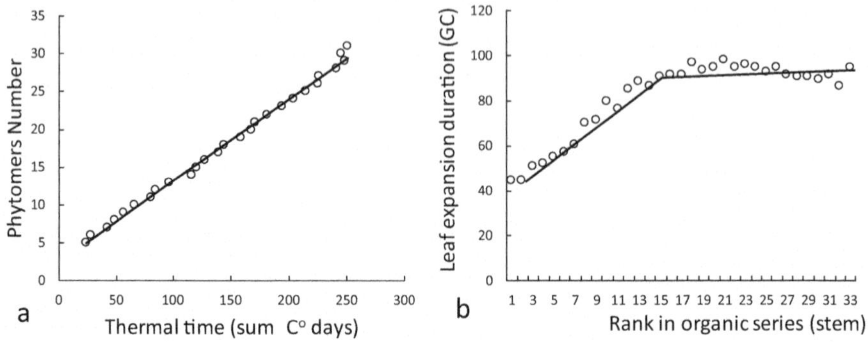

Fig. 9.24 Study of sunflower development. (**a**) Observations (○) and fittings (—) of the link between the number of leaves on the stem and the sum of temperatures: it is linear. Development stops at 33 phytomers because of terminal flowering. This relationship can be extrapolated to assign several growth cycles beyond the cessation of the development (see Chap. 5). (**b**) Evolution of the expansion time (integer value) observed on the leaf blade according to the rank of the phytomer from the base of the sunflower stem increases linearly from 40 to 100 DCs within 15 phytomers. The internodes have a shorter expansion time, limited to 70 DCs, but evolves in a similar way to leaves. The capitulum has an estimated expansion duration of 90 cycles

9.5.5.4 Growth Scheme

The organ expansion times of the sunflower were empirically estimated and are particularly long (Fig. 9.24b) and reach almost 100 DCs for the leaves.

Measurements of leaf blades show that the first rows have an expansion time of 40 DCs. This duration increases linearly according to the rank, and stabilizes around 90 DCs from the 15th rank. The expansion times of petioles and internodes are subject to variation as blades reach a maximum of 90 DCs. It is also the estimated duration of the expansion for the capitulum.

It is essential to consider the variations in the observed expansion times of the organs to compute the demand correctly. A simple organic series is established by assigning weights to the organs according to their ranks on the stem, blades, petioles, internodes, and capitulum. The weight of organ compartments is also considered in the estimation of parameters.

9.5.5.5 Result on Plant Functioning

9.5.5.5.1 Empirically Measured Parameters

The specific leaf weight ε, which corresponds to leaf thickness, is deduced from the ratio of leaf fresh weight to its surface area, which is computed as 0.024 g/cm^2.

9.5.5.5.2 Computed Parameters

The five stages of measurements correspond to harvested plants (one plant per stage) which are, therefore, independent samples. These measurements produce about 625 weights of organs in the organic series to be fitted simultaneously during the growth process. There are 13 parameters to be estimated.

The fitting is good: $R^2 = 0.98$, number of data = 625, number of computed parameters = 13 (the parameter P_b is fixed to 1). The values of the 13 parameters are invariants that control well the fittings of the organic series during the long growth period (120 GCs).

In this experiment, all the parameters of organ source and sink functions (P_o, B_{o1}, B_{o2}) are computed. The sink variation in capitulum needs only two parameters to control (B_{f2} is fixed to 1). The production area Sp (2596 cm^2) is not far from the available area corresponding to the plantation density Sd (2000 cm^2), which makes a model validation.

Three parameters are imprecise: the seed volume Q_0, the sink strength of internode P_i, and of cap P_f. This is because seed and cap are present only once in the stem, while the vegetative organs of the phytomer (leaves, sheaths) are represented 33 times. For the seed, its effect dissipates during the growth; the cap captures all the demand at the end of the growth, which makes it difficult to be precise. The sink strength of the internodes P_i is also fuzzy, which maybe because the internodes on the stem are very variable.

Figure 9.25 shows the evolution of the number of phytomers, plant demand, individual LAI, and produced biomass per cycle as a function of GC.

To correctly fit the base of the organic series, it is necessary to empirically vary the expansion duration of organs, which is shown in Fig. 9.26. This makes it possible to refine the demand.

The triplets (P_o, B_{o1}, B_{o2}) are used to calculate the sink variation in the organ with the cycles (Fig. 9.27). In Fig. 9.27a, the variations in leaf expansion durations observed on the plant as a function of rank during growth monitoring are compared with those determined empirically to optimize the fit of the model to the organic

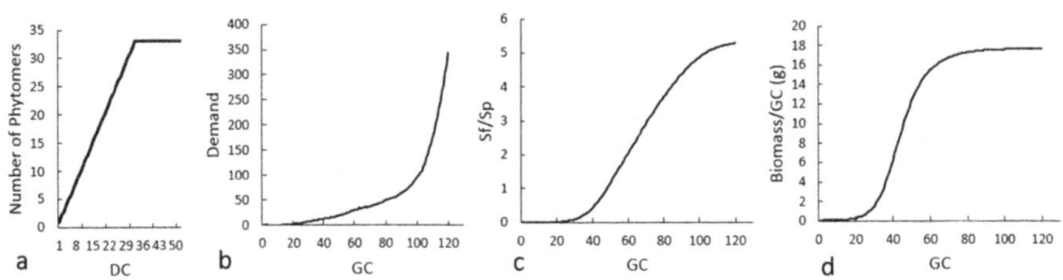

Fig. 9.25 History of the development and growth of Sunflower. (**a**) Number of phytomers per GC. Development stops at 33 DC (equals to 33 GC). (**b**) Evolution of demand per GC. The appearance of the capitulum of Sunflower at cycle 33 gradually captures all the demand. (**c**) Evolution of individual LAI. Production saturates when $Sf/Sp > 3$. (**d**) Evolution of biomass production per GC. After exponential growth, the Biomass/GC reaches saturation. The effect of leaf wither is not yet visible

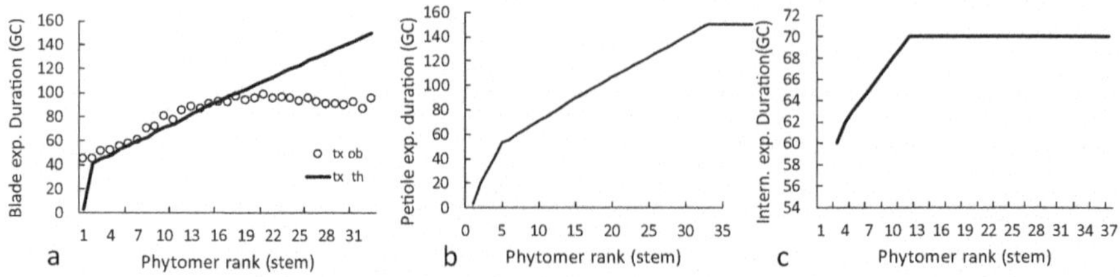

Fig. 9.26 Evolution of organ expansion duration from the basis of organ series on sunflower. (**a**) Blade observations (o) are compared with empirical simulations. (**b**) Petiole. (**c**) Internode

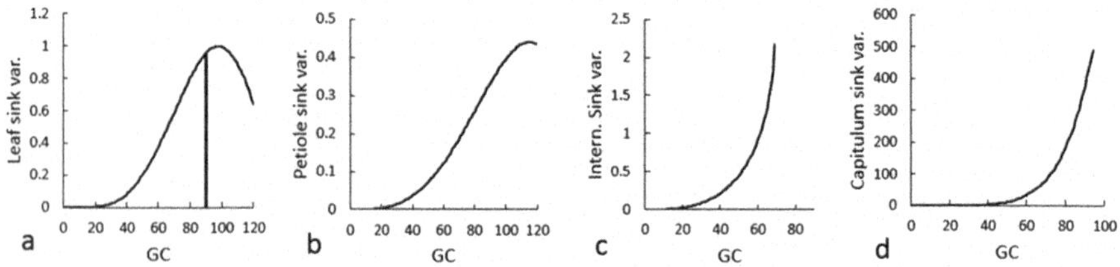

Fig. 9.27 Sink variation in each type of the organ during the expansion. (**a**) Blade. (**b**) Petiole. (**c**) Internode. (**d**) Cob

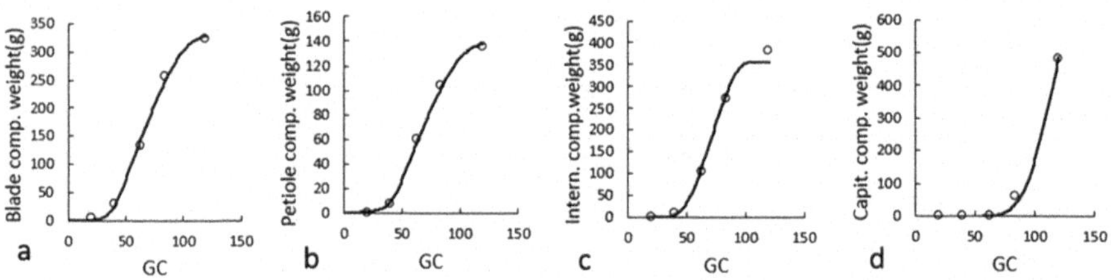

Fig. 9.28 Multi-fitting of the observed organ compartments (O) and computed organ compartments by the model (–) for the five measurement dates [20, 40, 63, 84, 120 GC]. (**a**) Blades. (**b**) Petioles. (**c**) Internodes. (**d**) Capitulum

series. A good agreement can be seen up to the 20th phytomer. Moreover, the Beta law of leaves requires 150 DCs, whereas the observation of the development gives only 120 DCs.

In reality, there is no disagreement. The last leaves around the 30th phytomer have an expansion lasting 120—30 = 90 DCs (Fig. 9.24 b), which corresponds well to the observations. The Beta law is exploited only up to 90 DCs (marked by a vertical line in Fig. 9.27a). However, it is necessary to stretch the Beta law up to 150 DCs to finalize a good fit of the organic series. Moreover, the contraction of the expansion durations at the base of the stem makes the Beta law fully exploited if the duration is smaller than 90 DCs.

The fitting of organ compartments along the CA is good (Fig. 9.28). Individual internode irregularities disappear under the cumulative effect of stem biomass.

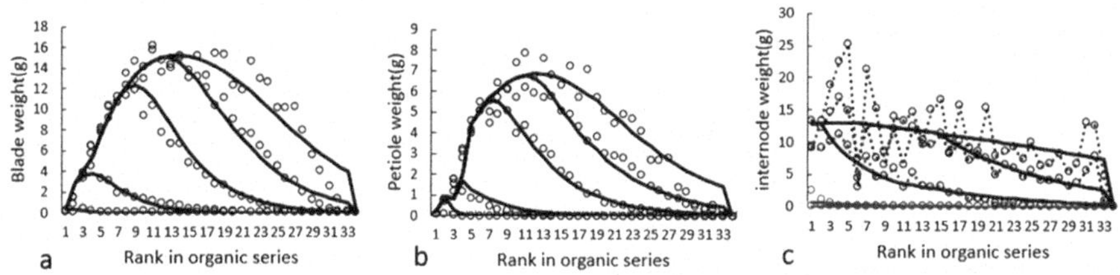

Fig. 9.29 Multi-fitting of the observed organic series (○) and computed by the model (−) for the five measurement dates (20, 40, 63, 84, 120 GCs) on sunflower. (**a**) Blades. (**b**) Petioles. (**c**) Internodes (Gloups software)

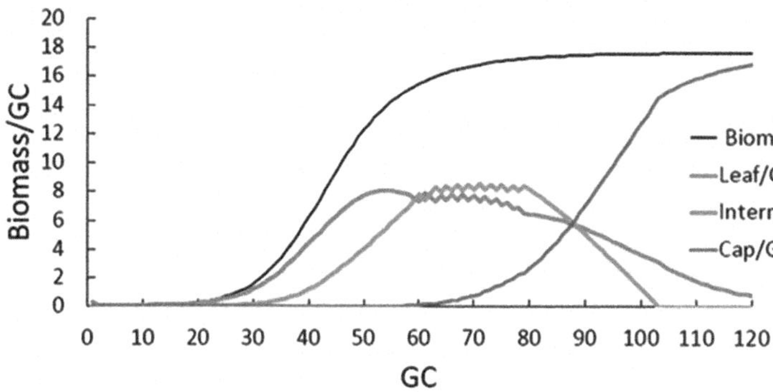

Fig. 9.30 Production and distribution of biomass with the GC in the different plant compartments of sunflower. Biomass production is saturated around the 60 GCs due to the production area *Sp*. At first, the leaves take up all the biomass, a rosette is made. From the 40th GC, the stem is formed, and beyond 100 GCs the capitulum takes all the biomass. The capitulum is a sink that grows until it matures

For the Capitulum, the organic series and compartments are identical because there is only one organ on the stem. The fitting of the organic series (Fig. 9.29) is satisfactory, but the internodes are very noisy. Indeed, the latter is not clearly defined on sunflower, as for a coffee tree. They result from the spiral arrangement of the leaves. The distance along the stem between the insertions of two leaves is very variable and can be almost null.

Knowledge of source–sink parameters allows us to simulate the growth and architecture of the plant. Biomass production over time and its distribution are displayed (Fig. 9.30). Growth is exponential at first, then production saturates: this is the effect of individual LAI and planting density. The internodes have a delay in their expansion and cause the rosette stage before the stem lengthening. Finally, the appearance of the capitulum quickly mobilizes all the available biomass. This limits the growth of leaves beneath the capitulum. If the capitulum is removed when it initiates, the expansion of the last formed leaves is no longer limited (personal communication, H. Rey).

Table 9.15 Source–sink parameters of sunflower

Param	r	Sp	Q_0	P_b	P_p	P_i	P_f	B_{b1}	B_{p1}	B_{i1}	B_{f1}	B_{b2}	B_{p2}	B_{i2}	B_{f2}
Mean	146	2596	0.3	1	0.4	2.9	487	5.9	4.5	3.8	6.7	3.6	2.1	0.7	1
SD	8.7	193	0.1	0	0.02	0.8	126	0.5	0.4	0.7	0.5	0.4	0.2	0.1	0
CV (%)	6	7	36	0	5	27	25	9	10	19	7	11	13	19	0

Table 9.16 Source–sink parameters of sunflower computed without cap (ghost sink)

Param	r	Sp	Q_0	P_b	P_p	P_i	P_f	B_{b1}	B_{p1}	B_{i1}	B_{f1}	B_{b2}	B_{p2}	B_{i2}	B_{f2}
Mean	144	2448	0.3	1	0.4	3.1	1027	6	4.6	3.4	8.7	3.7	2.1	0.7	1

9.5.5.6 Computing the Action of a Ghost Cap

The action of the capitulum can be detected indirectly, only by analyzing the organic series of the leaves and internodes. The capitulum becomes a ghost sink whose parameters are computed by the WLSM (Table 9.16).

The values of the parameters in the case of the ghost sink are close to those calculated in Table 9.15. The weight of the observed cap is 479 g and the one computed from the organic series without the cap is 443 g which is close.

9.5.5.7 Simulation of the Growth and Architecture of Sunflower

The fitting of the organic series allows the estimation of the parameters, and thus the simulations of the growth and architecture of the plant by Gloups and Digiplante software (Fig. 9.31).

On the first 40 growth cycles, we see only the leaves that form a rosette. However, the development is completed at the 33th cycle, and all the organs have been initiated. Beyond that, the organs only expand under the effect of the produced biomass. The internodes of the stem are pushed out of the rosette.

9.5.5.8 Discussion

In this study, each growth stage is represented by a single plant. This is made possible due to the deterministic growth and the low phenotypic variability of this studied sunflower variety.

Of all the plants modelled with the GreenLab model, the sunflower is the one with the longest organ expansion times. The laws of variation in the selected sinks with three parameters (P_o, B_{o1}, B_{o2}) are flexible enough to control the evolution of the expansion of the organs within a long duration. The expansion duration of the leaves according to their rank could be followed experimentally. It corresponds well to the empirical expansion times that allow to adjust the organic series. Note that the value of the expansion times t_{xo} may exceed the observed plant age, as is the case for the

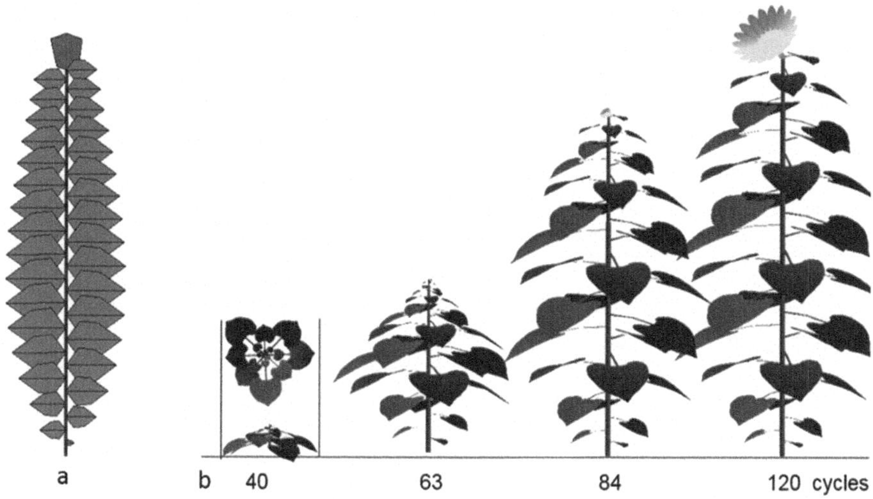

Fig. 9.31 Simulations of plant growth and architecture using the Gloups and Digiplante software. (**a**) 2D organic series of sunflower at 120 DCs (Gloups software). (**b**) 3D simulation of the growth and architecture of sunflower (Digiplante software). After a rosette stage (33 DCs), the stem expands. The capitulum from 80 DCs will gradually mobilize all the biomass produced, limiting the growth of the other organs

beet tape root and sunflower organs beyond the 20th rank. In this case, this value is imprecise, the smallest value should be used to obtain a good fit.

For sunflower, as for maize, the absence of root data does not prevent the modelling of the aerial part depending on 14 invariant parameters. It is deduced that the shoot–root balance is little disturbed during the growth.

9.5.6 Case of Cucumber Study of Zhang Bao-Gui (CAU)

9.5.6.1 Plant Description

Cucumber (*Cucumis sativus* L.) is an herbaceous plant whose stem, in the shape of a vine, has continuous development and growth. The basic phytomer consists of an internode, a leaf whose petiole must be considered because its biomass is not negligible, and a flower (hermaphrodite in the studied variety).

9.5.6.2 Cultivation Conditions and Sampling

The plants are grown in greenhouse with a single stem attached to a stake in separate containers. The weight of roots was measured. Four plants were destructively sampled at each developmental stage, as the location of the fruits on the stem varies slightly with the individual plants. Plant data were obtained from plants with similar fruit locations. This study concerns only one group, the others giving comparable results (Mathieu et al. 2007, 2012).

9.5.6.3 Development Scheme

The development of cucumber is deterministic and continuous. The relationship between the thermal time and the development cycle is supposed to be linear. The number of phytomers in the stem is presented as the number of development cycles. The three stages of development analyzed correspond to 9, 15, and 31 DCs. There is one flower per phytomer giving a fruit that may abort. The organic series of fruits on the stem is formed by alternating presence and absence of fruits. It is necessary to specify the position of the fruits on the stem in a mask file.

9.5.6.4 Growth Scheme

The functioning time of an individual leaf (t_a) is estimated to be 20 GCs. The duration of organ expansion varies with plant development. For leaf blades and petioles, we go from $t_{xa} = 4$ GCs to $t_{xa} = 20$ GCs in a linear way in four ranks along the stem. The variation and duration of the expansion time t_x are easily obtained empirically by adjusting the observed organic series to the computed ones. The expansion time of the internodes (t_{xi}) is 10 GCs. It varies from 8 to 10 GCs in four ranks. The observed fruit expansion time (t_{xf}) is 16 GCs. The expansion of leaves and internodes is immediate as soon as they are created, while that of fruits (t_d) is delayed by 4 GCs. The root whose weight (quite small) has been measured is assumed to have a constant sink over the entire growing period ($t_{xr} > 31$ GCs). These specificities are essential to consider, to calculate the demand and to make a correct fitting of the organic series by estimating the parameters.

The organic series of the leaves and internodes are continuous, but that of the fruits is discontinuous because of organ abortion. The position of each fruit must be specified in the form of a mask that constrains apparition where allowed. In the study, fresh weight was used as cucumber is produced for vegetable use. The rank of phytomers is numbered from the base of the stem. The fruits have developed in positions 3, 4, 5 and 15, 16, 17, which allows the mask to be defined for the fruit series. Elsewhere, they aborted at an early stage with a weight considered negligible.

9.5.6.5 Result on Plant Functioning

Parameter estimation with the WLSM is done by fitting the observed and computed simple organic series as closely as possible. The multi-fitting is multi-stage development (9, 15, 31 GCs). Only one plant is analyzed per stage.

Table 9.17 Source–sink parameters of cucumber (31 GCs) were computed with the organic series of fruits and without taking account of the fruit weights (ghost fruits)

Param	r	Sp	Q_0	B_{b1}	B_{p1}	B_{i1}	B_{f1}	Pr	P_{p1}	P_{i1}	P_{f1}
With fruits	11.2	1007	0.57	1.9	1.26	0.1	2	0.85	0.4	0.9	66
Ghost fruits	9.5	775	0.34	1.9	1.28	0.1	1.9	0.86	0.4	1	63
			Param	B_{b2}	B_{p2}	B_{i2}	B_{f2}				
			With fruits	6.9	1.55	0.01	1.1				
			Ghost fruits	6.9	1.57	0	1				

9.5.6.5.1 Empirically Measured Parameters

The specific leaf weight ε, deduced from the ratio of leaf fresh weight and its surface, is estimated as 0.03 g/cm^2. The other allometric parameters, concerning the shape of the internodes and fruits that are not involved, are fixed empirically to give a good looking at the 3D representation.

9.5.6.5.2 Computed Parameters

The three stages of measurements correspond to harvested plants (one plant per stage) which are, therefore, independent samples. These measurements produce 236 weights of organs in the organic series and compartments to be adjusted simultaneously during the growth process. There are 15 parameters (Table 9.17) to be estimated. The root system is assigned a constant sink P_r, since it represents barely 1% of the weight of the aerial part.

The fitting is good: $R^2 = 0.94$, number of data $=$ 236, number of computed parameters $=$ 15 (the parameter P_b is fixed to 1). The values of the 15 parameters (Table 9.17) are invariants that control well the fittings of the organic series during the growth period (31 GCs).

In this experiment, all the parameters of the source and sink functions of organs triplets (P_o, B_{o1}, B_{o2}) are computed. Three parameters have a big CV: the seed volume (Q_0), the two beta law parameters of the internode (B_{i1}, B_{i2}), since they have a small value.

Figure 9.32 shows the evolution of the number of phytomers, plant demand, individual LAI and produced biomass per cycle as a function of GC. It is noted that fluctuations in the demand due to the fruit sink do not affect the saturation of production due to LAI $>$ 3 per plant.

As mentioned above, to correctly fit the base of the organic series, it is necessary to empirically vary the expansion duration of organs, which is shown in Fig. 9.33. This variation occurs on the first four phytomers of the stem until the expansion time stabilizes for each type of organ.

The triplets (P_o, B_{o1}, B_{o2}) are used to calculate the sink variation in the organs with the cycles (Fig. 9.34).

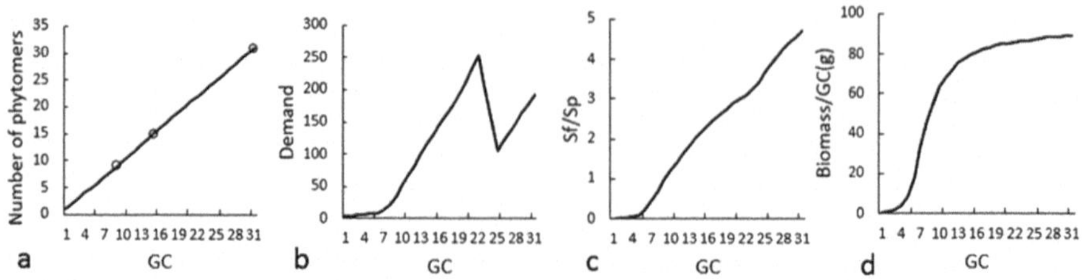

Fig. 9.32 History of the development and growth of cucumber. (**a**) Number of phytomers per GC. (**b**) Evolution of demand per GC. Fruits abortion creates the oscillations. (**c**) Evolution of individual LAI. Production saturates when $Sf/Sp > 3$. (**d**) Evolution of biomass production per GC. After an exponential growth, the plants reach saturation. The effect of leaf wither is not yet visible

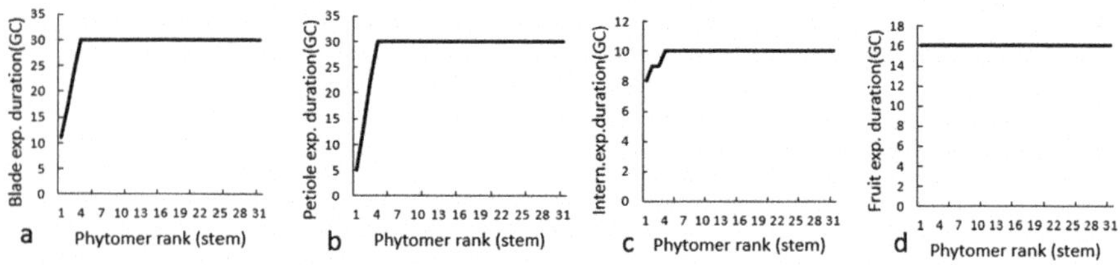

Fig. 9.33 Evolution of organ expansion from the basis of organ series of cucumber. (**a**) Blade. (**b**) Petiole. (**c**) Internode. (**d**) Fruit

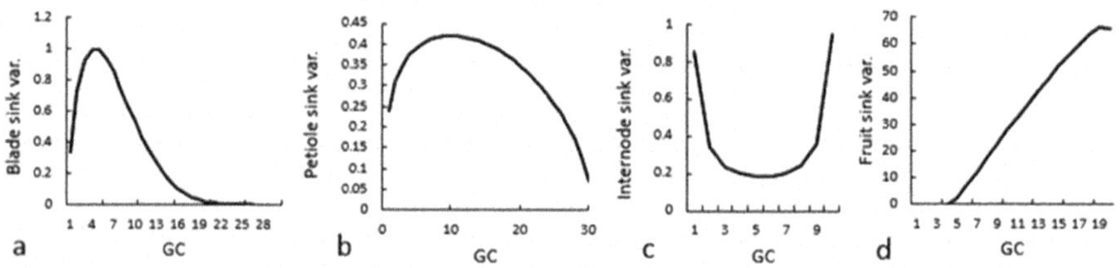

Fig. 9.34 Sink variation in each type of organ during the expansion. (**a**) Blade. (**b**) Petiole. (**c**) Internode. (**d**) Fruit

Figure 9.34 shows the great flexibility of the sink variation functions. The blade variation function covers 30 GCs. In reality, the expansion can only be visible up to 17 GCs, which is the observable expansion duration. But it is necessary to extend mathematically the expansion time up to 30 GCs to have a bell curve that best approximates the numerical values observed up to 17 GCs. Internodes have a constant sink between two and eight GCs. The fruits have a delay of four GCs after their initiation, before starting their expansion.

The hypothesis of a constant root sink during the growth looks relevant. The value is given in Table 9.17 is $P_r = 0.85$.

The multi-fitting of organ compartments along CA is quite good (Fig. 9.35).

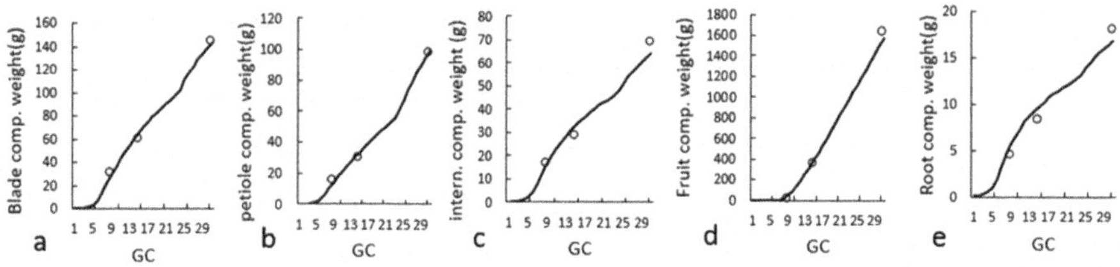

Fig. 9.35 Multi-fitting of the observed organ compartments (○) and computed organ compartments by the model (–) for the three measurement dates (9, 15, 31 GCs) of cucumber. (**a**) Blades. (**b**) Petioles. (**c**) Internodes. (**d**) Fruits. (**e**) Root

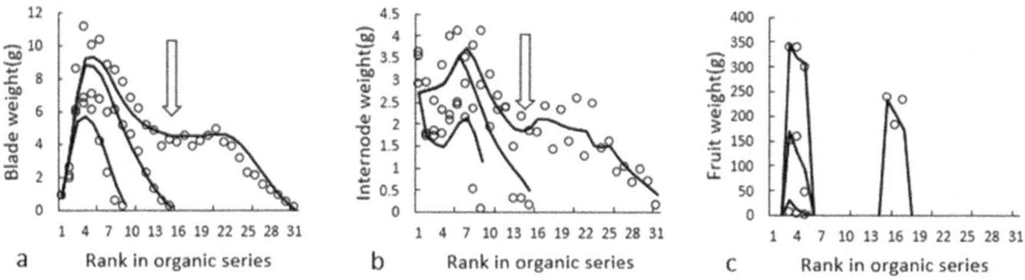

Fig. 9.36 Multi-fitting of the observed organic series (○) and computed organic series (–) by the model for the three measurement dates (9, 15, 31 GCs]. (**a**) Blades. (**b**) Internodes. (**c**) Fruits (Gloups software). The organic series of the petioles have similar shapes with that of the blades). Note that the shrinkage of the phytomers (arrow) is under the action of the demand for the fruits

The WLSM shows all its efficiency because it computes a triplet of sink parameters that fit compartments whose weights vary from 20 g to 1600 g (respectively for the root system and the fruit compartment). In cultivation with irrigated containers, the root systems develop very little.

For example, the root system of a tomato plant in a greenhouse represents only 5% of the plant weight, whereas it represents 30% in the field (personal communication Ep Heuvelink). The hypothesis of a constant sink for the root system under these growing conditions seems appropriate.

The good quality of the multi-fitting of organic series can be assessed in Fig. 9.36. The importance of fruit sink strength during their expansion explains the shrinkage observed on the organic series of blade, petiole and internodes. This is the action of the competition for biomass of the expanding fruits.

9.5.6.6 Computing the Action of Ghost Fruits

The organic series of the leaves and internodes of the cucumber is strongly modified by the fruit sink. The sink functions of the cucumber organs can be found by considering only the position of the fruits without knowing their weights. As in the case of beetroot and sunflower, the action of a ghost organ can be detected. This is shown in Table 9.17 which compares the values of the source–sink organs by considering successively the organic series of the fruits and then ignoring them. The sink function of the fruits is well recomputed.

The weight of the fruits observed is 1.62 kg. The weight of the fruits computed from the optimized parameters to fit all measurements that include the fruits is 1.57 kg. The weight of the fruits computed from the optimized parameters to fit the measurements with ghost fruits is 1.14 kg. The imprint of the fruits on the organic series of the leaves and internodes is so strong in the plant demand that the functioning of the fruit organ is rather well deduced.

9.5.6.7 Simulation of Cucumber Growth and Architecture

In Fig. 9.37, the development and growth of cucumber are depicted with its 2D organic series and its 3D architecture with all fruits at their observed locations.

9.5.6.8 Discussion

The high fruit sink greatly disrupts the organic series of phytomers that shrink in the middle of the stem. Normally, we should observe a fruit on each phytomer, but many

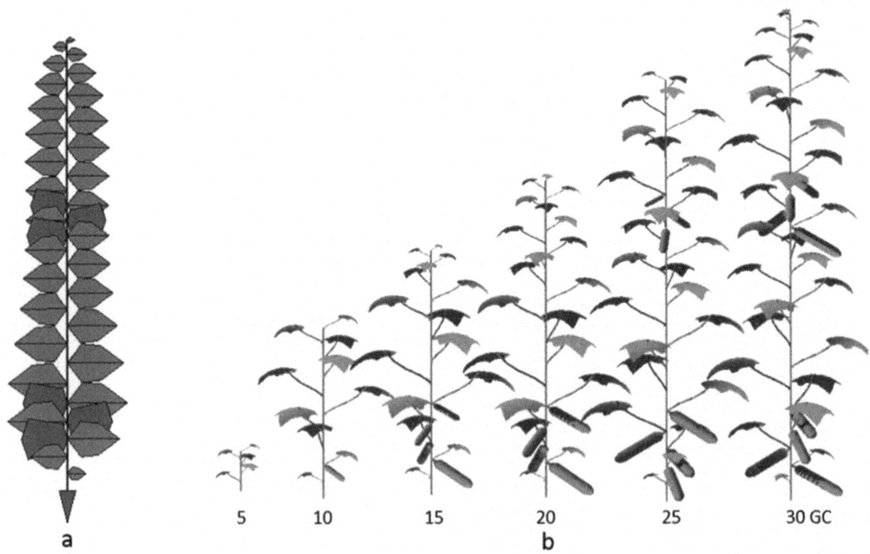

Fig. 9.37 Cucumber growth and architecture. (**a**) 2D organic series at 30th GC (Gloups software). (**b**) 3D growth simulation. We notice the grouping of fruits and the shrink of phytomers in the middle of the stem (Digiplante software)

of them abort. It should be noted that the new fruits abort in the area where the *Q*/*D* ratio is low. There is a possibility of a mechanical explanation of abortion here, which has been confirmed on sweet peppers and tomatoes. This will be discussed in Chap. 13.

9.5.7 Case of Tomato Study of Zhang Bao-Gui (CAU)

9.5.7.1 Plant Description

The tomato is an herbaceous plant whose stem has a continuous development at the beginning and after a rhythmic development, with a sympodial modular structure. The modules are made up of one or two vegetative phytomers bearing a leaf, followed by a phytomer bearing a leaf and a truss of fruits (Fig. 9.38). The leaf is composed of a multi-lobed blade and a petiole whose biomass is considered.

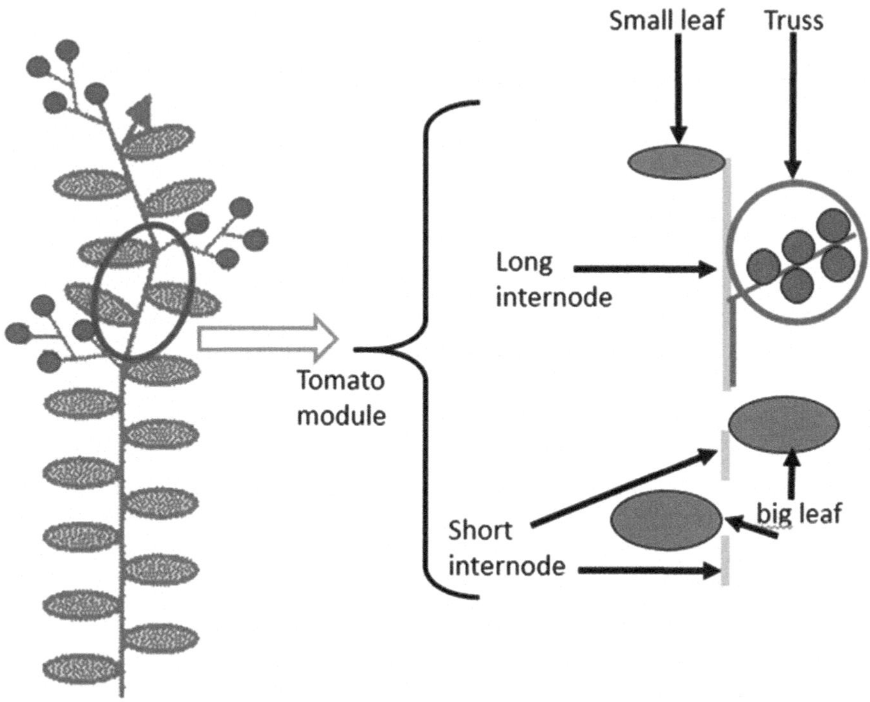

Fig. 9.38 Description of a module of Tomato stem (From Benlloch et al. 2007)

9.5.7.2 Cultivation Conditions

The plants are grown in greenhouse on a single stem attached to a stake and in separate containers. The root system was available and its weight was measured. Observations were made on a single plant and for three developmental stages with four repetitions. The fresh weights were measured.

9.5.7.3 Development Scheme

In the beginning, the stem meristem has a continuous development by repeating the same phytomer. A terminal flowering ends the juvenile phase, and the development becomes sympodial. The stem is produced by a succession of terminal flowering modules, each compound of two or three phytomers. There are two types of phytomers per module (Fig. 9.38). Vegetative phytomers have short internodes and big leaves, while fruiting phytomers have large internodes, small leaves and bear a truss of tomatoes at the end of the module. To further simplify the structure, the truss is considered as a compound fruit, with negligible petiole weight. The plant is pruned, removing the axillary branches at their birth, to keep a single-stemmed appearance. The plant was harvested after 28 GCs. The three stages of development analyzed correspond to 9, 17, 28 GCs.

9.5.7.4 Growth Scheme

The functioning time of leaves (t_a) is estimated as 15 GCs. There is a visible "base effect" on the organic series of leaves. For leaf blades and petioles, the duration of the expansion time (t_x) is easily obtained empirically by fitting the observed organic series to the computed ones. The expansion time of all vegetative organs (t_x) is 10 DCs. It varies from 5 to 10 DCs linearly in three ranks. The truss expansion time is set to 30 DCs. While the expansion of leaves and internodes is immediate as soon as they are created, that of fruits is delayed by $t_d = 4$ DCs. The organic series of the leaves and internodes are continuous, but with alternating organs of the same nature but of different weights corresponding to different sinks. Phytomers with the truss have long internode and small leaves. These series are, therefore, compound. That of the fruits on the stem is discontinuous and the position of each must be specified. The rank of phytomers is numbered from the base of the stem. Fruits developed at the end of the module on ranks 8, 11, 14, 17, 19, 22, 25. They are stored in a mask to force their positions. In this experiment, the root sink whose weight has been measured is assumed to be constant over the entire growing period ($t_{xr} > 28$ GCs).

9.5.7.5 Result on Plant Functioning

Parameter estimation with the WLSM is done by fitting the observed and computed compartments and simple organic series as closely as possible. The fitting is multi-stage development (9, 17, 28 DCs). Only one plant is analyzed per stage.

9.5.7.5.1 Empirically Measured Parameters

The specific leaf weight is computed from the ratio of leaf fresh weight and its surface, which is estimated as 0.025 g/cm^2. The other allometric parameters, concerning the shape of the internodes and fruits that are not involved, are fixed empirically to give a good resemblance in the 3D representation.

9.5.7.5.2 Computed Parameters

The three stages of measurements correspond to harvested plants (one plant per stage) which are, therefore, independent samples. These measurements produce 232 weights of organs in the organic series and compartments to be fitted simultaneously during the growth process. There are 15 parameters (Table 9.18) to be

Table 9.18 Source–sink parameters of tomato

param	mean	SD	CV (%)
r	23.1	2.3	10
Sp	6694	1841	27
Q_0	1.5	0	0
P_{b1}	1	0	0
P_{b2}	0.78		
P_{p1}	1.5	0.31	20
P_{p2}	1.2		
P_{i1}	1.27	0.31	24
P_{i2}	2.1		
P_f	48	9.1	19
P_r	0.54	0.04	8
B_{b1}	2.8	0.3	12
B_{p1}	6.4	2.538	39
B_{i1}	14	7.2	51
B_{f1}	2.6	0.16	6
B_{r1}	1	0	0
B_{b2}	1.8	0.2	14
B_{p2}	4.6	1.8	39
B_{i2}	9.2	4.6	50
B_{f2}	2	0	0
B_{r2}	1	0	0

estimated. In a container, the root system develops very little because it does not need to propagate to seek water. Here it is assigned a constant sink P_r, since it represents less than 1% of the weight of the aerial part.

The fitting is good: $R^2 = 0.94$ (the parameter P_b is fixed to 1). The values of the 15 parameters are invariants that control well the fittings of the organic series during the growth period (31 GCs).

As there are two types of phytomers (those without or with truss), two sinks are needed for the blades (P_{b1} and P_{b2}), the petioles (P_{p1} and P_{p2}) and the internodes (P_{i1} and P_{i2}). $P_{b2} = k_b{*}P_{b1}$; $P_{p2} = k_b{*}P_{p1}$ and $P_{i2} = k_i{*}P_{i1}$. ($k_b = k_p = 0.78$; $k_i = 1.71$).

The coefficient k_b is common to the blades and petioles. The parameters k_b and k_i are computed with the others by the WLSM. It is also assumed that the expansion laws of the second sink variation are the same as those of the first ones. There are thus only the two additional parameters k_b and k_i to be estimated.

In this experiment, not all the parameters of the source and sink functions of organs can be computed. The seed *Qo* is empirically set to 1.5 because the WLSM computes a too high value (21 g). This value decreases the global error significantly, but the root system is missed for the fitting. The parameter B_{f2} of the truss cannot be computed at the same time as the B_{f1}. It has a low sensitivity. The value $B_{f1}/(B_{f1} + B_{f2})$ is the important value for a good fitting (see maize study above). Setting the value of B_{f2} makes the WLSM computing a correct B_{f1}. Here we empirically chose the value $B_{f2} = 2$.

Figure 9.39 shows the evolution of the number of phytomers, plant demand, individual LAI and produced biomass per cycle as a function of GC. Here the LAI per plant does not reach saturation (LAI > 3). The plant demand is too big and the biomass production decrease after reaching a maximum.

As mentioned above, to correctly fit the base of the organic series, it is necessary to empirically vary the expansion duration of organs, which is shown in Fig. 9.40. This variation occurs on the first three phytomers of the stem until the expansion time stabilizes for each type of organ.

Figure 9.41 shows the bell shape of sink variation functions for all organs. The truss has an expansion delay of four GCs after its initiation, before starting their expansion.

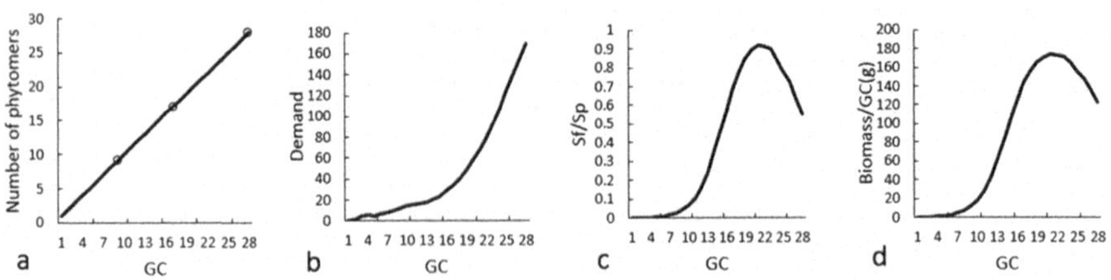

Fig. 9.39 History of the development and growth of tomatoes. (**a**) Number of phytomers per GC. (**b**) Evolution of demand per GC. (**c**) Evolution of individual LAI. (**d**) The produced biomass per GC. After exponential growth, the LAI and the produced biomass per GC decrease

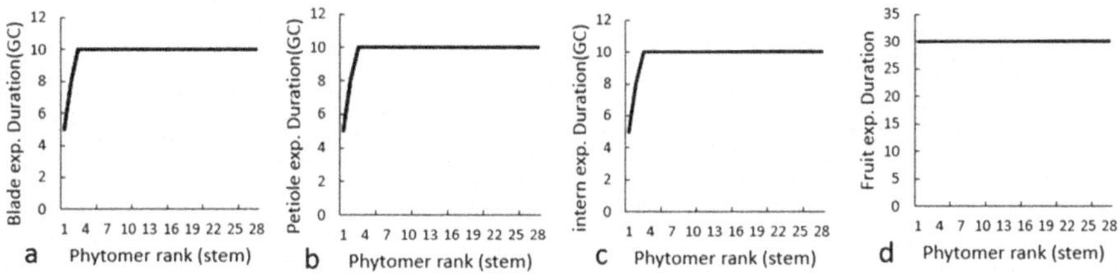

Fig. 9.40 Evolution of organ expansion time from the basis of organ series of tomato. (**a**) Blade. (**b**) Petiole. (**c**) Internode. (**d**) Fruit. The curves are the same for the organs of the blade, petiole and internode

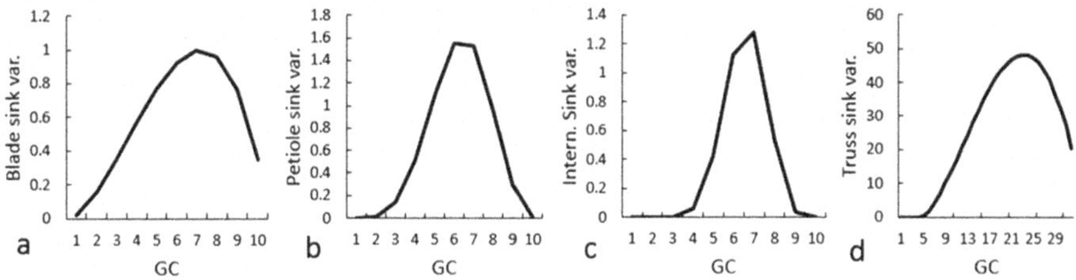

Fig. 9.41 Sink variation in each type of organs during the expansion. (**a**) Blade. (**b**) Petiole. (**c**) Internode. (**d**) Truss

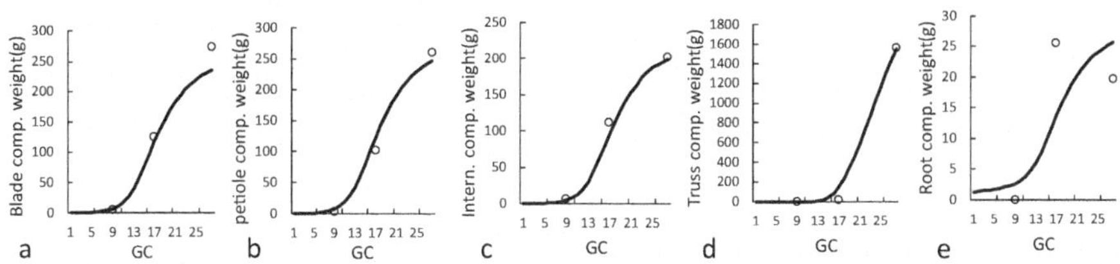

Fig. 9.42 Multi-fitting of the observed organ compartments (○) and computed organ compartments by the model (–) for the three measurement dates (9, 17, 28 GCs) of tomato. (**a**) Blades. (**b**) Petioles. (**c**) Internodes. (**d**) Fruits. (**e**) Root

The hypothesis of a constant root sink during the growth looks relevant. The value is given in Table 9.18, being $P_r = 0.54$.

The multi-fitting of organ compartments along CA is quite good (Fig. 9.42).

The good quality of the multi-fitting of organic series can be assessed in Fig. 9.43. The alternation of the two types of phytomers corresponding to two types of leaves and internodes in a module causes zigzags in the organic series. The WLSM correctly estimates the two sinks of the leaf and internode organs (Table 9.18).

9.5.7.6 Simulation of Tomato Growth and Architecture

In Fig. 9.44, the development and growth of tomatoes are depicted with its 2D organic series and its 3D architecture with all the fruits at their observed locations.

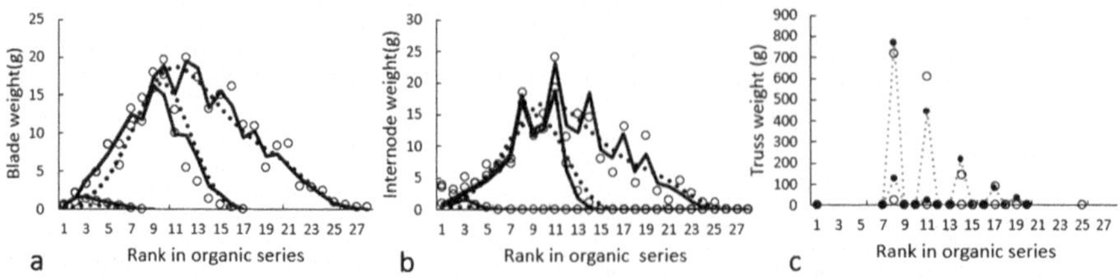

Fig. 9.43 Multi-fitting of the observed organic series (○) and computed organ compartments by the model (–) for the three measurement dates (9, 17, 28 GCs) of tomato. (**a**) Blades. (**b**) Internodes. (**c**) Fruits (Gloups software). The petioles organic series have similar shapes with that of the blades). Note the zigzags of the organic series are because the two kinds of blades and internodes organs have two different sinks. When ignoring the two kinds of phytomers (—), we can get the dotted line where zigzags are suppressed, but the fitting gets a lower quality

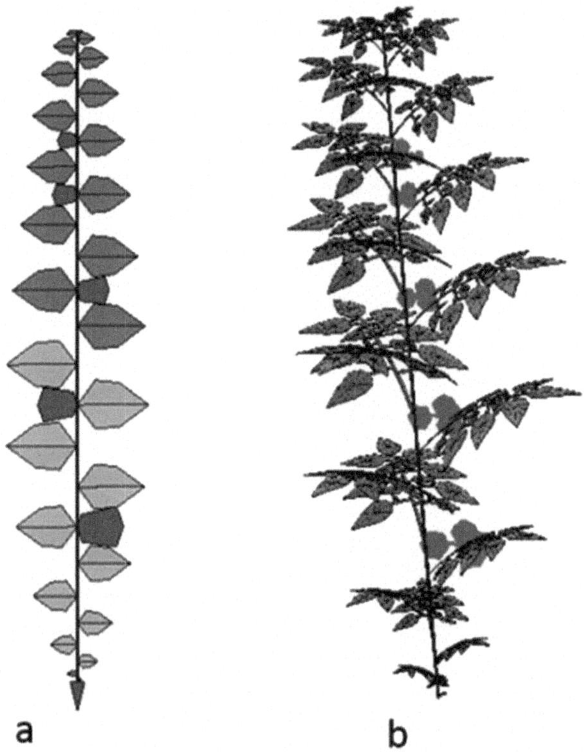

Fig. 9.44 Tomato growth and architecture at 28 GCs. (**a**) 2D organic series (Gloups software). (**b**) 3D growth simulation (GreenScilab software)

9.5.7.7 Discussion

To simplify the measurements, it is wise to transform the tomato truss into a single compound fruit whose triplet P_f, B_{f1}, B_{f2} controls the expansion. The peculiarity of having two types of phytomers for a single PA is considered in the calibration. Here again the source–sink parameters behave as invariants during the growth. The weight of the root system is 100 times smaller than that of the aerial system (because of the growth in the container), this is not a problem for the WLSM to estimate its sink.

9.5.8 Case of Sweet Pepper Study of Ma Yun-Tao (CAU)

In the Capsicum family, we studied sweet peppers and chili peppers. We present here a large-fruited sweet pepper of the Funcky variety.

9.5.8.1 Plant Description

After germination, the unbranched stem stops its development early because of an apical flowering that produces the first fruit. Two equal vigor branches are born immediately under the last two phytomers of the stem. Their continuous development is limited to a few phytomers, after which branching becomes sympodial. The most vigorous phytomer is the prolongation of the branch. Each node potentially bears fruit, but many early abortions are observed.

9.5.8.2 Cultivation Conditions

The plants were grown in greenhouses with stable climatic conditions in containers with a density of 5 plants/m^2. The root system was not available. For cultivation reasons, the twigs carried by the branches are pruned after the emergence of the first phytomer. The regular pruning of the twigs at the first phytomer limits the development of the plant architecture.

9.5.8.3 Development Scheme

The structure of the plant under study is, therefore, unnatural and it is necessary to adopt a mask to the botanical automaton to reproduce the pruning undergone by the plant during its development and the locations of fruit abortions.

It is relevant to distinguish three PAs: PA 1 for the stem, PA 2 for the branches and PA 3 for the twigs. The terminal flowering of the stem occurs at 11 phytomers. At the end of the stem, it gives rise to two similar branches which, at harvest, have developed 30 phytomers. The CA of development and growth is, therefore, 41 DCs.

It is useful to simplify the topological structure of a plant, when possible, to facilitate measurements and target preparation. In the GreenLab model, all plant architectures that produce identical cohorts also have the same organic series (even with different topological structures) and, therefore, the same source–sink functioning. Figure 9.45 shows that pruned twigs consisting of a single phytomer, placed end

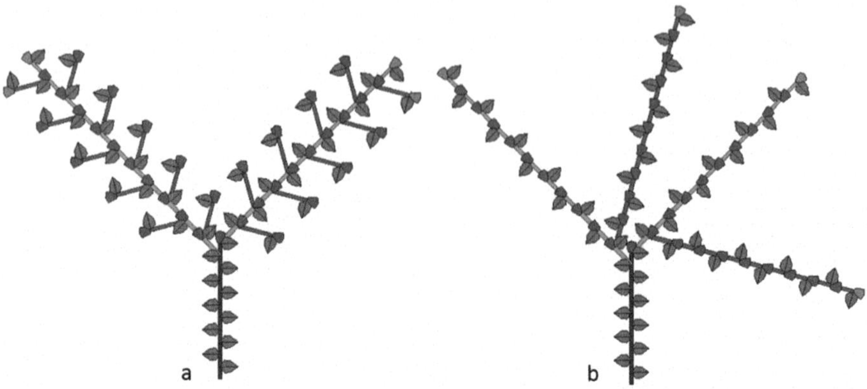

Fig. 9.45 Scheme of the development of the pruned sweet pepper. (**a**) Three PAs on the pruned sweet pepper. The stem (PA 1, blue) stops its development after 11 phytomers and transforms into flowers. Two similar branches (PA2, green) are born at the end of the stem. They branch off at each phytomer, which gives rise to a branch (PA3, red) pruned at the first phytomer. Each phytomer except on the stem potentially bears a fruit. (**b**) The pruned twigs placed end to end form an organic chronological series. From the aspect of the development and growth of the cohorts, Fig. 9.45b is not different (for cohorts) from Fig. 9.45a, but it is simpler to calibrate. Many fruits abort and the corresponding organic series are incomplete

to end, form an organic chronological series. Thus, the computational plant in Fig. 9.45b has the same number of organs as that of Fig. 9.45a, and the homologous organs with the same CAs and PAs have the same biomass.

Finally, the topological structure of the studied pepper can be broken down into an axis of PA 1 consisting of 11 phytomers, two axes of PA 2 consisting of 30 phytomers, and two axes of PA 3 consisting of 30 phytomers. This makes it possible to bypass the sympodial structure. The model fitting is multi-stage (11, 20, 28, 34, 41 DCs). The first four stages are fitted only on the organ compartments. The last stage is fitted in detail on the organ series. This way optimizes the number of measurements to be performed, while ensuring the stability of the source–sink parameters during the growth. Only one plant is analyzed per stage.

9.5.8.4 Growth Scheme

The functioning duration of the leaves t_a is estimated as 25 DCs. There is a visible "base effect" on the organic series of leaves. The duration of organ expansion t_x varies at the beginning of plant development. For leaf blades and petioles, the duration of the expansion time is easily obtained empirically by fitting the observed organic series to the computed ones. The expansion duration of leaf organs is 10 DCs. It varies from 5 to 10 DCs linearly in four ranks. The expansion duration of internodes looks stable with 6 DCs. The fruit expansion time is set to 20 DCs, while the expansion of fruits is delayed by $t_d = 3$ DCs.

9.5.8.5 Setting up Organic Series, Making a Mask

After the topological simplification of Fig. 9.45, as there are three PAs and four organ types (leaf blade, petiole, internode, fruit) per age, the target includes 12 organic series to be fitted in the same time by the model. While the organic series of leaves and internodes are continuous, those of fruits are discontinuous due to the abortion along with the rank of the organic series.

Setting a mask indicates which phytomer of an axis bears a branch, and which phytomer bears a fruit. The mask forces the simulation to realize the scheduled plant architecture. This method is mainly valid for simple plant topological structures in the absence of stochastic bud functioning. The mask indicates the positions of the branches in the topological structure (Table 9.19).

Table 9.19 Mask to simulate the sweet pepper architecture

	Axe PA 1	Axe PA 1	Axe PA 1	Axe PA 2	Axe PA 2	Axe PA 3
Rank phyto	Phytomer existence	Ramif number	Fruit number	Ramif number	Fruit number	Fruit number
1	1	0	0	1	1	0
2	1	0	0	0	0	0
3	1	0	0	0	0	0
4	1	0	0	0	0	0
5	1	0	0	0	0	0
6	1	0	0	0	0	0
7	1	0	0	0	0	0
8	1	0	0	0	0.5	0.5
9	1	0	0	0	0.5	0
10	1	0	0	0	0.5	0
11	1	2	1	0	0	0
12	0	0	0	0	0	0
13				0	0	0
14				0	0	0
15				0	0.5	0
16				0	0	0.5
17				0	1	0
18				0	0.5	0.5
19				0	0	0
20				0	0	0.5
21					0	0
22					0	0.5
23					0	0
24					1	1
25					0.5	1
26					0.5	1
27					1	1
28					1	0
29					1	0
30					0	

To not complicate the target, only one organic series per organ is available for each PA, so it represents here the average of the two branches repetitions.

The mask indicates the status of each phytomer according to its rank, whether it bears a ramification or a fruit. Thus, for an organic series, if both branches have no fruit at rank *k*, the rate is 0; if only one branch bears fruit at rank *k*, the rate is 0.5; if each branch bears a fruit at rank *k*, the rate is 1. Thus, if we have the total weight of fruits observed at rank *k*, in the computation, it will be multiplied by the presence rate in order to restore the average biomass loaded by the phytomer.

The target contains the biomass of the compartments of organs for the five dates 11, 20, 28, 34, 41 DCs, and the organ weights along with the ranks of the 12 organic series.

Table 9.19 shows that the stem (PA 1) is made up of 11 phytomers, and that the 11th bears one fruit and two branches (PA 2). Each branch (PA 2) bears a ramification (PA 3) and fruit on its first rank. The other ranks have a rate of fruiting of 0, 0.5 or 1. The branches (PA 3) remain unbranched and the rates of fruiting along the ranks are noted.

9.5.8.6 Results on Plant Functioning

Parameter estimation with the WLSM is done by fitting the observed and computed compartments and simple organic series. The fitting is multi-stage development (11, 20, 28, 34, 41 DCs).

9.5.8.6.1 Empirically Measured Parameters

The specific leaf weight ε, is computed from the ratio of leaf fresh weight and its surface. It is estimated as 0.036 g/cm^2. The other allometric parameters, concerning the shape of the internodes and fruits that are not involved, are fixed empirically to give a good resemblance in the 3D representation.

9.5.8.6.2 Computed Parameters

The five stages of measurements correspond to harvested plants (one plant per stage) which are, therefore, independent samples. These measurements produce 236 weights of organs in the organic series and compartments to be adjusted simultaneously during the growth process. There are 22 parameters (Table 9.20) to be estimated. But the second parameter of the sink variation law B_{i2} cannot be computed. Empirically $B_{i2} = 5$ gives a suitable value. Although the sweet pepper is an herbaceous plant, there is a girth growth on the stem, which must be estimated and is controlled by the sink P_c. The estimated parameter values (Table 9.20) allow a correct fitting of the organ compartments (Fig. 9.49) and the organic series of sweet pepper (Fig. 9.50).

Table 9.20 Source–sink parameters of sweet-pepper

Param	Mean	SD	CV (%)
r	11.82	2.04	17
Sp	1784	361	2
Q_0	0.65	0.66	101
P_c	3.88	2.16	56
P_{b1}	1	0	0
P_{b2}	1.5	0.32	21
P_{b3}	0.81	0.18	22
P_{p1}	0.42	0.15	35
P_{p2}	0.45	0.19	42
P_{p3}	0.26	0.11	42
P_{i1}	1.59	1.03	65
P_{i2}	8.47	3.9	46
P_{i3}	1.24	0.58	47
P_{f1}	18	5.93	33
P_{f2}	26.8	7.87	29
P_{f3}	28.26	8.2	29
B_{b1}	2.84	1.35	48
B_{b2}	2.92	1.52	52
B_{p1}	1.69	0.85	50
B_{p2}	1.5	0.32	21
B_{i1}	11.28	4.03	36
B_{i2}	5	0	0
B_{f1}	1.86	0.41	22
B_{f2}	2.83	0.82	29

The sink P_c of the girth, is not here computed according to Pressler's law. The demand for the girth growth is calculated by the equation:

$$D(i) = D_b(i) + D_p(i) + D_f(i) + P_c \cdot Q(i-1)/D(i) \tag{9.23}$$

D_b, D_p, D_f are the demand of blades, petioles, and fruits. D is the solution of an equation of the second degree, and it depends of the unknown parameter P_c.

The fitting is good: $R^2 = 0.92$, number of data $= 236$, number of computed parameters $= 22$ (the parameter P_b is fixed to 1). The values of the 22 parameters are invariants that control well the fittings of the organic series during the growth period (41 GCs).

Figure 9.46 shows the evolution of the number of phytomers per PA, plant demand, individual LAI (*Sf*/*Sp*) and produced biomass per cycle as a function of GC. Here the LAI per plant reaches a saturation (LAI > 3). The biomass production is saturated.

To correctly fit the base of the organic series of the stem, it is necessary to empirically vary the expansion duration of organs, which is shown in Fig. 9.47. This

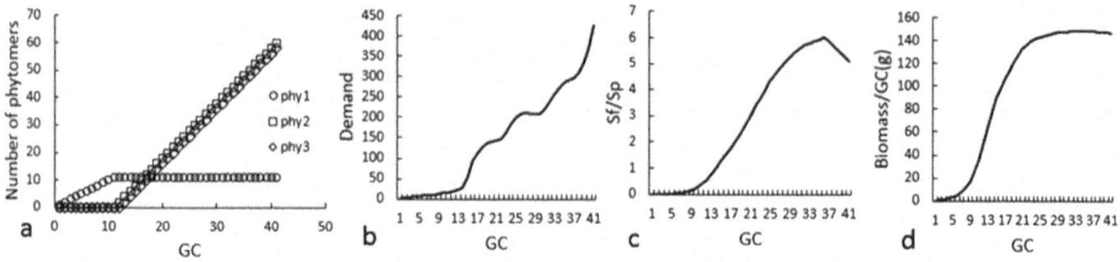

Fig. 9.46 History of the development and growth of sweet-pepper. (**a**) Number of phytomers per GC and PA. (**b**) Evolution of demand per GC. Fruit cohorts have group abortions which cause waves in demand (**c**) Evolution of individual LAI. (**d**) The biomass production per GC

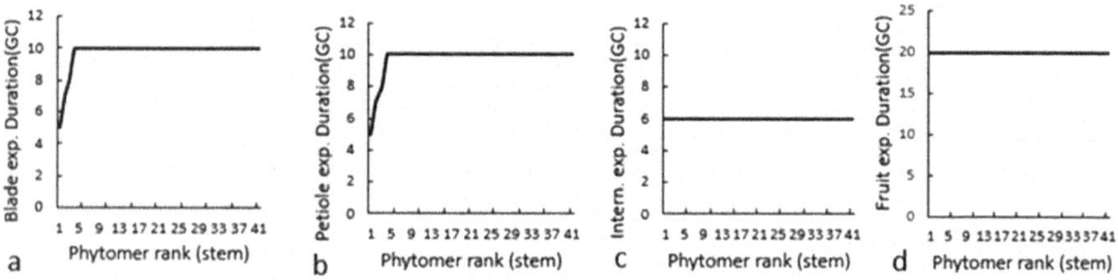

Fig. 9.47 Evolution of organ expansion time from the basis of organ series of sweet-pepper. (**a**) Blade. (**b**) Petiole. (**c**) Internode. (**d**) Fruit

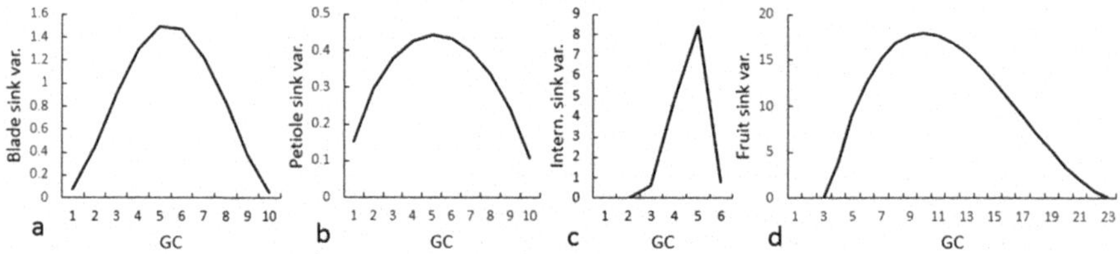

Fig. 9.48 Sink variation in each type of organs during the expansion. (**a**) Blade. (**b**) Petiole. (**c**) Internode. (**d**) Fruit. The functions corresponding to physiological age (PA) 2

variation occurs on the first five phytomers of the stem (only for leaf organs) until the expansion time stabilizes for each type of organ.

Figure 9.48 shows the bell shape of the sink variation functions for all organs. The internode has an expansion delay of one GC after its initiation. The fruit has an expansion delay of three GCs after its initiation, before starting their expansion. As there are three PAs, there are as many functions of sink variation in an organ as there are of PA. Let us recall that in the triplet P_o, B_{o1}, B_{o2}, the parameter P_o depends on the PA, but the parameters B_{o1}, B_{o2} do not change as explained previously. The functions corresponding to PA 2 are displayed here.

In Fig. 5.13 of Chap. 5, the measurements of daily increases of sweet pepper fruit are compared with those derived from the sink variation law computed from the

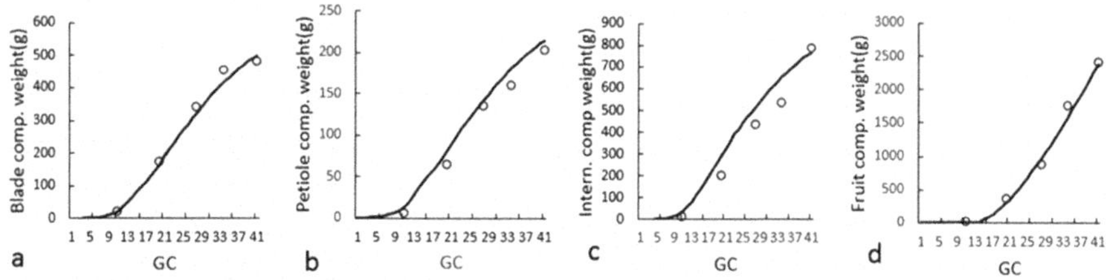

Fig. 9.49 Multi-fitting of the observed organ compartments (○) and computed organ compartments by the model (–) for the five measurement dates (11, 20, 28, 34, 41 GCs]. (**a**) Blades. (**b**) Petioles. (**c**) Internodes. (**d**) Fruits

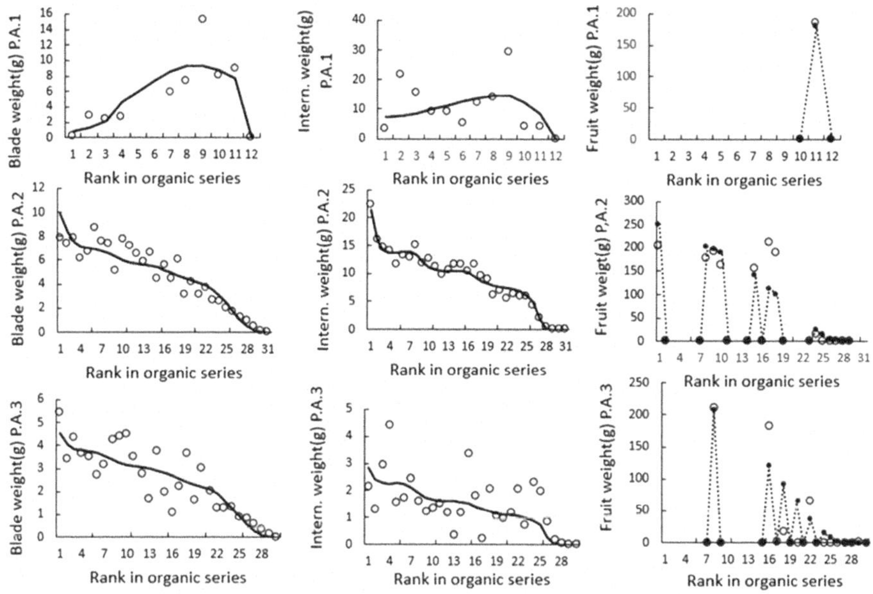

Fig. 9.50 Multi-fitting of the observed organic series (○) and computed organic series by the model (—) for the three PAs at the date 41 GCs, for leaf blades (petioles are not displayed), internodes and fruits

organic series. The two methods give similar results, and this brings validation for the model.

The multi-fitting of organ compartments (sum of all PAs) along CA is good (Fig. 9.49).

Figure 9.50 shows the organic series of leaf blades, internodes and fruits. The petiole series are not shown because they are like those of the leaf blades. The fitting of the organic series is satisfactory. The stem organs (PA 1) are noisy (leaves are senescent and two are missing.) The main branches (PA 2) have very homogeneous organic series. The organic series of PA 3 reconstructed from the first internodes of

Fig. 9.51 Sweet pepper growth and architecture at 41 GCs. (**a**) 2D organic series at 41 GCs in the simplified structure (Gloups software). (**b**) and (**c**): 3D growth simulation at 16 and 41 GCs (Digiplante software)

the axes three are also noisy. This is the result of the trauma related to pruning. Considering the compartments over the five growth stages shows that the parameters are invariants.

9.5.8.7 Simulation of Sweet Pepper Growth and Architecture

In Fig. 9.51, the development and growth of sweet pepper is depicted with its 2D organic series and its 3D architecture with all fruits at their observed locations.

9.5.8.8 Discussion

In this chapter, the sweet pepper is the first branched plant with several PAs. The WLSM easily computes the organ sinks of the different PAs. With branched plants, the simulation considers the topological structure that describes the architecture. If the plant has a deterministic development (*Bernoulli* process, parameter $b = 1$), then the appearance of events in the structure can be precisely located. It is possible to guide the simulation thanks to a mask that allows the development only on the existing branches. With deterministic plants, a single individual is often sufficient to perform the analysis of organic series (e.g., sunflower). If the structure remains small, the production of a mask is easy, but one quickly reaches a limit of the method when the branched system becomes more complex.

9.5.9 Case of Madagascar Watercress Study of Marie-Élodie Vavitsara (Mahajanga University)

This work was carried out as a part of Marie-Élodie Vavitsara's Ph.D. from Mahajanga University (Vavitsara 2014; Vavitsara et al. 2017). *Madagascar watercress* (*Spilanthes oleracea*) is an herbaceous crop grown in Madagascar, which is used as a vegetable for consumption.

9.5.9.1 Plant Description

The basic phytomer produces one internode, two leaves and potentially two twigs, up to the third order of branching. Development is stochastic and continuous; the stem is short and the shoots spread. All axes end with a terminal flower. The flowering of the stem forces the flowering of twigs that have not yet flowered, showing a dependence linked to the ontogenic age of the meristems (Chap. 7) and the growth of the twigs is delayed compared to the development pattern according to acropetal development (Fig. 9.52).

After the establishment of the first phytomers of the stem, apical flowering is observed, followed by simultaneous expansion of the preformed branches (S. Sabatier, CIRAD drawing).

9.5.9.2 Cultivation Conditions

Stable climatic conditions in tropical climates mean that calendar time can be equated with thermal time (in days). The root system was not available.

The plants were grown in pots large enough and spaced apart to avoid stress, and 65 plants were retained. Every 15 days, five plants were sampled for growth monitoring. Measurements were made in fresh biomass. Due to the delayed

Fig. 9.52 Botanical drawings illustrate the development and growth of *Spilanthes*

expansion of branches, simple organic series on them are not observable. The organs have been grouped into compartments per branch. There are, therefore, two types of organic series: simple organic series (stem single organs) and compound organic series (branch compartment organs).

9.5.9.3 Development Scheme

It is relevant to distinguish three PAs: PA 1 for the stem, PA 2 for the branches and PA 3 for the twigs. After a minimal rosette stage, the stem elongates and places a small number of phytomers (six to nine) because of the apical flowering that stops development. In a second stage, the preformed lateral branches then begin their expansion to order two with acropetal development. In a third step, if conditions allow, the preformed third-order branches begin their expansion with a basipetal development. The stochastic aspect of the development is visible: branches are missing and twig pairs do not always have the same number of internodes.

We can write: number of phytomers = 0.14 × number of days (Fig. 9.53a). The number of phytomers produced during development follows a distribution, characterized by an average and a variance from which the *Bernoulli* process parameter $b_1 = 0.94$ is derived (Fig. 9.53b).

To transform the number of days into the number of cycles, we have: number of development cycles = (0.14/0.94) × number of days. Stopping the development by synchronous terminal flowering of the axes does not mean stopping the growth, which continues over four cycles. Below is a detail of 4 of the 20 observed crowns (Fig. 9.54).

The aspect of stochastic development appears. The branches reach a limit of their development to seven DCs.

The analysis of the crowns provides a rhythm ratio $w = 1$, and a value $b_1 = 0.87$ and $b_2 = 0.94$ for the *Bernoulli* processes of the stems and branches. We can choose an average $b = 0.9$ for all the axes. The parameter b_1 of the stems also benefits from another close estimate obtained by the link of mean and variance on the development follow-up with $b_1 = 0.94$ (Fig. 9.53b).

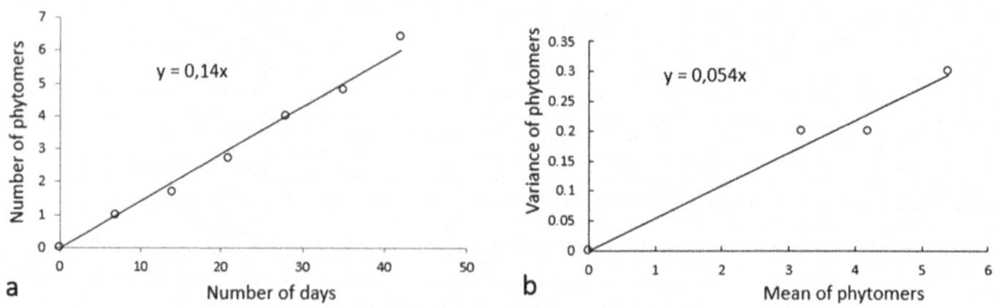

Fig. 9.53 *Spilanthes* development scheme. (**a**) Linear relationship between the development of the *Spilanthes* stem and the number of days. (**b**) The linear relationship between the mean and the variance of the number of phytomers of the stem

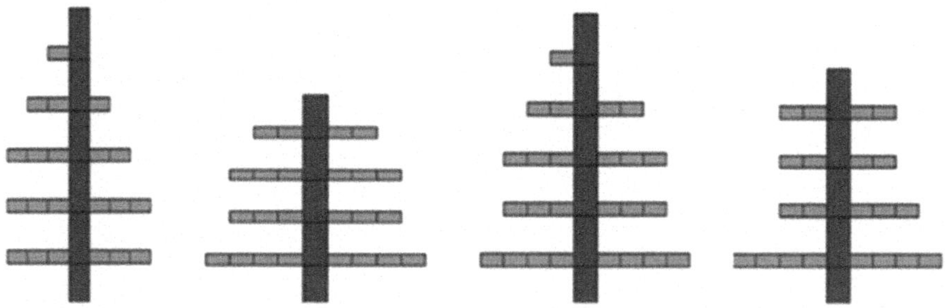

Fig. 9.54 Details of four *Spilanthes* stem crowns, with observations on branches and stems. The stochastic aspect of the structures is clear, not all pairs of branches have the same development; some branches are missing

The branching rate (a_2) is estimated as 0.97 for each axillary meristem of the stem and does not vary along the stem.

Five stages were selected to perform the measurements (4, 6, 8, 10, 12 DCs) during plant growth.

9.5.9.4 Growth Scheme

If the development of the architecture stops after the apical flowering of the stem, the growth continues, and the preformed organs in the lateral embryonic structures continue to expand up to 12 DCs. Based on the observation of leaf growth from the apex to the base of the stem, the expansion time for all organs (t_x) is set to 7 DCs. It remains constant along the stem. The leaves remain green on the plant until the end of the growth estimated at 12 DCs. The functioning time of the leaves (t_a) is, therefore, 12 DCs.

The development scheme is simple and the potential structure can, therefore, be easily established (Fig. 9.55a). But the growth of *Spilanthes* does not follow the development pattern; it is indeed a good illustration of the acropetal-basipetal model of herbaceous plants (Chap. 7). The complication comes from the fact that the beginning of the expansion of an organ is not only dependent on the notion of the cohort, but also on its position in the structure.

The leaves are not displayed. Until DC 8, the development of the architecture remains hidden in embryonic form. We can only observe a rosette of leaves. The first branches flower in their termination, immediately followed by a general apical flowering at DC 9, of all the axes, which completes the development. The growth pattern is different. Thus, the stem emitted by the rosette remains unbranched until DC 8. The expansion branching of the preformed branches then spreads rapidly in an acropetal manner on the stem. Over the PAs 2 and 3, they appear basipetal. The simulation is in accordance with the botanical diagrams in situ shown in Fig. 9.52.

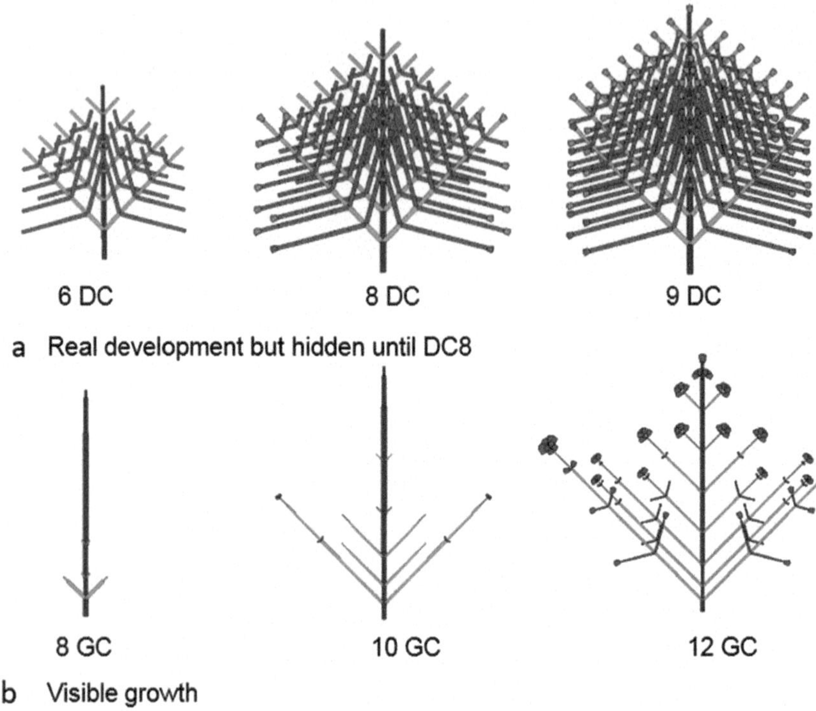

Fig. 9.55 Potential development structure and growth pattern of *Spilanthes* (Gloups software) (**a**) Real development but hidden until DC 8. (**b**) Visible growth

Let us recall the equations established in Chap. 7, which specify the To_g delay in the expansion of an organ according to its position in the structure:

$$T_{o_g}(i,j,k) = g_{o2}(T_1 - i) + g_{o3}(T_2(i) - j) + g_{o4}(T_3(i,j) - k) \tag{9.24}$$

T_1, $T_2(i)$, $T_3(i, j)$ refer to the development times of the axes according to their branching order and their positions on the carrier axis.

Parameters i, j, k refer to the position of the phytomer in the structure marked in coordinates that gives the distance counted in DC from the path to the seed (SDS) to order 3 (Chap. 7 Fig. 7.7). The delay in expansion (expressed in DC) of the organ o of the phytomer related to its position is designated by $To_g(i, j, k)$. If the expansion of a phytomer immediately follows its creation, this delay is null.

The length of the path in DC which leads from the base of the plant to the phytomer considered following the branching orders of the successive branches is: $i + j + k$. The number of phytomers of the axes of the structure once development is completed is T_1 for the stem, $T_2(i)$ for the second-order shoots in position i on the stem, etc. The parameters go_i control the intensity of the expansion delay according to the order of branching. The observed architecture of the plant, therefore, expresses only part of the development at a given age (Fig. 9.55b).

These parameters are estimated empirically for *Spilanthes* by simulation, to reproduce the appearance of the growth stages observed on the plant (Figs. 9.52 and 9.55b).

9.5.9.5 Setting up Organic Series

Measurements are expressed in fresh weight. Only well-developed main stem organs can be weighed individually. On branches that are expanding, some organs are still in the embryonic stage and it is not practically possible to establish organic series. In addition, the laws of delay in expansion, linked to the position in the structure, break down the notions of cohorts of organs. The organs of the same type of branch are weighed separately by compartments (leaves, internodes, fruits) according to the rank of the branch that carries them along the stem. This results in "compound" organic series (see Chap. 7 Fig. 7.17 for the way of sampling.)

Five growth stages were selected to perform the measurements (4, 6, 8, 10, 12 DCs). At each stage, five plants were collected to measure the fresh weight of organs in the simple and compound organic series whose averages are calculated and which were constructed to serve as a target for parameter estimation.

9.5.9.6 Results on Plant Functioning

9.5.9.6.1 Empirically Measured Parameters

A single leaf organ is assimilated to the whole leaf blade and petiole. The specific leaf weight ε, is deduced from the ratio of leaf fresh weight and its surface. It is estimated at $\varepsilon = 0.02$ g/cm^2 fresh weight.

Allometries and functioning times of the organs are estimated based on observations. The parameters g_o that control expansion delay laws have been experimentally set to best mimic botanical observations of growth stages. Thus, for leaves and internodes of axes two, the parameter is set to $g_o = 0.5$ which ensures an acropetal delayed branching. For the terminal flower, the parameter is set to $g_o = 0$ which ensures immediate expansion of the flower when the axis has finished its development.

For leaves and internodes of axes 3 the parameter is set to $g_o = 2$ which ensures a basipetal delayed branching. For the terminal flower the parameter is still set to $g_o = 0$.

9.5.9.6.2 Computed Parameters

The stages (4, 6, 8, 10, 12 DCs.) correspond to harvested plants (five plants per stage) which are, therefore, independent samples. These measurements produce 174 weights of organs to be adjusted simultaneously during the growth process. There are 17 parameters to be estimated. But the second parameter of the sink variation law B_{f2} cannot be computed. Empirically $B_{f2} = 2.5$ gives suitable values for fitting. Remember that the ratio $B_{o1}/(B_{o1} + B_{o2})$, produces a couple of values that produces generally a good fitting. Although the *Spilanthes* is an herbaceous plant, there is girth growth on the stem, which must be estimated and is controlled by the P_c sink. The Pressler law is available here.

The estimated parameter values (Table 9.21) allow a correct fitting of the organ compartments (Fig. 9.50) and the simple and compound organic series of *Spilanthes* (Fig. 9.60).

The fitting is good: ($R^2 = 0.97$; number of Data = 174; number of computed parameters =17 (parameter P_b is fixed to 1). The values of the 17 parameters are invariants that control well the fittings of the organic series during the growth period (12 GCs).

Note that organ sink values of the PA 3 are imprecise (high CV). This is due to their late appearance at the end of the axes and their few contributions to plant demand.

Figure 9.56 shows the evolution of the number of phytomers per PA, plant demand, individual LAI (*Sf*/*Sp*) and biomass produced per cycle as a function of

Table 9.21 Source–sink parameters of *Spilanthes*

Param	Mean	SD	CV (%)
r	18.6	2.5	14
Sp	54.1	22	21
Q_0	0.005	0.004	76
P_c	0.05	0.01	31
P_{b1}	1	0	0
P_{b2}	0.4	0.1	10
P_{b3}	16.5	20.4	124
P_{i1}	1.2	0.2	17
P_{i2}	0.5	0.1	24
P_{i3}	4.9	4.1	83
P_{f1}	2.4	0.6	25
P_{f2}	2.8	0.8	28
P_{f3}	0.6	0.5	79
B_{b1}	3.1	0.9	30
B_{b2}	3.7	1.1	31
B_{i1}	3.2	1	32
B_{i2}	2.5	1	41
B_{f1}	2.8	0.3	12
B_{f2}	2.5	0	0

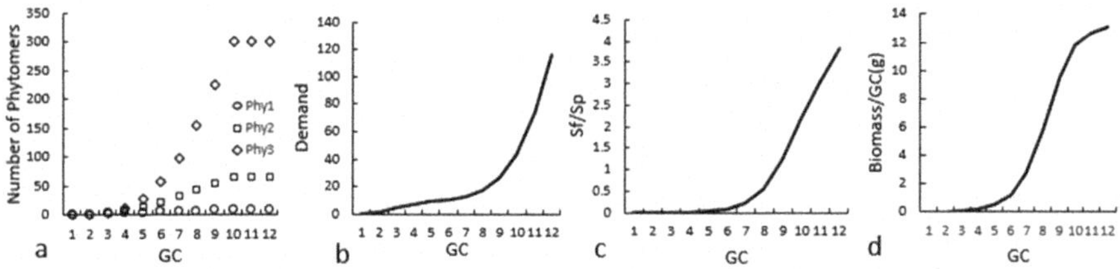

Fig. 9.56 History of the development and growth of *Spilanthes*. (**a**) Number of phytomers per GC and PA. (**b**) Evolution of demand per GC. (**c**) Evolution of individual LAI. The LAI per plant reaches a high value (> 3) which saturates the biomass production. (**d**) The biomass production per GC

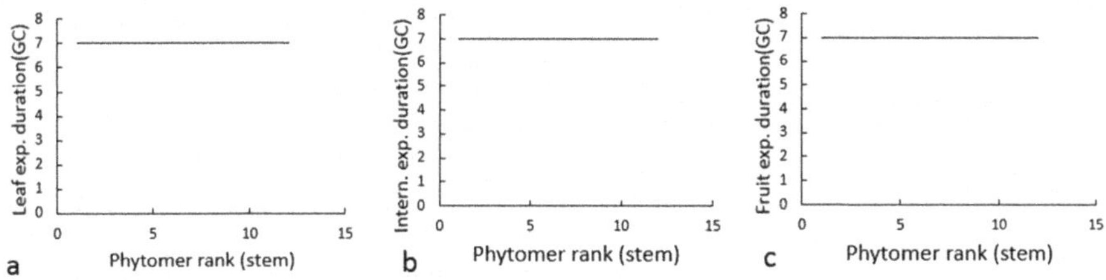

Fig. 9.57 Constant expansion time of *Spilanthes*. (**a**) Leaf. (**b**) Internode. (**c**) Fruit. Expansion times are the same for the three organs

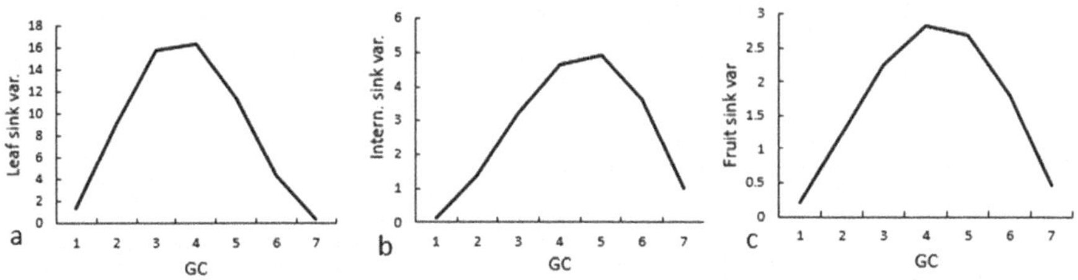

Fig. 9.58 Sink variation in each type of organ during organ expansion. (a) Sink variation in the blade (PA 3). (**b**) Sink variation in internode (PA 3). (**c**) Sink variation in fruit (PA 2)

GC. Here the LAI per plant reaches a saturation (LAI > 3). The biomass production is saturated.

The fact that there are potentially two branches by phytomers and that there are three PAs means that the number of potential phytomers in PA 3 is high (299). But in reality, most of them are still in the embryonic stage, and they are not important for the plant biomass.

For this plant the expansion duration of organs is constant (Fig. 9.57).

Figure 9.58 shows the bell shape of sink variation functions for all organs. As there are three PAs, there are many functions of sink variation in an organ. Let us recall that in the triplet P_o, B_{o1}, B_{o2}, the parameter P_o depends on the PA, but the

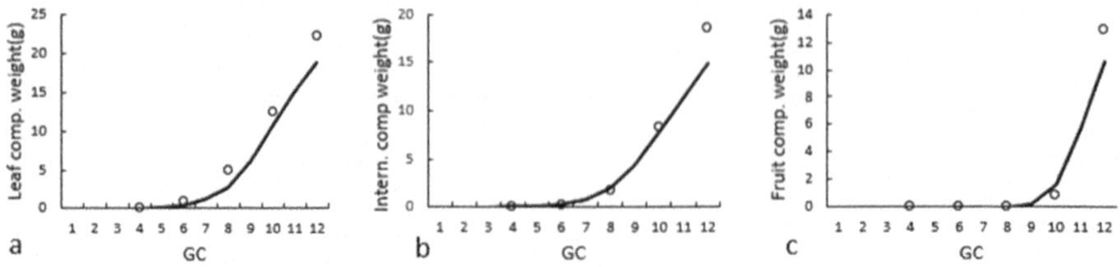

Fig. 9.59 Evolution of the weights of organ compartments of *Spilanthes*. Multi-fitting of the observed (○) and computed organ compartments by the model (—) for the five measurement dates [4, 6, 8, 10, 12 GCs]. (**a**) Blades. (**b**) Internodes. (**c**) Fruits

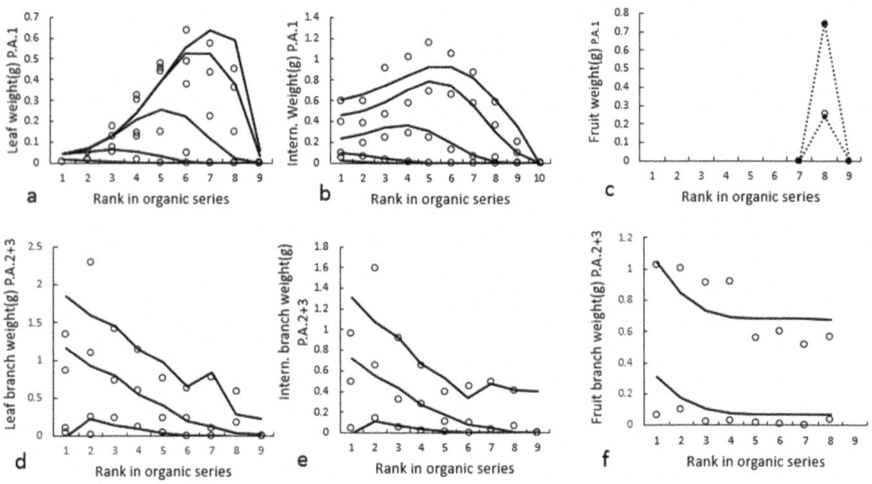

Fig. 9.60 Multi-stage growth adjustment of *Spilanthes* (4, 6, 8, 10, 12 DCs). (**a**, **b**, **c**) Simple organic series of leaves, internodes and fruits on the stem. At the base (**b**), the secondary growth of the stem internodes is visible compared to the evolution of the leaves. (**d**, **e**, **f**) Compound organic series of leaves, internodes, fruits on the stem belonging to branches (mixed PA 2 and PA 3). On the branches, the first fruits are larger because they appear slightly before the generalized flowering of the plant. "○" represents the observations, "—" represents the calculations.

parameters B_{o1}, B_{o2} do not change. The functions corresponding to PA 3 are displayed here for leaves and internodes, and PA 2 for fruits. The WLSM cannot compute B_{f2} so it is set empirically to 2.5

The multi-fitting of organ compartments (sum of all PA) along CA is good (Fig. 9.59).

Simple (stem) and compound (branches) organic series are well fitted by the model (Fig. 9.60). Fig. 8.60 shows the advantage of compound organic series composed of branches along the stem. The simple organic series cannot be identified separately because the delays in expansion suppress the cohorts and part of the phytomers remain embryonic. Each branch is, therefore, only known by its number

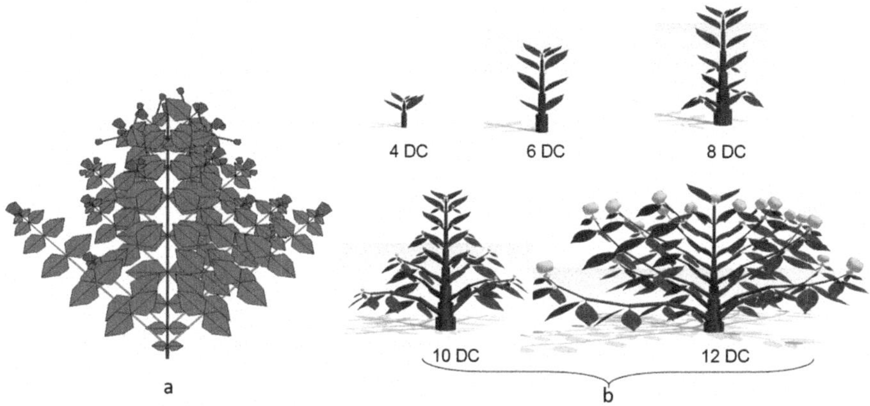

Fig. 9.61 Growth and architecture of the *Spilanthes* simulated at 12 DCs. (**a**) 2D organic series (Gloups software). (**b**) 3D growth simulation at 4, 6, 8, 10, 12 DCs showing the acropetal pattern by the QingYuan software

of phytomers, the total weights of leaves, internodes and fruits for all the PAs. It is satisfactory that the WLSM can find the sink values according to these PAs while these organs are mixed in the branch.

Considering the compartments over the five growth stages shows that the parameters are invariants.

9.5.9.7 Simulation of the Growth and Architecture of Spilanthes

Growth and development using the computed parameters and various allometries by the QingYuan software (CASIA) allow reproducing the plant architecture (Fig. 9.61) at the different developmental stages, which is the synthesis of this study.

9.5.9.8 Discussion

Spilanthes is certainly one of the most complex plants modelled using the GreenLab model. Development is stochastic and growth is delayed with the acropetal expansion of the branches. Compound organic series allows the estimation of source–sink parameters of branches. The method has been applied to other similar growth patterns of herbaceous plants such as rapeseed (Jullien et al. 2011; Wang et al. 2018) or chrysanthemum (Kang et al. 2012).

9.5.10 *Case of Rice Study of Zheng Bang-You (CAU)*

This study comes from the Ph.D. thesis of Zheng Bang-You at Chinese Agriculture University (Zheng et al.2008). The main results obtained are presented here.

9.5.10.1 Plant Description

Rice (*Oriza sativa* L.) is an herbaceous plant. Its development is characterized by an initial rosette-like stage, followed by an elongation of the stalk ending in a spike. Stems called tillers, in variable numbers, are issued at the same time as the main stem at the rosette stage. Although botanically the tillers can be considered as reiterations in terms of the development, the phytomers have smaller biomass and the spikes have larger biomass. It is, therefore, considered that there are two PAs, the main stalk being PA 1 and the tillers being PA 2. A characteristic of grasses is that the leaf is supported by a sheath attached to the node of the phytomer, it acts as a petiole (as for Maize). This sheath coats the stem and its weight is equivalent to that of the leaf it carries, and its sink must, therefore, be considered. The stalk, whose development is deterministic ($b = 1$), consists of 16 phytomers.

The first internodes of the stems are very short and do not expand until the 11th phytomer. Their elongation is allowed only from this rank.

In this study, we followed the growth of the plant until the phenomenon of remobilization. The mature phytomers reallocate part of the biomass of their organs into the common pool to provide the biomass necessary for the spikes when photosynthesis declines due to leaf senescence.

The architecture of rice is highly variable for branching, which consists of the emission of secondary stems (called tillers) at the rosette level. The number of tillers follows a distribution that can reach several tens per plant.

9.5.10.2 Cultivation Condition

The experiment was carried out in the field. The planting density was about 20 plants/m^2. The stochastic aspect has been eliminated by choosing plants with eight tillers from the population to focus on source–sink relationships in a deterministic development framework.

9.5.10.3 Development Scheme

If one wants to describe the topological structure of the plant in detail for the plant with eight tillers, the operation is complicated, as is the associated botanical automaton. It is observed that tillers have a delay of three DCs. Analysis of the chronological structure shows that it can be simplified by moving the tillers and sticking them in the right places on the stem to eliminate delays in appearance. The tillers that

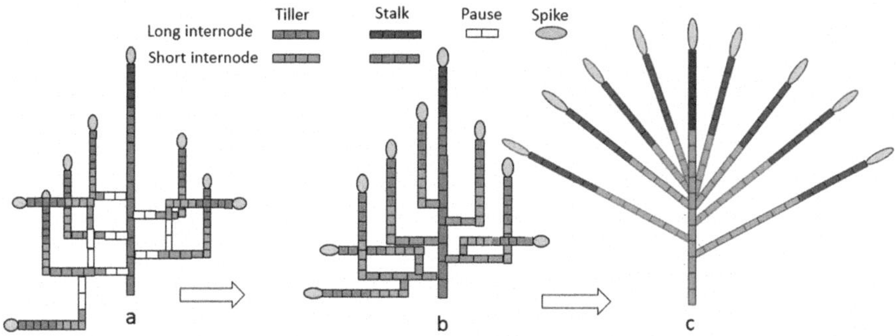

Fig. 9.62 Rice development for eight tillers. The stalk is PA 1 (blue). The tiller is PA2 (green). (**a**) Chronological structure showing the pauses before the tiller appearance. (**b**) Topological structure (pauses are removed). (**c**) Simplified topological structure with whorls of tillers on the main stem giving same cohorts and organic series with that of (**a**) (From Zheng et al. 2008)

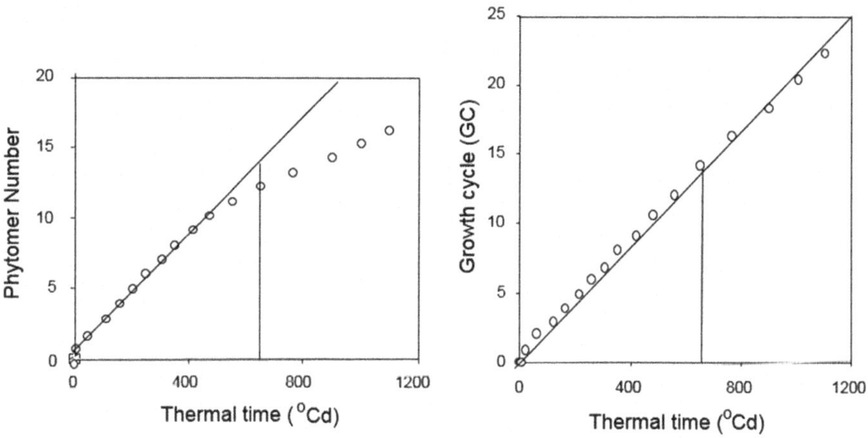

Fig. 9.63 Development of the rice stalk as a function of thermal time. Transformation of DCs into GCs beyond the 14th phytomer (From Zheng et al. 2008)

emerge at the same time are placed in whorls on the stem (Fig. 9.62). This simplified structure has the same pattern of development because it produces cohorts of organs identical to the original, and, therefore, the same organic series as the complete structure. The botanical automaton is simpler to build and the simulation is faster.

All organs from the tips of the axis are the same and synchronous because they belong to the same cohort. The spikes, therefore, appear together at the 16th DC.

Finally, it should be noted that the last internode of the stems carries a terminal spike but does not carry a leaf.

The development of rice is linear as a function of thermal time up to 800 °C·d, then oblique to slow down. The duration of the development cycle is longer. It is, therefore, advisable to reason in GCs to follow the growth by giving the same value of GCs in thermal time as that of the DCs of the first part of the development. It is estimated that the total duration of the observed growth is 29 GCs (Fig. 9.63).

Eight stages were selected to perform the measurements (7, 10, 12, 15, 18, 21, 25, 29 GCs) during the growth. Only stage 18 has been measured for organic series; organ compartments are considered for all stages.

9.5.10.4 Setting up Organic Series

Each of the two PAs at stage 16 DC, corresponding to Fig. 9.62 has four organic series: blade, sheath, internode and terminal spike. Measurements are expressed in dry weight (g).

For all eight stages the weights of organ compartments are measured.

The root system is not taken into account, which is assumed to not disturb the above-ground part of the plant architecture.

9.5.10.5 Growth Scheme

The particularity of the study of rice is to consider the phenomenon of biomass remobilization, a common phenomenon in grasses. Once the expansion period is over, the organs partly empty into the common pool, which will allow the growth of the spikes, relaying photosynthesis, which decreases with the senescence of the leaves. Thus, each organ has a maximum weight q_{max}, a functioning time t_f greater than the expansion time t_x ($t_f > t_x$) during which it does not lose weight, and a remobilization coefficient k_m ($0 < k_m < 1$). Beyond t_f DCs, the organs lose weight. The fraction $F\%$ ($0 < F\% < 1$) of biomass lost is empirical data measured after the end of remobilization. The damping parameter k_m expresses the speed with which the organ empties.

Remobilization is modeled by the function:

$$q(n) = q_{max}\left(1 - F\%\left(1 - (1 - k_m)^{n - t_f}\right)\right) \tag{9.25}$$

Only the k_m parameters of the organs under consideration are to be estimated by the inverse method. At the end of remobilization, the organ can be weighed and $F\%$ can be deduced from this. Figure 9.64 shows the form of variation in the function of remobilization. A particularity of grasses is short organ expansion durations ($t_x = 4$ DCs), except for the first 11 phytomers whose short internodes have no expansion. The growth process lasts 29 GCs. The spike expansion duration is set to 13 DCs. The first part is due to the photosynthesis with the functioning time of leaf $t_f = 5$ DCs. The remaining time is occupied by the remobilization process.

If the development of the architecture stops after the apical flowering of the axes (at 16 GCs), the growth continues until 29 GCs.

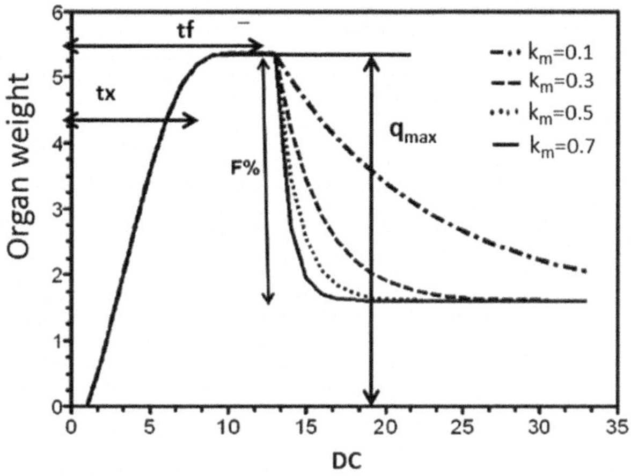

Fig. 9.64 Effect of the parameters $F\%$ and k_m on the decrease in weight of the organs subject to remobilization

9.5.10.6 Results on Plant Functioning

It is considered that at the growth stage of 18 GCs, organ weights are at their maximum, and the phenomenon of remobilization is still negligible. This stage, therefore, provides an architecture free of remobilization for the organic series and compartments.

9.5.10.6.1 Empirically Measured Parameters

The specific leaf weight ε, is deduced from the ratio of leaf fresh weight and its surface, which is estimated as 0.0035 g/cm^2. Looking at the organ compartments, it is empirically observed that the blades and sheaths lose 75% of their biomass through remobilization ($F\% = 0.75$) and the internodes 20% ($F\% = 0.2$). The functioning time of blades is estimated to be 5 DCs. It is also the starting time for remobilization. The remobilization times of sheaths and internodes are also set to five DCs. The expansion times of blades, sheaths and internodes are estimated as 4 DCs, and 13 DCs for the spike.

An important issue is the k-value of the cultural coefficient, which describes the average angle θ of inclination of the normal to the surface of a leaf concerning the vertical ($k = \cos(\theta)$). This coefficient is given empirically in Beer Lambert's law used in crop models. It is the same coefficient that appears in the plant growth equation of GreenLab (Chap. 5, Eq. 9.18).

When the leaves have a good average horizontality, (case of the plants treated earlier in this chapter) the value $k = 1$ is appropriate. The agreement between *Sp* calculated and *Sd* (available surface per plant) allows us to verify it. But the rice leaves are erect. We chose $\theta = 30°$ that gives $k = 0.5$, for computing the parameters with the WLSM.

The quality of the fitting does not depend on sensitivity on the value of k, which means that it does not influence the parameters of the sink functions. But the parameter values r and Sp depend strongly on it.

Thus with $k = 1$, the WLSM gives $r = 98.3$ and $Sp = 1185\ \mathrm{cm}^2$. But the value of Sd of this experiment corresponding to 20 plant/m^2 is $Sd = 500\ \mathrm{cm}^2$. The value $k = 1$ is, therefore, not suitable. It means also that the parameters r and Sp are no longer functional parameters, but only calibration coefficients.

Taking $k = 0.5$, which is more suitable for rice, the WLSM gives $r = 49.1$ and $Sp = 591\ \mathrm{cm}^2$.

This time the value of Sp, is close to that of Sd, the parameters r and Sp retrieve their biological functionality. ($1/r$ is the resource use efficiency and Sp is the inverse of the planting density).

9.5.10.6.2 Computed Parameters

There are eight measurement stages at 7, 10, 12, 15, 18, 21, 25, 29 GCs. These stages correspond to harvested plants (measurements are the average 2–4 plants per stage), which are independent samples. These measurements produce 143 weights of organs to be fitted simultaneously during the growth process. There are 20 parameters theoretically to compute. Qo and the second parameters of the four sink variation laws parameters B_{o2} of all organs cannot be computed together with the other 15 parameters. Fixing the sink parameter values, it is possible to compute in this reduce set, r, Sp, Q_0. Empirically $Q_O = 0.02$ g, $B_{b2} = 7$, $B_{p2} = 5$, $B_{i2} = 1$, $B_{f2} = 19$, give suitable values for fitting. Remember that the ratio $B_{o1}/(B_{o1} + B_{o2})$, is sufficient to produce generally a good fitting.

After 16 GCs, with remobilization, we observe a decrease in the weight of the vegetative organs for organ compartments in favor of the spike. It is mainly the blades and sheaths that are affected, the internodes being little affected. The remobilization coefficients for blade, sheath and internode are described by K_{rb}, K_{rp} and K_{ri} according to Fig. 9.64.

Although remobilization takes place in the organs of the phytomer, it would be too tedious to measure at this level. We, therefore, choose to calibrate the coefficients on the evolution of the weights of the compartments.

The estimated parameter values (Table 9.22) allow a correct fitting of the organ compartments (Fig. 9.68) and the organic series of rice (Fig. 9.69).

The fitting is good: $R^2 = 0.99$, number of data = 143, number of computed parameters = 15 (the parameter P_b is fixed to 1). The values of the 15 parameters are invariants that control well the fittings of the organic series during the growth period (29 GCs).

Figure 9.65 shows the evolution of the number of phytomers per PA, plant demand, individual LAI (Sf/Sp) and biomass produced per cycle as a function of GC. Here the LAI per plant reaches a saturation (LAI > 3). The biomass production is saturated.

Table 9.22 Source–sink parameters of rice

Param	r	Sp	Q_0	P_{b1}	P_{b2}	P_{p1}	P_{p2}	P_{i1}	P_{i2}	P_{f1}	P_{f2}	k_{rb}	k_{rp}	B_{b1}	B_{p1}	B_{i1}	B_{f1}
Mean	49	591	0.02	1	0.76	1.13	0.77	0.52	0.37	2.01	2.22	0.18	0.19	10.2	7.30	0.40	2.79
SD	5	103	0.01	0	0.03	0.07	0.05	0.08	0.06	0.43	0.49	0.02	0.02	1.29	1.26	0.17	0.31
CV (%)	11	17	55	0	4	6	6	16	16	22	22	12	12	13	17	44	11

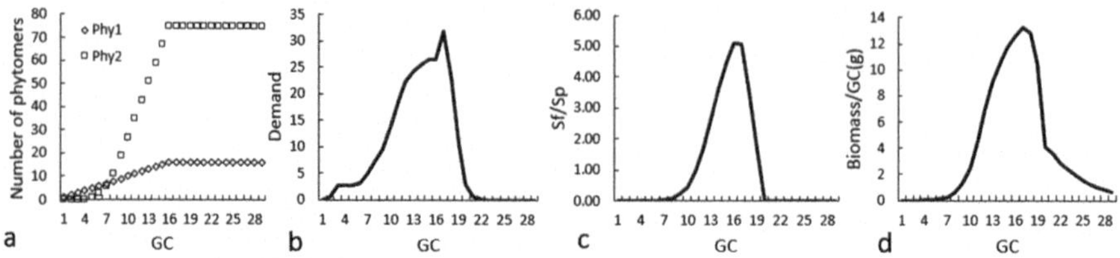

Fig. 9.65 History of the development and growth of rice. (**a**) Number of phytomers per GC and PA. (**b**) Evolution of demand per GC. (**c**) Evolution of individual LAI. The LAI per plant reaches a high value (> 3) which saturates. (**d**) The biomass production per GC

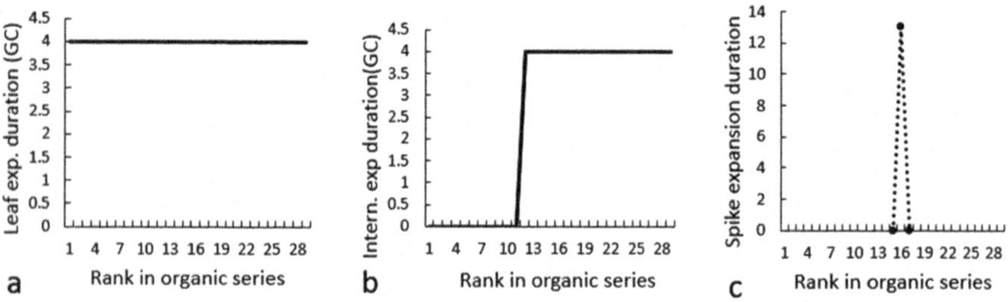

Fig. 9.66 Organ expansion durations of rice. (**a**) Leaf. (**b**) Internode. (**c**) Spike

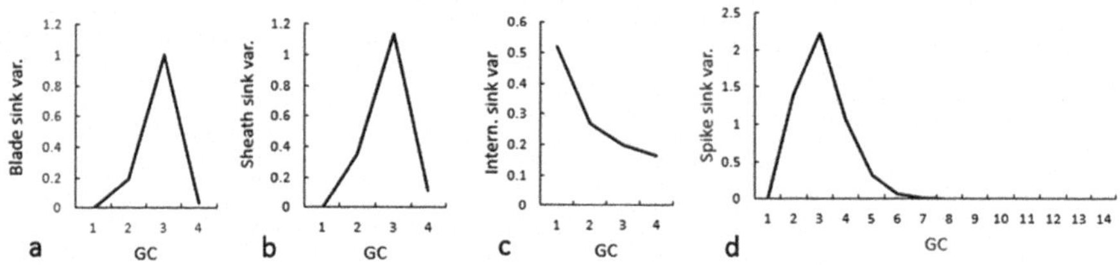

Fig. 9.67 Sink variation in each type of organs during the expansion. (**a**) Blade (PA 1). (**b**) Sheath (PA 1). (**c**) Internode (PA 1). (**d**) Spike (PA 2)

The demand of the plant (Fig. 9.65b) ceases after 16 + 4 GCs which corresponds to the demand of the last leaf of the stem. LAI per plant (*Sf*/*Sp*) (c) ceases after 16 + 5 GCs which corresponds to the functioning of the last leaf of the stem. However, the expansion of the spikes continues as it takes advantage of the remobilization of the phytomers. Biomass produced by photosynthesis ceases after 20 GCs. But the biomass provided by the remobilization of phytomers takes over. This causes a discontinuity in the biomass available per cycle (Fig. 9.65d).

For this plant, the expansion duration of organs is constant ($t_x = 4$ DCs) for leaf (blade and sheath), it is null for internodes until rank 11. For the spike that appears on the 16th internode, it is $t_x = 13$ DCs (Fig. 9.66).

Variations in the organ sinks occur over a short duration for phytomers ($t_x = 4$ GCs; Fig. 9.67). A pause of one GC precedes the expansion of the spike which lasts 13 GCs. In reality, this expansion is decomposed into two parts.

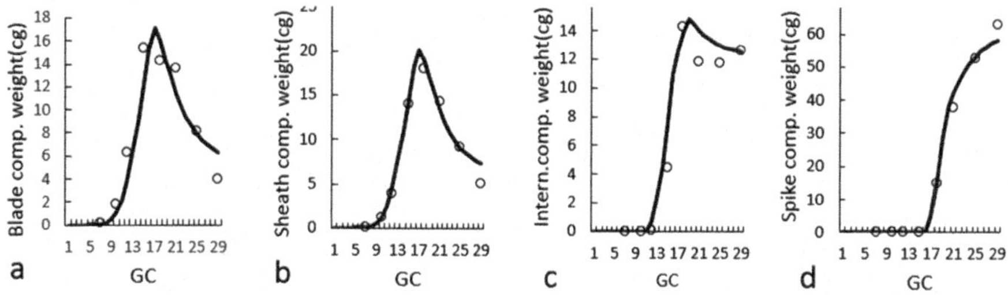

Fig. 9.68 Multi-fitting of the observed (○) and computed organ compartments by the model (—) for the five measurement dates [7, 10, 12, 15, 18, 21, 25, 29 GCs]. (**a**) Blades. (**b**) Petioles. (**c**) Internodes. (**d**) Spike

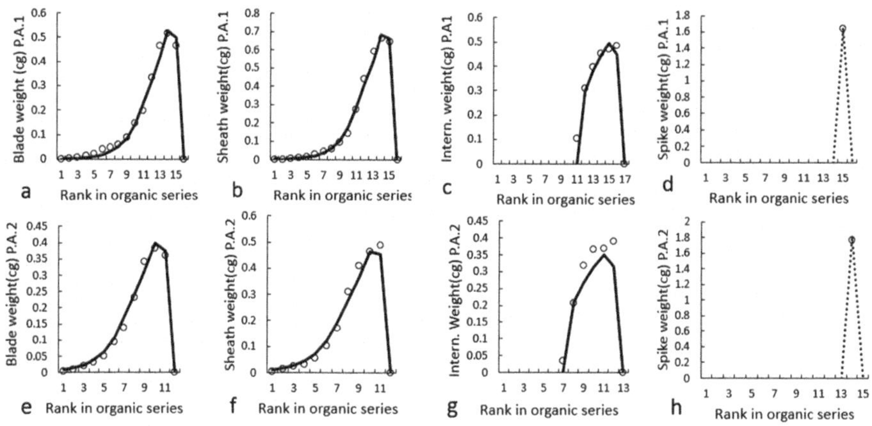

Fig. 9.69 Fitting of single-stage growth on rice (18 GCs). (**a**, **b**, **c**, **d**) Simple organic series of blades, sheaths, internodes and spike on the stem (PA 1). (**e**, **f**, **g**, **h**) Simple organic series of blades, sheaths, internodes and spike on tillers (PA 2). "○" represent the observations, "—" represents the computations.

The first one, which lasts 6 GCs, corresponds to the actual law of organ demand. In the second part, the demand is null for the whole plant and the expansion of the spike is done passively by remobilization, because it is the only active organ of the plant.

The multi-fitting of organ compartments (sum of two PAs) along CA is good (Fig. 9.68).

In Fig. 9.68, the remobilization is particularly visible in the compartments after 18 GCs. The model of remobilization is quite efficient. The loss of biomass ($F\%$) is 75% for the leaves. The damping coefficients are respectively $K_{rb} = 0.18$ and $K_{rp} = 0.19$. For the internodes, the loss is estimated at $F\% = 20\%$. The coefficient is fixed empirically at $K_{ri} = 0.2$ because the WLSM cannot calculate it due to the weak remobilization of the internodes.

Simple (stalk and tillers) organic series are well fitted by the model (Fig. 9.69) at the stage of 18 GCs, where the remobilization is still negligible. Considering the compartments over the eight growth stages shows that the parameters are invariants.

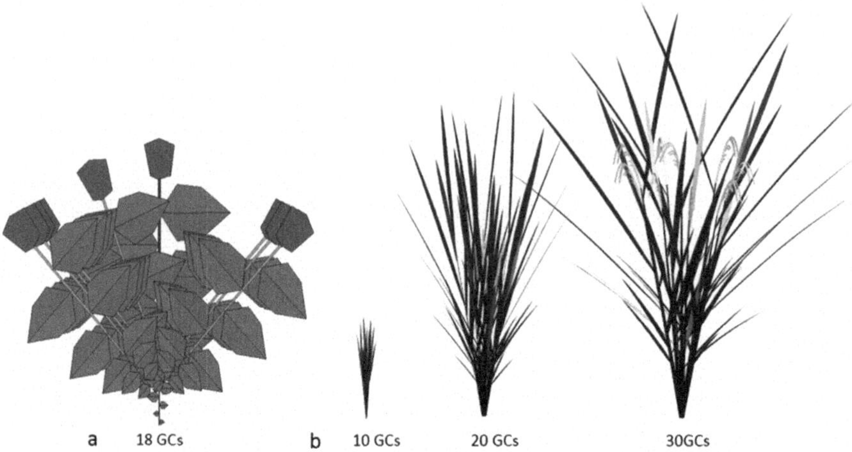

Fig. 9.70 Growth and architecture of rice. (**a**) 2D structure at 18 GCs showing organic series (Gloups software) (**b**) 3D growth simulation of architecture at 10, 20, 30 GCs simulated by Digiplant software

If we analyze only the main stem instead of the whole plant with its eight tillers, this is equivalent to dividing the production area *Sp* by 9. The WLSM fits the main stem with $Sp = 60.1$ cm^2. This is close to the value (Table 9.22) (591/9 = 65.7 cm^2. The consistency of the model is once again demonstrated. It is, therefore, possible to compute the source–sink relationships from the stem only, which simplifies the measurements.

9.5.10.7 Simulation of the Growth and Architecture of Rice

Organic series using the computed parameters and various allometries are displayed in 2D with Gloups software (Fig. 9.70a). Growth and development of rice are simulated by the Digiplant software to reproduce the plant architecture (Fig. 9.70b) at the different growth stages, which is the synthesis of this study.

9.5.10.8 Discussion

Of all the plants modelled with the GreenLab model, the rice is the one with the shortest organ expansion times. This makes the three parameters of the sink triplet p_o, B_{o1}, B_{o2}, not computable together, and B_{o2} must be set empirically. The role of the cultural parameter k is firstly efficient due to the erected leaves. As with sweet peppers, we can simplify the topological structure and eliminate the stochastic aspect by choosing plants with the same architecture, allowing us to focus solely on source–sink relationships. Modelling the remobilization of organs seems simple and on the dependency of a single damping coefficient.

9.5.11 Case of Canola Study of Wang Xiu-Juan (CASIA)

This study comes from the project of Xiujuan Wang at the Institute of Automation, Chinese Academy of Sciences (CASIA).

9.5.11.1 Plant Description

Canola (or oilseed rape) is an annual plant with inflorescences of yellow flowers. Seeds are sown in the autumn, and before winter, the plant develops a rosette of 10–15 leaves (Jullien et al. 2011). Stem extension begins with the return of the growing season in the spring. Flowering begins before stem extension has finished and continues for more than one month. The number of ramifications is pre-determined during the organ initiation early in the growth cycles in autumn. The meristem produces leaves, which bear axillary buds that can produce a ramification (Tittonel 1990).

The growth of Canola is continuous. At the growth end, the apical meristem of the stems transforms into an inflorescence which is a raceme (a raceme is a single-stemmed inflorescence consisting of an axis of indefinite development whose terminal meristem sequentially produces flowers one by one). After germination, the apical meristem of the main stem produces about 30 phytomers before the appearance of the inflorescence. Each lateral meristem can develop into a ramification (Fig. 9.71). The ramifications in distal positions elongate earlier than those in basal positions, which results in a basipetal flowering sequence, while the flowers within the ramification have an acropetal flowering sequence (Wang et al. 2011, 2014). The

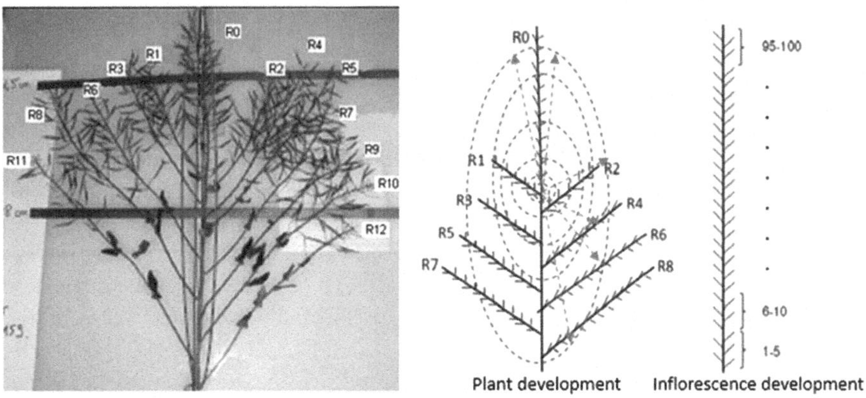

Fig. 9.71 Schematic diagram of winter Canola. Plant development: the inflorescences initiate acropetally but expand basipetally. Flowering starts on the main stem (R0) and is followed by lateral inflorescences from top to bottom (basipetal). Inflorescence development: flowering and pod setting on the inflorescence occur from bottom to top (acropetal)

time lag between the appearance and elongation of a ramification, called "the growth delay," needs to be determined when the internodes and flowers become significant sinks.

9.5.11.2 Cultivation Conditions

The experiment was carried out during the 2011–2012 and 2012–2013 growing seasons at the experimental station of the Zhejiang Academy of Agricultural Sciences. The planting density was 13 plants per square meter.

9.5.11.3 Measurements

Both destructive measurements and continuous observations were made. Continuous observations were made on 10 plants, twice per week. Detailed topological observations included the number of leaves on the main stem, and the number and position of ramifications on the main stem, and the number of leaves for each ramification.

In destructive measurements, four plants were selected to measure the fresh weights of organs (internode, blade, petiole, and ramification) for each position on the main stem, and the total fresh weights of ramifications for each kind of organ, at six stages along the growing period. The measured data of the last four stages were used to estimate the source–sink parameters of the model.

The organs have been grouped into compartments by ramification. There are, therefore, two types of organic series on the stem: simple organic series (organs) and compound organic series (ramifications).

9.5.11.4 Development Scheme

Following the emergence of the leaves, internodes of the main stem begin to elongate. The expansion of the ramifications is delayed compared to the main stem. Lateral inflorescences expand along the main stem from the top to the bottom. Flower emergence starts on the main inflorescence and develops basipetally to the lateral inflorescences (Tittonel 1990), causing the basal and oldest ramifications to bear the youngest inflorescences. Pods are set after all of the leaves of the main stem have emerged. The first pods are initiated on the main stem and then on the ramifications from apical to basal. Within a ramification, pod setting remains acropetal (Jullien et al. 2011). The stochastic aspect of the development is visible: ramifications are missing and do not always have the same number of internodes.

For the Canola, it is relevant to distinguish two PAs: PA 1 for the stem, PA 2 for the ramifications. Compound organic series are made of compartments of organs on axis 2 bore by the main axis. The development scheme is simple and the potential structure can, therefore, be easily established. But the growth of Canola does not

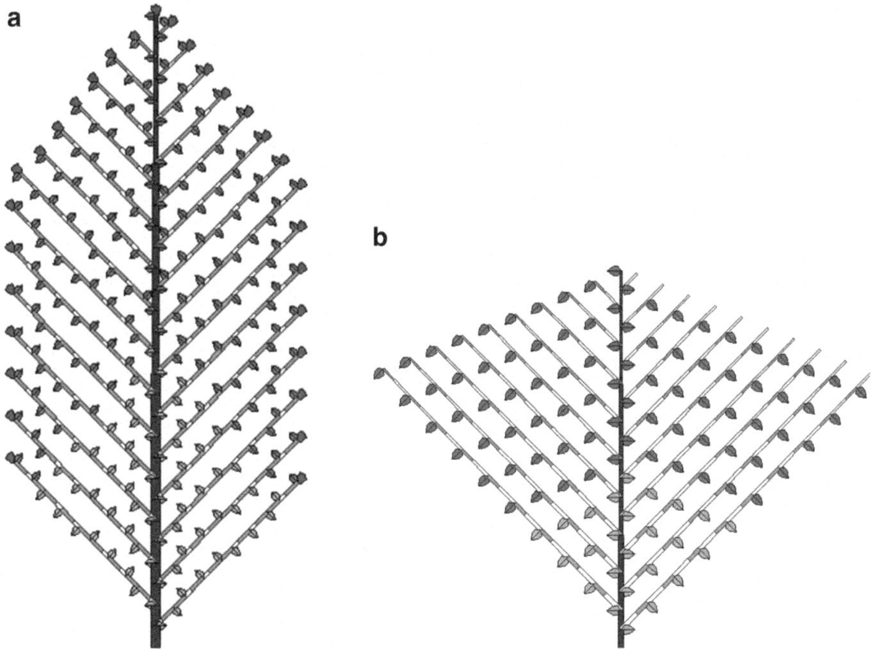

Fig. 9.72 Topological structure of Canola. (**a**) The development completes at 30 DCs, but the growth continues until 40 DCs. (**b**) Pauses are visible inside the axes 2 corresponding to $w = 0.6$ (20 DCs). Inflorescences are at the tip of the axis (Gloups software)

follow the development pattern. The complication comes from the fact that the beginning of the expansion of an organ is not only dependent on the notion of the cohort but also on its position in the structure (Chap. 7). The vegetative development cannot exceed 30 DCs for the main stem and axes 2 (Fig. 9.72). Stopping the vegetative development by synchronous terminal flowering of the axes does not mean stopping the growth, which continues over ten cycles with the growth of racemes doing pods. Five stages were selected to perform the measurements (5,18,31, 35, 40 DCs) during plant growth.

The crown analysis (Chap. 3) provides a rhythm ratio $w = 0.6$, and a value $b_1 = 0.86$ and $b_2 = 0.97$ for the *Bernoulli* processes of the stems and ramifications. We can choose an average $b = 0.9$ for all the axes. The branching rate (a_2) is estimated as 0.97 for each axillary meristem of the stem and does not vary along the stem. Fig. 9.73 is a detail of four of the 15 observed crowns.

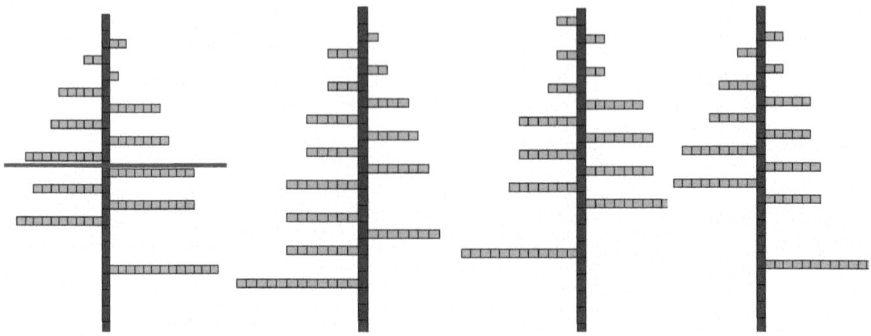

Fig. 9.73 Details of four Canola stems, observations of ramifications and stems. The stochastic aspect of the structures is clear; some ramifications are missing

Fig. 9.74 Development of the Canola as a function of thermal time. Transformation of DCs into GCs beyond the 30th phytomer

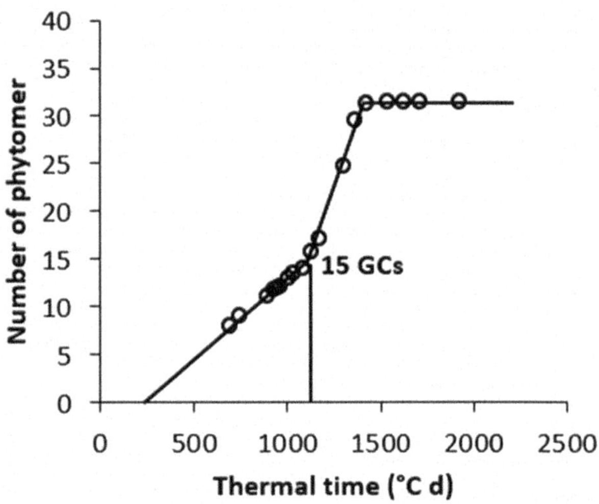

9.5.11.5 Growth Scheme

9.5.11.5.1 Variation in the Environment (Parameter E)

We set $E = 1$ for a constant environment. But when the production of phytomer according to thermal time (that is proportional to the number of DC) is not linear, this means that the radiation/DC change, so E must be adapted. A case encountered several times is that there are two slopes, one is a_1 if $T < T_1$, otherwise, a_2 if $T \geq T_1$, which can link the number of phytomers to the thermal times. For the first period ($T < T_1$), we set $E(T) = 1$, and for the second period ($T \geq T_1$), we have $E(T) = a_2/a_1$.

For the Canola, from non-destructive measurements, we displayed the number of emerged phytomers along the main stem according to the thermal time to compute the phyllochron (Fig. 9.74). It shows a linear piecewise relationship between the number of organs and the thermal time, which highlights a modification of the phyllochron.

The study of the development in thermal time shows that the slopes pass from $a_1 = 0.87$ to $a_2 = 1.02$, then we find $a_1/a_2 = 0.85$ growth cycle for a development cycle and $T_1 = 15$. Thus, $E(T) = 1$ when $T < 15$ and $E(T) = 0.85$ when $T \geq 15$.

9.5.11.5.2 Parameters that Control the Growth

If the development of the architecture stops after the apical flowering of the stem, the growth continues, and the preformed organs in the lateral embryonic structures continue to expand up to 40 DCs. Based on the observation of leaf growth from the apex to the base of the stem, variation in expansion time is a major point for the fitting. To fit correctly the beginning of the organic series on the main stem in most cases, the expansion time of organs must vary. Here, we set that between 1 DC and 10 DC, the expansion time increase linearly from 10 DC to 19 DC and after it remains constant for leaves and internodes. The photosynthetic time of the leaves (t_a) is set to 26 DCs for leaves and remobilization times at 30 DCs and 35 DCs respectively for leaves and internodes.

9.5.11.5.3 Parameters Controlling the Delays of Organ Expansion

Parameters named g_o control the delays of organ expansion (Chap. 7, Kang et al. 2009). Let see again their effects on the growth of plants. For simplicity, we suppose parameters are the same for all the organs (leaves, internodes and fruits). $g_o = 0$ means the axis 2 immediately expand (no delay), $g_o < 1$ means expand with an acropetal development, $g_o > 1$ means expand with a basipetal development, and $g_o = 1$ means all the axis 2 expand at the same cycle.

Negative values of g_o (> -1) (mesopetally) will program distribution centered on N*abs(g_o). This particular case is programmed to give bell shapes to the compound series of organs centered from a phytomer of the main stem. $g_o = -0.3$ means all the axis 2 are at the basis of the stem, $g_o = -0.5$ means all the axis 2 are at the middle of the stem, $g_o = -0.8$ means all the axis 2 are at the top of the stem. The Canola is the pattern of $g_o = -0.8$ (Fig. 9.75). See Chap. 7 divergent branching.

9.5.11.5.4 Remobilization

The phenomenon of biomass remobilization exists for the study of Canola, a common phenomenon in grasses (see rice in this chapter). Once the expansion period is over, the organs partly empty into the common pool, which will allow the growth of the inflorescences, relaying photosynthesis, which decreases with the senescence of the leaves. Thus, each organ has a maximum weight q_{max}, a functioning time t_f greater than the expansion time t_x ($t_f > t_x$) during which it does not lose

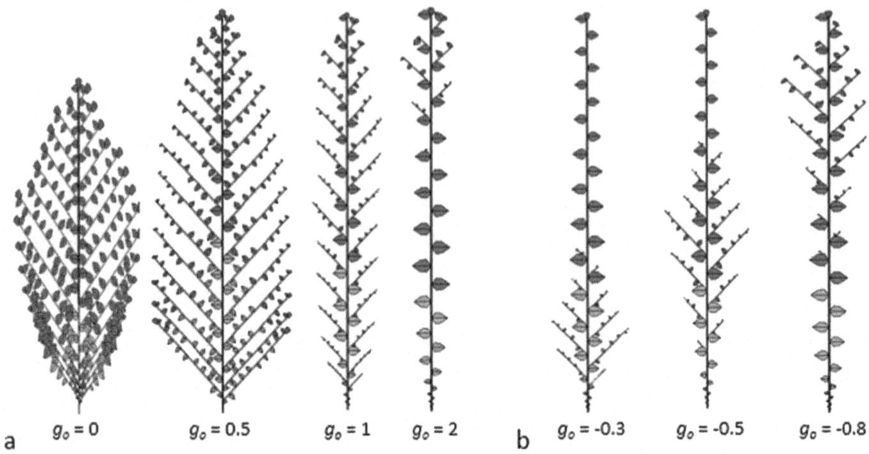

Fig. 9.75 Topological structure of Canola for the positive g and negative g (divergent series). (**a**) Topological structure when g is positive ($g_o = 0$, 0.5, 1 and 2). (**b**) Topological structure when g is negative ($g_o = -0.3$, -0.5 and $-$ 0.8) (Gloups software)

weight, and a remobilization coefficient k_m. Beyond t_f DCs, the organs lose weight. The fraction $F\%$ of biomass lost is empirical data measured after the end of remobilization. The damping parameter k_m expresses the speed with which the organ empties.

Remobilization is modeled by the function (25):

$$q(n) = q_{max}\left(1 - F\%\left(1 - (1 - k_m)^{n - t_f}\right)\right)$$

Only the k_m parameters of the organs under consideration are to be estimated by the inverse method. At the end of remobilization, the organ can be weighed and $F\%$ can be deduced. Figure 9.64 shows the form of variation in the function of remobilization.

For the Canola, it appears that the histograms of the weights of leaves decrease for a given rank of the organic series along the 31, 35 and 40 DCs. For the internodes, it increases from 31 to 35 and decreases from 35 to 40. This is controlled by two parameters. One is the percent ($F\%$) of remobilized biomass, one is the speed of remobilization (k_m) that makes exponential decreasing. Starting remobilization begins after time functioning of organs.

Two ways are possible, the remobilization starts sequentially after the age of each organ reaches its time functioning t_x; or, all organs make remobilization synchronously when the plant age exceeds the time of functioning. The Canola is the second way. No remobilization was observed for leaves. Here we set 80% of biomass is remobilized beginning at 36 DC at the damping coefficient of 0.0.25 for internodes.

9.5.11.6 Setting up Organic Series

Measurements are expressed in fresh weight. Only well-developed main stem components can be weighed individually. On ramifications that are expanding, some organs are still in the embryonic stage and it is not practically possible to establish organic series. In addition, the laws of delay in expansion, linked to the position in the structure, break down the notions of cohorts of organs. The organs of the same type of ramification are weighed separately by compartments (leaves, internodes, fruits) according to the rank of the ramification that carries them along the stem. This results in "compound" organic series (see Chap. 7, Fig. 7.17 for the way of sampling.)

Five growth stages were selected to perform the measurements (5,18,31, 35, 40 DCs). At each stage, four plants were collected to measure the fresh weight of organs in the simple and compound organic series whose averages are calculated and which were constructed to serve as a target for parameter estimation.

To study compartments at the same time, organic series needs that there is no missing data for the compartments. But, for the Canola, the compartments are not considered when fitting. The inflorescences at the tip of the axis gather the weights of internodes and pods to give the biomass of the raceme that is transformed into a compound fruit as for the tomato trusses.

9.5.11.7 Results on Plant Functioning

9.5.11.7.1 Empirically Measured Parameters

A single leaf organ is assimilated to the whole leaf blade and petiole. The specific leaf weight ε is deduced from the ratio of leaf fresh weight and its surface. It is estimated at $\varepsilon = 0.02$ g/cm^2 fresh weight.

Allometries and functioning times of the organs are estimated based on observations. The parameters g_o that control expansion delay laws have been experimentally set to best mimic botanical observations of growth stages. Thus, the parameter is set to $g_o = -0.87$ which ensures a divergent delayed branching for leaves (mesopetally), internodes of axes two. For the terminal flower, the parameter is still set to $g_o = 0$.

9.5.11.7.2 Computed Parameters

There are five measurement stages (5, 18, 31, 35 and 40 DCs). These stages correspond to harvested plants (four plants per stage) which are, therefore, independent samples. These measurements produce 451 weights of organs to be adjusted simultaneously during the growth process. There are 12 parameters to be estimated.

Table 9.23 Source–sink parameters of Canola

Param	Mean	SD	CV (%)
r	60.1	3.4	6
Sp	964	102	11
P_{b1}	1	0	0
P_{b2}	1.2	0.37	21
P_{i1}	1.4	0.09	7
P_{i2}	15.9	4.4	29
P_{f1}	193	55	29
P_{f2}	306	108	35
K_{ri}	0.25	0.02	7
B_{b1}	1.7	0.17	5
B_{i1}	5.4	0.16	6
B_{f1}	5.4	0.54	10

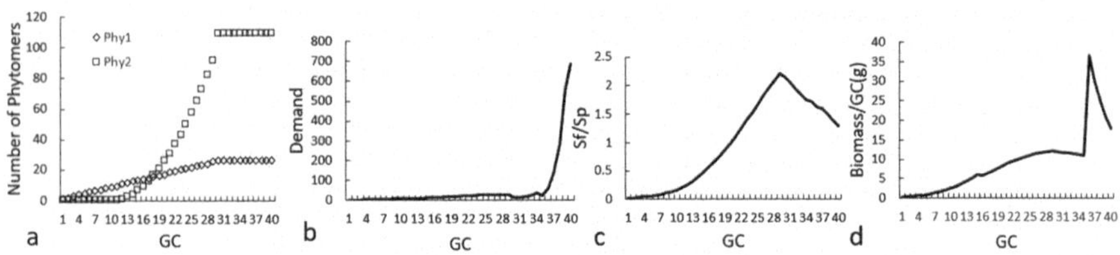

Fig. 9.76 History of the development and growth of Canola. (**a**) Number of phytomers per GC and PA. (**b**) Evolution of demand per GC. (**c**) Evolution of individual LAI. The LAI per plant reaches a high value (> 3) which saturates. (**d**) The biomass production per GC

The estimated parameter values (Table 9.23) allow a correct fitting of the simple and compound organic series of Canola (Fig. 9.78).

The fitting is good: ($R^2 = 0.97$; number of data = 451; number of computed parameters = 11 (parameter P_b is fixed to 1). The values of the 11 parameters are invariants that control well the fittings of the organic series during the growth period (40 GCs).

The available surface/plant Sd according to planting density is $Sd = 10{,}000/13 = 769$ cm$^{2.}$ The WLSM computes $Sp = 964$ cm^2 which is not so far from Sd.

Figure 9.76 shows the evolution of the number of phytomers per PA, plant demand, individual LAI (*Sf*/*Sp*), and biomass produced per cycle as a function of GC.

LAI per plant (*Sf*/*Sp*) (c) ceases after 35 + 5 GCs which corresponds to the functioning of the last leaf of the stem. However, the expansion of the pods continues as it takes advantage of the remobilization of the phytomers. Biomass produced by photosynthesis for a leaf ceases after 26 GCs. But the biomass provided by the remobilization of phytomers takes over. This causes a discontinuity in the biomass available per cycle (Fig. 9.76d).

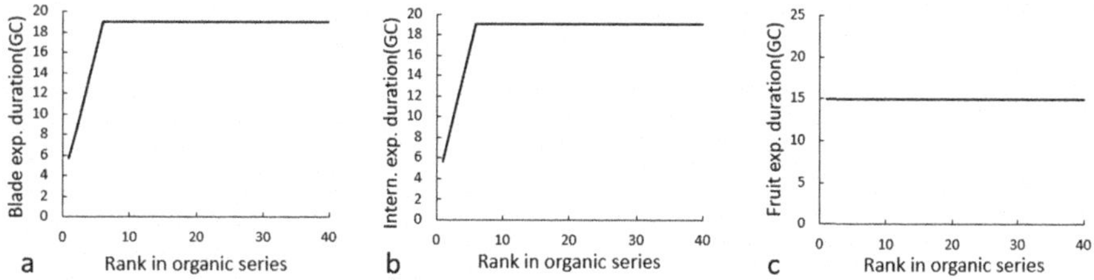

Fig. 9.77 Expansion time variation in Canola. (**a**) Leaf. (**b**) Internode. (**c**) Fruit. Expansion times are the same for the leaves and internode

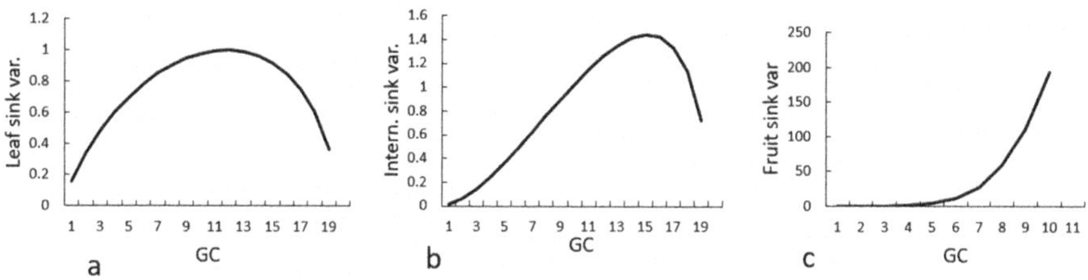

Fig. 9.78 Sink variation in each type of organ during organ expansion. (**a**) Sink variation in blade (PA 1). (**b**) Sink variation in internode (PA 1). (**c**) Sink variation in fruit (PA 1)

For this plant, the expansion duration of leaves and internodes increases until six DCs for the stem and ramifications, then remains constant until 40 DCs. For the inflorescences, it is set to $t_x = 15$ DCs for the stem and ramifications (Fig. 9.77).

Figure 9.78 shows the shape of sink variation functions for all organs. As there are two PAs, there are several functions of sink variation in an organ. Let us recall that in the triplet P_o, B_{o1}, B_{o2}, the parameter P_o depends on the PA, but the parameters B_{o1}, B_{o2} do not change. The functions corresponding to PA 1 are displayed here for leaves, internodes, and fruits.

On Fig. 9.79 a and b, it is shown that remobilization does not affect the leaves of the stem, but only its internodes whose weights decrease at the end.

Simple (stem) and compound (ramifications) organic series are well fitted by the model (Fig. 9.79). This figure shows the advantage of compound organic series composed of ramifications along the stem. The compound organic series cannot be identified at phytomer level because the delays in expansion suppress the cohorts and part of the phytomers remain embryonic. Each ramification is, therefore, only known by its number of phytomers, the total weight of leaves, internodes and fruits. It is satisfactory that the WLSM can find the sink values, while these organs are mixed in the ramification.

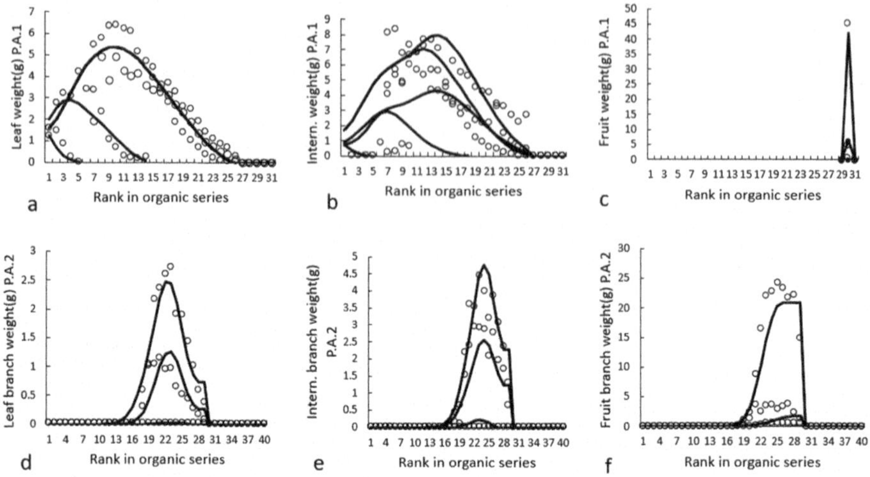

Fig. 9.79 Multi-stage growth adjustment of Canola (31, 35 and 40 DCs) for the stem and ramifications. (**a**, **b**, **c**) Simple organic series of leaves, internodes and fruits on the stem (PA 1). (**d**, **e**, **f**) Compound organic series of leaves, internodes, fruits on the stem belonging to ramifications (PA 2). "o" represents the observations, "—" represents the computations

9.5.11.8 Simulation of the Growth and Architecture of Canola

For needs of 3D computer graphics, it is possible to reconstitute a raceme from the growth of the compound fruit. It is enough to give it as many phytomers as expansion cycles, and at each cycle of development to give to the phytomer the sink computed by the beta law of the fruit.

Growth and development using the computed parameters and various allometries by the Gloups software allow reproducing the plant architecture in 2D and 3D (Fig. 9.80).

9.5.11.9 Discussion

The values $G = -0.87$ gives good result on compound series leaves, internodes. The values of k_m can change a lot together with r and Sp (they are linked), so it should be relevant to force $Sp = 1/d$ if d is known.

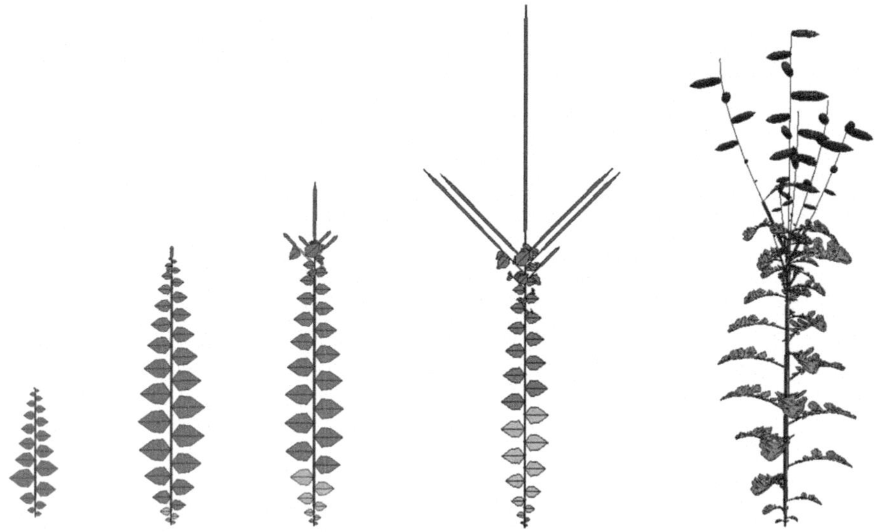

Fig. 9.80 2D organic series of the Canola simulated at 20, 30, 35, 40 DCs. The racemes are displayed not in detail but as a simple growing fruit (Gloups software). Last figure is a 3D display with a raceme reconstitution at 40th DCs

9.5.12 Case of Wheat Study of Feng Lu (CIRAD)

9.5.12.1 Plant Description

Wheat is a grass-like herb. Its development is characterized by an initial rosette-like stage, followed by an elongation of the stem which ends in an ear. A variable number of secondary stems, called tillers, are issued at the same time as the main stem at the rosette stage. As for Rice the leaf is supported by a sheath attached to the node of the phytomer, it acts as a petiole. This sheath envelops the stem and its weight is equivalent to that of the leaf it carries, so its sink must be considered. The vegetative phytomers are in a constant number of eight on the main stem. The first five phytomers have very short internodes and are responsible for the rosette. The last three vegetative phytomers have long internodes, the last of which bears an ear. The ear is at the end of a leafless internode that is a special phytomer. The weight of the internode and the ear are combined to form a fruit organ.

9.5.12.2 Cultivation Condition

In this experiment, the plant is cultivated in the field with a high density of 200 plants/m^2, leaving an average available surface area $Sd = 50$ cm^2 per plant. At this density, the potential structure is only partially explored. Tillers with a branching order of 3 do not appear. Only a proportion of second-order tillers reach maturity, the others either do not branch or abort before the ear creation. Measurements were made on 4 developmental stages (8, 10, 14, 18 DCs) and 8 plants per stage were sacrificed to establish the organic series of leaves, sheaths, internodes and spikes.

9.5.12.3 Scheme and Calibration of Development

The potential structure has only second order tillers. They appear from rank three to seven on the stem. As development is stochastic, each element of this structure has a probability of existence that depends on Bernoulli, branching and mortality processes. A detailed analysis shows that these values depend on the order of insertion of the tillers on the stem.

- The crown analysis gives:

Bernoulli process:
$b_1 = 1$ for the stem, $b_2 = 0.91$ for the second-order tillers.
Frequency "$a(k)$" of branching of the tillers on the stem as a function of the rank k from the base:
$a(k) = [0\ 0\ 0.87, 0.91, 0.50, 0.13, 0.2]$ (tillers locate at rank three to seven from the base of the stem).
Tiller abortion:
There is no mortality on the main stem (viability $c = 1$).
At rank k from the base of the stem the branched tiller is x DC aged. Recall that the viability (supposed constant) of a x DC aged meristem is c^x. To establish the average viability, it is sufficient to know the proportion $F(k)$ of aborted tillers at rank k on the stem, which is $1\text{-}c^x$. This gives the average value of viability c_k: $c_k = (1-F(k))^{1/x}$.
The average viability of second order tiller development (abort before ear creation) is calculated as a function of rank k from the base:
$c(k) = [0\ 0\ 0.95, 0.94, 0.86, 0.74, 0.52]$ (tiller at rank three to seven from the base of the stem).
These values are used to calculate the probability of the existence of a phytomer in the potential structure. In rank k from the base of the stem and rank x from the base of a tiller, the probability of the existence of the phytomer is $b_2\, c^x$. The scalar product between the organ sinks and their probability of existence (which is that of the carrier phytomer) gives the plant demand. If we calculate the average number of phytomers by the probabilities provided by the crown analysis during development and compare it with the averages of the observations, we can see a good agreement (Fig. 9.81a). A visualization of the potential structure with color intensities of the phytomers related to the probabilities of existence is given in Fig. 9.81b. The last tillers rarely occur.

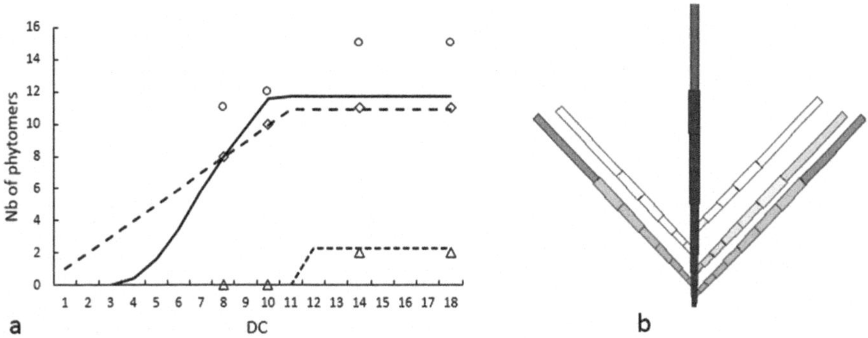

Fig. 9.81 Number of phytomers and potential structure of wheat. (**a**) Number of phytomers computed by the stochastic botanical automaton as a function of the number of DCs (Circles: main stem, squares: tillers, triangles: petiole of the ear). There is a slight underestimation for the number of phytomers in the tillers (12 against 15), but the small sample size allows for this. The potential total number of phytomers is 30. (**b**) Display of the potential structure with color intensities corresponding to the probabilities of existence of the phytomers

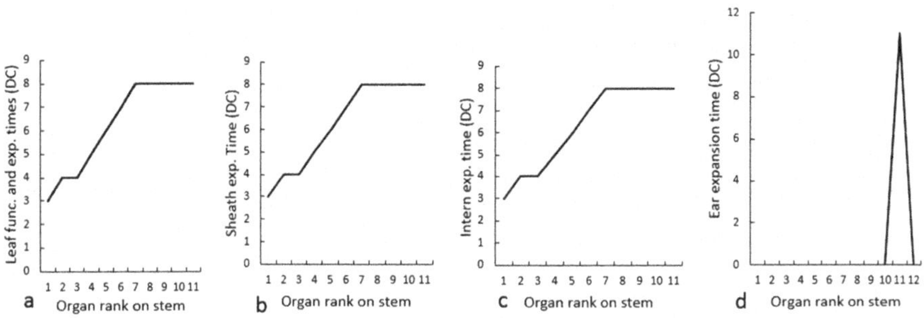

Fig. 9.82 Functioning time and expansion time of wheat. (**a**) Evolutions of functioning time of (solid line) and expansion time (dot line) of blades. (**b**) The expansion time of sheaths. (**c**) Expansion time of internodes. (**d**) the expansion time of ears

9.5.12.4 Scheme of Growth

The scheme of growth immediately follows the scheme of development. The functioning time of the leaves t_a is estimated in DC and corresponds to the interval in DC between the date of creation and the date of observation of yellowing. It varies according to the row of insertion on the stem. On short phytomers, the observation gives $t_a = 3$ DCs and on long phytomers $t_a = 8$ DCs.

The organs have an immediate expansion, whose duration t_x varies along with rank on the stem. Empirically it varies from $t_x = 3$ DCs for blades, sheaths, and internodes to eight DCs. On other hand, the internode expansion varies from 3 DCs for short internodes to 8 DCs for long ones (Fig. 9.82). For the ear expansion time, it is estimated 11 DCs. from the termination of the axes. The weight of the root system is not considered. It is assumed that there is proportionality between the weights of the aerial and underground parts. Beyond 18 DCs leaves and internodes will undergo biomass remobilization that is not studied here.

9.5.12.5 Setting up Organic Series

The main stems (PA1) are described in simple organic series (leaf, sheath, internode, ear). Order 2 tillers (PA2) are sub-structures organized in compound organic series composed of compartments of organs as for the *Spilanthes* and Canola, in order to simplify measurements. In addition, on these tillers the blades and sheaths are assembled into a single leaf organ, as they have a similar expanding behavior. At the end of the axis, the ear and the internode carrying it are a terminal structure of physiological age 3 (PA 3) resulting from a physiological age transition on the stem or tillers. This highlights the great flexibility in organizing the targets for model calibration into organ compartments, simple, or compound series in the GreenLab model. The weight of the root system is not measured; it is assumed either negligible or to be proportional to the weight of the plant.

9.5.12.6 Model Calibration

Calibration of the parameters on the organic series of the target is carried out by the WLSM over 4 growth stages (8, 10, 14, 18 DCs).

9.5.12.7 Parameters Measured Empirically

The specific leaf weight ε which corresponds to a thickness is deducted from the ratio of fresh weight/surface area. It is estimated at $\varepsilon = 0.008$ g/cm^2. The cultural coefficient is set at $k = 0.52$, because leaves are erected. The seed biomass that cannot be estimated by the WLSM in this experiment is set at $Q_O = 0.013$ g as in a previous experiment (Kang et al. 2008). The available surface area/ plant according planting density (200 plants/m^2) is $Sd = 50$ cm^2 to be compared with production surface computed by the WLSM.

9.5.12.8 Computed Parameters

There are four stages of measurement: 8, 10, 14, 18 DCs. There are 14 parameters from 139 Data, coming from measurements, to be estimated (11 sinks, 4 sink variation parameters, resistance r, surface Sp) (Table 9.24).

Figure 9.83 shows the evolution, plant demand, individual LAI (Sf/Sp), and biomass produced per cycle as a function of GC. Here the LAI per plant reaches a saturation (LAI > 3). The biomass production is saturated.

Table 9.24 Source–sink parameters of wheat

Param	r	Sp	Q_0	B_{b1}	B_{p1}	B_{i1}	B_{f1}	P_{b1}	P_{b2}	P_{p1}	P_{p2}	P_{i1}	P_{i2}	P_{i3}	P_{f3}
Mean	37.2	41.9	0.005	0.77	0.34	2.95	1.26	1	1.09	1.32	0.3	0.56	0.24	0.43	1.05
SD	6.51	13.5	0.00	0.20	0.23	0.43	0.32	0	0.13	0.43	0.0	0.15	0.07	0.15	0.33
CV (%)	18.0	32.0	69.00	25.0	66.0	15.0	25.0	0	12.0	33.0	0.0	27.0	29.0	34.0	31.0

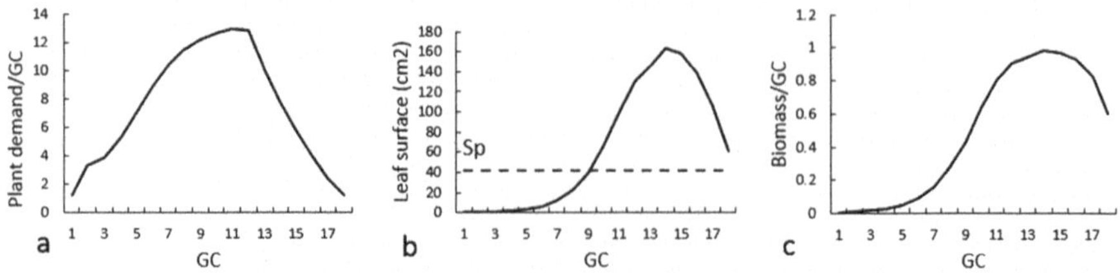

Fig. 9.83 History of the development and growth of wheat. (**a**) Evolution of demand per GC. (**b**) Evolution of leaf surface. (**c**) The produced biomass per GC. The value of *Sp* is the dashed line. The LAI per plant reaches a high value (> 3) which saturates

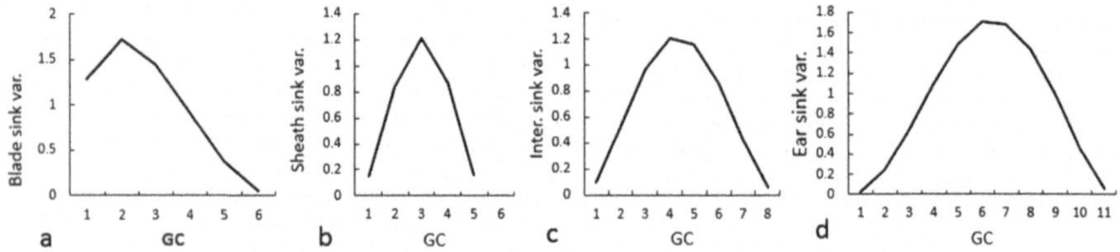

Fig. 9.84 Sink variation in each type of organs during the expansion. (**a**) Blade (PA 1). (**b**) Sheath (PA 1). (**c**) Internode (PA 1). (**d**) Spike (PA 2)

The second parameters of Beta laws are fixed to 3. Variations in the organ sinks occur over a short duration for phytomers ($t_x = 4$ GCs; Fig. 9.84).

The observed simple and compound organic series is correctly fitted by the model (Fig. 9.85).

The computed production surface area ($Sp = 41.9$ cm^2) is close to the value ($Sd =$ 50 cm^2) depending on the density (200 seeds/m^2), which confirms the suitability of the GreenLab model. Note that by choosing a value $k = 1$ for the crop coefficient, we find $r = 83$ and $Sp = 124$ cm^2 the other sink coefficients being unchanged. The values of *Sp* and *r* double and lose their meaning without changing the quality of the fitting. It is, therefore, important to consider the orientation of the leaves (cultural coefficient *k*) in relation to the verticality to estimate correctly the resource use efficiency $1/r$ and the production area of *Sp*.

On physiological age 2 (PA 2), the computed sinks of the sheaths and leaf blades are very fuzzy; in fact, the two organs have been added together to make the leaf meta-organ to simplify the measurements. What counts is the sum of the sinks PA 2 (blade +sheath = 1.86) which is to be compared with that of PA 1 (= 2.21) slightly higher as expected. The computed sink variation parameters are the first parameters of the beta law. Recall that the two parameters cannot always be estimated at the same time. The quality of the fit is not very sensitive to the sum of the two parameters and depends mainly on their ratio.

Figure 9.86 shows the good match between the plant biomass computed by the model and that observed on observations at the different stages of development

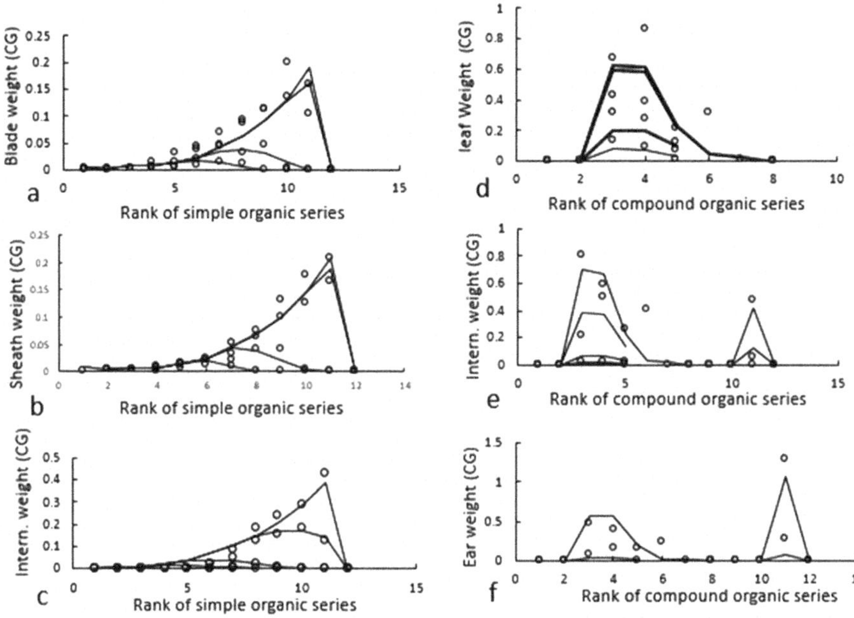

Fig. 9.85 Multi-stage fittings of the simple and compound organ series of wheat. For simple series, they concern the leaves (**a**), the sheaths (**b**) and the internode (**c**). For the compound series, they concern the leaves of the tillers (assembly blade+sheath) at the bottom of the stem (**d**), the internodes (**e**), and the ears (**f**). The ears with their internodes of the stem and the stem internodes of the tillers at the bottom and the end of the stem constitute the physiological age 3. The data from the observations (8, 10, 14, 18 DCs) fit correctly with the model (Gloups software)

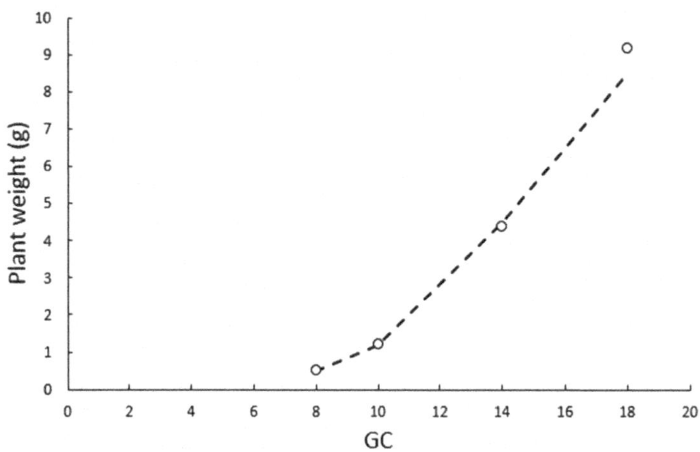

Fig. 9.86 Average observed plant weight (circles) and plant weight computed (dash line) by the model for the four development dates 8, 10, 14, 18 DCs. The functioning parameters of the meristems and organ sink sources give a good representation of plant growth. From the tenth DC onwards, growth is linear because the *Sp* value is constant and the foliar index per plant is high

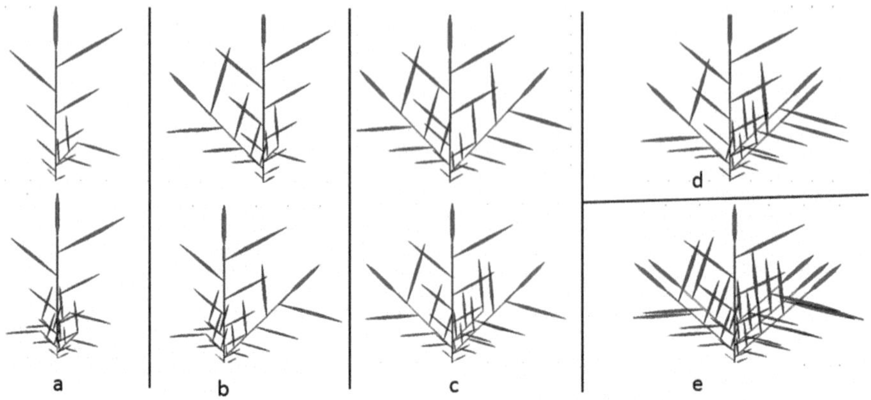

Fig. 9.87 Stochastic simulations of wheat tillering, obtained from the meristem functioning computed from the observations. The aborted tillers without ears are visible at the stem base. Two simulations per number of non-aborted tillers produced are displayed. (**a**) One ear. (**b**) One ear, one tiller. (**c**) One ear, two tillers. (**d**) One single simulation one ear, three tillers. (**e**) The potential structure with 5 tillers. All the simulations are contained in this structure (Gloups software)

(8, 10, 14, 18 DCs). This is a verification, because the weight of the plant was not a data to be fitted.

9.5.12.9 Stochastic Simulation of Wheat Tillering

Figure 9.87 shows the wheat's stochastic architecture with the tillering in all its variability, thanks to the accurate measurements made on the branching and abortions of the tillers which vary along the main stem.

In Chap. 11, we will see that the numerical values of the probabilities that control the functioning of the tillers also depend strongly on density.

9.5.12.10 Discussion

It is interesting to note that by the play of supply/demand alone, the dimension of the phytomers is well reproduced. Thus, the internodes at the base of the stem are short and those under the ear are long.

9.5.13 Case of Yam study of Adji Beda (Daloa University)

9.5.13.1 Plant Description

Yam is a herbaceous plant that exhibits climbing tendencies. It belongs to the *Dioscoreaceae* family and is characterized by its lianascent (monocotyledonous) nature. Yams typically grow as vines, and they are often dioecious, meaning there are separate male and female plants. These plants are known for their tuberous, starch-rich roots.

The leaves of yam plants are petiolate, cordate (heart-shaped), and generally simple. They are entirely smooth, oval in shape, and can be arranged alternately or oppositely depending on the specific yam species. The growth of yam vines is continuous and typically unbranched for a period of up to two months.

At the base of the yam stem, there are cataphylls, which are often spiny. The distance of these cataphylls from the collar of the stem can vary, ranging from 2 to 16 cm, depending on the yam species. In the case of this study (*Dioscorea rotundata*), the phyllotaxy is alternate, with spiral leaves at the base of the stem and opposite leaves at the top.

9.5.13.2 Cultivation Condition

The study was conducted by taking a yam tuber and cutting it into small, equal-sized square pieces. These pieces were then planted in a nursery using pots filled with local potting soil (ferrallitic soil) at the University of Daloa in Côte d'Ivoire in 2023. The climate in this region is humid tropical, with temperatures ranging from 21 to 34 °C, and an annual rainfall ranging from 1000 to 1900 mm. To ensure optimal growth conditions and minimize competition among the plants (to facilitate free growth), a plant density of 4 plants/m^2 was maintained (1 m spacing both within rows and between rows).

Maintenance of the yam plants consisted of daily watering and manual weeding. The first observation date was selected after a period of two months from germination of the yam plants, with measurements of the simple organic series of leaves, petioles and internodes, and the root system as a compartment. Destructive measurements were carried out on five plants to establish the organic series (leaves, petioles, internodes and root compartment). The fresh weight of each internode, petiole, leaf blade and root (Q_0 and Q_r), as well as the dimensions of these organs (diameter, length, width, leaf area and thickness) were recorded on the selected stems.

9.5.13.3 Development Scheme

Measurements were taken at the two months from five stems of variable development, with a maximum of 17 phytomers. Development was estimated to be controlled by a *Bernoulli* process (17 DCs, $b = 0.9$).

Yam sprouts from the reserves stored within a piece of the tuber. The basal part of the resulting stem bears a few embryonic leaves known as cataphylls, which are not photosynthetic. As the stem continues to grow, subsequent phytomers on the stem bear alternate or opposite leaves. At the young stage (two months), the stem has no branches or flowers and initiates new tubers. Consequently, the study, therefore, focuses on the growth of a single-stemmed yam plant.

9.5.13.4 Growth Scheme

The growth scheme immediately follows the development scheme. Specifically, the functioning time and the expansion time of the leaves t_a are estimated to 15 DCs, which correspond to the interval in DC between the date of leaf creation and the date when the leaves were observed to turn yellow.

During the initial stages of growth in yam plants, the first leaves are not photosynthetic. As a result, the plant relies primarily on the reserves stored within the tuber to support its early growth. However, as the growth continues, there is a transition. Photosynthesis gradually becomes the primary source of energy and nutrients for the plant as the functional leaves replace the cataphylls.

This is the first case in this chapter where the influence of reserves on germination is propagated over several growth cycles. Typically, in the previous instances, plants were grown from seeds (Q_0), and the reserves within the seeds were entirely transferred to the seedling during the first growth cycle. In contrast, in the current case, the reserves stored within the tuber play an ongoing role in supporting growth and development beyond the initial germination stage. This distinction highlights the significance of these reserves in sustaining the growth of the yam plant over an extended period.

As photosynthesis takes over, the root system and the newly formed tuber are considered to play the role of the root compartment. Initially, this compartment is assigned the value of Q_0 (assimilated to a seed), and is gradually emptied as these reserves are used for sustaining the growth and development of the plant. Subsequently, it is refilled by the production of a new tuber.

To provide a clear understanding of this process, the weight of the tuber piece is measured at the time of planting. This tuber piece forms a part of the reserves that support the initial growth of the seedling. At harvesting, the root system, including the newly formed tuber, is weighed.

9.5.13.5 Setting up Organic Series

The organic series considered for the stem, which is an average of five stems, includes measurements related to the internodes, petioles, leaf blades, and the weight of the root system. It is important to note that the organic series with leaves only begins at phytomer five, as the first four leaves are non-photosynthetic cataphylls with negligible weight. Consequently, at the initial stages of growth, only the reserves stored in the tuber can contribute to the growth. Moreover, the weight of the tuber piece is 34.3 g. This is the maximum value that can correspond to the seed Q_0 in the GreenLab model. However, the WLSM should compute a smaller value because only a part of the tuber can be used for the growth.

There is a total of 17 phytomers in the plant. The first four phytomers have no leaves. The leaves are further categorized into two parts: the leaf blade and the petiole. Consequently, we have three distinct organic series. The weight of the root system, along with the newly formed tuber Q_r is 39.8 g.

9.5.13.6 Result on Plant Functioning

9.5.13.6.1 Parameters Measured Empirically

The leaf thickness (SLW) is estimated at $\varepsilon = 0.06$ g/cm^2. Given the young age of the plant, since we are in the case of free growth, we cannot estimate the production accurately. Therefore, the production area is set at a high value to maintain free growth: $Sp = 100{,}000$ cm^2.

9.5.13.6.2 Computed Parameters

The initial tuber is assumed to deplete its biomass in a damped manner as the plant grows. If t is the number of DCs since the sprouting of the tuber, α as the damping-off coefficient and Q_0 as the initial biomass used by the tuber to support seedling growth, then the amount of biomass contributed to the seedling at DC t is:

$$\Delta q(t) = Q_o \cdot \alpha \cdot (1 - \alpha)^{(t-1)}$$

The Gloups software is used to calculate the initial biomass Q_0. The parameter α is determined empirically through simulation in order to best fit the organic series. In this case, a suitable value for α is found to be 0.3, which optimally fits the data and accounts for the damped manner in which the initial tuber biomass is used as the plant grows (Fig. 9.88a).

We have a dataset consisting of 44 data, and nine parameters need to be calculated. Fig. 9.88 displays the development and growth of Yam.

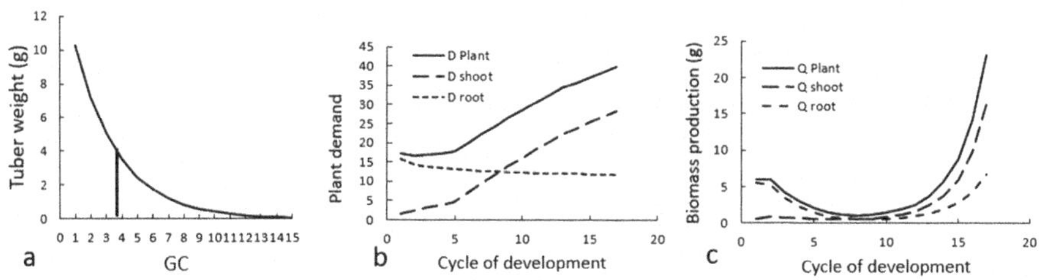

Fig. 9.88 History of the development and growth of Yam. (**a**) Weight loss of the tuber which supplies biomass to support the seedling. Photosynthesis becomes the primary energy source after the fourth DC. (**b**) Development. (**c**) Growth**.** Data for the whole plant, shoot and root system are displayed

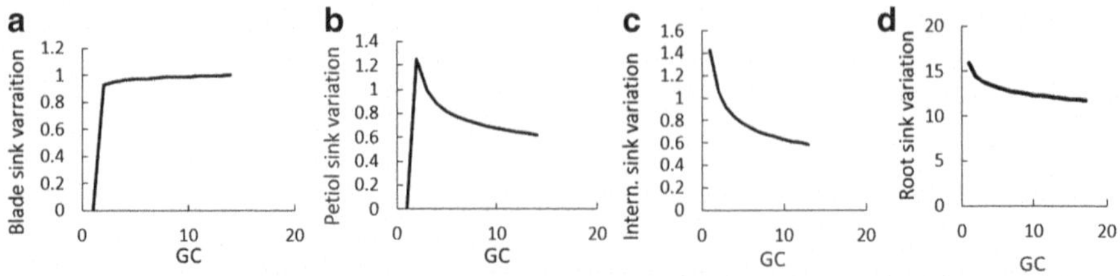

Fig. 9.89 Sink variation in each type of organs during the expansion. (**a**) Blade. (**b**) Petiole. (**c**) Internode. (**d**) Root system

Table 9.25 Source–Sink Parameters of Yam

Param	r	Q_0	B_{b1}	B_{p1}	B_{i1}	B_{r1}	P_r	P_{p1}	P_{i1}
Mean	6.4	19.9	1.02	0.78	0.72	0.91	31.9	1.2	2.8
SD	1.2	4.1	0.14	0.10	0.06	0.21	18.5	0.3	0.64
CV (%)	19	21	14	13	8	24	58	28	23

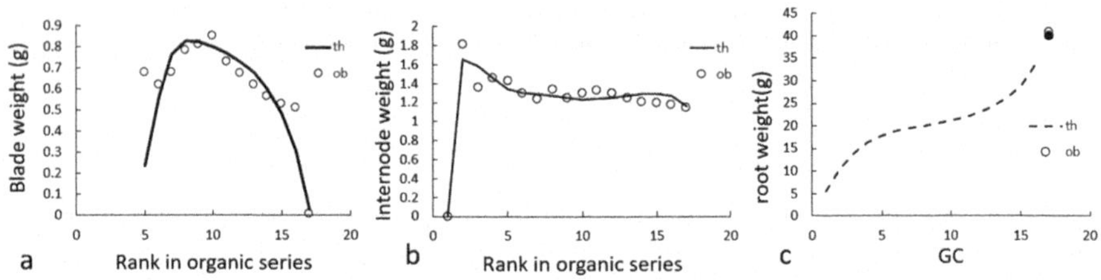

Fig. 9.90 Single-fitting of the organic series observed (O) and computed by the model (–) in Yam. (**a**) Blade. (**b**) Internode. (**c**) Root system. Petioles are not displayed and similar to blades

Figure 9.89 shows the sink variation laws. The second parameters of Beta laws of the leaf blade, petiole, internode, and root expansion functions are set to 1.

Table 9.25 shows the parameter values that have been computed with the WLSM based on the organic series.

The sink functions for yam are stable, and differ significantly from those of cotton plants, which show a bell-shaped curve. This indicates that the resource allocation patterns for these two plant types are distinct.

Furthermore, the weight Q_0 of the tuber piece that acts as the seed in the model is $Q_0 = 19.9$ g. Consequently, the tuber contributes 60% (19.9/34.3) of its biomass to germinate the seedling. The petiole biomass is not negligible, and its sink strength (1.2) is similar to that of the blade (1).

Furthermore, the model effectively fits the observed simple organic series and the weight of the root system, which with the new tuber. This is illustrated in Fig. 9.90, indicating the model's accuracy in simulating and describing the growth and development of the yam plant based on the available data.

Table 9.26 Comparison of parameter considering values with and without root system

	r	Q_0	B_{b1}	B_{p1}	B_{i1}	P_{p1}	P_{i1}
Without root	7.3	2.9	1.1	0.9	0.8	1.0	2.4
With root	6.5	20.0	1.0	0.8	0.7	1.3	2.8

9.5.13.6.3 Analysis Without Considering Root System

Similar to beetroot, plant analysis can be carried out without considering the root system. However, it is important to note that the adjustment of the organic series becomes less correct and, notably, the parameter for Q_0 are no longer consistent (Table 9.26).

The parameters of the organ sinks remain stable in both cases. However, the difference lies in the value of Q_0, which is ten times smaller when the root system is not considered. The weight of the initial tuber is, therefore, greatly underestimated.

9.5.13.6.4 Ghost Sink for Root System

When attempting to calculate root system parameters solely from shoot system data, it can be a viable approach as long as the weight Q_0 of the tuber is predetermined and set in advance. This predetermined weight of the tuber plays a pivotal role in such calculations, providing a necessary reference point for estimating root system parameters based on shoot system data.

9.5.13.7 Growth Simulation

Figure 9.91 shows the yam simulation in both 2D and 3D. Notably, it visually represents the absence of leaves on the first four phytomers situated at the base of the stem is visible. Additionally, the last leaf generated by the stem meristem is relatively small, and its expansion does not really begin until the second DC.

9.6 Conclusion

The 13 plants studied are correctly modelled by the GreenLab model which is well suited to the study of plant growth. The Gloups software is generic and fit correctly all the target files with regression coefficients characterizing the statistical adjustment to the model all more than 0.92.

Plant architecture is the result of the accumulation of thermal time and radiation along with the steps that are growth cycles, so the architectural methods gain certain independence from eco-physiology when the environmental conditions are not limiting. In this way, the driving factor can be normalized and temperature and light can be ignored. The resource use efficiency $1/r$ is set using the normalized

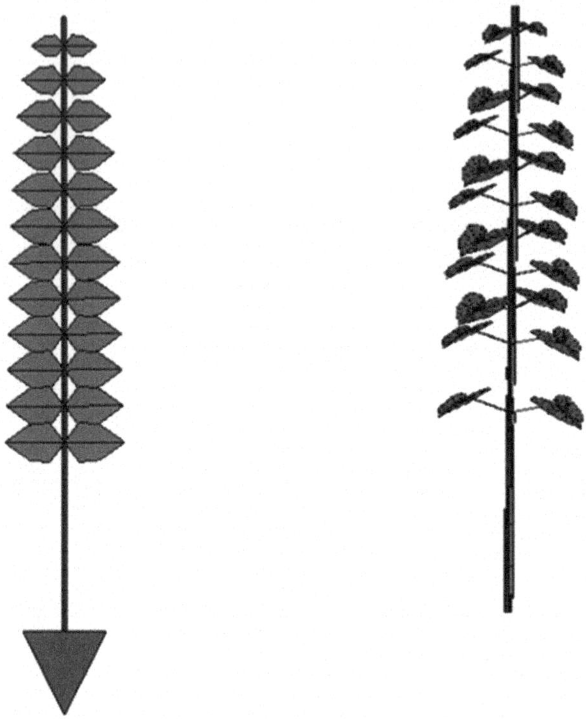

Fig. 9.91 Simulations of 2D and 3D architectures in Yam with 17 DCs

climate value $E = 1$. All other source–sink parameters (Sp, P_o, B_{o1}, B_{o2}) are not affected by the normalization. This makes it possible to study the growth of a harvested plant without knowing its climatic environment.

The organic series makes it possible to calculate the source–sink parameters. The use of organic series linked to physiological ages allows for adapted sampling, which considerably reduces the volume of measurements. Often it is possible to simplify the topological structure if the cohorts produced remain unchanged (rice, sweet pepper, etc.). The missing data are well managed by the model.

Another validation of the model is given by the good correspondence between the calculated production surface Sp and that Sd given by the planting density.

The production surface Sp, used in the Beer-Lambert law, associated with the resource use efficiency $1/r$, bypasses the heavy calculation of light interception by the plant, which requires the geometrical reconstruction of its architecture and the use of radiative transfer methods (radiosity, ray tracing).

The invariance of the parameters during the growth under given conditions makes the model particularly robust, and these parameters characterize the development and growth of the studied plant. They are, therefore, potentially useful for the selection or optimization of crop experiments (breeding, pruning, planting density, irrigation, etc.).

References

Arbenz K, Wohlhauser A (1986) Analyse numérique. PPUR, Presses polytechniques

Benlloch R, Berbel A, Serrano-Mislata A, Madueño F (2007) Floral initiation and inflorescence architecture: a comparative view. Ann Bot. 100(7):659–676

Brisson N, Mary B, Ripoche D, Jeuffroy M-H, Ruget F, Nicoullaud B, Richard G, Beaudoin N, Recous S, Tayot X, Plénet D, Cellier P, Machet J-M, Meynard J-M, Delécolle R (1998) STICS: a generic model for the simulation of crops and their water and nitrogen balances. I. Theory and parameterization applied to wheat and corn. Agronomie 18(5–6):311–346

Christophe A, Letort V, Hummel I, Cournède P-H, de Reffye P, Lecœur J (2008) A model-based analysis of the dynamics of carbon balance at the whole-plant level in Arabidopsis thaliana. Funct Plant Biol 35(11):1147–1162

Chew Y-H, Wenden B, Flis A, Mengin V, Taylor J, Davey C-L, Tindal C, Thomas H, Ougham H-J, de Reffye P, Stitt M, Williams M, Muetzelfeldt R, Halliday K-J, Millar A-J (2014) Multiscale digital Arabidopsis predicts individual organ and whole-organism growth. PNAS 111(39): E4127–E4136. https://doi.org/10.1073/pnas.1410238111

Cournède P-H, Letort V, Mathieu A, Kang M-Z, Lemaire S, Trevezas S, Houllier F, de Reffye P (2011) Some parameter estimation issues in functional-structural plant modelling. Math Modell Nat Phenom 6(02):133–159

de Reffye P, Lemaire S, Srivastava N, Maupas F, Cournède P-H (2009) Modeling inter-individual variability in sugar beet populations. In: 2009 third international symposium on plant growth modeling, simulation, visualization and applications. IEEE, Beijing, 9-13 November, 2009, pp 270–276

Feng L, Rey H, Mailhol J-C, Kang M-Z, de Reffye P (2013) Concept and calibration of virtual wheat including stochastic tillering. In: Sievänen R, Nikinmaa E, Godin C, Lintunen A, Nygren P (eds) Proceedings of the 7th international conference on functional-structural plant models (FSPM2013). Finnish Society of Forest Science, Vantaa, pp 279–281), Saariselkä, Finland, 9–14 June 2013

Feng L, Mailhol J-C, Rey H, Griffon S, Auclair D, de Reffye P (2014) Comparing an empirical crop model with a functional structural plant model to account for individual variability. Eur J Agron 53:16–27

Guo Y, Ma Y-T, Zhan Z-G, Li B-G, Dingkuhn M, Luquet D, de Reffye P (2006) Parameter optimization and field validation of the functional–structural model GreenLab for maize. Ann Bot 97(2):217–230

Heuvelink E (1999) Evaluation of a dynamic simulation model for tomato crop growth and development. Ann Bot 83(4):413–422

Jullien A, Mathieu A, Allirand J-M, Pinet A, de Reffye P, Cournède P-H, Ney B (2011) Characterization of the interactions between architecture and source-sink relationships in winter oilseed rape (Brassica napus) using the GreenLab model. Ann Bot 107(5):765–779

Kang M-Z, Evers J-B, Vos J, de Reffye P (2008) The derivation of sink functions of wheat organs using the GREENLAB model. Ann Bot 101(8):1099–1108

Kang M-Z, de Reffye P, Heuvelink E (2009) Modeling the growth of inflorescence. In: 2009 Third international symposium on plant growth modeling, simulation, visualization and applications, Beijing, China, pp 303–310. https://doi.org/10.1109/PMA.2009.52

Kang M-Z, Yang L-L, Zhang B-G, de Reffye P (2011) Correlation between dynamic tomato fruit-set and source-sink ratio: a common relationship for different plant densities and seasons? Ann Bot 107(5):805–815

Kang M-Z, Heuvelink E, Carvalho S-M, de Reffye P (2012) A virtual plant that responds to the environment like a real one: the case for chrysanthemum. New Phytol 195(2):384–395

Kang M-Z, Hua J, Wang X-J, de Reffye P, Jaeger M, Akaffou S (2018) Estimating sink parameters of stochastic functional-structural plant models using organic series-continuous and rhythmic development. Front Plant Sci 9:1688. https://doi.org/10.3389/fpls.2018.01688

Lemaire S, Maupas F, Cournède P-H, Allirand J-M, de Reffye P, Ney B (2009a) Analysis of the density effects on the source-sink dynamics in sugar-beet growth. In: 2009 third international symposium on plant growth modeling, simulation, visualization and applications. IEEE, Beijing, 9-13 November, 2009, pp 285–292

Lemaire S, Maupas F, Cournède P-H, de Reffye P (2009b) A morphogenetic crop model for sugar-beet (Beta vulgaris L.). In: Crop modeling and decision support. Springer, Berlin/Heidelberg, pp 116–129

Ma Y-T, Li B-G, Zhan Z-G, Guo Y, Luquet D, de Reffye P, Dingkuhn M (2007) Parameter stability of the functional–structural plant model GREENLAB as affected by variation within populations, among seasons and among growth stages. Ann Bot 99(1):61–73

Ma Y-T, Wen M-P, Guo Y, Li B-G, Cournède P-H, de Reffye P (2008) Parameter optimization and field validation of the functional–structural model GREENLAB for maize at different population densities. Ann Bot 101(8):1185–1194

Mailhol J-C, Ruelle P, Revol P, Delage L, Lescot J-M (1996) Operative modelling for actual evapotranspiration assessment: calibration methodology. In: Procedings of international conference on evapotranspiration and irrigation scheduling (ASAE 1996). China Computer Federation, San Antonio, pp 474–479

Mathieu A, Zhang B-G, Heuvelink E, Liu S-J, Cournède P-H, de Reffye P (2007) Calibration of fruit cyclic patterns in cucumber plants as a function of source-sink ratio with the GreenLab model. In: The 5th international workshop on functional structural plant models (FSPM 2007), Napier, pp 1–4, 4-5 November, 2007

Mathieu A, Letort V, Cournède P-H, Zhang B-G, Heuret P, de Reffye P (2012) Oscillations in functional structural plant growth models math. Model Nat Phenom 7(6):47–66

Okoma P, Akaffou S, de Reffye P, Hamon P, Hamon S, Konan O, Kouassi K-H, Legnate H, Letort V, Sabatier S (2016) Practical method for non-destructive measurement of leaf and stem dry biomass in African Coffea species. Agrofor Syst 92(3):667–675. https://doi.org/10.1007/s10457-016-0031-z

Rey H (2003) Utilisation de la modélisation 3D pour l'analyse et la simulation du développement et de la croissance végétative d'une plante de tournesol en conditions environnementales fluctuantes (température et rayonnement). Doctoral dissertation, Montpellier, ENSA

Rey H, Cournède P-H, de Reffye P, Lecoeur J (2006) Sunflower (Helianthus annuus sp.) growth analysis by source-sink relationships study with GreenLab model. In: The second international symposium on plant growth modeling, simulation, visualization and applications (PMA06). IEEE, Beijing., 13–17 November, 2006

Song Y-H, Guo Y, Li B-G, de Reffye P (2003) Virtual maize model I. biomass partitioning based on plant topological structure. Acta Ecologica Sinica 23(11):2333–2341

Tittonel E-D (1990) Evènements liés à l'évolution florale chez le colza Brassica napus L. var Oleifera Metzg. Thèse de doctorat thesis. Université Paris 6, Paris

Vavitsara M-E (2014) Modélisation de l'architecture et de la production végétale des herbacées. Application à l'amélioration de la production du Spilanthes acmelea L. Thèse es Science. Université de Mahajanga, Madagascar

Vavitsara M-E, Sabatier S-A, Kang M-Z, Ranarijaona H-L, de Reffye P (2017) Yield analysis as a function of stochastic plant architecture: case of Spilanthes acmella in the wet and dry season. Comput Electron Agric 138(1):105–116

Wang X-J, Mathieu A, Cournède P-H, Allirand J-M, Jullien A, de Reffye P, Zhang B-G (2011) Variability and regulation of the number of ovules, seeds and pods according to assimilate availability in winter oilseed rape (Brassica napus L.). Field Crop Res 122(1):60–69

Wang X-J et al (2014) Application of a probabilistic model for analysing the abortion of seeds and pods in winter oilseed rape (Brassica napus). Ann Appl Biol 165(3):414–428

Wang X-J, Li D, Lin B-G, Hua J, Kang M-Z, Zhang D-Q, de Reffye P, Wang F-Y (2018) Stochastic simulation of branch morphological structure in oilseed rape. Scientia Sinica Vitae 49(1):67–76

Yang L-L, Wang Y, Kang M-Z, Louran G, Dong Q-X, de Reffye P (2009) Relationship between tomato fruit set and trophic competition-a modelling approach. In: 2009 third international symposium on plant growth modeling, simulation, visualization and applications. IEEE, Beijing, 9-13 November 2009, pp 198–205

Zhan Z-G, de Reffye P, Houllier F, Hu B-G (2003) Fitting a functional-structural growth model with plant architectural data. In: Hu B-G, Jaeger M (eds) Proceedings of 1st international symposium on plant growth Modeling, simulation, visualization and applications (PMA03). Springer and Tsinghua University Press, Beijing, pp 108–117

Zhang B-G, Kang M-Z, Letort V, Wang X-J, de Reffye P (2009) Comparison between empirical or functional sinks of organs-application on tomato plant. In: 2009 third international symposium on plant growth modeling, simulation, visualization and applications. IEEE, Beijing, 9-13 November 2009, pp 191–197

Zheng B-Y, Shi L-J, Ma Y-T, Deng Q-Y, Li B-G, Guo Y (2008) Comparison of architecture among different cultivars of hybrid rice using spatial light model based on 3D digitizing. Funct Plant Biol 35(9/10):900–910

Chapter 10
Validation of the GreenLab Model on Woody Plants

Philippe De Reffye, Véronique Letort, and Mengzhen Kang

Abstract This chapter explains that internode growth includes primary and secondary phases, with the latter being determined by the presence of functional leaves. The model adopts Pressler's law and the common pool concept to compute the biomass allocation for ring formation. It employs the concept of physiological age to effectively manage the plants' sub-structures. Despite the architectural complexity and sampling challenges of trees, the 'crown analysis' provides functional parameters to reconstruct the potential structure, focusing on the organic series of the living axes. This chapter focuses on the method of parameter estimation in the GreenLab model. Validation of the GreenLab model is established on woody plants, including cotton, coffee, pine, guayule, maple, teak and Khaya.

Keywords Woody plants · Inverse method · Application on cultivated plants · Model validation

P. De Reffye
CIRAD, UMR AMAP, Montpellier, France

AMAP, Univ Montpellier, CIRAD, CNRS, INRAE, IRD, Montpellier, France

V. Letort
Mathématiques et Informatique pour la Complexité et les Systèmes, CentraleSupélec-Université Paris-Saclay, Gif-sur-Yvette, France
e-mail: veronique.lechevalier@centralesupelec.fr

M. Kang (✉)
State Key Laboratory for Multimodal Artificial Intelligence Systems, Institute of Automation, Chinese Academy of Sciences (CASIA), Beijing, China

School of Artificial Intelligence, University of Chinese Academy of Sciences, Beijing, China
e-mail: mengzhen.kang@ia.ac.cn

P. De Reffye et al. (eds.), *Plant Model Greenlab for Botany and Agronomy*,
https://doi.org/10.1007/978-981-97-7541-5_10

10.1 Introduction

The modelling of secondary growth according to the GreenLab model is described in Chap. 8. This mainly concerns woody plants, although some herbaceous plants also have girth growth, such as *Spilanthes* (Chap. 9). Secondary growth results in the girth growth of the axes in the form of concentric rings, which may or may not be visible depending on the type of woody plant.

The growth of an internode can be decomposed into two phases. The first phase, beginning at the internode emergence, is that of primary growth: it lasts for a limited time in the development cycle (DC) and ceases when the internode reaches its final length. The pith is then formed. The second phase is that of secondary growth and corresponds to the increase in diameter at each cycle, as long as the internode is fed by at least one functional leaf upstream.

The GreenLab model adopts Pressler's law from forestry and the common pool concept from agronomy. Chapter 8 shows that the internodes of the axes under the leaf crown see the common pool partially through the different paths that lead to it. Each node can be associated with a 'window' on the common pool, which can be computed thanks to the breakdown of the architecture into sub-structures (Chap. 4) depending on the notion of physiological age (PA).

At each DC, each leaf produces a ring element that can be seen as a 'pipe' element which travels the path from the leaf to the collar and is provided with a sink. The sum of the sinks of these elements constitutes the demand of the secondary growth, with which the biomass of the produced rings is computed. The assembling of the pipes in bundles distributes the biomass in the rings of the whole plant.

The GreenLab model does not simulate step by step the diffusion of biomass from leaves in the architecture, which would be too costly in computation. Insead, at each DC, algorithms directly ensure the allocation of the biomass to the internodes.

Besides the problem of modelling secondary growth, trees are also characterized by the complexity of their architecture, which raises the problem of sampling design. In most application cases, the stochastic aspect of architecture cannot be neglected. Moreover, the increasing mortality of meristems with their age causes missing data since branches have fallen. Thus, the notion of organ compartment is no longer available.

But the 'crown analysis' (Chap. 3) remains operational and provides the functioning parameters of the meristems, which allows us to reconstruct the potential structure without considering the missing data. Then the organic series of living axes, even with small sample size, is of great interest. Indeed, thanks to the notion of physiological age, the tree is seen as an assembly of multi-scale sub-structures. The notions of tree crowns and organic series are separated parts of an often extensive structure that does not need to be described in detail. They are sufficient to constitute the target to compute the source–sink parameters of the plant.

Nevertheless, there are additional difficulties for trees compared to herbaceous plants regarding sampling.

- The trees have a long life span, so it is difficult to obtain homogeneous plantations of different ages suitable for multi-fitting. Usually, only a given age can be found.

- Trees such as elms or maples have a rigid structure and a regular bearing, but some trees, like fruit trees, undergo a strong interaction between the bending of the branches and the branching process. While it is possible to simulate this interaction, calibrating such a model remains challenging.
- Finally, data on fruiting are only available for the current year for all measured trees (if there was no specific follow-up for these trees).

10.2 Studies of Plant Growth and Plant Architecture of Woody Plants

The duration of the DC (phyllochron) in calendar time is generally longer in woody plants than in herbaceous plants. For example, the duration of a DC in *robusta* coffee is around two or three weeks, whereas it is less than one day for sunflowers.

The main issue is the handling of tree structures that may contain thousands of phytomers. Exploring the topological structure or building the 3D architecture can quickly become time-consuming and memory-intensive for a standard PC. But the use of Matrix mode and Phyto-Facto mode (Chap. 8) is still working for the source–sink method via the organic series and bypass managing the potential structure directly.

10.2.1 Case Study 1: Cotton Plant (Li Dong, Zhan Zhi-Gang, China Agricultural University (CAU))

Cotton is one of the plants that has been most studied by the AMAP and GreenLab teams in France and China. Experiments of pruning and defoliation allowed to specify the parameters of development and growth. The first studies focused on stochastic development (see Chap. 3) (de Reffye et al. 1988).

Growth was first studied on single-stemmed individual plants to understand the functioning of secondary growth (de Reffye et al. 1999; Guo et al. 2003).

The study of the growth for the branched plant was performed with a deterministic mode (Li et al. 2009), as algorithms using the potential structure were not yet available. It is possible to ignore the stochasticity in the absence of meristem mortality.

To establish the organic series, different instruments are required (Fig. 10.1).

10.2.1.1 Plant Description

The cotton plant is a shrub with different types of axes, each associated with different PAs. The stem bears vegetative branches that can be considered as reiterations. Stem and vegetative branches do not bear fruits; instead, they have monopodial branching and bear fruiting branches. Fruiting branches have a sympodial structure, consisting of modules that include a phytomer with an internode, a leaf, a fruit and a terminal axillary bud. These modules extend the branch by apposition (Petit Model from

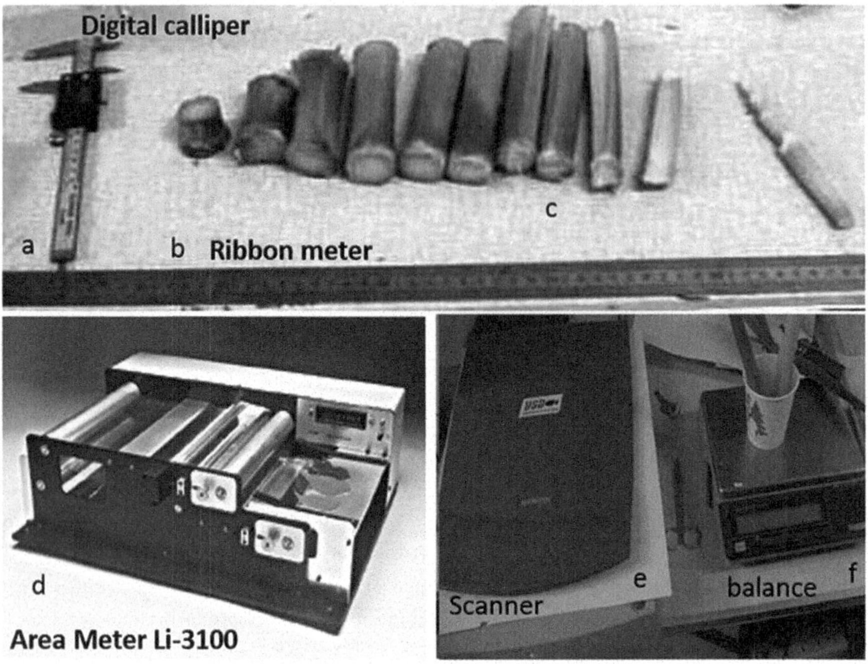

Fig. 10.1 Instruments for measuring organic series. (**a**) Digital calliper for internode length and diameter. (**b**) Ribbon meter for leaf length and diameter. (**c**) Organic series of internodes. (**d** and **e**) Scanner for leaf surface area. (**f**) Digital balance for organ weights

Hallé 1978). To simplify, this modular branch structure is often ignored and replaced by a continuously growing axis, which includes the same number of phytomers as modules. But the emission of a module from an axillary bud takes longer than the emission of a phytomer from a terminal meristem. So, the transformation of modular growth into continuous growth induces a rhythm ratio $w < 1$.

From a botanical point of view, the cotton plant, therefore, has two PAs: the stem and its reiterations (vegetative branches) are of PA 1 and the fruiting branches are of PA 2. The vegetative branches are not numerous (from 0 to 4 branches) and are located at the base of the stem. They appear a little less vigorous than the stems. Their branching is immediate or delayed.

In the model, for more convenience to detail the architecture and functioning of the cotton plant, three PAs are created:

- The stem: PA 1. A phytomer on the stem bears potentially two types of branches in the leaf axils (one vegetative and one fruiting).
- The vegetative branches: PA 2. A phytomer on a vegetative branch bears a fruiting branch in the leaf axils).
- The fruiting branches: PA 3. A phytomer on a fruiting branch bears a fruit (called a capsule) in the leaf axils.

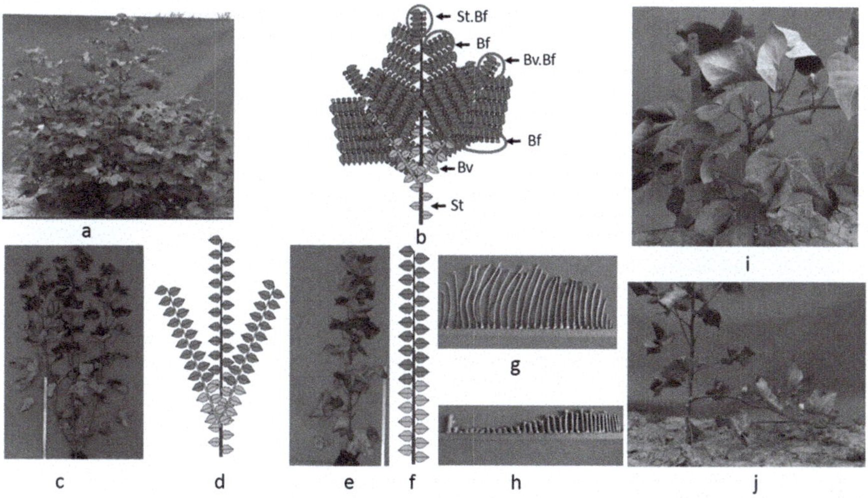

Fig. 10.2 Photos of the plant and associated topological structures. (**a** and **b**) Complete cotton plant with its stem (St, blue), vegetative branches (Bv, green), fruiting branches (Bf, red). Note the terminal transformation of the stem and the vegetative branches into fruiting branches (St.Bf and Bv.Bf). (**c** and **d**) Vegetative cotton plant (fruiting branches have been removed). (**e** and **f**) Single-stemmed cotton plant (vegetative and fruiting branches have been removed). (**g** and **h**) Organic series of petiole and internodes of single-stemmed cotton (**e**). (**i**) Top of a stem transformed into a fruiting branch. (**j**) Isolated fruiting branch showing capsules and its modular structure (Photos from Li Dong, CAU and simulations from Gloups software) (Color figure online)

The stems and vegetative branches of the plant have continuous development. They do not bear branches normally at the beginning. After a certain time, the terminal meristem may undergo a transition in PA, and the development becomes modular. The stem or vegetative branches turn into fruiting branches.

Different pruning protocols have been carried out to study plant behaviour.

- No pruning (free growth): the normal growth of the cotton plant with its vegetative and fruiting branches is kept (Fig. 10.2a, b).
- The vegetative cotton plant: axillary meristems giving fruiting branches are removed (Fig. 10.2c, d).
- The single-stem cotton plant: all axillary meristems are removed (Fig. 10.2e, f).

Other combinations of prunings have been carried out (Fig. 10.2j). In the simulations, a mask file is adapted to each case to mimic the associated pruning pattern.

10.2.1.2 Growth of Cotton Plant: Single-Stemmed Case

It is possible to obtain a single-stemmed plant, by removing all the axillary buds from the leaves of the stem at their birth. The resulting cotton plant has lost its stochastic aspects, and its development becomes quasi-deterministic ($b \approx 1$). The pruned plant offers an aspect of gigantism compared to the control branched plant (see Chap. 5). Thus, the manifestation of secondary growth can be studied under the optimal conditions on this single stem.

10.2.1.2.1 Plant Description

The plant is made of a single stem whose phytomers are constituted of an internode and a leaf. The leaf consists of a petiole and a blade.

10.2.1.2.2 Cultivation Condition

A field experiment on cotton was conducted at CAU in 2003. The planting density was about 3 plants/m^2 ($Sd = 3333$ cm^2), to minimize the competition. Irrigation and nutrients were applied at non-limiting rates. Five observation dates were selected between the germination and the maturation with a single plant per sample to form the simple organic series and organ compartments. No data on the root system was available.

Destructive measurements of plants were performed every week. The fresh weight of each internode, petiole and blade, the diameter and length of each internode and petiole, and the leaf area of each blade were measured to build the organic series. Meteorological data were collected at a nearby weather station, and plant transpiration was computed from potential evapotranspiration and plant leaf area index (LAI) (Fig. 10.3a, Jones and Kiniry 1986).

10.2.1.2.3 Development Scheme

The development of a single stem cotton plant is deterministic: the number of phytomers is in a good linear relationship with thermal time (Fig. 10.3b). The

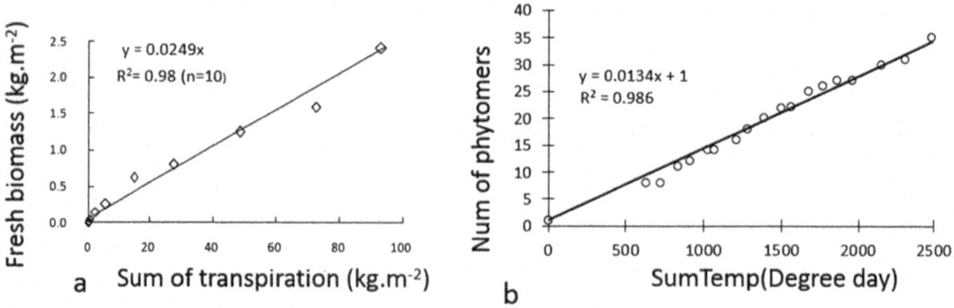

Fig. 10.3 Cotton development. (**a**) Relationship between plant fresh biomass production and the sum of plant transpiration. (**b**) Relationship between plant organogenesis and the sum of temperature

observations stop at 27 phytomers. The relationship between the development cycle (DC) and thermal time makes it possible to set the five dates in chronological age (CA) expressed in GCs: 8, 12, 18, 22 and 27.

10.2.1.2.4 Growth Scheme

The functioning duration of the leaves (t_a) is estimated to be 15 DCs. There is no visible 'base effect' on the organic series of leaves. The constant duration of organ expansion (t_x) is set to 15 DCs for blades, petioles and internodes.

10.2.1.2.5 Results on Plant Functioning

The target contains the biomass of the compartments of organs for the five observation dates, and the organ weights along with the ranks of the 3*5 = 15 organic series (three types of organs multiplied by five dates).

10.2.1.2.5.1 Empirically Measured Parameters

The specific leaf weight ε is computed from the ratio of leaf fresh weight to its leaf surface. It is estimated to be 0.03 g/cm^2 (Fig. 10.4a).

For a generic geometrical model of an internode, we need to consider two cases: primary growth (the pith) and, in the case of woody plants, secondary growth (the rings). Unfortunately, the total weights of the rings of the internodes on the stem cannot be experimentally separated, because the internal boundary of the pith is not clear. The effect of secondary growth on the internode diameter is negligible for the last created phytomers. The pith geometry is necessary to construct the length of the stem.

The pith's geometric shape is regarded as cylindric. For secondary growth, a new ring is added around the pith, thereby increasing the diameter of the internode. We regard the internode as a perfect cylinder with a length h and a cross-section area s (see Chap. 5). Its volume v ($v = h{\cdot}s$) depends on the quantity of biomass q and the density ds: $v = q/ds$; the shape coefficient β characterizes the elongation of the cylinder, and the coefficient γ is the allometry between s and h according to the relationship: $h/s = \beta{\cdot}v^{\gamma}$. The dimensions of the cylinder are then linked to the volume using allometry (see Chap. 5, Eq. 5.2):

$$
\begin{aligned}
h &= \sqrt{\beta} \cdot (v)^{\frac{1+\gamma}{2}} \\
s &= \sqrt{\frac{1}{\beta}} \cdot (v)^{\frac{1-\gamma}{2}}
\end{aligned}
\qquad (10.1)
$$

But the coefficients β and γ cannot be estimated directly.

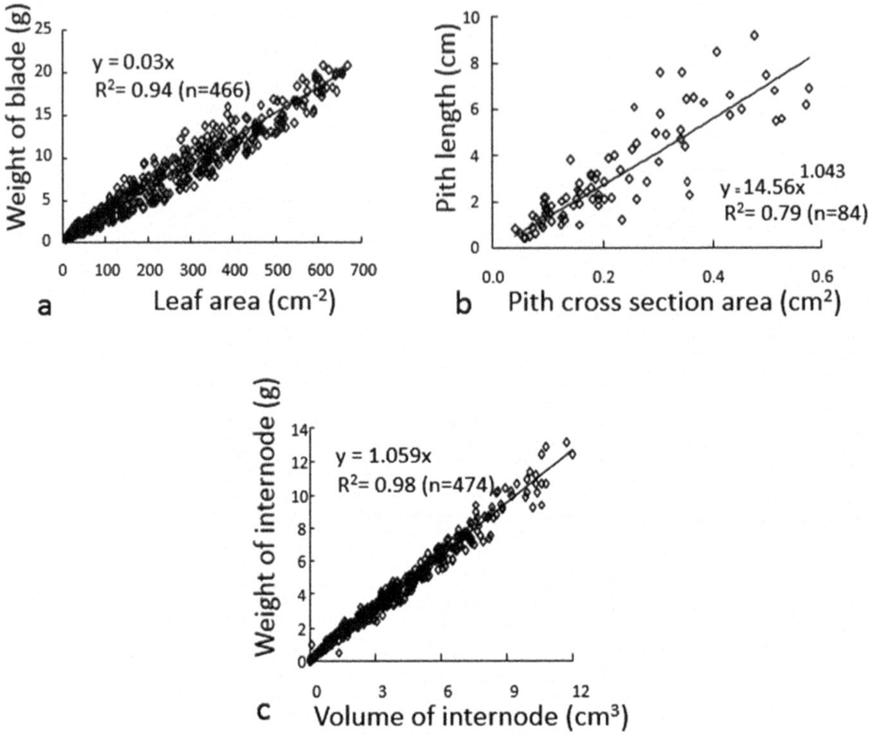

Fig. 10.4 Relationships of cotton growth. (**a**) Fresh weight and area for leaf blades. (**b**) Length and cross-section area for pith. (**c**) Fresh weight and volume for internodes

From the experimental data, we directly established an allometric relationship of Eq. 10.2 between the length h and the area of cross-section s. This relationship must be established only on the youngest internodes of the stem, where the effect of secondary growth on the diameter is still negligible. If d is the measured cylinder diameter, we have $s = \pi.d^2/4$.

The shape of allometry is determined using the last five internodes on the stem with Eq. 10.2 (Fig. 10.4b).

$$h = K \cdot (s)^{\alpha} \tag{10.2}$$

The least-square method gives the values of K and α. By comparing Eqs. 10.1 and 10.2, we can deduce the values of β and γ.

$$\begin{cases} \gamma = \dfrac{\alpha - 1}{\alpha + 1} \\ \beta = K^{1-\gamma} \end{cases} \tag{10.3}$$

Figure 10.4c shows that the density of the internode is close to 1. We got the same kind of results on the geometry of petioles. The interest of allometry is not only to obtain the organ geometry from its biomass. It is also a way to estimate the organ biomass directly from the measurements on the plant, without destructive samplings to get the organ biomass. With the measured data for the last five internodes, where the contribution of the ring to the total diameter is still small, the allometric relationship (Eq. 10.2) for the pith between its length h and cross-section area s was established as $h = 14.6 \cdot s^{1.043}$ (Fig. 10.4b), which gives $\beta = 13.7$ and $\gamma = -0.021$.

10.2.1.2.5.2 Computed Parameters

There are three kinds of organic series (blades, petioles and internodes). The five stages of measurements correspond to harvested plants (1 plant per stage) which are, therefore, independent samples. These measurements produce 272 weights of organs in the organic series and compartments to be fitted simultaneously during the growth process.

The fitting is good: $R^2 = 0.98$; number of computed parameters = 12 (the parameter P_b is fixed to 1). The values of the 12 parameters (Table 10.1) are invariants that control well the fittings of the organic series during the growth period (27 GCs).

Here the question of secondary growth arises. It concerns two parameters, P_c and λ (Chap. 8). For $\lambda = 1$ (the error of the WLSM = 23.7), we obtain the Pressler's mode, which gives the parameters of Table 10.1. The value $\lambda = 0$ does not give a correct fitting and is, therefore, excluded.

In this experiment, all the parameters of the sink functions of organs, the triplets (P_o, B_{o1}, B_{o2}), are computed. Four parameters are imprecise, having a large standard deviation: the seed volume (Q_0), the sink variation parameters of the internode (P_i, B_{i1}, B_{i2}). The sink parameters of the internodes are fuzzy because two processes (the expansions of pith and rings) are inextricably mixed in the biomass observations.

The computed surface of production $Sp = 2840$ cm^2 is close to the available surface $Sd = 3333$ cm^2 given by the planting density. This is again a validation of the GreenLab model for crops.

Figure 10.5 shows the evolution of the number of phytomers, plant demand, individual LAI (Sf/Sp) and produced biomass per cycle as a function of GC.

As mentioned above, the expansion duration of organs is constant, as shown in Fig. 10.6.

Table 10.1 Source–sink parameters of the single-stem cotton plant

Param	r	Sp	Q_0	P_b	P_p	P_i	P_c	B_{b1}	B_{p1}	B_{i1}	B_{b2}	B_{p2}	B_{i2}
Mean	54.4	2840	8.4	1	0.13	0.36	0.2	2.5	1.3	1.1	3.9	2.6	1.7
SD	2.7	285	2.1	0	0.01	0.03	0.02	0.16	0.16	0.32	0.33	0.37	2.83
CV (%)	5	10	26	0	6	16	10	6	12	36	8	14	49

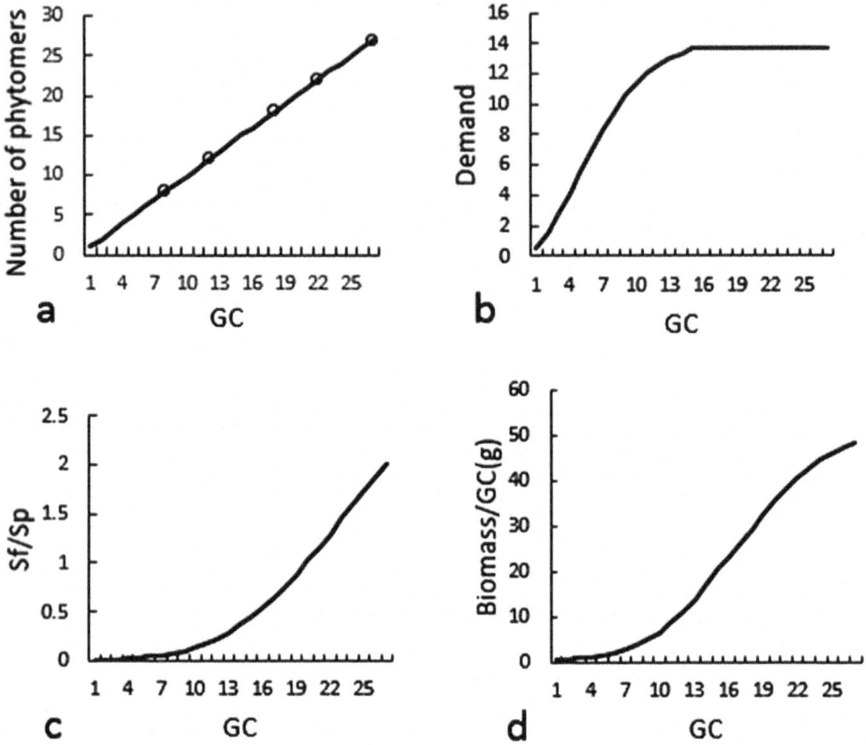

Fig. 10.5 History of the development and growth of single–stemmed cotton plant. (**a**) Number of phytomers per GC. (**b**) Evolution of demand per GC. The death of leaves saturates the demand. (**c**) Evolution of individual LAI. Production is not yet saturated ($Sf/Sp < 3$). (**d**) Evolution of biomass production per GC. After exponential growth, the plant production reaches saturation. The effect of leaf senescence is visible

The triplets (P_o, B_{o1}, B_{o2}) are used to compute the sink variation in the organs with the cycles (Fig. 10.7).

The multi-fitting of organ compartments along the CA is good (Fig. 10.8).

The good quality of the multi-fitting of organic series can be assessed in Fig. 10.9.

10.2.1.2.6 Simulation of Single-Stem Cotton

In Fig. 10.10, the development and growth of single stem cotton plants are represented with the longitudinal and radial rings at the bottom of the stem, its two-dimensional (2D) organic series at 27 GCs and its 3D architecture at 8, 18 and 27 GCs.

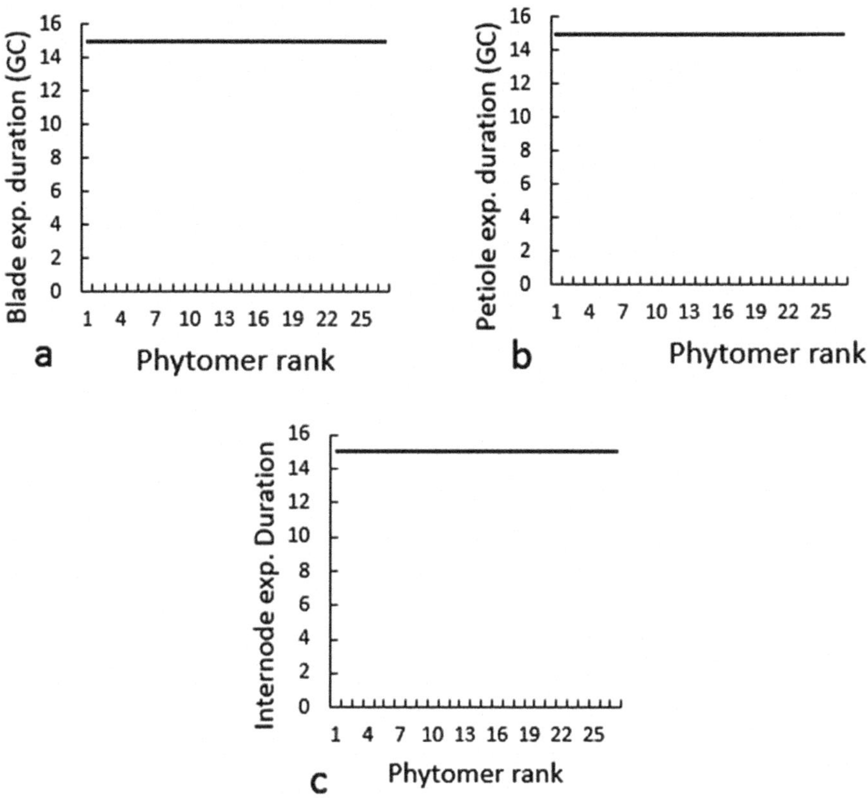

Fig. 10.6 Evolution of organ expansion from the basis of organ series of single stem cotton plant. (**a**) Blade. (**b**) Petiole. (**c**) Internode. Expansion durations of organs are constant on this cotton plant

10.2.1.2.7 Discussion

The WLSM proposes a partitioning of the wood biomass between the pith and the rings in the constitution of the internode cross-section, through the triplet sink parameters of the pith and the sink parameter of the ring element (P_c). The use of Pressler's law (see Chap. 8) gives a good solution for this partitioning.

To obtain the geometrical profile of the stem with its stacks of rings, it is necessary to know the allometric relationships between the volume of the pith, its length and the surface area of its section, as depicted in Fig. 10.4b. We obtain the stem length (180 cm) and the diameter at the base (2 cm). All the internodes along the stem have different dimensions. The volume of the internode is the sum of the pith volume and the ring's volume (Eq. 8.12, Chap. 8).

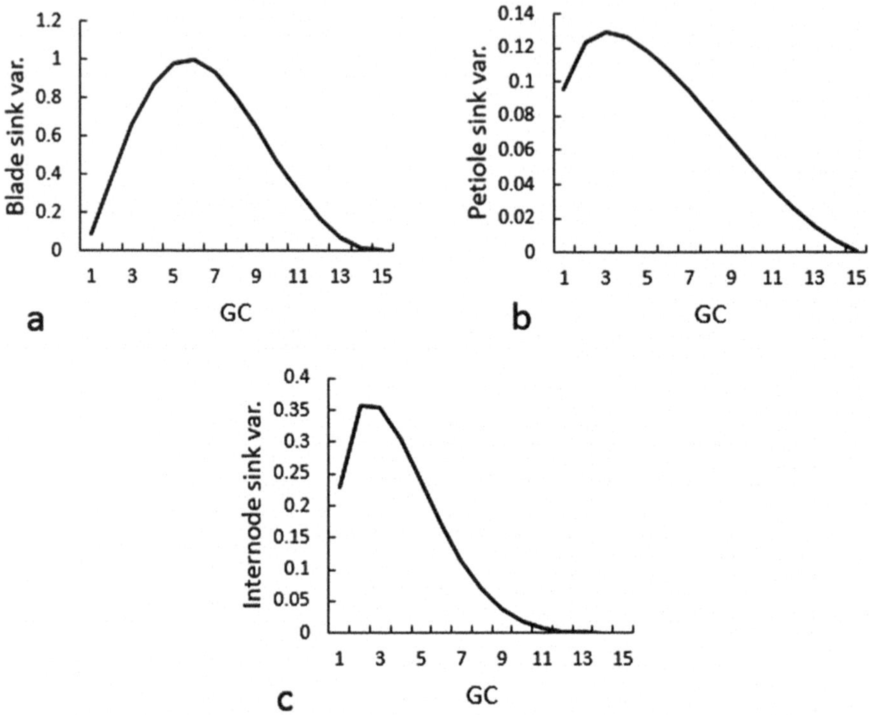

Fig. 10.7 Sink variation in each type of organ during the expansion. (**a**) Blade. (**b**) Petiole. (**c**) Internode

Note that the sink of the internode is almost zero after 10 GCs, so the visible expansion stops. But to obtain the best shape of sink variation, we set $T_{xi} = 15$ GCs. This shows that the apparent expansion time can be different from the optimized expansion time

10.2.1.2.8 The Search for the Root Sink Without Observations of the Root System

In the demand of the natural plant, all the sinks of the organs are involved and the root has also taken their part. Therefore, the compartment and the organic series implicitly contain the action of the root demand. So, it is possible to compute the root sink function as a kind of 'ghost' sink, without using any root system data. In the case of beetroot, the procedure works well (Chap. 9). It is shown that the simulation of cotton stem growth does not require knowledge of the competition of the root system (Fig. 10.9). The source–sink functioning of the aboveground organs is sufficient to correctly reproduce the architecture of the cotton plant stem at all stages of growth.

The question is whether there is still enough information in the target file to assess a root system. Depending on the conditions, two modes of shoot–root biomass partitioning can be proposed.

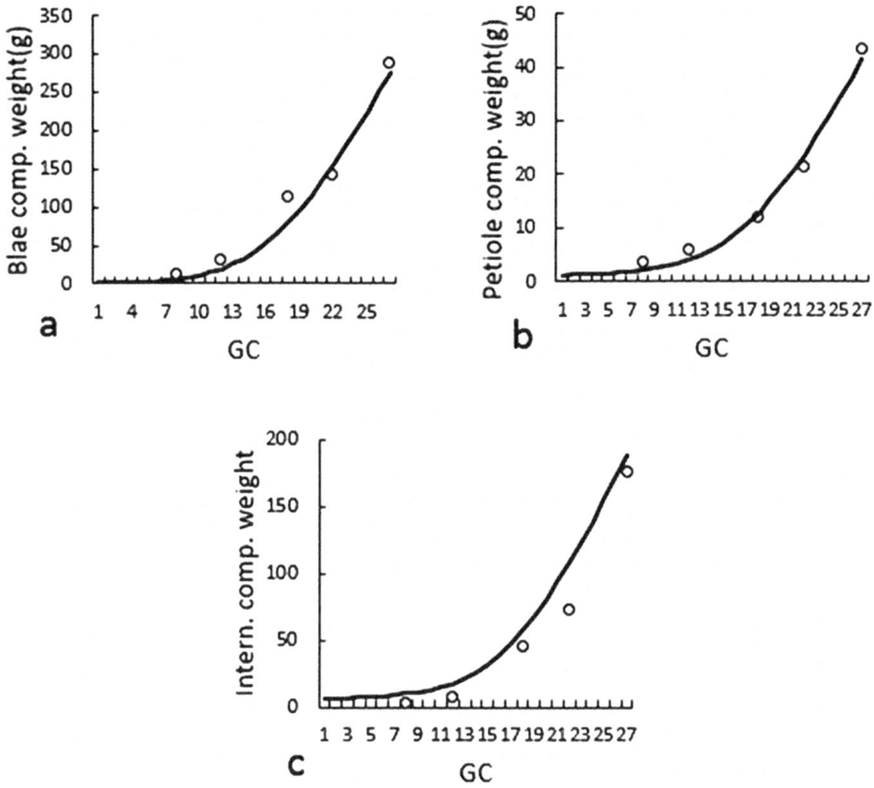

Fig. 10.8 Multi-fitting of the observed organ compartments (○) and computed organ compartments by the model (–) for the five measurement dates (8, 12, 18, 22, 27 GCs) of cotton. (**a**) Blades. (**b**) Petioles. (**c**) Internodes

1. If the aboveground part using WLSM cannot fit the model without the underground part, then the root system is an organ like any other that uptakes biomass from a common pool (Fig. 10.11a). This is the beetroot case.
2. If the aboveground part can fit the model without the underground part, then the plant distributes the biomass from the common pool into two parts, with the ratio p for the aboveground system and $1- p$ for the underground system. This ratio may vary with time (Fig. 10.11b). This is the cotton case.

The two systems are not equivalent. The second mode allows separating the aboveground and underground systems, which is not possible with the first mode. As the two root–shoot systems are independent, it should be considered that the organ sink parameters of the aboveground part (Q_o, Sp and the triplets P_o, B_{o1}, B_{o2}) are already known from the organic series analysis for the stand-alone aboveground part.

On the other hand, it is necessary to recompute parameter $1/r$ (resource use efficiency) as well as the triplet of parameters (P_r, B_{r1}, B_{r2}) of the root system. It is

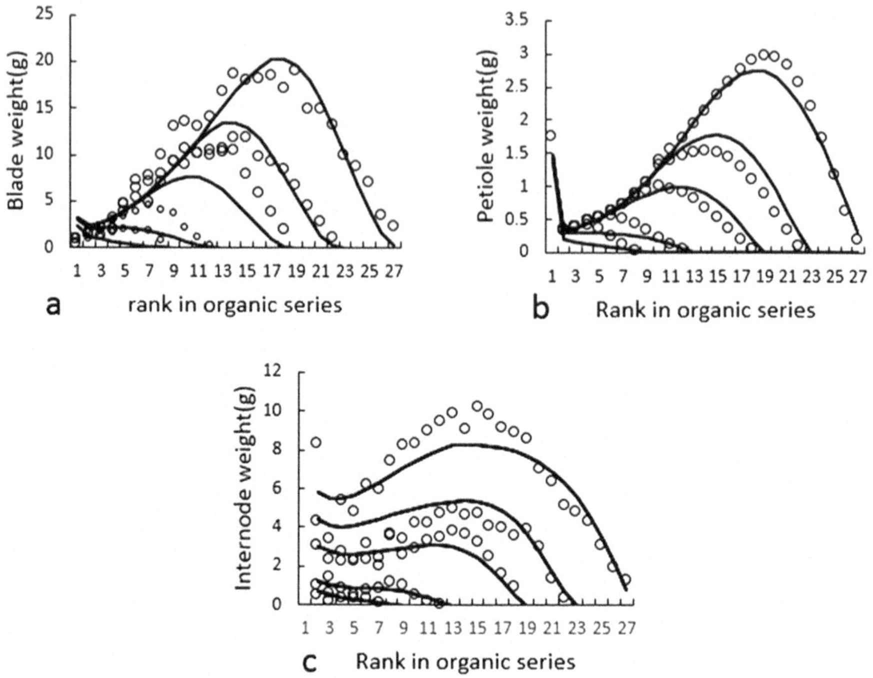

Fig. 10.9 Multi-fitting of the observed organic series (○) and computed organic series (–) by the model for the five measurement dates (8, 12, 18, 22, 27 GCs). (**a**) Blades. (**b**) Petioles. (**c**) Internodes. Secondary growth is clearly visible at the beginning of the organic series of internodes (Gloups software)

also necessary to determine the expansion time T_{xr} of the root system, which may exceed or not the current age of the observed plant. We know the demand for the stem at DC i is $Dshoot(i)$. The demand of the root system at DC i is $Droot\ (i)$, which can be written according to the triplet of the beta law as:

$$Droot(i) = Dshoot(i) \cdot p_r \cdot \left(\frac{i+0.5}{T_{xr}}\right)^{Br1-1} \cdot \left(1 - \frac{i+0.5}{T_{xr}}\right)^{Br2-1} \cdot \left(\frac{1}{T_{xr}}\right) \quad (10.4)$$

Noting $Dr(i) = Droot(i)/Dshoot(i)$, the total plant demand is then:

$$Dplant(i) = Dshoot(i) + Droot(i) = (1 + Dr(i)) \cdot Dshoot(i) \quad (10.5)$$

In this case, it seems reasonable to give a value for T_{xr} that exceeds 27 DCs. As an example, we arbitrarily choose $T_{xr} = 35$ DCs. The parameter B_{r2} must also be set empirically to the value $B_{r2} = 2.5$, because the WLSM cannot calculate the triplet of the sink parameters of the root system (they are not identifiable together). While keeping the values for the sinks of the aboveground system, we have to compute r, Sp, P_r, B_{r1}.

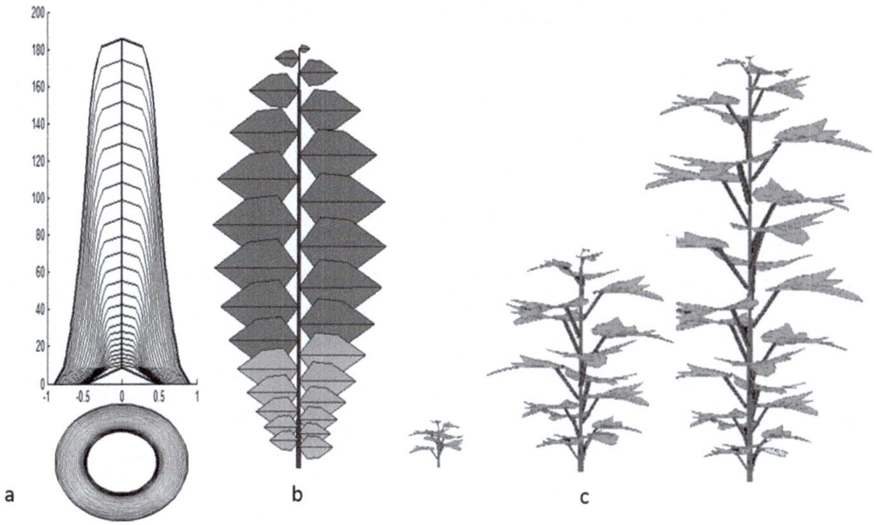

Fig. 10.10 Single-stemmed cotton growth and architecture. (**a**) Secondary growth with longitudinal and radial rings (enlarged) at the bottom of the stem. (**b**) 2D organic series at 27 GCs. Dead leaves are yellow (Gloups software). (**c**) 3D growth simulation at 8, 18, 27 GCs (Cornerfit software from CAU) (Color figure online)

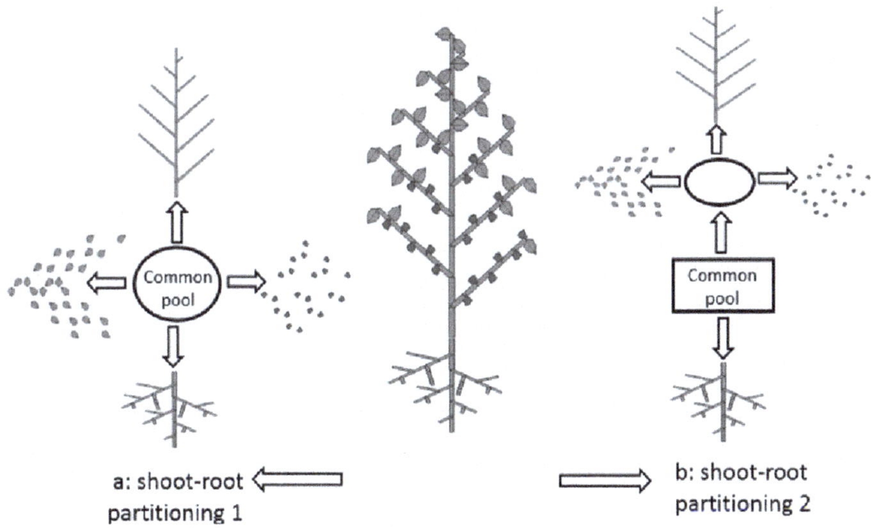

Fig. 10.11 Two modes of biomass partitioning between shoots and roots

10.2.1.2.8.1 Using Mode 1

If we search with WLSM for the root sink in mode 1, we find $P_r = 0$ and the other parameters take the values of the previous fitting with the aboveground part alone. This means the root system would be almost negligible, which is not reasonable. So, mode 1 must be discarded as it does not accurately represent the contribution of the root system to the overall plant biomass allocation.

10.2.1.2.8.2 Using Mode 2

If we search with WLSM for the root sink in mode 2, we find again $P_r = 0$ and the other parameters take the values of the previous fitting with the aboveground part alone. This means that the stem/root balance is too little modified to extract meaningful data from the root system, suggesting that mode 2 also fails to provide a reliable assessment of the root system's contribution.

However, it is possible to go further. Let us assume that experiments have been carried out taking into account the root system and that the fitting to mode 2 is satisfactory. This gives a reference value for the resource use efficiency *1/r (1/r* is expressed in $g{\cdot}cm^{-2}$ per DC). Such values are known for crops like tomatoes in the greenhouse (see Chap. 5). This value can be used to compute the evolution of the hidden root system, as *r* and *Sp* are given.

Let us choose arbitrarily $r = 40$ and keep $Sp = 2855\ cm^2$. Then only the triplet parameters (P_r, B_{r1}, B_{r2}) need to be estimated. The WLSM can easily find the triplet values (Table 10.2).

Figure 10.12 gives the result of the plant simulation with its root system.

The WSLM should find $B_{r1} = B_{r2} = 1$ in the case of a constant shoot ratio. For this case, the WLSM error = 26 and $P_r = 0.3$. Figure 10.12c shows that a slight

Table 10.2 Triplet of the sink parameters of root system using mode 2

Param	P_r	B_{r1}	B_{r2}
Mean	0.38	1.3	1.3
SD	0.03	0.2	0.4
CV (%)	8	17	30

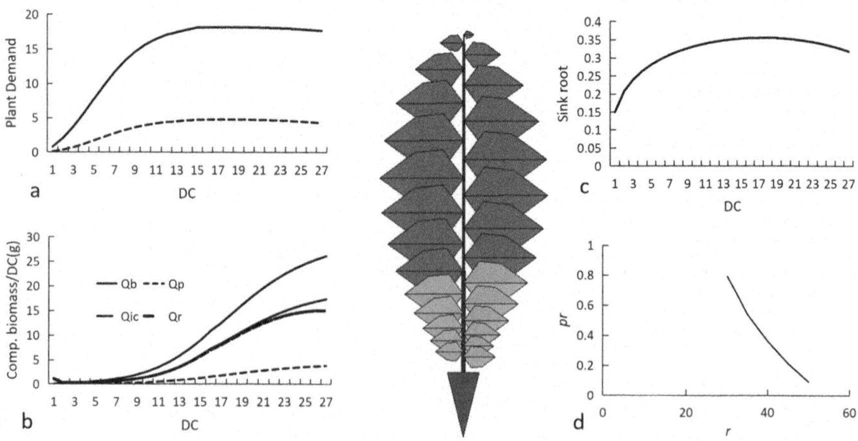

Fig. 10.12 Root system evolution for $r = 40$. (**a**) Plant demand. (**b**) Biomass compartment (Q_b = blade; Q_p = petiole; Q_{ic} = internode; Q_r = root. (**c**) Root sink variation. (**d**) Sink strength of root and *r*

variation in the root sink is found depending on $B_{r1} = B_{r2} = 1.3$, with the WLSM error $= 25$ and $P_r = 0.36$.

By adjusting the value of r, it is possible to compute the proportion of biomass P_r that is allocated to the roots. This relationship is not linear (Fig. 10.12d).

10.2.1.2.9 Discussion

This study shows that it is possible to obtain information on the root system even in the absence of direct data. For cotton, in contrast to the case of beetroot (Chap. 9), the model can fit well the organic series of the stem, even in the absence of the root system.

To know the evolution of the root system, the resource use efficiency ($1/r$) of the plant must first be computed using data containing the weight of the root system.

Once this value is known, the evolution of the root system can be computed using the data taken from the stem.

10.2.1.2.10 Effect of Defoliation

As a functional–structural plant model, the GreenLab model combines the dynamic simulation of plant architecture, plant biomass production of source organs and biomass partitioning among the sink organs, resulting in resource-dependent growth (Zhan et al. 2006). The objective of this study is to provide better information on how cotton plants respond to defoliation and how they adapt to their trophic dynamics.

A field experiment was conducted at China Agricultural University in Beijing in 2003. The cotton (*Gossypium hirsutum* L. cv. G33B) seeds were sown at a spacing of 0.6 m × 0.6 m. Plants emerged on May 26, 2003, and their growth was not limited by water or nutrition. All the axillary buds were removed as they appeared, resulting in a single–stemmed plants with purely vegetative growth. The control plants (also single-stemmed) were not subject to the removal of leaves. Four leaf-pruning treatments were implemented, which differed in the number of leaves retained on the top of plants through the growing season after regular defoliation started: 6, 9, 12 and 15 leaves were kept for the four treatments, respectively. Each treatment is identified with T_x, where x is the number of leaves retained on the treated plants.

Leaf removing began when plants had developed 11, 12, 17, 20 cycles for the treatment T_6, T_9, T_{12} and $T_{15,}$ respectively, corresponding to the first intervention in the field. The oldest leaf was removed once a new phytomer appeared at the top of each plant. As a result, plants subject to the same treatment always carried a fixed number of leaves: 6, 9, 12 or 15. The experiment aimed to have simple plant architectures to compare results and test the capability of the theoretical model for plants responding to leaf-pruning interventions.

The experiment had five blocks, one for each treatment. Five plants per treatment were chosen for in situ measurement in random order. These plants were destructively measured on the last date, allowing for the establishment of the total biomass when combined with previously collected leaf data. Destructive samples were taken weekly. One plant was collected per replication and sampling date. The root system was not considered.

For each sampling date, the observed organ fresh weight and dimension data were input in a target file, which subsequently served as a reference for statistical optimization of model parameters.

10.2.1.2.10.1 Effect of Defoliation on Development

No significant effect of leaf-pruning was found on the number of phytomers developed at the end of the experiment, indicating that the cotton plant phyllochron did not differ among treatments. The average number of phytomers developed by the plant was 36.5 ± 0.5 (with a sample size of 30 plants), as shown in Fig. 10.13.

10.2.1.2.10.2 Effect of Defoliation on the Growth

The weight of the cotton plant depends on the number of functional leaves after their removal. The removed leaves are weighed to compute the weight of the leaf compartment. Surprisingly, the control plants, which retain all their leaves, produce significantly less biomass than the treatment with 12 functional leaves (Fig. 10.14).

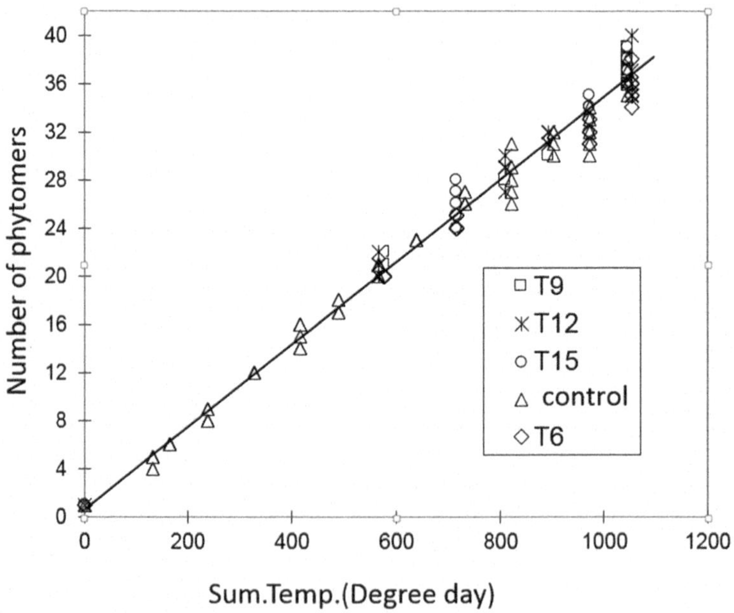

Fig. 10.13 Number of phytomers of single-stemmed cotton plants developed as a function of thermal time. Leaf pruning does not affect the development

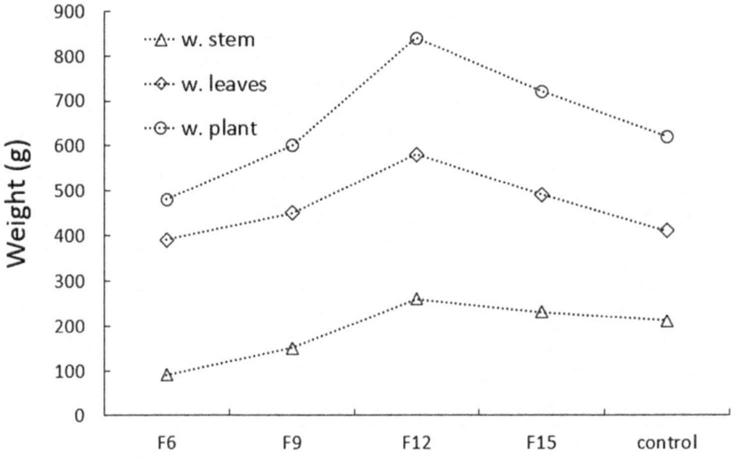

Fig. 10.14 Fresh weight of the cotton plant, including both leaves and stem, varies depending on the number of functional leaves after pruning (6, 9, 12, 15, control). The weights of removed leaves have been weighed

10.2.1.2.10.3 Pictures of the Treatments

Pruning the branches increases the size of the stem organs (see Chap. 5). Removing the number of functional leaves changes the growth of the stem. The control plant is smaller than the treatment plant with 12 functional leaves (Fig. 10.15).

10.2.1.2.10.4 Discussion

We modified the plant structure by removing the source organs that altered both the plant production and the plant demand. Regular leaf pruning had no effect on the organ production rate per unit of thermal time, whereas strong effects were observed on plant organ sizes, biomass acquisition, and organic series along the stem. The total fresh aboveground biomass production per plant was observed to be a bell-shaped curve concerning different defoliation treatments and the peak appeared at the treatment plants with 12 leaves, which were significantly greater than the control plants. The plants with six leaves were the smallest plants, which is logical. The fact that the control plant with all the leaves is smaller than the plant with 12 leaves on the stem is a reliable result have been obtained repeatedly.

It should not be forgotten that leaves are not only source organs but also sink organs like internodes or fruits

There are cases of branches with chlorophyll-free leaves, especially in ivy (Fig. 10.16). They are, therefore, not photosynthetic. These branches develop normally like normal branches. They are fed from the common pool.

One interpretation of the observed plant behavior is that the leaves could remain more as sinks than sources, after 12 GCs, because the older leaves become shaded within the canopy and receive less light. In the simulation, the expansion time and

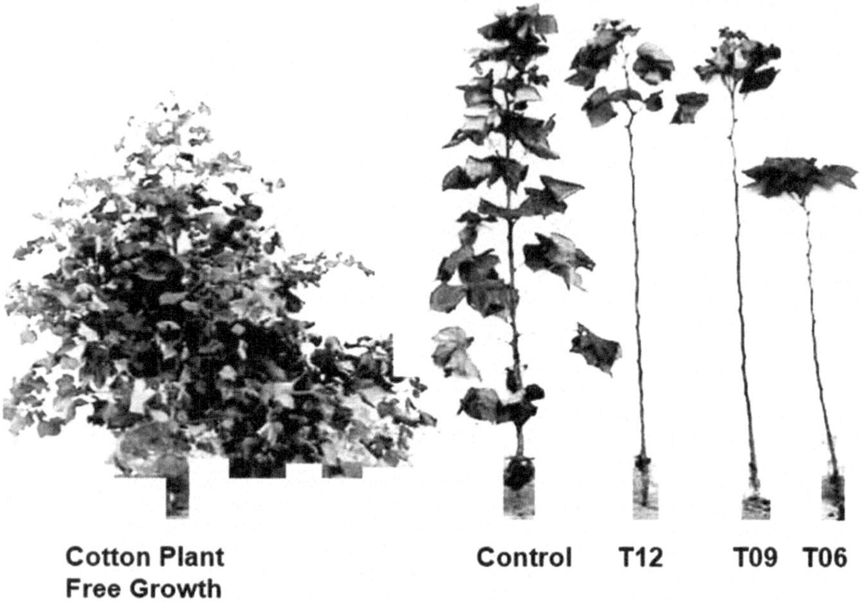

Fig. 10.15 Pictures (same scale) of the effect of pruning branches and leaves on the growth of the cotton plants (the removed leaves have been weighed). Plants T_{12} weighs more than control plants (Li Dong, Zhan Zhi-Gang, CAU)

Fig. 10.16 Ivy branch with chlorophyll-free leaves

the functioning time were fixed to 15 GCs for mature leaves, because the leaves do not show degeneration until this stage. Another possibility is linked to secondary growth: the more it is important, the more it depresses growth. The balance between shoot and root could also be involved.

Analysis of variance on the functional parameters optimized using the multi-fitting approach showed that most of the parameters keep stable. But the weight/surface area ratio, called specific leaf weight (SLW), increases with the number of functional leaves, and also the sink of the petiole. This kind of effect is not predictable by the GreenLab model.

10.2.1.3 Growth of Cotton Plant: Vegetative Case

It is possible to obtain a vegetative plant, by removing all the axillary buds that give fruiting branches at their birth from the leaves of the stem. The resulting cotton plant has lost its stochastic aspect: the development becomes quasi-deterministic ($b \approx 1$). The pruned vegetative plant is taller than the control branched plant (see Chap. 5).

10.2.1.3.1 Plant Description

The plant has a stem that bears two vegetative branches at the 5th to 6th internodes.

10.2.1.3.2 Cultivation Condition

The cultivation conditions are the same as for the single-stemmed cotton plant, but the planting density is: 0.8 m × 0.8 m ($Sd = 6400$ cm^2).

10.2.1.3.3 Development Scheme

Vegetative cotton plant development is deterministic. The vegetative branches had immediate branching with a rhythm ratio of $w = 1$. The observed plants stop at 37 phytomers on the stem, which means 37 DCs of the development.

10.2.1.3.4 Growth Scheme

The functioning duration of the leaves (t_a) is estimated as 15 DCs. The constant duration of organ expansion (t_x) is set to 15 DCs for blades, petioles and internodes.

10.2.1.3.5 Results on Plant Functioning

The target contains the biomass of the compartments of organs for only one date and of the organ weights along with the ranks of the $3*2 = 6$ organic series (because we assess two PAs and there are three kinds of organs: blades, petioles and internodes).

10.2.1.3.5.1 *Empirically Measured Parameters*

We keep the same allometries as those obtained from the single-stemmed case. Some functioning parameters resulting from the study of the single-stemmed cotton are included here: the second parameters of variations in the beta laws, which is to minimize the number of parameters that need to be estimated.

10.2.1.3.5.2 *Model-Based Estimated Parameters*

The stage of measurements corresponds to the only harvested plant. In total, the model is confronted with 189 experimental data collected at organ and compartment levels.

The fitting is good: $R^2 = 0.94$, number of computed parameters = 12 (the parameter P_b is fixed to 1), see Table 10.3.

As it is a single fitting (one stage), the invariance of parameters during the growth process cannot be established and can only be assumed.

Tables 10.1 (see above) and 10.3 allow comparing the parameters obtained from the single-stem and vegetative cases. We find $r = 54.8$ in Table 10.1 and $r = 44.5$ in Table 10.3. This is probably due to the SLW. We have not measured the SLW in this case, and ε should be smaller than 0.03 because the sizes of the organs of the vegetative cotton plant are smaller than those of the single-stemmed one. The value of $Sp = 4861$ cm^2 is in the same range as Sd (6400 cm^2). The values of Q_0 are not statistically different.

The sink strengths of the vegetative branches are smaller than the ones of the stem. It justifies our choice of considering them as a different PA and not as pure

Table 10.3 Source–sink parameters of the vegetative cotton plant

Param	Mean	SD	CV (%)
r	44.5	4.0	9
Sp	4861	702	14
Q_0	4.5	1.5	33
P_{b1}	1	0	0
P_{b2}	0.64	0.04	6
P_{p1}	0.28	0.02	6
P_{p2}	0.17	0.01	6
P_{i1}	0.59	0.08	13
P_{i2}	0.47	0.06	14
P_c	0.23	0.04	17
λ	0.15	0.13	83
B_{b1}	2.8	0.2	9
B_{p1}	1.8	0.2	10
B_{i1}	2.4	0.4	18
B_{b2}	3.9	0.0	0
B_{p2}	2.7	0.0	0
B_{i2}	5.7	0.0	0

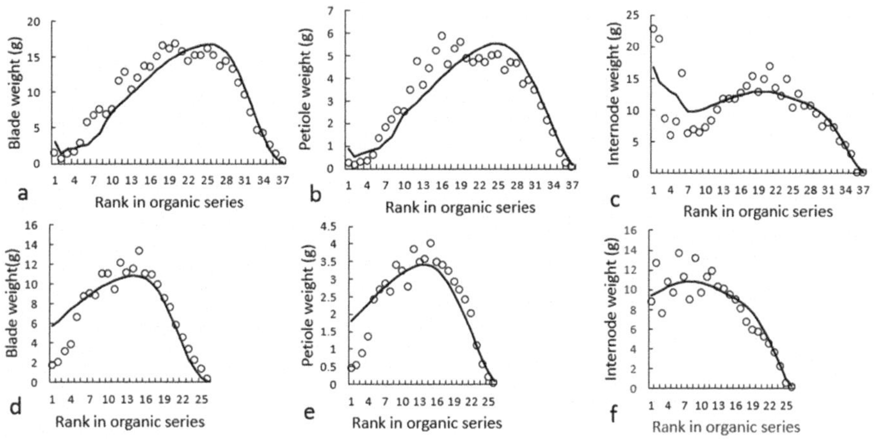

Fig. 10.17 Single fitting of the organic series of the vegetative cotton plant at 37 DCs. (**a**–**c**) Organic series of leaves, petioles, internodes on the stem. (**d**–**f**) Organic series of leaves, petioles, internodes on the vegetative branches

reiterations of the stem: their developments are indeed similar to the stem's one, but not their growth.

The secondary growth is also different. The value of $\lambda = 0.15$ is close to a uniform propagation for the biomass allocated to the rings. So, Pressler's law does not apply here (see Chap. 8). The parameters of the beta laws are not statistically different from those of the single-stemmed cotton plant.

The good quality of the single-fitting of organic series can be assessed in Fig. 10.17. There are two sets because there are two PAs (stem and vegetative branches).

It can be seen (Fig. 10.17d, e) that there is a base effect for the sizes of leaves at the bottom of the branches, i.e. branches at the base of the stem have different behaviour than the other ones. It is too negligible to be considered in Gloups software.

10.2.1.3.6 Simulation of Vegetative Cotton

In Fig. 10.18, the development and growth of the vegetative cotton plant are depicted with the longitudinal and radial rings at the bottom of the stem, its 2D organic series and its 3D architecture at 37 GCs.

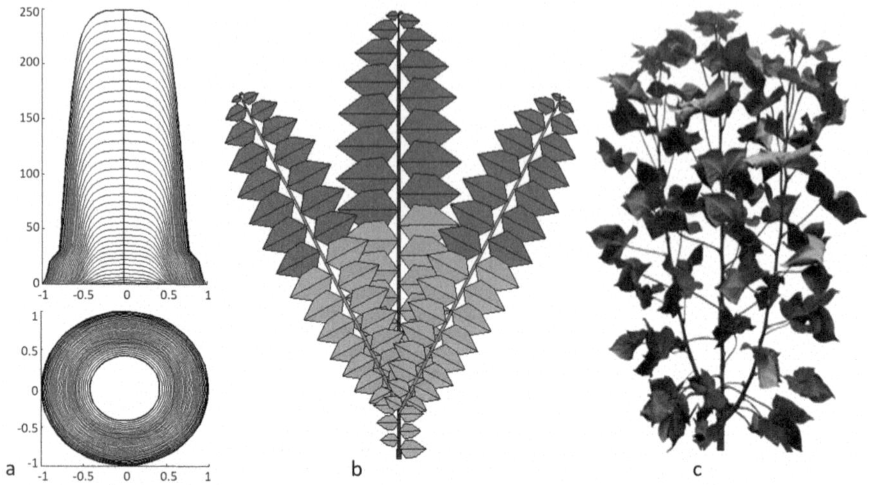

Fig. 10.18 Vegetative cotton growth and architecture. (**a**) Main stem secondary growth (longitudinal and radial rings at the bottom). (**b**) 2D organic series at 37 GCs. Dead leaves are yellow (Gloups software). (**c**) 3D photo of cotton (Li Dong, CAU) (Color figure online)

10.2.1.4 Growth of Cotton: Control Growth Case

10.2.1.4.1 Description of the Plant

Free growth cotton is a woody plant with three types of axes (physiological ages or PAs): the main stem, vegetative branches and fruiting branches. The fruiting branches that appear after the vegetative branches are composed of a series of terminal flowering modules that alternate. The development of cotton has been described in its stochastic aspect on a variety grown at CIRAD in Montpellier. For the experiments at the CAU, the work was simplified with the notion of the average plant, because the source–sink analysis in the stochastic case was not yet available at this time. It is obtained by replacing *Bernoulli* parameters with rhythm ratios of the same value, so the potential structure is the deterministic and average plant. The pauses induced by the rhythm ratios are kept. This transformation introduces a bias in the profiles of the organic series, but the approximation is acceptable if the axis development remains limited (see Chap. 6, Fig. 6.5b).

Later, the implementation of Gloups with its stochastic component makes it possible to resume fittings taking into account the *Bernoulli* process. The two modes are exposed and compared.

10.2.1.4.2 Cultivation Condition

The planting density is lower than 1 plant/ m^2. The observations here have been made on a single cotton plant and complete with its vegetative and fruiting branches, for a single stage of development (the terminal one).

10.2.1.4.3 Development Scheme

The analysed cotton plant had a vegetative branch on the phytomer ranks 5, 6, 7 of the stem and a fruiting branch on each phytomer of the stem and vegetative branches from rank 5. The phytomers 5, 6, 7, therefore, bear two different types of axes. Figure 10.19 shows the relationships between the numbers of phytomers in the different branched axes according to the rank on the carrier axis. The noise in the data is considered low enough to allow working on the average plant. The average rhythm ratio is $w = 1$ between a vegetative branch and stem, and $w = 0.5$ between a fruiting branch and stem. The rhythm ratio of the Asian variety ($w = 0.5$) is twice as high as that of the Mediterranean variety (Chap. 3).

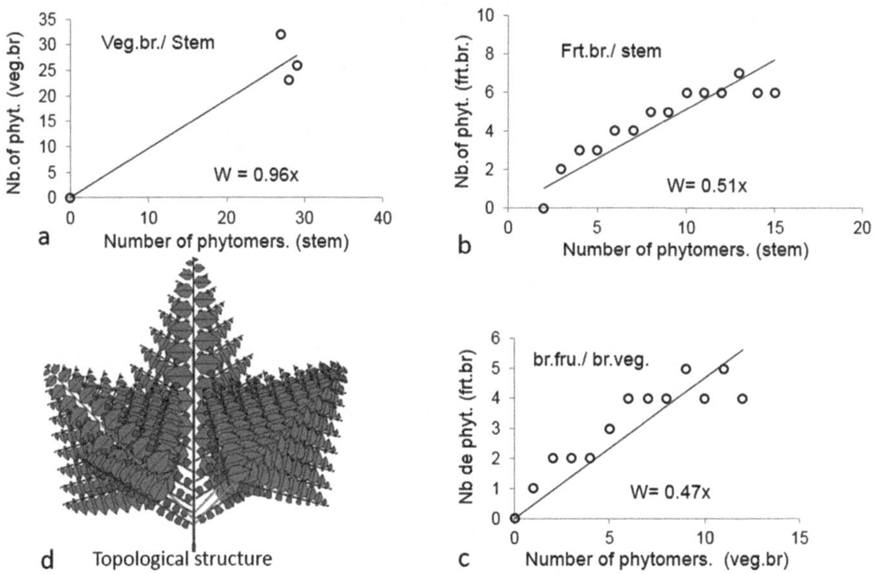

Fig. 10.19 Number of phytomers of the axes, and rhythm ratios of the branches in cotton. (**a**) Vegetative branch/stem. (**b**) Fruit branch/stem. (**c**) Fruit branch/vegetative. (**d**) Deterministic topological structure (after Li Dong, China Agricultural University (CAU)). The stochastic aspect of development is obvious in the scattering of dots, but the strong linear trend simplifies the development to a mere rhythm ratio ($w = 1$ vegetative branch/stem, $w = 0.5$ fruit branch/stem). Three vegetative branches are similar to reiterations at the base of the stem. We can distinguish the location of the three vegetative branches on the stem and that of the fruits on the fruiting branches. The dead leaves have fallen

This makes it possible to analyse individual plants with deterministic development. A mask is applied to the botanical automaton to remove the missing branches in the observed topological structure.

Even for this shrub, the number of phytomers is very large. The stem, vegetative and fruit branches have 34, 84 and 508 phytomers, respectively, giving a total of more than 2000 organs.

10.2.1.4.4 Growth Scheme

The growth and development of the axes are synchronous. The phytomers have immediate expansion.

10.2.1.4.5 Results on Plant Functioning

There is only one analysed development stage (the stem is made of 34 phytomers at this stage). The measurements are expressed in fresh weight. There are three PAs and, therefore, three types of axes (stem, vegetative branch, fruit branch). Each type of axis has three organic series with blades, petioles, internodes. The axes of PA 3 bear fruits and carry, in addition, a series of capsules (cotton fruits). So, there are 10 organic series to be fitted.

10.2.1.4.5.1 Empirically Measured Parameters

Allometries, functioning and expansion times of the organs are estimated from the observations. We keep some values from the single-stemmed study: Q_0 and the second parameter of the beta laws of sink variations. The thickness of the leaf corresponding to the SLW is measured as $\varepsilon = 0.02\ \text{g/cm}^2$. It is smaller than that of the single-stemmed cotton plant.

The single-stemmed cotton case gives the functioning and expansion times of organs at 15 DCs.

For the deterministic case, the age of the plant in DCs is equal to the number of phytomers in the stem: $T = 34$ DCs.

But for the stochastic case, there is a dilatation of the time scale. According to crown analysis on the cotton plant, we set the *Bernoulli* parameters as $b \approx 0.85$. Thus, the age of the plant is $T = 34/0.85 \approx 40$ DCs. The functioning and the expansion times of organs remain 15 DCs.

10.2.1.4.5.2 Computed Parameters

There are four kinds of organic series (blades, petioles, internodes and fruits). The stage of measurements corresponds to the only harvested plant. These measurements produce 236 weights of organs in the organic series to be adjusted simultaneously during the growth process.

Table 10.4 Source–sink parameters of the complete cotton plant in free growth in the case of deterministic and stochastic analysis

	Deterministic			Stochastic		
Param	Mean	SD	CV (%)	Mean	SD	CV (%)
Q_0	7.6			7.6		
r	43.4	3.4	8	38.3	3.2	8
Sp	55,344	12,374	22	33,598	5953	18
P_{b1}	1.00	0	0	1	0	0
P_{b2}	0.90	0.04	4	0.89	0.04	4
P_{b3}	0.64	0.04	6	0.63	0.04	6
P_{p1}	0.23	0.01	5	0.22	0.01	5
P_{p2}	0.20	0.01	5	0.19	0.01	5
P_{p3}	0.16	0.01	6	0.16	0.01	6
P_{i1}	0.53	0.09	18	0.25	0.08	34
P_{i2}	0.36	0.07	18	0.15	0.05	35
P_{i3}	0.23	0.04	19	0.10	0.04	35
P_{f3}	4.12	0.54	13	4.47	0.77	17
P_c	0.24	0.02	7	0.30	0.02	5
λ	0.10	0.00	0	0.1	0.00	0
B_{b1}	2.96	0.16	5	3.41	0.20	6
B_{p1}	1.66	0.15	9	1.88	0.18	9
B_{i1}	2.47	0.20	8	1.39	0.23	17
B_{f1}	6.53	1.04	16	7.84	1.57	20
B_{b2}	3.80	0.00	0	3.8	0.00	0
B_{p2}	2.60	0.00	0	2.6	0.00	0
B_{i2}	5.70	0.00	0	5.7	0.00	0
B_{f2}	5.00	0.00	0	5	0.00	0

In both deterministic and stochastic cases, the fitting is good: $R^2 = 0.98$; the number of computed parameters = 16 (the parameter P_b is fixed to 1).

The values of the 16 parameters are given in Table 10.4. Comparing the values for the same parameters, it is found that the sink strengths of internodes are essentially different. The values in the stochastic case are two times less than those in the deterministic case. The parameter λ cannot be estimated, but empirically, the error in the fitting is minimum for $\lambda = 0.1$. It is close to the previous value for the vegetative cotton plant; the biomass for rings follows roughly a uniform propagation.

Although the correct parameters come from the stochastic model, it is interesting to note that the deterministic model reconstructs the plant architecture as well.

Figure 10.20 shows the evolution of the number of phytomers, plant demand, individual LAI and produced biomass per cycle as a function of GC.

Sink variation laws are similar to the ones of the single-stemmed case (Fig. 10.21).

The good quality of single-fitting of organic series in both deterministic and stochastic modes can be assessed in Fig. 10.22.

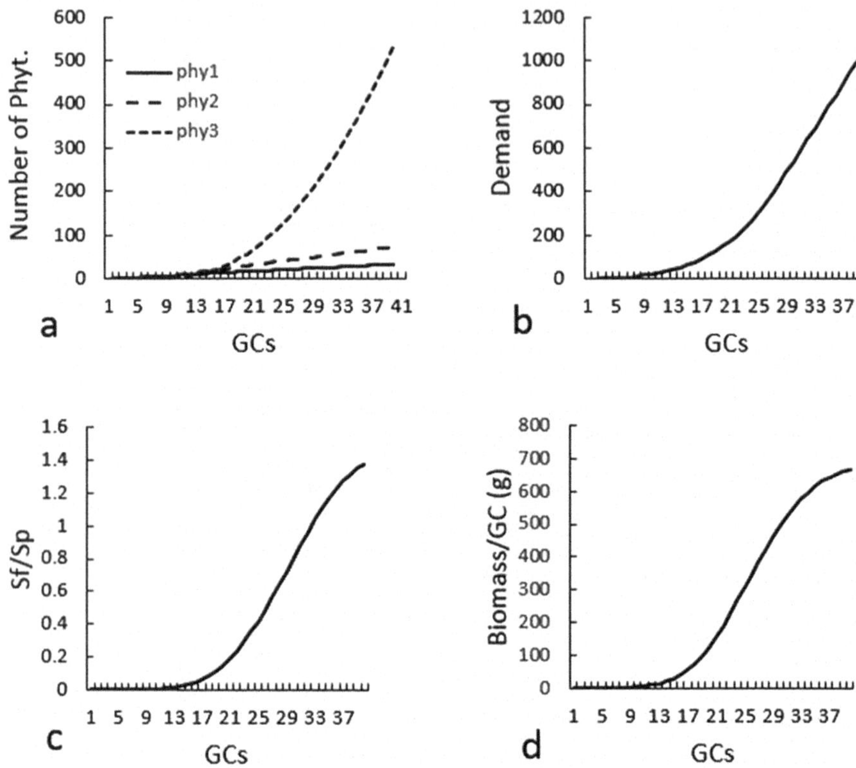

Fig. 10.20 History of the development and growth of the complete cotton plant. (**a**) Number of phytomers per GC. (**b**) Evolution of demand per GC. (**c**) Evolution of individual LAI. Production not yet saturated (*Sf/Sp* < 3). (**d**) Evolution of biomass production per GC

There are about 200 measured organs belonging to the organic series instead of the 2000 organs into the plant architecture. This shows the interest in using the concept of organic series for reducing the sampling. Moreover, as shown in Chap. 9, it is not necessary to measure each rank of phytomer.

10.2.1.4.6 Simulation of Cotton Architecture

The development and growth parameters allow simulating the 3D architecture of the cotton plants whose average plant is represented (Fig. 10.23).

The simulation reproduces the studied plant with three vegetative branches carried by ranks 5th, 6th, 7th of the stem (Gloups software).

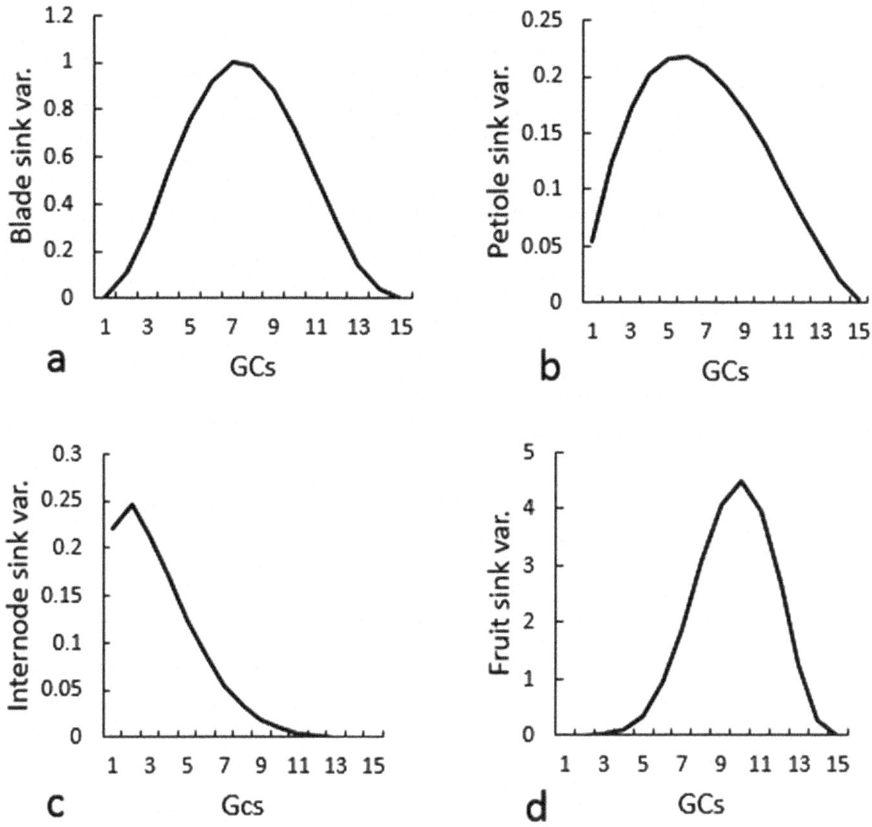

Fig. 10.21 Sink variation in each type of organ during the expansion. (**a**) Blade. (**b**) petiole. (**c**) Internode. (**d**) Fruit. Note that the sink of the internode is almost zero after 10 GCs, so the visible expansion stops. The expansion of fruits needs four GCs before starting

10.2.1.4.7 Discussion

This example of a complete cotton plant shows that the GreenLab model correctly adjusts the organic series taken from a single plant. The three PAs provide six organic series and 16 parameters are computed. By setting the second sink variation parameters of the beta laws to the values of the stem cotton plant, the coefficients of variation can be reduced without affecting the quality of the fittings.

In this example, the Matrix mode can compute the parameters 16 times quicker than the List mode. The List mode is necessary only to draw the 2D plant structure or the 3D plant architecture.

The absence of mortality makes it possible to define an average cotton plant with the use of a mask to fix the position of the branches. If the fruit branches are always presented from a certain rank onwards, the vegetative branches are only presented in small numbers at the base of the stem. They must, therefore, be positioned with a

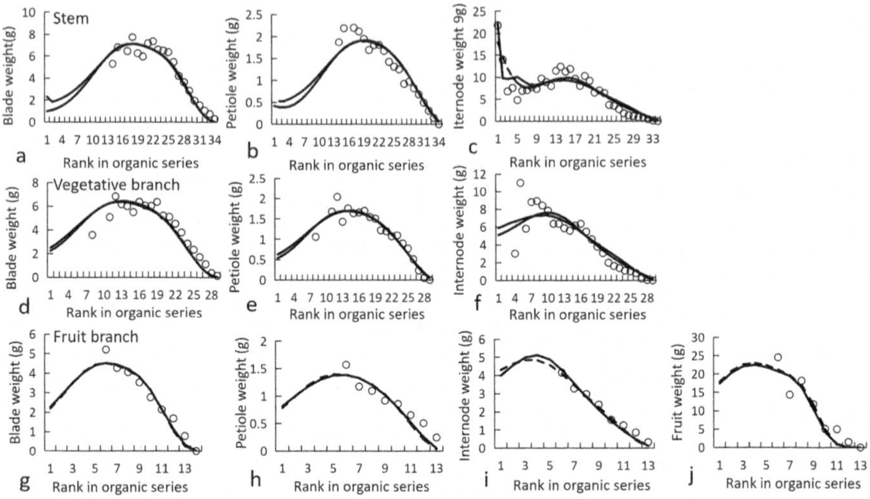

Fig. 10.22 Single-fitting of the observed organic series (○) and computed organic series (deterministic – and stochastic --) by the model for one measurement date (34 or 40 GCs). (**a**) Blades. (**b**) Petioles. (**c**) Internodes (stem). (**d**) Blades. (**e**) Petioles. (**f**) Internodes (vegetative branches). (**g**) Blades. (**h**) Petioles. (**i**) Internodes. (**j**) Fruit branches (Gloups software)

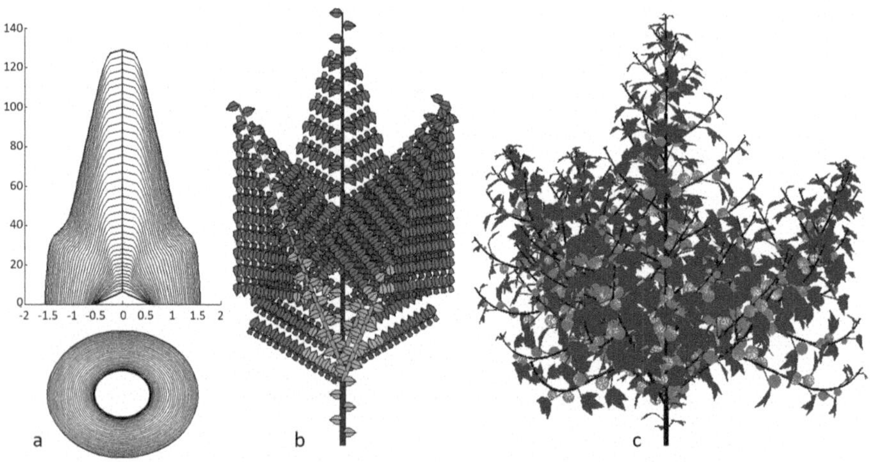

Fig. 10.23 Simulation of complete cotton architecture. (**a**) Main stem secondary growth (longitudinal and radial rings at the bottom). (**b**) 2D organic series at 40 GCs. Dead leaves are yellow. (**c**) 3D simulation (Gloups software) (Color figure online)

mask. For the chosen plant, the branching is immediate, which allows the plant to remain within a deterministic framework.

The fittings between the deterministic and the stochastic cases can be considered as of equal qualities. But the sinks of internodes differ. The use of the average plant should be reserved for cases where stochastic algorithms are not implemented to analyse the plant architecture.

10.2.2 Case Study 2: Coffee Plant Study (S. Sabatier, P. Okoma, S. Akaffou, CIRAD_RCI)

The coffee tree was the first plant studied with the AMAP model for its computational architecture (de Reffye 1981a, b). This plant is perfect to study meristem behaviour and makes it possible to develop the crown analysis (Chap. 3). But this analysis only concerned plant development.

For plant growth, the source–sink analysis using organic series will be undertaken later. The first result on young coffee trees and various species showed that the biomass production was independent of plant architecture (see Chap. 5, Fig. 5.22) (Letort et al. 2012). Single-stemmed (pruned) young coffee trees have the same fresh biomass as the branched ones for a given species. But the organs of the pruned coffee plant are larger. This trait is well predicted by the GreenLab model.

The computation of source–sink relationships in the case of stochastic plants was taken into account once the notion of the potential structure had been validated (see Chap. 6). This structure derives directly from the bud parameters computed by the 'crown analysis'. The potential structure allows to compute the potential demand of the plant at any DC and the profiles of the organic series. The fittings with the WLSM of the observed and computed organic series give the parameters of the source–sink functions of the plant.

The notion of the mask becomes useless and is not adapted to the *Bernoulli* process of bud functioning.

In addition, the effects of pruning and genetic variability have been studied on several coffee species (Ph.D. of Okoma 2018).

This section presents successively: the study of the growth of a young coffee tree, the effect of pruning on development and growth, and the genetic variability of coffee trees. The study concerns only young plants (two years) with the flowering and the mortality of the axes have not yet installed.

10.2.2.1 Plant Description

Coffee is a woody shrub plant with stochastic development (Chap. 3) and continuous growth. The shrub offers three types of stochastic organic series (leaves, internodes and fruits). The architecture follows Roux architectural model: the stem is orthotropic, and each phytomer has two leaves with potentially two plagiotropic branches. So, there are two PAs. As the petiole is reduced, the leaf is considered to be the blade only. At the young stage, there is no mortality or flowering. We observe the gradual implementation of the branching from the base of the trunk. In the adult stage, the reiterations appear.

10.2.2.2 The Use of Allometry to Compute Organ Weight

Measuring the volume of an organ requires its dimensions, fresh or dry weight and density. The density is always close to 1.

For leaves, the SLW gives the constant ε. The weight of a leaf is easy to measure with a balance, but its surface area requires to be scanned, which is a destructive and time-consuming process. In order to avoid it, an allometric relationship was first established between the surface area measured by the scanner and the product length L by width D of the leaf. The weight of the leaf is obtained from the allometry coefficient k, its length L, its width D and the SLW ε. Weight $= k \cdot \varepsilon \cdot L \cdot D$ (density $= 1$) (Fig. 10.24a, c).

For the internode, the volume V is computed as a cylinder by measuring its length L and its diameter D ($V = \pi \cdot L \cdot D^2/4$). The volume gives the weight in fresh weight or dry weight, taking into account the desiccation coefficient (Fig. 10.24b, d).

When working with dry weight, the desiccation reduces the weight of the organs by 80–90%. Knowing this relationship, it is, therefore, possible to estimate the dry weight of the organs from the measurements carried out on the plant. When the allometry works properly, the advantages are the non-destructive measurements: only a meter and a digital calliper are sufficient to obtain the organic series (Fig. 10.1). No need for additional work in the laboratory using the devices like balance, scanner and oven for desiccation.

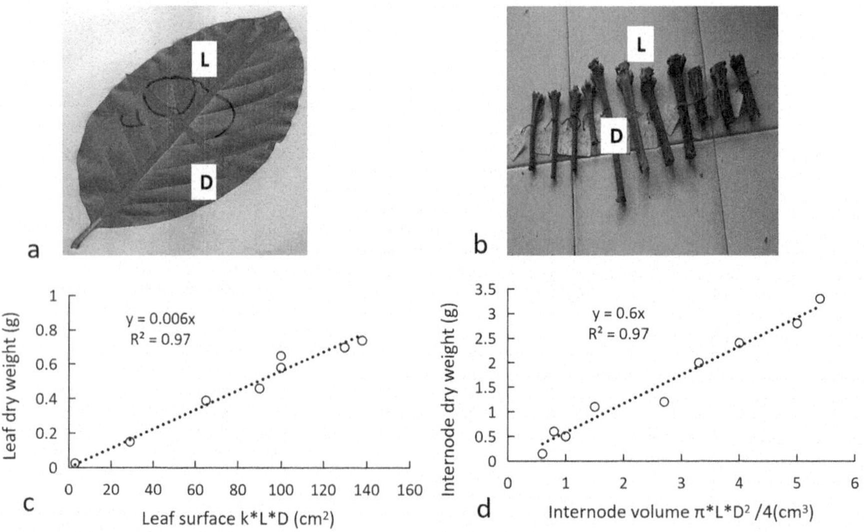

Fig. 10.24 Allometry between the weight of the organ and its dimension. (**a**) Coffee leave. (**b**) Organic series of internodes of a stem. (**c**) Allometry between the dry weight of leaf and the product L·D. (**d**) Allometry between the dry weight of internode and its estimated volume (P. Okoma)

10.2.2.3 Growth of a Young Deterministic Coffea Liberica

On a small coffee tree, all the details of the architecture can be seen. In particular, it is evident that all the branches represent the same series of phytomers from their extremities, either wholly or partly, depending on their chronological age (CA). Here we consider a young *Coffea liberica*. The architecture is deterministic ($b = 1$). The rhythm ratio is $w = 1$. The branching starts at the 4th node of the stem consisting of eight phytomers. There are, therefore, four pairs of branches, each with 20 phytomers. The plant grown in the greenhouse is abnormally stretched due to a lack of light.

10.2.2.3.1 Growth Scheme

The functioning duration of the leaves (t_a) is estimated to be 12 GCs (i.e. more than the CA of the plant). There is no clearly visible 'base effect' on the organic series of leaves. The constant duration of organ expansion (t_x) is set to four GCs for leaf blades and internodes. The organ expansion time in GC is short (four GCs), which translates to more than eight weeks in calendar time (the phyllochron is at least two weeks).

10.2.2.3.2 Results on Plant Functioning

The target contains the biomass for one stage of growth (eight GCs), and the organ weights along the ranks make four organic series.

10.2.2.3.2.1 Empirically Measured Parameters

The SLW ε, is computed as 0.016 g/cm^2 from the ratio of leaf fresh weight and its surface area.

10.2.2.3.2.2 Estimated Parameters

There are two kinds of organic series (blades and internodes) and two kinds of PAs (stems and branches), but only one stage of growth (eight GCs) and a single plant. These measurements produce 24 organ weights in the organic series to be adjusted simultaneously during the growth process.

The fitting is good: $R^2 = 0.95$; number of computed parameters $= 9$ (the parameter P_b is fixed to 1). The second parameter of the sink variation is set empirically as $B_{b2} = B_{i2} = 2$. The empirical value of $\lambda = 0$ means a uniform propagation for the biomass allocated to the rings. The values of the nine parameters (Table 10.5) are displayed for the growth period (eight GCs).

Table 10.5 Source–sink parameters of young *Coffea liberica*

Param	r	Sp	Q_o	P_{b1}	P_{b2}	P_{i1}	P_{i2}	P_c	B_{b1}	B_{i1}	B_{b2}	B_{i2}
Mean	33.9	5599	2.73	1.00	0.57	0.47	0.02	0.09	2.89	2.54	2.00	2.00
SD	8.2	3869	1.56	0	0.087	0.213	0.032	0.028	0.47	0.68	0.00	0.00
CV (%)	24	69	57	0	15	45	146	32	16	27	0	0

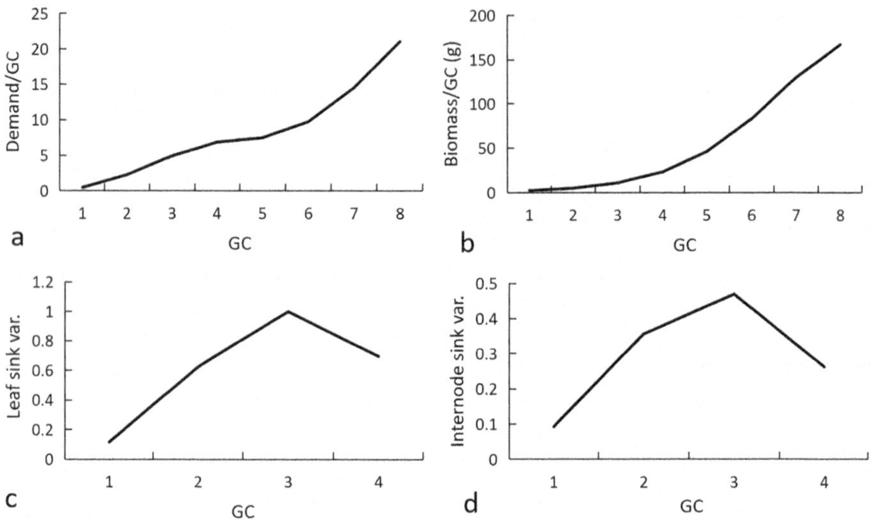

Fig. 10.25 Fitting of *Coffea liberica*. (**a**) Evolution of demand. (**b**) Evolution of biomass production. (**c**) Leaf sink variation. (**d**) Internode sink variation

Figure 10.25 shows the evolutions of demand and biomass production per GC. The triplets (P_o, B_{o1}, B_{o2}) are used to compute the sink variations in leaves and internodes.

The CVs are high because there is only one stage for the fitting. The parameter values of branches are smaller than those of the stem.

The good quality of single-fitting of organic series can be assessed in Fig. 10.26.

10.2.2.3.3 Simulation of the *Coffea liberica*

The development and growth parameters allow simulating the 3D architecture of the *Coffea liberica* (Fig. 10.27).

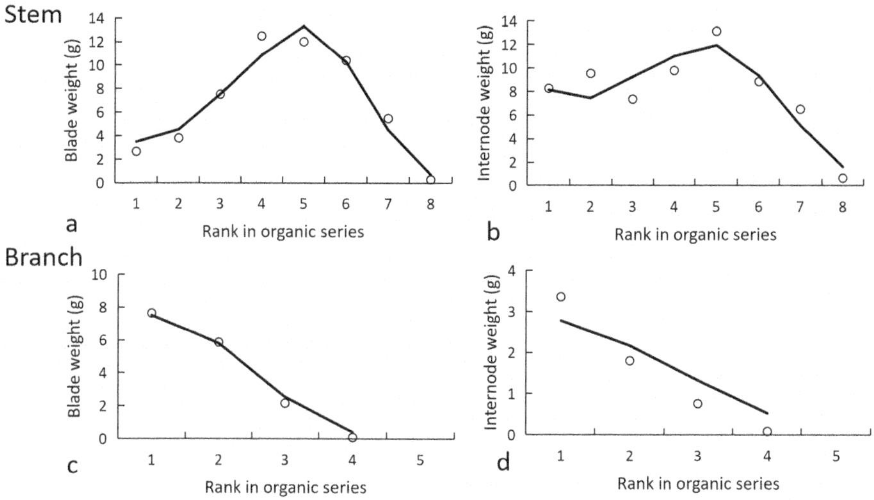

Fig. 10.26 Single-fitting of the observed organic series (○) and computed organic series (–) by the model for the one measurement date (eight GCs] of *Coffea liberica*. (**a**) Blades (stem). (**b**) Internodes (stem). (**c**) Blades (branch). (**d**) Internodes (branch) (Gloups software)

Fig. 10.27 Young *Coffea liberica*. (**a**) Main stem secondary growth (longitudinal and radial rings at the bottom). (**b**) 3D growth simulation at eight GCs (Digiplant software from Ecole Centrale Paris (ECP)). (**c**) The photography of the plant

10.2.2.4 Growth of a Young Stochastic *Coffea pseudozanguebarie*

The study was carried out in Côte d'Ivoire by Sélastique Akaffou (Okoma et al. 2018) in a collection of coffee trees.

10.2.2.4.1 Cultivation Conditions

Here we are interested in the architecture of 14 young *Coffea pseudozanguebarie* of the same age (two years old) and cultivated under the same conditions in the field. Mortality and flowering have not yet occurred, so there are only four organic series: the leaves and internodes of the stem and branches. The organs are measured in dry weight.

10.2.2.4.2 Development Scheme

Figure 10.28 displays the structure of five young coffee tree crowns, as documented from the survey. Their variability illustrates the stochasticity of their development. For each rank K from the top of the stem, the number of phytomers is variable and represented in Fig. 10.29a.

Table 10.6 shows the details of the crowns of the 14 coffee trees. These have different numbers of phytomers. Starting from the top, each row of phytomers has

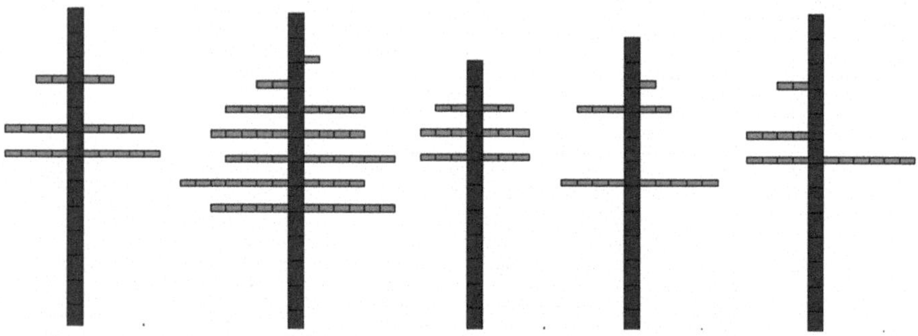

Fig. 10.28 Five displays of young coffee tree crowns (*C. pseudozanguebarie*) among the 14 observed crowns

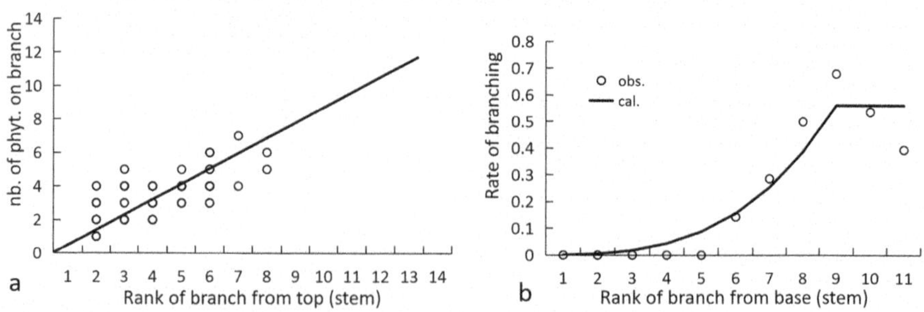

Fig. 10.29 Statistical analysis of the crowns of *Coffea pseudozanguebarie*. (**a**) Number of phytomers of the branches according to their insertion rank on the stem under the top. The slope of the line is $r = w \cdot b_2/b_1$. (**b**) Evolution of the branching rate from the base of the stem

Table 10.6 Details on the crown structures for 14 *C. pseudozanguebarie*

		Coffee tree number																											
		1	1	2	2	3	3	4	4	5	5	6	6	7	7	8	8	9	9	10	10	11	11	12	12	13	13	14	14
	Stem size and branch number																												
	0	14	14	14	14	13	13	13	13	13	13	11	11	9	9	10	10	12	12	13	13	11	11	13	13	13	13	9	9
	0	1	2	1	2	1	2	1	2	1	2	1	2	1	2	1	2	1	2	1	2	1	2	1	2	1	2	1	2
		Nb of phyt. on branches along ranks from top																											
Rank from top	1	0	0	0	0	0	0	0	0	0	0	0	0	0	0	0	0	0	0	0	0	0	0	0	0	0	0	0	0
	2	1	1	0	0	0	0	0	0	0	0	2	2	0	0	3	2	1	0	0	0	3	2	0	0	1	0	4	4
	3	0	0	0	0	2	0	2	0	2	2	3	3	2	2	4	4	3	2	2	2	4	4	2	2	2	0	5	5
	4	3	0	0	0	0	0	2	2	0	0	3	3	3	2	0	0	0	0	0	0	0	0	4	3	4	4	0	0
	5	3	3	4	0	4	0	4	3	4	4	0	0	0	0	0	0	0	0	4	4	0	0	4	3	5	4	0	0
	6	0	0	4	4	6	4	0	0	5	4	0	0	0	0	0	0	5	4	3	3	0	0	0	0	6	4	0	0
	7	0	0	0	0	0	0	0	0	0	0	0	0	0	0	0	0	0	0	0	0	0	0	0	0	7	4	0	0
	8	0	0	0	0	0	0	0	0	0	0	0	0	0	0	0	0	0	0	0	0	0	0	0	0	6	5	0	0
	9	0	0	0	0	0	0	0	0	0	0	0	0	0	0	0	0	0	0	0	0	0	0	0	0	0	0	0	0
	10	0	0	0	0	0	0	0	0	0	0	0	0			0	0	0	0	0	0	0	0	0	0	0	0		
	11	0	0	0	0	0	0	0	0	0	0	0	0					0	0	0	0	0	0	0	0	0	0		
	12	0	0	0	0	0	0	0	0	0	0							0	0	0	0			0	0	0	0		
	13	0	0	0	0	0	0	0	0	0	0									0	0			0	0	0	0		
	14	0	0	0	0																								

potentially two branches. We note the number of phytomers of these branches. If the branch is absent, the number is 0.

The branching is immediate, but there is a proportion of axillary buds that remain dormant. A base effect is observed, resulting in a gradual increase in the branching rate on the stem at the beginning and then remaining stable (Fig. 10.29b). This is modelled according to the method in Chap. 3, by the function of the branching probability a_i, where i is the position on the stem.

We get:

$$a_i = 0.62 \cdot \left(\frac{i}{9}\right)^{3.45} \quad \text{if } i \leq 9$$

$$a_i = 0.62 \qquad \text{if } i > 9$$

After the ninth phytomers, the branching probability stabilizes at $a = 0.62$.

The crown analysis gives the values of the *Bernoulli* parameters and the rhythm ratio between the stem and the branches: $b_1 = 0.8$, $b_2 = 0.9$ and $w = 0.75$.

These values, which numerically describe the activity of meristems at each DC, allow to simulate stochastic development and to compute the potential structure of the coffee tree. The probability of the existence of a phytomer on the trunk and the branch are b_1 and $a_i \cdot b_1 \cdot w \cdot b_2$, respectively. In the potential structure, the existence of phytomer takes the value 1 or 0 within the periodic sequence generated by the rhythm ratio w (see Chap. 3).

10.2.2.4.3 Growth Scheme

The functioning duration of the leaves (t_a) is estimated as 11 GCs. There is no clearly visible 'base effect' on the organic series of leaves. The duration of organ expansion (t_x) is short and constant, set to five GCs for leaf and internodes. The phyllochron is at least two weeks.

10.2.2.4.4 Results on Plant Functioning

The target contains the biomass for one date (16 GCs) with four organic series.

10.2.2.4.4.1 Empirically Measured Parameters

The SLW ε is computed as 0.016 g/cm^2 from the ratio of leaf fresh weight and its surface area.

Table 10.7 Source–sink parameters of *Coffea pseudozanguebarie*

Param	r	Sp	Q_o	P_{b1}	P_{b2}	P_{i1}	P_{i2}	P_c	B_{b1}	B_{i1}	B_{b2}	B_{i2}
Mean	180.8	100,000	0.61	1.00	0.67	0.49	0.64	0.01	0.09	1.45	2.00	2.00
SD	12.5	0	0.13	0.00	0.05	0.10	0.09	0.01	0.25	0.48	0.00	0.00
CV (%)	7	0	21	0	7	19	15	200	266	33	0	0

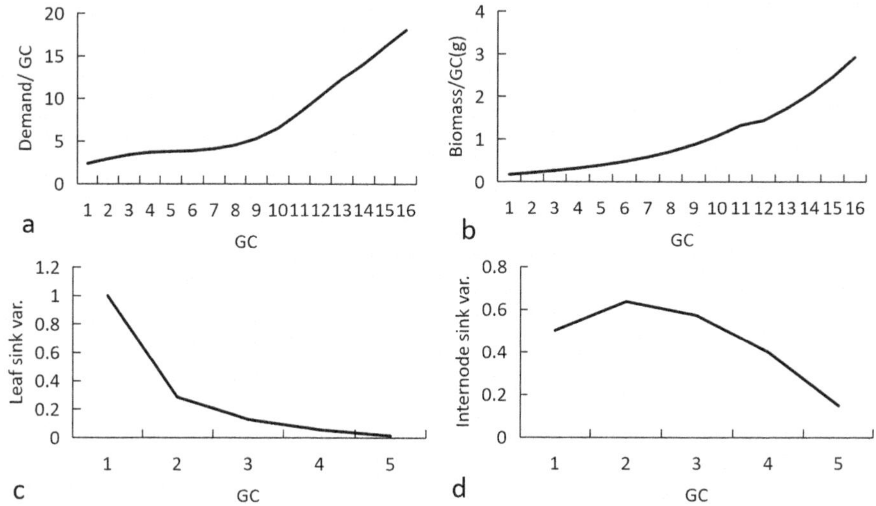

Fig. 10.30 *Coffea pseudozanguebarie*. (**a**) Evolution of demand. (**b**) Evolution of biomass production. (**c**) Leaf sink variation. (**d**) Internode sink variation

10.2.2.4.4.2 Computed Parameters

There are two kinds of organic series (blades and internodes) and two kinds of PAs (stems and branches) for 14 plants. These measurements produce 35 organ weights in the organic series to be fitted simultaneously.

The fitting is good: $R^2 = 0.95$; number of computed parameters = 8 (parameter P_b is fixed to 1). The Sp is set to a high value ($Sp = 100{,}000$ cm^2 means free growth) and the second parameter of the sink variation is set empirically to $B_{b2} = B_{i2} = 2$. The empirical value of $\lambda = 0$ means a uniform propagation for the biomass allocated to the rings. The values of the eight parameters (Table 10.7) are displayed for the growth period (16 GCs).

Figure 10.30 shows the evolutions of demand and biomass production by GCs. The triplets (P_o, B_{o1}, B_{o2}) are used to compute the sink variation in leaves and internodes.

The girth growth (P_c) is negligible at this stage of growth.

The sink of leaves is maximum for the first GCs. The good quality of single-fitting of organic series can be assessed in Fig. 10.31.

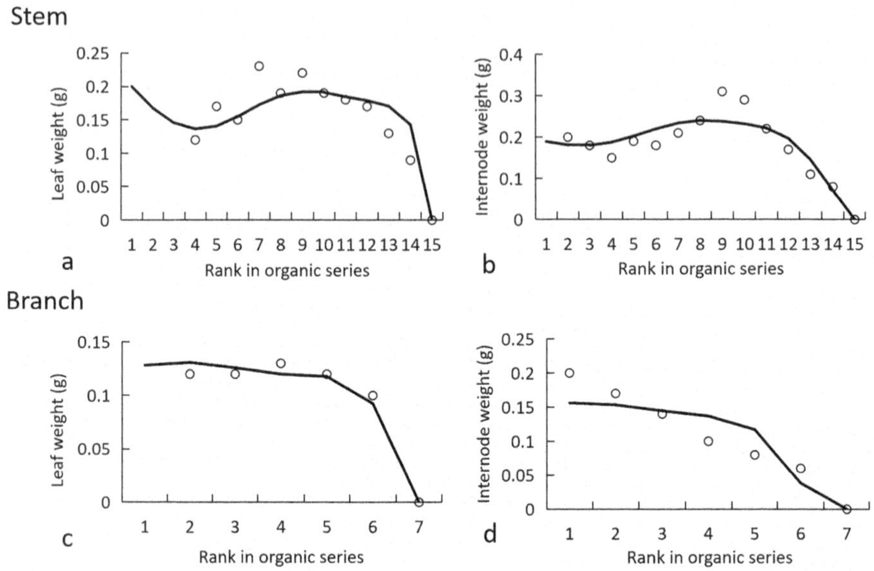

Fig. 10.31 Single-fitting of the observed organic series (O) and computed organic series (–) by the model for the one measurement date (16 GCs) of *Coffea pseudozanguebarie*. (**a**) Blades (stem). (**b**) Internodes (stem). (**c**) Blades (branch). (**d**) Internodes (branch) (Gloups software)

10.2.2.4.5 Simulation of the *Coffea pseudozanguebarie*

The development and growth parameters allow simulating the 3D architecture of the *Coffea pseudozanguebarie*. Four stochastic simulations are represented (Fig. 10.32).

The secondary growth observed on the stem is very limited.

10.2.2.4.6 Discussion

The development of the method for sampling measurements on plants with stochastic architecture was a long process. In a first step, the crown analysis method provided the values of the parameters that control the development. In a second step, the organic series method, coupled with the notion of potential structure, provided the values of the source–sink parameters that control growth. The method is particularly effective on shrubs such as coffee trees.

Woody plants (especially trees) have a much slower development than herbaceous plants. In most cases, measurements are available for only one date. It is rare to be able to benefit from multi-fitting to ensure the stability of the parameters.

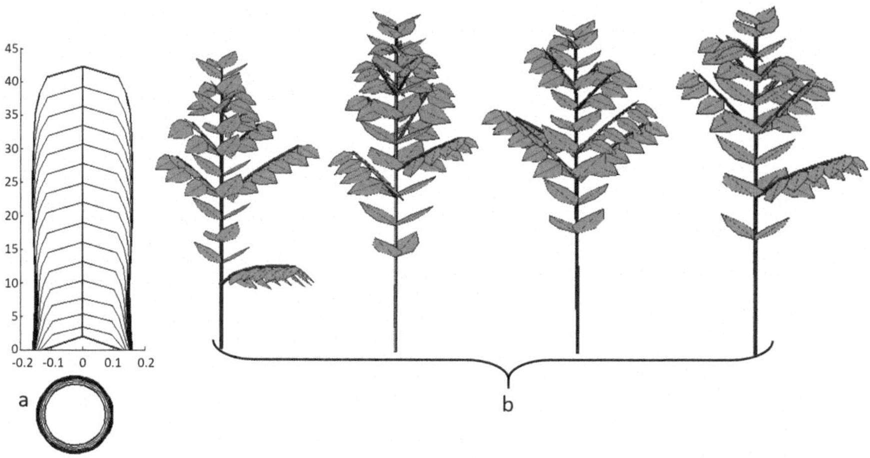

Fig. 10.32 *Coffea pseudozanguebarie*. (**a**) Main stem secondary growth (longitudinal and radial rings at the bottom). (**b**) Four 3D stochastic simulations at 16 GCs (Gloups software). Secondary growth is very low at this stage of growth

10.2.2.5 Effect of Pruning on the Functioning of Meristems

A theoretical study of young coffee trees has shown that biomass production does not depend on the architecture (Chap. 5). So, in this case, the removal of axillary meristem does not influence the final weight of the plant, which is true only for young development stages where secondary growth is negligible. The organs of the single-stemmed plant are simply larger than those of the branched control plant.

However, actions on the architecture can influence the development. Pruned plants have a more regular development, which increases the value of parameter b controlling the activity of the meristems by the *Bernoulli* process (Chap. 3). Thus, for the same age, pruned plants may produce more phytomers on the main stem than branched control plants.

For branched coffee trees, the 'crown analysis' gives the value of the parameter b of the stem. However, for the single-stemmed coffee plant, it is necessary to use the distribution of the number of phytomers per stem (a binomial law) for a given date to obtain the parameter b as a function of the mean m and the variance v, according to the equation $b = 1 - v/m$ (see Fig. 3.11, Chap. 3). The possible effects of delayed and variable germination must be eliminated.

Table 10.8 gives the measured values of the stem parameter b for single-stemmed and control coffee plants for six different species.

Pruning has a regulating effect on development: it becomes more deterministic. This leads to the increase in organ volume, probably due to the increase in trophic pressure Q/D. This point will be further developed in Chap. 13.

Table 10.8 Effect of pruning on the development parameter *b* for one-year-old trees

	Single-stemmed	Control (branched)	Single-stemmed	Control (branched)
Species	*b*	*b*	Phytomer number	Phytomer number
Coffea canephora	0.91	0.88	11	8
Coffea dewevrei	0.92	0.77	*	*
Coffea liberica	0.95	0.77	11	10
Coffea stenophylla	0.89	0.78	13	11
Coffea pseudozanguebarie	0.79	0.8	17	12
Coffea racemosa	0.91	0.7	26	19
Average	0.89	0.78	15.6	12

10.2.2.6 Study of the Inter-species Variability

The study of different species of coffee makes it possible to explore the genetic variability of development and growth parameters. Two experiments were carried out, one in a greenhouse and the other in the field on free-growing coffee plants.

10.2.2.6.1 Single-Stemmed Case (Sylvie Sabatier, CIRAD)

An experiment on nine species of coffee trees with different behaviours for development was conducted in a greenhouse. They grew for the same period and under the same conditions in containers. The axillary meristems of these plants were removed to obtain a single stem. Each species was represented by three or four individual plants. The studied species are *C. arabica* (ARA), *C. brevipes* (BRE), *C. canephora* (CAN), *C. dewevrei* (DEW), *C. humilis* (HUM), *C. liberica* (LIB), *C. pseudozanguebarie* (PSE), *C. racemosa* (RAC), *C. stenophylla* (STE). Great genetic variability is observed. Coffee trees with small leaves (*C. racemosa, Pseudozanguebarie*) develop quickly, while large-leaved coffee trees (*C. canephora, C. dewevrei, C. liberica*) develop slowly. It takes eight days for the stem to make a new phytomer for *C. racemose* and 24 days for *C. liberica*.

These coffee trees have experienced free growth, so the production area *Sp* is large and not defined. The growth equation is written (see Chap. 5, Eq. 5.33):

$$Q(t) = \frac{E}{\varepsilon \cdot r} \cdot \left(\sum\nolimits_{i=\max(1,\, t-t_a+1)}^{t} N_a(i) \cdot \sum\nolimits_{j=i}^{t} \frac{p_a \cdot Q(j-1)}{D(j)} \right) \cdot \Delta t \qquad (10.6)$$

Table 10.9 Source–sink parameters of nine species of coffee trees

Species	DC	r	Q_0	P_b	P_i	P_c	B_a	B_i
ARA	21	118	0.66	1.00	0.62	0.31	1.47	2.11
BRE	13	133.2	1.09	1.00	0.23	0.03	1.36	1.88
CAN	12	69.3	0.77	1.00	0.46	0.17	1.84	3.3
DEW	9	63	2	1.00	1.01	0.03	1.57	3.8
HUM	11	68	1.04	1.00	0.58	0.02	2.19	2.31
LIB	9	58.2	1.97	1.00	0.36	0.09	1.42	2.08
PSE	22	71	0.29	1.00	0.01	0.67	1.61	2.2
RAC	29	112.6	0.21	1.00	4.2	0.27	1.79	3.8
STE	15	160	0.63	1.00	0.21	0.08	1.24	1.41

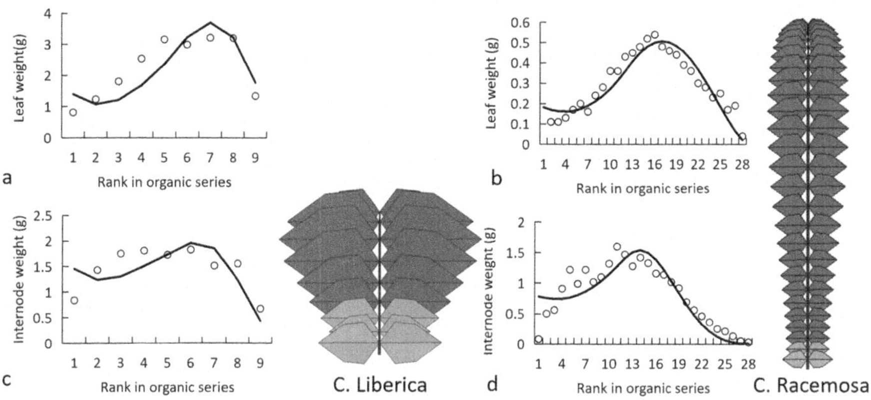

Fig. 10.33 Organic series of two species of coffee trees at the same age. These species have very different development rates. (**a**, **c**) *C. liberica*. (**b**, **d**) *C. racemosa*

10.2.2.6.1.1 Standard Analysis

By convention $E = 1$ and $\Delta t = 1$, because the cycle is time unit. But the numbers of cycles (DC) are different for the different species, although the plants are the same age in calendar time. The analysis of the organic series of leaves and internodes gives the source–sink parameters and the DC period computed from the mean and variances of the stem (Table 10.9). The WLSM computes six parameters. The second parameters of the sink variation law are fixed to 2. Figure 10.33 shows the organic series of a slow-growing, large-leaved (*C. liberica*) and a fast-growing, small-leaved (*C. racemosa*) coffee tree.

Table 10.10 Source–sink parameters of nine species of coffee trees transformed with rhythm ratio

Species	DC	w	r	Q_0	P_b	P_i	P_c	B_a	B_i
ARA	29	0.72	186.3	0.82	1	0.65	0.25	1.47	2.25
BRE	29	0.45	324.5	0.95	1	0.22	0.02	1.36	2.03
CAN	29	0.41	199	0.78	1	0.43	0.14	1.73	2.83
DEW	29	0.31	247.8	1.94	1	0.90	0.03	1.56	3.60
HUM	29	0.38	199	0.81	1	0.59	0.02	1.77	1.86
LIB	29	0.31	204.1	0.84	1	0.27	0.09	1.40	1.76
PSE	29	0.76	138.7	1.12	1	0.70	0.35	1.98	3.73
RAC	29	1	112.7	0.21	1	4.25	0.26	1.79	3.8
STE	29	0.51	397	1.02	1	0.19	0.03	0.79	1.9

10.2.2.6.1.2 Synchronized Analysis

One problem with the notion of the cycle (DC) is that it does not allow a direct comparison between the growths of different species, as the thermal times associated with the cycle (the phyllochron) differ from one species to another.

A solution is to use a global rhythm ratio w. The value 1 should be assigned to w for the species with the fastest development. Let N_1 be the associated number of cycles for this species during a given duration. During the same period, the other species produce N_2 cycles and will thus have a rhythm ratio $w_2 = N_2/N_1$. Then all the plants can grow together in N_1 cycles, with periodic pauses corresponding to their w.

Here, to synchronize the growth of different species of coffee trees, *C. racemosa* is chosen as a reference, as it grows on 29 DCs. By comparison, we can deduce the rhythm ratio of a species from it. For example, the *C. liberica* grows on 9 DCs. Its rhythm ratio is, therefore, $w = 9/29 = 0.31$. The age of all the plants is then 29 DCs.

It is necessary to recompute the functioning parameters of these plants from their targets. The organ functioning durations (t_{xo}) have to be replaced by the values t_{xo}/w. The synchronization of the development and growth of different species of coffee trees is then ensured. Table 10.10 gives the new values for the source–sink parameters.

A comparison of Tables 10.8 and 10.9 shows that the variable r is essentially increased by this transformation involving rhythm ratios.

With the transformation using the rhythm ratios, the different species have a synchronized growth because they share the same duration of the development cycle.

Figure 10.34 shows the 2D pictures of the organic series of the nine species at the age of 29 DCs for coffee trees.

10.2.2.6.1.3 Analysis of Some Influential Functional Parameters

Equation 10.6 can be written as:

$$Q(t) = \frac{E}{\varepsilon \cdot r} F(t) \cdot \Delta t \tag{10.7}$$

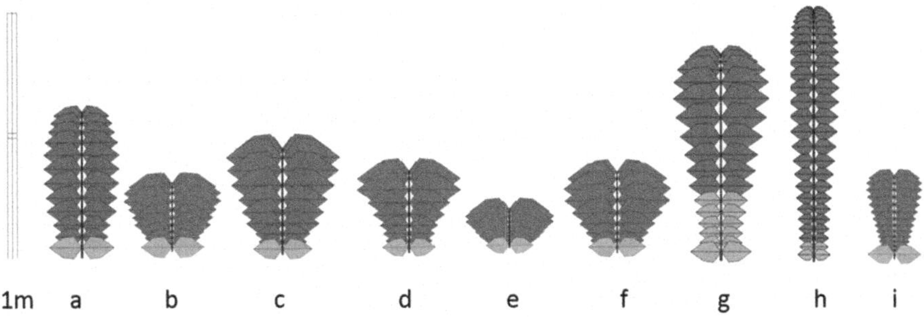

Fig. 10.34 Nine single-stemmed coffee trees (2D) of different species and rhythm ratios at the age of 29 DCs. (**a**) *C. arabica*. (**b**) *C. brevipes*. (**c**) *C. canephora*. (**d**) *C. dewevrei*. (**e**) *C. humilis*. (**f**) *C. liberica*. (**g**) *C. pseudozanguebarie*. (**h**) *C. racemosa*. (**i**) *C. stenophylla*

Table 10.11 Values of Q, Δt, ε, r for each species of the coffee tree

	$Q(g)$	Δt (days)	ε	r
ARA	2.2	18	0.02	118
BRE	1.4	20	0.021	133.2
CAN	3.3	25	0.022	69.3
DEW	4.5	26	0.024	63
HUM	1.8	26	0.029	68
LIB	3.7	24	0.025	58.2
PSE	1.7	14	0.029	71
RAC	1.1	8	0.023	112.6
STE	0.6	17	0.021	160

With:

$$F(t) = \sum\nolimits_{i=\max(1,\ t-t_a+1)}^{t} N_a(i) \cdot \sum\nolimits_{j=i}^{t} \frac{p_a \cdot Q(j-1)}{D(j)} \tag{10.8}$$

The four parameters E, ε, r, Δt are constant and play an important role in biomass production. They do not depend on the development and the sink values. Parameter E is the same for the nine species, but ε, r, Δt, differ from one species to another. It seems that there is a statistical relationship between the average biomass production (Q) and the three parameters. A linear relationship (10.9) is assumed (see Fig. 10.34):

$$Q = k \cdot \frac{\Delta t}{\varepsilon \cdot r} \tag{10.9}$$

Table 10.11 gives the values of Q, Δt, ε, r for each species of the coffee tree.

Table 10.12 gives the average biomasses Q per cycle produced by the plants and the associated values of the ratio $\frac{\Delta t}{\varepsilon \cdot r.}$

Figure 10.35 shows their linear relationships.

Table 10.12 Relationship between the biomass production and the functional parameters Δt, r, ε

	ARA	BRE	CAN	DEW	HUM	LIB	PSE	RAC	STE
Q	2.2	1.4	3.3	4.5	1.8	3.7	1.7	1.1	0.6
$\frac{\Delta t}{\varepsilon \cdot r}$	7.6	7.2	16.4	17.2	13.2	16.5	6.8	3.1	5.1

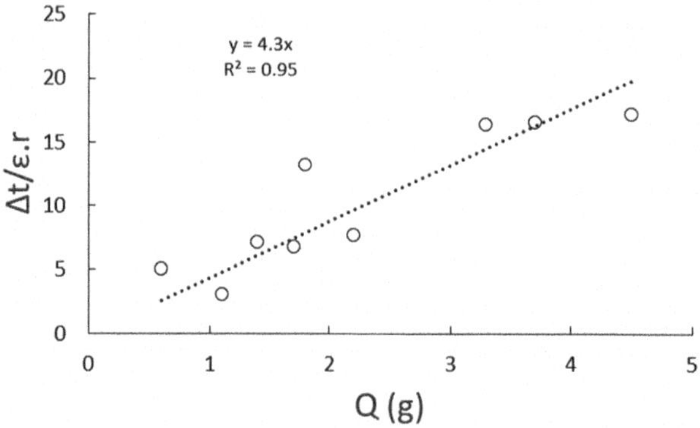

Fig. 10.35 Relationship between the biomass production Q and the functional parameters Δt, r, ε

It shows that the biomass production (Q) increases ($R^2_{Q,\Delta t/(\varepsilon \cdot r)} = 0.95$) with the increase in the light use efficiency ($1/r$) and the phyllochron (Δt). The statistical correlation can be tested with only one variable:

$$R^2_{Q,\Delta t} = 0.5; R^2_{Q,r} = 0.8.$$

When considering Δt and r together, it gives $R^2_{Q,\Delta t/r} = 0.85$.

So, the contribution of the SLW ε seems negligible. Resource use efficiency 1/r varies significantly by species.

In Fig. 10.35, we can see that there are two groups of plants. The coffee trees with small leaves on the left and big leaves on the right (*C. liberica*, *C. dewevrei*, *C. canephora*).

10.2.2.6.2 Control Case (Branched Trees: P. Okouma, Daloa University)

Here we are dealing with coffee trees that have grown in the field in Ivory Coast over two years. These young coffee trees have not yet bloomed, and the branches are free of any mortality. This study is extracted from the thesis of Pamelas Okoma (Daloa University).

The studied species are *C. canephora*, *C. dewevrei*, *C. liberica*, *C. pseudozanguebarie*, *C. racemosa*, *C. stenophylla*. The weights of organs were estimated directly from non-destructive measurements on the plants using the established allometric relationships. All young coffee trees have only two PAs (trunks and branches), except for *C. racemosa*, which is very bushy with branches at order 3.

Table 10.13 Parameters values for the development of six coffee species

	CAN	DEW	LIB	PSE	RAC	STE
t_a	10	8	8	12	24	14
t_x	4	4	5	4	8	6
w_{21}	0.9	0.71	0.67	0.9	0.7	0.78
w_{32}					0.7	
a_2	0.81	0.9	0.84	0.71	0.6	0.87
a_3					0.6	
na	8	12	9	5	6	12
c_a	3.03	4.03	3.15	0.46	0.27	0.68
b_1	0.88	0.77	0.77	0.8	0.7	0.78
b_2	0.93	0.97	0.95	0.8	0.7	0.92
b_3					0.7	

Table 10.14 Sources–sink parameters of the coffee species, computed by WLMS

	CAN	DEW	LIB	PSE	RAC	STE
r	122.2	63.4	162.6	109.7	300.0	292.0
Sp	3758	2179	18,587	1634	13,332	3252
Q_0	7.54	3.64	8.37	1.61	0.12	0.71
P_{b1}	1.00	1.00	1.00	1.00	1.00	1.00
P_{b2}	1.00	0.85	0.98	1.14	1.08	0.98
P_{b3}					0.93	
P_{i1}	9.07	2.11	0.84	1.09	0.44	0.40
P_{i2}	1.21	0.26	0.10	1.13	0.55	0.79
P_{i3}					0.22	
P_c	0.22	0.29	0.17	0.44	0.11	0.40
λ	1.00	1.00	0.20	1.00	0.62	1.00
B_{b1}	1.23	2.47	0.28	1.22	0.39	1.07
B_{b2}	2.00	2.00	2.00	2.00	2.00	2.00
B_{i1}	0.07	0.20	0.00	0.48	0.65	0.64
B_{i2}	2.00	2.00	2.00	2.00	2.00	2.00

The standard study deals with the crown analysis to determine the parameters of the buds (Table 10.13). It is followed by the analysis of the organic series to determine the source–sink parameters of the organs (Table 10.14). For *C. racemosa,* there are two rhythm ratios: w_{21} for order 2 branches and stem, and w_{32} for order 3 branches.

10.2.2.6.2.1 Results for Development and Growth

Table 10.13 shows that the stochastic behaviours, as well as the rhythm ratios, are very obvious. The values of functioning times for leaves (t_a) and expansion times for leaves and internodes (t_x) are assessed empirically. Changing t_a will change the value of parameter r without influencing the fitting of the organic series. As the CA is in DC, for the same duration of calendar time, t_a and t_x are much larger for *C. racemosa* than that for *C. canephora*, because of their respective phyllochrons.

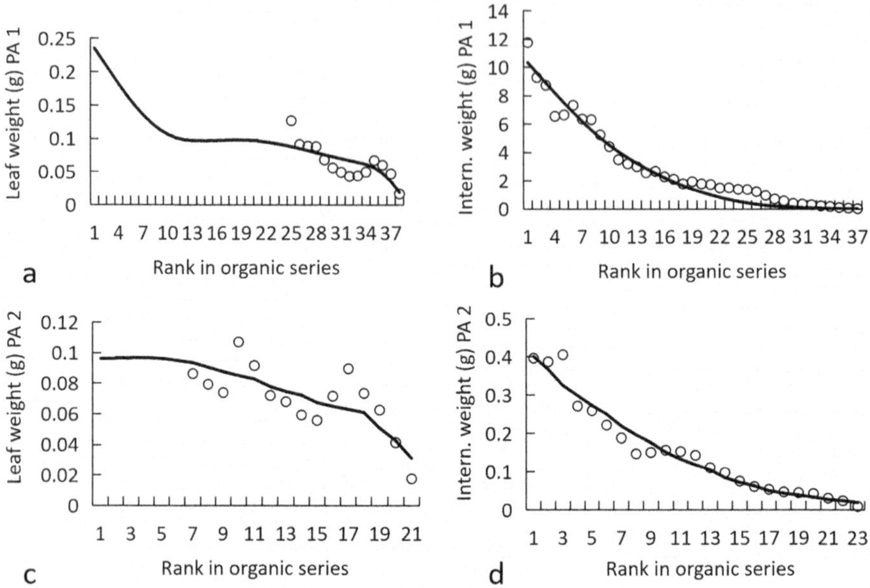

Fig. 10.36 Fitting the organic series of the two physiological ages (PAs) of *C. stenophylla*. (**a**, **c**) Leaves. (**b**, **d**) Internodes (the dead leaves are missing)

The *Bernoulli* parameters (b) are identical for all species. The rhythm ratios range from 0.7 to 0.9. Note that the rhythm ratio is also variable between clones of the same species. For example, in *C. canephora*, w is between 0.9 and 1.3. All coffee trees have a base effect for branching. The ratio of branched phytomers at the ith phytomers from the bottom (a_i) is given by the equation: $a_i = a \cdot \left(\frac{i}{na}\right)^{ca}, i \leq na$ and $a_i = asi,\ i > na$ (see Fig. 10.28b).

The branching probabilities range from 0.6 to 0.9 after stabilization.

The fittings of the organic series are of good quality (e.g. *C. stenophylla*, Fig. 10.36).

The values of r vary from a factor of 1 to 5. This fact is largely due to the variation in the phyllochron, which is ignored when the CA is expressed in cycles. The variation in Sp is mainly due to the species-specific geometric nature of the architecture of the plant that intercepts light in different ways. The weight of the seed is in accordance with the fact that *C. canephora*, *C. dewevrei*, *C. liberica* have large organs, while *C. pseudozanguebarie*, *C. racemosa*, *C. stenophylla* have small organs. Parameter λ is equal to one in three species of coffee trees which, therefore, follow Pressler's law (see Chap. 8). The *C. liberica* is close to a uniform distribution ($\lambda = 0$), and the *C. racemosa* ($\lambda = 0.62$) is in an intermediate situation.

Note that all the leaves on branches have a sink strength P_b close to 1 for all PAs, but the pith sinks of the internodes show great variability.

Fig. 10.37 Four pictures of coffee trees with three Monte Carlo simulations (Gloups software). (**a**) *C. canephora*. (**b**) *C. liberica*. (**c**) *Racemosa*. (**d**) *C. stenophylla* (Gloups software)

10.2.2.6.2.2 Simulations of the Different Coffee Species (Fig. 10.37)

The development and growth parameters, computed from the organic series, allow the simulation of the 3D architecture of the coffee tree species. Three stochastic simulations are represented for each.

10.2.3 Case Study 3: Mongolian Pine Study (Wang Feng, CAU)

10.2.3.1 Description of the Plant

The pine is a conifer whose architecture follows the Rauh model. The physiological ages are organized hierarchically in accordance with the orders of branching. The studied trees are young from one to six years old. Each year, the terminal meristems produce a new growth unit (no *Bernoulli* process in the sequence of GUs). Mortality is not yet a factor in these young trees, and the only variability comes from branching, which becomes noticeable from the third year. The numbers of branches follow a statistical distribution at the whorl level. This is the simplest case of stochastic development. On conifers, phytomers are extremely short, so they are impossible to distinguish in practice. The GU, whose expansion time is short in relation to its functioning time, can be considered a meta-phytomer, consisting of an internode and a leaf that represents the mass of the needle sleeve.

10.2.3.2 Observation Data

The study was carried out on plantations of young pines with trees aged between one and six years. The production area was estimated as $Sp = 2400\ \text{cm}^2$, corresponding to the available area per plant (Sd) according to planting density. The value of the climatic factor was monitored in detail to quantify the production, which is not taken into account here. Its normalized form ($E = 1$) is sufficient to properly approximate the source–sink relationships. The reduction in the GU to a meta-phytomer implies that the development cycle (DC) is the year. The expansion of the GU is immediate ($t_x = 1$ DC), and the needle functioning time is set to $t_a = 3$ DCs corresponding to the presence of active needles on the tree for three years. The statistics of the number of branches per whorl were based on 100 six-year-old individual plants. No branches were observed on the trunk before two years and no branches of order 3 before four years. In a young pine tree, the number of branches produced per axis category can be counted. The crown analysis is useless here because the development of the axes is deterministic without mortality (parameters $b = 1$ and $w = 1$).

Biomasses of GUs of internodes and needles were collected from four individual plants per year. The weight of the root system was not measured. Biomasses are measured in fresh weight. Allometries between the length and weight of the GU are computed using data from young GUs with negligible secondary growth.

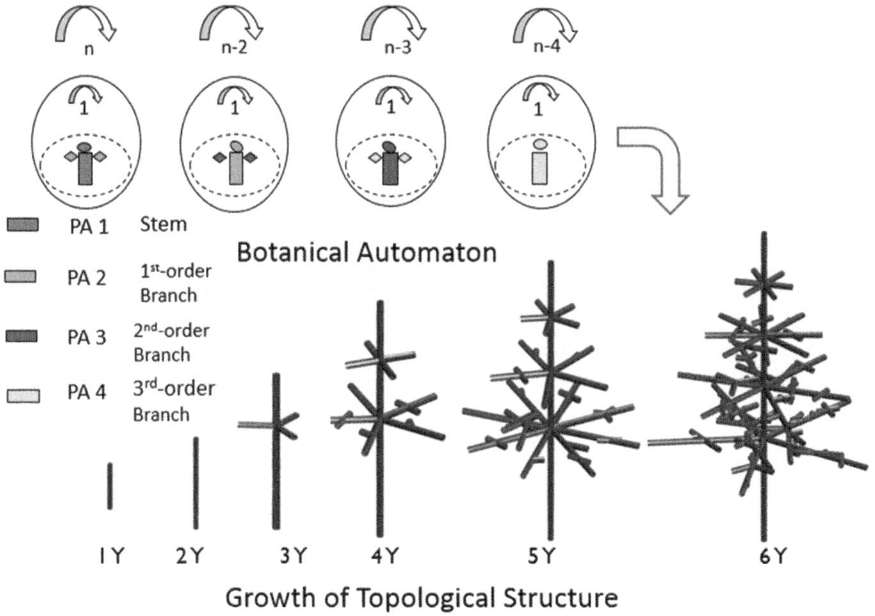

Fig. 10.38 Botanical automaton controlling the development of pine architecture. The growth unit (GU) is simplified into a 'meta-phytomer' with a variable number of branches per whorl. There are four physiological ages (PAs) that correspond to the order of branching

10.2.3.3 Development Scheme and Calibration

The architecture of the pine tree is considered here as an assembly of axes made up of GUs bearing whorls of branches. The botanical automaton that generates this development is simple (Fig. 10.38).

The stochastic aspect only exists in the distribution of the number of branches per whorl (Fig. 10.39). This distribution fits well with a binomial distribution, which makes it possible to compute the branching probability a_p of a branch belonging to a whorl. Potentially, one whorl has n_p branches of order p (or PA p) (Table 10.15).

The fitting of the distribution of the number of branches per meta-phytomer with binomial laws gives the potential number of axillary buds as well as the probability of branching. This makes it possible to compute the number of meta-phytomers per cohort and PA for the six years of growth thanks to the Botanical Automaton.

The fitting between the observed and computed phytomer production for the different CA and PA is shown in Fig. 10.40. The GUs only branch from the second GU of an axis. The increase in GU as a function of the age is linear for the trunk, quadratic for branches of order 2, cubic for branches of order 3 and biquadratic for branches of order 4.

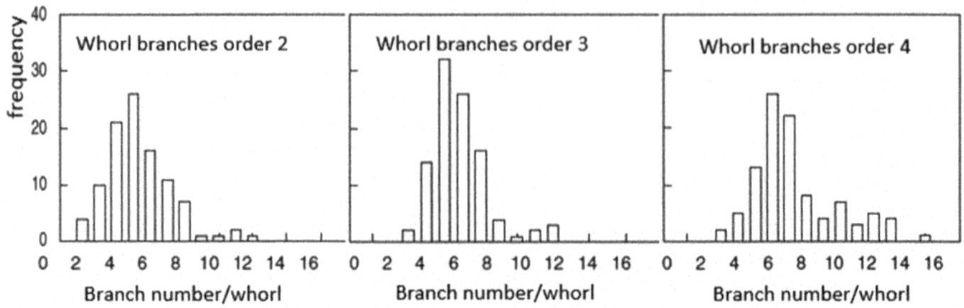

Fig. 10.39 Distributions of the number of branches per whorl along the axes with different PAs. These are fitted to binomial laws (n_p, a_p). n_p represents the number of potential branches and a_p represents the branching probability of physiological age (PA) p axes

Table 10.15 Number of branches per whorl as a function of the number of potential branches and the probability of branching (binomial law), depending on physiological age (PA) and several meta-phytomers per PA produced by the botanical automaton at chronological age of six DCs

PA	n_p	a_p	Number of GUs
1			6
2	15	0.27	40
3	6	0.22	53
4	4	0.13	13

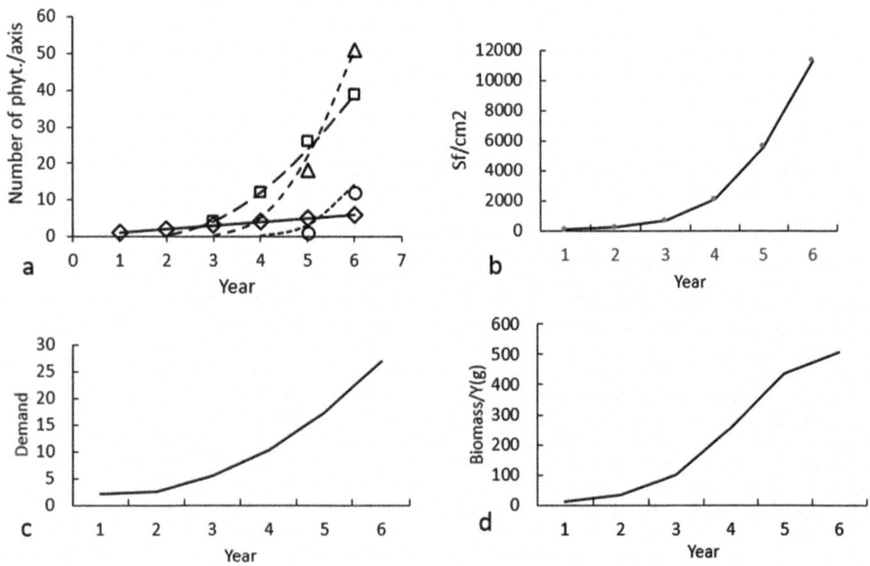

Fig. 10.40 (**a**) Fittings of the mean numbers of produced growth units (GUs) in order of branching and chronological age on the pine for a six years growth. The observations are represented by diamonds, squares, triangles and circles, respectively, for orders 1, 2, 3 and 4. The other graphs show (**b**) Evolution of leaf surface, (**c**) Plant Demand per GC, (**d**) Biomass production per GC

10.2.3.4 Growth Scheme

The growth pattern in pine tree is found to be identical to their development pattern. Simplified GUs have an immediate expansion. The growth cycle is the year. The needles are considered as a single leaf organ. The model of secondary growth associates each leaf with a ring element (Chap. 8).

10.2.3.5 Setting Up the Organic Series and Compartment of Organs

Organic series measurements are carried out on 6-year-old trees with four PAs and two types of organs (leaves and internodes), resulting in eight types of organic series. The leaves are only present over the last three years. The weights of the organ compartments are measured over the six years of growth to verify the stability of the source–sink parameters.

10.2.3.6 Calibration of the Growth Model

Calibration of the GreenLab model uses the WLSM to fit the observed organic series and compartments of organs by optimizing the source–sink parameters.

- Parameters measured empirically:

The SLW ε of the leaf, which corresponds to a thickness (although this notion is not directly related to a biophysical measure in the case of needles), is deduced from the needle sleeve. It is estimated as $\varepsilon = 0.035$ g/cm^2 in fresh weight. The light extinction coefficient k of Beer's law is set to $k = 1$, and the value of the available surface Sd per plant is set to 2400 cm^2. This can be compared with the computed production surface Sp. The allometry of the length–volume ratios of the GU is modelled. The GU is made up of a central pith, which corresponds to its shape at the time of emission, and which is subsequently surrounded by woody rings. To establish the allometry, only young GUs that do not yet have secondary growth should be considered.

This allometry is modelled by the equation (Chap. 5): $h(i,t) = \sqrt{\beta} \cdot (qm(i,t))^{\frac{1+\gamma}{2}}$.

h is the length of the GU and qm is its volume. Parameters β and γ (scale and shape coefficients) are computed. The fittings are done for each PA separately. Pith surface $\sigma(i,t)$ is computed by the following equation: $\sigma(i,t) = \sqrt{\frac{1}{\beta}} \cdot (qm(i,t))^{\frac{1-\gamma}{2}}$.

Figure 10.41 shows the form of the relationships that link the volume of the GU pith to its length and the quality of the fittings.

Values of β per PA (scaling coefficients) are: $\beta = [76, 163, 197, 358]$;
Values of γ per PA (shape coefficients) are: $\gamma = [-0.23, -0.30, -0.20, 0.14]$.

We can see that it is evident that the scale coefficient (β) is the parameter that changes the most according to the branching order. As the order increases, the GU

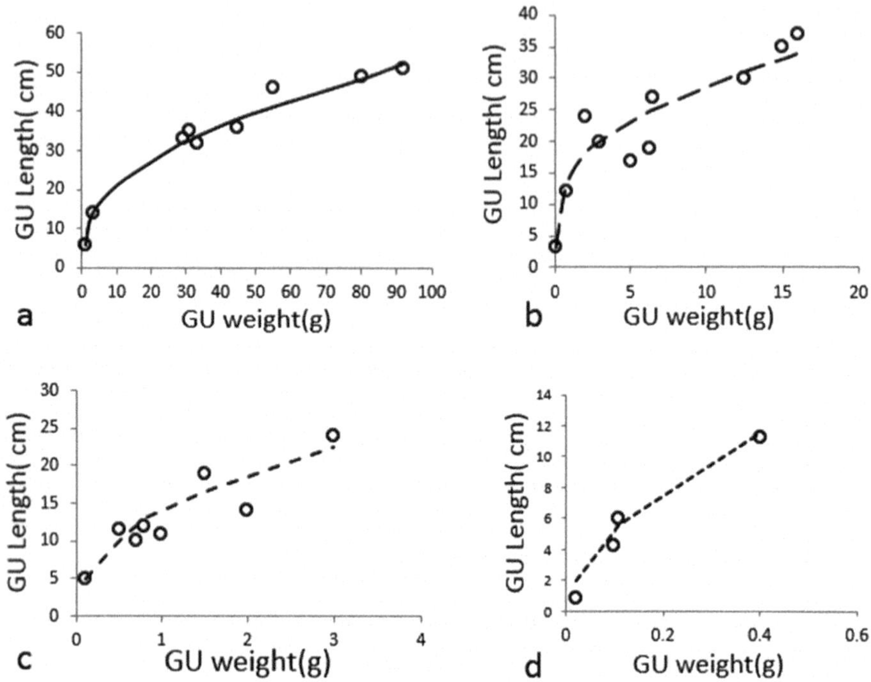

Fig. 10.41 Allometric relationships between the lengths h and the weight q of the growth unit (GU) pith in order of branching. (**a**) Order 1. (**b**) Order 2. (**c**) Order 3. (**d**) Order 4. These relations are established on the youngest GUs and are necessary to obtain the evolution of the lengths and the diameters of the axes

becomes more tapered. These parameters are crucial in ensuring the right length of the GU as a function of the pith volume.

- Estimated source–sink Parameters:

The WLSM provides the values of 10 hidden source–sink parameters for the 36 measurements on 6-year growth stages. The fitting is good: $R^2 = 0.99$.

Table 10.16 shows that the PAs (physiological ages = orders of branching) are very distinct. To be able to fit the secondary growth correctly for all PAs, the parameter P_c alone was not sufficient. Corrective terms cp had to be added: these parameters are indexed by the branching orders and represent secondary ring sinks, which distribute the allocated biomass to the rings differently according to PAs. The corrective term takes the value 0 on PA 3 and PA 4 axes, meaning that they have negligible secondary growth in comparison with that of PA 2 branches ($cp = 0.15$ for PA 2, which is few) and with that of the trunk ($cp = 1$). The biomass for secondary growth is thus preferentially directed to the trunk.

Table 10.16 Source–sink parameters of a pine tree with four physiological ages (PAs)

Parameters	PA 1	PA 2	PA 3	PA 4
Needle sink P_a	1	0.42	0.15	0.05
Internode sink P_i	0.76	0.11	0.05	0.02
Ring sink P_c	0.35			
	0.23			
Coef. rings *cp*	1	0.15	0	0
Resistance *r*	5.8			
Surface *Sp*	3036			
Q_0	5			

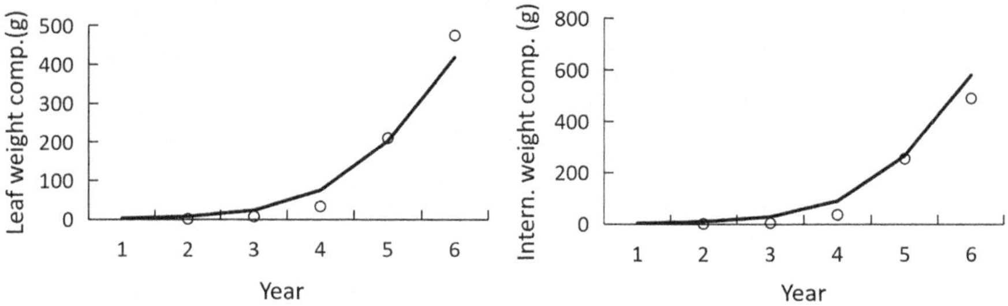

Fig. 10.42 Fitting pine biomass production over six years of growth for needles (leaves) and internodes compartments

The parameter *Sp* (3036 cm^2) is once more close to the available surface *Sd* (2400 cm^2) provided by the planting density, thus reinforcing the confidence in the model robustness.

Figures 10.42 and 10.43 show the evolution of the biomass of compartments (needles and internodes) and the organic series profiles at six years old of the branches for the four PAs.

10.2.3.7 Simulation of the Growth and Architecture of Young Pine Trees

In Fig. 10.44, one can observe the growth of the tree and the random effect produced by branching from one repetition to the next.

10.2.3.8 Discussion

In pine trees, the allometries between biomass and organ dimensions must be accounted for, because, in forestry applications, the shape and dimensions of the axes determine in part their economic value and their potential usage. The computation of the secondary growth needs to take into account the PA. The tree ring element method, which links the cross-sectional area of a tree ring to the number of

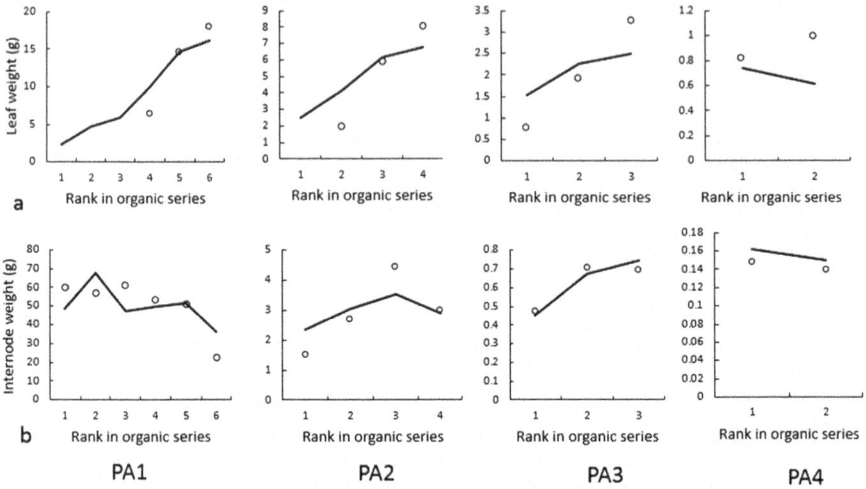

Fig. 10.43 Fitting the organic series of needles (**a**) and internodes (**b**) for the four PAs of pine at six years old (only the last three growth units (GUs) have needles)

leaves seen upstream, is not sufficient. Corrective terms (*cp*) are needed that modify the distribution of secondary growth according to the PA (here the order of branching). This is especially obvious in the short axes of trees with no secondary growth, and this is the case for orders 3 and 4 of branching in pine (Table 10.16).

10.2.4 Case Study 4: Guayule (Sylvie Sabbatier, Marc Jaeger, Nicolas Hemery, Ali Abed)

10.2.4.1 Description of the Plant

The Guayule (*Parthenium argentatum, Asteraceae*) is a shrub native from northern Mexico, attracting growing interest in research and agriculture (Ray 1993). The objective is to understand the interaction between the development of Guayule and natural rubber production. The Guayule is a shrub that forms abundantly branched clumps. Its architecture is of a sympodial type and is composed of undifferentiated modules with terminal flowering. It follows the Leeuwenburg model according to F. Hallé's classification. Each module bears new modules in its terminal position.

In one year, the plant produces axes made up of 8–9 successive modules. These modules are made up of internodes, the number of which follows a binomial law. To simplify, the module is considered to be a meta-phytomer (Phyto-Facto mode). All the paths that lead from the collar of the plant to the apexes are organic series made up of a chain of modules, ordered according to their rank from the base. The total weight of leaves and internodes per module is considered to constitute compound organic series made of meta-phytomers.

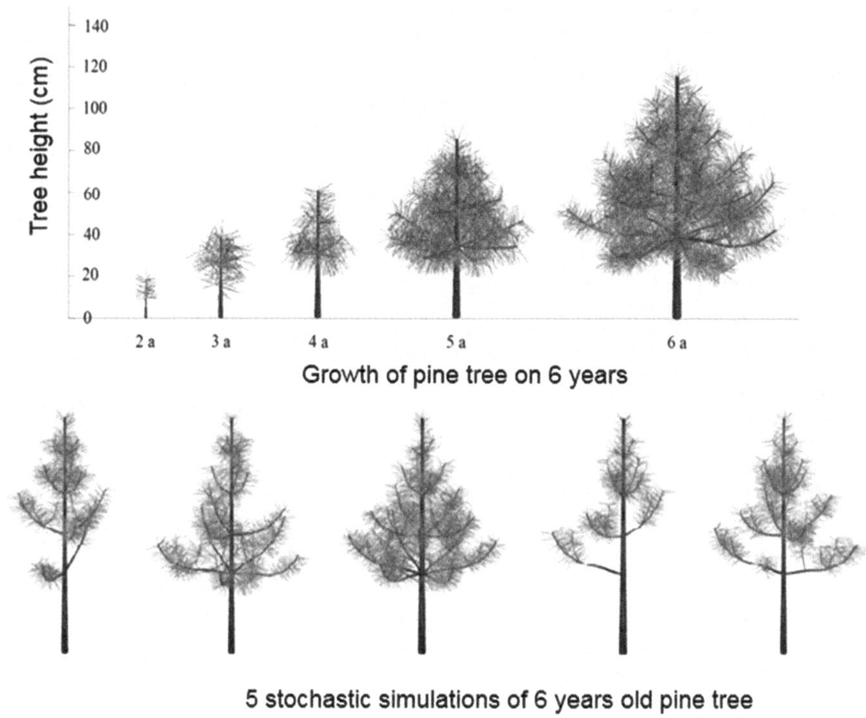

Fig. 10.44 Growth and architecture of a pine tree from two to six years with five stochastic repetitions of six years (Digiplant software). The simulation takes into account the allometries between the dimensions and the biomasses, the stochastic parameters of the development and the source–sink parameters calibrated on the measured organic series (Wang Feng, China Agricultural University (CAU))

10.2.4.2 Observation Data

Experimental plots were conducted on a variety named CL1 (parent USDA 11591 triploid). The individuals were planted in the field according to a planting density ($0.91/m^2$).

The measurements were carried out to count the numbers of phytomers per module as well as the numbers of relay sub terminal axes. Biomasses were measured in dry weight. Internodes and needles were collected from four individual plants. The weight of the root system was not measured. Allometries between the length and weight of the module were computed, using the data from new GUs with negligible secondary growth.

10.2.4.3 Development Scheme and Calibration

The architecture of the Guayule, which is sympodial, produces similar ramifications, which are considered as reiterations. There is, therefore, only one PA.

10.2.4.3.1 Development of a Module

On a sample of 100 modules, we observe a binomial distribution (10, 0.27) of the number of phytomers per module, with all ranks together. It means that we have a *Bernoulli* process with 10 DCs and a probability of success $b = 0.27$ (Fig. 10.45a). This distribution is stable according to the rank of the module, which has an average of 3 phytomers.

10.2.4.3.2 Branching Pattern

The number of branches per module follows a binomial distribution (Fig. 10.45b). However, this distribution evolves bottom up along the successive ranks. For instance, at the second rank, we get a distribution of (10, 0.34), which makes it possible to compute the branching rate $a = 0.34$ for a branch belonging to a whorl with potentially 10 branches. At rank 7, it is only (6, 0.21). This characterizes a damping process.

The product $N{\cdot}a$ gives the average number of branches per module. Thus, by setting the potential number of branches to 5, we obtain the following branching rates a(k) depending on the rank k (Table 10.17 and Fig. 10.45c).

The number of branches/module at the rank k is computed by $nb \cdot br(k) = 5 \cdot a(k)$.

Fitting the observed branching rate $a(k)$ according to the rank k of the sympodial axis (Table 10.17) can be satisfactorily achieved with the equation: $a(k) = 1.12.\ k^{-0.66}$.

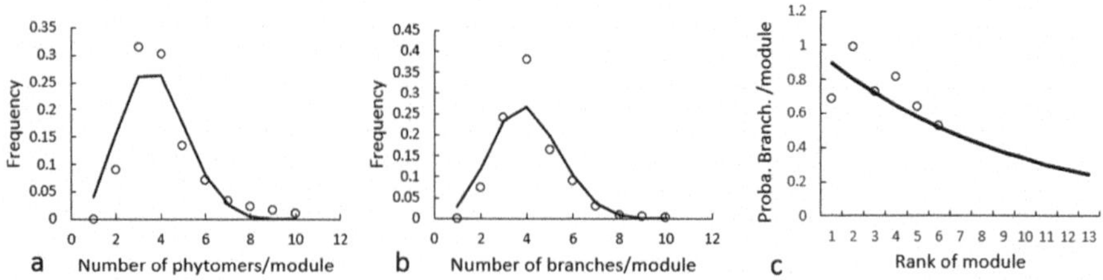

Fig. 10.45 Branching of Guayule. (**a**) Distributions of the number of phytomers per module along the ranks of modules inside plant architecture. (**b**) Distributions of the number of branches per whorl along with the ranks of modules. These are fitted to binomial laws. (**c**) The probability of branching per module along with the rank, the effect of damping

Table 10.17 Branching probabilities according to the rank (with potentially five branches per module)

Rank	1	2	3	4	5	6	7	8
Nb of branches per module	3.7	4.9	3.9	4.3	3.6	3.1		
Obs. rate $a(k)$	0.7	1.0	0.7	0.8	0.6	0.5	*	*
Comp. rate	0.90	0.80	0.72	0.65	0.58	0.52	0.47	0.42

*Represents the non-observed data

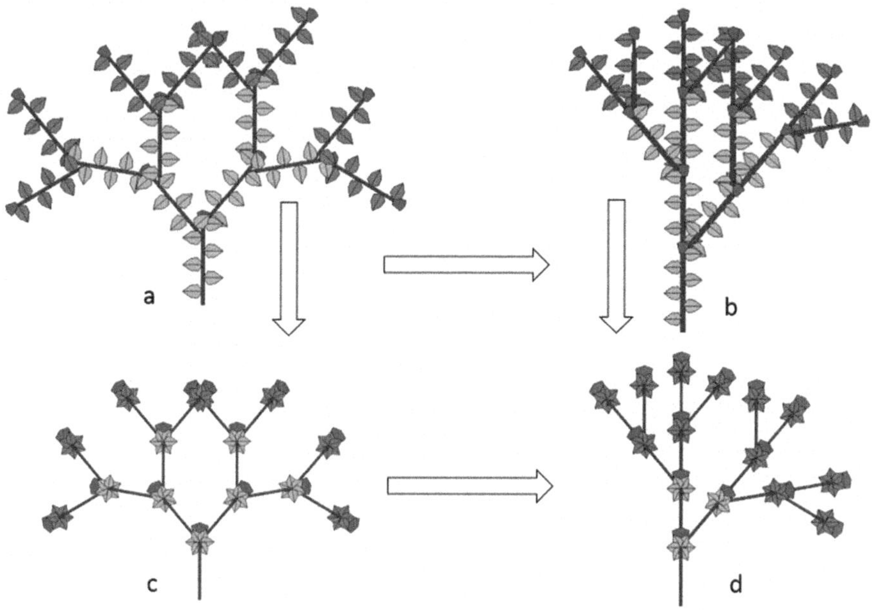

Fig. 10.46 Development of Guayule. (**a** and **b**) transformation of a sympodial model (**a**) into a monopodial model (**b**) that produces the same cohorts. (**c** and **d**) transformation of growth units (GUs) of (**a**) and (**b**) into meta-phytomers. Trees (**a**), (**b**), (**c**) and (**d**) produce the same cohorts of organs and the same organic series. They have the same growth in biomass

10.2.4.3.3 Simplification of the Development

The cycle corresponds to the time interval necessary for the establishment of a new module with its functioning time.

But the compound organic series is essentially a chronological organic series, because in a sympodial system the axes are not continuous but made of successive relays, which complicates data processing. It makes sense to transform the sympodial system into a monopodial system (Fig. 10.46), which produces the same cohorts of meta-phytomers with topological organic series (Chap. 5). Source–sink relationships become simpler to compute.

To switch from a sympodial system to a monopodial system, it must be assumed that there is always at least one terminal relay for each module. If $n_a(k)$ is the number of relays per sympodial module, the number of branches of the equivalent monopodial system is $n_a(k) - 1$.

However, it is estimated that there are on average five potential branches per module in the sympodial structure. This gives four potential branches in the equivalent monopodial system. In reality, we observe $n_a(k)$ branches at rank k. The branching rate is,therefore, $a(k) = (n_a(k) - 1)/4$.

Another problem to be solved is the maximum order of reiteration in the monopodial system. This is computed empirically by fitting the compound organic series of the phytomers. Indeed, the volume of the internodes depends on the secondary growth and, therefore, on the order of reiteration. Considering the monopodial scheme, the best fitting is obtained for three orders of reiteration.

10.2.4.4 Growth Scheme

The growth pattern is identical to the development pattern. Simplified modules in meta-phytomer forms have an immediate expansion. The sum of leaves is considered as a single leaf organ. The model of secondary growth associates each leaf with a ring element according to Pressler's law (Chap. 8).

10.2.4.5 Setting up the Organic Series

Organic series are compound organic series of modules representing meta-phytomers. Each meta-phytomers bears a terminal inflorescence. The state of development shows 14 ranks of modules, corresponding to one year and a half of growth.

10.2.4.6 Calibration of the Growth Model

Calibration of the model uses the WLSM to best fit the observed and computed compound organic series by optimizing the source–sink parameters of the GreenLab model.

- Parameters measured empirically:
  The SLW ε of the leaf is estimated as $\varepsilon = 0.005$ g/cm^2 in dry weight. The light extinction coefficient k, which occurs in Beer's law, is set to $k = 1$. Because of the Phyto-Facto mode, the organ expansion times of the module is 1 DC, the functioning time of the leaves is assessed to 4 DCs.
- Estimated source–sink parameters:
  There are three kinds of compound organic series (blades, internodes, inflorescences). The stage of measurements corresponds to 14 GCs. These measurements produce 31 organ weights in the compound organic series to be adjusted simultaneously during the growth process.

Table 10.18 Source–sink parameters of single-stem Guayule plant

Param	r	Sp	Q_0	P_c	P_{i1}	P_{f1}
Mean	35.3	7896	3.18	1.02	0.45	0.91
SD	22.7	2677	3.92	0.21	2.86	0.13
CV (%)	64	34	123	21	632	14

The fitting is good: $R^2 = 0.92$; number of computed parameters $= 6$ (the parameter P_b is fixed to 1). The values of the six parameters (Table 10.18) control well the fittings of the organic series for the growth period (14 GCs). As there is only one available stage of development, the stability of these parameters during the growth period is not established.

Even if the adjustment is satisfactory, the accuracy of the parameters given by the CV is poor.

Parameter Sp (7896 cm^2) is close to the available surface Sd (9000 cm^2) provided by the planting density, which is, again, satisfactory as to the adequacy of the GreenLab model to plant growth.

Figure 10.47 shows the evolution of the number of phytomers, plant demand, individual LAI, and produced biomass per cycle as a function of GC.

The quality of fitting of compound organic series can be assessed in Fig. 10.48.

Figure 10.48 shows a good fitting of the weight for internodes per module. The secondary growth is very important and is well fitted by the Pressler's law.

10.2.4.7 Simulation of the Growth and Architecture of Guayule Trees

In Fig. 10.49, one can visualize the growth of the potential structure of the bush.

10.2.4.8 Discussion

The Guayule is the first sympodial plant modelled by GreenLab. The bushy aspect of its architecture does not make it easy to measure. A necessary simplification is to transform the modules into meta-phytomers. Another simplification consists in transforming the sympodial structure into a monopodial structure, which gives the same organic cohorts and organic series, thus preserving the source–sink relationships. The stacking of axes in large branching orders leads to structures that are too large to handle for standard PCs. Accuracy on the evolution of the branching rate along the ranks of modules is an important factor to take into account, as it induces a great sensitivity on the growth simulation.

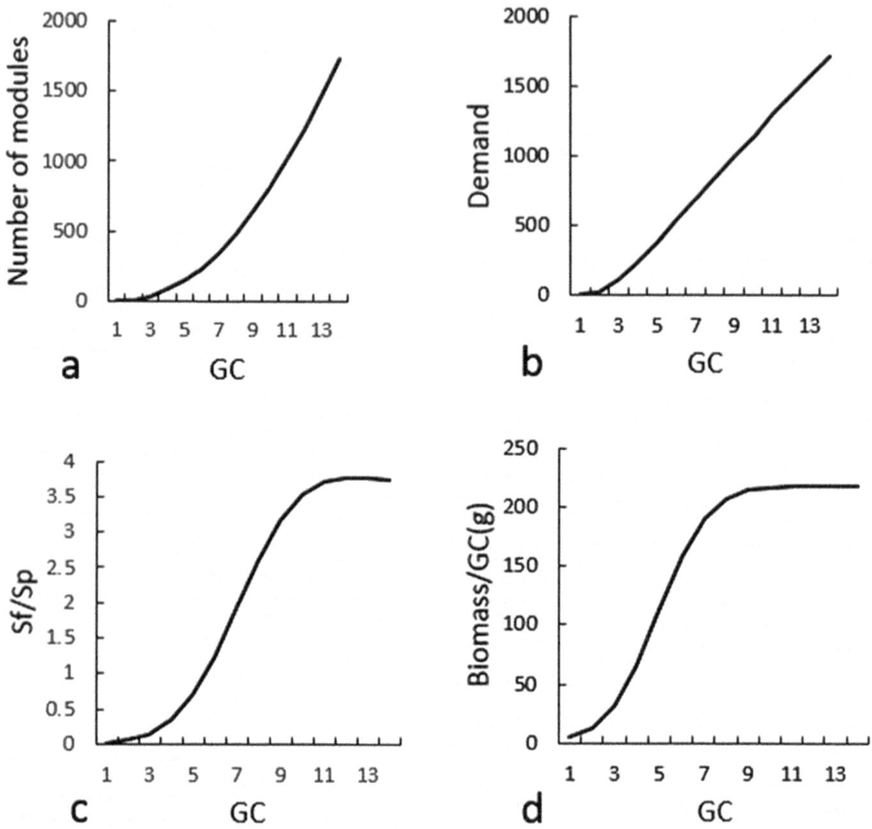

Fig. 10.47 History of the development and growth of Guayule. (**a**) Number of modules per GC. (**b**) Evolution of demand per GC. (**c**) Evolution of individual LAI. Production saturates $Sf/Sp > 3$. (**d**) Evolution of biomass production per GC. After an exponential growth phase, the plants reach saturation because $Sf/Sp > 3$

10.2.5 Case Study 5: Maple Tree Study (O. Togourdeau, S. Delagrange, CEF- UQAM)

Part of this study was published in (Taugourdeau et al., 2013).

10.2.5.1 Description of the Tree

The architecture of the sugar maple follows Rauh's architectural model. The structure is monopodial, and there are four categories of orthotropic axes, characterized by their vigour and length. They correspond to physiological ages (PAs). The phytomers carry two leaves, and phyllotaxis is opposite. The shoots are preformed

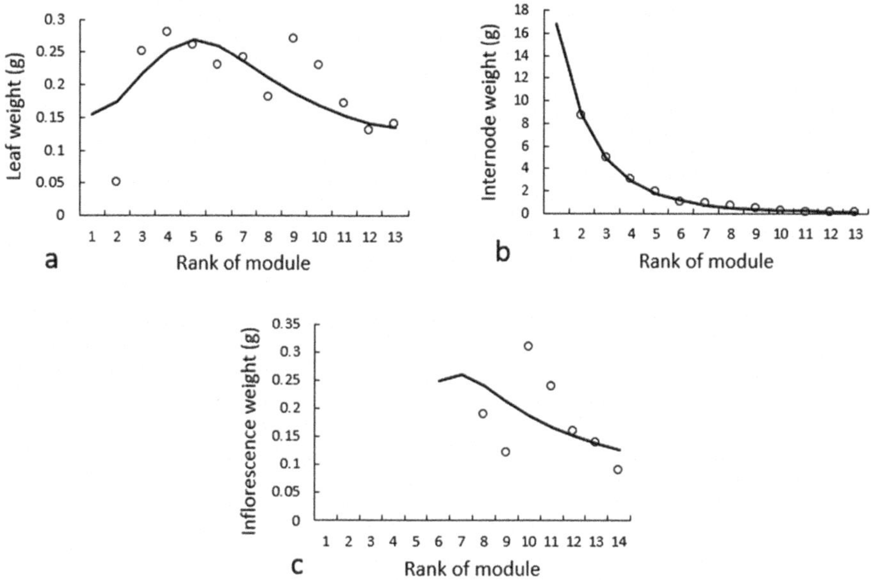

Fig. 10.48 Fitting of the observed compound organic series (○) and computed compound organic series (–) by the model for the measurement date (14 GCs). (**a**) Blades. (**b**) Internodes. (**c**) Inflorescences (data before rank 6 are missing) (Gloups software)

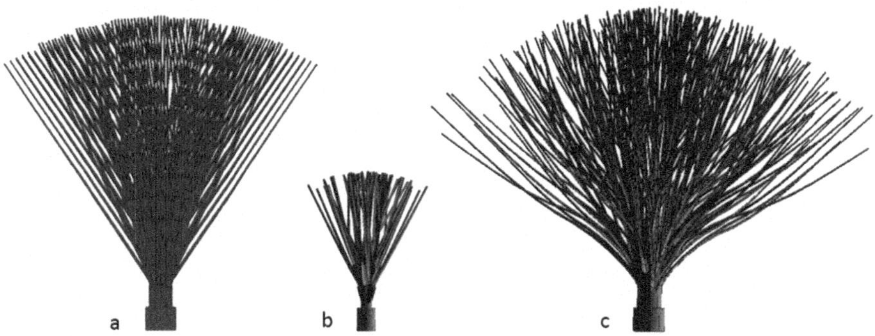

Fig. 10.49 Growth and architecture of Guayule (potential structure). (**a**) 2D representation at 14 GCs. (**b**) and (**c**) 3D representation at 7 and 14 GCs

and monocyclic. They are lengthened in spring over a short time period (a few weeks). All types of axes can reiterate, and the entire tree is a reiterated complex (see Chap. 4). Physiological ages are characterized by numbers of phytomers ranging from two per GU for short axes to 10 for long axes (trunk and its reiterations).

10.2.5.2 Observation Data

Sugar maples were studied in a set of forest plots in Quebec (Canada). All the measured trees were dominant. There were 31 selected developmental stages, with ages ranging from 3 to 78 years. Over these long growing times, axis mortality is a major component to be taken into account because the values of the meristems' functioning parameters are changing.

10.2.5.2.1 Production Surface Area

The trees studied are not in a regular planting pattern; the projection area *Sp* depends on the stage of development of the tree and no longer on the density of the plantation. The projection surface *Sd* of the crown on the ground, if its thickness is sufficient, should be a default estimate of the production surface *Sp* (Chap. 5). Under these conditions, the evolution of *Sp* is assumed to follow a U-shaped curve, capable of reflecting the initial phase of non-limited growth followed by the limited growth phase, with the minimum and maximum of which are empirically determined by ground measurements of crown projection surfaces. The shape parameters of this function are hidden and must be estimated using WLSM from the multi-stage evolution of the organic series. The projection surfaces were thus measured empirically for all individuals over 10 years old, so that they could be compared with those estimated by the model.

10.2.5.2.2 Determination of Axis Types

The four PAs were identified from architectural measurements such as the length of the GU, the number of internodes, the average length of an internode, and the type of carried branch.

10.2.5.2.3 Data for Development

The same dataset was used to calibrate the development in terms of distributions of the number of phytomers per GU and the branching rate of the different types of axes. The mortality evolution of the branch categories along the bearing axes was studied by the crown analysis method in Chap. 3 using a data set of about 50 branches collected from six trees of varying ages.

10.2.5.2.4 Data for Growth

Given the volume and complexity of adult maple architecture, it is not possible to consider it at the level of the phytomer. The total dry weights of the leaf and internode compartments of the living terminal GUs were weighed for the four PAs and 31 selected CAs, resulting in meta-phytomers. In addition, to estimate secondary growth, the diameters at the base of the trees were measured. This provides at least the standard variables used in forest models (height and diameter at the base of the tree, crown size). The compound organic series are thus reduced to their simplest expression: a terminal GU per type of axis and the diameters at the base of the trunk but repeated 31 times.

10.2.5.3 Scheme and Calibration of the Development Model

The four types of axes have significant effects on GU formation, meristem mortality and branching patterns. Each year, the apical meristems set up new preformed GUs if they do not abort.

Preformed tree GUs, such as beech or maple, usually have a small number of internodes (about 10) whose distribution follows a binomial distribution for a given cohort (see Chap. 3) but which changes significantly throughout development. Figure 10.50 shows the distribution of the number of phytomers per GU over the 22–23 years old cohorts corresponding to the largest observed GUs. The distribution adjusts to a binomial law $B(11, 0.75)$ whose mean is rounded to the value 8.

The GU size is not only subject to stochastic fluctuations in a given cohort (which precisely generates a binomial law) but also varies with the age of the cohorts. At first, on the trunk, the GUs consist of only two short phytomers. They gradually grow by increasing their number of phytomers to an average of eight and lengthening them. This is the classic base effect. To simplify the computations, given the low variability in the number of preformed phytomers, the average value of the number of internodes per GU and cohort was used to model development.

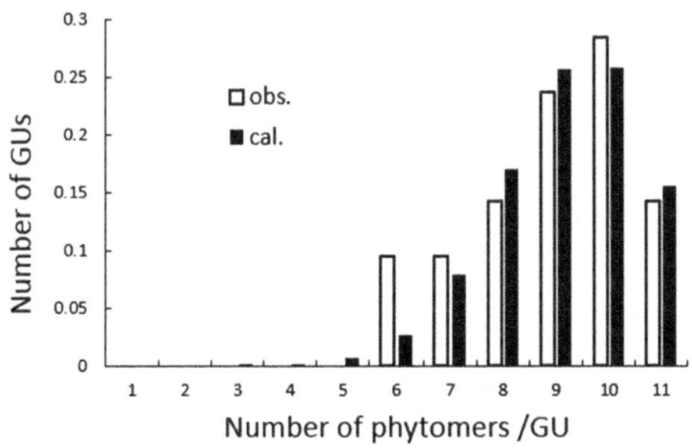

Fig. 10.50 Distribution of the number of phytomers per growth unit (GU) on the trunk (observed distribution, white histograms) over the grouped cohorts of about 20 years. This distribution adjusts to a binomial distribution (computed distribution, black histograms) $B(11, 0,75)$

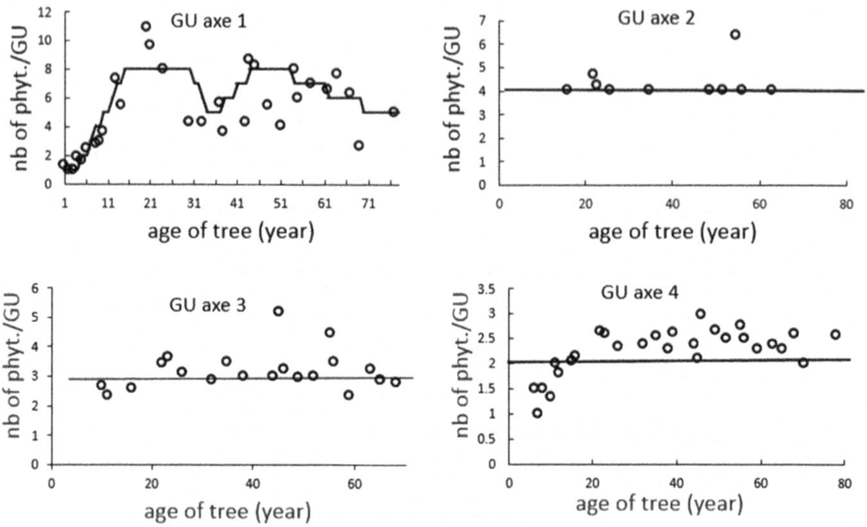

Fig. 10.51 Evolution of the number of phytomers per growth unit (GU) as a function of chronological ages (CAs) and physiological ages (PAs) in maple (axis 1 = PA 1 (trunk and reiterations). Axe 4 = PA 4 (short axis). This number can be considered constant for PA axes >1 and is highly variable on axes 1 (trunks and reiterations) with a marked base effect

In the GreenLab model, the parameters can be associated with a stress function that allows their variation according to CA to be continuously monitored, by interpolating between the values identified for different selected stages, similar to the reference axis (Chap. 2). Here, the parameter that varies is the average number of phytomers per GU of PA 1. Its evolution as a function of CA and PA is shown in Fig. 10.51. It can be seen that only the size of the GUs of axes 1 (trunks and reiterations) varies greatly according to CA. It is characterized by a marked base effect, followed by an 8-phytomer plateau, interrupted by a marked depression. The axes of types 2, 3 and 4 have GUs too small for the depression to be visible and branch too late (except type 4) for a basic chronological effect to be visible. A constant number of phytomers per GU of 4, 3 and 2 is established for axis types 2, 3 and 4, respectively.

10.2.5.4 Study of Axis Mortality

The trunk and its reiterations (type 1 axes) have no mortality, except for traumas followed by a relay that ensures the continuation of the axis and is, therefore, not taken into account in the model. The other types of axes experience mortality that increases with CA and PA. This mortality is measured by the crowns analysis (Chap. 3), which gives the mortality rate of the different types of axes as a function of their position relative to the apex of the bearing axis (Fig. 10.52).

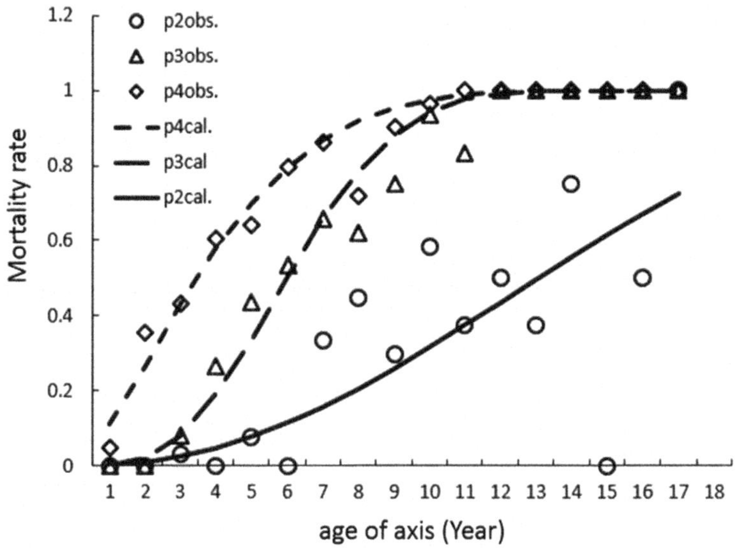

Fig. 10.52 Evolution of the mortality rate as a function of axis age in years and by physiological age (PA)
Axes 2 (○) have a maximum lifetime of 30 years. The rate is adjusted by the cumulative beta law of parameters (30, 2.45, 3.9). Axes 3 (Δ) and 4 (◊) have a maximum lifetime of 17 years and are adjusted, respectively, by the cumulative beta laws with parameters (17, 4.12, 7.3) and (17, 1.58, 5.22). Axes 4 (short axes) have the highest mortality rate

10.2.5.5 Study of Branching: Plasticity of the Growth Unit

The ramification of GUs depends on two factors. One is the existence of the type of phytomer that has a branch of a given PA (Chap. 4); the other is the associated branching rate.

The reference configuration of the potential GU is the maximum observable size (Fig. 10.53). All the others are obtained by contraction using a stress function. Among the possible contraction strategies, a good point of view that respects acrotony is to truncate the GU by displaying only the number of phytomers produced from the base (Chap. 6). Only PA 1 GUs have a variable dimension during the development. GU2, GU3, GU4 show astonishing stability.

The number of reiterations of the types of phytomer u_{ij} produced by the botanical automaton is measured, and so is the observed branching rate on their axillary meristems. The trunk is considered able to reiterate up to nine times, considering the number of bifurcations observed on the studied trees. Reiterations only occur at CAs when GU 1 are at their maximum size (resulting in two waves of reiterations). The maple's phytomers potentially have two branches per node. These may have different PAs. Their observed average number is fixed with their associated branching rates experimentally. The reiteration rate on GU 1 is estimated as

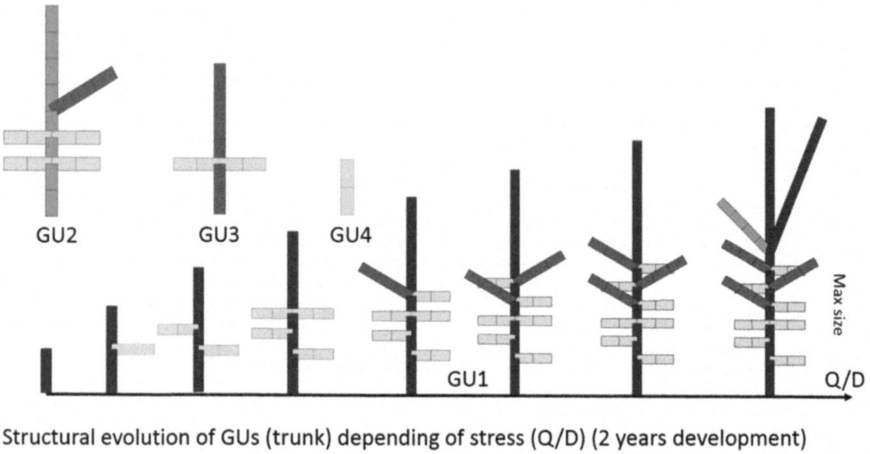

Fig. 10.53 Configuration of two-year-old growth units (GUs), physiological age (PA) 1 according to the number of produced phytomers. Smaller GUs have two phytomers and do not branch. Only GUs of maximum size can reiterate. The branches are arranged to reproduce the number of types of observed branches according to the size of the GU. GUs 2, 3 and 4 have stable configurations. GUs 1, 2, 3 and 4 can reiterate up to a high order (limited to 9 in the case of maple)

$a = 0.08$ when the phytomer exists. The branches of PA 2, 3 and 4 appear on GU 1 with the rates $a = 0.6$, 0.85 and 1, respectively.

On GU 2, branching rates are $a = 0.5$ and 1 for PAs 3 and 4; on GU 3, branching rate is $a = 1$ for PA 4.

GU 3 and GU 4, which have high mortality, can reiterate by forming small sympodial complexes whose reiteration rates are estimated at approximately $a = 0.15$. On maple, branching rates are stable.

10.2.5.6 Production of Phytomers during the Development

Once the botanical automaton has been built, it is possible to compute the tree's development up to 78 years, the date of the last observation. It is necessary to count nine development cycles (DCs) per year, considering that there is a maximum of eight phytomers to create (GU 1) and that an intermediate pause is required to separate the years.

The functioning of the preformed GU is simulated as follows (Chap. 6): during eight cycles, the phytomers are produced one by one in an embryonic state. In the eighth cycle, all the phytomers in the GU expand using the previous year's reserves. Then the leaves produce the annual biomass production during this cycle. This procedure considerably shortens the functioning time for the same result. The botanical automaton gives the following production (Table 10.19).

There are, therefore, about 800,000 phytomers produced at 78 years by the botanical automaton. Reiterations considerably increase the number of phytomers.

Table 10.19 Estimates of the number of phytomers produced by maple by physiological age (PA) at the age of 78 years

PA	Number of phytomers
1	3046
2	2394
3	116,511
4	655,374

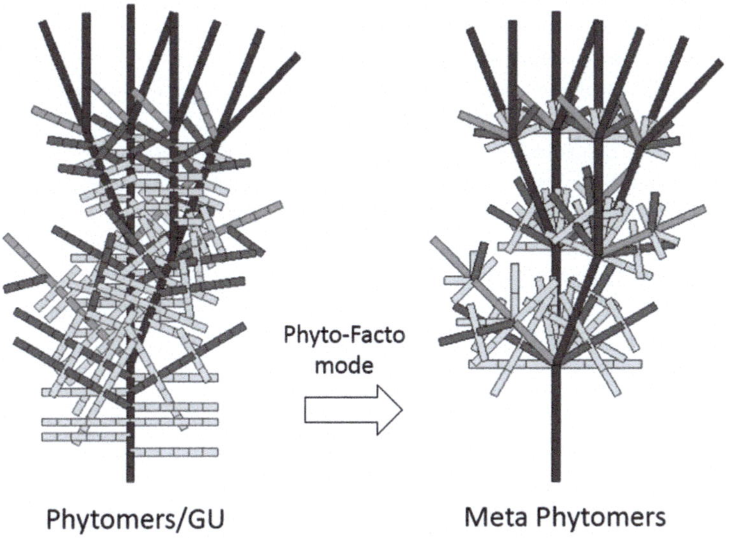

Fig. 10.54 Transformation of a growth unit (GU) in meta phytomer (Phyto-Facto mode). This operation reduces the development computation, but the growth and obtained shapes are unchanged

This limits the use of list mode on a standard PC that cannot manage such a complex topological structure. Only the matrix mode can perform computations to estimate cohorts and organic series production.

However, the matrix mode does not allow for drawing the plant. The Phyto-Facto mode transforms the GUs into simple meta-phytomers, which strongly reduces the computation of the development, without changing the growth and the dimensions of the architecture (Chap. 8, Fig. 8.30). Figure 10.54 shows the effect of the Phyto-Facto mode.

10.2.5.7 Use of Ontogenic Age for Reiteration Development

The reiterations of the trunk are its duplications, but this statement is not enough. They are a localized duplication of the carrier axis, initialized with an ontogenic age that corresponds to the time of their initiation (see Chap. 4).

10.2.5.8 Establishment of Organic Series

Due to a large number of measured trees, only a sampling of terminal GUs of the different types of living axes is measured (Chap. 8, Fig. 8.22). The compound organic series is made up of the weights of leaf and internode compartments per GU. There is, therefore, only one element per series for one of the 31 selected chronological ages.

In addition, the diameter at the base of the tree is measured (Chap. 8, Fig. 8.22). The basal section area multiplied by the length of the internode at this level gives the volume of the pith of the internode with its rings. The density is rounded to 1 to obtain the biomass from the volume of the phytomer.

There are 31 measured growth stages and four PAs, so there are 124 leaf compartment measurements and as many internodes on terminal GUs, and 31 measurements of trunk diameter, all of which must be fitted by the model.

Finally, the evolution of the crown projection surface was estimated, for each selected development stage, from the measurement of its extension to the ground in the four cardinal directions. This surface is to be compared to the production surface *Sp* optimized in its evolution by the model when adjusting the organic series.

10.2.5.9 Calibration of the Growth Model

There are 31 development stages available to adjust the organic series composed of the tops of the trunk axes and basal diameters by the respective outputs of the model. Over this long period, the evolution of the parameters of the functioning of meristems has been observed, in particular for the size of the GUs of axes 1 and the mortality of meristems. What about source–sink parameters? WLSMs can only converge if the evolution of the parameters is taken into account. In this case, to obtain a fitting, it must be assumed that the production area varies according to a U-shaped curve (Chap. 5), and that the sink of the ring element damps as the chronological age increases.

10.2.5.9.1 Equation for the U-Shape of *Sp* Evolution

It can be assumed that the variation in *Sp* with CA adjusts to an empirical function with a U-shape. This function of *Sp(t)* must depend on a small enough number of parameters to be estimated from equations of GreenLab model: a good candidate for such a function is a transformation of the beta law, already acknowledged for its flexibility.

Consider the beta function discretized in steps of 1 and defined over the interval [1,*T*] (see Chap. 5).

$$B(t) = \left(\frac{(t-0.5)}{T}\right)^{aSp}\left(1-\frac{t-0.5}{T}\right)^{bSp}\frac{1}{T} \tag{10.10}$$

This function has a minimum $Bmin = 0$ and a maximum $Bmax$, in the interval [1, T]. The shape of this curve is bell-shaped if $aSp > 1$ and $bSp > 1$.

By transformation of B, we create the function BF of the same nature as B with the same parameters a and b, and whose imposed minimum and maximum are *Smin* and *Smax*:

$$BF(t) = Smin\left(1-\frac{B(t)-Bmin}{Bmax-Bmin}\right) + Smax\left(\frac{B(t)-Bmin}{Bmax-Bmin}\right) \tag{10.11}$$

The bell curve can be transformed into a U-shaped curve by inverting its shape while keeping the maximum and minimum:

$$Sp(t) = Smax\left(1-\frac{BF(t)-Smin}{Smax-Smin}\right) + Smin\left(\frac{BF(t)-Smin}{Smax-Smin}\right) \tag{10.12}$$

In the case where a sufficient number of individuals at various stages of development are available, we can empirically estimate *Smin* and *Smax*, using the measurable ground projection surfaces of canopies whose foliage thickness is sufficient to obscure the light.

By the inverse method, parameters *aSp* and *bSp* can be estimated along with the other parameters by fitting the data using WLSM (i.e. organic series over several stages of development) to the model. If the fit is satisfactory, the numerical values identified for *Sp(t)* can then be compared with the respective readings of the *Sh(t)* ground projection areas of the canopies. If there is a match, the model works correctly. This method, which has the merit of being able to be calibrated on observations, gives good results on trees such as teak and maple (Fig. 10.56a).

10.2.5.9.2 Parameters Obtained Empirically

The allometry of the leaves (thickness ε) is estimated as $\varepsilon = 0.01$ g cm^{-2}; for the internodes, the shape coefficient $\beta^2 = L_{\varphi}(i)^2$ and $\alpha = -1 \cdot L_{\varphi}(i)$ is the average value of the length of the internode in GU of the PA φ at CA i. This method gives the height of the tree at all CAs by summing the lengths of the phytomers.

The minimum projection area *Sp*, $Smin = 3000$ cm^2, is reached at 11 years and the maximum area $Smax = 400{,}000$ cm^2 at 78 years (the last measured stage).

The ratio between the average weights of GU organs for different PAs is an estimate of the ratio of their sink strengths to the GU (because variations in sink functions are assumed to be the same by organ type for all PAs). The leaves of the trunk have, by definition, a sink normalized to the value 1, which serves as a reference. Since the internodes were measured at the end of the season, the diameter

Table 10.20 Relative sink values as a ratio of compartments

	Axe 1	Axe 2	Axe 3	Axe 4
Leaf sink	1	1	1	0.6
Internode sink	0.75	1.8	0.5	0.1

consists of the pith of the internode and the first ring of secondary growth. On the terminal GU, the contribution of the ring is negligible compared to the pith for the internodes of axes 1, 2, 3 and 4.

The sink functions of axes 2, 3, 4 are obtained by multiplying the ratios corresponding to the sink of axis 1 for the concerned PA (Table 10.20). These allometries, computed on the average of all growth stages, are, therefore, fixed.

The seed is empirically initialized to the value $Q_0 = 0.2$. It is known that it is impossible to estimate it by the inverse method beyond the very beginning of growth.

10.2.5.9.3 Hidden Source–Sink Parameters

There are only a few hidden source–sink parameters:

- The parameter r of resource use efficiency in the production equation;
- The coefficients aSp and bSp of the projection surface Sp in U shape, whose minimum $Spmin = 3000$ cm^2 and maximum $Spmax = 400{,}000$ cm^2 were estimated from the projection surfaces measured at 11 and 78 years;
- The damping of the sink of the ring element during the growth of the tree (Chap. 8, Fig. 8.10b). The WLSM does not converge if the p_c ring element is constant. It must decrease with chronological age. In DC i, the sink of the ring element is modelled according to the equation: $p_c(i) = p_{c1} \cdot (p_{c2})^i$. It is, therefore, necessary to identify two parameters for secondary growth: p_{c1} and p_{c2};
- The coefficient λ, an additional parameter that determines the balance of the biomass distribution mode in rings (Chap. 8);
- The sink of the pith of the internodes of axes 1, p_i.

There are, therefore, seven parameters to be estimated by the WLSM from the data.

The sink parameters of axes 2, 3 and 4 are obtained by the multiplicative coefficients presented in Table 10.20. The fitting gives a $R^2 = 0.9$ that is correct.

Table 10.21 gives the source–sink parameters of the maple computed by the WLSM.

The quality of the fittings can be assessed (Fig. 10.55) on the evolution of leaf weight and internodes of apical GUs throughout the 31 growth stages.

Despite the noise on the data, due to the heterogeneity of the measured population, the organic series is well restored. However, the leaves are better fitted than the internodes. At the beginning of growth, the internodes of axis 1 are overestimated. This is because the distribution of biomass between the internode pith and leaves is

Table 10.21 Source–sink parameters of maple (multi-stage calibration)

Parameters	Axe 1	SD	CV (%)	
Leaf sink P_a	1	0	0	Fixed
Internode sink P_i	0.95	0.06	6	Estimated
Ring sink P_{c1}	3.91	0.29	8	Estimated
Allocation coeff. λ	1	0	0	Estimated
Damping of the ring P_{c2}	0.975	0.002	0	Estimated
Resistance r	9.62	0.57	6	Estimated
Max *Sp*	400,000	0	0	Fixed
Min *Sp*	3000	0	0	Fixed
aSp	1.28	0.04	7	Estimated
bSp	2.68	0.18	6	Estimated
Q_0	0.2	0	0	Fixed

not stabilized before 10 years, an effect that the model does not take into account at this time.

The diameter growth at the base of the tree was transformed into the volume by the eq. $Q = \pi \cdot l \cdot d^2/4$, with l being the length of the internode and d being the diameter at this location. Figure 10.56b shows that the forecast of secondary growth at the base of the tree is correct. The simple secondary growth model is a good approximation: Pressler's law is well verified with $\lambda = 1$. Experimentally, to obtain a satisfactory fit, the sink of the ring element must decrease over the cycles by a damping coefficient $P_{c2} = 0.975$.

The shape coefficients *aSp* and *bSp* of the U curve, which controls the evolution of *Sp*, are well computed. The evolution of the computed production area *Sp* is quite comparable to the evolution of the measured projection surfaces (Fig. 10.56a). This last point, which is a good validation of the model, gives a tangible biological interpretation to the concept of production area since the projection surfaces were measured independently. Figure 10.56b shows that the prediction of the weight of the first GU of the trunk is well assessed.

The number of phytomers in the adult tree is large. The PA 4 axes (short) are those that produce the most and build the largest part of the leaf area (85%). These axes are dedicated to the biomass production and can be referred to as 'harvesting shoots'; around the age of 80 years, the botanical automaton generates 40,000 leaves per cycle (year). Figure 10.57a shows the different contributions of the four types of axes. The dimensions of the architecture are essentially ensured by axes 1 (trunk and reiterations). They can be referred to as 'exploration shoots' since they allow the tree to find and establish in lighted spaces. The trunk at the beginning of growth only produces short GUs and does not branch. It is necessary to wait for 6, 11 and 15 years to see the successive branches of axes 4, 3, 2 and 1 appear on the GUs of the trunk.

The particular structure of the evolution of the GU of axes 1 means that two waves of reiteration occurred during the growth. These waves induce a decrease in the *Q*/*D* ratio and, in turn, the shrinkage of GU 1. These waves do not appear in the

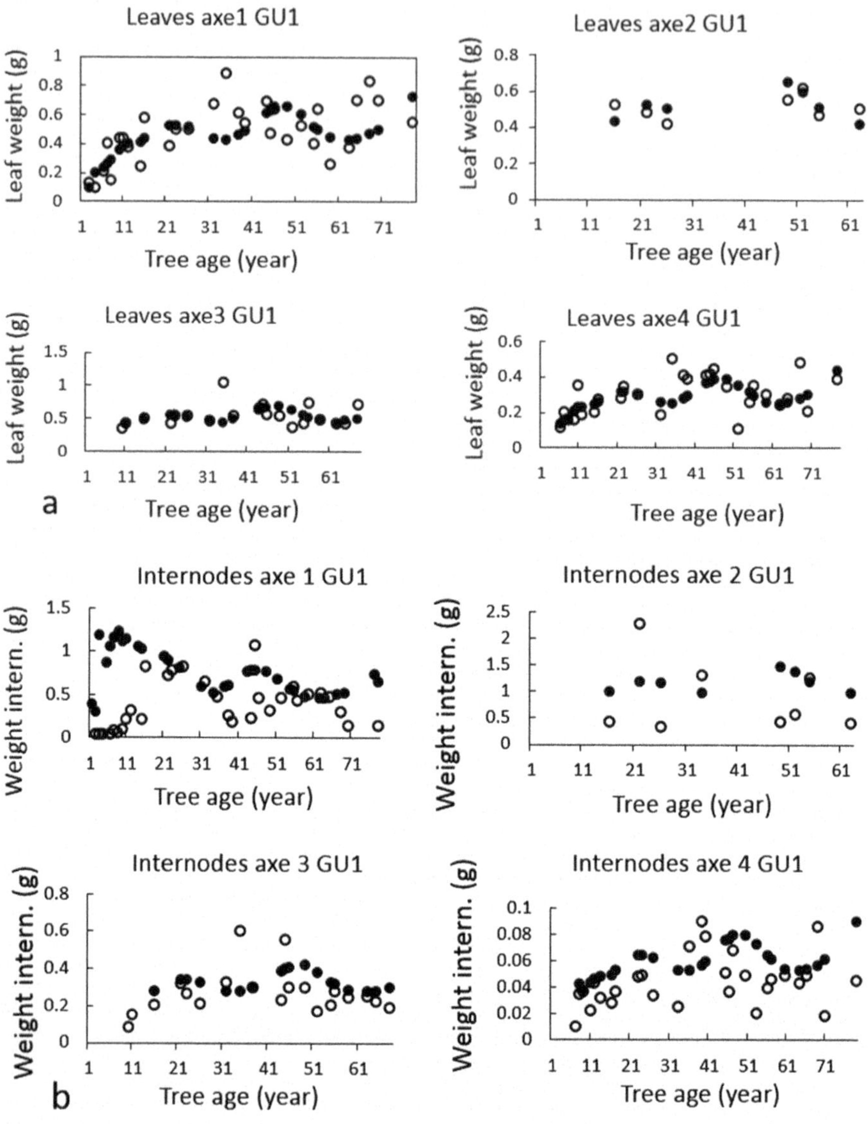

Fig. 10.55 Evolution of weights of leaves and internodes. (**a**) Fittings of observed leaf weights (○) and computed leaf weights (•) by the model for different chronological ages (CAs) and physiological ages (PAs), on apical growth units (GUs). (**b**) Fittings of observed internode weights (○) and computed internode weights (•) by the model

curve of biomass production (solid line in Fig. 10.57b) because the expansion of the production is linear, but they appear on the evolution of demand. On the other hand, *Q/D* oscillations (Fig. 10.57c) modify the number of GUs and the size of organs for the development of axis 1. The proportion of biomass allocated to the rings decreases

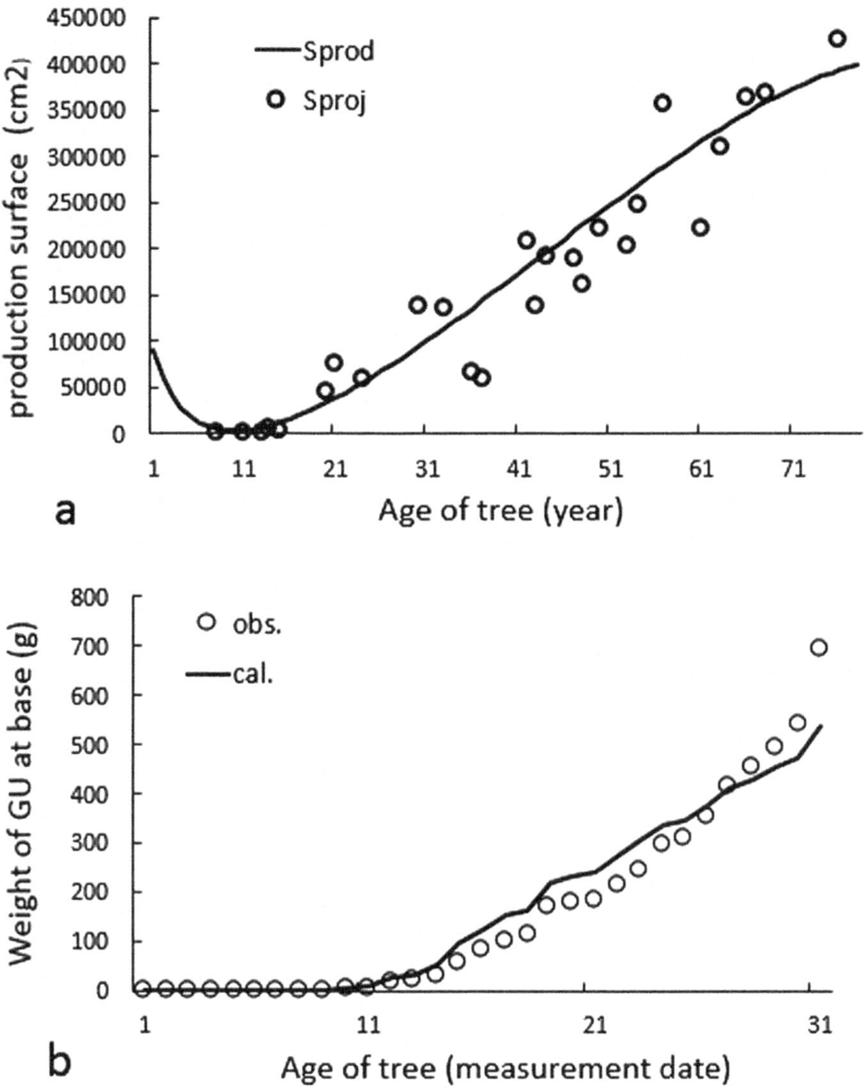

Fig. 10.56 Evolution of surfaces and weights of growth units (GUs) at the base of the trunk. (**a**) Comparative evolutions of the projection surfaces of the crown trees (○) and the computed production surfaces (curve), over the growth time. (**b**) Adjustments of the observed (○) and computed (-) GU weights at the base of the trunk for the observation dates

during the growth. Owing to this mechanism, the biomass can be preferentially allocated to primary growth that ensures the growth in height.

The Gloups software cannot simulate, in list mode, the architecture of the tree beyond a young age of around 10 years, as computation times become prohibitive

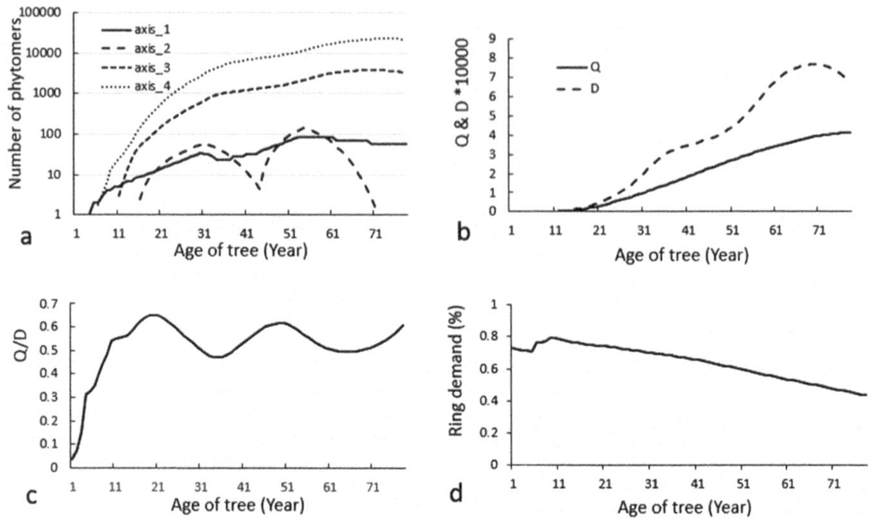

Fig. 10.57 Characteristics of Maple production. (**a**) Production of the number of phytomers in logarithmic coordinates according to the types of axes. Short axes represent 85% of production. Axes 3 and 2 correspond to the dashed lines. The solid line corresponds to axes 1 (trunk and reiterations). The trunk only branches in the 6th year. (**b**) The exponential increase in production at the beginning becomes linear after the 20th year, because it is controlled by the linear expansion of the crown projection surface, which is of the same order of magnitude as that of the production surface. The increase in demand is distorted by the appearance of reiterations. (**c**) Supply on demand oscillates under the effect of waves of reiterations. (**d**) The proportion of biomass going to the rings decreases with the age of the tree, which preserves its growth in height

for a PC. An 80-year-old Maple tree produces about 800 DCs. The memory required to store the topological structure is insufficient.

In matrix mode, it can easily compute the growth of the tree over a few hundred years and the profile of its axes (organic series), including the trunk with its rings (but not its architecture). The profile of the rings gives access to the study of the quality of the wood. The effect of the two waves of reiteration on the trunk profile can be distinguished (Fig. 10.58).

10.2.5.10 Simulation of the Development of Maple Tree Architecture

In list and Phyto-Facto mode, the software can display the primary structure of the tree (corresponding to PA 1), meaning it can show only the trunk and its reiterations beyond 100 years. Figure 10.59 shows five random simulations of the Maple tree without the branches at 78 years old.

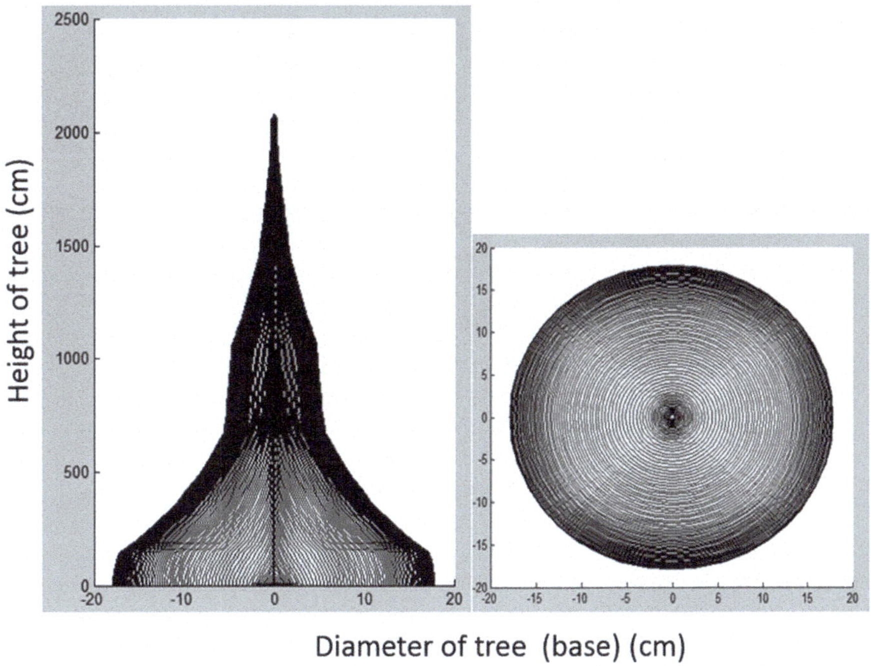

Fig. 10.58 Visualization of the rings (longitudinal and radial) obtained by the simulation of secondary growth. Rings are made up of an assembly of ring elements produced by the leaves and assembled in bundles towards the base of the tree. The more active leaves a phytomer sees upstream, the wider its ring is. The effect of the two waves of reiteration is visible. At 78 years old, the tree is about 22 m in height

10.2.5.11 Discussion

While the tree can be computed at 80 years old, and it can be disassembled and visualized in separate parts (details of the axes), its architecture cannot be simulated as its topological structure is too voluminous. But foresters are mainly interested in these axes as separate parts, including the profile of the rings in the trunk that can be provided by GreenLab modelling. Finally, this study does not include the root system and fruiting, which were not available. Assuming that the biomass allocated to the root system would be proportional to the total biomass produced, and that fruiting at the terminal GU level would be proportional to its weight, their neglection only implies an overestimation of the value of parameter r.

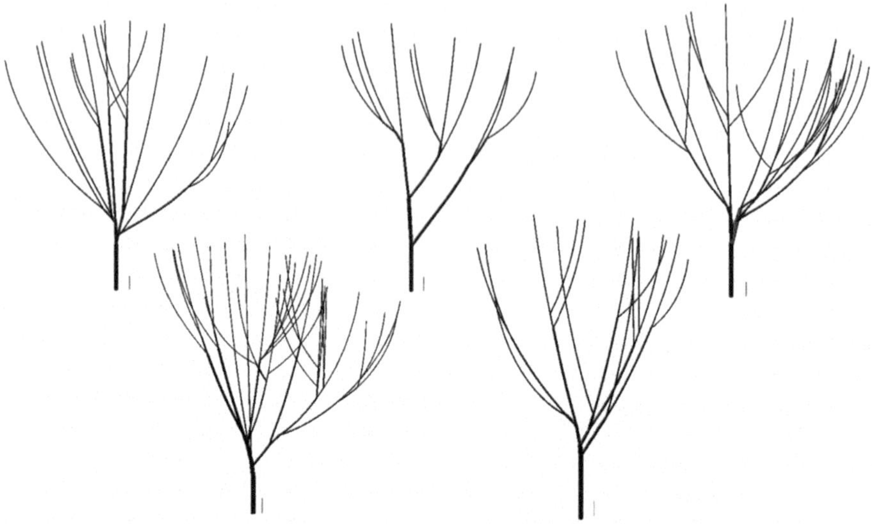

Fig. 10.59 Five random simulations of the primary architecture (trunk and reiterations without the branches) of a 78-year-old maple tree. We distinguish the two waves of reiterations, 1 m at the foot of the tree gives a scale for reference

10.2.6 Case Study 6: Teak Tree Study (Tondjo Kodjo, Sylvie Sabatier)

10.2.6.1 Description of the Tree

The teak tree, *Tectona grandis*, is a tropical tree known for its highly valued wood. The study took place in plantations in Togo (Tondjo et al 2018). The development of the shoots is subject to seasonality due to the alternation of dry and wet seasons. The tree has rhythmic growth and branching. Its axes consist of a succession of growth units. Its flowering is terminal and generally occurs during the 5th year of growth. At the same time, the development of the axes, initially monopodial, becomes sympodial. The studied trees, which are between one and five years old, have not yet bloomed, and all their axes are alive. There are four categories of axes: the trunk (PA 1), the main branches (PA 2), the long branches (PA 3) and the short branches (PA 4). The phyllotaxis is oppositely decayed. Each phytomer bears two leaves and can bear zero, one, or two branches. The trunk may bear branches, long and short twigs, preferably in the middle part of the GU (there is no acrotony).

Each year, the terminal meristems produce a new GU consisting of a preformed part followed by a neoformed part, whose number of phytomers varies greatly. Subsequently, a second GU is formed in the same year (during the short-wet season), giving rise to a bicyclic annual shoot. Only the first GU of the annual shoot is branched and has a greater number of internodes than the second GU if it exists. The architecture of teak has a highly stochastic appearance.

10.2.6.2 Observation Data

The study was carried out on plantations of young teak trees between one and five years old. The climatic factor is assumed to have a constant, normalized value $E = 1$, which is reasonable to represent the tropical conditions where the temperature varies little.

10.2.6.2.1 Data for Development

To establish the distribution laws of the number of phytomers per GU and branching, trees aged one, two, three, four and five years were analysed with 70, 30, 12, 7, 6 and 5 repetitions of individuals, respectively. The stochastic part of the development of the axes is found in the branching and the lengthening of the GU of the annual shoot. At the level of the GU, the development must be modelled according to a *Bernoulli* process, which is composed of the law of cessation in neoformation. All annual shoots have the same number of potential cycles, even if they stop their development early or later. The number of DCs in a year is at least equal to the number of phytomers produced by the largest annual shoot.

10.2.6.2.2 Data for Growth

Dry biomass measurements were carried out on trees aged one, two, and five years, with six replicates of individuals per age, respectively. Leaf and internode dry weights were measured at the phytomer level and above the GU level in cumulative weight. The GU becomes a compartment. Due to the large number of phytomers in a five-year-old tree, only the weights of the leaf and internode compartments per GU were measured to build compound organic series. Note that the level of phytomer remains, and GUs are not transformed into meta-phytomers. The leaf functioning time is one year. The organs simultaneously stop their expansion at the end of the growth. The last phytomers formed, therefore, have an unfinished expansion. The pause period following the cessation of functioning can be reduced to one cycle, as the increase in its duration has no impact on development and growth.

10.2.6.3 Scheme and Calibration of the Development Model

The maximum age of the analysed trees is five years. They begin to branch in the third year and carry a mixture of long and short branches and twigs. At five years, only the trunk bears branches. When GUs are grouped according to their PA and CA, and their rank in the annual shoot, the distribution of the number of phytomers per GU can be established, and thus the laws of meristem activity can be obtained. The year of formation of a GU has no significant influence on the distribution of its number of phytomers. GUs are grouped according to two criteria: PA or category of axis and the rank of the GU in the annual shoot (maximum polycyclism =2 macrocycles). The development study is divided into two parts: the analysis of the distribution of phytomers per GU and the analysis of branching, which only concerns the trunk for the studied trees.

10.2.6.3.1 Analysis of Phytomers Distribution in GUs

The stochastic GU model used here is described in Chap. 3. The distributions generated are similar to those of the cocoa tree (Chap. 3, Fig. 3.25d). The distributions of the number of phytomers per GU fit well with negative binomial laws shifted by three phytomers. This number represents a constant number of pre-formed phytomers. All GUs have a neoformed part which is controlled by the two parameters (K, c) of the law, which give the probability of cessation of activity of the meristem according to the DC.

It is known that the composition of the *Bernoulli* process with a binomial-style cessation law gives a law of the same nature as the cessation law (positive or negative). It is, therefore, impossible to extract the value of parameter b without growth monitoring. For the sake of simplicity and without notably lacking precision on the growth analysed, $b = 1$ is then set. The fitting of the distributions by the WLSM is done with software Gloups_dev and gives the following Table 10.22 for the distributions whose sampling sizes are displayed in brackets.

The quality of the fittings is shown in Fig. 10.60. Given the associated uncertainty on these values, the parameter c can be considered as constant ($c = 0.42$). The effect of PA becomes significant from PA 3 onwards, as does the rank of the GU in the annual shoot. The second GU of a shoot is twice as small. The short axes of PA 4 only have monocyclic shoots.

These distributions give the maximum number of DCs per annual shoot by suming the numbers of cycles of GU 1 and GU 2 truncated at the 5% threshold. The end of growth is estimated at 36 DCs. An additional pause must be added to mark the annual cessation of terminal meristem functioning for all PAs, which brings the total to 37 DCs. These cycles function synchronously with the resumption of development, which slows down or stops at varying rates depending on the PA of the terminal meristem and the negative binomial law that controls the cessation of activity.

Table 10.22 Parameters of the activity laws of the meristems on teak

	GU 1		GU 1	GU 1	GU 2		GU 2	GU 2
	phyt. pref.		phyt. neo. K	c	phyt. pref.		phyt. neo. K	c
PA 1	(210)	3	7	0.49	(150)	3	3	0.35
PA 2	(149)	3	7	0.45	(70)	3	3	0.44
PA 3	(220)	3	4	0.41	(77)	3	2	0.48
PA 4	(100)	3	3	0.3	*		*	*

*Represents no data available

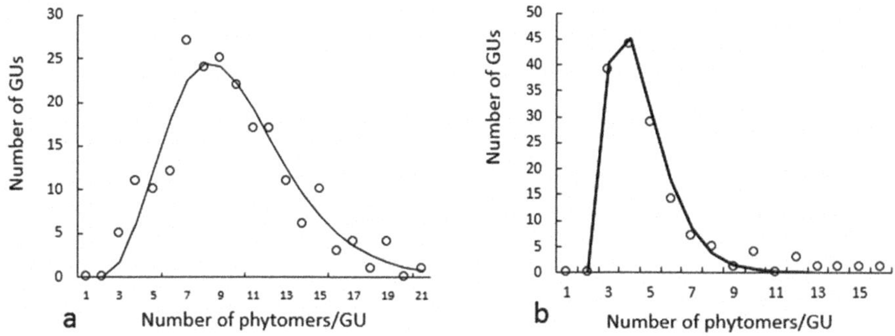

Fig. 10.60 The two consecutive growth unis (GUs) of a bi-cyclic annual shoot on the trunk of the teak tree are different. Fitting of the distributions of phytomers per GU on the trunk to negative binomial laws for GU 1 (**a**) and GU 2 (**b**) of the annual shoot is shown. The distributions of GU 2 are shorter. Similar distributions are observed for the other physiological ages (PAs)

Table 10.23 Polycyclism rate in teak according to the physiological age (PA) of the meristem

	Nb of GU 1	Nb of GU 2	Polycyclism rate
PA 1	210	150	71%
PA 2	149	70	46%
PA 3	220	77	35%
PA 4	100	0	0%

10.2.6.3.2 Rate of Polycyclism

The formation of a second GU is also dependent on the PA of the meristem (Table 10.23).

The frequent occurrence of polycyclism on the trunk (71%) is weaker on the branches and disappears on the short axes (PA 4).

Branching Modalities

Branching only concerns order 2 and is only observed at five years of age on the three and four-year-old shoots of the trunk. In order to increase the number of observations and enhance the accuracy, the measurements were completed on six- and seven-year-old trees. The branching probability is low. Each phytomer potentially carries two branches which may be of different PAs. Acrotony is not marked. Each type of branched axis can be located anywhere on the bearing GU.

Figure 10.61 shows the combinatorics generated by the three types of axes.

The distribution of the 739 types of twig pairs per observed phytomer is given in Table 10.24.

It is observed that 80% of the phytomers are unbranched. The probabilities of branching for a node are theoretically 0.054, 0.091 and 0.055 for the axes of type 2, 3, 4, respectively.

Table 10.25 shows the distribution that would result from random effect, using the axis type proportions observed in Table 10.24.

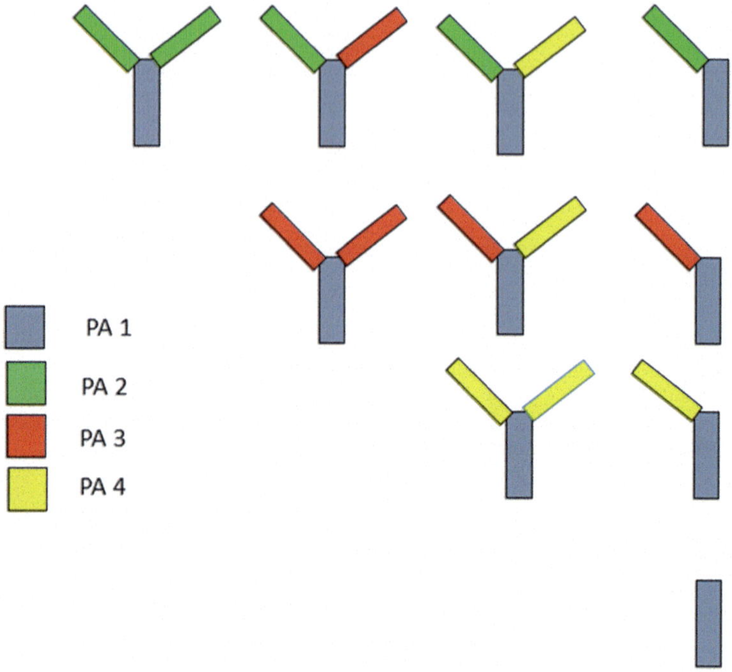

Fig. 10.61 Three distinct situations can be observed on each phytomer: the phytomer bears two branches (which gives six pairs of possible combinations with three physiological ages (PAs)), the phytomer carries a branch (which gives three possible combinations with three PAs), the phytomer is unbranched

Table 10.24 Observed numbers of different pairs of axes per phytomer

Observed	Axis 2	Axis 3	Axis 4	Void	Total	rate
Axis 2	20	6	3.5	10.5	40	0.054
Axis 3	6	42	4.5	14.5	67	0.091
Axis 4	3.5	4.5	16	16.5	40.5	0.055
Void	10.5	14.5	16.5	550	591.5	0.800
Total	40	67	40.5	591.5	739	1

Table 10.25 Numbers expected by the computation of different pairs of axes per phytomer

Theoretical	Axis 2	Axis 3	Axis 4	Void	Total
Axis 2	2.2	3.6	2.2	32.0	40.0
Axis 3	3.6	6.1	3.7	53.6	67.0
Axis 4	2.2	3.7	2.2	32.4	40.5
Void	32.0	53.6	32.4	473.4	591.5
Total	40.0	67.0	40.5	591.5	739.0

Table 10.26 Observed numbers of trunk phytomer bearing two axes, one axis and one void and two voids

Observed	Axis	Void	Total	Ratio
Axis	106	41.5	147.5	0.200
Void	41.5	550	591.5	0.800
Total	147.5	591.5	739	1.000

Table 10.27 Numbers are expected by the computation of two axes, one axis and one void and two voids per phytomer

	Axis	Void	Total
Axis	29.3	118.2	147.5
Void	118.2	473.3	591.5
Total	147.5	591.5	739

A comparison of Tables 10.24 and 10.25 shows that branches of the same PA have a strong tendency to cluster on the same phytomer, which increases the number of unbranched phytomers. The proportions of pairs of the same type are strongly higher compared to what would be expected in a random distribution of axis types. For example, there are 20 pairs of axis 2 when there should only be 2.2 (the rate of having 2.2 pairs is calculated as 739*0.054*0.054, which equals 2.2).

An overall assessment of branching can be given by considering phytomers with two axes, one axis and one void and two voids, with Table 10.26.

The probability of branching is $a = 0.2$. Using this value, one should obtain the branching pattern (Table 10.27).

A comparison of Tables 10.26 and 10.27 shows that there are more axis-axis and void-void pairs than the detriment of axis-void pairs. This pattern is typical of a coupling effect, as modelled in Chap. 3 and which has already been observed in detail on the coffee tree. Let us recall the equations of Chap. 3, which model the coupling between pairs of axes.

A simple coupling model is defined by assuming a branching probability a and a coupling r between the realizations according to the equations:

$$
\begin{aligned}
p(1,1) &= a(r + (1-r)a) \\
p(1,0) &= 2(1-r)a(1-a) \\
p(0,0) &= (1-a)(r + (1-r)(1-a))
\end{aligned}
\tag{10.13}
$$

These equations allow to compute the branching probability and its associated coupling from the proportions of observed pairs.

In this case, the pairs taken from Table 10.26 are:

Phytomers with 2 axes: AA = 106;
Phytomers with 1 axis: AV = 83;
Phytomers with 0 axis: VV = 550;
for a total of $N = 739$ phytomers.

By applying the previous Eq. 10.13, we can adjust these classes according to the number N and the parameters a and r. We find $r = 0.65$ (the parameter a is obviously unchanged at $a = 0.2$). The coupling is strong. Phytomers of the trunk tend to carry a pair of branches or to remain unbranched.

10.2.6.3.3 Number of Branches per Growth Unit (GU)

In the teak tree, the probability of branching of the trunk for each phytomer depends on the PA of the branched axis. It is observed that 80% of the phytomers are not branched. The branching probabilities (Table 10.24) for an axillary meristem are 0.054, 0.091 and 0.055 for axes of type 2, 3, 4, respectively. The same phytomer can carry two axes which differ according to PA, but it is observed that there is a 60% chance of having two-branched axes for the same PA. These branching probabilities mean that the number of branches carried by a GU is proportional to its degree of development, which is the number of phytomers per GU (Fig. 10.62).

The stochastic branching model for the teak tree can be simulated. The number of branches per GU increases linearly with the number of phytomers, with a low branching probability.

The data obtained on meristem activity are now sufficient for computing the potential structure of the tree. This process is data-intensive in list mode because all the phytomers of the trunk can potentially branch. However, this disadvantage is eliminated by using the matrix mode of the GreenLab model, which computes the production of this potential structure without going through the construction of the topological structure. In the list mode simulation, the topological structure is poorly

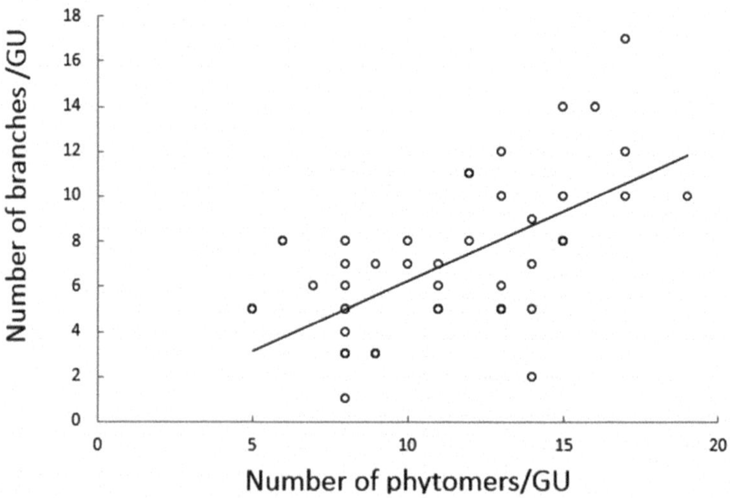

Fig. 10.62 Number of branches carried by a growth unit (GU) on the trunk as a function of its number of phytomers (o = observations). The trend is linear, but the low number of branches causes a strong dispersion

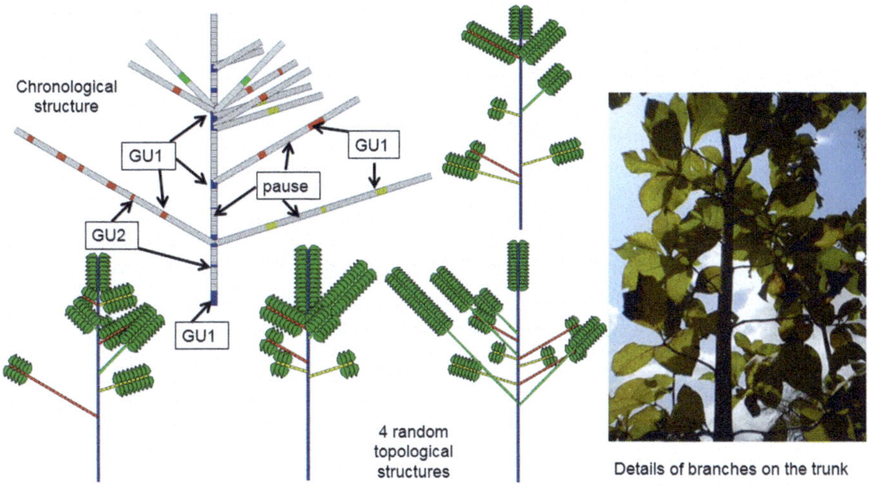

Fig. 10.63 Stochastic simulations in chronological and topological modes of the development of a teak tree at five years from the botanical automaton calibrated with the data. The physiological ages (PAs) 1, 2, 3, 4, are of the respective colours (blue, green, red and yellow). It is noticeable that the simulated chronological structure is mainly composed of pauses. If these are suppressed in the display, a topological structure is obtained, with four stochastic simulations displayed. The leaves are only present at the shoot of the year (Color figure online)

Table 10.28 Average number of phytomers produced by teak at five years according to the physiological age (PA) of the meristems

PA	1	2	3	4
Nb of phyt.	78.4	100.2	116.2	55.1

provided because of the scarcity of branches, but stochastic realizations of the development can be simulated and drawn (Fig. 10.63).

In Fig. 10.63, the display, in chronological mode, shows the importance of the meristem pauses in relation to the functioning cycles in which the phytomers are created. These pauses are used for the expansion of the organs, hence playing a crucial role in the correct computation of plant growth.

10.2.6.3.4 Number of Phytomers Produced by the five-Year Teak Tree

The theoretical numbers of phytomers produced by the stochastic botanical automaton categorized by PA at five years are given in Table 10.28.

10.2.6.4 Establishing Organic Series

There are two types of organs: leaves and internodes. Internodes can be measured along the axes thanks to leaf scars, but this is not the case for leaves, which are only present on the most recent annual shoot. They are, therefore, under-represented in comparison to the internodes. To simplify, the leaf organ comprises the petiole and the leaf blade. If the detail observation at the phytomer level is possible in the first years, in the fifth year, at a rate of 37 DCs per shoot, this gives a potential trunk close to 200 phytomers. At this level of detail, a follow-up of the growth of the teak is too time-consuming and complex for the observers.

Measurements are then taken at the assembled level of the GU, which becomes a phytomer compartment. Two types of organic series are used: the phytomers along with the GU (simple series) and the GUs along the axes (compound series). The phytomers (simple organic series) were measured on one-year-old teak trees. The cumulative weights of GUs in internodes and leaves (at the ends of the axes) were measured on two- and five-year-old trees (compound organic series). The samples for the organic series were taken from six trees.

10.2.6.5 Scheme of Growth

The growth pattern is identical to the development pattern. The annual shoots have a complex functioning, linked to the establishment of two GUs, one with a preformed part and the other with a neoformed part. Here the growth time step of the phytomer is the development cycle (DC). The beginnings of the growth for the first GUs of the annual shoots are synchronous. Due to the lack of follow-up observations, we also assumed the synchronism for the second GUs of the annual shoot. We will see later that the quality of the fittings on the second GUs is inferior to those on the first, which suggests that this synchronism might not be so strict in reality. The leaves all stop working after 36 DCs, which marks the end of the growing period. The 37th DC marks the annual pause. The parameters of meristem activity allow the potential structure to be computed (Chap. 3).

If a constant production area *Sp* is kept, the WLSM does not converge in multi-fitting. So, it is assumed *Sp* varies with age and follows a U-shaped curve with the increase in CA as for the Maple tree. This curve can be approximated by a function with two hidden parameters, which are computed by the WLSM to optimize the fit of the organic series.

10.2.6.6 Calibration of the Growth Model

10.2.6.6.1 Parameters Obtained Empirically

The SLW ε is estimated as $\varepsilon = 0.015$ g/cm^2 in dry weight. The light extinction coefficient k, which occurs in Beer's law, is set to 1.

Allometries between organ sinks: organ sinks of the same type are assumed to share the same variation function. So, only the sink strength is different from one PA to another one. The directly measurable ratio between the weights of the organs is, therefore, proportional to the ratio of the sinks. The master sink is that of the organ of

PA 1 (leaf or internode, respectively). The others can be deduced from the measured allometry, which has the advantage that the hidden parameters are not multiplied unnecessarily.

The allometry of the master sinks is 1 by definition. The other sinks are deduced from the value of the master sink, multiplied by the value of the allometry corresponding to that of the organ and the considered PA. Organ sinks decrease with PA. For the leaves, the values do not depend on the rank of the GU. For the internodes of the GU 2, the sinks decrease strongly compared to the GU 1, which gives shorter internodes visible on the annual shoot. The sink values from cycle two become the value of cycle one multiplied by the allometry coefficient measured between cycle two and cycle one.

10.2.6.6.2 Hidden Source–Sink Parameters

10.2.6.6.2.1 Fitting One-Year-Old Teak Tree

One-year-old teak trees do not branch and are described by a single PA (the stem). At this young stage, the measurements are taken at the phytomer level. The six selected stems of the same age have a fairly regular monocyclic distribution in the number of internodes: it can be assimilated to a shifted positive binomial law ($d = 13$, $N = 10$, $b = 0.5$). This distribution combines the *Bernoulli* process with the cessation of the activity of the meristem. It is important to note that this selection does not conform to the natural distribution observed due to the oriented selection of individuals, which results in a more clustered distribution. As the source–sink parameters are (by hypothesis) independent of the meristem activity parameters, this discrepancy is not important, but it is necessary to calibrate the source–sink parameters based on this distribution.

The GUs here comprise a maximum of 23 phytomers, in the selection that the second cycle is absent. The leaves have an estimated functioning life of 24 DCs. The organs have an estimated expansion time of 15 DCs for sink variation functions (Fig. 10.64).

The organic series of leaves and internodes represent 46 data adjusted by six parameters (Table 10.29). The second parameter of the beta law is set to 3 (this second parameter is often too insensitive to be estimated).

Figure 10.65 shows that the organic series are correctly adjusted by the WLSM. It is not possible to compute *Sp*, which is arbitrarily set to a large value.

The second coefficients of the beta laws for sink variation (B_{b2} and B_{i2}) are set to 3.

The fitting includes 40 measurements ($R^2 = 0.88$) with five hidden parameters (Table 10.30).

Sink variation parameters are consistent with observations. The low ratio $B_{i1}/(B_{i1}+B_{i2})$ indicates a rapid onset of expansion for the internode, which is rapidly damping, whereas it is slower for the leaf. Figure 10.64a shows the effect of the cessation of development on the size of the organs. The last leaves are large because

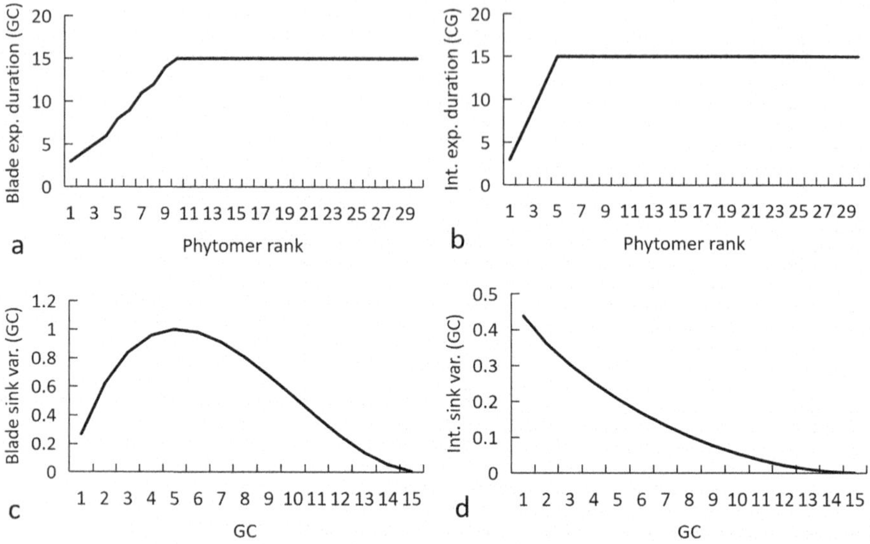

Fig. 10.64 Expansion durations and sink variation functions of leaves and internodes. (**a**) Expansion durations of leaves. (**b**) Expansion durations of internodes. (**c**) Sink variation functions of leaves. (**d**) Sink variation functions of internodes

Table 10.29 Allometries between leaf sinks and internodes according to physiological age (PA) and growth unit (GU) cycle in the annual shoot

	Leaf sink allometries			
GU type	P_{b1}	P_{b2}	P_{b3}	P_{b4}
GU 1	1	0.78	0.62	0.59
GU 2	1	0.78	0.62	0
	Internode sink allometries			
GU type	P_{i1}	P_{i2}	P_{i3}	P_{i4}
GU 1	1	0.14	0.36	0.14
GU 2	0.21	0.7	0.1	0

they have continued to expand after the meristem has stopped functioning. The last internodes formed remain small because of the shape of the sink variation, which limits the expansion. On the other hand, the effect of secondary growth is clear on the growth of the internode as a function of the rank from the top.

10.2.6.6.2.2 Multi-fitting Teak Tree of Two and Five Years Old

There are two axillary meristems per trunk phytomer and three possible physiological ages for branches. Branching probabilities are given in Table 10.24.

Complete measurements of organic series on GUs were carried out on individual plants at the age of two and five years. For the five-year-old trees, there are too many phytomers at this scale. Fittings are made to the cumulative organ weights at the level of the growth units (GU 1 and GU 2) as organ compartments (leaf or

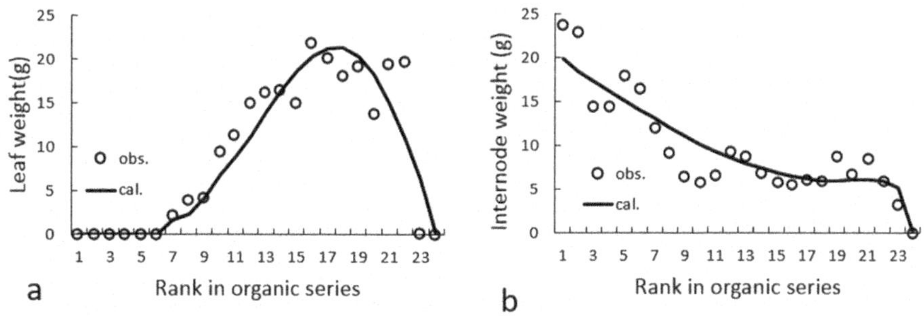

Fig. 10.65 Adjustments to the observed (circle) and computed (line) organic series of the leaves (**a**) and internodes (**b**) of the stems for the one-year-old teak tree. The first leaves are senescent. The following ones increase in weight by the increase in *Q*/*D*. The last leaves are large due to the prolongation of leaf expansion after stem elongation has stopped. This phenomenon is not visible on the internode, which has a shorter expansion time (Fig. 10.64d). Secondary growth is visible on the first internodes at the base of the growth unit (GU)

Table 10.30 Source–sink parameters of the one-year-old teak tree

Parameter	r	Sp	Q_0	B_{b1}	B_{i1}	P_c	P_i
Mean	136.8	100,000	1	1.89	0.95	0.18	0.44
SD	6.05	0	0	0.15	0.35	0.015	0.25
CV (%)	4	0	0	8	37	8	56

internode). We obtain compound organic series composed of GUs on the axes. However, because of the neoformation, one cannot, as for the maple, simplify the development using meta-phytomer by attributing a meta-cycle of development per GU.

We keep the DC of the phytomer and the laws of sink variation established over the first year at the phytomer scale. These cannot be deduced from the weight of the compartments alone. Branching only appears from the second year onwards. There are four types of axes (PAs) and only one order of branching (order 2). The unknown parameters are: the leaf sinks and the internode sinks P_i on GU 1 and GU 2 (15 unknowns), the ring sink P_c, the resistance r, the surface Sp (two unknowns). This gives 19 parameters to be estimated. Since some parameters can be directly estimated from allometry between PAs (Table 10.28), only nine parameters need to be estimated finally. As we are working at the GU compartment level, there are 29 measurements at the internode compartment level and nine at the leaf compartment level.

The WLSM does not converge towards a solution if a single value for Sp is sought, which is generally the case for high-density herbaceous field crops (Chap. 9). As in the case of maple, the production area, Sp expands with the expansion of the tree crown.

The Fitting includes 38 measurements ($R^2 = 0.99$) with seven hidden parameters (Table 10.31).

Table 10.31 Five-year source–sink parameters of the teak tree (multi-fitting)

Parameter	Axis 1	Axis 2	Axis 3	Axis 4	SD	CV (%)
Leaf sink P_a	1	0.78	0.62	0.59		
Internode sink P_i	7.81	4.2	5.5	1.5	1.5	20
Coef. Int. sink GU 2 *kpi*	0.16				0.13	84
Ring sink P_c	1.8				0.19	23
λ	1.16				0.01	1
Coef. Rings *cp*	1	0.7	0.7	0.7		
Resistance *r*	55				12	23
Surface *Sp*	*maxSp*	*minSp*	*aSp*	*bSp*		
	50,000	1000	1.01	1.24	0.01 0.14	1 11
Q_0	1					

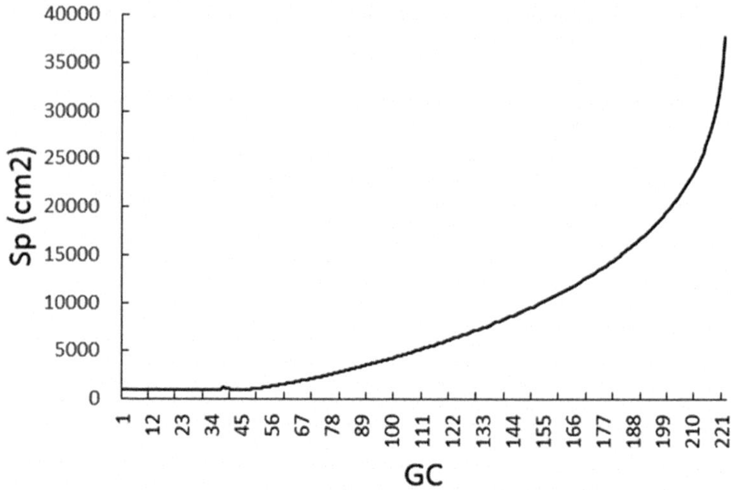

Fig. 10.66 The variation in the hidden production area *Sp* is due to the shape coefficients *aSp* and *bSp* of the *Sp* function and optimized by the WLSM. Its shape minimizes the error between the observations and the computations of the organic series of growth unit (GU)

As for the maple, the shape of the *Sp* curve (Fig. 10.66) depends on two shape parameters: *aSp* and *bSp*, which are computed at the same time with the other source–sink parameters. The areas $minSp = 1000\ \text{cm}^2$ and $maxSp = 50{,}000\ \text{cm}^2$ were fixed empirically, as they were not included as hidden parameters. The coefficient λ is close to 1, which shows that we are close to Pressler's law. It should be noted that the ring model is fitted with a coefficient $\lambda > 1$. Furthermore, there is no effect of polycyclism on the size of the leaves. On the other hand, the effect on the reduction of internodes is important. The internode sinks are multiplied by a factor $kpi = 0.16$ for the second cycle.

The ring elements are empirically reduced by a factor $cp = 0.7$ on the axes with PA > 1. The weight of the seed is empirically fixed at $Q_0 = 1$ g.

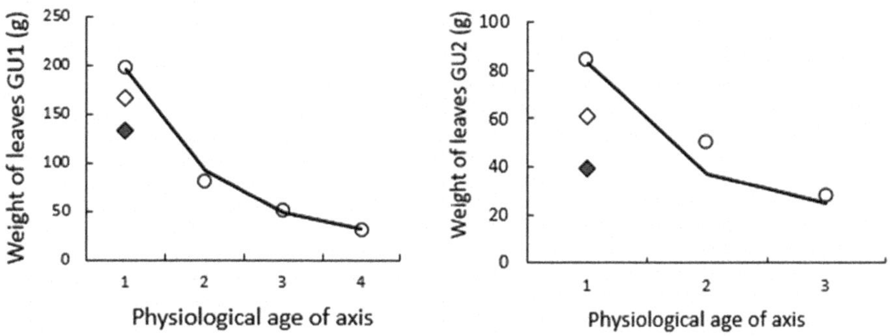

Fig. 10.67 Multi-fitting (two and five years) of the leaf weights of growth unit (GU) 1 and GU 2. The observed (◇, year two and ○, year five) and computed (◆ and —) values for the four PA of teak axes. The leaves are presented only in the last year. PA 4 has no GU 2. The leaf weight of the GU decreases with the axis category and according to the number of the GU cycle on the annual shoot

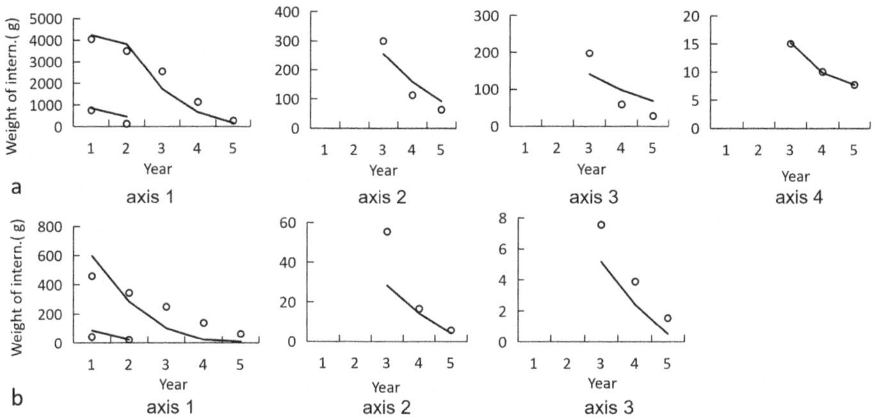

Fig. 10.68 Multi-fitting (two and five years). (**a**) Weights in the internodes of growth unit (GU) 1. (**b**) Observed GU 2 (circle) with the computed values (lines) for the four categories of teak axes. The weight of the internodes of GU decreases with the category of axis and according to the rank of the GU on the annual shoot

The fittings for the GU organic series are shown in Figs. 10.67 and 10.68 for leaves and internodes, respectively. The estimated parameters provide a good representation of the biomass distribution in the tree structure at the GU level. The trunk accounts for the majority of the weight due to the effect of secondary growth.

It is worth noting the efficiency of the WLSM, which manipulates very different order-of-magnitude numbers. Indeed, the weights of the organic compound series are of the order of 2000 g on axes 1, 200 g on axes 2 and 3, and 20 g on axes 4 for GU 1. For GU 2, they are about 10 times lower.

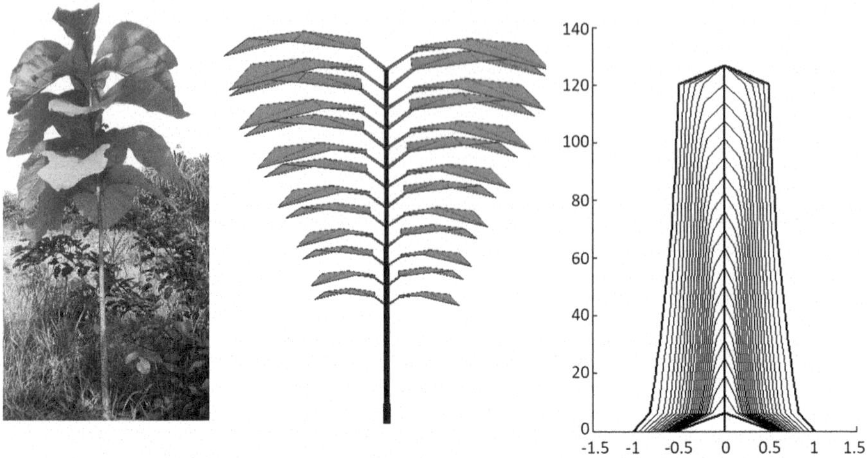

Fig. 10.69 A photo and simulation of the one-year-old teak tree. The leaves at the base of the stem fall off and increase in size as a function of their rank. The profile of the rings shows the parts due to pith and secondary growth increase with the increase in stem diameter (Gloups software)

10.2.6.7 Simulation of the Teak Tree

The parameter values of the development and growth found by calibration on the data allow simulating the architecture of teak in the interval one to five years. The allometries were set to obtain the right order of magnitude in trunk height and base diameter.

Figure 10.69 shows the simulation of a one-year-old teak tree and the stem profile with rings.

Figure 10.70 shows four stochastic simulations of teak trees at five years. Here the stochastic part in the development is at its maximum complexity. The numbers of phytomers per GU follow negative binomial laws, which are indexed by PA and the cycle rank of GU 1 or GU 2. The intensity of branching also depends on the type of axis. The analysis of the organic series has provided the source–sink parameters that drive the growth dynamics.

Secondary trunk growth is complex due to rhythmic development and polycyclism (Fig. 10.71).

Fig. 10.70 Four random simulations of teak trees at five years. The trees are derived from the stochastic functioning of the botanical automaton. This generates the different categories of axes along the stem in the right proportions. However, the coupling between meristems is not taken into account here, which does not influence the number of emitted twigs

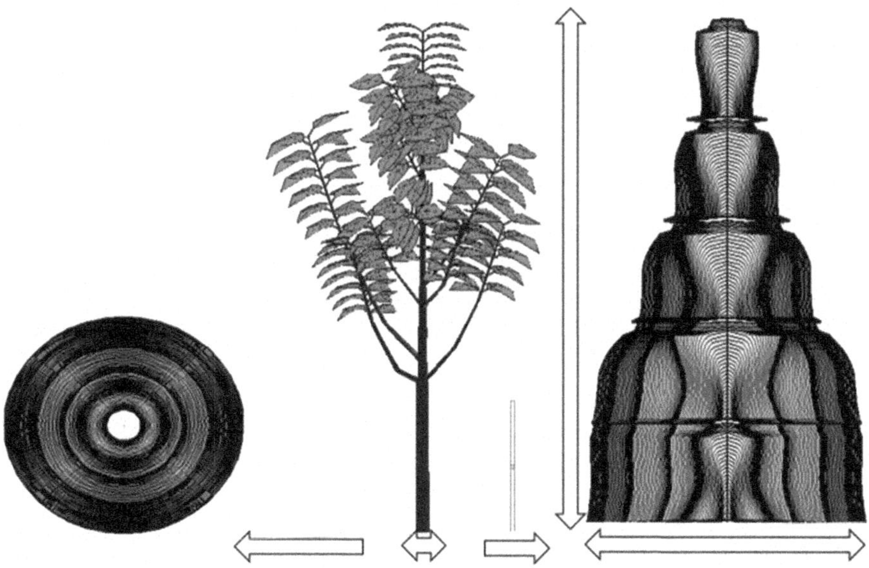

Fig. 10.71 Simulation of a teak tree and internal structure of the trunk. The profile of the stem diameter is the complex result of the distribution law of the biomass allocated to the rings in the stem. The effect of rhythmic growth is clearly visible. The diameters here are much enlarged

10.2.7 Case Study 7: Khaya Tree Study (Adji Beda, Selastique Akafou)

10.2.7.1 Description of the Tree

Khaya senegalensis (Mahogany tree) belongs to the Rauh model, characterized by an orthotropic trunk and branch, monopodial branching, rhythmic growth, lateral sexuality with five orders of branching and four axis categories. Each phytomer bears a single leaf with phyllotaxy arranged in an alternate spiral pattern. Development is asynchronous and polycyclic with two to six GUs per year on the trunk and one to three GUs per year on the branches. Preformed GUs are followed by minimal neoformation. The number of phytomers per GU varies from five to 25 in the young stage (up to two years).

10.2.7.2 Observation Data

The study was carried out on plantations of young Khaya trees aged 6, 12 and 24 months in nurseries and fields in Ivory Coast. The climatic factor is assumed to take a constant, normalized value $E = 1$, which is reasonable to represent the tropical conditions where the temperature varies little.

10.2.7.2.1 Data for Development

To establish the distribution laws of the number of phytomers per GU and branching, trees aged 6, 12 and 24 months were analysed with 30 repetitions of individuals, respectively. The stochastic part of the development is found in the branching and the lengthening of the axes at GU and phytomer levels (double-scale automaton). At the level of the axis, the development of the GUs is polycyclic, meristems produce GUs according to a *Bernoulli* process. This makes an asynchronous production of GUs. At the level of the GU, the development of the phytomers must be modelled according to a *Bernoulli* process, which is composed of the law of cessation in neoformation.

Only two types of axes were observed at this young stage (two physiological ages). Branches only appear after 10 GUs on the stem.

Same with the case of the teak tree, the meristem functioning of the Khaya tree is continuous at the beginning of development. The rhythmic functioning of the GUs does not appear until six months of age.

Given the number of phytomers in the architecture of Khaya at 24 months (more than 300), it is necessary to change the level of observation from the phytomer level to the GU (metaphytomer) level. The fresh weights of the leaf and internode compartments of the GUs were measured for the two physiological ages.

10.2.7.2.2 Data for Growth

Organ weights are measured in fresh matter (g). Six-month-old trees give simple organic series of phytomers. After six months, the phytomers are grouped into compartments at the GU level, which gives compound organic series. The time is then measured in macrocycles (MDC) rather than in microcycles (DC). The number of DCs in a GU is estimated to be 25 DCs. Leaves are functional over five consecutive GUs ($t_a = 5$ MDCs) that corresponds to 125 DCs, and organ expansion does not exceed five consecutive GUs ($t_x = 5$ MDCs) that corresponds to 125 DCs.

10.2.7.3 Scheme and Calibration of the Development Model

We have to consider two cases according to the double scale development automaton.

Before six months, the development is continuous and the *Bernoulli* process is expressed at the phytomer level. After six months, there is the formation of GUs, and the *Bernoulli* process is expressed at the level of the GU. But within the GU, we observe the distribution of phytomers characterized by a mean and variance.

10.2.7.3.1 The Bernoulli Process at the Young Development Stage

As the plants do not branch at six months, the plant consists of a single physiological age, and the average number of phytomers on the stem is $M = 18.14$ with a variance $V = 6.28$ (from a sample of 14 stems). The Bernoulli process parameter is given by the formula $p = 1 - V/M$ (Chap. 3), which gives $p = 0.65$.

After six months, the development becomes rhythmic and double-scale. The *Bernoulli* process is expressed in the stochastic production of GUs by the terminal meristem of the stem. The mean and variance of the number of GUs per stem at one and two years are shown in Table 10.32 The value of the parameter P is derived from this data.

The linear relationship between V and M is visualized in Fig. 10.72, as $V = 0.29{*}M$ which gives $p = 0.71$.

Crown analysis can be performed on two-year-old trees that have branches. This gives the parameter of *Bernoulli* processes $p = 0.6$ for the trunk and $b = 0.84$ for the branches, and the rhythm ratio $w = 1$. This value ($p = 0.6$) is smaller than the one calculated by the means and variances of Table 10.32 ($p = 0.71$). But we know that

Table 10.32 Means and variances of the number of growth units (GUs) per stem at one year (12 months) and two years (24 months)

Months	12	24
M	4.22	12.7
V	0.89	3.8

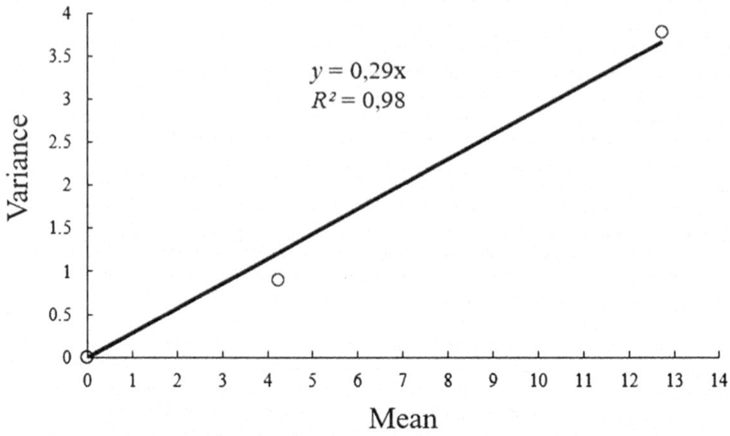

Fig. 10.72 Linear relationship between the mean and variance of the number of growth units (GUs) per stem over two development stages. This shows the stability of the *Bernoulli* process over time

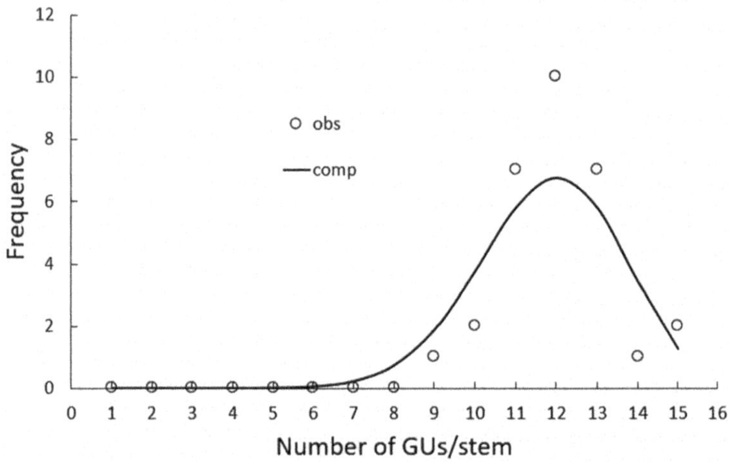

Fig. 10.73 The distribution of the number of growth units (GUs) on the stems of Khaya at two years old is a binomial distribution $B(15, 0.72)$. This shows the relevance of the *Bernoulli* process

the precision of p is less than that of b. In addition, the branches are poorly developed because they are only located at the top of the stem, which makes the crown analysis less reliable. We finally choose $p = 0.71$.

10.2.7.3.2 Distribution of GUs on Stems

The *Bernoulli* process on GUs generates binomial laws at a given age. This can be seen at the age of two years (Fig. 10.73). The parameters of the binomial distribution $B(15, 0.72)$ are in good agreement with the value of $p = 0.71$ in Fig. 10.72.

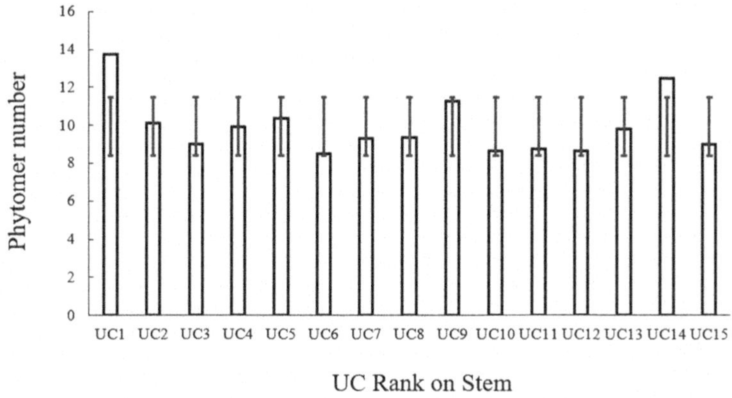

Fig. 10.74 Means and variances of the number of phytomers per growth unit (GU) along the stem. We can see no evident trend and small variability

10.2.7.3.3 Distribution of Phytomers inside GU

Figure 10.74 shows the evolution of the means and variances of the number of phytomers per GU as a function of the rank of the GU on the stem. We observed minimal variations and no evolutionary trends along the stem. It is, therefore, considered that the GUs can be grouped as coming from a single distribution. The mean is $M = 9.9$ and the variance is $V = 2.3$.

The distribution of the number of phytomers per GU fits well to a negative binomial distribution (Fig. 10.75). Although this distribution has both preformed and neoformed parts, the latter is poorly developed, as shown by the average number of phytomers per GU ($M = 9.9$). For simplicity, we consider that this distribution is only preformed and all the phytomers simultaneously elongate in the form of a flush.

There are two kinds of definitions for negative binomial laws: the first one counts successes and failures, and the second one counts only failures.

We choose here the second definition of the negative binomial law.

The computed negative binomial distribution of the phytomers within the GUs on the stem is: $BN(12, 0.56)$.

The computed negative binomial distribution of the phytomers within the GUs on the branches is: $BN(12, 0.66)$.

10.2.7.3.4 Distribution of Phytomers Inside the Stem

The composition of the laws of GU production on the stem (Fig. 10.72) and the production of the number of phytomers per GU (Fig. 10.75a) gives the distribution of the number of phytomers on the stem.

Since the distributions are binomial laws, it is proven that the addition of N binomial laws $B(K_1, b)$, gives the distribution $B(N \cdot K_1, b)$.

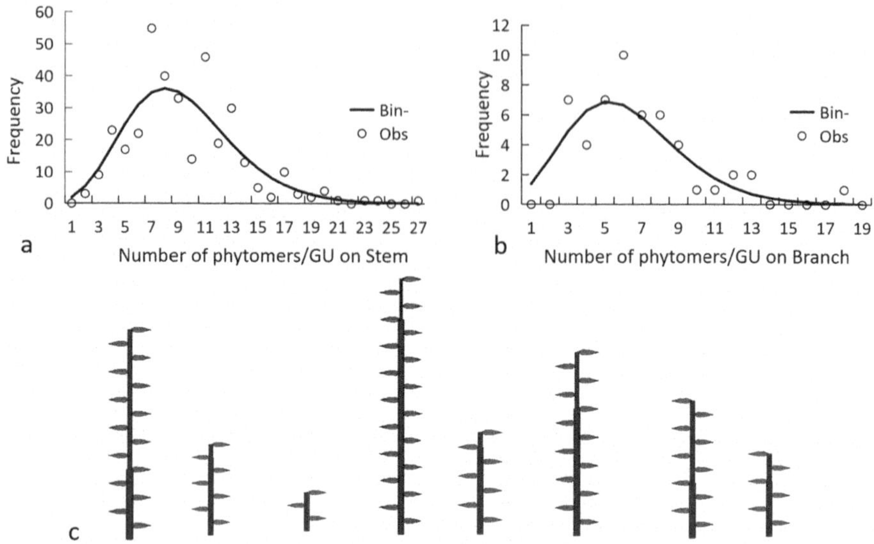

Fig. 10.75 Phytomer distributions within the growth units (GUs). (**a**) On the stem and (**b**) on the branch. They fit well to negative binomial laws of the second type with $B(12, 0.56)$ and $B(12, 0.66)$, respectively. (**c**) Sample of eight random GUs produced by the distribution of (***a***)

At one year, the distribution of the number of GUs per stem is the positive binomial distribution:

$F1 = B(6, 0.71)$.

The distribution of the number of phytomers per GU is the negative binomial distribution:

$F2 = B(12, 0.56)$.

The composition of the two laws gives the distribution F of the number of phytomers per stem according to the formula: $F(X) = \sum_{i=1}^{K1} F1(i) \cdot \sum_{j=1}^{i} F2(i, X)$

(10.13)

Figure 10.76 compares the measured and calculated data with a sample of 30 stems.

Table 10.33 shows the good agreement of the model prediction for the number of phytomers produced by stems for one- and two-year-old trees.

The number of phytomers per stem is a consequence of the functioning of the meristems as a double-scale automaton. This number is not directly useful.

10.2.7.3.5 Branching Study

Twenty-five out of the 30 observed individuals are branched (the branching rate is 83%) (Letort et al 2020). The branching of the stem appears only from the 10th GU

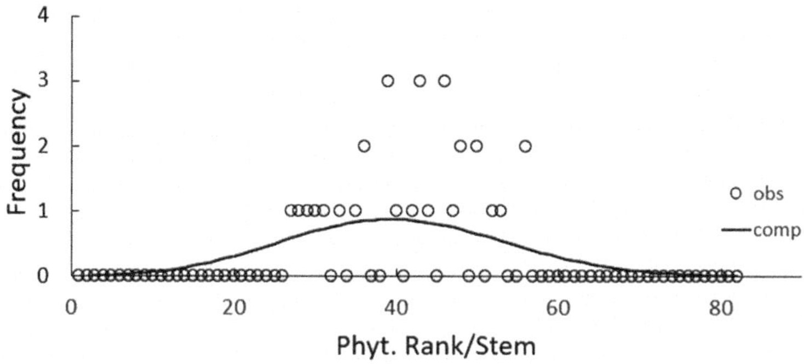

Fig. 10.76 Distribution of the number of phytomers per stem in one-year-old Khaya trees

Table 10.33 Observed (Obs) and computed (Comp) phytomer numbers on the stem for one- and two- year-old Khaya trees

	Obs	Comp
One year	41.8	40.1
Two years	102.3	100.4

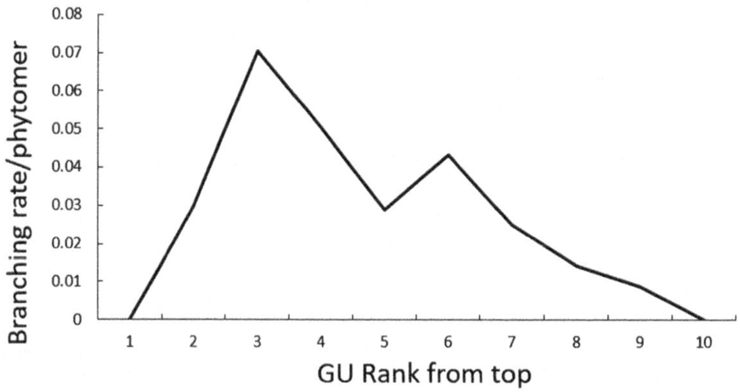

Fig. 10.77 Branching rate of stem phytomers as a function of rank from the top. This decreases after the 6th growth unit (GU) below the top of the stem

and intensifies along the axis (Fig. 10.77). This corresponds to the progressive increase in trophic pressure.

The average branching rate of a phytomer over the last five GUs on the stem is about $a = 0.05$. By giving a constant value $a = 0.07$ within a GU, we obtain 83% of branched stems, a value in good agreement with the previous one. This is the value that we finally retain for the development.

Table 10.34 Average theoretical number of phytomers produced by the botanical automaton at 6, 12 and 24 months according to the physiological age of the axes

Physiological ages		1	2
Number of phytomers	6 months	15.4	–
	12 months	43.4	–
	24 months	96.7	30.4

10.2.7.4 Average Number of Phytomers Produced by Khaya Tree at 6, 12 and 24 Months

Table 10.34 shows the numbers of phytomers produced by the stochastic botanical automaton by physiological age at different ages using 30 random Monte Carlo simulations.

10.2.7.4.1 Simulation of Khaya Senegalensis Development at Two Years

The software Gloups software simulates the development of the Khaya at two years old, taking into account the parameters of meristem functioning computed previously (Fig. 10.78).

10.2.7.5 Establishing Organic Series

There are two physiological ages and two types of organs, leaves and internodes, which produce four types of organic series.

The growth is continuous phytomer by phytomer until six months. At this stage, we obtain the simple organic series of leaves and internodes following the rank of phytomer along the stem. At the end of this period, the growth becomes rhythmic and only the sum of weights of the organs per GU is measured. Thus, following the rank of GU along the stem, the compound organic series consisting of the weights of the leaves and internodes compartments of the GU are obtained.

10.2.7.6 Scheme of Growth

During the continuous phase of development (< six months), we estimate the functioning and expansion times of the phytomers at 14 DCs. In the rhythmic phase (> six months), it is considered that the GUs are preformed in the bud and the expansion of the organs occurs in the last cycle at the end of the development within the GU. The growth step is the macrocycle (which is itself made up of microcycles, see Chaps. 2, 4, and 6). This mode of functioning is same with that of the Maple. But the Maple is monocyclic while the Khaya is polycyclic. For polycyclic, the GU can expand over several macrocycles. The variation in the sinks is no longer at the level of the phytomer but at the level of the GU. The expansion of organ sinks at the level of successive GUs is observed.

Fig. 10.78 Simulation of Khaya in list mode at two years old (375 DCs). Some stems are not branched (proportion 17%) shows the stochastic aspect of development. Branches are only established after one year of development

The functioning time $TFGU_{DC}$ and the expansion time $TXGU_{DC}$ of the organs are computed taking into account the GUs in DC. Thus, $TFGU_{DC} = TF_{GU}*TGU_{DC}$ for the photosynthesis and $TXGU_{DC} = TX_{GU}*TGU_{DC}$ for the expansion. $TFGU_{DC}$ is expressed in DC. TF_{GU} is the functioning time of GU (photosynthesis) in macrocycles (MDC) and TGU_{DC} is the duration of GU production in DC. $TXGU_{DC}$ is expressed in DC. TX_{GU} is the expansion time of GU in macrocycles (MDC) and TGU_{DC} is the duration of GU production in DC.

In the case of Khaya, we estimate that: $TXGU_{DC} = TFGU_{DC}$. We set $TF_{GU} = TX_{GU} = TGU_{DC} = 25$ DCs. We have, therefore, $TXGU_{DC} = TFGU_{DC} = 125$ DCs.

Moreover, the development time of the tree in DC at two years is $15*25 = 375$ DCs.

10.2.7.7 Calibration of the Growth Model

10.2.7.7.1 Parameters Obtained Empirically

The SLW ε is estimated as 0.013 g/cm^2 in fresh weight. The light extinction coefficient k, which occurs in Beer's law, is set to 1. Since the trees are young, the production area (*Sp*) is a priori very large ($Sp \approx 1{,}000{,}000\ \text{cm}^2$), which should be confirmed by the WLSM.

10.2.7.7.2 Hidden Source–Sink Parameters

10.2.7.7.2.1 Fitting Six-Month-Old Khaya Tree

The *Bernoulli* parameter of the meristem during the development up to six months was estimated as $p = 0.65$. This is to be taken into account for the fitting of the organic series of leaves and internodes. As for the teak tree, the last leaves formed are large, indicating that the development has stopped but the growth has continued. The growth is estimated at 30 DCs while the largest stem has formed 21 phytomers.

The organic series of leaves and internodes including 41 data are fitted by six parameters. The second parameter of the beta law is set to 2 (often too insensitive to be estimated).

Figure 10.79 shows the evolution of demand, biomass production per cycle. After 14 DCs, the demand stabilizes because the organs have completed their expansion. A plateau is formed. Nevertheless, the demand continues to increase slightly because of the demand of the ring elements, which are formed as long as the leaves are photosynthesizing ($t_f = 27$ DCs).

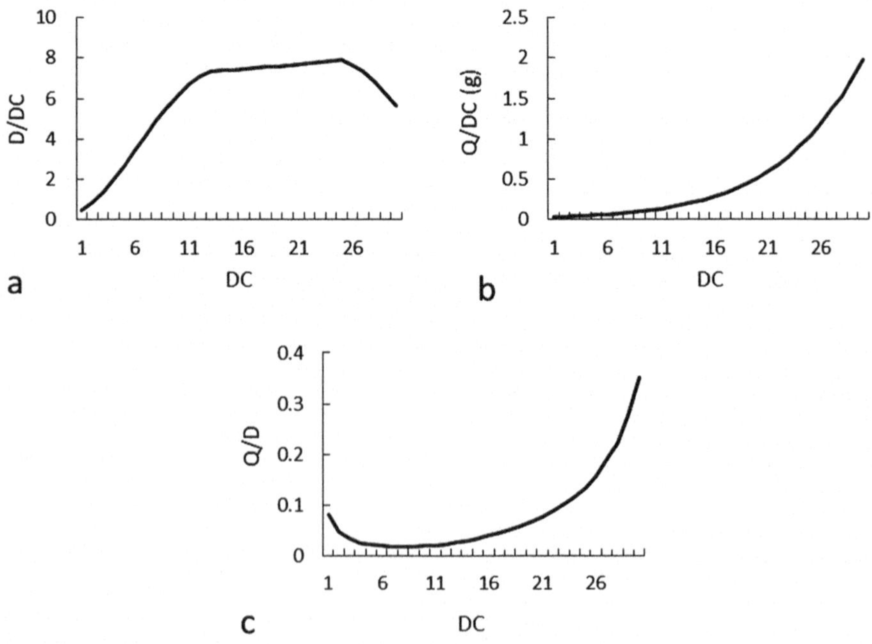

Fig. 10.79 Evolution of (**a**) demand, (**b**) biomass production, (**c**) trophic pressure during the first six months growth of Khaya tree

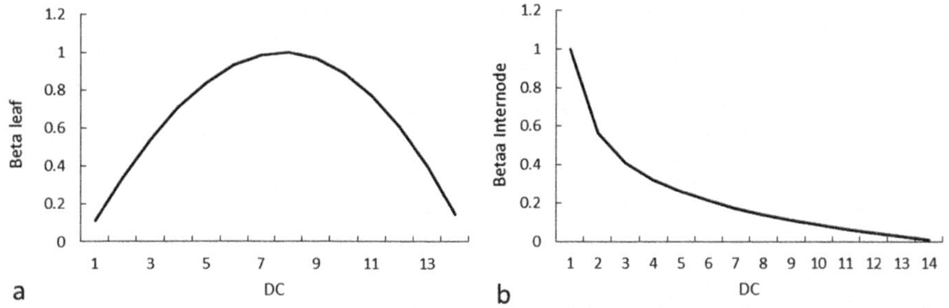

Fig. 10.80 Sink variation function of the leaf (**a**) and internodes (**b**) at six-month-old of Khaya tree

Table 10.35 Source–sink parameters for Khaya tree at six-month-old

Param	r	Q_0	Sp	B_{b1}	B_{i1}	P_c	P_{i1}
Value	401	0.8	10,000,000	2.3	0.82	0.06	0.17
SD	47	0.4		0.3	0.5	0.06	0.17
CV (%)	12	46		15	61	54	102

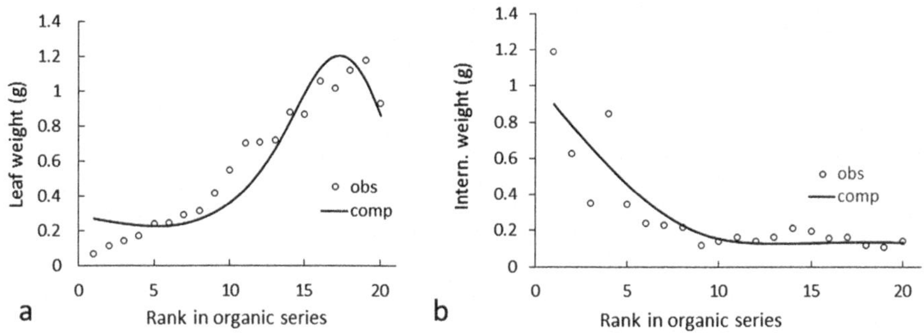

Fig. 10.81 Fittings to the observed (circle) and computed (line) organic series of the leaves (**a**) and internodes (**b**) of the stems for the six-month-old Khaya tree

The growth is exponential, which confirms that we are in free growth.

Figure 10.80 shows the shape of the organ sink variations. The shape of the sink variations in leaf and internode is similar to those that are produced with single-stemmed plants (see cotton, coffee, teak, etc.)

Table 10.35 gives the values of different hidden parameters. The second beta law coefficients for the sink variation in the leaf (B_{b2}) and internodes (B_{i2}) are set to 2.

The correlation coefficient between the computed and observed organ weights is $R^2 = 0.89$.

The Q_0 value (0.82 g) computed by the model is close to the average of seeds sown in the nursery (0.88 g).

Figure 10.81 shows the fitting of the observed and computed organic series. The last leaves are large due to the prolongation of leaf expansion after stem elongation has stopped (Fig. 10.81a). This phenomenon is not visible on the internode whose sink decreases rapidly (Fig. 10.81b).

10.2.7.7.2.2 Muti-Fitting of 1 and 2 Years Khaya Tree

The last stage of measurements is at 24 months (15 macrostates). There is an intermediate stage of 12 months (six macrostates). Only these two stages have the GUs. At one year, the branches are missing. Each stage is the result of the samples made on 30 trees.

At two years old, the age corresponds to 375 DCs, which gives too heavy a computational load for the Gloups software, especially for the WLSM. As we assume that the GUs are preformed with a one DC expansion, the Matrix and Phyto-Facto modes are operational without noticeable loss of accuracy. It is, therefore, necessary to operate with meta-phytomers corresponding to the macrostates.

The distributions of the phytomers, which follow negative binomial laws, require the assessment of parameters related to the meristem activity. This assessment require to compute a collection of potential structures simulated with the developmental parameters instead of a single theoretical potential structure (see paragraph 5 in Chap. 6). The fitting accuracy depends on the size of the simulated samples. The computational load obviously depends on this sample size, and a compromise is necessary. Here the size of 30 simulated random potential structures is chosen, which corresponds to the size of the observations.

The crown analysis gave *Bernoulli* parameters at the GU level ($p = 0.71$ for stems and $b = 0.84$ for branches).

Because of the transition to the meta-phytomer, the organic series become compound, and the sinks are expressed at the level of GU for the leaf and internode compartment. It is the GU that expands without going down to the level of the phytomer. The beta laws thus follow the evolution of the GU sinks for the leaf and internode compartments. Similarly, the seed weight becomes the reserve allocated to the first GU.

The expansion time of the GU is estimated at five macrocycles corresponding to 125 DCs.

The compound organic series of leaves and internodes, including 43 data, are fitted by nine parameters. The second parameter of the beta law is set to 3 (often too insensitive to be estimated).

Pressler's law does not correctly adjust the secondary growth of the Khaya axes. The base of the tree is exaggerated in diameter. From a modelling point of view, a good way is to assume that the cross-sectional area of the ring elements is not constant along the paths, leading from the leaves to the collar, as postulated by Pressler's law or the pipe model theory. To solve this problem, the volume of biomass allocated to the rings is computed in the same way, but at constant volume, the cross-sectional area of the ring element evolves along with the ring element (see Sect. Pressler's law in Chap. 8). This is again controlled by a beta law of parameters B_{c1} and B_{c2} (Fig. 10.82c). The length of the ring element increases at each time step (DC), but the coefficients of the beta law remain constant, which make it possible to estimate them with the WLSM during the fitting of the stem profile. We set $B_{c2} = 1$. Only B_{c1} is estimated. Figure 10.82 shows the variation in the sinks at the level of the organ compartments. It decreases rapidly.

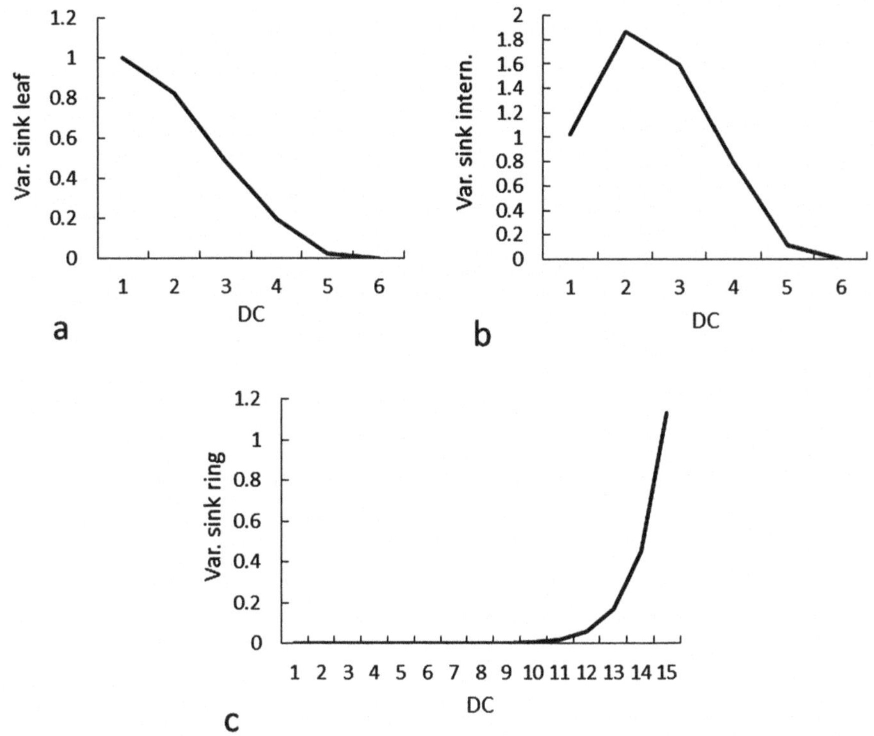

Fig. 10.82 Sink variations in the leaf compartment (**a**) and the internode compartment (**b**). Evolution of the sink of the section along with the ring element (**c**)

Table 10.36 Source–sink parameters for Khaya tree at 12 and 24 months (multi-fitting)

Param	r	Q_0	B_{b1}	B_{i1}	P_c	P_{b2}	B_{c1}	P_{i1}	P_{i2}
Value	32.2	71.3	1.3	2.0	1.1	0.5	13.8	1.9	0.9
SD	5.3	25.4	0.2	0.5	0.3	0.0	7.6	1.0	0.2
CV (%)	17	36	17	24	25	8	55	52	20

Table 10.36 gives the values of the source–sink parameters for the multi-fitting of the compound organic series of one- and two-year-old trees. Figure 10.83 shows the results of the fitting. The match with Pressler's law (dotted lines) clearly shows its inadequacy in taking into account the evolution of stem profiles and the effectiveness of the corrected Pressler's law.

The correlation coefficient between the computed and observed organ weights of the compartment (GU) is $R^2 = 0.99$.

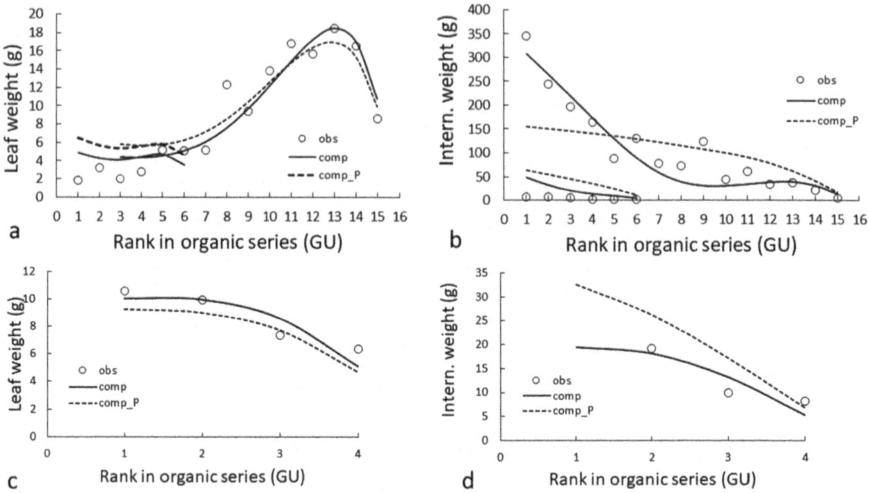

Fig. 10.83 Muti-fittings to the observed (o) and computed (–) compound organic series of the leaves (**a** and **c**) and internodes (**b** and **d**) for the stems and the branches for one- and two-year-old Khaya tree. Solid lines (comp) are the fittings with the new improved Pressler's law. Dotted lines (comp_P) are those for the standard Pressler's law. They do not match well with the observed data (obs) for the internode organic series

10.2.7.7.2.3 Simulation of Khaya Tree

The development, growth and geometry data allow the architecture of the Khaya tree at different growth stages to be visualized in 2D or 3D.

Figure 10.84 shows a 2D profile of the tree at six months with the ring structure within the stem.

Figure 10.85 shows the difference between secondary growth according to Pressler's enhanced law and standard Pressler's law. The base of the tree is expanded. The Phyto-Facto mode produces a geometry close to that of the list mode.

Figure 10.86 shows seven random 3D simulations of the two-year-old Khaya tree.

10.3 Conclusion

In this chapter, the behaviour of the GreenLab model towards woody plants, shrubs and trees is studied. The first result is that the secondary growth is generally well-calibrated by the model using Pressler's law, which stipulates that the surface area of the woody rings of a phytomer is proportional to the number of active leaves seen upstream. This important result justifies the use of the common pool, even in the case of trees. The common pool is located in the leafy crown, and the phytomers are located below. The thickness of a phytomer ring is proportional to the number of

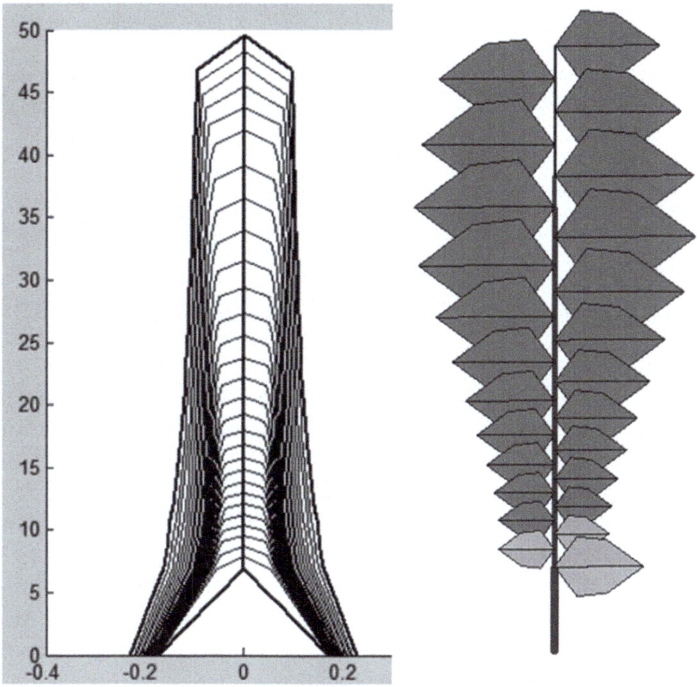

Fig. 10.84 Simulation of six-month-old Khaya tree. The leaves increase in size as a function of their rank. The profile of the rings shows the parts due to pith and secondary growth increase with the increase in stem diameter (Gloups software)

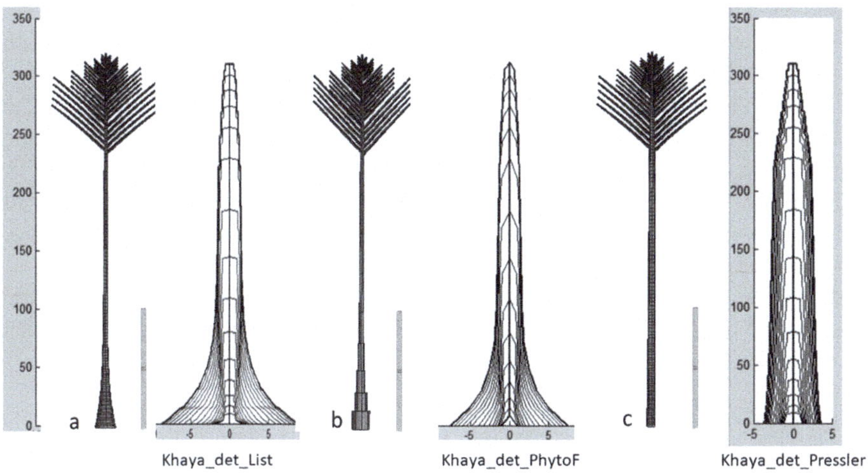

Fig. 10.85 Comparison of Pressler's laws enhanced in list mode (**a**) and Phyto-Facto mode (**b**) with the original Pressler's law (**c**), for the same amount of biomass attributed to the rings

Fig. 10.86 Seven random 3D simulations in list mode of Khaya tree at two years old (375 DCs)

photosynthetic leaves seen upstream in the leaf crown. However, adjustments are necessary for Pressler's law when considering extended growth duration concerning the ring sink decreasing with the age (the case of the Maple), and the variation in the cross-sectional area of the ring element along the axes (the case of the mahogany). These second-order corrections do not affect the robustness of Pressler's law and the calculation model of secondary growth.

The second result is the use of the meta-phytomer. On shrubs (cotton trees, coffee trees), it is possible to work at the level of the phytomer. However, this is not reasonable for woody plants with GUs, which can produce tens of thousands of phytomers. The simplification of modules or GUs into meta-phytomers (Phyto-Facto mode) greatly speeds up the computation of the growth. The use of the matrix mode allows solving the source–sink relationships while keeping the notion of the axis, based on the computation of cohorts and their orthogonal organic series, which decomposes the architecture into separate parts. The profile of the trunk is thus preserved, which is important for forestry.

Finally, we have to remember once again that the application of the GreenLab model requires mono-specific plantations that are uniform in terms of soil and density with even-aged plants. This is most often the case in agronomy. While the crown analysis to compute the development is easy to implement, the analysis of the organic series is more delicate.

Monocyclic trees with preformed GUs are the best candidates to be modelled (maples, poplars, etc.). Polycyclic trees with neoformed GUs (teaks, elms, mahogany, etc.) are much more difficult to deal with, mainly for sampling issues.

The most challenging case is undoubtedly fruit trees where the interaction between the development and the gravity is obvious and generates epitony, which makes it almost impossible to perform a representative sample of measurements.

References

de Reffye P (1981a) Modèle mathématique aléatoire et simulation de la croissance et de l'architecture du caféier Robusta. 1ère partie. Étude du fonctionnement des méristèmes et de la croissance des axes végétatifs. Café Cacao Thé 25(2):83–104

de Reffye P (1981b) Modèle mathématique aléatoire et simulation de la croissance et de l'architecture du caféier Robusta. 2ème partie. Étude de la mortalité des méristèmes plagiotropes. Café Cacao Thé 25(2):219–230

de Reffye P, Cognée M, Jaeger M, Traore B (1988) Modélisation stochastique de la croissance et de l'architecture du cotonnier. 1. Tiges principales et branches fructifères primaires. Coton et Fibres Tropicales 43(4):269–291

de Reffye P, Blaise F, Chemouny S, Jaffuel S, Fourcaud T, Houllier F (1999) Calibration of hydraulic growth model on the architecture of cotton plants. Agronomie-Paris 19(3):265–280

Guo Y, de Reffye P, Song Y-H, Zhan Z-G, Dingkuhn M, Li B-G (2003) Modeling of biomass acquisition and partitioning in the architecture of sunflower. In: Proceedings Of PMA03 : The First International symposium on plant growth modeling, simulation, visualization and their applications, Beijing, China, October 13-16, 2003. Tsinghua University Press

Hallé F, Oldeman R-A-A, Tomlinson P-B (1978) Tropical trees and forests. Springer Verlag, Berlin, Heidelberg, New-York, p 441

Jones CA, Kiniry JR (eds) (1986) CERES-Maize, a simulation model of maize growth and development. Texas A&M University Press, College Station, 194 p

Letort V, Sabatier S, Akaffou S, Hamon S, Hamon P, de Reffye P (2012) Interspecific variability of biomass production of young Coffea: no influence of branch pruning. Experimental evidence and theoretical analysis. In: 2012 IEEE 4th international symposium on plant growth modeling, simulation, visualization and applications. IEEE, pp 224–227

Letort V, Sabatier S, Okoma MP, Jaeger M, de Reffye P (2020) Internal trophic pressure, a regulator of plant development? Insights from a stochastic functional–structural plant growth model applied to Coffea trees. Ann Bot 126(4):687–699

Li D, Letort V, de Reffye P, Zhan Z-G (2009) Modeling branching effects on source-sink relationships of the cotton plant. In: Li B-G, Jaeger M, Guo Y (eds) The third international symposium on plant growth modeling, simulation, visualization and applications (PMA 09), 9-13 November 2009, Beijing, China. IEEE Computer Society, Los Alamitos, pp 293–302

Okoma P, Akaffou S, de Reffye P, Hamon P, Hamon S, Konan O et al (2018) Estimation of stem and leaf dry biomass using a non-destructive method applied to African Coffea species. Agrofor Syst 92(3):667–675

Ray D-T (1993) Guayule: a source of natural rubber. New Crop 338–343

Taugourdeau O, Delagrange S, de Reffye P, Messier C (2013) Modelling sugar maple development along its whole ontogeny: modelling hypotheses and calibration methodology. In : Proceedings of the 7th International Conference on Functional-Structural Plant Models (FSPM2013), Saariselkä, Finland, 9-14 June 2013. Eds. Risto Sievänen, Eero Nikinmaa, Christophe Godin, Anna Lintunen, Pekka Nygren. Vantaa : Finnish Society of Forest Science, pp. 234-236.

Tondjo K, Brancheriau L, Sabatier S, Kokutse AD, Kokou K, Jaeger M et al (2018) Stochastic modelling of tree architecture and biomass allocation: application to teak (Tectona grandis L. f.), a tree species with polycyclic growth and leaf neoformation. Ann Bot 121(7):1397–1410

Zhan Z, Rey H, Li D, Guo Y, Cournède P-H, de Reffye P (2006, November) Study on the effects of defoliation on the growth of cotton plants using the functional structural model GreenLab. In: 2006 Second international symposium on plant growth modeling and applications. IEEE, pp 194–201

Chapter 11
The Transition from the Single Plant to the Stand

Paul-Henry Cournède, Frédéric Blaise, Philippe De Reffye, and Mengzhen Kang

Abstract This chapter discusses the integration of environmental factors such as temperature, light, water, and planting density into an eco-physiological context to monitor and predict plant development and stand production in greenhouses or fields. It emphasizes the transition from individual plant growth to stand simulation, highlighting the use of the leaf area index (LAI) and harvest index (HI) in agronomic crop models to estimate biomass production and yield. However, these models often overlook the root system and face challenges in calculating LAI under non-uniform planting densities. Structural–functional plant models (FSPMs) focus on detailed physiological processes within individual plants but do not address stand production effectively, using complex and costly techniques. The GreenLab model stands out by bridging the gap between agronomic crop models and botanical architectural models, offering compatibility with crop models, and enriching them with a deeper understanding of plant components, thereby providing new criteria for plant breeding and optimization of crop systems.

P.-H. Cournède
CentraleSupélec Laboratory of Applied Mathematics and Systems (MAS), Châtenay-Malabry, France
e-mail: paul-henry.cournede@centralesupelec.fr

F. Blaise
CIRAD, Montpellier, France
e-mail: frederic.blaise@cirad.fr

P. De Reffye
CIRAD, UMR AMAP, Montpellier, France

AMAP, Univ Montpellier, CIRAD, CNRS, INRAE, IRD, Montpellier, France

M. Kang (✉)
State Key Laboratory for Multimodal Artificial Intelligence Systems, Institute of Automation, Chinese Academy of Sciences (CASIA), Beijing, China

School of Artificial Intelligence, University of Chinese Academy of Sciences, Beijing, China
e-mail: mengzhen.kang@ia.ac.cn

P. De Reffye et al. (eds.), *Plant Model Greenlab for Botany and Agronomy*,
https://doi.org/10.1007/978-981-97-7541-5_11

Keywords Environmental factories · Planting density · Crop and FSPM models · LAI · Production surface

11.1 Introduction

The aim here is to place the plant in an eco-physiological context where environmental parameters (temperature, light, water, planting density) monitor development and growth to access stand production under greenhouse or field conditions. The effect of these parameters has been defined in Chap. 5. The development of the plant is regulated by converting calendar time (days) into thermal time (sum of temperatures), which makes it possible to define the notion of the cycle. During a cycle, a meristem produces a phytomer and leaves receive an amount of energy proportional to the radiation accumulated during the cycle.

This chapter discusses the transition from single plants to stand simulation for crop production. In practice, it is necessary to establish a link between the GreenLab structural–functional model that works at the level of the plant's phytomers and the crop models that work at the level of the organ compartments produced per square meter, and predict harvests according to environmental parameters. It is, therefore, a question of inserting the individual plant into a stand that is presented as an agronomic plot in the field or the greenhouse.

Agronomic crop models such as CERES (Jones and Kiniry 1986), Stics (Brisson et al. 1998), Pilot (Mailhol et al. 1996, 2011), TomSim (Heuvelink 1996), Apsim (Keating et al. 2003), and Irribet (Jeuffroy et al. 2008) rely primarily on the use of the leaf area index (LAI) to predict biomass production per square meter, considering radiation, temperature, and essential inputs (irrigation, fertilizer, etc.). The useful fraction of production (leaves, fruits, roots, wood, etc.) called the harvest index (HI) gives the yield of the plot. Most of the crop models ignore the root system.

The biomass per square meter ratio, foliage and harvest indices are sufficient for a model such as Pilot to predict harvests according to a climate scenario that incorporates water stress. The LAI is measured directly in the field by a device (Li-Cor), which makes it possible to estimate the interception of radiation by the canopy to compute biomass production. The Pilot approach forces the functioning of the stand based on the monitoring of the measured LAI, which is verified to follow the temperature sums. Production is approached as kinetics. Some models, such as TomSim in greenhouses, detail the distribution of biomass per square meter in the different compartments and estimate LAI from foliar biomass. This also introduces feedback into the functioning and approaching production more dynamically.

If agronomists have modelled, with crop models, the action of climate and soil on yields, they have overlooked the yield components linked to the plant architecture that are involved in the development with the abortion of the meristems and the growth with the abortions of the competing organs in the source–sink relationships. These two essential components are not only accessible at the compartment of organ level but also at the phytomer level.

Difficulty also with the computation of LAI when the planting density is not homogeneous and Beer-Lambert's law (Chap. 5) does not apply.

Functional-Structural Plant Models (FSPMs), such as L_Peach (Allen et al. 2005) and Vica (Wernecke et al. 2000), do not address stand production; they focus on the integration and simulation of physiological processes in the functioning of the individual plant with a high level of detail. The interception of light by 3D models uses ray tracing or radiosity, which are complex and costly techniques. The synthesized biomass is diffused step by step in the virtual plant. These models, based on an academic simulation of knowledge, are not calibrated on observations. The physiology integrated into these functional–structural plant models is not compatible with the coarser physiology used by the process-based models and adapted to stand functioning.

The GreenLab model is an exception because it bridges the gap between crop models of agronomy and architectural models of botany. A crop model can be perceived "as the projection of a structural–functional model of plant" according to Charlotte Baey (2014) in her thesis. The GreenLab model is fully compatible with crop models. It aims to enrich them by allowing agronomists to detail, if necessary, the components of yield based on a better knowledge of the plant, to offer new criteria for plant breeding, phenotyping, and crop systems optimization.

11.2 Introduction to Stand

A stand is a collection of plants, a mixture of various species that can be natural (tropical forest, scrubland, etc.), or artificial (field). In both cases, it corresponds to the result of competition between individuals for access to resources (light, water). To be able to model the development and growth of a stand, based on architectural methods, it is necessary to find sufficient repetitions of the same plant in similar situations in situ to be able to estimate the parameters of its functioning; otherwise, modeling is not possible. Thus, a natural forest is out of reach because of the multitude of species involved and the lack of knowledge about the history of its establishment, both globally and individually, each plant having developed under different environmental conditions. Only agronomic plantations can be modelled today, primarily monospecific stands and, if possible, even-aged stands, with well-controlled cultivation trails. It should be noted that the case of associated or intercropped crops (maize with soybean, coffee with flemingia, etc.) can be reduced to locally homogeneous and monospecific stands in strip planting.

Simulation of a heterogeneous plantation is theoretically possible to test hypotheses, but we are leaving the framework of the rational and validated use of the model. We distinguish between large herbaceous crops such as beet, corn, sunflower, and forest stands (spruce, poplar).

For field crops (beet, maize, etc.), the development time is a few months, observations are at square meter level, and the ground is quickly covered by foliage. In the absence of stress, organ abortions are negligible except for fruits. In addition, there is no secondary growth. Architectural data are easily accessible, and experiments are repeated on small scales of time and space.

For forest stands (pine, eucalyptus, poplar, etc.), the case is more complicated. The growth period is decades-long and observations are at the hectare level.

Development is complex (growth units, polycyclism, etc.) and stochastic with ubiquitous meristem mortality and girth growth (rings). Frequently, replacements must be made during the establishment of the stand, which disrupts the homogeneity. Finally, architectural data are difficult to access, so it is necessary to operate by sampling. Forest systems that take decades to complete include thinning that significantly modifies plant–environment interactions at the level of individual development. Despite these difficulties, tree growth modelling is feasible for species whose architectures can be described by the botanical automaton (Chap. 4). It is easy to see that the architecture of a poplar tree shows a certain homogeneity in its stochastic development, while a pedunculated oak shows great complexity.

In this chapter, we will study the development and growth of monospecific and even age stands as a collection of individuals. Even at this level, the homogeneity of the stand is broken at the individual level for four reasons:

- The volume of the seed: at the beginning of growth, biomass production is proportional to the volume of the seed, the parameter Q_0 (Chap. 4). In this chapter, the seeds are assumed to have an equal volume, following calibration;
- The germination date: if the planting date is common to all plants, the germination dates are more or less scattered. However, a slight germination delay between two plants quickly leads to a growth difference with an exponential effect. Early germination will allow a plant to occupy the land and restrict its neighbors' access to resources (light, water, etc.) and thus take advantage. In forest plantations, there are often mortalities. Dead trees must be replaced with new ones, which leads to standing heterogeneity;
- The variability of the local environment: light, water, nutrients. In this chapter, the environment is considered without stress and homogeneous in the plantation;
- The local planting density around a plant induces a differentiated competition for access to light and soil resources in the environment of its neighbors, which modifies its growth. While the notion of planting density is global (it is the number of plants sown per square meter), it does not imply the regularity of spacing between individual plants. These are subject to germination defects and the inevitable irregularity of a seedling. Plants that have more space, regardless of germination times, will grow less constrained by competition.

Germination date and local density are two important factors of individual variability in monospecific and even-aged plantations. Here, architectural methods are essential because they provide crop production models with additional information to which they do not have access through their traditional measurement strategies, which are the harvest index (HI), the leaf index (LAI), and the biomass produced per compartment. Indeed, the harvest index can fluctuate greatly, for the same amount of biomass produced per unit area.

11.3 Study of Maize Growth in Its Environment

Here we continue the study on the maize variety modelled in Chap. 9, taking into account the influence of the environment. Figure 11.1 shows the evolution of the produced biomass and its distribution, as well as the variations in the different organ sinks in maize.

11.3.1 Influence of Climate on Maize Growth, Modelled with GreenLab

11.3.1.1 Clarification on Water Efficiency

The environmental factors that primarily drive development and growth are temperature, light, and water (Chap. 5). The components of the atmosphere (CO_2) and the mineral and organic elements of the soil are assumed to be non-limiting factors.

The GreenLab production equation (Chap. 5) is:

$$Q(t) = \frac{E(t)}{r} \cdot Sp \cdot \left(1 - \exp\left(-k \cdot \frac{Sf(t)}{Sp}\right)\right) \tag{11.1}$$

E(*t*) is the value of the climate data that drives the growth in each cycle, and the resource use efficiency (1/*r*) is for its conversion into plant biomass.

In greenhouses, radiation is the primary factor that drives the growth of crops throughout the year. Temperature and irrigation are controlled, light varies significantly with the seasons, as artificial lighting does not sufficiently compensate for the

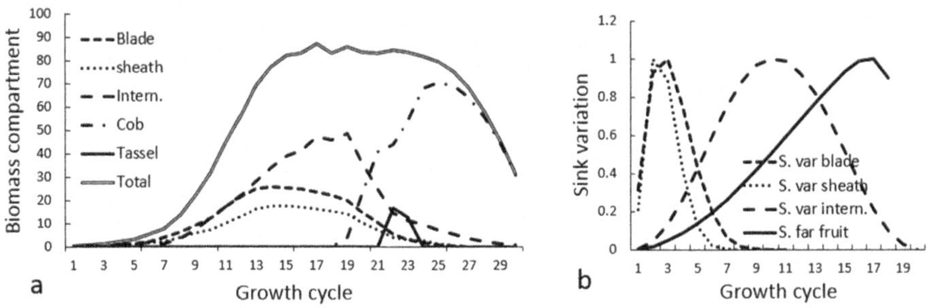

Fig. 11.1 Biomass production and sink variation during the maize growth. (**a**) Evolution of biomass production and its distribution in the different compartments. We observe an exponential increase in production at the beginning and stabilizes due to the limitation of the production surface *Sp*, then decreases due to the progressive senescence of the leaves. (**b**) The variations in the sinks are contrasted according to the type of organ (sink forces are normalized to the value 1 so that their variations can be compared)

decrease in brightness in winter. $E(t)$ is the measured radiation, and $1/r$ is the Light Use Efficiency (LUE).

In the field, crops grow at the same time of year, and the water is the limiting factor in two ways:

- By the water content in the soil: it is assumed not to be limited here because it is controlled by irrigation;
- By atmospheric humidity: agronomists have noted that crop production is proportional to canopy transpiration, especially when the atmosphere is dry and is null in saturated humidity, in the absence of sunlight. The latter can vaporize water on the surface of the leaves even in a water-saturated atmosphere. This defines the potential evapotranspiration (PET) that characterizes the daily drop in the level of an aquatic surface. This integrative parameter depends on radiation, atmospheric humidity, and wind. We, therefore, define $E(t) = \text{PET}(t)$, which is the average value of the PET at cycle t computed in thermal time and which corresponds exactly to the period associated with the calendar time. For example, in maize, it takes about 225 L of transpired water to produce 1 kg of dry matter corresponding to 10 kg of fresh matter. The Water Use Efficiency (WUE) is, therefore, 4.5 g/H_2O kg^{-1}. Equation 11.1 remains valid using the PET, and $1/r$ corresponds to the WUE.

The proportion of water retained in the plant is, therefore, very small compared to transpired water. Sixty tons of fresh matter per hectare corresponds to 50 tons of water retained in the stand. This corresponds to a water height of 5 mm for a consumption of about 110 mm of water, or 1100 tons/ha.

11.3.1.2 Field Experiments (China Agricultural University)

Between the year 2000 and 2005, repeated experiments were carried out to analyze the effect of climate on maize growth by Ma Yun-Tao in China (Ma et al. 2007, 2008). The planting density was 60 × 60 cm. Biomass measurements were made on organic series and compartments, which are the averages of four plants at each sampling in the plot. Climate data (daily temperatures and evapotranspiration) were recorded each year during the growth period to study inter-annual variability. In 2003, two experiments (sowing in May and July) were made to study intra-year variability. Since edaphic factors are not limiting, the effect of climate is, therefore, tested on temperature and PET, which replaces the parameter E in the GreenLab equation. The parameter $1/r$ is the efficiency of the PET that calibrates the equation.

For the five experiments carried out between 2000 and 2005, the relationships between the sum of temperatures and the number of developed phytomers align on the same line, which once again verify the link between thermal time and development (Fig. 11.2a). This makes it possible to know the cycle expressed in thermal time and to transform the PET per day into PET per development cycle (DC).

Table 11.1 shows that the five experiments can be divided into two groups due to the differences in the daily average temperature. Group A corresponds to the

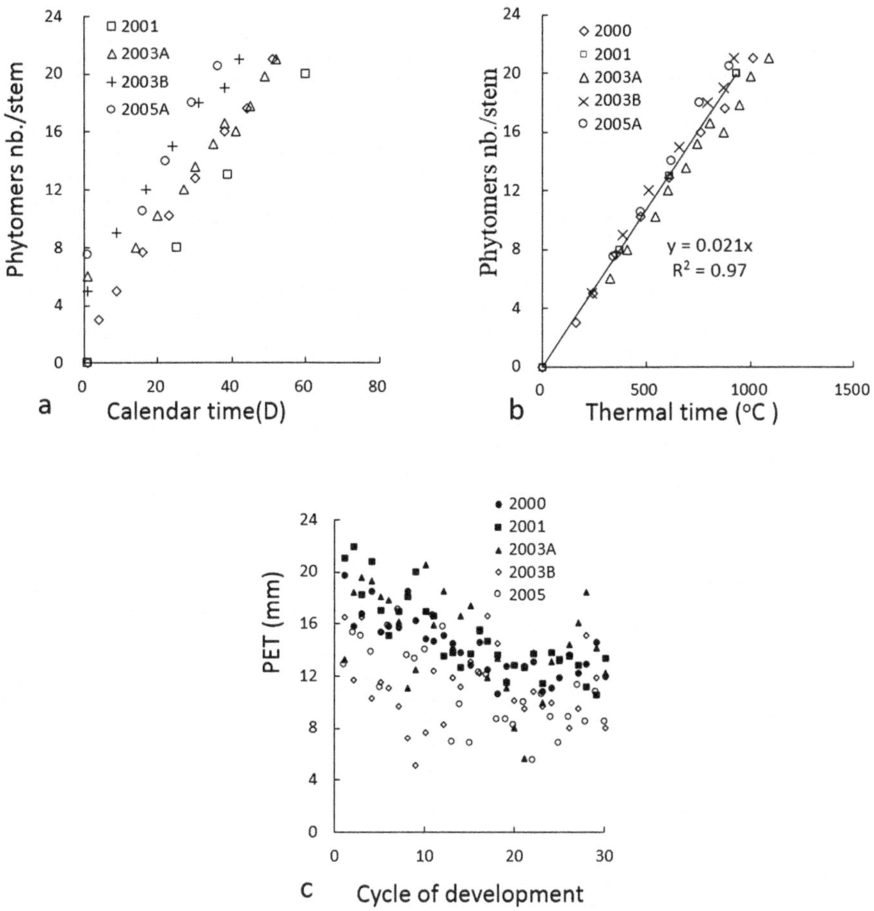

Fig. 11.2 Number of produced phytomers and PET in the experiments. (**a**) Number of phytomers produced in calendar time (days) in the different experiments (Plants develop faster in 2003B and 2005 because they are warmer). (**b**) The linear link between thermal time and the corresponding numbers of phytomers, where the climate effect disappears. (**c**) PET per experiment and development cycle (DC). The PETs per cycle were lower in 2003B and 2005

Table 11.1 Average temperature and PET characteristics per day or per cycle in the five experiments

Experiment	Culture (day)	Biomass (g)	Temp (°C)	ΣTemp (DC^{-1})	PET (day^{-1})	PET (DC^{-1})
Group A 2000	89	2063.3	16.8	47	5.5	15.2
Group A 2001	107	1755.2	14.6	45.3	5.2	16.1
Group A 2003A	86	1552.8	16.9	47.3	5.4	15
Group B 2003B	75	1042.4	18.3	42.1	4.9	11.3
Group B 2005A	77	1286.4	19.7	43.3	5.4	11.8

The duration of the crop, the biomass produced, the average daily temperature, the sum of temperature per cycle, the daily PET, and the PET per cycle are displayed experiments of 2000, 2001, and 2003A with a daily temperature below 17 °C. Group B corresponds to the experiments of 2003B and 2005 with a daily temperature above 18 °C. In addition, during this period, the average daily PET was approximately the same.

By the sum of the temperatures, group B grows faster than group A. With the number of DCs remaining constant, the duration of the cycle in calendar time is therefore reduced, leading to a shorter crop cycle. As a result, plants in group B receive less energy and produce less biomass than plants in group A. Given the minimal daily variation in PET, the PET per cycle (DC) is also lower in group B, resulting in reduced biomass production. This result is confirmed in the Table 11.1.

In group B, the daily temperature is higher than that in group A. The daily PETs have no differences in both groups, but the PETs per cycle are lower in group B because of the shorter cycle in calendar time.

11.3.1.3 Stability of GreenLab Source–Sink Parameters Concerning Climate

For each experiment, by substituting parameter $E = 1$ for the PET(t) value at each cycle t, we can compute the model parameters, concerning climate, by adjusting the organic series (the value $1/r$ becomes the use efficiency of the PET). This was satisfactory for all treatments. An analysis of variance was performed to test the effect of the seasons on both the weight of the compartments and the model parameters.

The experiments showed that the biomasses of the compartments (leaves, internodes, ears) varied significantly, while the model parameters remained stable. The percentage of biomass distribution in the compartments is also stable since the stability of sink functions, which will be specified later. It can be deduced that the significant variation in biomass production is entirely explained by variations in temperatures and daily PETs. These environmental variables, therefore, do not modify the source–sink and endogenous parameters of the model. The variation in PET per cycle has no significant effect on the source–sink parameters (Fig. 11.3), except for the parameter r. Indeed, fluctuations in PET are removed by the expansion of organs. This allows us to autonomously study the growth by normalizing $E = 1$ (Chap. 9). It is as if the adjustment is made with the average value $\mathrm{E} = \frac{\sum \mathrm{PET}(t)}{T}$ on the number of cycles T.

11.3.1.4 Production Prediction Based on Climate Data

Since the values and stability of the model parameters are established for maize, if a climatic scenario with daily temperatures and PETs is given, in the absence of limiting factors, it is possible to predict the potential harvest. Thus, the parameters calibrated during the year 2000 make it possible to compute the evolution of the

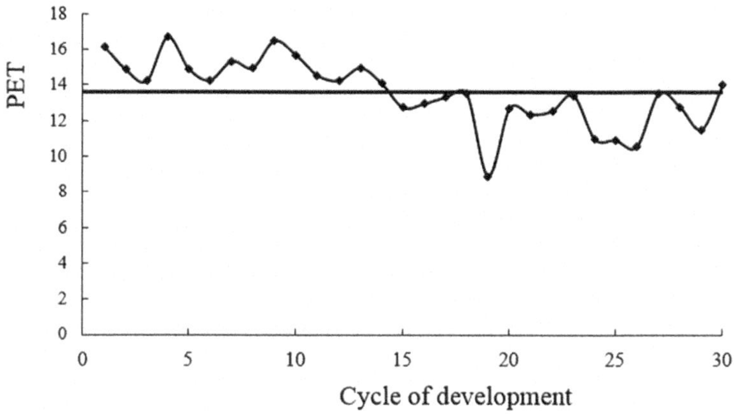

Fig. 11.3 Observed evolution of PET per cycle in an agronomic experiment on maize (China Agricultural University (CAU))

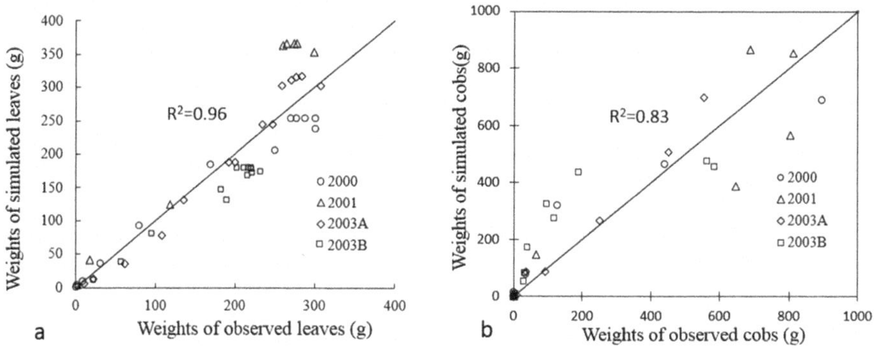

Fig. 11.4 Prediction of the compartment weights of the blade (**a**) and cob (**b**) according to the source–sink parameters computed in 2000 for different climate scenarios, compared to their realization. The compartments of the sheath and internode are as well fitted as the ones in the blades

biomass of the plant compartments according to the chronological age in cycles, with climatic data from other years.

Figure 11.4 shows the adequacy between the predictions based on parameters computed in 2000 and observations from subsequent years, for which climate data (temperature, PETs) are different. The compartments displayed are the blades and cobs. Dispersion in the cobs is because the expansion is subject to more complex mechanisms, such as grain filling.

The 2003A and 2003B treatments correspond to sowing dates in 2003. The 2003A treatment corresponds to sowing in May and 2003B in July. The second period is warmer, and the plant grows faster, but the daily PET has no difference. While plants develop in cycles, they have different calendar durations, which modifies the growth (Fig. 11.5). The plant sown in July is, therefore, smaller.

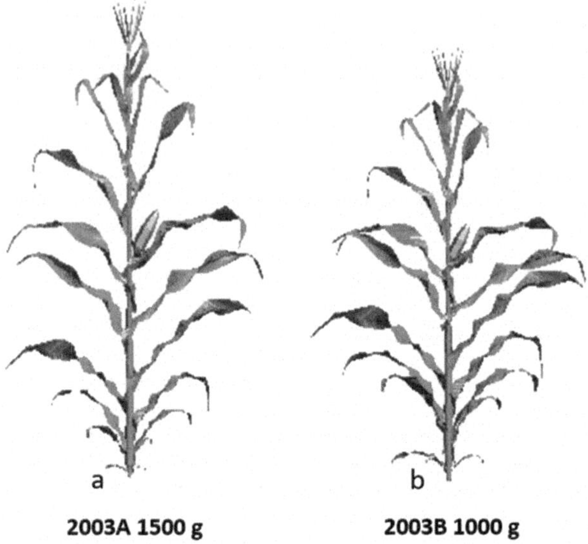

Fig. 11.5 Architectures of two maize plants planted in May (2003A) and July (2003B) respectively (Experiments by Ma Yun-Tao, China Agricultural University (CAU)). The source–sink parameters are the same, but the environmental parameters (temperature and daily PET) differ. The average daily temperature is 2 °C higher for the 2003B treatment, while the daily PET remains stable. This reduces the cycle time and, therefore, the PET per cycle. The temperature increased the rate of development over a limited time period, which reduced the growth by the same amount. Thus, the weight of plant (**a**) planted in May reaches 1500 g and that of plant (**b**) planted in July reaches 1000 g

11.3.2 Study of the Effect of Planting Density on Production Per Square Meter

In 2005–2006, a maize experiment was carried out at CAU to compare the effect of planting density on biomass production and distribution in the plant compartments. Three densities were tested: 2.77, 5.55, and 11.11 plants/m^2, corresponding to the spacing of 60 × 60, 60 × 30 and 30 × 30 cm^2, respectively. The fitting of the organic series provides the source–sink parameters and they are correctly fitted by the model over five growth stages. Individual plant weights are highly dependent on the density, and are half as low at the high density (750 g) compared to the lowest density (1500 g) (Fig. 11.6a), as is the weight of cobs (Fig. 11.6b).

The study showed that the number of produced phytomers as a function of thermal time (development) is not affected by the planting density (Fig. 11.7a). This result is consistent with the normal relationship between development and thermal time. If the sink–source parameters are stable concerning climate, as shown in Sect. 3.1, there is variability concerning density obviously for *Sp* but also slightly for parameter *r*, with the other parameters being stable. As a result, the

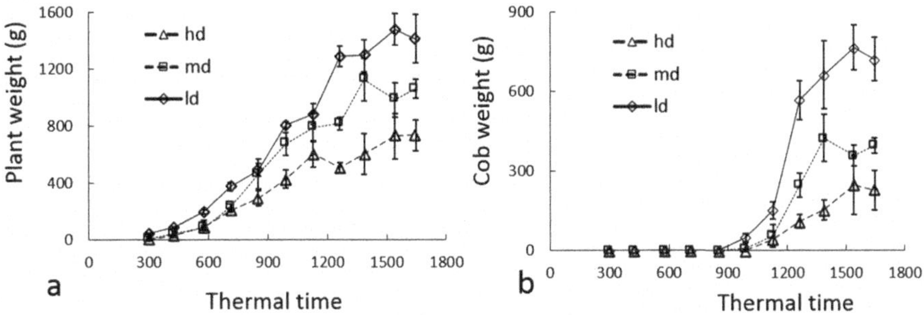

Fig. 11.6 Evolution for the maize of biomass production as a function of low density (ld), medium density (md), and high density (hd). (**a**) Weight of the plant. (**b**) Weight of the cob. The increase in density d reduces the surface area Sp and, therefore, the production. We have $Sp \approx Sd = 1/d$

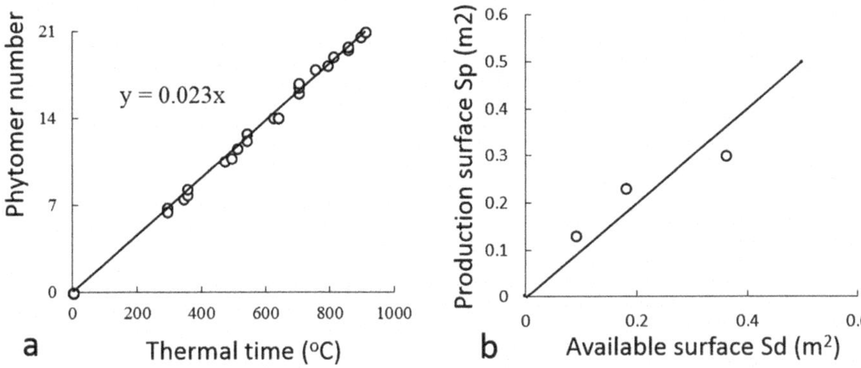

Fig. 11.7 Effects of the density on the number of phytomers and production surface. (**a**) The relationship between thermal time and the development (phytomer numbers) is not modified by the planting density (three densities together in the observations (○)). (**b**) The computed production surface Sp is close to the available surface Sd per plant, the inverse of planting density, as predicted in the model, $Sp \approx Sd = 1/d$

distribution of biomass by the compartment does not depend on the density (Fig. 11.8a). The variation in Sp among planting densities shows that the projection area Sp is close to the available area Sd offered by the density ($Sd = 1/d$), which is a characteristic of the prediction of GreenLab (Fig. 11.7b).

Under these conditions, we move from the individual plants to the stand defined per square meter. The parameters keep the same values, except for Sp, which is adjusted according to the planting density. To predict the harvest per square meter, simply multiply the weight of the individual plant by the planting density. This is illustrated in Fig. 11.8b. The experiment was repeated for two consecutive years. In this figure, the total produced biomass per density and square meter is displayed, as well as the useful harvest, i.e., the biomass of the cobs per density and square meter. It can be seen that the production in the second year is correctly predicted on the three planting densities by the parameters computed in the first year.

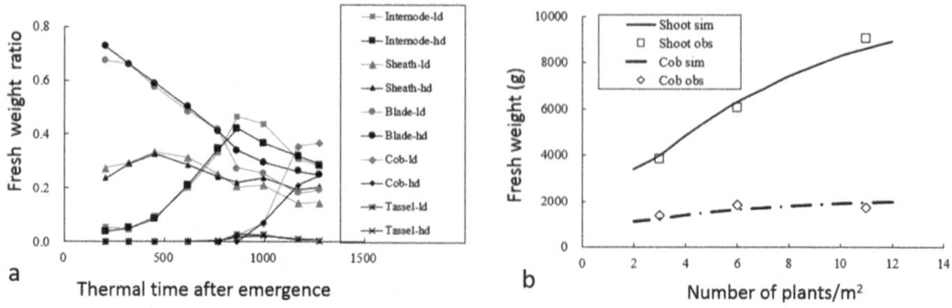

Fig. 11.8 Evolution of biomass production during the growth. (**a**) Comparative evolutions of the percentages of biomass distribution in compartments for low (ld) and high density (hd) in maize. These are stable although the produced biomass varies from simple to double. (**b**) The total biomass production of plants per square meter increases with the density, but that of cobs does not change

The harvest index decreases from about 50 to 25% as density increases from low to high. It can be seen that the total produced biomass per square meter doubles from low to high density (4–8 kg), while the harvest by weight of cobs remains relatively constant (2 kg).

This result is consistent with the model prediction. Indeed, for total biomass, it depends strongly on the planting density, because at high density, all the ground is already covered as soon as the plants germinate and the LAI is high from the beginning. Biomass production is, therefore, already saturated, whereas for low density, it takes more time for the plants to cover the ground and come into contact, which explains the difference in cumulative production of total biomass per square meter among the different densities. However, the situation is different with regard to biomass production per square meter of cobs. The cobs appear late in DC 15 according to the development pattern that is independent of density. At this stage, all plants are in contact for all densities; the LAI is high and production per square meter no longer depends on density. However, the distribution of biomass in the compartments per plant is also independent of density. It is, therefore, the same cumulative biomass that is assigned to the cobs per square meter, when they appear.

This example shows the interest of functional–structural models in understanding the HI (harvest index) and their predictive efficiency; this is particularly true for the GreenLab model (Fig. 11.9).

11.3.3 Transition from the Individual Plant to the Plant Stand

In the previous section, the harvested plants were selected at the same development stage, so they are cohorts, and they have also been sown individually with controlled distances between plants. In a field, the plant stand is subjected to at least two additional factors of variation. First, there is variability in germination dates that causes the development of individuals at a given age to follow a distribution law, and

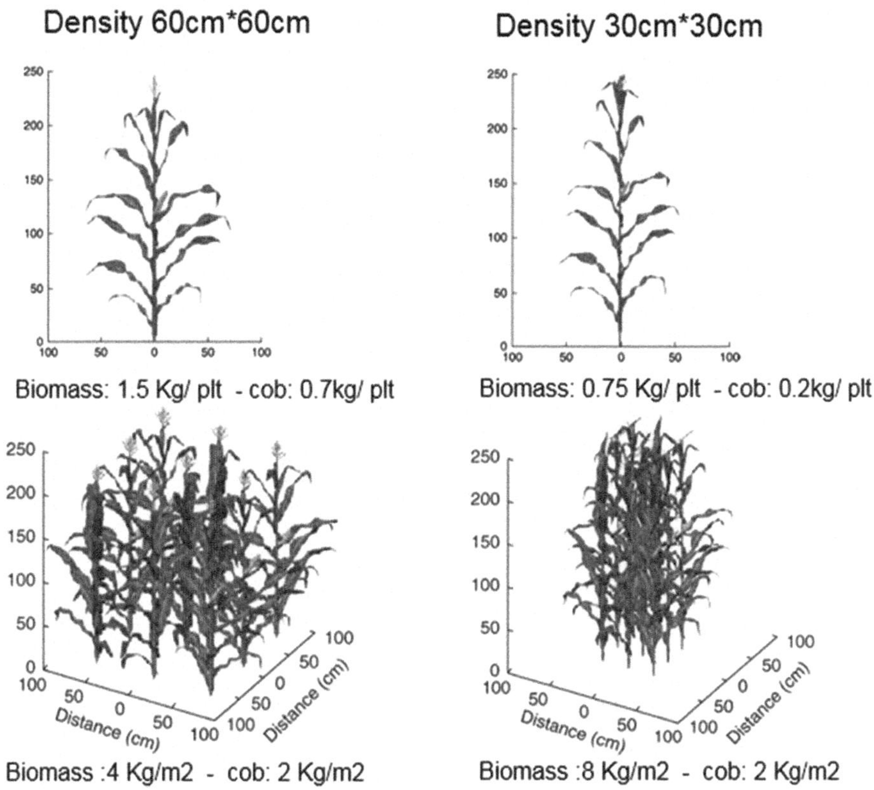

Fig. 11.9 Representation of an individual plant and the corresponding stand per square meter under low and high density. The weights of individuals and corresponding stands vary in the opposite direction to the increase in density. The production per square meter in kg of cobs does not change

second, the seed drill produces irregularities in plant spacing within a row. The local density per plant is, therefore, variable, not to mention the shortages due to seed mortality.

There is, therefore, a gap between the artificial stand generated by manual operations and the natural stand generated by automatic seeding. The methods of computing production in the two cases differ.

The principle of field crop models is to convert the light intercepted at the level of the square meter into the biomass at each time step of the calendar (day). Most of them use the equation:

$$Q(t) = \frac{E(t)}{r} \cdot (1 - \exp(-k \cdot \mathrm{LAI}(t))) \tag{11.2}$$

where Q(t) is the biomass production at time t; $1/r$ is the resource use efficiency (RUE); $E(t)$ is the climate data; LAI(t) is the LAI measured at time t by optical instruments immersed in the canopy. The coefficient k is relative to the orientation of the leaves. By summing over time we obtain the total biomass production, i.e., the weight of the harvest. It should be noted that the notion of the cultivated area is implicit in Eq. 11.2 (often 1 m^2).

11.3.3.1 Comparison Between the Pilot Model and the GreenLab Model

The Pilot program, developed by J.-C. Maihol (IRSTEA) (Mailhol et al. 1996), works according to the principle of Eq. 11.2. The harvest corresponds to the biomass production per square meter, multiplied by an empirical harvest index. The particularity of the model lies in a powerful water module that estimates the quantity of water available in the soil according to irrigation and crop evapotranspiration. The model combines radiation and water resource efficiencies.

The LAI is measured directly in the plantation by an instrument (Li-Cor) that evaluates the light intercepted by the canopy, regardless of any reference to architecture or density.

In the absence of water stress, biomass production is totally controlled by the LAI. As a first approximation, the evolution of the LAI depends only on the sum of the temperatures from one year to the next, which stabilizes according to the equation:

$$\mathrm{LAI}(J) = \mathrm{LAI}_{\max}\left(\left(\frac{TT(J) - T_e}{T_{\max}}\right)^{\beta} \cdot \exp\left(\frac{\beta}{\alpha} \cdot \left(1 - \left(\frac{TT(J) - T_e}{T_{\max}}\right)^{\alpha}\right)\right)\right) \tag{11.3}$$

On day J the sum of the temperatures is $TT(J)$, it takes a time T_e for the germination to start the following planting, and the maximum LAI is reached at time $T_{\max}$. The parameters α and β are calibrated on the measurements and serve as a reference for a stress-free plantation. Each year we check whether the development of culture is in line with the theoretical model. In general, the performance of predictions by the Pilot model is satisfactory.

In 2008, a maize experiment was carried out in the IRSTEA fields in Montpellier to compare the performance of the Pilot crop model and the GreenLab model.

This variety of French maize is different from the Chinese one studied in Chap. 9. The stems have 19 phytomers instead of 21.

The planting density was 7.5 plants per square meter, and the plot was irrigated to remove water stress. The plants were measured in two types of measurements in the same field to provide data for both models. The Pilot measured data are LAI and biomass production per unit area at different growth stages. The data measured and predicted by the model are in good agreement (Fig. 11.10).

GreenLab data are the biomass of organic series taken from plants at the same development stage for different dates. We average 10 plants each time. The fitting of these series with the WLSM gives the set of source–sink parameters that control the growth (see Table 11.2 and Fig. 11.11).

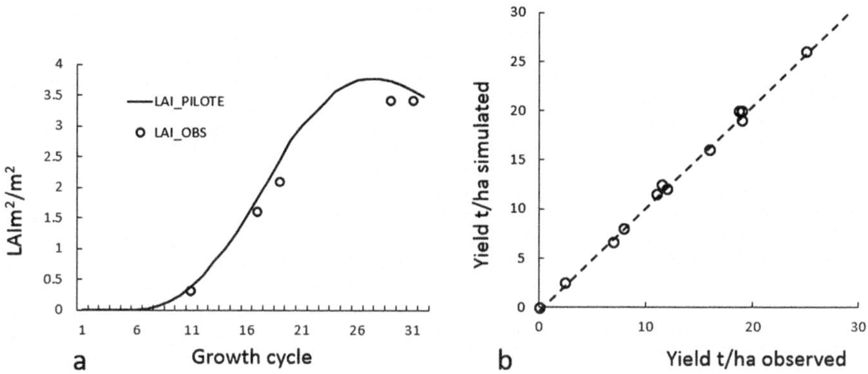

Fig. 11.10 LAI and biomass were measured and predicted by the Pilot model. (**a**) Evolution of the LAI measured with Li-Cor (○) on maize experiment compared to the theoretical curve (Eq. 11.3) computed by thermal time. (**b**) Predictions of the Pilot model in biomass production per hectare compared to observations (○) during the experiment

Table 11.2 Computed source–sink parameters of maize (IRSTEA)

Param	r	Sp	sSp	Q_o	B_{b1}	B_{p1}	B_{i1}	B_{f1}	P_{p1}	P_{i1}	P_f	P_m
Mean	18.3	1139	159.4	4.1	2.8	2.3	18.6	7.2	0.9	6.3	2854	3.6
SD	1.6	117	77.4	1.5	0.2	0.1	0.6	0.2	0.0	0.3	0	0.3
CV (%)	9	10	49	38	7	6	3	2	4	4	0	7

In particular, GreenLab correctly computes the planting density from the data. $Sp = 1139$ cm^2/plant is computed, while the surface per plant is $Sd = 1333$ cm^2/plant associated with the actual planting density (7.5 plants/m^2). The individual LAI of GreenLab is in correspondence with the leaf area measured per square meter for each sampled development stage.

This is done from individual plants to the stand level for LAI and biomass by multiplying the individual data by the density ($d = 7.5$ plants/m^2) and comparing them with the respective data obtained by Pilote at the square meter level. There is a discrepancy in curves between the LAI (Fig. 11.12a) and biomass production (Fig. 11.12c). This is because the measurements for Pilot were made from the plant stand with a mixture of germination dates, while those for GreenLab were made at the same development stage, i.e., on the cohorts of individuals.

The distribution of the number of leaves per plant after sprouting can be used to deduce the germination delay law (in DC). This theoretical law (a negative binomial law, solid line in Fig. 11.12e) fits the observations (○). Germination lags vary between one and six DCs. The composition of cohorts according to their frequency in germination times correctly reflects stand observations for the LAI (Fig. 11.12b) and total biomass production (Fig. 11.12c).

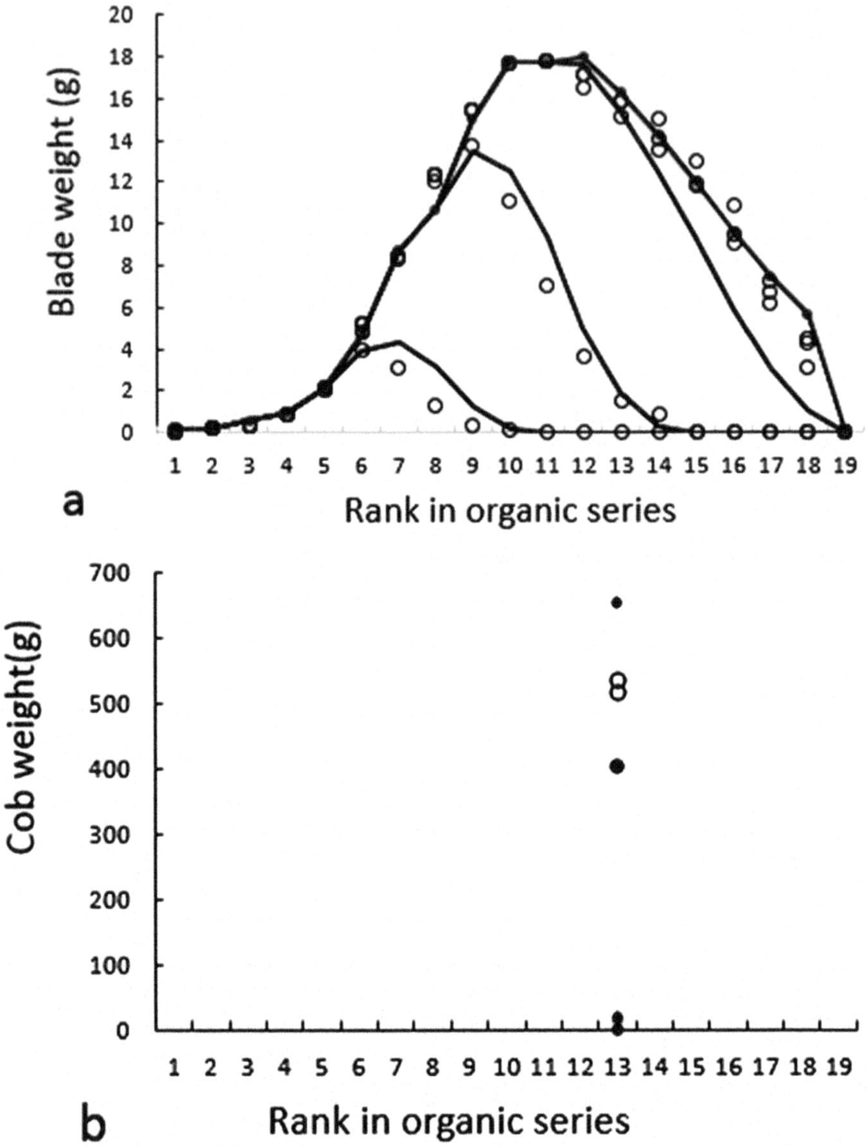

Fig. 11.11 Observed and computed weights of blade and cob for maize. (**a**) Blade weights of the observed leaf blades (○) and computed (curves) over several development dates with the same set of parameters. (**b**) Weight of the cob (○ = observed, ● = computed) over the same dates. The other organic series (sheaths, internodes, male ear) are also well fitted

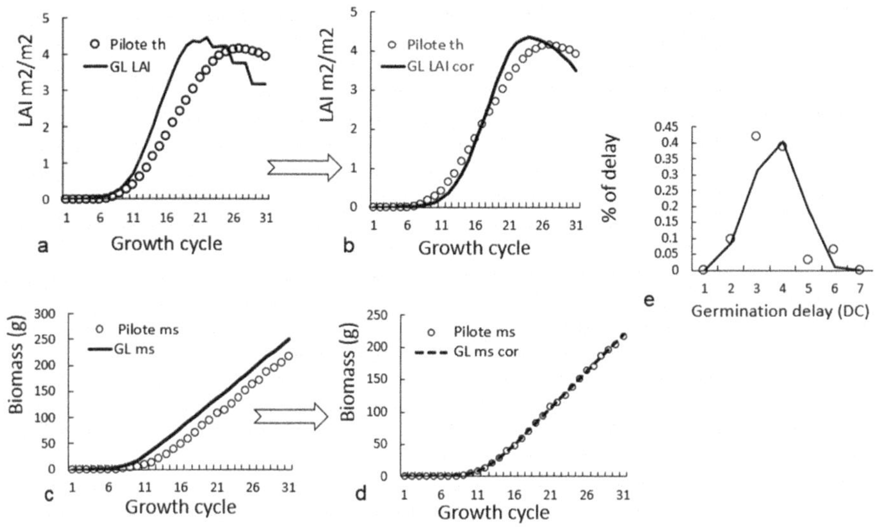

Fig. 11.12 Comparison of outputs for the Pilot and GreenLab models. (**a**, **c**) The solid curves are computed by GreenLab from the cohorts and the values computed by Pilot (○) for LAI and cumulative biomass per plant, respectively. The curves are shifted. If we take into account the law of germination delay (**e**), which produces a mixture of cohorts to simulate natural population, the lag is eliminated for LAI (**b**) and biomass and (**d**)

11.3.3.2 Individual Biomass Variability: The Use of Differential Statistics

The average biomass per plant is not sufficient to describe the functioning of the stand. There is individual variability whose origins are defined at the beginning of this chapter. It is not possible for reasons of the complexity of measurements to go down to the level of the organ or organic series to take this reality into account. On the other hand, it can be easily integrated into the compartments that represent the sum of the biomass for the same type of organ (leaves, internodes, fruits). In statistical terms, therefore, the mean and variance of total biomass per plant are associated with the means and variances of organ compartments. A statistical method, called "statistical differentials" (Johnson and Kotz 1970), makes it possible to compute the means and variances of these compartments from the source–sink parameters and one or more input source variables (E, Sp, Q_0) defined by their means and variances.

By focusing on the cohort concept, the influence of the time to germination on biomass production is eliminated. It is also assumed that the seeds (parameter Q_0) are graded in biomass and that the environment is homogeneous (constant parameter E). Under these conditions, only the local planting density around the individual plant should be taken into account.

The produced biomass $Q(t)$ in cycle t is a function of the values $Q(i)$ from the seed ($i < t$) and the variable Sp. The other parameters (sink functions, etc.) are constants. The production function of GreenLab is written in Chap. 5:

$$Q(t) = \frac{E(t) \cdot Sp}{r} \cdot \left(1 - \exp\left(-\frac{k}{\varepsilon \cdot Sp} \cdot \sum\nolimits_{i=\max(1,\, t-t_a+1)}^{t} N_a(i) \sum\nolimits_{j=i}^{t} \frac{P_a(j-i+1) \cdot Q(j-1)}{D(j)}\right)\right) \quad (11.4)$$

It can be written as a function G depending on Sp and variables $Q(i)$:

$$Q(t) = G(Sp, Q_0, Q(1), \ldots, Q(t-1)) \quad (11.5)$$

The variance $vQ(t)$ of $Q(t)$ is derived from the derivatives of function G and the variances of the variables by the following approximation in the so-called differential statistics method:

$$vQ(t) \approx (G'_{Sp})^2 vSp + \sum (G'_{Q(i)})^2 \cdot vQ(i) + 2\sum G'_{Sp} \cdot G'_{Q(i)} \cdot r_{SpQ(i)} \sqrt{vSp \cdot vQ(i)} + 2\sum\sum G'_{Q(i)} \cdot G'_{Q(j)} \cdot r_{Q(i)Q(j)} \sqrt{vQ(i) \cdot vQ(j)} \quad (11.6)$$

The variance of Sp noted vSp represents the variation in production area per plant due to the local planting density. It is established that the correlations $r_{SpQ(i)}$ between Sp and $Q(i)$ are 1 as well as the correlations $r_{Q(i)Q(j)}$ between $Q(i)$ and $Q(j)$. The derivatives of G concerning $Q(i)$ and Sp are easily computed numerically. The values of $Q(i)$ and their variances $vQ(i)$ are, therefore, computed by recurrence from the seed Q_0. In this system, vSp appears as a parameter at the same level as Sp, so it can be computed by the inverse method. If biomass production per cycle is not observable, cumulative weights are observed by weighing. The means and standard deviations of the compartments are deduced from Eq. 11.5 from their partial demands and, therefore, the standard deviations.

The method for the organ compartment noted o (o = leaf, internode, fruit, …) is established here.

If $D_o(i)$ is the demand of the sum of the organ sinks of compartment o, and $D(i)$ is the demand of the plant in cycle i, we have:

$$D(i) = \Sigma_o \cdot D_o(i) \quad (11.7)$$

The biomass that goes to the organ compartment o is:

$$Q_o(i) = \frac{D_o(i)}{D(i)} \cdot Q(i) \quad (11.8)$$

The standard deviation of the biomass of the compartment is:

$$sQ_o(i) = \left(\frac{D_o(i)}{D(i)}\right) \cdot sQ(i) \tag{11.9}$$

The weight of the plant obtained by the sum of the biomass production in each cycle is:

$$TQ(t) = \sum\nolimits_{i=1,t} Q(i) \tag{11.10}$$

The weight of compartment o in cycle t is:

$$TQ_o(t) = \sum\nolimits_{i=1,t} Q_o(i) \tag{11.11}$$

The variance of weights of the compartment o is:

$$\text{var}(TQ_o(t)) = \left(\sum\nolimits_i sQ_o(i)\right)^2 \tag{11.12}$$

The influence of the variation in *Sp* on the variation in the weights of the plant compartments is quantified using the variances.

The computation of the plant system parameters is, therefore, based on both the organic series and the means and the variances of the weights of the organ compartments which make it possible to estimate the variance *vSp* associated with *Sp*.

11.3.3.2.1 Application on Maize

When applied to the maize experiment, the system works properly (Fig. 11.13). During the growth, the simultaneous evolution of the compartment biomass for the mean and standard deviation can be monitored. The means and variances of the compartments can be measured directly from the samples taken at each growth stage. The computation of the parameters of the plant system is, therefore, based on both organic series and on the means and variances of the weights of the organ compartments that make it possible to estimate the variance *vSp* associated with *Sp*. In practice, for the computations, we take the standard deviation *sSp* instead of *vSp* ($sSp = (vSp)^{1/2}$), because it has the same dimension as *Sp*. Table 11.2 shows the sink–source parameters computed for this maize.

The fitting is good: $R^2 = 0.98$, number of data = 444, number of parameters = 11.

Sheaths have an expansion delay of two GCs and cobs of three GCs. The second parameters of Beta laws are: 5, 5, 10, 3 for blades, sheaths, internodes, and cobs, respectively. For cobs, the sink strength and the first parameter of the sink variation law cannot be computed together in this example. All the organic series are well fitted (see Fig. 11.11). For tassels, it is not possible to compute the variation law of the sink. The two parameters are fixed to 2.

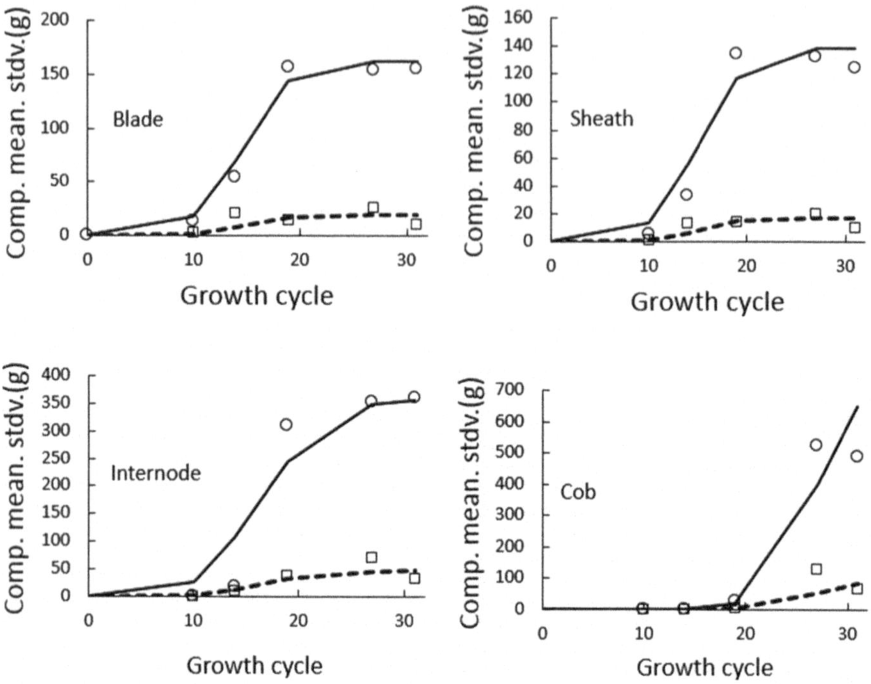

Fig. 11.13 Evolutions in the means (lines) and standard deviations (dashes) of the weights of the organ compartments and their adjustments by the model. (○) and (□) represent the mean and standard deviation values of the measurements on maize

The computed standard deviation *sSp* of *Sp* allows to control of the statistical variations in all the organ compartments (Fig. 11.13).

11.3.3.2.2 Application on Beetroot (S. Lemaire, ITB)

Differential statistics have also been successfully applied to beetroot (Lemaire 2010).

The measurements are simplified when the functions of sink variations have been computed beforehand, which are assumed to be stable. Thus, in the case of beet, they have been computed for a density of 10 plants/m^2, as well as the seed Q_0 (see Chap. 9). For other planting densities, only the compartments can be used. The computed parameters are then the resource use efficiency $1/r$, the production area *Sp*, the standard deviation *sSp*, and the taproot sink p_r. The experiment was carried out on four planting densities: 1, 5, 10, 20 plants/m^2.

The model correctly computes the production surface *Sp* that matches the available surface *Sd* deduced from the planting density. The means and the standard deviations of the biomasses of compartments were computed together for the four densities, which work properly (Fig. 11.14). For low density, $Sd > Sp$, because there is no more competition for the space between plants, only the self-shadowing plays a role.

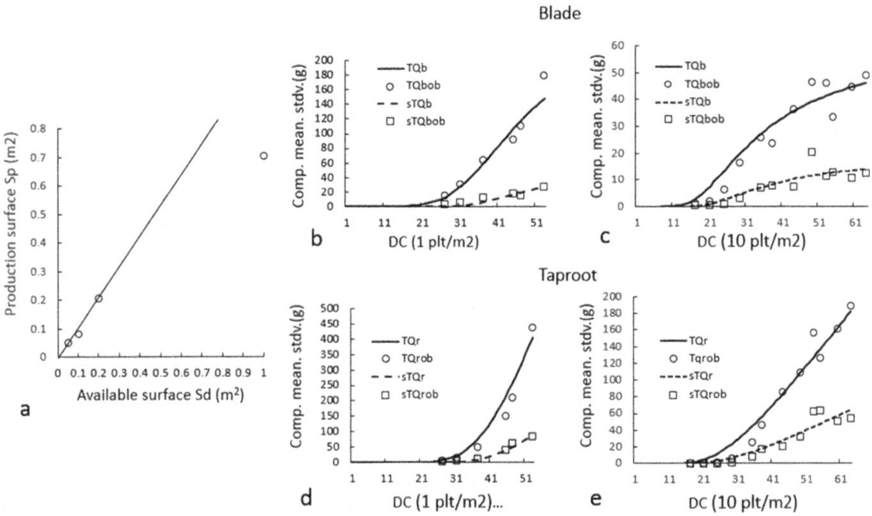

Fig. 11.14 The means and the standard deviations of the biomasses of compartments computed for the four densities (1, 5, 10, 20 plants/m^2). (**a**) Prediction of the production area *Sp* according to the planting density (available area *Sd*). For low density (1 plant/m^2), we have $Sd > Sp$. (**b**, **c**, **d**, **e**) fitting of biomass means and standard deviation for leaf and root compartments, for two different planting densities (1 plant/m^2 and 10 plant/m^2). A strong decrease in the weight of the compartments is observed with the increase in density

11.3.4 Simulation of a Maize Field

To simulate a field of maize from individual plants whose source–sink parameters are known, it is, therefore, necessary to take into account the germination delay law and the variability of the available surface area per plant *Sd* (here $Sp = Sd$) taking into account the average density *d*: ($Sd = 1/d$). The variability of *Sd* is characterized by its standard deviation *sSd* and the area per plant is simulated by a normal distribution with parameters *Sd*, *sSd*. The delay of germination law requires two parameters, while the variability of biomass requires only one (*sSd*). The transition from the individual plants to the stand is, therefore, very efficient (three additional parameters), and can also be computed from observations using the inverse method (WLSM).

Using the Simeo software of CIRAD (Griffon and de Coligny 2012), a field is constructed whose mesh size varies according to the normal law (*Sp*, *sSp*). The mesh size defines individual areas per plant. To reproduce the field, for each plant, a germination delay and a production area are randomly selected. Foliar and harvest indices are then consistent with those provided by stand measurements (Fig. 11.15).

The case of maize thus shows the link between a functional–structural model (GreenLab) and a crop model (Pilot) and their close compatibility.

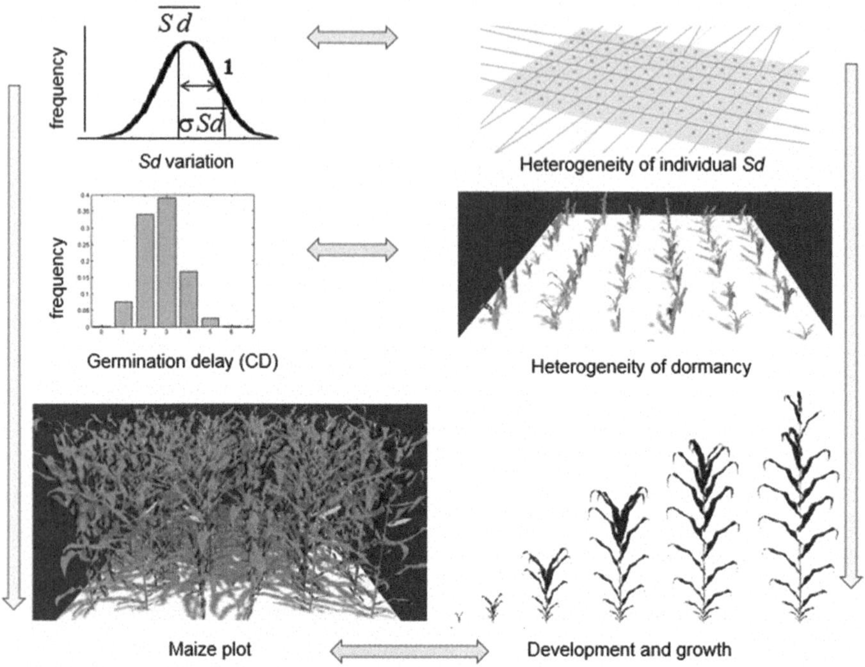

Fig. 11.15 Simulation of a maize field. Each maize plant grows with a germination delay and a production area randomly selected according to the distribution laws calibrated on the individual data. The reproduced plot has the same foliar and harvest indices as the original plot. The population is reproduced

11.4 Effects of Climate on the Stability of Source–Sink Parameters: The Case of Chrysanthemum (Kang Meng-Zhen, Ep Heuvelink)

11.4.1 Description of the Study

While the action of the climate has little visible effect on the values of the source–sink parameters of maize, it is not the case for chrysanthemum, an herbaceous plant marked by a strong basipetal development.

Chrysanthemums are grown for ornamental purposes. The architecture of the plant is, therefore, of particular economic importance. Unlike maize, where the weight of the organs is crucial, for chrysanthemums, it is the dimensions and numbers of organs (leaves, internodes (height of the plant), flowers) that are of primary concern. Light and temperature significantly modify the growth and architecture of the plant, especially their basipetal development (Chap. 7). This involves measuring their influences on plant growth, and adjusting and optimizing their values in a greenhouse.

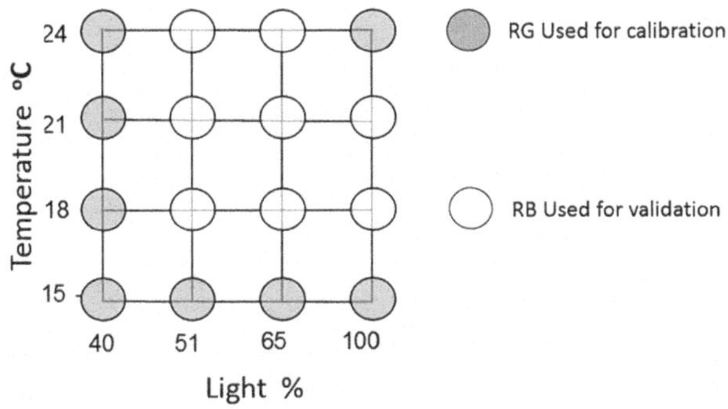

Fig. 11.16 Light–temperature combinations were used to study chrysanthemum growth. RG represents the combinations used for calibration and RB represents those used for validation

It should be noted that while the parameters should remain stable during growth under various environmental conditions, for the model, to be robust, nothing prevents the values of these parameters from being dependent on these conditions. Thus, in electricity, resistances depend on temperatures, but Ohm's law remains valid.

This experiment aims to study the phenotypic variation in the plant architecture as a function of the combination of environmental parameters. Four levels of light intensity and four levels of temperature have been selected; they define a grid of controlled cultivation conditions for 16 plots in the greenhouse. Eight combinations were selected for calibration and eight combinations were used for validation (Fig. 11.16). These studies were published in Kang et al. (2006a, b, 2012).

Source–sink parameters were computed by fitting the simple and compound organic series of chrysanthemums during the growth under different temperature–light combinations.

If the temperature controls the rate of development as expected, we can see that light influences the number of phytomers on the stem. Flowering occurs earlier with the increase in light.

11.4.2 Method for Estimating Parameters by Interpolation

The source–sink parameter values of the GreenLab model are computed through the 16 controlled experiments for light and temperature according to the experiment grid (Fig. 11.16). It is proposed to predict the values of the parameters within the grid by regression, from those computed on its edges, using a quadratic form that computes the effect of light and temperature and mimics a limited development of the unknown second-order functions of these parameters.

$$f = a_1T^2 + a_2L^2 + a_3T + a_4L + a_5T \cdot L + a_6 \qquad (11.13)$$

The coefficients a_i are computed by regression only on the data from the eight experiments corresponding to the edges (RG) of the grid in Fig. 11.16.

11.4.3 Results

The eight combinations (RG) on the edge of the grid allow us to compute the coefficients a_i by regression. They are used as a reference to estimate the parameters of the GreenLab model for the other eight temperature–light (T*L) combinations (RB). This predicts plant growth by running the model and comparing it with the actual growth achieved in each combination. Finally, a good agreement is obtained between the predictions and the actual data (here the weights of the organ compartments on the stems and branches). Chrysanthemum growth is, therefore, linked to the combination of the two environmental factors (light and temperature).

The combinations of T*L do not affect the phytomers of the stem for the primary growth of internodes. Sink values of blades and internodes are stable. However, the increase in light intensity increases the sink value of the flowers. It should be noted that the temperature has no significant effect on sink values. The remaining parameter values, except for the light use efficiency, increased with light intensity.

The combinations of T*L influence girth growth, light inerception per cycle, branch organ sinks, and parameters that control basipetal development (Chap. 7).

Figure 11.17 shows that for the nine growth stages and eight T*L combinations of the RB experiments, the weights of the observed organs of all types are well predicted, which is a validation of the model.

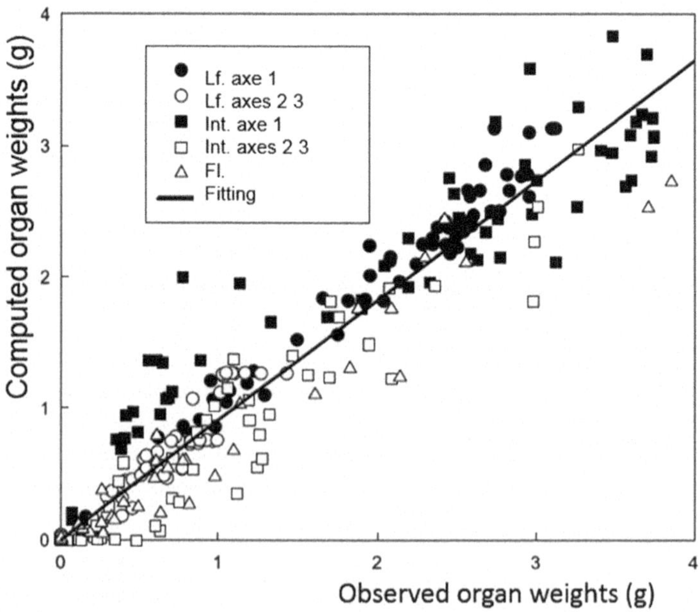

Fig. 11.17 Prediction of the weight of the different stem and branch organs for eight combinations of light and temperature over nine growth stages

The source–sink parameters for the RB combinations are estimated from the function 11.13.

The parameters g_i of basipetal development (Chap. 7) of the expansion delay laws were estimated together with the source–sink parameters. The interpolation procedure is validated by comparing the observations with the predicted values. The leaves expand immediately throughout the plant. The leaves carried by the branches form rosettes beneath the stem leaves, even before the expansion of the internodes has pushed them out of the stem. The indices refer to the branching order. It can be seen that the values of the parameter vary with the increase of temperature but the light intensity has no significant effect.

These parameters, which modify the appearance of organs on the branches, play an essential role in the computation of demand and in the architecture of the plant. A decrease in temperature increases the basipetal development on the stem. The increase in light reduces the development of the stem, which accelerates flowering and increases the sink strength of the flowers. As a result, the plant is stronger and the flowers are larger.

The observed numbers of flowers are in good agreement with the predictions computed with the light and temperature combinations and the decrease in light reduces the number of flowers produced while delaying their appearance (Fig. 11.18).

The model also reproduces well in the grid at the edges, leaf area, and height of the plants (Fig. 11.19). This prediction is important for assessing the visual quality of flower production.

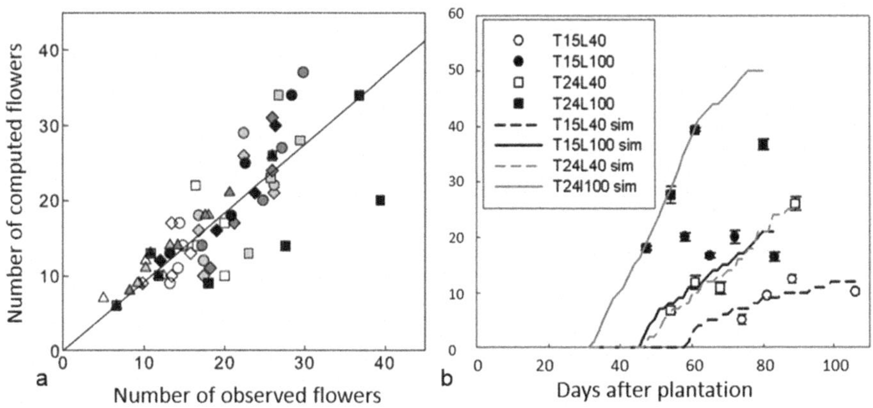

Fig. 11.18 Flower production. (**a**) Comparison of the number of flowers observed and predicted by the laws of delay in expansion obtained by interpolation. The light intensities L (40, 51, 65, 100) are represented by the forms Δ, ◊, ○, □ respectively. Temperatures T (15, 18, 21, 24) are represented by the filling colors white, light gray, dark gray, and black. (**b**) Number of flowers produced according to the age of the plant (observed: markers, computed lines), according to the light and temperature combinations (Color figure online)

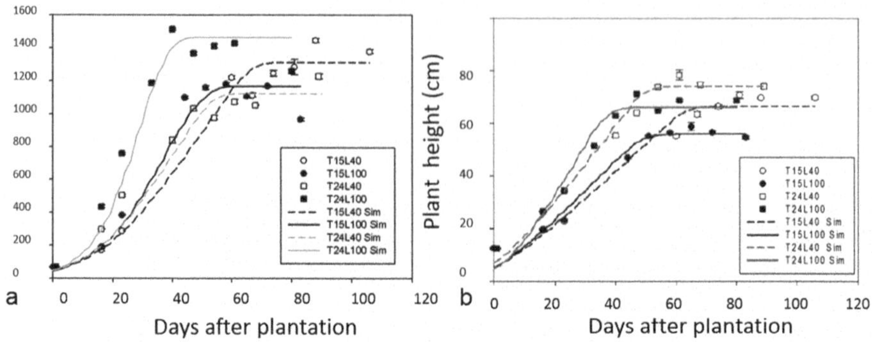

Fig. 11.19 Observed and computed foliar surfaces and heights of plants. (**a**) Observed and computed foliar areas for different combinations of light and temperature. (**b**) Plant height growth for the same combinations. Model validation is satisfactory

11.4.4 Visualization of the Plants Obtained by the Different Temperature–Light Combinations

Figure 11.20 shows photographs of the plants grown under various environmental combinations, for the same duration in calendar time. These are arranged in four blocks, each maintaining a constant temperature of 15, 18, 21, and 24 °C, respectively. In each block, the light intensity decreases by percentages of 100%, 64%, 54%, 40% per plant. It is visible that low temperatures delay flowering and promote vegetative development. The effect of increasing light intensity reduces the duration of development and shortens the length of the stem before flowering. It also increases the sink strength of the flowers and their volumes.

11.4.5 3D Visualization of Plants Simulated by GreenScilab Software

The simulation of chrysanthemum growth and the visualization of its architecture according to climatic parameters are finally carried out by the GreenScilab software (Kang et al. 2006a, b, 2009). Each combination of temperature–light values is matched by a set of parameters that control the development and growth of the plant. The visual results of these simulations are close to the photographs of the plants subjected to the corresponding combinations (Fig. 11.21).

Fig. 11.20 Observations on chrysanthemum growth and flowering for different temperature–light combinations. The increase in light triggers earlier apical flowering and reduces development. The sink of flowers is stronger. (Photos from Heuvelink, Wageningen University)

11.4.6 Conclusion on the Growth of Chrysanthemums

For stabilized values of temperature and light (T*L), the parameters of chrysanthemum development and growth are constants. These constants can take different values if the climatic conditions T*L of the crop vary without changing the model formulation. The parameter values are obtained from a set of combinations T*L on the edge of a grid that constitutes a learning curve, by optimizing a quadratic function. This is predicted by interpolating the value of the parameters for other

Fig. 11.21 Simulations of chrysanthemums grown in different climatic combinations of temperature and light (GreenScilab software, Kang et al. 2006a, b, 2009)

Table 11.3 Differences between the five models

	Modelling scale	
	Compartment	Organ
Empirical distribution	Pilote	CERES
Biomass allocation	STICS LNAS	GreenLab

combinations T*L within the grid. This function has been proven effective in providing a good prediction of plant growth and architecture as a function of temperature and light, which validates the model.

11.5 Predictive Value of the GreenLab Model Compared to Crop Models, Case of Beet (Charlotte Baey)

This study was conducted by Charlotte Baey. Several plant crop models have been developed since the 1970s, each with its specificities and objectives. They do not have the same scales for modelling and taking into account the plant, and can integrate a variable number of physiological mechanisms. These different levels of complexity generally imply an increase in the number of parameters to be estimated. However, when a model is used as a predictive tool, it must be as robust as possible, in particular to be able to adapt to different situations from those in which it was calibrated.

A comparative study of the predictive capacities of five plant growth models for sugar beet was carried out (Table 11.3). The models concerned are CERES (Jones

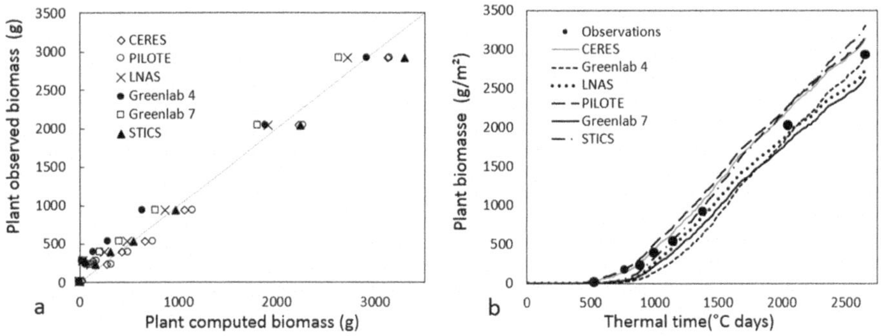

Fig. 11.22 Prediction of total biomass. (**a**) Observations *versus* predictions. (**b**) Changes in total biomass production as a function of thermal time. The dots correspond to the observations and the lines to the predictions of the different models

and Kiniry 1986), STICS (Brisson et al. 1998), Pilot (Mailhol et al. 1996, 2011), LNAS (Baey et al. 2014) and GreenLab (in two variations). This work follows the thesis work of Sébastien Lemaire (2010).

In the PBM (*Process-based model*) CERES and Pilot, the LAI (leaf area index) evolves empirically, independently of the created biomass, and a harvest index makes it possible to distribute the biomass production between leaves and roots at the end of the growing period. In the LNAS and GreenLab models, biomass distribution is dynamic throughout the plant growth period, using biomass allocation functions. The difference between these two models is the modeling scale, which is done at the individual organ level in GreenLab and the organ compartment level in LNAS. Thus, the calibration of the GreenLab model is done here at the level of the compartments, not at the level of the organic series (leaves). In this case, it has been shown that the plant architecture can no longer be reconstructed if the sink functions of the organs have not been previously estimated.

The approach used in Stics can be seen as an intermediary between the two previous ones, as it is based on an empirical evolution of LAI coupled with a source–sink mechanism inducing the feedback of allocation by a trophic stress index.

The five models correctly predict the total biomass production and the fraction going to the taproot (Figs. 11.22 and 11.23).

The study showed good compatibility between the GreenLab structural–functional model and conventional crop models (or PBM). For GreenLab in this study, only the compartments are used. The involved high available planting densities, make the production area *Sp* of the model equal to the available area per plant (*Sd* = 1/density, see section above about differential statistics on Beetroot). We, therefore, have a good approximation in the context of Beer-Lambert's law. The GreenLab model behaves as a "projection" of crop models and, normally, it gives comparable results. By staying at this level, it does not bring anything more. The interest of the model is to provide more in-depth information on the functioning of the crop while remaining within the same agronomic framework. In conventional PBM growth monitoring, measurements on plant compartments are in the shape of

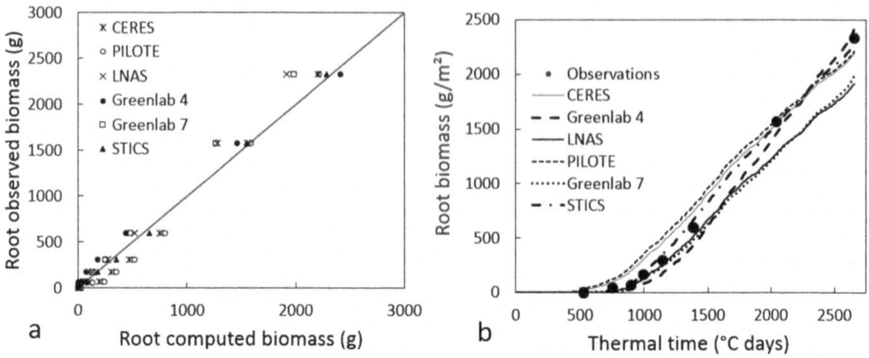

Fig. 11.23 Prediction of root biomass. (**a**) Observations *versus* predictions. (**b**) Changes in root biomass production as a function of thermal time. The dots correspond to the observations and the lines to the predictions of the different models

dots. The organ cohorts are mixed. In the FSPM GreenLab, organic series also provide a memory of growth processes, as organ cohorts are identified. Ideally, a single plant alone provides the entire history of its development and growth. The optimization of source–sink relationships, which essentially plays a role in the harvest index to increase yields, is certainly a new approach that deserves to be explored as its links with enhanced genetics.

11.6 Passage from the Isolated Tree to the Forest Stand

Trees most often have a rhythmic development and a complex structure that evolves according to a development sequence that presents reiterations of the architectural unit. The GreenLab model, using the crown analysis, makes it possible to compute the effect of mortality on growth. Another characteristic is the secondary growth that produces most of the accumulated biomass on the tree. The data are difficult to access, and only mono-specific and even-aged plots allow the modelling to be carried out under the best possible conditions, thanks to individual replicates. A distinction must be made between the development approach based on dendrometric measurements of length (heights, diameters), crown volume characteristic of empirical forest models (EFM), and the growth approach which considers the biomass production and its distribution in the tree structure and which is only in its early stages in FSPMs, such as Lignum (Perttunen et al. 1996, 1998). Indeed, taking into account source–sink relationships significantly increases the degree of complexity of the modelling. In the case of the GreenLab system, the sampling-based measurement method has already shown that it is applicable to model the development of adult trees (Japanese elm, cherry tree, etc., Chap. 3) and potentially applicable to model the growth of adult trees (sugar maple, teak, mahogany, etc., Chap. 8).

11.6.1 Simulation and Visualization of Empirical Forest Models: A Black Pine Model (PNN Feng Lu, Philippe Dreyfus)

Empirical forest models (EFMs) are used to estimate and predict the production of wood in a stand. In each region and for a given tree species, growth tables in height and diameter of trunks and tree crowns are recorded for different densities and sylvicultural scenarios (pruning, thinning), which gives access to the estimation of wood production. The measurements made within the framework of the EFMs are too brief to be able to reproduce the architecture of the tree. They even do not explicitly include the notion of a growth unit (GU). However, it is interesting to visualize the simulated plantations, and to communicate with the various organizations involved in ecology and environmental protection. An additional architectural study can then complete the collected basic dendrometric data. It concerns phyllotaxis, the structure of GUs, and branching patterns. The 3D simulated tree then becomes realistic, while implicitly containing the dendrometric data (height, trunk diameter, crown volume) of the production tables. The stand generated by an EFM can then be realistically visualized.

Such work was carried out by Feng Lu (2012) who coupled the GreenLab model with the EFM empirical model of PNN black pine, developed by Philippe Dreyfus (1993).

The latter model is based on a large database of forest plots that have undergone various sylvicultural scenarios. The main data recorded are:

- the annual growth in height and diameter of the trunk;
- the length and width of the tree crown (which includes branch mortality);
- the variable number of branches per whorl on the trunk.

In addition, the black pine has been the subject of a precise architectural study at the AMAP laboratory (Yves Caraglio), which makes it possible to specify the functioning of the botanical automaton concerning development and geometric parameters. The merging of the two approaches allows simulating the architecture of the tree. For each sylvicultural scenario, the GreenLab automaton is modified to restore the architecture of the individual tree to its correct dimensions (Fig. 11.24). With Simeo software (Griffon and de Coligny 2012) trees are planted at the right density and the plot is visualized in a computer graphic image at any stage of development (Fig. 11.25).

In this type of simulation, there is no spatial interaction between individuals during the growth, but only restitution of the architecture by the model from individual dendrometric measurements made on plots of known density. Biomass is not considered.

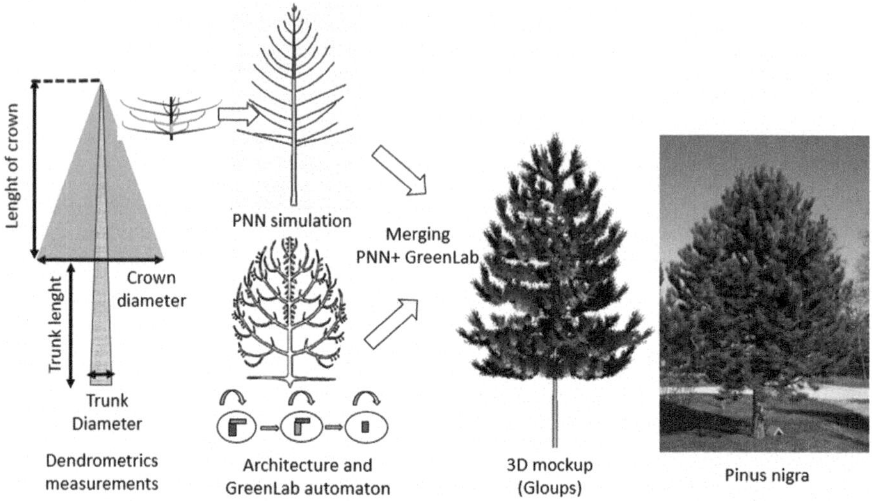

Fig. 11.24 Coupling of the dendrometric model of the black pine PNN (Dreyfus 1993). It provides the dimensions of the tree with a botanical model (GreenLab) by simulating the detail of the architectural model with a botanical automaton. The tree architecture is restored both in detail and in volume (Gloups software, in Java, Feng Lu)

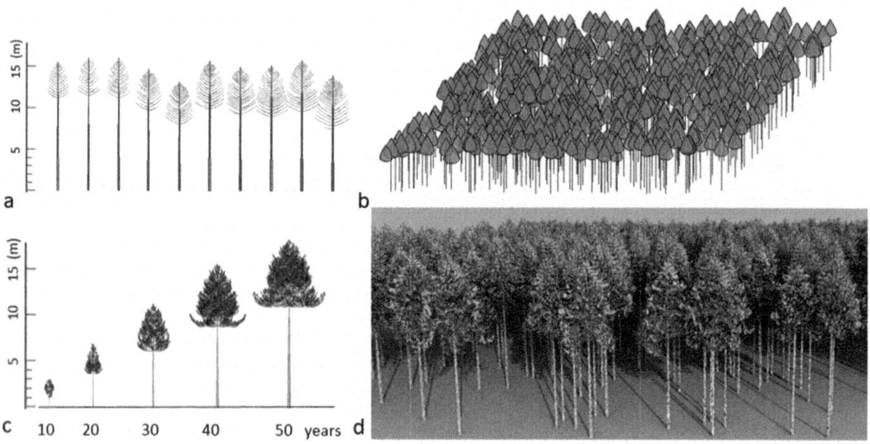

Fig. 11.25 Simulation of a 50-year-old black pine plot by Capsis software (Dufour-Kowalski et al. 2012) based on the PNN model (Dreyfus 1993). (**a**) Individual variability is restored. (**b**) The plot is restored. (**c**) The growth of the tree is simulated in 3D by the implementation of GreenLab in the Xplo editor. (**d**) Visualization of the plot by computer graphic image in the Simeo platform. Xplo and Simeo are AmapStudio tools (Griffon and de Coligny 2012)

11.6.2 Simulation of the Growth of a Forest Stand Based on Spatial Interactions

In woody plants, the growth is indefinite and takes decades in trees. It has two components: a primary one (the internodes) and a secondary one (the rings). The development is most affected by the density because of the changes it causes in the functioning of meristems (see Chap. 8). This is mainly related to the size of the GUs, their branching potential and the abortion of terminal meristems according to physiological ages. Branch mortality is a major component of development that shapes architecture through continuous pruning. The production area *Sp* is not constant. It evolves according to a U-shaped curve (Chap. 8, maple), which can be determined by an inverse method from growth monitoring. In the case of trees, data acquisition is difficult. Nevertheless, the studied large trees (Japanese elm, Canadian sugar maple) show that the botanical automaton can be informed by measurements on tree crowns and organic series taken by targeted sampling, based on architectural knowledge of botany.

Here the objective is to show that it is possible to accurately simulate the growth of a forest stand, from a perspective point of view, based on the knowledge acquired in the previous chapters. To simplify, it is assumed that the stand is mono-specific and even-aged (the trees were planted on the same date) and the local available surface area per tree is variable and does not evolve with age. It results from the mesh size generated by the positions of individuals in the plantation. The main concept used is the potential structure (Chap. 3). At a given age, each individual produces a simulated architecture that is a subset of the potential structure, because of developmental failures (branching, mortality, etc.).

There are several points of view to address the problem:

- A first spatialized geometric simulation model implemented in AmapHydro that manages local interactions such as space competition. The crown of a tree is truncated when in contact with the crown of neighboring trees;
- A second procedural model with GreenLab which globally manages growth–development interactions considering the available surface area *Sd* for each tree, which limits the volume of the crown (Digiplante, Gloups, QingYuan software);
- The two models can be combined (QingYuan software).

11.6.2.1 AmapHydro Growth Model and Stand Simulation (Frédéric Blaise)

AmapHydro growth model has played a pioneering role in the field of plant growth simulation, despite its limited functionalities. The AmapHydro software developed by Frédéric Blaise (Blaise et al. 1996; de Reffye et al. 1995) shapes the envelope of the tree crowns according to the contacts between neighboring trees. When two apical branch meristems, whether from the same or neighboring trees, are found in

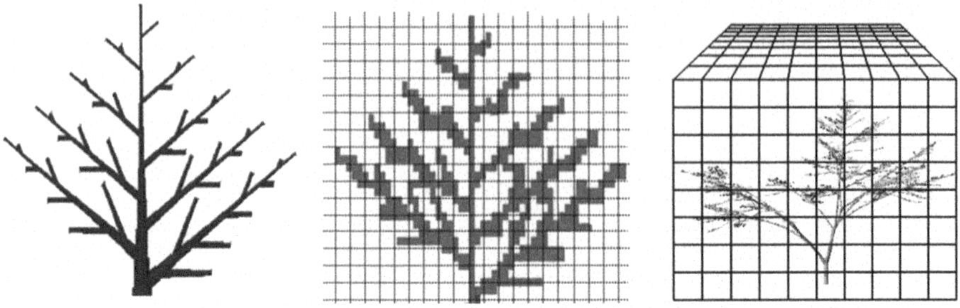

Fig. 11.26 Plants are placed in a voxel space (AmapHydro software)

the vicinity in a cube located in space and referred to as "voxel" (Fig. 11.26), their abortions can be decided, resulting in the death of the respective axes. The volume of the voxel is crucial for the intensity of the interaction; a smaller voxel results in fewer interactions, and vice versa. By using this system, during tree growth, spatial competition between branches of the same tree or those of neighboring trees at the voxel level will produce realistic pruning. It is an adaptation of Ken Mitchell's model on Douglas (Mitchell 1975) that truncates the envelopes of toupee trees in contact, thus reducing the trunk rings downstream (Chap. 8). But here we go down to the scale of the phytomer in architecture. The size of the tree crown will obviously decrease with the planting density, which will automatically modify the secondary growth by applying Pressler's law. It should be noted that the tree crowns explore the available space and are, therefore, asymmetrical in nature.

In the AmapHydro model, the architectural development is very simple, and is not stochastic. The GU is reduced to the meta-phytomer (see Maple in Chap. 10). Photosynthesis is proportional to the number of functional leaves. This simplification remains acceptable for describing architectures with monopodial branches such as those of conifers (spruces, pines), and it is sufficient for the intended exploratory purpose. A grove's growth is simulated to visualize space competition and its consequences on secondary growth (Fig. 11.27). The results are qualitatively realistic. The isolated trees do not prune and the trees surrounded by its neighbors has a small, high crown. The trees on the periphery of the grove have asymmetric pruning.

Simulation of plots to compute the volume of wood produced as a function of planting density becomes possible (Fig. 11.28) and sylvicultural scenarios including thinning can be implemented (Fig. 11.29).

This simulation mode is time-consuming in terms of computation time and memory and the system will quickly reach its limits.

11.6.2.2 GreenLab Growth Model and Stand Simulation

In the previous chapters, the production area *Sp* in the GreenLab equation plays the role of a parameter that inserts the plant into its space competition environment. When the LAI is sufficient (>3), it is the available ground area (*Sd*) per plant in the

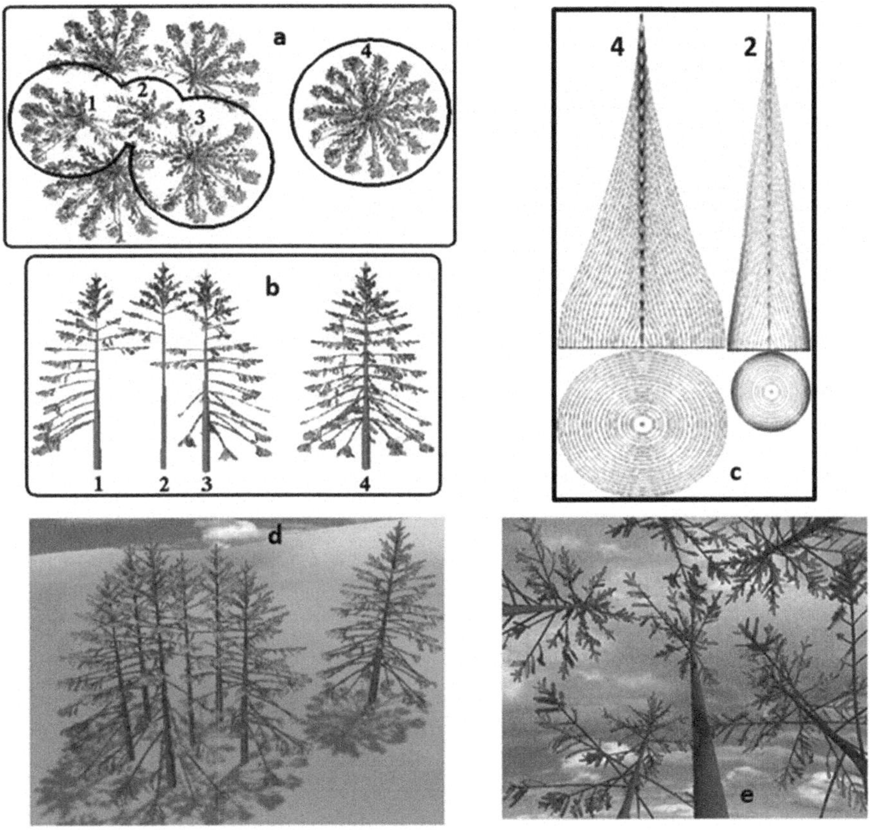

Fig. 11.27 Simulation of a spruce grove. Spatial competition during growth and its consequences on the dimensions of tree crowns and ring stacks. (**a**) Top view. (**b**) Side view: A grove of five trees and an isolated tree. Tree 2 experiences intense competition from all sides, resulting in heavy pruning and almost symmetrical crown, and the diameter of the trunk at the base is thin. Tree 4 is isolated, symmetrical, and not pruned. The trunk has a large diameter at the base. Trees 1 and 3 are peripheral and, therefore, asymmetrical because there is no competition for branches outside the grove. (**c**) Stacking of tree rings 2 and 4. (**d**) and (**e**) 3D visualization external and internal to the grove

case of high-density herbaceous crops ($Sd = 1/d$) that limits photosynthesis. It is still the projection surface on the ground of the tree crown during the growth of the tree (Chap. 10, maple). This evolves according to a U-shaped curve but is bounded by the available surface area *Sd* per tree (considering its neighbors) and which results locally from the mesh size generated by the positions of the individuals in the plantation. The agronomic experiences of the previous chapter validate this approach.

To simplify and without qualitatively modifying the result, the value of *Sp* is set directly to its final value *Sd*. A plantation can, therefore, be considered as a set of plants inserted in individual *Sd* that form a mesh, whose extrusion forms tubes that

Fig. 11.28 Spruce plantation with densities (**a**) (625), (**b**) (1600) and (**c**) (10,000) trees per hectare. (**I**) 3D view of the plantations. (**II**) Top view of the density (**a**) showing the competition. (**III**) diameters at the base of the trunks of the three densities

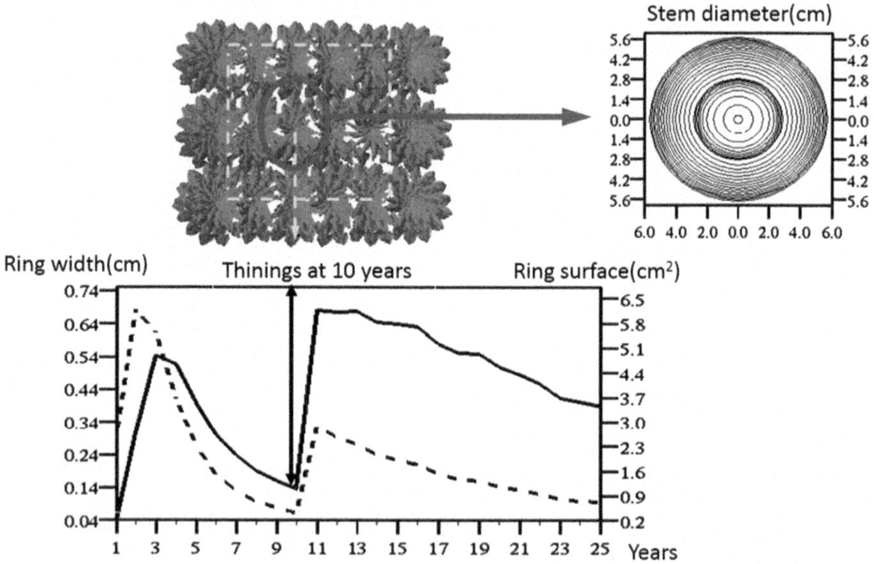

Fig. 11.29 Sylvicultural scenario on a spruce plot with a density of 625 trees/ha, viewed from the above. The plantation is 25 years old. When it was 10 years old, one thinning was conducted, removing trees from every other row. As a result, the canopies of the remaining trees have invaded the liberated area and regain volume. The effect of thinning is visible in dark circles

limit the architecture of the trees in their respective volumes. This *Sd* value can be estimated according to the nearest neighbors by a Delaunay tessellation; it paves the plane with polygons according to an algorithm that delimits the contours around the base of the trees according to their neighbors.

We use here the fact that the development has a high degree of plasticity related to the supply/demand ratio (*Q*/*D*), which focuses in particular on the size of GUs, branching, and mortality of axes (see Chap. 13). This introduces a feedback loop between growth and development.

Each of the three parameters that control development, mortality, and branching, depends on a function of form $F = k_1 + k_2 \cdot Q/D$.

The *F* value is compared to a threshold whose value is empirically fixed. When the threshold is exceeded, a branching is allowed, while abortion may occur.

The abortion of buds is no longer caused by proximity between neighboring buds, but by the limitation of the development of an axis that depends (linearly) on the *Q*/*D* ratio at its creation, which determines its pruning (Mathieu 2006).

The *Q*/*D* ratio increases at the beginning of growth, leading to a base effect with consequences of an increase in the number of phytomers of GUs and a progressive appearance of branching. The life span of the branches depends on the *Q*/*D* ratio at their births.

Then, due to saturation of biomass production due to *Sd,* the *Q*/*D* ratio will stabilize. This limits the development of new axes, increases pruning, and the *Q*/*D* ratio eventually stabilizes. Thus, the tree crowns tend towards limits in their respective available volumes, limited by the individual *Sd* projection surfaces. Depending on the setting of the mortality of the main branches (axes 2) as a function of *Q*/*D*, the projections of the tree crown will remain more or less contained within their respective *Sd*. Empirically, it is easy to obtain realistic simulations.

In Fig. 11.30, the isolated tree can reiterate because a high *Sd* corresponds to high *Q/D* values that produce sufficiently developed GUs on the trunk to trigger reiteration. On the contrary, the tree surrounded by its neighbors with a low *Sd* in the grove is associated with low *Q/D* values, which limits branching and increases branch mortality. The tree does not reiterate and its crown is reduced. Branch mortality is no longer the result of contact with a neighborhood, but the limitation of development caused by the *Q/D* values under the dependence of *Sd d*uring growth. For a visual result equivalent to that in Fig. 11.20, the simulation is greatly accelerated because there is no need to consider the position of the organs in space. A difference nevertheless appears by the fact that the crowns remain symmetrical because the competition is considered globally and not locally. It should be noted in Fig. 11.30 that the value of *Sd* has little effect on the height of the simulated tree, which corresponds well to the observation that the height of trees is little affected by the planting density (Eichorn's Law).

To scale up the plot with a heterogeneous density, a good viewpoint is to divide the plot into polygons according to the Delaunay triangulation method. The *Sd* values are then polygons that delimit the territory around each tree according to the planting scheme. Thus, the isolated tree that corresponds to high *Sd* values will reiterate and have a spherical crown, while the tree surrounded by its neighbors will have an upright growth form with a reduced crown. All *Sd* are different, which result in many different architectures. However, they can be grouped by class to limit the number of growth simulations to be performed. Each *Sd* is assigned the growth simulation results of the tree corresponding to the nearest class. Stands with

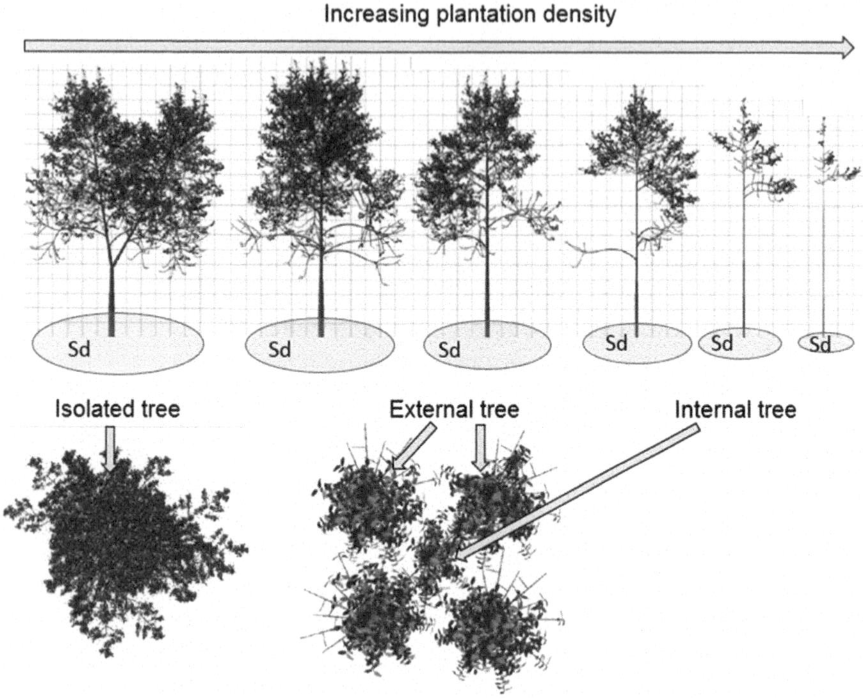

Fig. 11.30 Tree growth and architecture as a function of plantation density. The isolated tree reiterates. When placed in a grove, the intersections of its projected crowns define individual *Sd* surfaces that condition and limit the growth of the tree. As a result, the tree clamped by its neighbors has a reduced crown cover due to an increase in branch mortality (Digiplante software, ECP)

heterogeneous densities can then be simulated (Fig. 11.31). This system is very efficient in simulating the growth of a plantation. Overall, the local growth and development of each tree as well as the production of biomass from the stand, are accurately represented. But there is no local interaction, and the tree crowns remain symmetrical. To simulate the production of a stand, without favoring the computer graphics image aspect, the method seems to be effective (Cournède et al. 2008, 2010).

11.6.2.3 QingYuan Growth Model and Stand Simulation

To manage local interactions and crown asymmetry during growth, photon mapping is a good way. The light comes from one direction and is intercepted by the foliage. Each leaf receives its portion of radiation according to the local extinction coefficient due to the simulated shading. The results for biomass production between photon mapping and *Sd* use are similar in the case of homogeneous density (Wang et al. 2012, 2013). But the first can simulate local interactions between the axes during the growth. Abortions of the axes will depend on the extinction of light in their

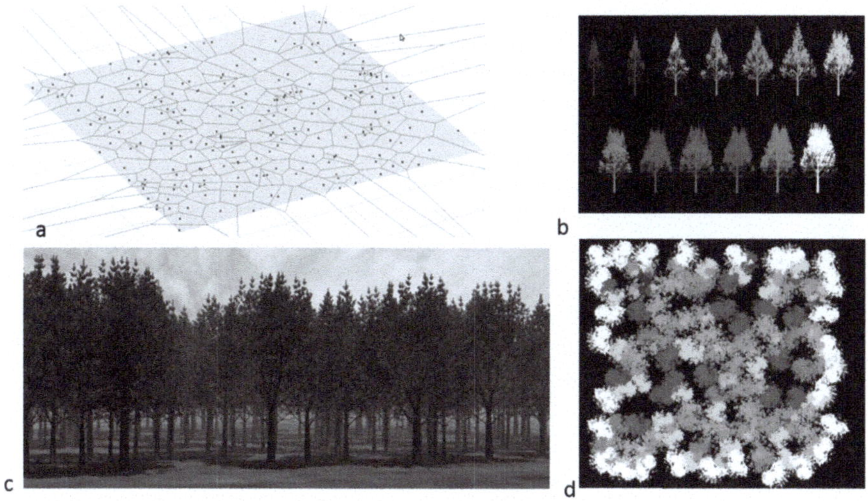

Fig. 11.31 Simulation of tree growth in a mono-specific and even-aged plot with a heterogeneous planting density. (**a**) A Delaunay mesh gives the available area *Sd* for each tree according to its neighbors. (**b**) The areas *Sd* are grouped into classes whose tree growths are pre-computed. (**c**) A plot is thus very quickly reconstructed according to its planting scheme. The local planting density is represented by a color. The white color corresponds to a low density (case of peripheral trees that reiterate) and the blue color to a high local density that gives trees with small tapered crowns. (**d**) The plantation can then be visualized in computer graphics (Digiplante and Simeo software) (Color figure online)

environment under the effect of shading. The results of the simulations are visually realistic. The asymmetric aspect of the crowns obtained by geometric methods based on voxels is also presented (Fig. 11.32). Obviously, the method is more expensive in simulation and limited to small plots; it is more dedicated to computer graphics.

11.7 Conclusion

The behavior of the GreenLab structure–function model concerning climate data is consistent with the knowledge acquired on crop (or PBM) models. The passage from the isolated plant to the stand level is possible with this model, provided that the specific monoculture stands are considered.

In fields and under uncontrolled conditions, the climatic effects of the season do not significantly affect the values of the source–sink parameters, and therefore, the distribution of biomass in the compartments (especially in the case of herbaceous). The effect of density affects the parameter *Sp* (production area), which tends towards the inverse of planting density when the leaf area index is high.

In greenhouses, the strong variations in light and temperature significantly modify the values of the source–sink parameters of basipetal herbaceous plants.

Fig. 11.32 Effect of light on the interactions between the canopies of two trees during the growth (Hua Jing, CASIA, QingYuan software)

However, for a given combination, these parameters are constant and the GreenLab model works correctly. It is possible by regression to estimate the values of each parameter with a quadratic shape as a function of light and temperature from the data. Therefore, the model can correctly predict their effect on plant development, growth and ultimately architecture, on other temperature-light-density combinations. Recently, the use of neural networks is a new approach to simulate interactions between plant growth and environmental parameters (see Chap. 12).

The source–sink parameters that determine the production and distribution of biomass in the plant are of particular interest for phenotyping related to the selection or optimization of cropping systems (density, irrigation, etc.) where simulation can play an essential role.

References

Allen M-T, Prusinkiewicz P, DeJong T-M (2005) Using L-systems for modeling source-sink interactions, architecture and physiology of growing trees: the L-PEACH model. New Phytol 166(3):869–880

Baey C, Didier A, Lemaire S, Maupas F, Cournède P-H (2014) Parametrization of five classical plant growth models applied to sugar beet and comparison of their predictive capacities on root yield and total biomass. Ecol Model 290:1120

Blaise F, Houllier F, de Reffye P (1996) Simulation of tree architecture and growth in a forest stand: Amappara software. In: Nepveu G (ed) Proceedings of connection between silviculture and wood quality through modelling approaches and simulation software, 1st IUFRO WP S5.01-04 Workshop. Hook, Suède, pp 46–55

Brisson N, Mary B, Ripoche D, Jeuffroy M-H, Ruget F, Nicoullaud B, Richard G, Beaudoin N, Recous S, Tayot X, Plénet D, Cellier P, Machet J-M, Meynard J-M, Delécolle R (1998) STICS: a generic model for the simulation of crops and their water and nitrogen balances. I. Theory and parameterization applied to wheat and corn. Agronomie 18(5–6):311–346

Cournède P-H, Mathieu A, Houllier F, Barthélémy D, de Reffye P (2008) Computing competition for light in the GreenLab model of plant growth: a contribution to the study of the effects of density on resource acquisition and architectural development. Ann Bot 101(8):1207–1219

Cournède P-H, Guyard T, Bayol B, Griffon S, de Coligny F, Borianne P, de Reffye P (2010) A forest growth simulator based on functional-structural modelling of individual trees. In: Li B-G, Jaeger M, Guo Y (eds) Proceedings of plant growth modeling, and their applications (PMA09), IEEE CPS, Beijing, China, pp 34–41, November 9–13

de Reffye P, Houllier F, Blaise F, Barthélémy D, Dauzat J, Auclair D (1995) A model simulating above- and below-ground tree architecture with agroforestry applications. IUFRO World Congress. 20, 1995-08-06/1995-08-12, Tampere, Finland. Agroforest Syst 30(1–2):175–197

Dreyfus P (1993) Modelling Austrian black pine response to silvicultural practices in the South East of France. In: Burkhart HE, Gregoire TG, Smith JL (eds) Proc IUFRO S4.01 conf "Modelling stand response to silvicultural practices", Publ FWS-1-93. Virginia Polytech Inst and State Univ, Blacksburg, pp 5–18

Dufour-Kowalski S, Courbaud B. Dreyfus P. Meredieu C, de Coligny F (2012) Capsis: an open software framework and community for forest growth modelling. Ann For Sci 69(2):221–233

Feng L, de Reffye P, Dreyfus P, Auclair D (2012) Connecting an architectural plant model to a forest stand dynamics model—application to Austrian black pine stand visualization. Ann Forest Sci 69(2):245–255

Griffon S, de Coligny F (2012) Amapstudio: a Software Suite for Plants Architecture Modelling. In: Proceedings of plant growth modeling, simulation, visualization and applications (PMA12). IEEE Press, Shanghaï, pp 141–147, 31 Oct–3 Nov 2012

Heuvelink E (1996) Dry matter partitioning in tomato: validation of a dynamic simulation model. Ann Bot 77(1):71–80

Jeuffroy M-H, Bergez JE, David C, Flénet F, Gate P, Loyce C, Surleau-Chambenoit C (2008) Utilisation des modèles pour l'aide à la conception et à l'évaluation d'innovations techniques en production végétale: bilan et perspectives. Agronomie 18:311–346

Johnson N-L, Kotz S (1970) Distributions in statistics: continuous unvariate distributions-1. Houghton Mifflin

Jones C-A, Kiniry J-R (eds) (1986) CERES-Maize, a simulation model of maize growth and development. Texas A&M University Press, College Station. 194 p

Kang M-Z, Heuvelink E, de Reffye P (2006a) Building virtual chrysanthemum based on sink-source relationships: preliminary results. In: III international symposium on models for plant growth, environmental control and farm management in protected cultivation, vol 718, pp 129–136

Kang M-Z, Qi R, de Reffye P, Hu B-G (2006b) GreenScilab: a toolbox simulating plant growth in the Scilab environment. In: 8th middle eastern simulation multiconference, Alexandria, Egypt, pp 174–178

Kang M-Z, Qi R, de Reffye P, Hu B-G (2009) GreenScilab: a toolbox simulating virtual plants in the Scilab Environment. In: 2009 IEEE international workshop on open-source software for scientific computation (OSSC 2009)

Kang M-Z, Heuvelink E, Carvalho S-M, de Reffye P (2012) A virtual plant that responds to the environment like a real one: the case for chrysanthemum. New Phytol 195(2):384–395

Keating B-A, Carberry P-S, Hammer G-L, Probert M-E, Robertson M-J, Holzworth D, McLean G (2003) An overview of APSIM, a model designed for farming systems simulation. Eur J Agron 18(3):267–288

Lemaire S (2010) Système dynamique de la croissance et du développement de la betterave sucrière (*Beta vulgaris* L.). AgroParisTech (Ph. D. Thesis)

Ma Y-T, Li B-G, Zhan Z-G, Guo Y, Luquet D, de Reffye P, Dingkuhn M (2007) Parameter stability of the functional–structural plant model GREENLAB as affected by variation within populations, among seasons and among growth stages. Ann Bot 99(1):61–73

Ma Y-T, Wen M-P, Guo Y, Li B-G, Cournède P-H, de Reffye P (2008) Parameter optimization and field validation of the functional–structural model GREENLAB for maize at different population densities. Ann Bot 101(8):1185–1194

Mailhol J-C, Ruelle P, Revol P, Delage L, Lescot J-M (1996) Operative modelling for actual evapotranspiration assessment: calibration methodology. In: Proceedings of international conference on evapotranspiration and irrigation scheduling (ASAE 1996). China Computer Federation, San Antonio, pp 474–479

Mailhol J-C, Ruelle P, Walser S, Schütze N, Dejean C (2011) Analysis of AET and yield prediction under surface and buried drip irrigation systems using the crop model PILOTE and Hydrus-2D. Agric Water Manage 98:1033–1044

Mathieu A (2006) Essai sur la modélisation des interactions entre la croissance et le développement d'une plante: cas du modèle GreenLab. Thèse de doctorat Mathématiques appliquées. École centrale de Paris

Mitchell K-J (1975) Dynamics and simulated yield of Douglas fir. Forest Sciences, Monograph 17, 39 p

Perttunen J, Sievänen R, Nikinmaa E, Salminen H, Saarenmaa H, Väkevä J (1996) LIGNUM: a tree model based on simple structural units. Ann Bot 77:87–98

Perttunen J, Sievänen R, Nikinmaa E (1998) LIGNUM: a model combining the structure and the functioning of trees. Ecol Model 108(1–3):189–198

Wang H-Y, Kang M-Z, Hua J (2012) Simulating plant plasticity under light environment: A source-sink approach. In: 2012 IEEE fourth international symposium on plant growth modeling, simulation, visualization and applications (PMA), pp 431–438

Wang H-Y, Kang M-Z, Hua J, Wang X-J (2013) Modeling plant plasticity from a biophysical model: biomechanics. In: Proceedings of the 12th ACM SIGGRAPH international conference on virtual-reality continuum and its applications in industry. ACM, pp 115–122

Wernecke P, Buck-Sorlin G, Diepenbrock W (2000) Combining process- with architectural models: the simulation tool VICA. Syst Anal Model Simulat 39(2):235–277

Chapter 12
Applications of Plant Architecture Modelling

Philippe De Reffye, Marc Jaeger, and Mengzhen Kang

Abstract This chapter highlights the significance of plant architecture across various plant-related disciplines, with a focus on the AMAP and GreenLab models. It introduces the use of three-dimensional (3D) plant mock-ups for computer graphics in landscaping and urban planning, as well as their utility in remote sensing for assessing plant stand health and productivity. This chapter emphasizes the potential of plant breeding using source–sink dynamics, offering a more stable and functional approach in comparison to traditional measurements techniques. It also discusses the application of functional–structural models in optimizing farming systems to reduce resource waste and pollution, and in improving pruning and thinning techniques through simulation. Additionally, the text points to the relevance of such models in forestry for analysing tree mechanics and wood quality. Heuristic methods using neural networks are also mentioned as a promising tool for estimating environmental impacts on model parameters. Agronomy should make increasing use of these structural–functional models which require the consideration of the plant structure.

Keywords Computer graphics applications · QTL · Remote sensing · Optimization technics · Optimal control

P. De Reffye · M. Jaeger
CIRAD, UMR AMAP, Montpellier, France

AMAP, Univ Montpellier, CIRAD, CNRS, INRAE, IRD, Montpellier, France
e-mail: marc.jaeger@cirad.fr

M. Kang (✉)
State Key Laboratory for Multimodal Artificial Intelligence Systems, Institute of Automation, Chinese Academy of Sciences (CASIA), Beijing, China

School of Artificial Intelligence, University of Chinese Academy of Sciences, Beijing, China
e-mail: mengzhen.kang@ia.ac.cn

P. De Reffye et al. (eds.), *Plant Model Greenlab for Botany and Agronomy*,
https://doi.org/10.1007/978-981-97-7541-5_12

12.1 Various Applications of Computer Graphic Models

Visualization needs are vast and are expressed in a wide range of applications. In our experience, for more than 30 years, since the appearance of plant simulation models, the representation of models and the three-dimensional (3D) visualization of plants, cultures, and landscapes have shown their interests, covering various and often related aspects.

12.1.1 Plant Models for Botany

Models (as AMAP-GreenLab) integrate most of the concepts developed in plant architecture by the Botanic Institute of Montpellier, France, and in particular the concept of 'architectural model' by botanist Francis Hallé. Thanks to the computer tools, simulating the development of plant architecture is a synthesis and a capitalization of knowledge on the subject. Photosynthesis aspects are unnecessary: the coupling of development and geometry are sufficient to obtain realistic 3D plants (Chap. 4). The advent of computer-generated imagery, an exceptional promotion tool for modelling, has generated a demand for the computer-aided production of plant models. Digital representation makes it possible if it is realistic enough to recognize a species or even a clone. Thus, the digital 3D plant becomes a learning tool. This property has been used several times. Examples include virtual begonia collections (Lecoustre et al. 1992), funded by the City of Rochefort for communication purposes (Fig. 12.1), or representations of rubber tree clones whose simulations were distributed on colour plates to help producers recognize their clones in real environments (Dinouard et al. 1987).

The Tobacco Institute (Seita) commissioned a study on the growth of ornamental tobaccos to optimize their cultivation. Growth modelling of five species (Fig. 12.2) was carried out by Hervé Rey (Poisson et al. 1997).

Detailed studies on the architecture of date palms and oil palms (Elhoumaizi et al. 2002) have been carried out and have focused specifically on palm phyllotaxis and morphology. They are of a geometric nature (Fig. 12.3).

Representations for entertainment and multimedia applications where the plants serve as a key element of contextual realism, represent an important historical channel in the field of 3D plant growth simulation and visualization (de Reffye et al. 1988; Jaeger 1987, 2012). For instance, the Japanese firm Dentsu commissioned a 3D growth simulation of Japanese elm for a film showcased at the Osaka World Expo in 1990 (Fig. 12.4).

Visualizations play a crucial role in educating people about the evolution of structures as modeled by dynamic models. The 3D models facilitate understanding the underlying principles (Jaeger and de Reffye 1991; de Reffye et al. 2008).

The reconstruction of fossil flora or landscapes is also an unexpected field of application for the use of virtual plant models. With the paleobotany laboratory of

Fig. 12.1 Begonia Sophie Cécile. From the plant collection of the municipality of Rochefort. Simulation of anatomical details and synthetic images in a greenhouse (Lecoustre et al. 1992)

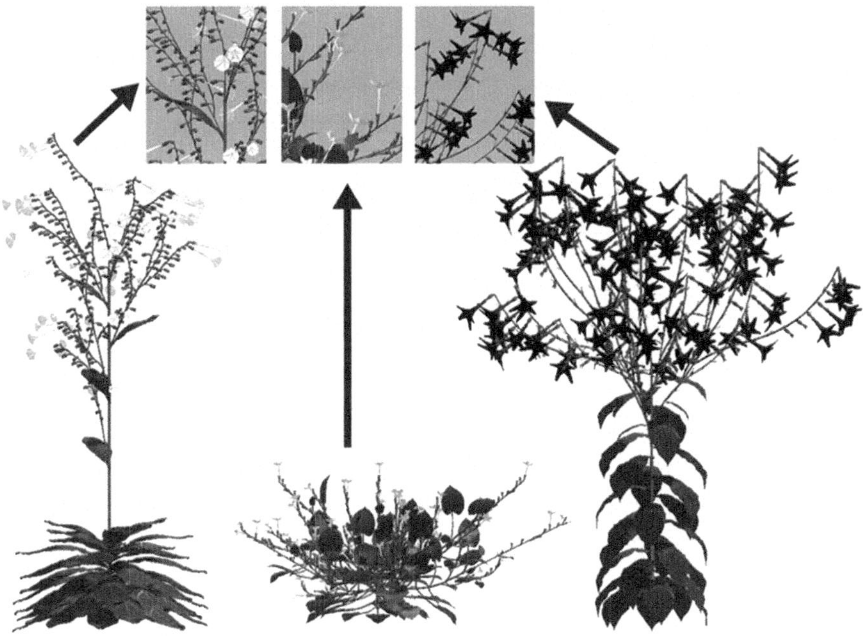

Fig. 12.2 Three ornamental tobaccos, *Nicotiana excelcior*, *umbricata* and *sanderae*. Collaboration between Seita and CIRAD, Hervé Rey, AmapSim software

Fig. 12.3 Palm tree study. (**a**) Date palm, Hervé Rey and René Lecoustre. Simulation and image of Sébastien Griffon (CIRAD). (**b**) Oil palm. Jean Dauzat and René Lecoustre. Simulation and image: Marc Jaeger, (CIRAD)

Fig. 12.4 Simulation of a 100-year-old Japanese elm. Developed by Philippe de Reffye, Daniel Barthélémy, P. Dinouard using the AmapSim software

the University of Montpellier, reconstructions have been modelled to visualize carboniferous fossil plants and landscapes (Fig. 12.5) (Daviero et al. 1996, 2000).

In the 1990s, these achievements were new and original. They provided additional insights not only through the visual representation but also through the botanical content of the simulated plants.

Fig. 12.5 Fossil landscape of the carboniferous: calamities, lepidodendrons, sigillars, etc. (Daviero et al. 1996)

12.1.2 Plant Models for Urban Planning and Landscaping

On a larger scale (from plot to landscape), applications of plant visualization contribute to dialogue and communication with decision-makers and the general public on development projects involving landscape aspects (Edelin et al. 1989; de Reffye et al. 1987), regulatory aspects (Griffon and Auclair 2009) or environmental aspects (Jaeger et al. 2010).

The advent of architecture and computer graphics software has made it possible to realistically visualize buildings or structures. A demand for a 3D plant collection has emerged to simulate urban planning and landscaping projects where plants coexist with buildings.

This demand has been met by software such as Onyxtree (Bosanac and Zanchi 2005, http://www.onyxtree.com), Xfrog (Lintermann and Deussen 1997, 1999, http://xfrog.com), which provide a three-dimensional plant database. These softwares aim at the realism of appearance but integrate little architectural botanical knowledge. In contrast, based on AMAP's work, a plant database was developed (Fig. 12.6) based on the botanical knowledge of more than 600 plants on five continents. This database has been integrated into computer-generated image software suites and distributed to many users by Bionatics SA (http://www.bionatics.com). This software publisher and distributor offers plant generation tools for

Fig. 12.6 Trees extracted from the tree library of Bionatics. From left to right: frangipani, weeping willow, fir, elm, Lawson cypress, beech, chestnut

Fig. 12.7 Synthesis image of a landscape (Bionatics)

multimedia and flight simulators and uses plant collections in its Landsim 3D landscape design software. The latter is used by cities, communities, companies or architectural firms to carry out urban or landscape development projects, on small or large scales (Fig. 12.7).

12.1.3 Plant Models for Remote Sensing and Radiative Transfers

In the context of these applications, the aim is no longer to visualize plants, but to compute the interception of light by the plant canopy to access photosynthesis or the biomass production of a plant stand.

12.1.3.1 Use of Models in Remote Sensing

This work was conducted by Thierry Castel (Castel et al. 2001). Satellites capture images of land use by vegetation that must be interpreted in terms of crop production. This is one of the challenges of remote sensing. 3D models of plants calibrated on cultivated species are used to analyse images and estimate the condition of a plantation and its biomass.

The growth of Austrian black pine has been modelled and simulated by the AmapSim software to obtain computer models with an architecture that is as close as possible to reality. The height of the trunk, its diameter profile, the size of the crown, the length and inclination of the branches are dendrometric characteristics that allow us to know the stage of development and the production of wood. The Cesbio laboratory in Toulouse has developed software that simulates radar echoes of virtual geometric objects. The pine model adapted to the appropriate format (via a computer module, AmapSar, developed by Barczi) allows the Cesbio software to recognize the radar echo of the simulated tree models (Fig. 12.8a). The application of black pine plantations in Lozère to radar images has made it possible to correctly estimate tree development and woody biomass produced in plots of different ages (Fig. 12.8c). These can be visualized in place in a geographic information system (GIS). Conversely, GIS data can be used to represent landscapes and crop systems (Fig. 12.8b). The first applications in remote sensing are the exploitation of GIS forestry data with the visualization of large plant scenes and various reforestation strategies. These studies were produced in collaboration with the University of Tokyo (Saito et al. 1993) (Fig. 12.8d).

From satellite images, it is possible to estimate the LAI (leaf area index) of the vegetation cover of a field crop. With plant growth and architecture modelling, the value of the simulated LAI is related to the stage of development of the plantation and, therefore, to the biomass production, as well as its distribution in plant compartments (leaves, stems, fruits). And the evolution of LAI as a function of thermal time (or in several cycles) is generally stable from one year to the next. The LAI estimated from the satellite image, therefore, makes it possible to identify the stage of development and growth of the plantation. This approach (Fig. 12.9) was tested on the plant model of maize (Kang et al. 2012).

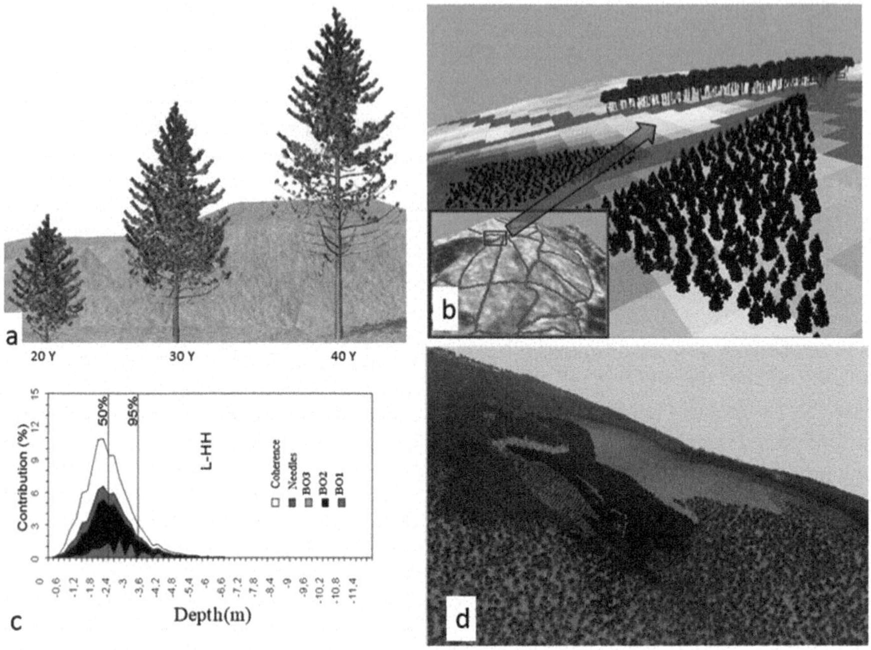

Fig. 12.8 Use of simulated plants in remote sensing applications. (**a**) Growth simulation of black pine at 20, 30, 40 years (modelled by Caraglio Y., CIRAD AMAP). (**b**) GIS including plots of different stands of black pines at different ages of plantation (zoom of a plot). (**c**) Simulation of a radar echo. The received signal depends on the dendrometric characteristics of the tree (AmapSar and AmapSim software, Barczi, CIRAD-AMAP). (**d**) The Chichibu forest, the first entire landscape generated from virtual plants and a GIS (Saito et al. 1993), collaboration University of Tokyo-CIRAD

12.1.3.2 Use of Models to Compute Light Interception

The photosynthesis of a plant canopy is directly proportional to the light intercepted by the foliage. At the plot level, Beer-Lambert's law (Chap. 5) is generally well adapted. At the scale of detail in the plant, computer algorithms that compute the interception of light by models are required. The models are supposed to be calibrated to reality, which implies that the spatial distribution, orientation and size of the leaves are accurately reproduced. In addition, the optical properties of the leaves (transparency, Lambertian reflection) must be known. This being assumed to be achieved, different algorithms were used around the 3D models.

12.1.3.2.1 Ray Tracing

Jean Dauzat (Dauzat et al. 2008) developed the Archimedes software, which computes the light intercepted by the cover of a virtual plantation with 3D models of

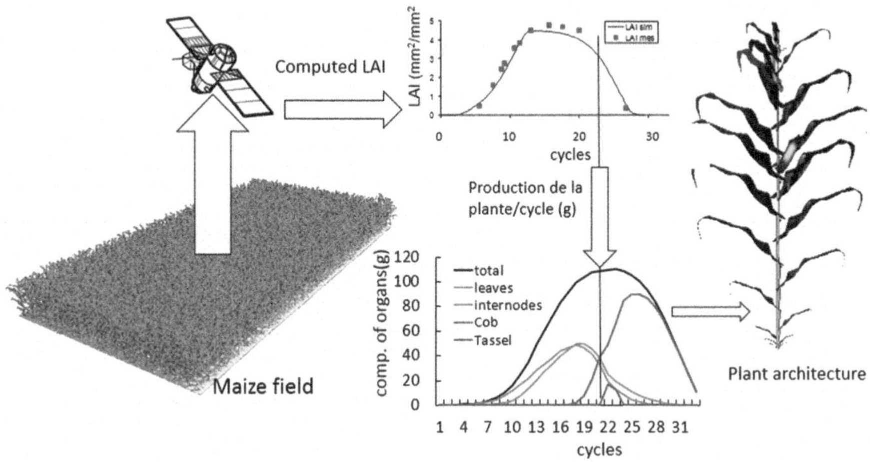

Fig. 12.9 Computation of the production of a maize field, based on satellite images that estimate LAI (Corpetti et al. 2018). We deduce from this the state of development and growth of the individual plant and by extension that of the stand through the planting density

Fig. 12.10 3D plant models and computation of light interception. (**a**) 3D model of a coconut tree with the computation of the light intercepted (by ray tracing) by the palms. (**b**) Mapping of the cumulated light on the ground in a coconut plantation, after six months. The intensity decreases from red to blue (Color figure online)

coconut, cotton, sunflower, etc. simulated by AmapSim software (Barczi et al. 2008). An algorithmic approach, the technique of ray tracing, is used. Rays intercept the polygons whose leaves are made up (Fig. 12.10a). The aim is to optimize the planting density according to solar radiation; but in these virtual plantations, plants are static objects that do not function (Mialet-Serra et al. 2001). This type of scene can be simulated at any stage of development. A mapping of the cumulated light on the ground is carried out (Fig. 12.10b). It considers the daily path of the sun and the seasons, and can be used to optimize planting density, including the case of intercropping.

12.1.3.2.2 Radiosity

This work was conducted by Cyril Soler and François Sillion at INRIA. Radiosity is another technique, with more complex formalism and implementation than ray tracing, based on the multi-directional reflection of light in a virtual scene composed of facets. This technique allows a realistic rendering of the lighting, with a good approximation of the light energy received on the surfaces present in the scene. Applied to plants, the radiosity computation provides each leaf with its lighting. We can thus estimate the radiative transfer in a maize field (Chelle and Andrieu 1998). The light intercepted by the foliage of a model, combined with the notion of light use efficiency for photosynthesis, makes it possible to compute biomass production and distribution in the plant. A growth-interception feedback loop of light is thus formed. The PlantRad radiosity software developed at INRIA (Soler et al. 2003) made it possible to grow and visualize virtual plots simulated by the AmapHydro software. Light plays a dual role; it illuminates the virtual plant scene and provides the energy captured by the leaves to ensure stand growth (Fig. 12.11). The same

Fig. 12.11 Planting of 233 poplars and acacias simulated by AmapHydro and growing using light scattered by the PlantRad radiosity software (Soler et al. 2003)

principle was applied to tomatoes modelled and simulated by AmapSim software (Dong et al. 2007), coupled with the GreenLab model.

While the principle is attractive because the stand function is simulated in detail, the time involved in computing the interception of light on detailed model sets is too long for these methods to be usable. This software has been adapted for large cultivated areas where plants are no longer individualized: the plant stand is then described as a turbid environment.

12.1.3.2.3 Photon Launching

Other methods can be used such as photon launching (Wang et al. 2012), which can also be used to define a local lighting environment (Chap. 11, Fig. 11.32).

In photon mapping (Jensen 2001), light propagation is simulated by tracing light particles (i.e. photons) originated from light source. When a photon strikes on an organ, it can be absorbed, reflected, transmitted or diffused according to the material property of the organ. To accelerate the collision detection between a photon and an organ surface, bounding volume hierarchy (BVH) method (Shirley and Morley 2003) is used: instead of detecting whether a photon collides with every object in the complex scene, a regular bounding volume (e.g. a cubic) of this object is used for collision detection. Each individual organ is wrapped in a bounding volume, which is leafy node in a data structure of binary tree. These nodes are then grouped as small sets and enclosed within larger bounding volumes. With such hierarchy, during collision testing, a small set does not need be examined if its parent volume does not intersect with the rays, which reduces the time complexity.

By replacing the Beer-Law using the light interception model (photon mapping) for biomass production in the GreenLab model, the integrated model was able to simulate the effect of light condition on plant structure through source–sink regulation (Wang et al. 2012).

In order to evaluate the algorithm, different schemes of ray emission are tested using different number of light sources. The results of biomass production are closed with different number of photons emitted, while the simulation time can be dramatically different. However, bigger neighbourhood size leads to more computational time. The search radius of neighbourhood has little effect on the biomass production. For the plant stand, it is useful to enlarge search radius in order to consider the occlusion by surrounding trees.

The simulations of tree competition under isotropic and anisotropic light conditions were studied using the same parameter file. Under the isotropic light condition, the morphological structure of the trees is affected by the local environment, especially neighbourhood competition even under isotropic light environment. In Fig. 12.12, it is observed that the tree in the middle grows less and has a smaller crown than the surrounding trees as it receives less light. Planting density, which affect local light environment, plays a role on plant morphology. The greater the density, the greater effect the trees have with each other.

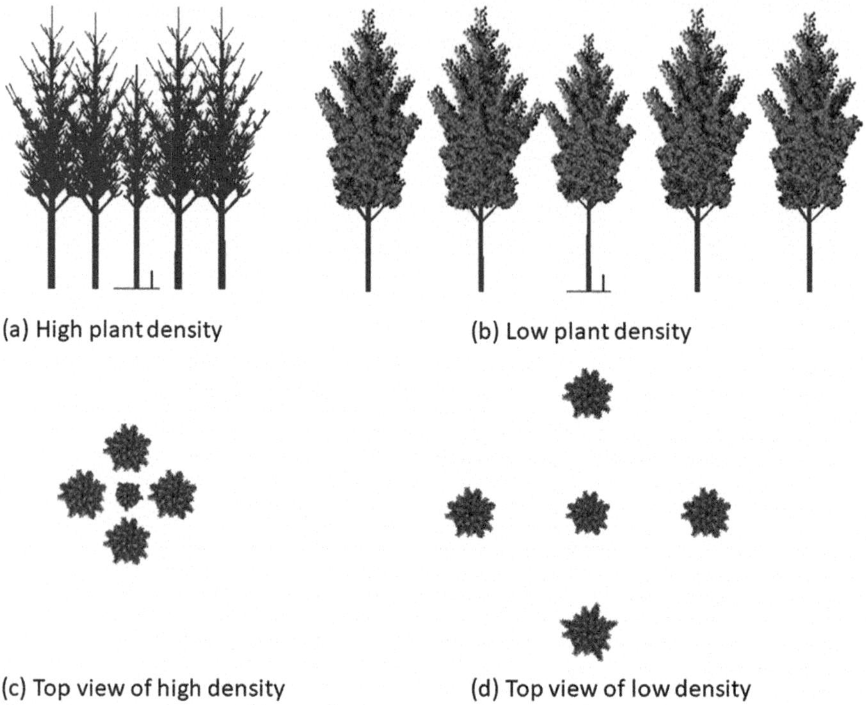

Fig. 12.12 Trees grown with high plant density (**a**, **c**) and low plant density (**b**, **d**) under isotropic light source

Figure 12.13 shows the dynamical growth process of two neighbouring trees under the anisotropic light condition. The point light locates at the top left-hand side and the tree in the left receives more light. The change of the number of organs, e.g. via variable branching, which induced by ratio of *Q*/*D* affected under different light photon flux density, result in structural plasticity.

By integrating light interception and photosynthesis model into GreenLab model, plant plasticity can be simulated under different light environment, through a source–sink approach. This method can simulate the light interception and reflect the adjustment between plant morphology and local environment, which provides an alternative way of computing biomass production aside Beer-Law approach. The integrated model is useful when the aim is to simulate the effect of geometrical structure on light interception and plant growth, or to produce unsymmetrical tree crown for landscape design. It is helpful for both the graphics applications and tree growth modelling.

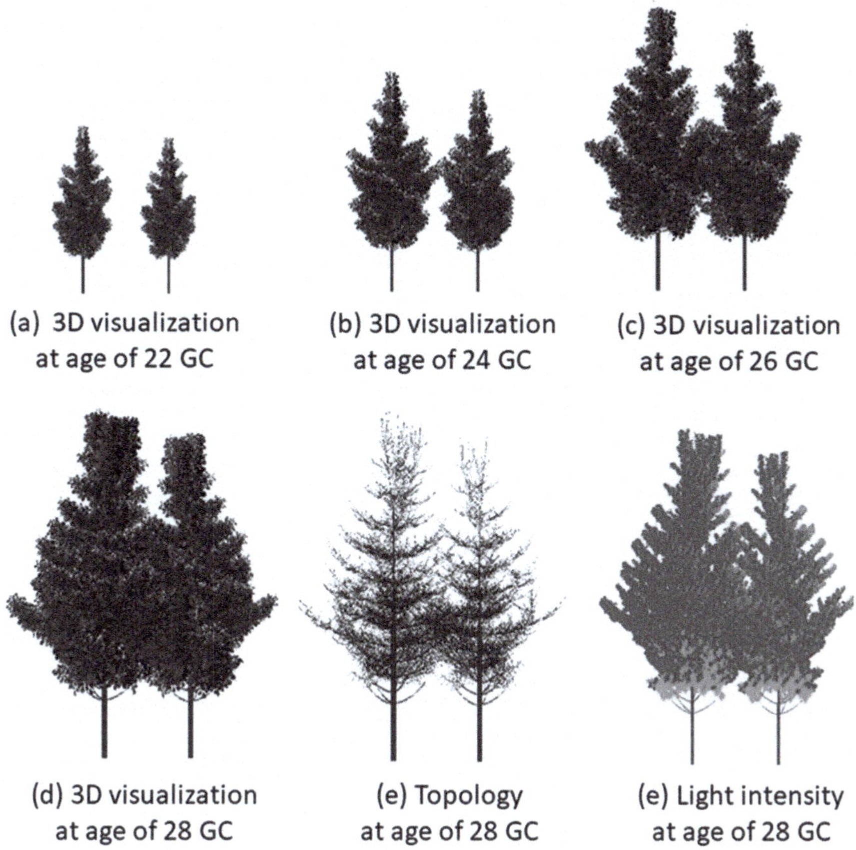

Fig. 12.13 Two neighbouring trees competing for light. 7(**a**)—Fig. 7(**d**) the 3D visualization of two neighbouring trees competing for light at 22, 24, 26, 28 GCa, respectively. Fig. 7(**e**) and Fig. 7 (**f**) show the topology and light intensity at plant age of 28 GCs, respectively

12.1.4 Tree Models for Wood Quality

Secondary growth produces rings in trees that pile up each year in trunks and branches (Chaps. 8 and 10). Trunk slicing and unrolling reveal the rings in the form of veins with the nodes that come from the beginning of the branches remaining on the tree and covered by the annual increases in diameter (see Chap. 8). The AmapHydro and Gloups software simulate the construction of rings based on the distribution of biomass along woody axes. Secondary growth has a double influence on both mechanical properties and wood quality. On the one hand, there are mechanical stresses that are released after slicing and that may deform the boards, and on the other hand, the visual appearance of the boards depends greatly on the conditions in which the tree has grown.

12.1.4.1 Mechanical Balance of the Standing Tree

This study was conducted by Thierry Fourcaud at INRA-LRBB. When the tree grows, its architecture is subjected to the forces of gravity that cause it to bend. To maintain its balance, it is formed in the rings of reaction wood, denser and harder, whose appearance differs from ordinary wood and which allows the axes to be straightened. Mechanical stresses are thus balanced throughout the growth process thanks to reaction wood: tension wood in deciduous trees (which are then as guyed) and compression wood in gymnosperms. Thierry Fourcaud has developed an external module (AmapMeca) that models the formation of reaction wood according to gravity constraints, and its action on axis straightening (Fourcaud 1995, Fourcaud et al. 1997, 2003). The growth simulation produces models on which the action of gravity is visible externally. They also contain, internally in the axes, the rings with the location of the reaction wood (Fig. 12.14).

12.1.4.2 Trunk Slicing and Trunk Unrolling of Computational Trees

Secondary growth modelling with the description of rings and nodes, which complements the primary growth of the woody axes in forming the tree's architecture, provides computer models on which it is possible to simulate slicing to obtain boards, or unrolling to obtain veneers. AmapHydro software provides virtual trunks from simulated plantations. Geometric algorithms allow to simulate the cuts in any part of the trunk and visualize the appearance of the cross-sections with their rings and nodes (due to branch insertions).

Fig. 12.14 Mechanical balance of a growing tree (Fourcaud 1995). (**a**) Location of tension wood in a growing tree. (**b**) A grove of pines whose trunks are deformed by reactions due to mechanical stresses generated by interactions between individuals during growth

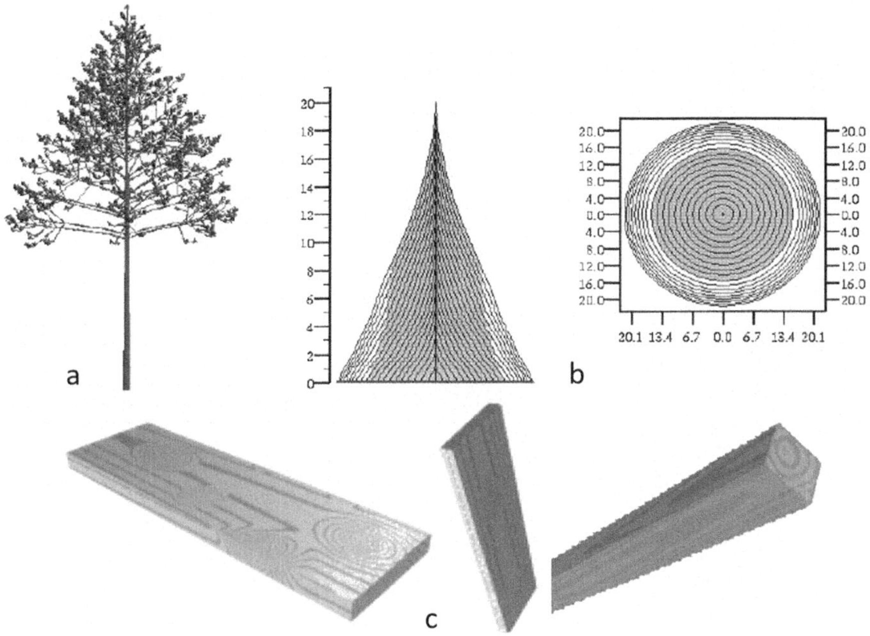

Fig. 12.15 Example of the application of 3D simulated plants in wood quality. (**a**) Architecture of a tree simulated by the AmapHydro software (Frédéric Blaise, CIRAD-AMAP). (**b**) Internal description of the trunk with the rings in longitudinal and radial sections. We distinguish between juvenile and adult wood. (**c**) Extraction of virtual boards from the simulated trunk (Nicolas Szafran, IM^{2}AG)

For slicing, the virtual cross-section is positioned inside the trunk. In Fig. 12.15, a tree is simulated from which the trunk is extracted with its stacks of rings (Fig. 12.15b). We distinguish between juvenile wood (dark) which still contains traces of the branch inserts forming the knots and adult wood (light) which no longer contains them (pruning the branches removes them).

Sawing software has been developed at the modelling and computation laboratory and the IM^{2}AG training and research unit of the Joseph Fourier University in Grenoble by Nicolas Szafran (Szafran et al. 2003). Figure 12.15c shows an example of a board extracted from the simulated tree. A similar study with another formalism was carried out by Guimberteau et al. (2004).

For unrolling, the trunk is virtually centred on a rotating axis. A saw cuts it according to an Archimedes spiral and generates wooden blades that are used to make veneers. Depending on the thickness of the blade, the drawings obtained vary. Frédéric Blaise has developed a computerized veneering module to generate veneers from virtual trunks (Fig. 12.16). Boards and veneers in turn allow for the assembly of virtual furniture.

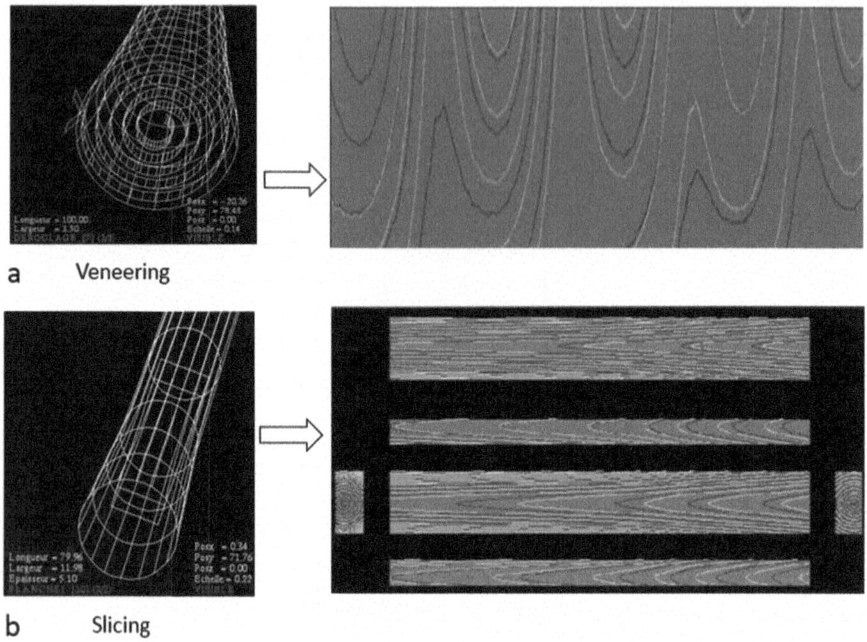

Fig. 12.16 Virtual slicing and unrolling. (**a**) Virtual slicing of cross-sections positioned inside the trunk. (**b**) Virtual unrolling of a trunk (Frédéric Blaise, CIRAD-AMAP)

12.2 Genetics and Plant Structural–Functional Model

This study was conducted by Véronique Letort and Brigitte Courtois (Letort et al. 2008). The characteristics that can be identified on a phenotype have long been the subject of selection processes, first natural and then guided by humans. Their heritability has been studied since Mendel's early work, which laid the theoretical foundations of genetics. These characteristics can be qualitative (flower colour) or quantitative (plant height). In the latter case, the problem is more complex because these characteristics result from functioning rather than being fixed only by the alleles present in certain genes. For example, the height of a stem depends on the number of phytomers and the length of the internodes. The number of phytomers in turn depends on the activity of apical meristem, which is in the development domain, and the length of the internode depends on allometries and source–sink relationships, which are in the growth domain. Thus, the phenotypic characteristics commonly used in selection, such as plant height and leaf surface under the ear, cannot reasonably be linked directly to a dominant gene and are certainly the result of the action of several genes involved in the functioning of the plant, whose expression is conditioned by the action of the environment.

On the contrary, the source–sink parameters of a structural–functional model such as GreenLab are more likely to be related to the action of a small number of particular genes because they are at the origin of the functioning and not of its result.

Using quantitative genetics and marker-assisted selection methods, based on phenotypic traits measured on a plant and its progeny, it is possible to locate in its chromosome map the areas that affect the expression of these traits. They have been called QTL (*quantitative trait locus*). Software such as QTL cartographer (Basten et al. 2005) gives the most probable positions of the significant QTLs on the chromosome map.

To study the interest of using growth models in these marker-assisted selection methods, the simulation is a prospective study that is rich in teaching, before carrying out experiments on field crops. The QTL identification procedure is simulated here to illustrate the results to be expected when the traits considered are model parameters rather than conventional phenotypic traits.

It is assumed here that the parameters of the GreenLab model are determined by linear combinations of genes that have several alleles represented by actual values varying around nominal value 1. To simplify, it is assumed that the markers are identified to the QTLs directly (and even to the gene). A theoretical chromosome is defined as a vector in which each component represents a gene: the different values taken by a given component represent the different alleles of the corresponding gene. An individual will be defined by a pair of these vector-chromosomes (Fig. 12.17), from which a virtual functional chromosome is derived by rules of dominance and additivity. The same gene can contribute to the values of several parameters of the

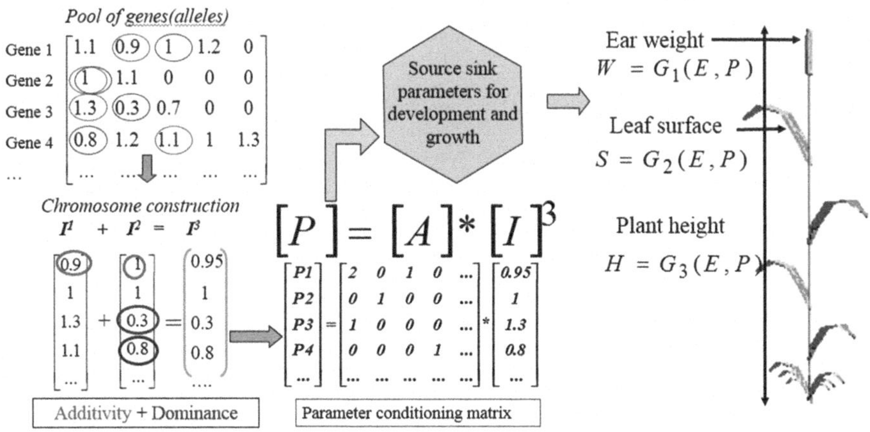

Fig. 12.17 A chromosome is constructed from a gene pool. In matrix I of the gene pool, the two strands I1 and I2 define an expressed strand I3 computed according to the additivity and dominance rules chosen. The parameters of the GreenLab model are linear functions of the values associated with the genes. A matrix [A] defines the contribution of each gene to the expression of the parameter. The set [P] of parameters is injected into GreenLab to grow the model in its physical environment E. The growth and architecture of the plant are then simulated and the phenotypic characters (W, S, H) which are outputs of the model are computed

Table 12.1 The values of simulated parameters of the GreenLab model deduced from studies on wheat

GreenLab endogenous parameters	Values
Leaf thickness *e*	0.025
Inverse of resourse use efficiency *r*	30,000
Leaf sink	1
Sheath sink	1
Short internodes number	6
Long internode allometry	200
Short internode allometry	1
Internode sink	1
Number of stem phytomers	10
Ear sink	230
Seed	0.05

model (pleiotropic gene); similarly, several genes can combine linearly to fix the values of a parameter. A matrix formulation makes it possible to simulate the action of genes on parameter values, which is assumed to be linear. These in turn enable the GreenLab model to construct the phenotype derived from them and on which the characters are expressed (e.g. organ weight).

In the following example, we simulate grass, and to simplify, the matrix A is diagonal (one gene one trait) and we define Table 12.1 the values of the parameters of the GreenLab model deduced from studies on wheat.

These parameters correspond to those of the diagonal matrix A. Three non-coding markers are inserted between two genes, making a chromosome with 57 genes. The distance between the two markers is fixed at 10 centi-morgans. All genes follow the dominance/recessivity rule. In matrix I, there are two alleles whose values are fixed and multiplied by the homologous parameter of matrix A, which gives the value of the gene. Two homozygous parents are defined where the values of the alleles are defined. These parents are crossed by the computer program and provide the F1. This is done on populations of 500 virtual individuals. We then simulate self-fertilization up to the sixth generation F6, which gives almost homozygous individuals. The parameters of the GreenLab model of each phenotype obtained are estimated by inverse method and provide a database. This is analysed by the QTL cartographer software, which allows the location of marker genes on the chromosome that influence exogenous traits such as the weight of the ear of wheat or the total height of the plant to be located.

This virtual experiment showed that the chromosomal map of the QTLs (Fig. 12.18) can be easily found using statistical software tools (such as QTL cartographer) based on the numerical values of the parameters associated with the phenotypes obtained by segregation. On the other hand, based on the numerical values of phenotypic traits (leaf area, height), only major QTLs such as those associated with the GreenLab parameter $1/r$ (resource use efficiency) can be identified.

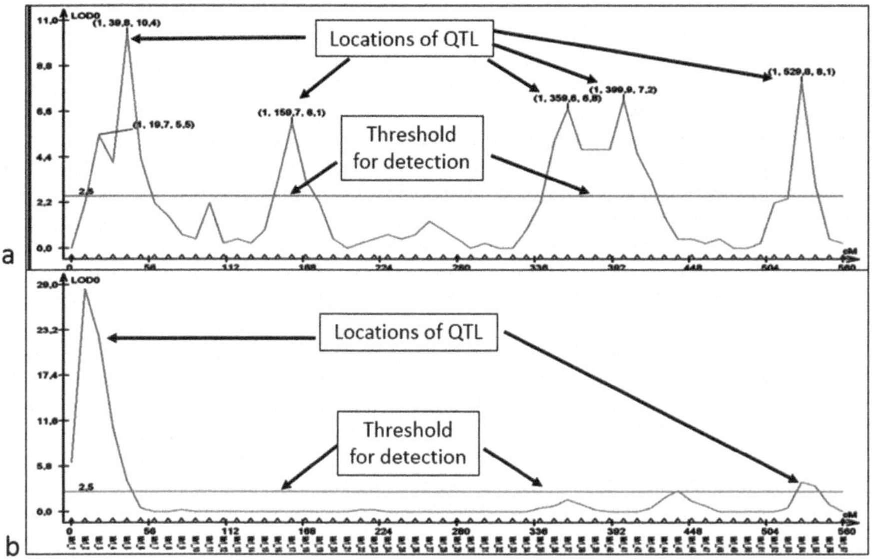

Fig. 12.18 Effects of the QTLs by virtual experiments (**a**) With a diagonal A matrix each gene corresponds to a parameter. The QTL_cartographer software detects the locations of the markers that contribute significantly to the weight of the ear. The detected peaks correspond to the genes that give the resource use efficiency, the number of short internodes, the number of phytomers in the stem, the spike sink and the leaf thickness. (**b**) The matrix A is no longer diagonal. GreenLab's parameters depend on several genes, which can contribute to several parameters. Only markers corresponding to resource use efficiency and leaf thickness are detected

Furthermore, if we try to see directly the QTLs that control phenotypic traits such as plant height and ear weight, the likelihood of identifying a specific trait—gene relationship is lower than in the case of a parameter—gene relationship.

Once the values of the genes that contribute to the identification of the parameters are known with the domain of their allelic variation, as well as the matrix [A] that gives the contributions of each QTL, it is possible to determine by a genetic algorithm which is the best combination to optimize the weight of the ear.

By simulating the generations of crosses and selecting a trait such as an ear weight, we can observe the progress obtained by the selection over successive generations (Fig. 12.19). In short, the in-silico experiment is conclusive. Genetic selection based on the hidden parameters of the model, identified by the inverse method, could provide new QTLs of interest, potentially more stable under different environments than those obtained from the phenotypic traits directly. It opens new perspectives on plant improvement.

All in all, the *in-silico* experience is conclusive. Genetic selection based on the hidden parameters of the model, identified by inverse method, could provide new QTLs of interest, potentially more stable under different environments than those obtained from the phenotypic traits directly. This opens new perspectives on plant breeding.

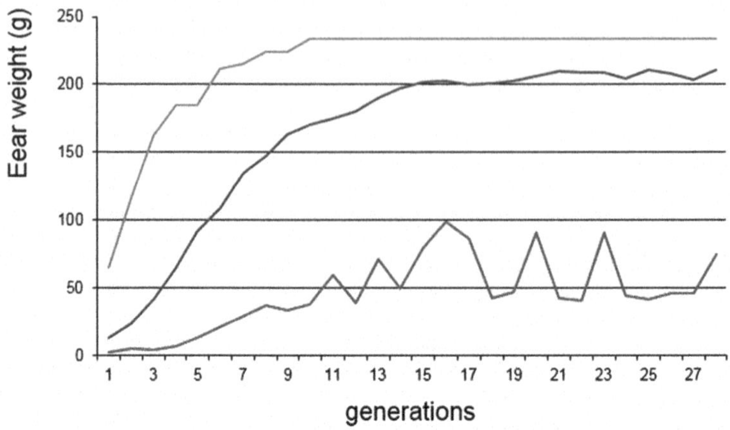

Fig. 12.19 Improvement in ear weight over successive generations (green curve the best individual, black curve the population average, red curve the worst individual) (Color figure online)

12.3 Optimization and Functional-Structural Plant Model

Depending on the type of plant and crop, the characteristics of the plant's development play a role in optimizing yield. The problem is discussed in the thesis of Qi (2009).

Thus, for beetroot, the objective is to have the largest possible taproot by optimizing source–sink relationships. In the case of maize, a compromise is being sought between the two conflicting objectives of having as much forage as possible and the best possible cob harvest. The solutions are distributed over a curve called the Pareto front.

When the organ of economic value is the leaf, the problem is complex. This is the case for tea. The leaf is both source and sink. Taking too many leaves reduces biomass production. The aim here is to find an optimal harvesting strategy without hindering too much growth.

The parameters of plant development and growth are the components of biomass production and its distribution in the plant, i.e. yield. In the previous paragraph, the first optimization of yield is made by the best possible allelic combinations of genes considering genetic variability. But there are other forms of optimization linked to the crop system, such as harvest index, irrigation and pruning strategy. In general, it is necessary to define an objective, in the form of a function that depends on the parameters of the system. The parameters generally have an individual range of variation.

When the functions of the model are derivable, mathematical tools, such as the Newton method, make it possible to compute the values of the parameters taken when the optimum function is reached.

12.3.1 Numerical Method for Optimization (Newton)

12.3.1.1 Cases with One Parameter

We consider the plant growth programme Gloups as a function $G(a_1)$ which depends on a parameter to be optimized a_1 among others.

The maximum of the function is reached for: $\left[\frac{\partial G}{\partial a_1}\right] = 0$.

We start from an initial value $a_{1o.}$ The solution is given by the recurrence:

$$[a_1]_{k+1} = [a_1]_k - \left[\frac{\partial^2 G}{\partial a_1^2}\right]_k^{-1} \cdot \frac{\partial G}{\partial a_1}$$

It is practically difficult to find an analytical formulation of the derivative. As for least squares method (Chap. 9), the numerical values of the first and second derivatives are obtained directly by assuming a small h and by doing:

$$\left[\frac{\partial G}{\partial a_1}\right] = \frac{G(a_1+h) - G(a_1-h)}{2h} \quad \left[\frac{\partial^2 G}{\partial^2 a_1}\right] = \frac{G(a_1+h) - 2G(a_1)+G(a_1-h)}{h^2}$$

12.3.1.2 Cases with Two Parameters

The maximum of the function is reached for: $\left[\frac{\partial G}{\partial a_1}\right] = 0$ and $\left[\frac{\partial G}{\partial a_2}\right] = 0$

We start from two initial values a_{1o} et $a_{2o.}$ The solution is given by the recurrence:

$$\begin{bmatrix} a_1 \\ a_2 \end{bmatrix}_{k+1} = \begin{bmatrix} a_1 \\ a_2 \end{bmatrix}_k - \begin{bmatrix} \frac{\partial^2 G}{\partial a_1^2} & \frac{\partial^2 G}{\partial a_1 \partial a_2} \\ \frac{\partial^2 G}{\partial a_2 \partial a_1} & \frac{\partial^2 G}{\partial a_2^2} \end{bmatrix}_k^{-1} \cdot \begin{bmatrix} \frac{\partial G}{\partial a_1} \\ \frac{\partial G}{\partial a_2} \end{bmatrix}$$

The first numerical derivatives are computed by the formulas:

$$\frac{\partial G}{\partial a_1} = \frac{G(a_1+h, a_2) - G(a_1-h, a_2)}{2h} \quad \frac{\partial G}{\partial a_2} = \frac{G(a_1, a_2+h) - G(a_1, a_2-h)}{2h}$$

The second numerical derivatives are computed by the formulas:

$$\frac{\partial^2 G}{\partial a_1^2} = \frac{G(a_1+h, a_2) - 2G(a_1, a_2) + G(a_1-h, a_2)}{h^2}$$

$$\frac{\partial^2 G}{\partial a_2^2} = \frac{G(a_1, a_2+h) - 2G(a_1, a_2) + G(a_1, a_2-h)}{h^2}$$

$$\frac{\partial^2 G}{\partial a_1 \partial a_2} = \frac{G(a_1+h, a_2+h) - G(a_1+h, a_2-h) - G(a_1-h, a_2+h) + G(a_1-h, a_2-h)}{4h^2}$$

$$\frac{\partial^2 G}{\partial a_2 \partial a_1} = \frac{\partial^2 G}{\partial a_1 \partial a_2}$$

This method can easily be extended to more than two parameters thanks to its matrix writing.

12.3.2 Optimization of the Harvest Index

A good example of the optimization of the relationships between source–sink parameters is provided by the search for the optimal value of the organ sink that gives the best harvest. By feedback, the increase in the sink decreases the leaf area and can even stop the plant growth. Assuming sufficient allelic variability, it is possible to determine mathematically on a computational plant and deriving the model, the optimum of the sink that gives the best production, all other source–sink parameters remaining unchanged, in particular the source parameters of photosynthesis, i.e. resource use efficiency *1/r,* production area *Sp* and seed Q_0. As an example, we take a computational single-stemmed plant with lateral flowering (Fig. 12.20). (Nosenzo et al. 2001)

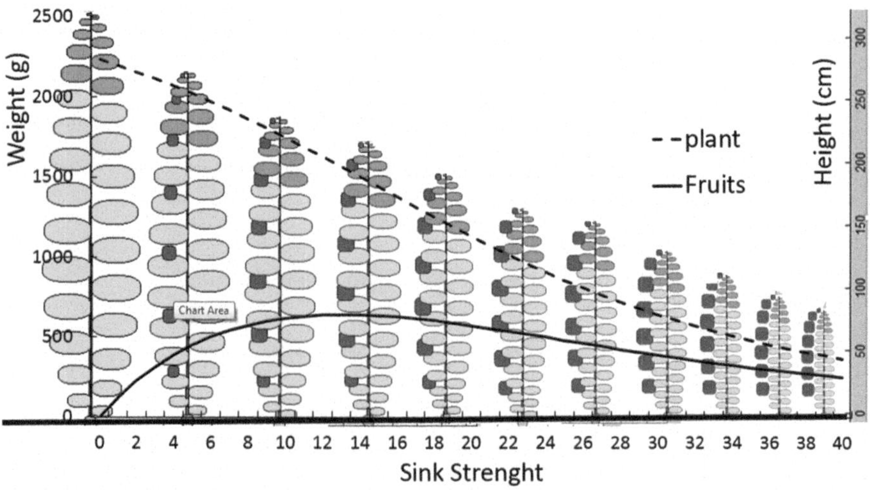

Fig. 12.20 Effect of increasing the fruit sink on plant growth. By feedback the leaf area decreases which decreases growth. The fruit harvest reaches a maximum

One can be left or right away from the optimum in two ways. Starting from the left you increase the fruit production while decreasing the weight of the plant. Starting from the right you increase the fruit production while increasing the weight of the plant. This is obviously the first condition that is the most favourable because the nutrient resources (water, fertilizers) are saved.

In nature, the feedback of fruit removal on growth is well observed. The removal of a sunflower cap strongly increases the volume of the stem and leaves. Sterile triploid hybrid coffee trees (without fruit) have increased vigour compared to their parents' diploid canephora and tetraploid arabica. In addition, if the fruit load is too high, abortions often occur to rebalance demand. However, in varieties such as arabicas caturra amarello, the absence of fruit abortion can cause the plant to die from an overly abundant first flowering. On sweet peppers, chilli peppers and tomatoes, the size of the fruit is very variable according to the variety. In chilli peppers that have small fruits, these are uniformly distributed along the stem. In sweet peppers with big fruits, as in cucumbers, waves of fruiting can be observed along the stems, alternating with areas of abortion, which are caused by feedback from excessive demand for the fruit.

It is said that if the allelic variability is sufficient, the fruit sink can vary considerably in cultivated species. The question arises as to whether the fruit sink of the observed variety is close to its optimum and whether a gain in yield is possible given the genetic variability.

12.3.2.1 Soybean Study

We have a variety of single-stemmed soybean grown at the CAU. In the first step, the source–sink parameters of the variety must be computed. The growth is deterministic on 32 DC. Weights are measured in grams and fresh matter. The root compartment is measured. The data comes from a single plant.

12.3.2.1.1 Empirically Measured Parameters

The leaf thickness (SLW) is $\varepsilon = 0.019$ g/cm^2.

On this plant the durations of expansion of the organs are constant. The leaves have a duration of functioning and expansion of $t_a = 15$ DCs, that of the petioles $t_p = 15$ DCs, that of the internodes $t_i = 5$ DCs, that of the fruits $t_f = 15$ DCs, that of the roots $t_r = 32$ DCs. Fruits do not appear before the 15th internode rank on the stem.

12.3.2.1.2 Computed Parameters

Measurements are available at stage 32 GCs. These measurements produce 99 data: weight of organs in the organic series and compartments to be adjusted

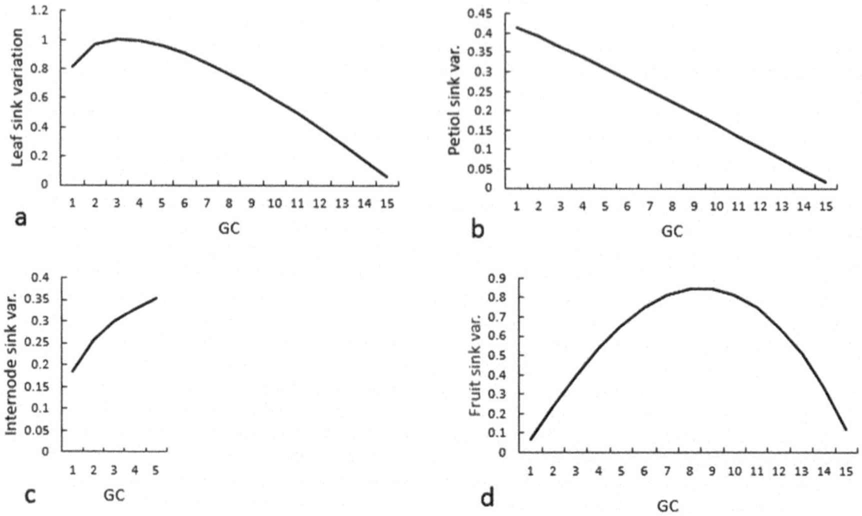

Fig. 12.21 Sink variation during organ expansion. (**a**) Sink variation in blade. (**b**) Sink variation in petiole. (**c**) Sink variation in internode (the expansion duration is short). (**d**) Sink variation in fruit

Table 12.2 Soybean source–sink parameters

Qo	r	Sp	p_a	p_p	p_i	p_c	p_f	p_t	B_a	B_p	B_i	B_f	B_t
0.51	57.7	1593	1	0.41	0.35	0.47	0.84	2	1.22	1.01	1.29	2.15	2
									2	2	1	2	1

simultaneously during the growth process. Eleven parameters are estimated according to a single-stage calibration (single-fitting).

The fitting is good: $R^2 = 0.98$, number of data = 99, number of computed parameters = 11. The second parameters of the beta laws are set.

There is a girth growth on the stem with a parameter value $p_c = 0.47$.

Two parameters (P_o, B_{o1}) are used to compute the sink variation in the organs with the cycles (Fig. 12.21). Parameter B_{o2} is fixed (Table 12.2)

12.3.2.1.3 Comparison Between the Observed and Computed Data

The observed organic series are used as the target data for the WLSM, which can optimize the values of source–sink parameters by the model. The fitting results are good on organic series (Fig. 12.22) and the regression coefficient is high ($R^2 = 0.98$). The weight of the observed root system is 24 g and the computed one 21.8 g.

The weight of the observed plant is 380 g, and the fruit yield is 59 g, with a fruit sink $p_f = 0.84$, which is relatively low.

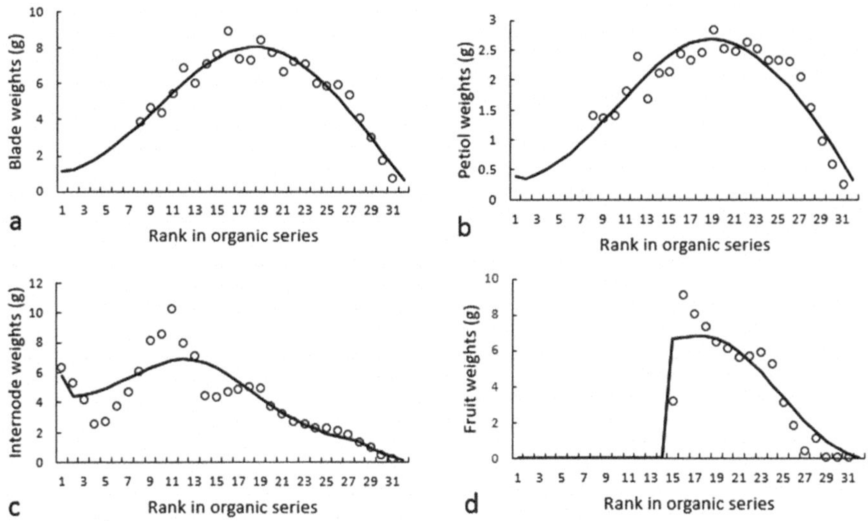

Fig. 12.22 Soybean: single-fitting of the observed (○) organic series and computed organic series by the model (—) for the measurement dates (32 GCs). (**a**) Blades. (**b**) Sheaths. (**c**) Internodes. (**d**) Fruits (no fruit before the rank 14 on the stem)

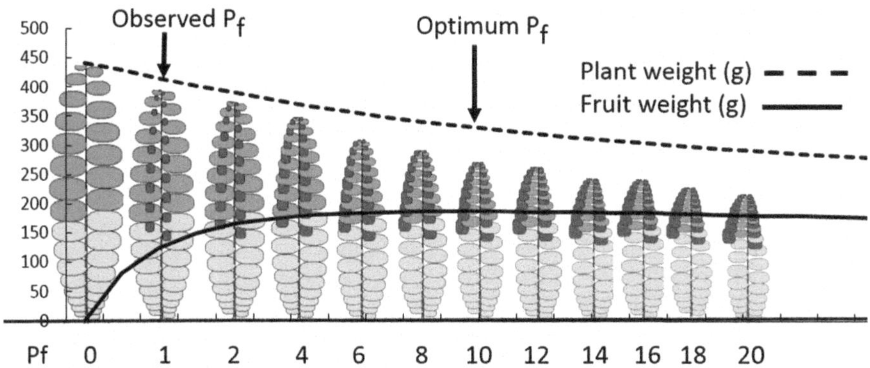

Fig. 12.23 Determination of the sink p_f that gives the plant the best harvest. For the same density, the maximum possible biomass produced by the plant is in the absence of fruit. In the presence of fruit, increasing the fruit sink will reduce the total biomass. There is an optimum that ensures the best harvest index. Beyond that, the fruit production will decrease in parallel with the total biomass

12.3.2.1.4 The Search of Optimal Fruit Sink

Searching for the optimal yield and keeping all the parameters the same (except p_f the fruit sink), we find $p_f = 9.9$. The weight of the optimal plant is 333 g and the fruit yield 187 g (Fig. 12.23). In this case, the potential gain in yield (68%) is quite significant.

The fact that the fruits do not appear until the 15th internode rank means that growth will only be slowed down by the fruits from the 16th DC onwards. This is the reason why this growth cannot be slow down before the 16th DC. The weight of the plant tends towards a minimum limit.

12.3.2.2 Beetroot Study

We return to the beetroot studied in Chap. 9.

Table 12.3 gives the values of the source–sink parameters of the plant computed on 64 DCs of growth.

The weight of the observed plant is 254.7 g and the fruit yield 197.7 g with a tape root sink $P_r = 417$.

12.3.2.2.1 The Search of Optimal Tape Sink

Searching for the optimal yield and keeping all the parameters the same (except p_r the tape root sink), we find $P_r = 378$. The weight of the optimal plant is 262.4 g and the fruit yield 198.9 g (Fig. 12.24). In this case, the potential gain in yield is quite negligible. The plant functioning is optimal.

Table 12.3 Computed source–sink parameters of shoot and root on beetroot

Param	r	S_p	P_b	P_r	B_{b1}	B_{r1}	B_{b2}	B_{r2}
Mean	170	1261	1	417	1.79	5.1	3	2

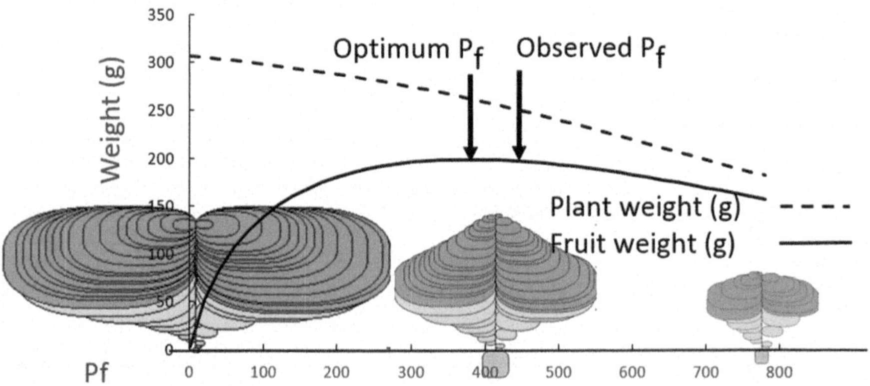

Fig. 12.24 Determination of the sink p_r that gives the plant the best taproot harvest. Increasing the taproot sink will reduce the total biomass. There is an optimum that ensures the best harvest index. In this case, the observed sink is close to the optimum

12.3.3 Optimization of Wood Production

In the case of trees, it is the secondary growth that is concerned. Secondary growth is concurrent with the primary growth of the tree. There is an optimum for the sink of the rings to produce the maximum amount of wood. The tree that produces the most biomass is the one that does not have rings. When the ring sink increases the ring biomass increases at first, goes through a maximum and then decreases with the total tree biomass (Fig. 12.25). However, an additional (mechanical) constraint is that the tree must not bend, so a secondary growth is necessary (Qi et al. 2009).

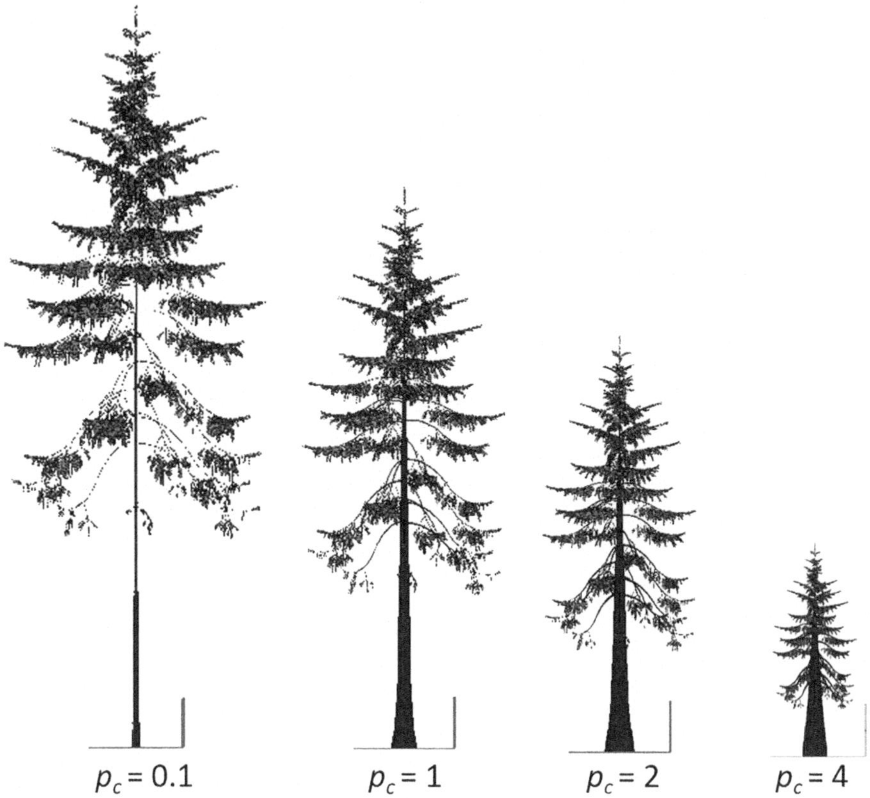

Fig. 12.25 Optimization of the ring sink for wood production. A conifer is simulated at 40 years of age by varying the sink of the ring upwards, which reduces the biomass of the tree accordingly. The maximum wood production is reached here for a sink of the rings $P_c = 1$. The 1 m scale at the foot of the trees allows us to compare their sizes

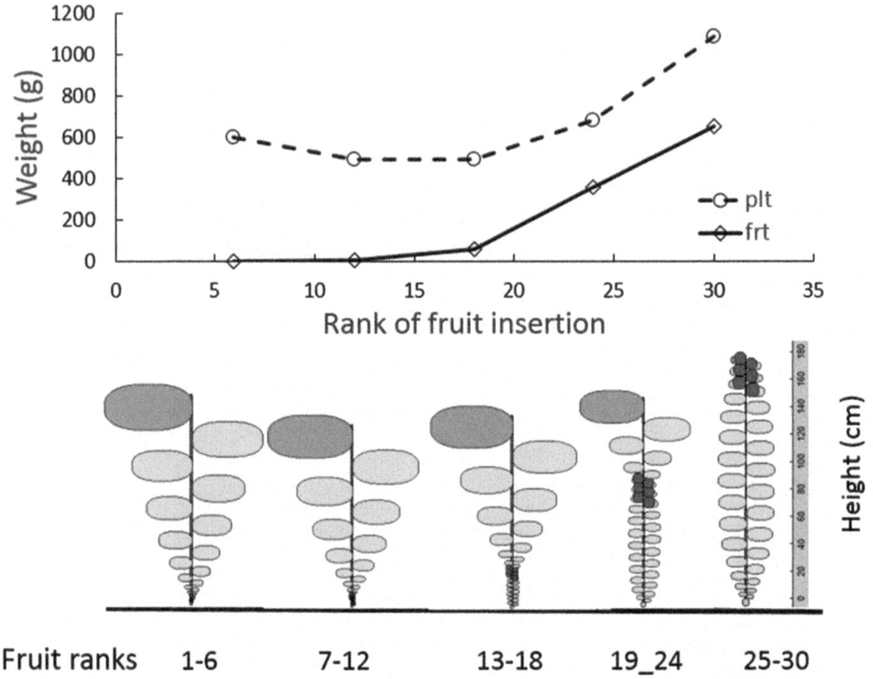

Fig. 12.26 The plant whose development stops at 30th DC grows up to 45th DC. The fruits in a series of six along the stem are positioned successively at rows 1, 7, 13, 19, 25. The higher the series of fruits is positioned, the better the harvest. In this case, the size of the last leaves of the stem decreases because the demand for the fruit slows down its growth. This phenomenon is commonly observed in herbaceous plants (e.g. sunflower) (Qi et al. 2010)

12.3.4 Optimization of Fruit Location on the Stem

The location of fruit organs in the plant structure, which is determined by the development programme, is also an important factor in biomass production. The functioning times of the organs and the shapes of their sink functions must be considered. The study was carried out by Qi et al. (2010) on the single-stem plant with variable fruit locations (Fig. 12.26). Here we choose a single-stemmed plant whose development stops after 30th DC. The functioning times of the organs whose sinks are constant are 15 DCs. The growth thus stops at 45th DC. The organic series of fruits are distributed along six phytomers. The rank of the first fruit of a series is successively at 1, 7, 13, 19 and 25.

In the example chosen, if the fruits are positioned at the bottom of the stem, they remain small because the *Q*/*D* ratio is low at the origin. As soon as the demand for the fruit ceases, the growth starts again. Maximum fruit production is reached when the fruits are positioned at the end of the stem.

In all cases, on a stem whose development is limited to a given number of phytomers, the series of fruit positions must reach its tip to obtain a maximum harvest. This solution has been chosen most of the time by nature for herbaceous plants (e.g., sunflowers, grasses) whose flowering is terminal. Depending on the sink strength, the appearance of the first fruit must be delayed, so as not to hinder growth. The harvest is optimal for a date of appearance corresponding to several associated fruits.

12.3.5 Biological Balances

Plant growth is dependent not only on environmental parameters, but also on other types of interactions such as those related to insect pest attacks. In the case of oil palm, insects of the genus *Coelaenomenodera* can defoliate a plantation. These same insect pests are themselves parasitized by microscopic wasps that infest their eggs. We are, therefore, dealing with dynamic systems with three interacting actors. The agronomic problem to be solved is, therefore, to treat plants effectively, i.e. to optimize the applications of insecticide treatments while sparing as much as possible the parasite population that attacks the pest population. The development of plants and insects also depends on the sum of temperatures. This allows development cycles to be synchronized between plants and insects. In insects, these correspond to the different larval stages of nymphs and adults. Insects are only vulnerable to insecticides at the adult stage when they fly to reproduce.

There are mathematical models specialized in parasitic host relationships, of which the Lotka–Volterra equations are an example. These must be adapted to describe precisely the plant–pest interactions, parasites in the different stages of development of the antagonists. This has been done for oil palm (Lecoustre and de Reffye 1984, 1986; Qi et al. 2009).

Although it is possible to simulate the developments and interactions of the three populations, plants, pests and parasites, the practical aspect of the study should be put into perspective. The model has a high sensitivity to initial conditions. A small change in parameters (reproduction or mortality rates) has major consequences on its evolution. Finally, the processing dates are imperative to be effective. In practice, logistics still must be able to keep up!

12.3.6 Optimal Control of Environmental Conditions

If there is a possibility of improving production based on the genetic variability observed in the endogenous source–sink parameters, another possibility is to optimize the environmental conditions (light, temperature, water). In greenhouses, water is never the limiting factor. We can act on light, temperature and CO_2. On the contrary, in the field, we cannot act on these climatic parameters except on water, which is the main limiting factor but can be controlled by irrigation.

With a limited water resource, the question is to distribute irrigation as well as possible during the duration of the crop growth to optimize production.

Crop models such as Pilot (Mailhol et al. 2011) have water modules that compute soil water content according to the dates of water supply (rainfall, irrigation), their respective intensities and the evapotranspiration of the canopy. These modules are generally mathematically sophisticated (Le Chevalier et al. 2007). In general, for a soil type, a *maximum* water saturation rate $Q_{H2O}\ max$ called 'field capacity' is considered beyond which additional water supply flows and does not penetrate. Symmetrically, we consider a minimum moisture content $Q_{H2O}\ min$ called 'wilting point', below which the plant cannot extract water from the soil for its functioning. Without an additional water supply, the water content decreases exponentially by evapotranspiration of the canopy.

A water module is interfaced with the GreenLab model by modifying at each cycle the climatic factor *Eo* according to the formula:

$$E(t) = Eo\frac{Q_{H_2O}(t) - Q_{H_2O}\ min}{Q_{H_2O}max - Q_{H_2O}\ min} \quad (12.1)$$

Climate factor is at its highest when the soil is saturated with water (field capacity) and zero when the wilting threshold is reached. In each cycle, a fraction of the transpired water is transformed into biomass, while the other evaporates. The system has been adapted to sunflower to optimize irrigation under conditions of limited water resources.

12.3.6.1 Soil Water Evaporation Model as a Function of Irrigation

Consider a tank with a square cross-section *Sd* ($Sd = 1\ m^2$ by default). The height of water inside at cycle t is $H(t)$. The volume of water in the tank is, therefore, $V(t) = Sd*H(t)$.

At time $t = 1$ the reserve is initialized to $H(1)$.

In soil science, the minimum height of water *Hmn* below which water can no longer be extracted by the plant is called the wilting point. If the water demand for evaporation is normalized to 1 ($E = 1$), the decrease in the water level $\Delta H(t)$ during the time Δt is given by the finite-difference equation:

$$\Delta H(t) = -c_1 \cdot (H(t) - Hmn) \cdot \Delta t \quad (12.2)$$

The parameter c_l characterizes the intensity of the drainage. This can also be written as a finite difference equation by making $\Delta t = 1$:

$$H(t) = H(t-1)(1-c_1) + c_1 Hmn \quad (12.3)$$

If this equation is integrated, we obtain:

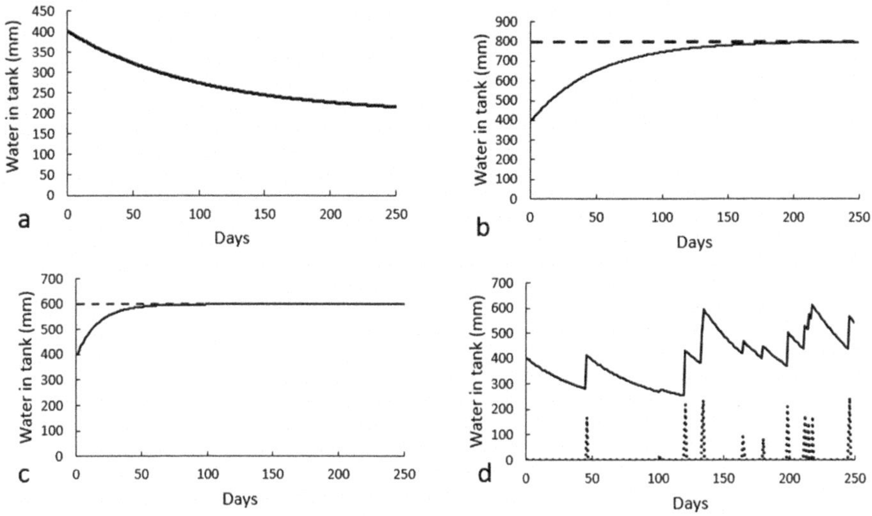

Fig. 12.27 (**a**) Daily water loss by evaporation in the tank from an initial height of 400 mm. (**b**) Filling the tank with a water supply of 20 mm/day from an initial height of 400 mm. (**c**) Filling the tank with evaporation and a water supply of 40 mm/day from an initial height of 400 mm. The system reaches an equilibrium. (**d**) Random water supply in the tank. The exponential decrease in the water height between two inlets can be observed, which constitutes the 'drainage'. If the water height is high, the water supply is limited by the 'field capacity' effect

$$H(t) = Hmn + (H(1) - Hmn)(1 - c_1)^{t-1} \tag{12.4}$$

Example is given for $c_1 = 0.02$, $H(1) = 400\ mm$, $Hmn = 200\ mm$. Figure 12.27a shows the daily water loss through evaporation in the tank from an initial height $H(1)$ of 400 mm.

12.3.6.2 Irrigation Model

There is a maximum height of water *Hmx* that the tank can store, which is called field capacity. Additional water runs off and is not absorbed by the soil. Let $\Delta HI(t)$ be the water input at cycle t that the soil partially absorbs according to its degree of saturation. It is assumed that this amount is small and constant at first.

The increase in water level $\Delta H(t)$ as a result of a constant water supply $\Delta HI(t)$ during time Δt is given by the differential equation:

$$\Delta H(t) = c_2(Hmx - H(t-1)) \cdot \Delta HI(t) \cdot \Delta t \tag{12.5}$$

If the water input $\Delta HI(t)$ is constant at each cycle, the solution of Eq. 12.5 is obtained as a finite difference with ($\Delta t = 1$).

$$H(t) = (1 - c_2\Delta HI(t))H(t-1) + Hmx \cdot c_2\Delta HI(t) \tag{12.6}$$

This equation is integrated and we obtain:

$$H(t) = Hmx + (H(1) - Hmx)(1 - c_2\Delta HI(t))^{t-1} \tag{12.7}$$

Example is given for: $\Delta H(t) = 20$ *mm/cycle*, $Hmx = 800$ *mm*, $H(1) = 400$ *mm*, $c_2 = 0.001$. Figure 12.27b shows the effect of the daily water input of 20 mm in the tank from an initial height of 400 mm.

12.3.6.3 Irrigation and Evaporation Model

12.3.6.3.1 Regular and Constant Water Supply in Each Cycle

Equations 12.2 and 12.5, which quantify the daily constant water loss and inflow $\Delta HI(t)$, must be combined, and a finite difference equation is obtained:

$$H(t) = (1 - c_1 - c_2\Delta HI(t))H(t-1) + c_1 Hmn + c_2\Delta HI(t)Hmx \tag{12.8}$$

This is a classical recurrence equation of the type:

$$H(t) = A \cdot H(t-1) + B \tag{12.9}$$

The solution to which is:

$$H(t) = A^{t-1}\left(H(1) - \frac{B}{1-A}\right) + \frac{B}{1-A} \tag{12.10}$$

The coefficients A and B in Eq. 12.7 are identified:
$A = 1 - c_1 - c_2\Delta H(t)$ and $B = c_1 Hmn + c_2\Delta H(t)Hmx$
The system from the initial value $H(1)$ tends to a limit H_L.

$$H_L = \frac{c_1 Hmn + c_2 Hmx\Delta H(t)}{c_1 + c_2\Delta H(t)} \tag{12.11}$$

This limit can be higher or lower than $H(1)$.

The value $H(L) = H(1)$ gives a constant level in the tank. It is obtained with the water inputs:

$$\Delta HI(t) = \frac{c_1(H(1) - Hmn))}{c_2(Hmx - H(1))} \tag{12.12}$$

Example is given for: $\Delta H(t) = 40\ mm/cycle$, $c_1 = 0.02$, $c_2 = 0.001$, $Hmx = 800\ mm$, $Hmn = 200\ mm$, $H(1) = 400\ mm$. This gives by Eq. 12.11 a stabilization of the water input–output for a water height in the tank: $H_L = 600$ mm. Figure 12.27c shows the behaviour of the evaporation and irrigation system.

12.3.6.3.2 Irregular Irrigation in each Cycle

Equation (12.8) is solved step by step according to the amount of water supplied $\Delta HI(t)$ at cycle t. An irrigation probability can be assigned to a given cycle and an arbitrary amount of water height $QH_20{*}rand$ ($rand$ is a random number between 0 and 1).

Figure 12.27d simulates random irrigation. In each cycle, the probability of water supply is 0.05. The height of water randomly added is between 0 and 250 mm.

The system evolves randomly between $Hmx = 800$ mm and $Hmn = 200$ mm, depending on the distribution of irrigations.

Example is given for: $\Delta H_I(t) = 250{*}rand(1)$, proba irrigation = 0.05/cycle, $c_1 = 0.02$, $c_2 = 0.001$, $Hmx = 800\ mm$, $Hmn = 200\ mm$, $H(1) = 400\ mm$. Figure 12.27d shows the effect of random water supply in the tank from an initial height of 400 mm.

Height of Water in the Tank and Quantity of Water

If we denote the unit of the soil surface by Σo (for crops $\Sigma o = 1\ m^2$) and by d the planting density, the average available surface per plant Sd is $Sd = \Sigma o/d$.

To get the amount of water in the tank, simply multiply the height of water $H(t)$ in the tank by the base area of the tank Σo.

$$Q(t) = (1 - c_1 - c_2 \Delta HI(t))Q(t-1) + Q_{mn}c_1 + Q_{mx}c_2 \Delta HI(t) \tag{12.13}$$

Equation 12.13 is the same as Eq. 12.8; it gives the available water stock in the tank as a function of irrigation and transpiration, instead of the water height.

12.3.6.4 Validation of the Water Balance Model on Rainfall Monitoring of a Cocoa Stand

In 1977, Pierre Jadin (CIRAD) monitored the water content in the soil of cocoa stand daily for one and a half years, measured with a neutron detector. This content depends on the transpiration of the cocoa stand and the rainfall. It is considered that the canopy is closed and that only its transpiration dries the soil. Moreover, in a humid climate over a long period, it can be considered that the PET varies a little around an average.

Under these conditions, the water balance depends on the rainfall and the parameters c_1 and c_2, which must be estimated from the measurements given by the neutron detector. The WLSM is well adapted to this problem

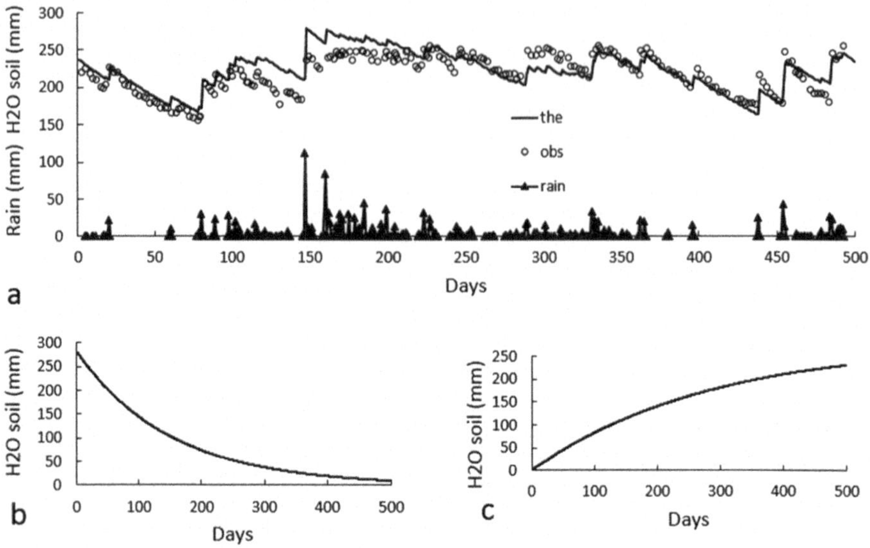

Fig. 12.28 Water balance of cocoa stands as a function of rainfall throughout one and a half years. (**a**) Curve of water height in the soil as a function of rainfall. The small deviations of the observations from the theory can be attributed to variations in PET. (**b**) Theoretical evaporation curve without irrigation. (**c**) Theoretical soil irrigation curve in the absence of evaporation and with a daily supply of 1 mm/day

Figure 12.28 shows the good fitting of the soil water content measurements (250 days) by the model ($R^2 = 0.92$). This value is obtained by empirically adjusting the field capacity and the wilting point: $Hmx = 280$ mm; $Hmn = 0$ mm.

The estimated parameter values are $c_1 = 0.0066$ and $c_2 = 0.011$ (Coefficient of Variation (CV) = 4%)

12.3.7 Model of Plant Transpiration as a Function of Irrigation

The plant is in a container with a surface area *Sd* and a water height *Hmx*. If the water height exceeds this value, the water will runoff. *Hmn* is the minimum height of water in the container for the plant to collect. *Sd* is also the area occupied by the plant in the stand. Based on a plot size of 1 m^2 and a planting density *d*, *Sd* is the inverse of plant density ($Sd = 1/d$). For simplicity, it is assumed that there is no soil evaporation. In agronomy, it is established that foliage transpiration follows Beer-Lambert's law, as does light capture.

Sf(*t*) is the leaf area of the plant. The leaf area that transpires according to Beer-Lambert's law is:

$$SA(t) = Sd\left(1 - \exp\left(-k \cdot \frac{Sf(t-1)}{Sd}\right)\right) \tag{12.14}$$

The amount of water transpired by the leaf surface of the plant in the container is:

$$\Delta Q_1(t) = c_1(H(t) - Hmn) \cdot SA(t) \tag{12.15}$$

But as the plant canopy is closed, we can pose $SA(t) = Sd$.

With photosynthesis, the water use efficiency (*WUE*) transforms this water into the quantity of matter $Q_m(t)$:

$$Q_m(t) = WUE \cdot \Delta Q_1(t) \tag{12.16}$$

The amount of water transpiration by the canopy is proportional to *Sd*.

The water evaporated by the soil and transpired by the plant is:

$$\Delta Q_2(t) = c_1(H(t) - Hmn) \cdot Sd \tag{12.17}$$

Moreover, at each cycle t a quantity of water $Sd * \Delta HI(t)$ is supplied, of which the container retains only $\Delta Q_3(t)$:

$$\Delta Q_3(t) = c_2(Hmx - H(t-1))Sd \cdot \Delta HI(t) \tag{12.18}$$

The water balance, i.e. the quantity of water $Q(t)$ in the container is written as a finite difference as a function of irrigation, and transpiration is given by Eq. 12.13.

12.3.7.1 Effect of Irrigation on Plant Architecture

To transition from the stand level to the individual plant level, the Beer-Lambert law must be used to compute the transpiration of the foliage. The value of *Sd* represents the surface area of the top of the container on which the plant is installed. The individual LAI and the ratio *Sf*/*Sd*, where *Sf* is the transpiring surface of the plant, are considered. It is assumed that only the container does not contribute to evaporate. The plant is irrigated according to a specific schedule. The container fills with water during each irrigation event and empties at each cycle by the transpiration of the plant. The parameters that control the height of water in the container are the same as those of the stand described in the previous paragraph: *Hmx*, *Hmn*, c_1, and c_2.

If the expansion time is 1 DC, organ volume depends on the *Q*/*D* ratio at irrigation date, and the organ series are irregular. If the expansion time is over several cycles, it smoothes the effect of irrigation, which is no longer visible on the organic series (Fig. 12.29).

For monocyclic temperate trees, the cycle represents one year. The effect of the climate is very visible, especially on the tree rings.

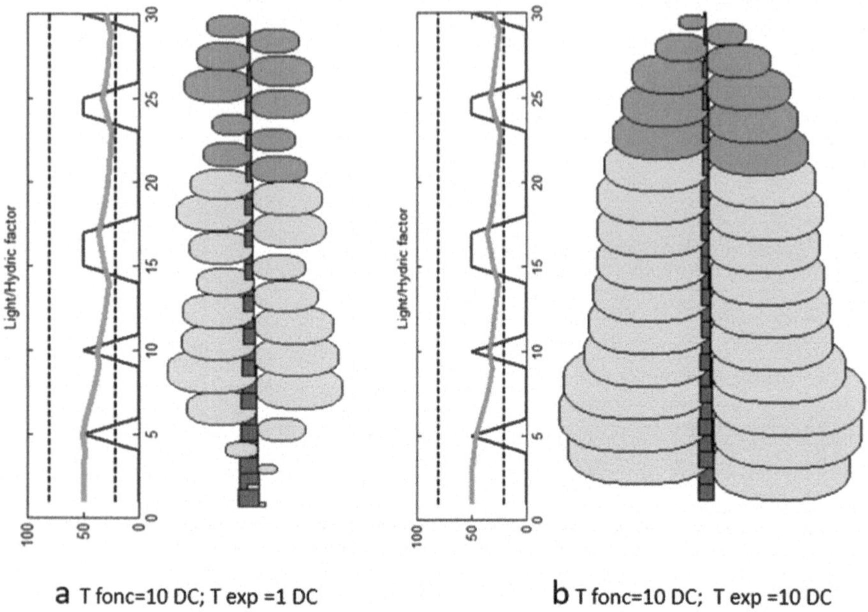

Fig. 12.29 Effect of organ expansion time on the growth of a plant subjected to spaced irrigation dates. (**a**) Expansion time is 1 DC. (**b**) Expansion time is 10 DCs. Graphics: blue line irrigation dates; green line water height in the container (Color figure online)

Table 12.4 Source–sink parameters of the computational plant

Parameter	r	S_d	Q_0	P_b	P_i	P_c	P_t	B_{b1}	B_{t1}	B_{i1}	B_{b2}	B_{i2}	B_{t2}
Value	6.7	1000	0.5	1	0.5	0.5	10	3	1	3	3	2	1

12.3.7.2 A Theoretical Case Study

A monoculture plant with sink parameters close to the cotton plant is chosen.

The plant grows in 90 GCs. The expansion and functioning times of the leaf and internode organs are: $t_a = t_{xa} = t_{xi} = 20$ GCs, for the leaves and internodes and $t_{xt} = 90$ GC for the root system. The thickness of the leaf (SLW) is: $\varepsilon = 0.05$ g/cm^2.

The value $WUE = 0.05$, which means that 5% of the transpired water becomes matter through photosynthesis. It is combined with an equal amount of carbon dioxide to make up 90% of the plant fresh matter, which is rounded off here to 100% for simplicity. The available surface area per plant is $Sd = 1000$ cm^2.

The source–sink parameters are given in Table 12.4

The hydraulic parameters are set to $c_1 = 0.017$ and $c_2 = 0.006$. The maximum water height in the container is $Hmx = 200$ mm; the minimum water height is $Hmn = 20$ mm. The initial water reserve is $H(1) = 150$ mm.

Compared to the standard model when all the plant foliage covers the area of occupation *Sd,* the water model gives:

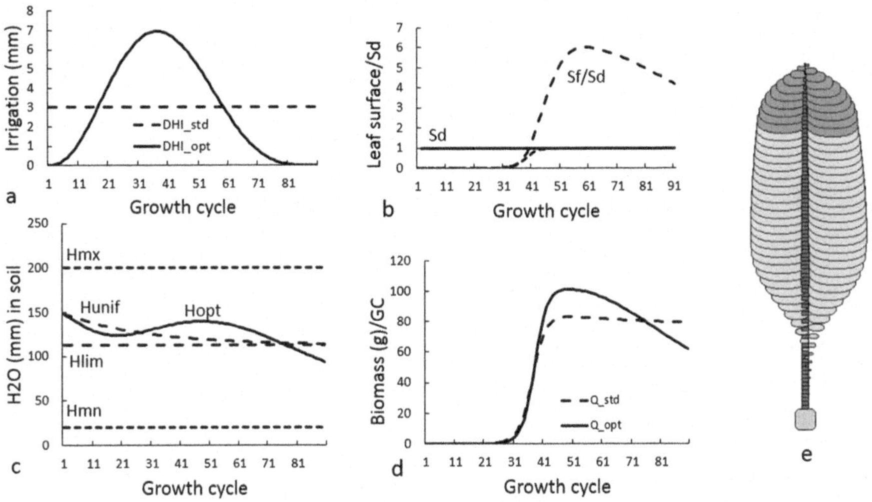

Fig. 12.30 Comparison of uniform irrigation with optimized irrigation. (**a**) Curve of optimized irrigation (solid line) and uniform irrigation (dotted line). (**b**) Evolution of individual LAI. From 40 GCs on the transpiring leaf area is the same as *Sd*. (**c**) Evolution of water in the soil; (dotted line = uniform irrigation; solid line optimized irrigation). In the case of uniform irrigation, the water height in the soil *Hunif* reaches the value *Hlim* that balances the inputs and outputs in the case of uniform irrigation and is given by formula (Eq. 12.10). (**d**) Biomass production/GC for uniform irrigation (dotted line) and optimized irrigation (solid line). (**e**) Optimized plant architecture (90 GCs)

$$1/r = WUE \cdot c_1 \cdot (Hmx - Hmn) \tag{12.19}$$

First, we assume uniform irrigation with a constant height of water/cm^2 supplied in each cycle: $\Delta HI(t) = 3$ mm/GC.

The water input at 90 GCs is, therefore, $QH_2O = 90{*}Sd{*}\Delta HI$, $QH_2O = 270$ kg.

A growth simulation of 90 GCs produces a total plant weight: $Qplt = 4.30$ kg.

From the point of view of optimization, we are looking for the best water input per GC to obtain the maximum weight of the plant, considering the limited amount of water given by the standard irrigation (here 270 kg).

A good point of view is to find a flexible parametric form of the irrigation curve that best approaches the theoretical optimum. This is again a Beta law with parameters h_1 and h_2 which are optimized by the Newtown algorithm. The two-parameter case is explained above in Sect. 4.3.

The uniform irrigation is given by h_1 and $h_2 = 1$.

Optimization gives: $h_1 = 3.69$; $h_2 = 4.98$.

The optimized plant weight is $Q\_opt = 4.62$ kg.

The optimization of irrigation in this theoretical case, therefore, ensures a gain of 7%. Figure 12.30 compares the effects of the uniform and the optimal irrigations on plant growth

Note: by spacing the successive irrigations by a few GCs, the yield is further improved. Thus, with a spacing of five GCs between the irrigations, the weight of the plant obtained is $Q\_plt = 4.74$ Kg. This is a gain of 13% compared to uniform irrigation. This is shown next section.

12.3.7.3 Application to Sunflower (Wu L, F-X Ledimet)

The system has been adapted to sunflower to optimize irrigation under water limitation conditions.

A first basic approach assumes that the supplies are provided periodically and that their intensities are distributed according to a flexible function (a Beta law) whose parameters are to be optimized. Figure 12.31 shows that a uniform distribution per cycle of the limited water supply does not give the best production for the capitulum. Using an optimized beta function, irrigation is better with a periodic water supply every five GCs than with a supply at each GC. This is because the ($H(t)$—Hmn) value is on average larger.

An even better approach uses the sophisticated optimal control method to obtain the sequence of water inputs that maximizes the objective function (plant production) (see Wu et al. 2012). The gain reaches 100%!

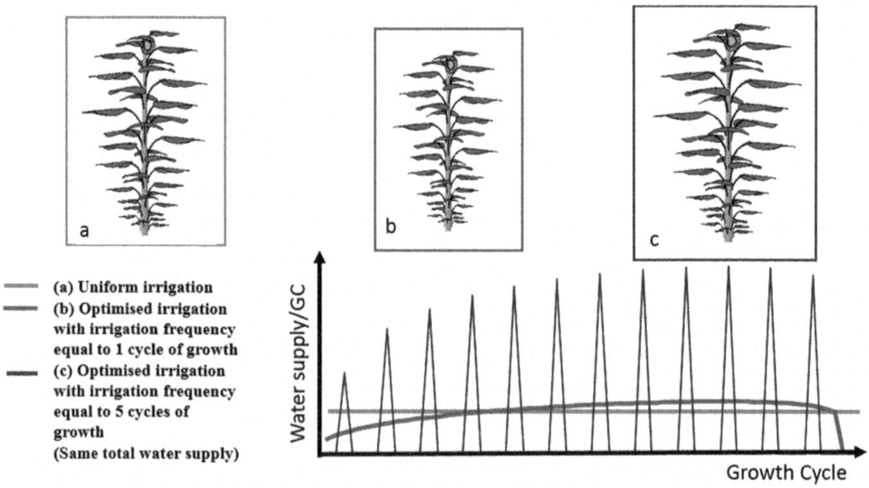

Fig. 12.31 Comparison of the effect of irrigation distribution on sunflower growth and production under limited water supply (after Wu et al. 2012). Given the limited water resources, a constant daily water supply does not give the best growth. By optimizing the irrigation with a supply every five days according to a beta law intensity profile, the optimal irrigation schedule is found. The cost function of optimization is the cap weight. (**a**) Sunflower height is 106 cm, fruit weight is 623 g, and total weight is 1587 g. (**b**) Sunflower height is 98 cm, fruit weight is 656 g, and total weight is 1510 g. (**c**) Sunflower height is 112 cm, fruit weight is 882 g, and total weight is 2000 g. Compared to case (**a**) with uniform distribution, case (**b**) results in a 5% increase in production, and case (**c**) results in a 41% increase

The optimized irrigation schemes are in good agreement with the empirical schemes recommended in agriculture (personal communication from J. Lecoeur (INRAE)).

12.3.8 Using Neural Networks to Control Environmental Effects (Fan X-R, Hu B-G)

The effects of environmental parameters (temperature, light, etc.) on growth and development were studied particularly in the case of maize and chrysanthemum in Chap. 11. The method consists of computing the values taken by the model parameters from measurements (organic series) on plants in various experiments. Some of the GreenLab parameters, such as resource use efficiency (parameter $1/r$) or leaf thickness (parameter ε, were found to be sensitive to light. The prediction of production requires knowledge of the changes in the numerical values taken by these parameters when the environment is subject to variation.

In the greenhouse, for example, with chrysanthemum (Chap. 11), a grid of light–temperature variation in the experiments has been established and in each cell at the edges of the grid, the model parameters are recomputed by the inverse method. Validation is obtained by predicting the behaviour of the model in the cells inside the grid. This allows computing the value of the parameters under the effect of the light–temperature combinations by a quadratic regression.

This approach corresponds to knowledge-driven models (e.g. the source–sink relationship in GreenLab), as opposed to data-driven models. In the latter, the quantity to be computed is expressed directly as a function of environmental parameters (e.g. biomass directly as a function of light, temperature) without any prior knowledge of how the system works. Neural networks are a very efficient method for making predictive models based on data. They are based on learning from databases. Their advantage is that they work without knowledge of the functioning (black box). A data-driven model often proves to be efficient or even better than a knowledge-based model, if the latter is approximate. For example, the control of climatic parameters in greenhouses is often managed by computer programs based on neural networks. Their disadvantage is that the functioning model with its parameters is ignored. Furthermore, a situation unforeseen by the training database can lead to aberrant responses.

In the case of the GreenLab model, it seemed interesting to mix the two types of models. Indeed, while the knowledge of the functioning of the plant on the sink parameters seems to be well established, the knowledge on the source parameters (resource use efficiency $1/r$, for example, water use efficiency) which depends on the environment) is more uncertain. The regression computation of the value of r using interpolation on a grid of the variations in the environmental parameters is not necessarily the most efficient.

The case of tomatoes in greenhouses is a good illustration. Heuvelink conducted a series of agronomic experiments in the greenhouses of Wageningen University on

Table 12.5 GreenLab parameters estimated for tomato according to the 12 planting dates

Experiments	P_p	P_i	P_f	Sp cm^2	r $mg{\cdot}cm^2/mm$
1	0.425	0.633	3.469	1158.711	0.166
2	0.428	0.638	3.508	1157.971	0.166
3	0.423	0.631	3.547	1132.218	0.168
4	0.435	0.648	3.517	1172.124	0.168
5	0.435	0.648	3.511	1151.715	0.167
6	0.427	0.637	3.522	1183.146	0.164
7	0.424	0.63	3.567	1166.277	0.164
8	0.446	0.666	3.615	1194.565	0.17
9	0.436	0.65	3.473	1165.134	0.167
10	0.419	0.62	3.351	1130.649	0.171
11	0.449	0.669	3.598	1196.886	0.164
12	0.439	0.652	3.642	1264.283	0.158
Mean	0.432	0.643	3.527	1172.807	0.166
SD	0.009	0.015	0.077	35.571	0.003
CV (%)	2.083	2.333	2.183	3.033	1.947

the growth of tomatoes at various times of the year (Chap. 5). The growing conditions changed each time. The production per organ compartment (leaf, stem, fruit) is monitored as a function of changing climatic parameters. The parameters of the Tomsim model are adjusted on each experiment, which allows the evolution of the biomass production of the crop per square meter to be followed with good accuracy.

It should be remembered that Tomsim and GreenLab are very similar models from an eco-physiological point of view. The main difference is that Tomsim works at the level of organ compartments and GreenLab at the level of the individual organs. If we are given a priori the sink variation functions of the organs, then Greenlab can operate at the aggregate level of the compartments and becomes equivalent to Tomsim. Tomatoes have been modelled by GreenLab in various experiments (Dong et al. 2008). The functions of the organ sinks are, therefore, known.

The Wageningen data were thus re-analysed with the GreenLab model. Over the 12 planting dates, relatively stable values were obtained for the model parameters (Table 12.5). The parameter $1/r$ is defined as the water use efficiency relative to the potential evapotranspiration (*PET*) measured at each cycle and identified by the parameter $E(t) = PET(t)$.

The PET is measured and computed at each cycle by the formula:

$$PET = a + b\left(\frac{\Delta}{\Delta + \gamma}\right) \cdot Rs$$

Rs is the solar radiation, Δ the slope of the water vapour pressure relative to the temperature per cycle, a, b and γ constants whose values are fixed.

The growth equation is written (Chap. 5):

$$Q(t) = E(t) \cdot r \cdot Sp \cdot \left(1 - exp\left(-k\frac{Sf(t)}{Sp}\right)\right)$$

Knowing that the value $E(t) = PET(t)$ is not necessarily the best possible contribution of the climate for the adjustment of the model, although the system gives satisfactory results, we look for a better expression of $E(t)$, according to the measurements of the parameters of the environment (light, temperature, CO_2) measured at each growth cycle in the greenhouse.

This time the data are based on radiation, planting date, temperature and CO_2. The mean values of the parameters in the table are taken as constants of the GreenLab model and only the climate represented by $E(t)$ varies and should account for the production and distribution of biomass. Neural networks are used to provide the best possible numerical value of $E(t)$ at each cycle t (Fan et al. 2015).

A training database is available on which to calibrate this network and validate its predictive ability on additional data.

The system works well and significantly improves the prediction of the GreenLab model in a variable climate environment. The model is mixed (Fig. 12.32), using GreenLab's knowledge of source–sink parameters and climate data to optimize their effect at each t-cycle on growth via a neural network. Figure 12.33 shows the good agreement between the data and the model. The improvement over the Tomsim model is not visible in the prediction of total biomass (see Chap. 5, Fig. 5.10). It can only be seen in the decrease in the difference between the measured and computed values. On the other hand, all organ compartments are well represented separately, whereas Tomsim only gives the total biomass.

It is interesting to compare the PET values actually measured in the greenhouse throughout the crop with the value computed by the neural network according to the environmental parameters (Fig. 12.34).

There is good agreement on average between the measured PET values and those computed by the neural network.

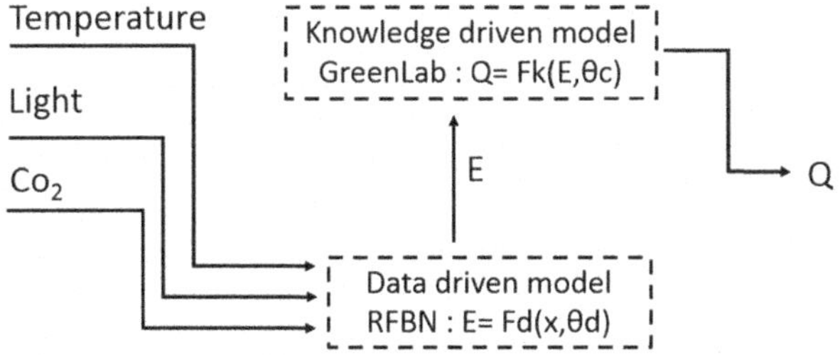

Fig. 12.32 Mixed model for computing the biomass production Q, based on the optimized value of the E factor as a function of the environmental parameters with a neural network and injected into the GreenLab model whose parameters are identified, to control growth

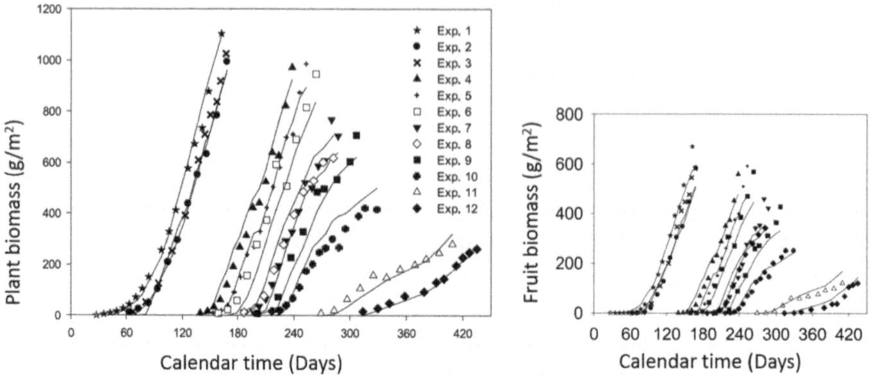

Fig. 12.33 Modelling of total biomass and fruit production per square meter for the tomato crop according to planting date (data Heuvelink, Wageningen University)

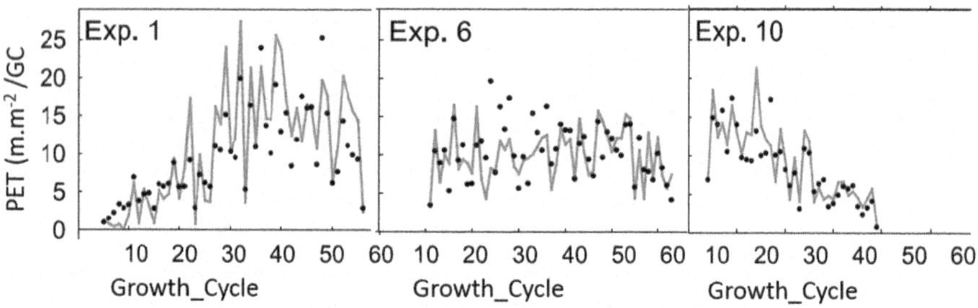

Fig. 12.34 Comparison of the measured PET values per cycle (dots) with the values provided by the neural network (green lines) for January, June and October

The use of a mixed model is, therefore, interesting and improves accuracy and prediction. It separates the source–sink parameters of the GreenLab model from the climate parameters and identifies them in different ways. The former by the least squares method, the latter by neural networks. The latter makes better use of the effect of climate on growth.

12.4 Conclusion

This chapter shows that the applications of a functional–structural model are potentially numerous, but they must be implemented experimentally to be validated.

During the development of the AMAP and GreenLab models, new possible applications of plant modelling in agronomy have emerged. The use of 3D models of plants is obviously of great interest for computer graphics, particularly for the needs of landscaping and town-scaping. For example, Bionatics offers a database of more than 600 plants from the Amap program for all five continents.

But 3D plant mock-ups are also interesting as a backup to help remote sensing interprets the images provided by satellites, to access the production or health status of plant stands (fields, plantations).

Plant breeding based on the identification and estimation of source–sink parameters rather than direct measurements (sizes, weights...) on plants seems very promising, because parameters are less subject to variability and closer to the plant functioning. It should be possible to increase the harvest by favouring the useful compartment, by optimizing source–sink relationships. The model provides also new parameters that are components of yield. Thus, for each plant, we can define a parametric map like a chromosomal map, which allows to reconstruct the plant by simulation. The search for QTL attached to these parameters, less sensitive to the environment than the usual and empirical measurements carried out on the plants, would make them good candidates for the breeding and phenotyping methods (Zhang et al. 2023).

The management of farming systems with the optimization of climatic parameters in greenhouses or fields provides numerous applications for a dynamic model such as GreenLab. The aim is to limit resource waste and pollution rates.

The growth–development interactions make it possible to consider further improvements in pruning or thinning techniques. Simulation becomes an effective method.

In forestry, this type of model is indicated for studying the mechanics of the standing tree or the quality of the wood according to forestry scenarios.

Heuristic methods using neural networks seem also promising to better estimate the effects of environmental parameters on the variation in model parameters.

Finally, agronomy should make increasing use of these structure–function models which require the consideration of the plant structure.

References

Barczi J-F, Rey H, Caraglio Y, de Reffye P, Barthélémy D, Dong Q, Fourcaud T (2008) AmapSim: an integrative whole-plant architecture simulator based on botanical knowledge. Ann Bot 101(8):1125–1138

Basten C-J, Weir B-S, Zeng Z-B (2005) QTL cartographer v1.17. North Carolina State University

Bosanac B, Zanchi P (2005) Onyx TREE professional. http://www.onyxtree.com/

Castel T, Baudoin A, Floury N, Le Toan T, Caraglio Y, Barczi J-F (2001) Deriving forest canopy parameters for backscatter models using the Amap architectural plant model. IEEE Trans Geosci Remote Sensing 39(3):571–583

Chelle M, Andrieu B (1998) The nested radiosity model for the distribution of light within plant canopies. Ecol Model 111(1):75–91

Corpetti T, Gong X, Kang M-Z, Hu B-G, Hubert-Moy L (2018) Time-consistent estimation of LAI by assimilation in greenlab plant growth model. Computers & Geosciences 130:57–68

Daviero V, Meyer-Berthaud B, Lecoustre R (1996) A morphometric approach to the architecture and ontogeny of the extant sphenopsid Equisetum telmateia Ehrh. Int J Plant Sci 157(5): 567–581
Daviero V, Meyer-Berthaud B, Lecoustre R (2000) Computer simulation of sphenopsids architecture. I.Principles and methodology. Rev Palaeobot Palynol 109:121–134
Dauzat J, Clouvel P, Luquet D, Martin P (2008) Using virtual plants to analyse the light-foraging efficiency of a lowdensity cotton crop. Ann Bot 101(8):1153–1166
de Reffye P, Cognée M, Jaeger M, Traore B (1988) Modélisation stochastique de la croissance et de l'architecture du cotonnier. 1. Tiges principales et branches fructifères primaires. Coton et Fibres Tropicales 43(4):269–291
de Reffye P, Edelin C, Jaeger M, Blaise F, Fournie G, Bree Lefebvre V (1987) L'arbre et l'ordinateur. In: pleasure of landscape, IFLA world congress. 14, 1987/09/01-04, Paris, France IFLA, pp. 198–201
de Reffye P, Jaeger M, Cournède P-H (2008) Le monde végétal. In: Comité International des Jeux Mathématiques (ed) Maths Nature Express. CILM, Paris, pp 26–35
Dinouard P, de Reffye P, Jaeger M (1987) Modélisation et simulation de l'architecture de l'hévéa. Revue Générale des Caoutchoucs et Plastiques 64(673):53–55
Dong Q-X, Louarn G, Wang Y-M, Barczi J-F, de Reffye P (2008) Does the structure function model GreenLab deal with crop phenotypic plasticity induced by plant spacing? A case study on tomato. Ann Bot 101(8):1195–1206
Dong Q-X, Wang Y-M, Yang L, Barczi J-F, de Reffye P (2007) GreenLab-tomato: a 3D architectural model for tomato development. N Z J Agric Res 50:1229–1233
Edelin C, de Reffye P, Jaeger M, Dinouard P (1989) La simulation de l'architecture des arbres et son rôle potentiel dans la conception et la gestion des paysages urbains. Revue Forestière Française, numéro spécial 41:143–153
Elhoumaizi MA, Lecoustre R, Oihabi A (2002) Phyllotaxis and handedness in date palm (Phoenix dactilifera L.). Fruits 57(5–6):297–303
Fan X-R, Kang M-Z, Heuvelink E, de Reffye P, Hu B-G (2015) A knowledge-and-data-driven modeling approach for simulating plant growth: a case study on tomato growth. Ecol Model 312:363–373
Fourcaud T (1995) Analyse du comportement mécanique d'une plante en croissance par la méthode des éléments finis. Thèse de 3e cycle en mathématiques appliquées, Université de Bordeaux, 148 p
Fourcaud T, Blaise F, Barthélémy D, Houllier F, de Reffye P (1997) A physiological approach for tree growth modelling in the software Amappara. Nepveu, G. (ed) Proceedings of 2nd workshop connection between sivliculture and wood quality through modelling approaches and simulation software (WP S5.01.04), 25-30/08/1996, berg-en dal, Afrique du Sud, pp. 321–334
Fourcaud T, Blaise F, Lac P, Castera P, de Reffye P (2003) Numerical modelling of shape regulation and growth stresses in trees. II. Implement Amappara Software Simulat Tree Growth Trees Struct Funct 17(1):31–39
Griffon S, Auclair D (2009) Visualising changes in agricultural landscape. In: Van Ittersum M-K., wolf J, Van Laar H-H (ed) Proceedings of AgSAP, Egmond aan zee, The Netherlands. Wageningen University and Research Centre, 10–12 march 2009, pp. 296–297
Guimberteau G, Mérillou S, Terraz O (2004) Simulation de la croissance interne de bois basée sur des subdivisions de volumes. In: Journées de l'Association Française d'Informatique Graphique. AFIG 2004
Jaeger M (1987) Représentation et simulation de croissance des végétaux. Thèse doctorat en informatique. Strasbourg; Université Louis Pasteur, 156 p
Jaeger M (2012) Enhancing computer generated natural scenes using quick and dirty image based recipes. In: proceedings of plant growth modeling, simulation, visualization and applications (PMA12), Shanghaï, 31 Oct—3 Nov 2012. IEEE press, pp. 164–171
Jaeger M, de Reffye P (1991) Le logiciel Amap: un outil de simulation et de représentation des végétaux. In: de Reffye P, Jaeger M (eds) L'informatique scientifique dans l'enseignement de la

biologie et de la géologie au lycée: actes du colloque. Institut National de la Recherche Pédagogique, Paris, pp 243–248. (Colloque Technologies Nouvelles et Education)

Jaeger M, Jia J-Y, Zhang X-P, Griffon S (2010) VR web based tree and vegetation representations for environmental applications and studies, an example on ChongMing Island. In: LandMod 2010: international conference on integrative landscape modelling, Montpellier, February 3–5, 2010, 8 p

Jensen H-W (2001) Realistic image synthesis using photon mapping. Ohmsha Ltd, 2001

Le Chevalier V, Jaeger M, Mei X, Cournède P-H (2007) Simulation and visualisation of functional landscapes: effects of the water resource competition between plants. J Comput Sci Technol 22(6):835–845

Lecoustre R, de Reffye P (1984) Contribution to an integrated control of Coelaenomenodera minuta uh., main pest of Elaeis guineensis in West Africa. Modelling of Coelaenomenodera population dynamics and practical applications. Oléagineux 39(10):461–469

Lecoustre R, de Reffye P (1986) Contribution to the development of an integrated control method against Coelaenomenodera minuta uh., main pest of Elaeis guineensis in West Africa. Oléagineux 41(8–9):367–376

Lecoustre R, de Reffye P, Jaeger M, Dinouard P (1992) Controlling the architectural geometry of a plant's growth—application to the begonia genus. In: Thalmann D, Thalmann NM (eds) Creating and animating the virtual world. S. L. Springer Verlag, pp 199–214

Letort V, Mahe P, Cournède P-H, de Reffye P, Courtois B (2008) Quantitative genetics and functional-structural plant growth models: simulation of quantitative trait loci detection for model parameters and application to potential yield optimization. Ann Bot 101(8):1243–1254

Lintermann B, Deussen O (1997) A modelling method and user Interface for creating plants. Graphics Interface:189–198

Lintermann B, Deussen O (1999) Interactive structural and geometrical modeling of plants. IEEE Comput Graph Appl 19(1):56–65

Mailhol J-C, Ruelle P, Walser S, Schütze N, Dejean C (2011) Analysis of AET and yield prediction under surface and buried drip irrigation systems using the crop model PILOTE and Hydrus-2D. Agric Water Manag 98:1033–1044

Mialet-Serra I, Dauzat J, Auclair D (2001) Using plant architectural models for estimation of radiation transfer in a coconut-based agroforestry system. Agrofor Syst 53(2):141–149

Nosenzo R, de Reffye P, Blaise F, Le Dimet F-X (2001) Modèles mathématiques de conduite culturale. In: Malézieux É, Trébuil G, Jaeger M (eds) Modélisation des agroécosystèmes et aide à la décision. CIRAD, Montpellier, pp 145–172. (Repères)

Poisson C, Rey H, Caraglio Y, Rabaste S, Barczi J-F, Guedon Y (1997) Modélisation de l'architecture et de la croissance de 5 espèces du genre Nicotiana. Annales du Tabac, France. Section 2(29): 37–54

Qi R (2009) Optimization and optimal control of plant growth: application of the GreenLab model for the decision help in agriculture. Doctorat École centrale de Paris

Qi R, Cournède P-H, de Reffye P (2009) Tri-trophic ecosystem oil palm-pests-auxiliaries: II. Sensitivity analysis, parameter identification and control. In: plant growth modeling, simulation, visualization and applications (PMA), 2009 third international symposium on. IEEE, pp. 327–333

Qi R, Letort V, Kang M-Z, Cournède P-H, de Reffye P, Fourcaud T (2009) Application of the GreenLab model to simulate and optimize tree stability and wood production: a theoretical study. Silva Fennica 43(3):465–487

Qi R, Ma Y-T, Hu B-G, de Reffye P, Cournède P-H (2010) Optimization of source-sink dynamics in plant growth for ideotype breeding: a case study on maize. Comput Electron Agric 71(1): 96–105

Saito K, Kumagai Y, Honjo T, Ishida Y, Lecoustre R, de Reffye P (1993) Photo-realistic forest landscape simulation—application of GIS and plant modeling technique. In: proceedings of NICOGRAPH'93. Tokyo, Japon, NICOGRAPH'93, 11/1993, pp. 226–236

Shirley P, Morley R (2003) Realistic ray tracing. Natick: AK Peters, Ltd.

Soler C, Sillion F-X, Blaise F, de Reffye P (2003) An efficient instantiation algorithm for simulating radiant energy transfer in plant models. ACM Trans Graph 22(2):204–233

Szafran N, Despreaux S, Biard L, Blaise F (2003) Sawing of logs in virtual trees using 3D intersections algorithms. In: Hu B-G, Jaeger M (ed) Plant growth modelling and applications (PMA03), proceedings of the 2003 international symposium on plant growth modeling, simulation, visualization and their applications, Beijing, China, 13–16 Octobre 2003. Beijing: Tsinghua University press, Springer, pp. 372–384

Wang H-Y, Kang M-Z, Hua J (2012) Simulating plant plasticity under light environment: a source-sink approach. In: proceedings of 2012 IEEE fourth international symposium on plant growth modeling, simulation, visualization and applications (PMA). IEEE, pp. 431–438

Wu L, Le Dimet F-X, de Reffye P, Hu B-G, Cournède P-H, Kang M-Z (2012) An optimal control methodology for plant growth—case study of a water supply problem of sunflower. Math Comput Simul 82(5):909–923

Zhang P-P, Huang J-Y, Ma Y-T, Wang X-J, Kang M-Z, Song Y-H (2023) Crop/ Plant Modeling Supports Plant Breeding:II. Guidance of Functional Plant Phenotyping for Trait Discovery. Plant Phenomics 5:Article 0091. https://doi.org/10.34133/plantphenomics.0091

Chapter 13
Growth and Development Interaction: The Use of Internal Trophic Pressure (*Q*/*D*)

Amélie Mathieu, Véronique Letort, Philippe De Reffye, and Mengzhen Kang

Abstract This chapter presents a possible improvement of the GreenLab model that aims to reflect the feedback between architectural development and growth, which are traditionally considered as separate processes in the Standard model. The novel approach integrates the activity of meristems and the expansion of organs, both of which are influenced by plant vigor (represented by the ratio biomass supply divided by demand, called trophic pressure), into a unified framework. The "Crown Analysis" and "Organic Series Analysis" provide numerical data for the botanical automaton, allowing for an empirical description of meristem function and organ expansion over thermal time. The model applies integrative biology to account for environmental and stochastic variability in plant development. The chapter also discusses how different environmental conditions can lead to diverse phenotypes from the same genotype, highlighting the phenotypic plasticity in response to environmental stress. It also notes that planting density significantly affects tree

Electronic supplementary material The online version of this chapter (https://doi.org/10.1007/978-981-97-7541-5_13) contains supplementary material, which is available to authorized users.

A. Mathieu
AgroParisTech, UMR EGC (1091), Grignon, France
e-mail: amelie.mathieu@agroparistech.fr

V. Letort
Mathématiques et Informatique pour la Complexité et les Systèmes, CentraleSupélec-Université Paris-Saclay, Gif-sur-Yvette, France
e-mail: veronique.lechevalier@centralesupelec.fr

P. De Reffye
CIRAD, UMR AMAP, Montpellier, France

AMAP, Univ Montpellier, CIRAD, CNRS, INRAE, IRD, Montpellier, France

M. Kang (✉)
State Key Laboratory for Multimodal Artificial Intelligence Systems, Institute of Automation, Chinese Academy of Sciences (CASIA), Beijing, China

School of Artificial Intelligence, University of Chinese Academy of Sciences, Beijing, China
e-mail: mengzhen.kang@ia.ac.cn

P. De Reffye et al. (eds.), *Plant Model Greenlab for Botany and Agronomy*,
https://doi.org/10.1007/978-981-97-7541-5_13

architecture. This new model emphasizes the importance of considering environmental factors and the dynamic nature of plant growth in understanding and predicting plant architecture.

Keywords Feedback growth–development · Trophic pressure · Architectural plasticity · Meristem and source-sink parameters new formulation

13.1 Aim of the Chapter

This chapter aims to introduce a novel model of functioning that is closer to reality. In the Standard model, the architectural development, which is the result of the activity of the meristems, is modeled as a process independent of the growth, which arises from photosynthetic relationships between source and sink organs. By contrast, the proposed model endeavors to unify these processes in a more relevant manner.

The "Crown Analysis" provides the numerical probabilities that serve as input for the botanical automaton, while "Organic Series Analysis" offers the numerical values for the evolution of organ sinks according to development cycles (DCs) and thus thermal time. In this way, meristem function and organ expansion can be described numerically in an empirical manner. Such fundamental operations enable the explanation of the entire architecture's functioning, with the principles of integrative biology by changing the scale (Chauvet 2006).

In this chapter, we try to establish a connection between the stochastic activity of meristems and the functioning of organ expansion through a variable that represents plant vigor. For development, the branching process increases with vigor, and the mortality of axes and organs decreases. For growth, the volume of organs is obviously linked to this vigor. Hence, it is essential to model the meristem activity and organ expansion as dependent on this vigor variable, which requires a formal definition to obtain a more mechanistic model.

The plant's architecture results from a morphogenetic program that is expressed in a specific environment. The same genotype will give rise to different phenotypes if the environment varies. Even with a highly homogeneous environment, it is impossible to generate identical individuals due to an inherent stochastic component (Chap. 6).

Over short growth periods (in the case of herbaceous plants), meristem activity can be considered stable, while over long growth periods (in the case of woody plants), significant variations occur. Notably, the volumes of the organs, dimensions of the growth units (GUs), branching, and mortality rates observed along vegetative axes evolve or are considerably modified in response to environment stress.

For instance, a beech tree growing under a canopy may only grow 1 cm per year, with short internodes (0.5 cm) and no secondary growth. It does not branch, and the stem meristem produces only a pair of phytomers. At 15 years of age, the tree is only 15 cm high. However, the same tree grown in an open field can reach a height of

15 m in the same period, with abundant branching and GUs that can exceed 1 m in length, being composed of more than 12 phytomers whose internodes can measure up to 10 cm, with strong secondary growth. This extreme case illustrates the enormous plasticity of the tree phenotypes from the same genotype in response to environmental effects.

The planting density also plays a crucial role in the development of a tree's architecture. When planted alone in open space, a tree will branch out and reiterate abundantly near its base, giving it a spherical shape and a large trunk diameter. In contrast, in a dense plantation, the tree will have a columnar shape, with a thin and pruned trunk.

To take account of time-dependent changes in meristem functioning, the Reference Axis of the AMAPsim model (Chap. 4) simulates the evolution of architecture development over long periods by progressively modifying GU functioning. Likewise, with the Botanical Automaton of the GreenLab model, the parameters of meristem activity can also evolve along the development axes. But these changes are forced, not the result of an internal feedback mechanism.

13.2 Introduction of Trophic Pressure

The mathematical formalism of the model (Chaps. 5 and 6) brings out in its equations a variable that plays a key role and appears as an integrative notion characterizing the course of growth: the ratio of biomass supply to demand, denoted *Q*/*D*. This ratio corresponds to a state variable of the model, and therefore offers the possibility of integrating a feedback effect of growth on meristem activity and thus on the development of architecture. The *Q*/*D* ratio can be considered as an "internal trophic pressure" exerted by biomass availability on meristems, directly influencing their functioning. This trophic pressure is a numerical expression of the qualitative notion of vigor presented at the beginning of the chapter. This choice, resulting from the model equations, is consistent with many observations reported on plants (Chaps. 7, 8, 9 and 10).

The main idea is to use the *Q*/*D* ratio as a regulatory variable that drives the functioning of meristems and organs. The evolution of *Q*/*D* value throughout plant growth often follows a sigmoid curve and is closely correlated with the observed evolution of the successive phenological stages during architecture development:

- The volume of the buds has a significant impact on the development of future shoots (Sabatier and Barthélémy 2001) and the size of the GUs, which controls the appearance of the lateral axes (Nicolini and Chanson 1999). However, carbon resources and the level of trophic competition in the plant partially control these volumes. The *Q*/*D* ratio plays a primordial role, and its evolution during tree growth strongly correlates with the evolution of GU morphology.
- The "base effect" is observed as branches gradually appear on the stem, becoming increasingly vigorous until the formation of reiterations. Branch life span gradually increases along the trunk from the base, and GUs are minimal in size at the

start and progressively lengthen along the trunk. These observations are consistent with the evolution of the *Q*/*D* ratio (whose value increases) in the juvenile growth phase.

- After a certain time, the growth stabilizes due to the saturation of production *Q* under the effect of the production surface *Sp* (the crown's shadow on the ground), reaching the limit *Sd* of available surface for the plant imposed by the planting density. Correspondingly, branching probability, GU size, and branch life span also stabilize, forming a crown that is approximately invariant in size, which stabilizes demand. This phase coincides with the stabilization of the *Q*/*D* ratio once the linear part of the sigmoidal growth is reached.
- Finally, senescence occurs, giving the crowns of an old tree a fractal appearance. The *Q*/*D* ratio decreases during this phase, which coincides with a complex form of senescence not discussed here.

Thus, the observations suggest a functional correlation between growth–development interactions and variations in the *Q*/*D* ratio. Therefore, utilizing internal trophic pressure as a regulatory variable to model the complex interaction loops between organogenesis, biomass production, and its distribution in structural–functional models of plant growth shows promise. However, this approach is hampered by the internal trophic pressure being a non-measurable quantity, which can only be assessed by estimating the model's parameters. An interesting aspect of the GreenLab model is that it reconstructs the history of demand and biomass supply by inverting the model from the static data of the plant architecture, and the quotient of these variables gives the history of the trophic pressure. Nevertheless, this methodology is not straightforward, especially when dealing with a stochastic model.

13.3 Methodology for the Use of Trophic Pressure

In the GreenLab Standard model, the evolution of demand is computed through iterations in an autonomous way, and environmental interactions cannot modify the trajectory of development. The growth is updated at each step of development without feedback between growth and development. The stochastic component (Chap. 6) does not affect the behavior of development (Fig. 13.1a).

Therefore, to provide greater flexibility and adapt the architecture development to environmental variations, a feedback loop between growth and development was introduced in the GreenLab model. This is done at each step of growth (Fig. 13.1b), resulting in the new version of GreenLab called "GreenLab Loop model".

The assumptions about development and source–sink functioning are retained from the Standard version. However, the parameters of the system are no longer the values of the probabilities and the sinks, but the coefficients of the functions depending on (*Q*/*D*) that allow computing them. This implies that the parameters of development and sinks are hidden in the organic series and must be retrieved by the weighted least squares method (WLSM).

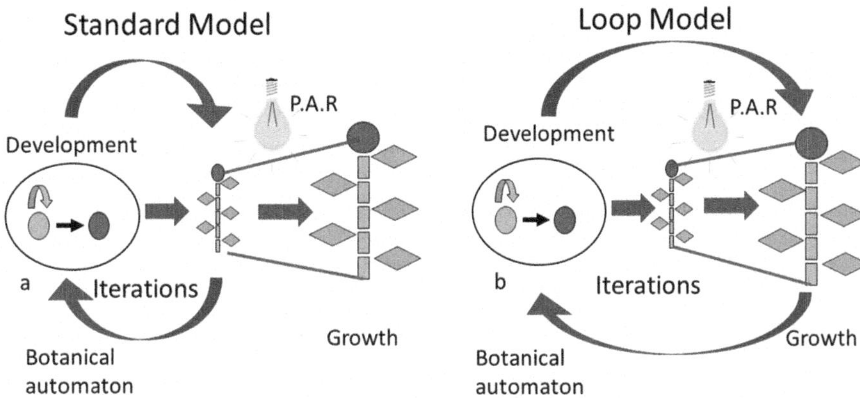

Fig. 13.1 (**a**) GreenLab Standard model with no feedback between growth and development. This one has a standalone functioning. At each step, growth is updated. (**b**) GreenLab Loop model introduces the feedback between growth and development. At each current step of functioning (DC), the growth induced by the photosynthetic active radiation (PAR) modifies the development of the next step

This chapter proposes to adapt the type of retroaction between growth and development by reflecting on the different types of development encountered in the previous chapters:

- Plants with continuous or rhythmic deterministic development.
- Plants with continuous or rhythmic stochastic development.

The aim is to propose a new formulation of the parameters of meristem activity and organ expansion, including their dependence on trophic pressure. This new formulation enhances the plasticity of the plant architecture in response to environmental variations.

To achieve this, the chapter places special emphasis on numerical simulation of plant growth in silico, which enables the study of various emergent properties of the GreenLab Loop model. The results obtained through these simulations are compared to classical observations in agronomic experiments. In a few available experimental cases, the WLSM allows calibration of the new source–sink parameters, thereby confirming the relevance of this new functional approach to plant growth.

13.4 Modeling of Deterministic Development with Feedback

Deterministic organogenesis is a unique phenomenon that is applied mainly to single-stemmed plants (maize, sunflower, etc.). However, it serves as a fundamental theoretical framework to understand the behavior of models with growth–development feedback. As demonstrated in Chap. 6, the findings from the deterministic case in Chap. 5 can be generalized to the stochastic case. The feedback between growth and development was initially analyzed deterministically on synchronous trees whose architecture is simple (architectural model of Rauh from Hallé et al. 1978). In this context, the GU is considered preformed and built within a single cycle (as in the Phyto_Facto mode in Chap. 10), with an annual time step. The botanical automaton generates GUs with acrotony according to the rules defined in Chap. 4. The Digiplante and QingYuan software tools have been used to investigate the model's behavior. The primary research findings have been published in several works by Mathieu et al. (2004, 2007a, b, 2008, 2009, 2012).

13.4.1 Assumptions for a Simplified Development–Growth Feedback Model

To express the dependence of various parameters on the *Q*/*D* variable, a simple linear formulation is used. The specified constraints are as follows:

- The production time of the GUs corresponds to 1 DC. All phytomers are made at the same time.
- The number of phytomers produced depends on *Q*/*D* at the time of GU initiation.
- The number of phytomers produced is bounded by an upper and a lower limit.
- The branching rate depends on *Q*/*D*.
- The maximal number of GUs of a branch is set as a function of *Q*/*D* at the time of the creation of the axillary meristem.

A simple linear formulation is chosen for expressing how each parameter depends on the *Q*/*D* variable:

$$par\left(\frac{Q}{D}\right) = A + B \cdot \frac{Q}{D} \tag{13.1}$$

Equation (13.1) gives a generic expression that can be applied to model the parameters in the system, such as the parameter *b* of the *Bernoulli* process, the branching rate *a*, or the viability *c*.

This formulation is based on a limited development (Taylor series) to order one of an unknown function that depend on the *Q*/*D* by considering higher-order terms negligible. As *Q*/*D* can only take positive values, the parameter *par* has a minimum bound, $par_{\min} = A$. We also impose a maximum limit $par_{\max}$ such that if $p(Q/D) > par_{\max}$, then we set $par(Q/D) = par_{\max}$. For example, the number of phytomers produced in a GU (n_Φ) can be expressed as:

$$n_\phi\left(\frac{Q}{D}\right) = n_a + n_b \cdot \frac{Q}{D} \tag{13.2}$$

Where n_a is the minimum number of phytomers per GU and n_b is a coefficient that increases n_Φ according to *Q*/*D*. We also modify the branching of the GUs according to the method of Chap. 4 by truncating the maximum GU from the top.

Similarly, we compute the limit number of GUs in a branch N_{ax} at the time of their creation using the equation:

$$N_{ax}\left(\frac{Q}{D}\right) = N_a + N_b \cdot \frac{Q}{D} \tag{13.3}$$

Where N_a is the minimum number of GU per axis and N_b is a coefficient that increases N_{ax} according to *Q*/*D*. This equation simulates the duration of functioning of the apical meristem before its abortion or transformation into a terminal inflorescence.

The functioning of the meristem is given by Eq. (13.1), which includes phytomer creation, branching, and mortality, and is triggered if the parameter *par* exceeds a threshold value.

The GUs are formed as a single block, and the size limit of the branches is determined at their birth. Furthermore, all axillary meristems of the same type are synchronized. Despite these constraints, we believe that our model provides a realistic simulation of growth–development interactions, and has produced results consistent with our empirical knowledge of the plasticity of tree architecture in a variable environment. This has been demonstrated using software packages such as Digiplante and QingYuan, which operate based on the principles of our model.

13.4.2 Example of a Simulation of the Growth–Development Interaction During the Growth

A tree with a branching structure of order 2 is chosen. The source–sink parameters of the organs are constant and identical on both the trunk and branches. The botanical automaton produces a maximum of four phytomers per GU. Each phytomer on the trunk can carry a branch that produces a maximum of five GUs, and branching only occurs if the computed branching rate exceeds a threshold value.

Figure 13.2 shows the progressive phases of the establishment of tree architecture, with a transient regime (base effect in the juvenile period) and a stationary regime in adulthood. These different regimes reproduce well the phenological stages of architecture during growth and depend on *Q*/*D* according to the equations used in the Standard model. As we said, it is possible to obtain the same effects with the standard deterministic model by assigning stress functions indexed to chronological age that constrain the developmental parameters to take the numerical values corresponding to the observations, but these effects are then forced.

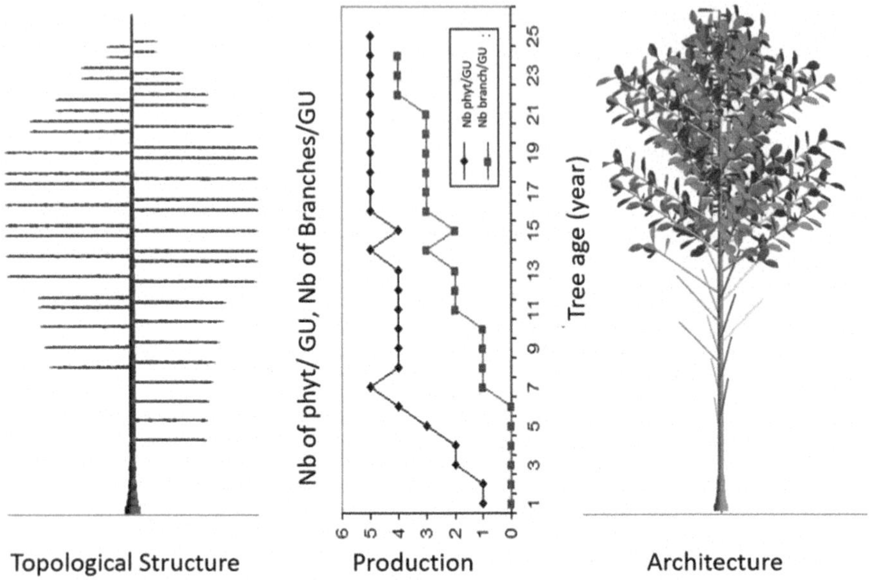

Fig. 13.2 Growth–development interactions in architecture. The first growth units (GUs) of the trunk have a minimum number of phytomers (here 1), and the branching probability is below the threshold. Then the GUs lengthen, branching is triggered, and the size of the branches increases until a limit is reached. Development is then stabilized (according to Mathieu, Ecole Centrale Paris, Digiplante software)

13.4.3 Occurrences of Reiterations

In the case of continuous growth, it is assumed that reiterations occur when the branching threshold is exceeded, while in the case of rhythmic growth, reiterations occur when phytomers distal to the GU that can carry reiterations are created, which occur above a certain threshold value of the *Q/D* ratio. The appearance of reiterations increases the demand and thus causes the *Q/D* ratio to fall below the threshold that triggers the reiterations. The reiterations can only appear again after a certain number of cycles, once *Q/D* exceeds the threshold again. This alternation of phases, in which *Q/D* successively takes values above and below the threshold for triggering reiterations, gives successive waves of reiterations (Fig. 13.3), as observed on the maple tree (Chap. 10).

13.4.4 Evolution of Branching in the Growth of Sympodial Systems

The number of axes produced by sympodial systems is exponential by nature, which leads to an exponential increase in demand. However, growth is limited by the

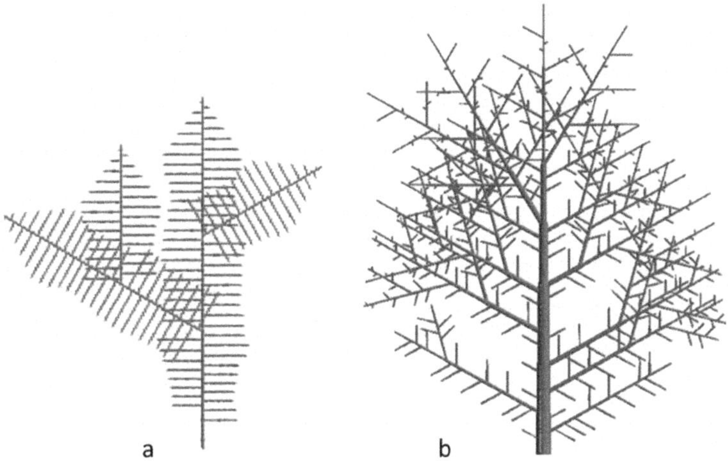

Fig. 13.3 Triggers for reiteration. (**a**) Case of continuous growth with two physiological ages. After a base effect, branches appear, followed by two waves of synchronous reiterations. The size of the branches decreases after these waves, because *Q*/*D* decreases. It takes some time for the second wave to appear, indicating that the branching probability has again exceeded the threshold. (**b**) Case of rhythmic growth. The tree has three physiological ages and reiterates on the first two. The growth unit (GU) lengthens during growth due to the increase in *Q*/*D* and eventually creates the phytomers that carry reiterations according to the acrotonic rule (Digiplante ECP software)

production surface (*Sp*). Without feedback between growth and development, the *Q*/*D* ratio decreases with increasing chronological age, resulting in a purely fractal appearance. To prevent excessive decrease in phytomer size, the tree often automatically reduces its branching rate by lowering *Q*/*D*, which brings the branching rate below the trigger level (Fig. 13.4). In the steady state, the demand is stabilized, and GUs produce only one terminal relay, resulting in a constant number of sympodial axes.

13.4.5 Influence of Planting Density on Growth–Development Feedback

Figure 13.5 shows the effect of planting density on tree development. The competition between neighboring trees is simulated by the available surface *Sd* per tree, considering the nearest neighbors. This surface limits the production *Q* per tree. A high tree density reduces the production *Q*, and therefore *Q*/*D*, which subsequently reduces the branching intensity and shortens the life span of the branches. As a result, short and high crowns are formed. The effect of pruning is limited, as the branches can hardly penetrate the neighboring crowns. This modeling approach is more efficient in simulation (memory and speed) than simulators that take the

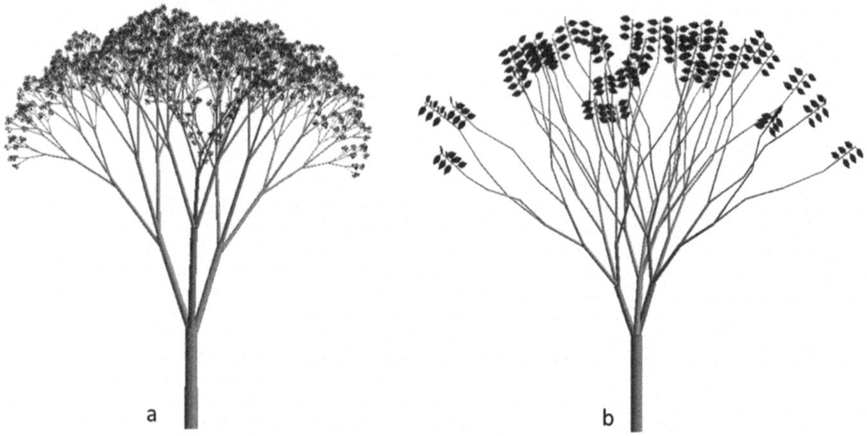

Fig. 13.4 Decreasing branching rate during the evolution of a sympodial system. (**a**) Fixed branching rate (three relay axes per growth unit [GU]). Limited growth leads to fractal behavior. (**b**) Branching rate depends on *Q*/*D*. The tree progressively transitions from three relay axes to two relay axes, then to one relay per GU (Ecole Centrale Paris, Digiplante software)

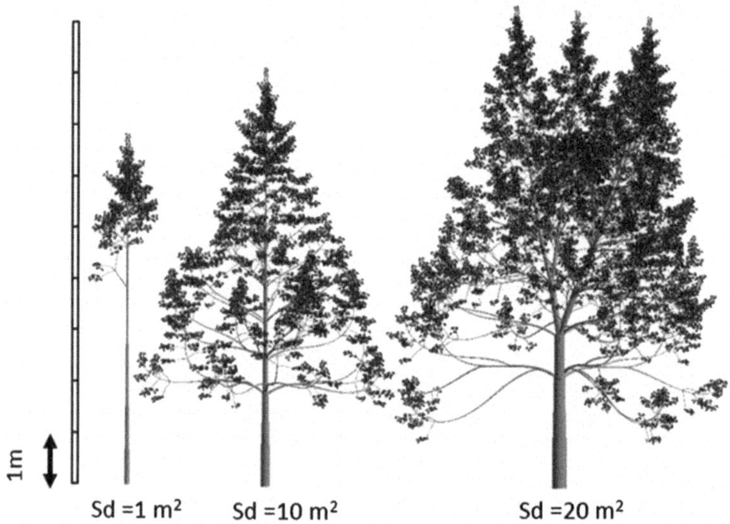

Fig. 13.5 Influence of planting density on crown development. The simulated tree is a Rauh architectural model with acrotony of the growth units (GUs) varying with *Q*/*D*. Reiterations can only occur if the available surface area *Sd* on the ground and per tree is sufficient. Conversely, a small surface area *Sd* results in narrow, heavily pruned crowns. The tree height is not very sensitive to the *Sd* parameter (Ecole Centrale Paris, Digiplante software)

approach of identifying branches to be pruned based on contacts between neighboring trees (Chap. 11). On the other hand, with low planting density, growth is stimulated and allows for high *Q*/*D*, resulting in the occurrence of reiterations.

However, it should be noted that density has little influence on tree height. If *Q*/*D* decreases, pruning increases, which subsequently reduces *D*. As a result, Eichorn's law is an emergent property in these simulations, stating that in a stand, the height of the dominant tree is minimally influenced by planting density.

13.4.6 Light Influence

Figure 13.6 shows the effect of light on growth–development feedback, illustrating the remarkable architectural plasticity that can be generated by this environmental parameter noted as *E* in the GreenLab model. The biomass produced (*Q*) is directly proportional to *E*. As the *Q*/*D* ratio increases, it has a direct effect on the intensity of development. Under low light level, such as in a densely shaded forest canopy, the *Q*/*D* is typically low, resulting in a tree that tends to vegetate and produce short GUs with only a few internodes. Such a tree would have little branches, if any. In contrast, a tree growing under optimal environmental conditions, with abundant light, will experience accelerated development, resulting in abundant branching or even reiteration. We can simulate the contrasting phenotypes obtained in a beech tree grown under two different conditions. A tree grown under a dense forest canopy will typically reach a height of only 15 cm in 15 years, with GUs consisting of only two internodes, and with internode lengths of 0.5 cm. In contrast, a tree grown

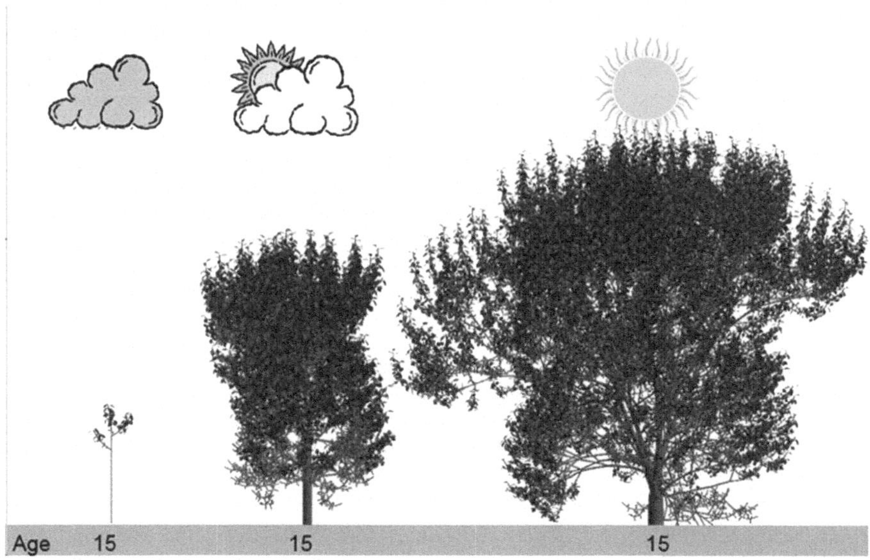

Fig. 13.6 Plasticity of tree architecture under the effect of light. The three simulated trees are identical in their functioning parameters and have the same age (15 years) but grow under different light conditions (factor *E* in GreenLab). The planting density here is not limiting (Digiplante software, Ecole Centrale Paris (ECP))

isolated in an open field with plenty of light can reach a height of 15 m, with GUs consisting of up to 12 internodes, and with internode lengths reaching up to 10 cm.

13.4.7 Interaction Between Growth and Fruiting

Fruiting introduces a major sink in the plant's demand. In terminal flowering plants like sunflowers, fruiting accounts for the entire demand toward the end of their growth, while in shrubs like coffee, the weight of berries on branches can be up to five times that of internodes and leaves. The growth–fruiting interaction plays a crucial role in altering the profile of organic series on the axes, with the volume of organs along the stem primarily depending on the fruit positions and the shape of their sink function, as discussed in Chap. 9 on cucumber.

While fruiting limits growth in most cases, it does not entirely halt it, as observed in coffee, beetroot, or cucumber, where it leads to a reduction in the expansion of vegetative organs. Physiological regulations induce abortions when the fruit load is too high, resulting in alternating fruitless and fruit-bearing zones on the stem. This occurs at the flowering stage in a cucumber or sweet pepper, where the newly formed and fertilized flowers may abort during expansion if the plant bears too many fruits. In fruit trees, abortions can occur during all stages of development due to an excessive fruit load, even in the meristems of the vegetative axes. Without these physiological regulations, the plant may die from sustaining the growth of too many fruits, leading to a decline in productive leaves. In coffee, many abortions are observed on the terminal meristems of branches when the fruit load is too high, a phenomenon known as "die back". In the *Caturra* variety of *Arabica* coffee, flowers must be removed at first flowering; otherwise, many vegetative meristems abort, and the plant may die.

We repeat hereafter a previous study (Mathieu et al. 2012) on a computational single-stemmed plant with lateral flowering. In this study, a flower can only give rise to fruit if it exceeds a threshold value T_f, expressed as:

$$f_1 + \frac{f_2 \cdot Q}{D} > T_f (\text{we normalize the value } T_f \text{ to } 1) \tag{13.4}$$

When fruit appears on a plant, it increases the demand for resources, which can cause the expression (13.4) to fall below a certain threshold, leading to the abortion of the next fruit. As a result, the plant can exhibit periodic alternations of phytomers with fruiting (F) and without fruiting (V). In Fig. 13.7, limited mode growth is used, with the functioning time of the leaves being 20 DCs, and the expansion time of the organs being 15 DCs. The leaves and internodes have the same constant sink function. Plant P1 is used as a control, with a constant sink function and no fruiting threshold, resulting in one fruit per phytomer. However, for plants P2, P3, and P4, the beta laws of the sink functions are modified by adjusting the shape of the sink

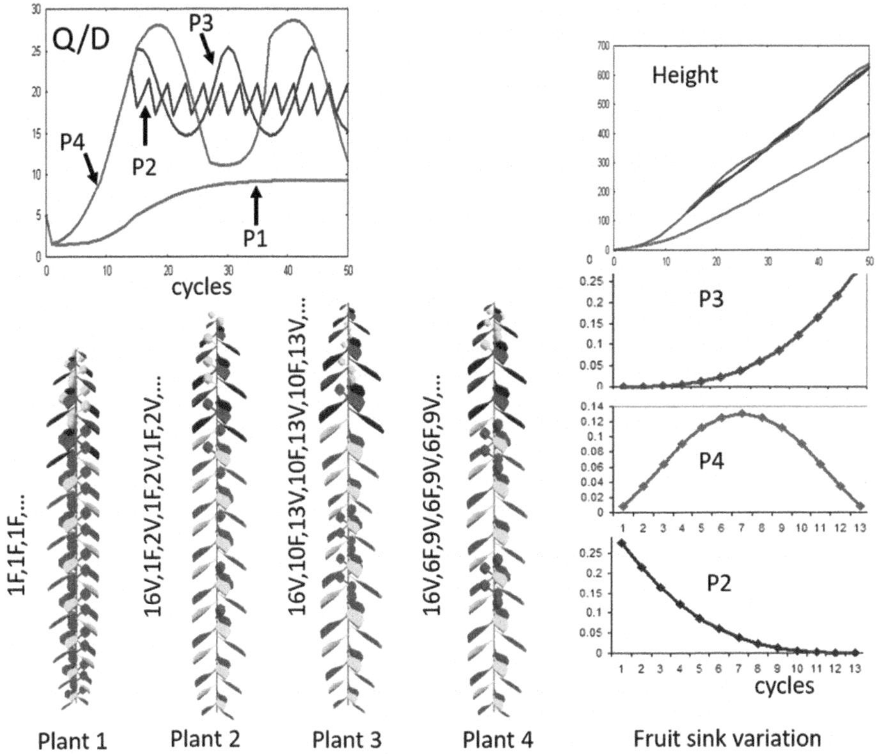

Fig. 13.7 Periodic alternations of fruiting (F) and non-fruiting (V) areas depending on sink functions in a single-stemmed computational plant. The lists with or without fruits are described in association with the simulated stems. The *Q*/*D* ratio oscillates, periodically crossing a threshold for fruiting. The first control plant produces fruits on each phytomer, while the other plants have periodic fruiting that depends on the fruiting threshold T_f and the shapes of the variations in sink functions (from Amélie Mathieu)

function through parameters that control its variation, while keeping the sink strength constant.

As expected, plant P1 is the smallest because of its higher fruit load that limits biomass allocation to leaves and thereby hinders future biomass production. For the other plants, simulation results reveal that a certain number of phytomers are necessary before the first fruit appears, which is also observed in real plants. Subsequently, periodic sequences of fruiting (F) and non-fruiting (V) zones are observed, and the nature of these sequences depends on the parameters of the fruit sink variation law.

In fruit trees, there is often an alternation between fruiting and non-fruiting years, which can be achieved by creating different *Q*/*D* thresholds for the functioning of branch and fruit meristems (Fig. 13.8). A high fruit load reduces the supply of biomass for the following year and can cause the abortion of vegetative buds.

Fig. 13.8 Simulation of alternating vegetative and fruiting periods in a computational branched tree. The crosses represent the abortion of axillary buds. The preformed growth units (GUs) of the current year can either produce branches or fruits, depending on the reserves stored at the end of the previous year (from Amélie Mathieu)

13.5 Modeling of Stochastic Development with Feedback

In Chap. 6, the standard stochastic model is presented, where probabilities are either constant or depend on functions indexed on chronological age in DCs, and this model is forced. In this case, demand does not depend on biomass supply. To compute the evolution of demand, which in turn allows the computation of biomass production, the parameters of meristem activity (*Bernoulli* process, viability, branching rate) must first be identified using the "plant crown analysis" (Chap. 3) from measurements on plant architecture.

In the stochastic model with feedback, the activity of the meristems depends on the trophic pressure (*Q*/*D* ratio) and, therefore, changes with each cycle. However, the similarity with the standard stochastic model is strong, and some model properties remain valid. The methods developed to identify its parameters can be adapted here. The organic series result from the stochastic activity of meristems associated with source–sink relationships. If the probabilities that control the meristem activity are expressed as a function depending on *Q*/*D*, then the parameters of this function can theoretically be computed by the inverse method on the organic series, as along with the parameters of the sink functions. As a result, the crown analysis may no more be necessary.

The model is once again based on the existence of a potential structure (Chap. 3) that contains all the possibilities of development, and which is only partially realized during a stochastic simulation depending on the activity of the meristems according to the values taken by *Q*/*D*.

13.5.1 Case of Continuous Growth

The GreenLab model highlights the important notion of free growth (Chaps. 5 and 6). This occurs when the whole leaf surface intercepts the light, corresponding to a very high value of *Sp*. The growth then follows an exponential or damped pattern, depending on whether we have:

$$\frac{(t_a+1)}{2}\cdot\frac{E}{A} > 1 \; or < 1.\text{It is balanced if } \frac{(t_a+1)}{2}\cdot\frac{E}{A} = 1.$$

It is important to note that t_a is here both the expansion time of organs and photosynthetic functioning time of leaves. Additionally, ε is the specific leaf weight (SLW, leaf thickness), $1/r$ is the resource use efficiency (RUE), while P_a, P_i, and P_f correspond to the constant sinks of leaves, internodes, and fruits, respectively. Assuming one leaf and one fruit per phytomer, we have:

$$A = \varepsilon \cdot r \cdot (P_a + P_i + P_f)$$

The formula $\frac{(t_a+1)}{2}\cdot\frac{E}{A}$ is important mainly for two reasons:

- The first is that it compactly links the source–sink parameters with deterministic or stochastic functioning to determine the type of growth behavior (exponential, damped, or constant). This formula is applicable irrespective of the architectural model and has been validated in the case of the Standard model, but it must also be operative in the case of the model with feedback (Loop model).
- The second is that the formula serves to verify the proper functioning of the Gloups software, especially by using the case $\frac{(t_a+1)}{2}\cdot\frac{E}{A} = 1$, which must produce constant growth for both the average of the simulated plants and the potential plant.

The following sections consider the cases of free growth and limited growth.

13.5.1.1 Stochastic Development with Feedback on Growth

In Chap. 3, we assume that meristem activity follows a *Bernoulli* process: a meristem can either produce a phytomer at each cycle i with probability b_i or be in a pause with probability $1 - b_i$. Although this variable is mostly stable, it can vary during the plant's life, particularly when the meristem ages (as in the case of coffee tree). Initially, we assume no mortality and branching failure, and the stochastic aspect is therefore solely concentrated on the *Bernoulli* process. We assume that b_i depends linearly on the *Q*/*D* ratio when it is low, and it tends toward a maximum (<1) when it is high. Consequently, we empirically select a mono-parametric homographic function:

$$b_i = \frac{K_b \cdot \left(\frac{Q_{i-1}}{D_i}\right)}{1 + K_b \cdot \left(\frac{Q_{i-1}}{D_i}\right)} \tag{13.5}$$

We write to simplify:

$$b_{\left(\frac{Q}{D}\right)} = \frac{K_b \cdot \left(\frac{Q}{D}\right)}{1 + K_b \cdot \left(\frac{Q}{D}\right)} \tag{13.6}$$

The parameter K_b is a constant used solely for testing the growth–development interactions under the effect of the trophic pressure in a particular case. Note that there could be other function forms that could produce the same curve shape, such as:

$$b_{\left(\frac{Q}{D}\right)} = 1 - \exp(-K_b \cdot (Q/D))$$

Let us assume, by recurrence from the seed, the demand of the organs in cycle $t - 1$.

13.5.1.1.1 Single-Stemmed Plant Case

In the case of a single-stemmed plant, and given the biomass available and the formulation of meristem activity, the updated demand at the plant's DC t can be written as follows (knowing that the phytomer includes an internode and a leaf, and they share the same sink function in this example):

$$D(t) = 2p \cdot \frac{K_b \cdot \frac{Q(t-1)}{D(t)}}{1 + K_b \cdot \frac{Q(t-1)}{D(t)}} + D_R(t-1) \tag{13.7}$$

P is the constant sink of the leaf organ or internode. With one leaf per phytomer, the sink of the phytomer is 2*P* during its expansion period. $D_R(t - 1)$ is the known residual demand of the axis organs at age $t - 1$ (after removal of organs that have exceeded their expansion time), and still a sink (biomass consumer) at age *t*.

The updated demand equation is second-degree equation, and the variable *D*(*t*) appears on both sides. Solving this equation gives the appropriate positive root, which is the value of *D*(*t*). The GreenLab model works by recurrence, and this equation ensures its proper functioning.

13.5.1.1.2 Case of Branched Structures

Equation (13.7) is a generalization of the computation for branched structures for all organ types *o* and physiological ages *p*. At any given age *t*, we can compute the residual demand of the phytomers of the structure at age $t - 1$, which are still active sinks. This is done using the list mode, which was defined in Chap. 4 and involves exploring the simulated topological structure.

In the standard stochastic version of the GreenLab model, all the existence probabilities of the phytomers in the potential structure are pre-computed (Chap. 6). The demand of the plant is then deduced from the sums of the existence probabilities of the phytomers and multiplying them by the respective sinks of their organs according to the following equation:

$$D(t) = \sum_{o,p} \left(\sum_{i=1}^{t} N_o^p(t - i + 1) \cdot P_o^p(i) \right) \qquad (13.8)$$

o is the type of organ; $N_o^p(t)$ is the number of new phytomers of physiological age *p* in the new cohort of the potential structure, given the stochastic functioning; and $P_o^p(i)$ is the sink of the phytomer organ. At age *t*, the supply $Q(t - 1)$ is known, while the demand *D*(*t*) is computed from formula (13.8). In this formula, the numbers of phytomers are determined by applying the Botanical Automaton with its probabilities, regardless of the available biomass supply.

But in the Loop model, the demand cannot be computed using the same approach as in the standard stochastic version. Instead, we start with the potential structure and step through it at each DC to determine the probability of whether a phytomer will or will not be set up, according to the ratio of *Q*/*D*.

At each transition from a cycle $t - 1$ to a cycle *t*, the following actions are performed:

- Eliminate organs that are older than their expansion duration.
- Compute the residual demand of the organs in the structure at age $t - 1$, which are still active sinks at cycle *t*.
- Add the demand of the possible new organs in number $N_o(t)$ in the new cohort set up by the meristems.

For simplicity, let us first assume that all organs have a constant sink P and that there is only one physiological age. Additionally, we assume that each expanding phytomer has a sink equal to $2P$ (one internode + one leaf).

When a new phytomer is created, there are two cases to distinguish: either it originates from an apical bud that extends the axis, or it develops from an axillary meristem that generates a new branch. In either case, the demand equation can be written as follows:

$$D(t) = \sum\nolimits_{s=1,N_o(t)} (\Omega(s) \cdot \Theta(s) \cdot 2P) + D_R(t-1) \tag{13.9}$$

Where $\Omega(s)$ represents the probability of existence of the phytomer of the axis carrying the child phytomer, and $\Theta(s)$ represents the probability of creating the new phytomer given by Eq. (13.6). $N_o(t)$ is the number of phytomers produced by the structure at cycle t and numbered in the associated list.

This equation, with the unknown $D(t)$ at cycle t, is solved numerically at each DC using the Newtown algorithm. Since the value of $Q(t-1)$ is known, the value of the parameter $b_{\left(\frac{Q}{D}\right)}$ can be deduced from Eq. (13.6), which then allows us to compute the phytomer production of the last created cohort.

For the theoretical computation of the demand, the probability of the phytomer's existence is stored in each botanical entity of a new cohort of the potential structure, as in the Standard model (Chap. 6). The sum of the probabilities of every botanical entity associated with the sinks gives the potential demand.

In the stochastic simulation, the number of new organs simulated in the new cohort at age t will be a binomial distribution sample ($N_o(t)$, $b_{\left(\frac{Q}{D}\right)}$). For each botanical entity (phytomer) in the new cohort, the value of $b\left(\frac{Q}{D}\right)$ is compared to a random number denoted *rand* (*rand* is between 0 and 1). If rand $< b\left(\frac{Q}{D}\right)$, the meristem will create a phytomer, otherwise, it will pause.

Therefore, we can compute the demand of a simulated or computed plant by following the steps described above.

13.5.1.2 Study on Growth–Development Feedback (Case of Free Growth)

Free growth is an interesting case because it allows testing the correct functioning of the simulation program with a simple mathematical formula: $\frac{(t_a+1)}{2} \cdot \frac{E}{A} = 1$.

13.5.1.2.1 Study of the Stochastic Model with Feedback in Free Growth

The plant simulation is set to operate in the free growth mode, which allows testing with the simple mathematical formula to ensure that the simulation program is functioning correctly. In the standard stochastic version of the GreenLab model

(Chap. 6), it was observed that the parameter b of the *Bernoulli* process has no effect on the conditions for obtaining an equilibrium in production. We confirm that this remains true when b depends on Q/D. Moreover, for the following text, the parameter b is supposed to be identical for all physiological ages of meristems.

In addition, the organs are assumed to have identical functioning and expansion time ($t_a = 10$ DCs), and the sink value is constant ($P = 1$). The resistance is set at $r = 55$, leaf thickness at $\varepsilon = 0.05$, and branch lifetime at $t_2 = 20$ DCs. The environment factor E is normalized to $E = 1$, and Sp is set to a very large value (free growth). These parameters verify the equilibrium growth condition of Chaps. 5 and 6 for the standard model: $\frac{(t_a+1)}{2} \cdot \frac{E}{A} = 1$. The value of K_b (Eq. 13.5) is set to 500.

A simulation of 20 random plants over 40 DCs is carried out. Figure 13.9 shows the evolution of biomass Q, demand D, Q/D ratio, and *Bernoulli* parameter b. The four curves stabilize, and the parameter b becomes constant ($b \approx 0.72$). The functioning of the plant becomes like that of the Standard model.

13.5.1.2.1.1 Inverse Method for Computing Development and Growth Parameters: Case of Bernoulli *Process*

The source–sink parameters and the b parameter depend on the Q/D ratio, which is implicitly contained in the organic series. In order to reconstruct the theoretical organic series from the potential structure, Eq. (13.8) in Chap. 6 must be used:

Fig. 13.9 Free growth of a branching structure (Roux model) over $T = 40$ DCs, with a *Bernoulli* parameter b depending on the trophic pressure: $b_{\left(\frac{Q}{D}\right)} = \frac{K_b \cdot \left(\frac{Q}{D}\right)}{1+K_b \cdot \left(\frac{Q}{D}\right)}$. The life span of the branches is $t_2 = 20$ DCs, and the growth and biomass reach equilibrium beyond $T > t_2$. (**a**) Evolution of the demand D. (**b**) Evolution of the produced biomass Q. (**c**) Evolution of the ratio Q/D. (**d**) Evolution of the *Bernoulli* parameter b, stabilizes at the value $b = 0.72$. (**e, f**) Two random simulations of the structure

$$\overline{q_o(k,t)_{\text{topo}}} \rightarrow \frac{\sum_{x=0}^{t-k+1} \binom{k+x}{x} \cdot b^k \cdot (1-b)^x \cdot q_o^\theta(k+x,t)}{1 - \sum_{x=t-k}^{\infty} \binom{k+x}{x} \cdot b^k \cdot (1-b)^x} \tag{13.10}$$

This formula is only valid if the parameter b is constant, which is not the case if it depends on Q/D. However, if the ratio Q/D stabilizes, the formula can be applied, but only from the top of the organic series. Thus, it is necessary to truncate the organic series from the base until the moment of stabilization is reached, and the value of parameter b is the one reached at that moment.

Now the source–sink parameters from the previous section (free growth) are used with an unlimited t_2 branch expansion. To obtain the average organic series of stems and branches, a simulation of 50 random plants at age $T = 40$ DCs is carried out. It is observed that the b parameter stabilizes at $b = 0.7$ from 25 DCs, which is kept for Eq. (13.10). Therefore, the simulated organic series are adjusted by the model on the 15 phytomers from the top.

The least squares method (WLSM in Chap. 9) can be used to correctly retrieve the values of K_b, r, and P. Table 13.1 compares the "true" parameters used for random simulations with those estimated from the WLSM. It is observed that the development parameter K_b and the source–sink parameters r and P are well retrieved by the WLSM from the fitting of the simulated and computed organic series.

Figure 13.10 shows the fitting of the computed organic series with the average of 30 random organic series. Equation (13.10) is effective due to the stabilization of the *Bernoulli* parameter. However, it is important to note that this stabilization is only applicable to the terminals of the organic series.

13.5.1.2.1.2 Estimation of Development and Growth Parameters: A Complete Case Study

To obtain a more realistic development, it is necessary to include branching and mortality in the *Bernoulli* process. The source–sink parameters from the previous section (free growth) are used, along with non-limiting t_2 branch development ($\varepsilon = 0.05$, $t_a = 10$, $t_x = 10, \ldots, r = 55, P = 1, \ldots$).

The development probabilities are assumed to evolve in a homographic form depending on Q/D according to a function of Eq. (13.5). This defines three parameters that control the development: K_a, K_b, K_c. The following values are set: $K_a = 250$, $K_b = 500$, $K_c = 2000$.

K_a controls the branching rate $a_{(Q/D)}$ according to the function:

Table 13.1 Parameters used for the plant simulation and recomputed parameters from the WLSM

	K_b	r	P
Param simul	500	55	1
Param comput	563.6	54.6	1.02
Standard deviation (SD)	7.1	0.08	0.01
Coefficient of Variation (CV)	1.2	0.12	0.8

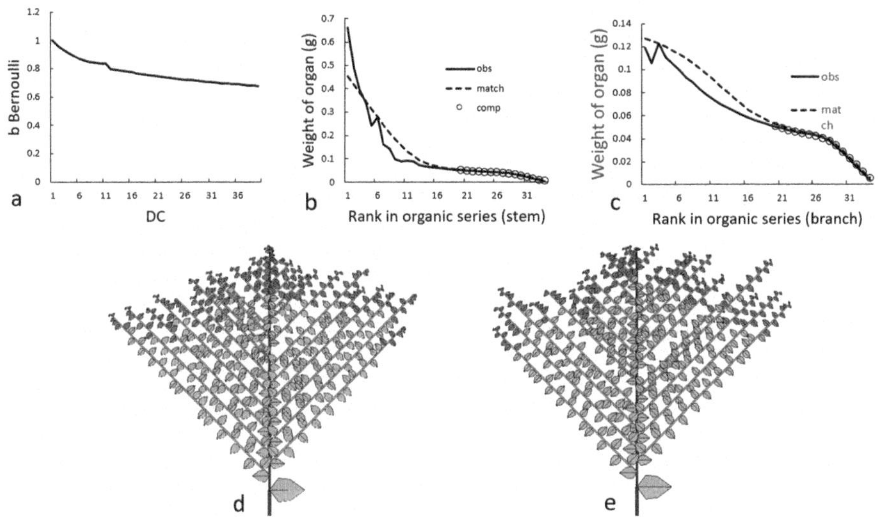

Fig. 13.10 Computation of the parameters K_b, *r*, *P* from the simulated organic series of stems and branches. (**a**) Evolution of the *Bernoulli* parameter depending on *Q*/*D*. (**b**) Solid line indicates the values from the simulation. (**c**) Dotted line indicates the values obtained from the computation using WLSM. In the beginning, the differences between the two lines come from the variation of *b*. However, toward the end, the simulated and calculated organic series converge as *b* stabilizes at 0.7. The WLSM method used to generate Table 13.1 is only applied to the last 15 phytomers (o). (**d**, **e**) Two random plant simulations at 40 DCs

$$a\left(\frac{Q}{D}\right) = \frac{K_a \cdot \left(\frac{Q}{D}\right)}{1 + K_a \cdot \left(\frac{Q}{D}\right)} \tag{13.11}$$

K_b controls the development rate (*Bernoulli*) $b_{(Q/D)}$ according to the function:

$$b_{\left(\frac{Q}{D}\right)} = \frac{K_b \cdot \left(\frac{Q}{D}\right)}{1 + K_b \cdot \left(\frac{Q}{D}\right)} \tag{13.12}$$

K_c controls the viability rate $c_{(Q/D)}$ according to the function:

$$c_{\left(\frac{Q}{D}\right)} = \frac{K_c \cdot \left(\frac{Q}{D}\right)}{1 + K_c \cdot \left(\frac{Q}{D}\right)} \tag{13.13}$$

To validate the mode, we simulate 50 random plants at 40 DCs to generate the organic series that will serve as the target for the model.

Figure 13.11 shows the evolutions of demand and biomass production over time (DC). The demand stabilizes as a result of the mortality of the branch meristems. In free growth conditions, the biomass production also stabilizes due to feedback, since: $\frac{(t_a+1)}{2} \cdot \frac{E}{A} = 1$.

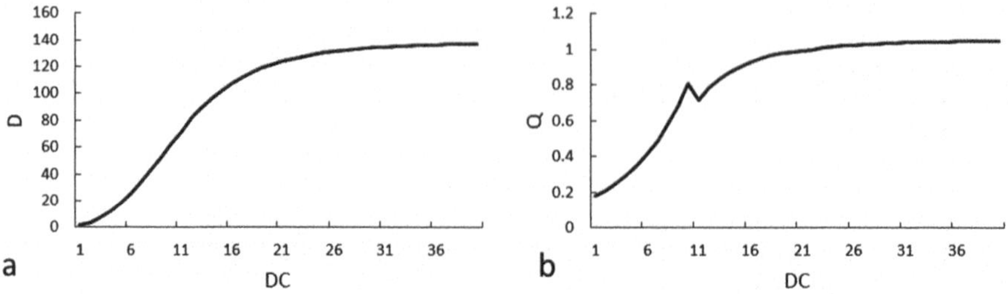

Fig. 13.11 Evolutions of demand and biomass produced as a function of time (DC). (**a**) Evolution of the demand *D* during the growth. (**b**) Evolution of biomass production *Q* during free growth

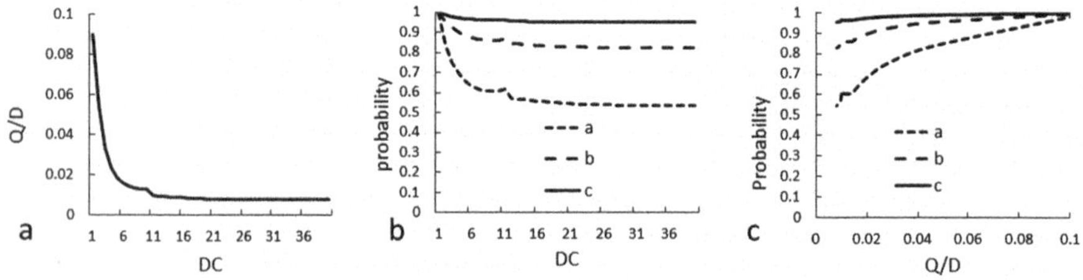

Fig. 13.12 (**a**) Evolution of the *Q*/*D* ratio during growth. (**b**) Evolution of these probabilities during growth. (**c**) Relationships between branching, development, and viability rates of meristems (*a*, *b*, *c*) as a function of *Q*/*D*

As a result, the *Q*/*D* ratio also stabilizes during the growth.

Figure 13.12a shows the evolution of the *Q*/*D* trophic pressure during 40 DCs of plant growth. This variable stabilizes rapidly. Similarly, Fig. 13.12b shows that the functioning probabilities attached to the meristems also stabilize. Figure 13.12c displays the values taken by functions of Eqs. (13.11, 13.12, and 13.13), which compute the probabilities associated with meristem activity as a function of the values taken by the *Q*/*D* ratio during growth, ranging between 0.005 and 0.1.

The branching rate stabilizes at $a = 0.60$, while the development rate (*Bernoulli*) stabilizes at $b = 0.82$ and the viability probability stabilizes at $c = 0.94$, respectively. The WLSM optimizes the five parameters K_a, K_b, K_c, r, P, so that the organic series computed from the potential structure best fit the simulated organic series (Fig. 13.13). The stabilized value $b = 0.82$ is used to compute the theoretical organic series from the potential structure.

The source–sink coefficients are well recomputed from the simulated organic series (Table 13.2). Additionally, the K_a, K_b, K_c coefficients are also accurately recovered by the WLSM, within their order of magnitude; this demonstrates that the stochastic functioning of meristems is intricately imprinted in the organic series and can be extracted without the need for crown analysis.

The standard deviations of the variables provided by WLSM are only indicative, as the conditions for BLUE (Best Linear Unbiased Estimator) estimators are not strictly met in cases where the variables are not statistically independent.

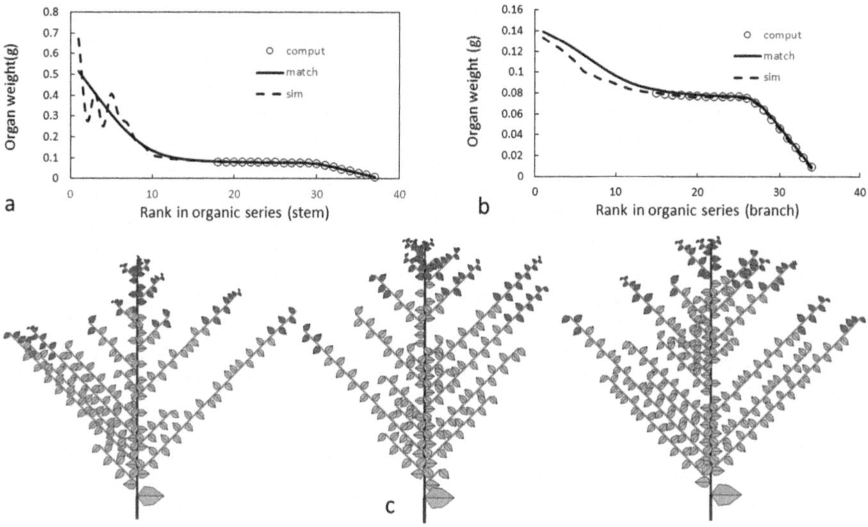

Fig. 13.13 The simulated organic series of stems and branches computed using the parameters K_a, K_b, K_c, r, P. (**a**) Organic series of the stem. (**b**) Organic series of the branch: --, simulated data: —, computed data that matches the simulations without optimization; and o, optimized data from the WLSM at the top of the organic series. The WLSM used to give Table 13.2 is applied only to the last 15 phytomers scored (o). (**c**) Three random plant simulations at 40 DCs, and the effects of branching and mortality are clearly visible

Table 13.2 Parameters used for the plant simulation in free growth and recomputed parameters from WLSM

Param	K_a	K_b	K_c	r	P
Simul	250	500	2000	55	1
Comput	204	595	2079	54.8	1.02
SD	13	18	159	0.05	0
CV (%)	6	3.6	8	0.01	0.05

We compared the performance of the Loop model, using stabilized values for the probabilities $a = 0.6$, $b = 0.82$, and $c = 0.94$ computed from the Q/D trophic pressure, with the Standard model.

At the 40th DC, the Loop model produced 33 stem phytomers and 198 branch phytomers, while the Standard model produced 33 stem phytomers and 162 branch phytomers.

The Loop model (with trophic pressure) has a decrease in the probabilities of meristem activity until stabilization, leading to a higher production of phytomers compared to the Standard model. As a result, computational plants generated using the Loop model are more abundant at the base. Although the organic series generated by both models are similar, the Standard model produces slightly larger phytomers due to lower demand (Fig. 13.14).

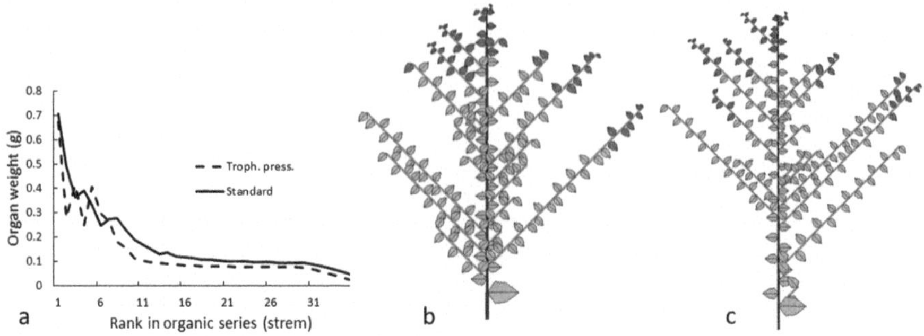

Fig. 13.14 Comparison of plants simulated with trophic pressure and those simulated by the Standard model for stabilized values: $a = 0.6$; $b = 0.82$; $c = 0.94$. (**a**) Organic series. (**b**) Plant simulated with trophic pressure. (**c**) Plant simulated by the Standard model

13.5.1.3 Study on Growth–Development Feedback (Case of Limited Growth)

This case is more accurate representative of plant growth and highlights the base effect, where branches gradually appear on the stem and their initial short life span increases to a medium value. Eventually, demand stabilizes due to branch mortality, and biomass production also stabilizes as the production area *Sp* limits growth.

The source–sink parameters used in this example, assuming non-limiting t_2 branch development, are $\varepsilon = 0.05$, $t_a = 10$, $t_x = 10$, $r = 30$, $Sp = 1000\ \text{cm}^2$, and $P = 1$. The first phytomer of the stem has no branches. The sigmoid dependence of meristem probabilities on Q/D is controlled by three parameters: $K_a = 50$, $K_b = 100$, $K_c = 200$. The functioning probabilities $a_{(Q/D)}$, $b_{(Q/D)}$, and $c_{(Q/D)}$ are computed from Eqs. (13.11, 13.12, and 13.13).

To generate the organic series, 50 random plants at 50 DCs are used as a target for the model.

Figure 13.15 shows the evolutions of demand and biomass production over time (DC). The demand stabilizes over time due to the mortality of the branch meristems. In limited growth conditions caused by *Sp*, feedback causes the biomass production to also stabilize.

In the early stages, the Q/D ratio is low, resulting in a low intensity of branching and increased branch mortality, which establishes the commonly observed base effect in young plants. However, the functioning rates of meristems stabilize beyond the 20th DC, with $a_{(Q/D)} \approx 0.89$, $b_{(Q/D)} \approx 0.91$, and $c_{(Q/D)} \approx 0.94$. At this stage, the Loop model functions like the Standard model, and the organic series from the potential structure can be processed with the negative binomial distribution (c.f. *Bernoulli* process) (Fig. 13.16).

The source–sink coefficients (r, Sp, P) are well recomputed from the simulated organic series (Table 13.3). The K_a, K_b, K_c, coefficients, which are of the same order of magnitude as those for free growth, were also recovered by the WLSM. This

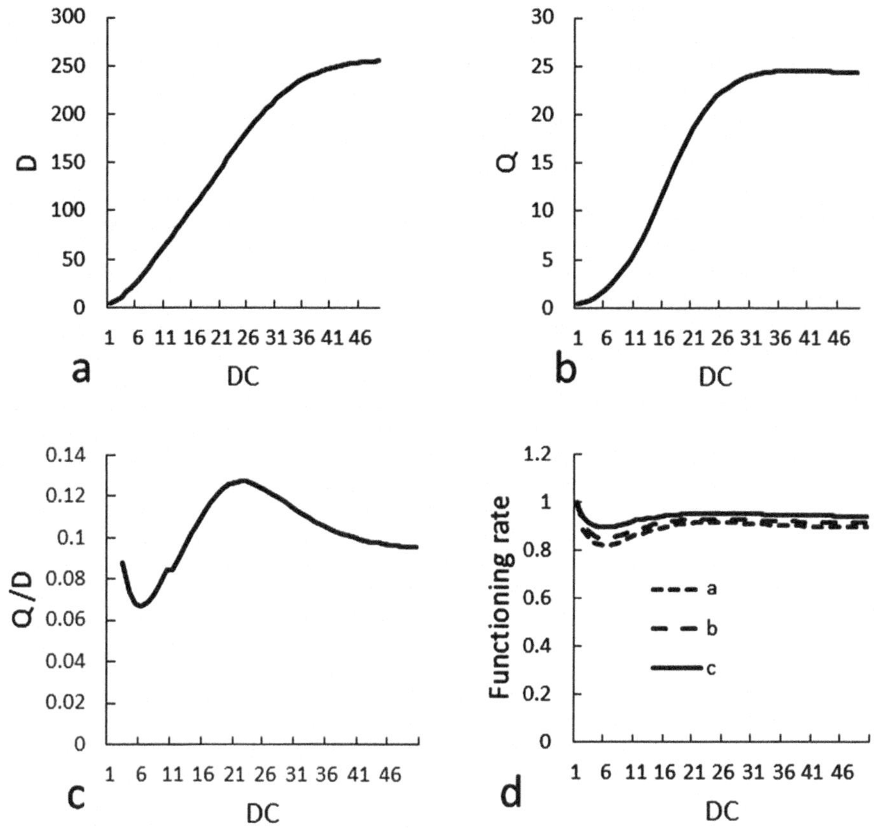

Fig. 13.15 Evolutions of demand, biomass production, trophic pressure *Q*/*D*, and the functioning rates of meristems over time (DC). (**a**) Evolution of the demand *D* during growth. (**b**) Evolution of biomass production *Q* during limited growth. (**c**) Evolution of trophic pressure *Q*/*D* during limited growth. (**d**) Evolution of the functioning rates of meristems for branching (*a*), development (*b*), and viability (*c*). After the 20 DCs, these rates tend to stabilize

confirms that the stochastic functioning of meristems is clearly reflected in the organic series and can be extracted without crown analysis.

13.5.1.4 Pruning Effect on Plant Development

The purpose of this paragraph is to compare the growth of two virtual plants, one single-stemmed and one branched, which share the same functional parameters defined in the previous paragraph. The single-stemmed model was created by pruning the axillary meristems at their creation in the branched model. Observations made on the pruning in coffee trees (Chap. 10) and cotton plants (Chap. 3) have shown that pruning not only increases the expansion of stem organs but also

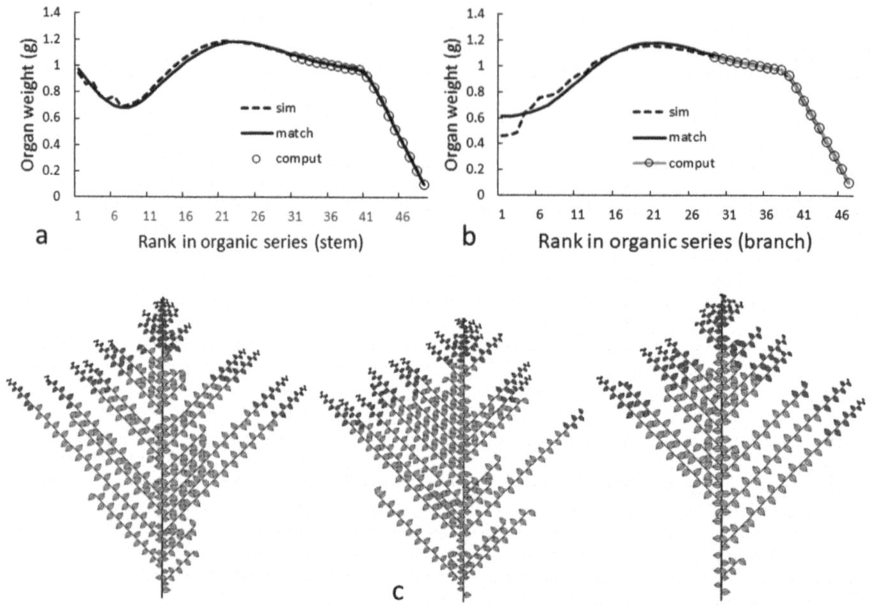

Fig. 13.16 The simulated organic series of stems and branches from the computation of the parameters K_a, K_b, K_c, r, Sp, P. (**a**) Organic series of the stem. (**b**) Organic series of the branch: "--" represents the simulated data, and "—" represents the computed data that match the simulations without optimization; "o" represents the optimized data obtained from the WLSM applied only to the top 15 phytomers scored (o) in the organic series, which is presented in Table 13.3. (**c**) Three random plant simulations at the 50th DCs. The effects of branching and mortality are clearly visible

Table 13.3 Parameters used for the plant simulation in limited growth and recomputed parameters from WLSM

Param	K_b	K_a	K_c	r	Sp	P
Simul	100	50	200	30	1000	1
WLSM	111	80	180	30.04	1007	1.01
SD	3.8	26.2	8.9	0.2	16.2	0
CV (%)	0.03	0.33	0.05	0.01	0.02	0.00

regularizes development (Chaps. 5 and 10). Therefore, removing axillary meristems of some shrubs to make them single-stemmed has a regulating effect. Conversely, if the branches are allowed to develop, the stems will exhibit irregular development, and the number of phytomers produced over this period will follow a binomial law (Chap. 3).

Thus, for several species of coffee trees (Chap. 10), the *Bernoulli*'s parameter b of the stem meristem is higher (on average $b = 0.89$) for single-stemmed plants than for branched stems ($b = 0.78$). A similar trend was obtained for cotton plants, where single-stemmed plants have a higher value ($b = 0.88$) compared to branched plants ($b = 0.80$). One possible explanation for this observation is that the removal of axillary meristems decreases the demand and trophic competition, allowing for organ expansion and an increase in the value of *Bernoulli*'s *b* parameter, as discussed in Chap. 6.

The aim here is to test the ability of the GreenLab Loop model to simulate this pruning effect. The removal of axillary meristems should trigger the double effect as observed in reality: a reduction in the pauses of the terminal meristem and an increase of the organic series of the single stem.

In our example, the numerical value of b_i as a function of Q/D is expressed as Eq. (13.4):

$$b_i = \frac{500 \cdot \left(\frac{Q_{i-1}}{D_i}\right)}{1 + 500 \cdot \left(\frac{Q_{i-1}}{D_i}\right)} \tag{13.14}$$

Here, it is assumed that the functioning duration of the organs is equal to their expansion duration, which is set at $t_a = 10$ DCs. The value of the sinks is also set at a constant ($P = 1$), while the resistance is set at $r = 55$, and the thickness of the leaf at $\varepsilon = 0.05$. K_b is assigned a value of 500, and branch development is considered unlimited. Since *Sp* takes a large value, the free growth case applies, with: $\frac{(t_a+1)}{2} \cdot \frac{E}{A} = 1$.

To investigate the pruning effect, 30 stochastic simulations are carried out at the age of 30 DCs, and the average values of b_i per DC are computed.

The b_i parameter rapidly decreases and stabilizes in both pruned and branched plants and stabilizes at $b = 0.94$ and $b = 0.70$, respectively (Fig. 13.17a, b). The removal of axillary meristems leads to a significant increase in the *Bernoulli* parameter (almost to the point where the development becomes deterministic) and the size of the organs. In contrast, the stochastic behavior of branched plants may be attributed to a decrease in the biomass supply and available trophic pressure per individual meristem (*Q*/*D*).

This study also shows that the gigantism of the organs resulting from the removal of axillary meristems is consistent with observed phenomena in various plants (Chap. 5, Fig. 5.27) and is reproduced in the stochastic mode of the Loop model.

13.5.1.5 Fruiting Effect on Plant Development and Growth

In the stochastic case, the function $f_{\left(\frac{Q}{D}\right)} = \frac{K_f \cdot \left(\frac{Q}{D}\right)}{1+K_f\left(\frac{Q}{D}\right)}$ represents the probability of triggering fruit expansion at its birth (equivalent to a non-self-pruning) as a function of trophic pressure. This demand for the fruit is expressed as the product of this probability multiplied by the sink strength.

This section explores the possible combinations between fruiting and growth, free or limited. For simplicity, it is assumed here that fruiting does not affect the *Bernoulli* process that controls the activity of vegetative meristems ($b = 1$). The stochastic aspect only concerns here the presence of the fruits.

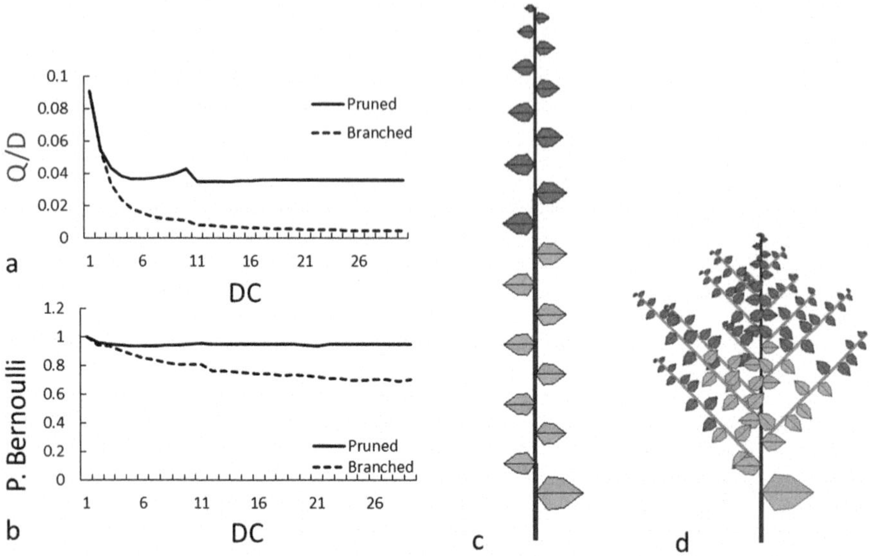

Fig. 13.17 Growth–development feedback in free growth: two stochastic simulations, drawn from a population of 30 simulated plants, of the growth over 30 DCs of a branched (**c**) and pruned (**d**) plant, respectively. The pauses are observed to be almost absent in the pruned plants. (**a**) Evolution of the average *Q*/*D* ratio during the growth of branched and pruned plants. This ratio stabilizes because: $\frac{(t_a+1)}{2}\frac{E}{A}=1$ and is ten times lower for the branched plant. (**b**) Evolution of the average *Bernoulli* parameter *b*, which also stabilizes during growth of branched and pruned plants. For pruned plants, *b* is close to 1. Simulated plants (**d**, **c**) (at the same scale) show the gigantism of organs produced by axillary meristems' ablation

13.5.1.5.1 Case of the Free Growth of a Single-Stemmed Plant

In this example of free growth mode, the growth parameters are:

$$Q_0=1,\ \ E=1,\ \ r=55,\ \ \varepsilon=0.05,\ \ t_x=t_a=10\ (Sp\ \text{ very large}).$$

The sinks for leaves, internodes, and fruits are constant and equal to $P_a=1$, $P_i=0.5$, and $P_f=1$ during the expansion period. However, the system is unbalanced because the expression $\frac{(t_a+1)}{2}\cdot\frac{E}{A}=1$ is not verified. If all the phytomers bore a fruit (K_f very large), the growth would be damped and eventually stop. To achieve constant growth, the sink fruit value for fruit (P_f) should be 0.5 to satisfy $\frac{(t_a+1)}{2}\cdot\frac{E}{A}=1$. In the stochastic case, if $f(Q/D)$ is the probability of a fruit appearing on a phytomer, then growth can be stable if the value of the product $P_f\cdot f(Q/D)=0.5$ (here we have $P_f=1$, so $f(Q/D)=0.5$). On the other hand, without any fruit ($K_f=0$), the growth would be exponentially accelerated.

Fifty random simulations are run over a period of 100 DCs, with the values of K_f as: [0, 5, 10, 20, 40, 80, 160, 9999]. Figure 13.18 displays the averages of the

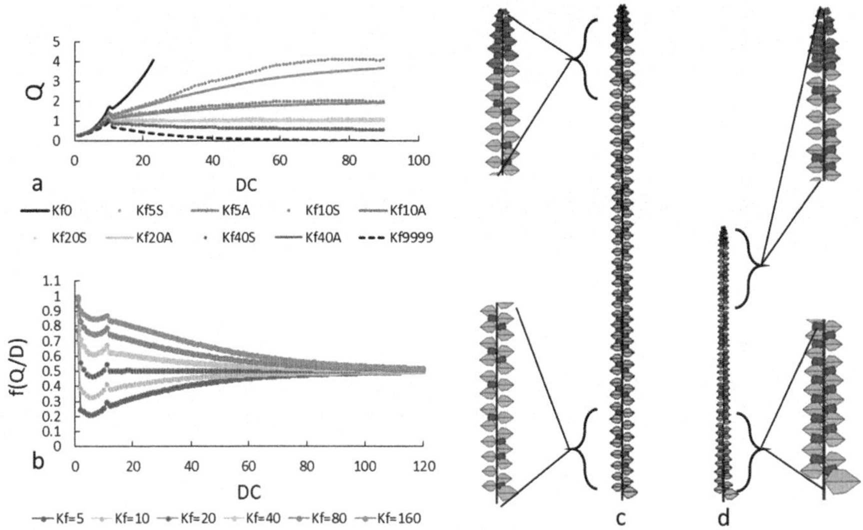

Fig. 13.18 Simulations of plants for different K_f [0, 5, 10, 20, 40, 80, 160, 9999]. The probability $f(Q/D)$ tends to the value 0.5 in an increasing or decreasing manner (**b**), and the biomass production stabilizes (**a**). For $K_f = 0$ ($P_f = 0$), the growth is exponential because $\frac{(t_a+1)}{2}\frac{E}{A} > 1$. For $K_f = 9999$ ($P_f = 1$), the growth is damped because $\frac{(t_a+1)}{2}\frac{E}{A} < 1$. The plant c ($K_f = 10$) is bigger than the plant d ($K_f = 80$) because the fruit load is low at the base

stochastic simulations and the potential structure calculations that give the output Q and the probability f with respect to the values of K_f (Fig. 13.18a).

Up to ten DCs, all growth trajectories are identical (see Chap. 5). We can observe the exponential growth of Q for the value $K_f = 0$ and the damped growth for $K_f = 9999$.

For the intermediate values of K_f, we can see that the production Q stabilizes and that the probability $f(Q/D)$ tends toward the value $f(Q/D) = 0.5$ (Fig. 13.18b).

This phenomenon can be understood by analyzing the dynamics of the different variables: according to the value of K_f,

(i) The biomass production increases or decreases.
(ii) The ratio Q/D will increase or decrease.
(iii) The value of the probability function $f_{\left(\frac{Q}{D}\right)} = \frac{K_f \cdot \left(\frac{Q}{D}\right)}{1+K_f \cdot \left(\frac{Q}{D}\right)}$ will approach the value $f_{\left(\frac{Q}{D}\right)} = 0.5$.

Once this value is reached, the free growth reaches equilibrium because the expression $\frac{(t_a+1)}{2} \cdot \frac{E}{A} = 1$ is then verified (we recall that $A = \varepsilon \cdot r \cdot (P_a + P_i + f \cdot P_f) = 0.05 \cdot 55 \cdot (1 + 0.5 + 0.5)$ and $t_a = 10$). The free growth becomes stable, and the demand in all cases reaches the value: $D = t_x \cdot \varepsilon \cdot r \cdot (P_a + P_i + f \cdot P_f) = 20$.

Starting with $K_f = 10$, the plant has low fruit production at the base and will experience accelerated growth until the value $f(Q/D) = 0.5$ is reached. Starting with $K_f = 80$, the plant has high fruit production at the base, and the growth will slow down until the value $f(Q/D) = 0.5$ is reached. As a result, the first plant obtained (Fig. 13.18c) is tall, and the second is small (Fig. 13.18d). At equilibrium, 50% of the phytomers bear fruit due to feedback between growth and demand. A notable difference between Fig. 13.7 (deterministic growth driven by threshold) and 12.18 (stochastic growth) is periodic fruit series in the former and stochastic fruit series in the latter.

13.5.1.5.2 Case of the Limited Growth of a Single-Stemmed Plant That Fruits

In this paragraph, we use the same parameters as in the previous paragraph except for $r = 25$, $Sp = 1000$, and $K_f = 5$ to perform 50 random simulations, which are averaged. The system is simpler than that of free growth because demand necessarily saturates due to the limited expansion time of the organs, and biomass supply is limited by the value of Sp (Fig. 13.19a). As a result, the Q/D ratio tends to a limit, and the function $f_{\left(\frac{Q}{D}\right)} = \frac{K_f \cdot \left(\frac{Q}{D}\right)}{1+K_f\left(\frac{Q}{D}\right)}$ stabilizes, giving the expected percentage of phytomers bearing fruit (Fig. 13.19b). The matching between the simulated organic series and those from the potential structure is good (Fig. 13.19c–e). The average of the 50 simulations allows us to appreciate the stochastic effect of the function $f_{\left(\frac{Q}{D}\right)}$. Three simulations of the plant are displayed (Fig. 13.19f), showing the bottom-up increase in the appearance of fruits.

13.5.1.6 Study of the Fruiting of a Branched Plant with Continuous Growth

For a branched plant, free growth can be balanced under the condition $\frac{(t_a+1)}{2} \cdot \frac{E}{A} = 1$if the number of phytomers per branch is limited because the demand must stabilize (Chap. 5). Under the standard stochastic condition (Chap. 6), it is sufficient that the branch meristems have a limited lifetime because branch mortality stabilizes the number of functional phytomers produced. Here we study the feedback induced by fruiting on branch development.

13.5.1.6.1 Case of the Free Growth of a Branched Plant That Fruits

We are interested here in studying shrubs, such as the coffee tree, which bear fruits on their stems and branches. These species often exhibit a phenomenon called “die back,” characterized by the abortion of branch meristems, which is linked to an

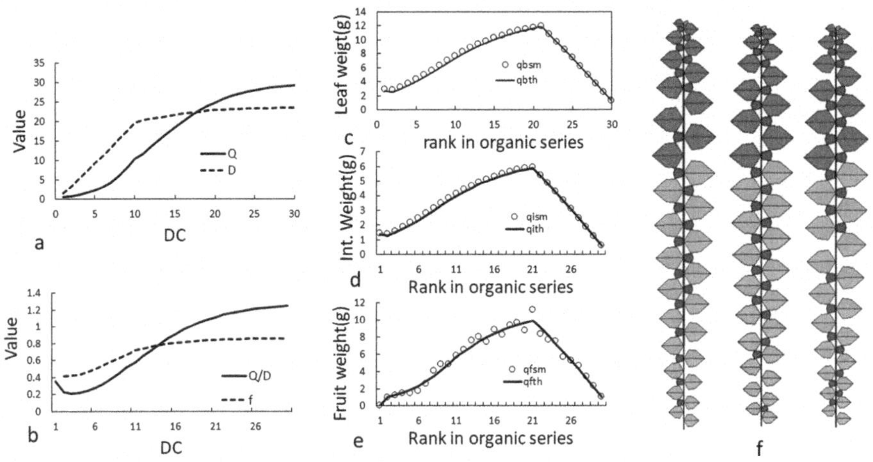

Fig. 13.19 Modeling of fruiting in limited growth of a single-stemmed plant with feedback of fruiting demand on growth. (**a**) Stabilization of supply and demand over 30 growth cycles. The demand cannot exceed the value imposed by the expansion time of the phytomers, and the biomass supply reaches a limit imposed by *Sp*. (**b**) As a result, the Q/D ratio stabilizes, and so does the probability of bearing fruit for a phytomer. (**c**, **d**, **e**) Comparison of the simulated (dots) and computed (lines) organic series by the potential structure. (**f**) Three random simulations at 30 cycles. The presence of fruits increases bottom-up along the stem

overload in fruits. To simulate the phenomenon, we introduce a viability function to the branch meristems: $c_{\left(\frac{Q}{D}\right)} = \frac{K_c\cdot\left(\frac{Q}{D}\right)}{1+K_c\cdot\left(\frac{Q}{D}\right)}$. It is important to note that only the organic series of the living branches are analyzed and compared to their counterparts in the potential structure. Dead branches are not included in the sampling for calculating sink–source parameters since the top-down description of the organic series is no longer valid (Chap. 7). Moreover, the stem has no mortality.

For the stochastic simulation of 30 coffee tree shrubs, the following parameters are chosen: $r = 55$, Sp is set to a large value, $\varepsilon = 0.05$, $Q_0 = 1$, $t_a = t_x = 10$, and organ sinks P_a, P_i, and P_f are, respectively, set to 1, 0.5, and 1 for leaves, internodes, and fruits of stem and branch phytomers. The mortality coefficient for the branch meristem is set to $K_c = 500$, and the fruit pruning coefficient is $K_f = 100$.

However, it is important to note that, given the values chosen for the functional parameters, the equilibrium condition $\frac{(t_a+1)}{2}\cdot\frac{E}{A} = 1$ is not achieved.

In the previous example of a single-stem, if the phytomers of the branches either do not bear any fruit ($K_f = 0 -> f = 0$, it grows exponentially) or deterministically bear fruit ($K_f = 99{,}999 -> f = 1$, it decreases exponentially), the growth cannot be in equilibrium. Removing the fruits will break the equilibrium, causing growth to accelerate, and the abortions of the meristems tend to disappear (Fig. 13.20a, plant without fruit). On the other hand, imposing a fruit on each phytomer will cause growth to vanish in a geometrical decrease, accompanied by a high mortality of the branches (Fig. 13.20b). A comparison between the two simulations with and without

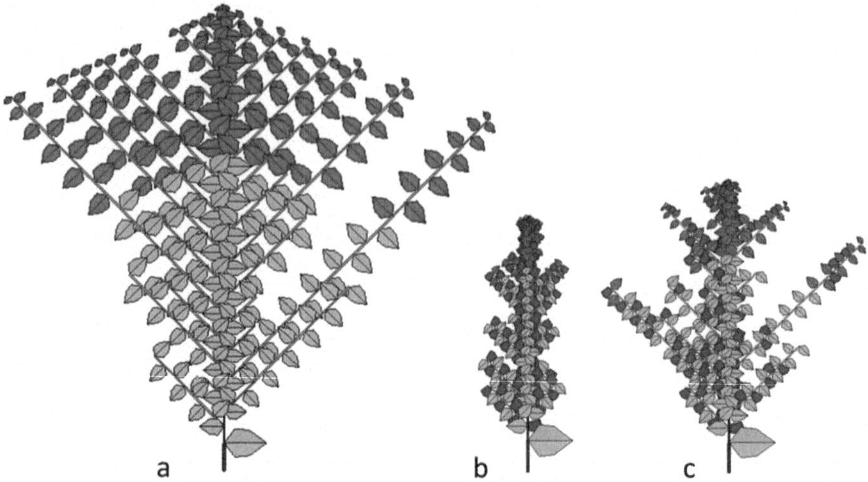

Fig. 13.20 Free growth–fruiting interactions in the branched plant simulating the die-back phenomenon. The simulated plants grow until 30 DCs. Parameter K_c that computes the viability c is fixed to $K_c = 500$. (**a**) Stochastic simulation without fruit with $K_f = 0$. In the absence of fruit at each phytomer, the growth increases out of control and meristem mortality disappears. (**b**) Simulation in which we impose a fruit at each node with $K_f = 99{,}999$. The growth will decrease indefinitely, and branch mortality will increase due to an overload of fruiting. (**c**) Simulation with $K_f = 100$, the coefficient that gives the probability of appearance of a fruit on a node according to Q/D. The system reaches a stabilized free growth

fruit clearly shows the influence of fruit demand on branch meristem abortion, referred to as "die back".

However, if we set $K_f = 100$ (Fig. 13.20c), the free growth will stabilize because Q/D stabilizes due to meristem mortality. The number of phytomers for a branch will reach a limit, with an average number equal to: $\overline{Nint} = \frac{1}{1 - c_{\left(\frac{Q}{D}\right)}}$.

On the other hand, the parameter f will converge to the value given for equilibrium by the formula $\frac{(t_a+1)}{2} \cdot \frac{E}{A} = 1$ with $A = \varepsilon \cdot r \cdot (P_a + P_i + f \cdot P_f)$. Solving this equation gives the limit value of $f = 0.5$.

By manipulating meristem and fruit abortions, we can control a fruit load to ensure free growth. This leads to the stabilization of Q/D ratio, resulting in the averages of the simulated organic series fit those derived from the potential structure. As a result, the Loop model with feedback approaches the Standard model.

In Fig. 13.21a, we observe that when the shrub does not produce fruit ($K_f = 0$), the biomass production diverges. Conversely, when the shrub produces a fruit on each phytomer ($K_f = 99{,}999$), we see a damping of the production. If K_f is an intermediate value (e.g., $K_f = 100$), production equilibrium can be achieved if the probability f reaches the value $f = 0.5$. The evolution of demand (Fig. 13.21b) follows the same pattern as that of the biomass produced.

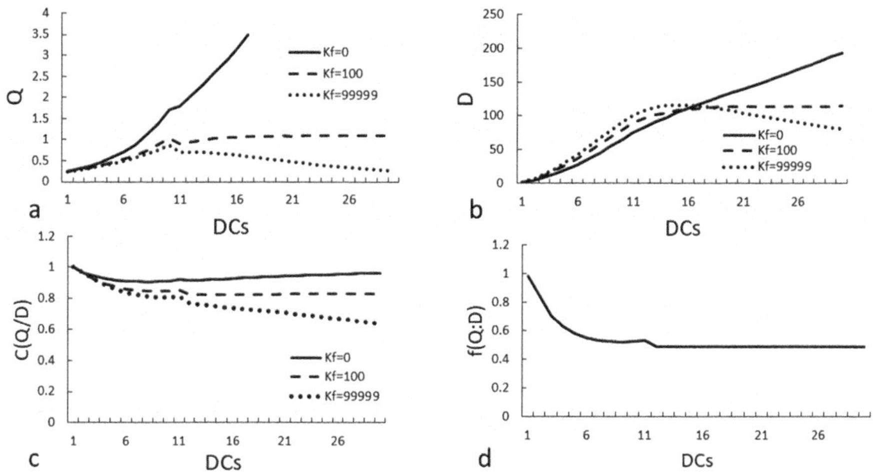

Fig. 13.21 Evolution of the biomass, the demand, viability, and probability of fructification for the different values of K_f. (**a** and **b**) Evolution of the biomass and the demand produced for the different values of $K_f = [0, 100, 99{,}999]$. We observe the stabilization for $K_f = 100$ and $K_c = 500$. (**c**) Evolution of the viability for the different values of K_f. This one stabilizes for $K_f = 100$. (**d**) Evolution of the probability f of fructification, until the stable value calculated $f = 0.5$. (Solid line computation, dotted line simulation average of 30 random plants)

Figure 13.21c shows the evolution of the viability c of branch meristems. In the absence of fruit, branch mortality ceases ($c \rightarrow 1$). However, when fruit is present on each phytomer, viability declines, and branch meristems abort at an increasing rate. Equilibrium is reached when viability stabilizes; in this case, with a value $K_f = 100$ and $c = 0.80$, branches typically die after producing on average five phytomers.

Figure 13.21d shows the evolution of the probability f and compares the results obtained from the potential structure with the average of 30 simulations. The latter converges to 0.5, which is consistent with theoretical expectations.

Figure 13.22 shows the goodness of fit between the averages of the 30 simulated organic series and their corresponding potential structures, with growth stabilizing at $K_f = 100$ and $K_c = 500$. The organic series of the fruits exhibit some disturbances due to the stochastic aspect. During the construction of the organic series, many dead branches are removed, resulting in a lower representation of still-alive long branches. This can lead to a significant dispersion of values for organ weights at the beginning ranks.

13.5.1.6.2 Case of Limited Growth of a Fruiting Branching Plant

This case is a representation of the growth of woody plants. During the early growth stages, the formed branches often die off, followed by growth stabilization. In this

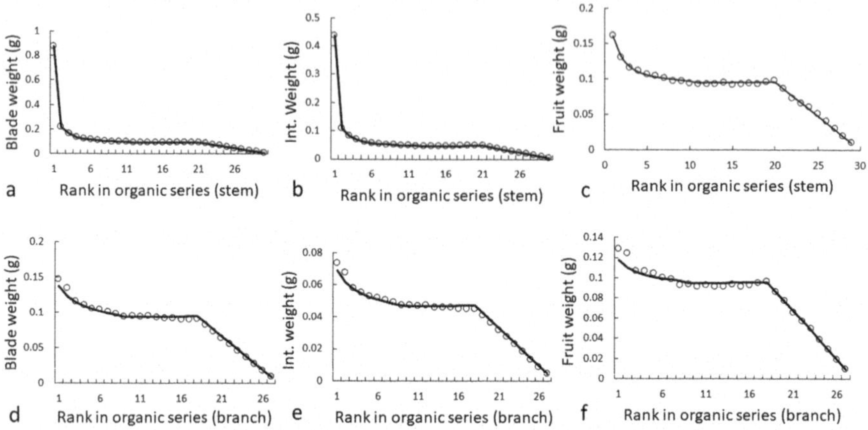

Fig. 13.22 Case of free growth. Comparisons between simulated (o) and computed (—) organic series of stem (**a**, **b**, **c**) and branch (**d**, **e**, **f**) for leaves, internodes, and fruits

study, 50 stochastic simulations of shrubs were performed until 40 DCs using the following parameter values.

For source–sink parameters: $r = 20$, $Sp = 1000\ \text{cm}^2$, $\varepsilon = 0.05$, $Q_0 = 1$, and $t_a = t_x = 10$; organ sinks: $P_a = 1$, $P_i = 0.5$, and $P_f = 1$. The parameter r is relatively low, resulting in a largeresource use efficiency (RUE) of $1/r$.

For meristem parameters: $K_c = 50$ and $K_f = 10$ ($K_b = 10{,}000 \rightarrow b = 1$), with no stochastic *Bernoulli* process and branching process ($a = 1$).

The low value of Sp leads to biomass production saturation. However, when biomass production saturates, demand continues to increase, which ultimately results in a decrease in Q/D trophic pressure, leading to a reduction in the viability of branch meristems and the fruiting rate.

Figure 13.23 shows the goodness of fit between the averages of the 30 simulated organic series and their corresponding potential structures. Table 13.4 shows that the WLSM correctly computes the source–sink parameters.

For the same RUE $1/r$, two parameters K_f and K_c can decrease the growth: the fruiting rate and mortality rate of branches. For example:

For source–sink parameters: $r = 20$, $Sp = 1000\ \text{cm}^2$, $\varepsilon = 0.05$, $Q_0 = 1$, and $t_a = t_x = 10$. Organ sinks $P_a = 1$, $P_i = 0.5$, and $P_f = 3$ (the sink of fruit is increased on purpose). Parameter r is low, resulting in a large RUE of $1/r$.

For meristem parameters: $K_c = 50$ and $K_f = 10$. No stochastic *Bernoulli* process ($b = 1$) and branching process ($a = 1$).

Figure 13.24 (f, e, g) shows three simulations of the computational plant at the age of 40 DCs, with values $K_f = [0, 10, 99{,}999]$. As K_f increases, the fruiting rate f also increases, leading to a decrease in growth and an increase in branch mortality, which depends on viability c, through the retroaction of trophic pressure.

The feedback of trophic pressure allows to regulate the growth. The model gives a good explanation for the observed effect of fruiting on the growth of coffee trees. For

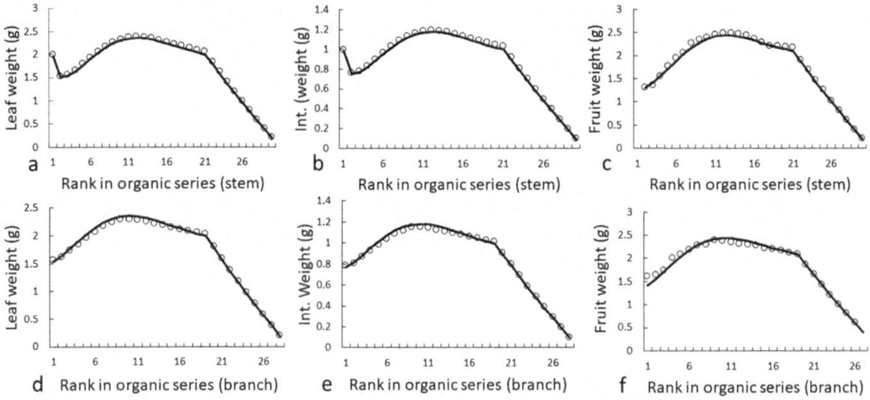

Fig. 13.23 Case of limited growth. Comparisons between simulated (o) and calculated (—) organic series of stem (**a**, **b**, **c**) and branch (**d**, **e**, **f**) for leaves, internodes, and fruits. Only the simulated fruit series are perturbed

Table 13.4 Parameters used for the plant simulation in limited growth and recomputed parameters from the WLSM

Param	K_c	K_f	r	Sp	P_i	P_f
Simul	50	10	20	1000	0.5	1
WLSM	48.7	10.4	19.6	963	0.5	1
SD	0.36	1.2	0.09	7.5	0	0
CV (%)	0.03	0.05	0.01	0.02	0	0

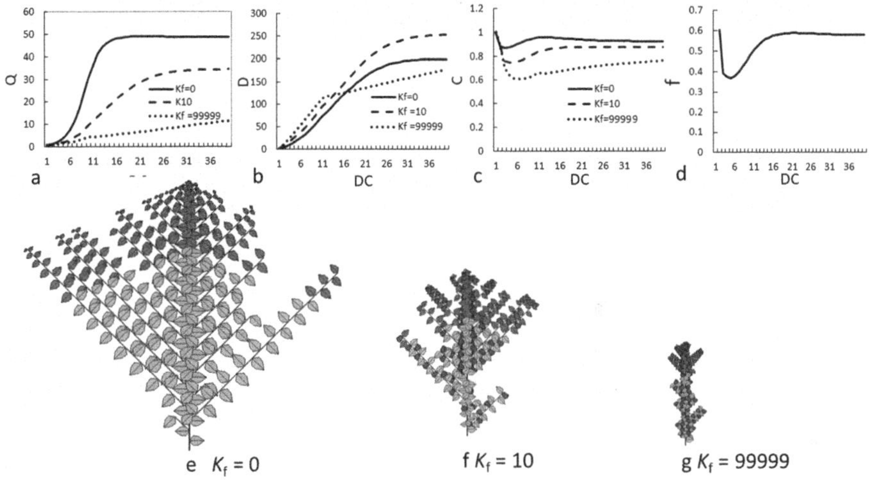

Fig. 13.24 Influence of fruiting rate on computational plant growth for $K_f = [0, 10, 99{,}999]$. (**a**) Evolution of biomass production. (**b**) Evolution of demand *D*. (**c**) Evolution of viability *c*. (**d**) Evolution of fruiting rate *f*. In the three cases, the biomass *Q* and the demand *D* reach a limit. For $K_f = 0$, there is no fruit ($f = 0$). Biomass production is the highest. The resulting plant (**e**) is the largest. For $K_f = 99{,}999$, there is one fruit per phytomer ($f = 1$), resulting in the lowest biomass production leading to a high mortality rate of the branches. The resulting plant (**g**) is the smallest. For $K_f = 10$, the fruiting rate stabilizes at $f = 0.58$, resulting in a medium-sized plant (**f**)

example, the triploid *canephora*arabica* hybrids, which are sterile and therefore produce very few fruits, exhibit vigorous growth and a highly developed architecture. In contrast, the *Caturra* variety of *Arabica* produces abundant fruits, resulting in weak growth and a packed architecture. Even young plants can abort if the fruit load is too high.

It is worth noting that increasing the sink strength of the fruit, P_f, has a similar effect on the viability of meristems to that of K_f.

13.5.1.7 Growth and Development Controlled by Production Surface *Sp*

The planting density, expressed by the production area *Sp*, can affect branch mortality and the viability *c* of meristems. For this example, we set $K_c = 50$, and the *Bernoulli* parameter and branching rate are both set to 1. We consider five different production areas: $Sp = [625, 1250, 2500, 5000, 10{,}000]$ (cm^2), and simulate 30 random plants at the age of 40 DCs.

For source–sink parameters, we set $r = 30$, $\varepsilon = 0.05$, $Q_0 = 1$, and $t_a = t_x = 10$; organ sinks: $P_a = 1$, $P_i = 0.5$, and $P_c = 0.5$.

Table 13.5 gives the values obtained at 40 DCs for the number of phytomers produced on the stem (N_1), on the branches (N_2), the synthesized biomass (Q), the demand (D), the height of the plant (H), and the viability (c), for the different values of *Sp*.

We find that the viability of meristem (c) decreases from 0.93 to 0.81 when *Sp* varies from 10,000 cm^2 to 625 cm^2. This corresponds to an average limited number of phytomers per branch of 14 and 5, respectively. This reduction in viability indicates that planting density reduces branch development, as observed in forest plots. Moreover, the height of the plant (H) is not very sensitive to *Sp*, which is consistent with Eichorn's law mentioned above. Additionally, the simulation results show the "base effect" that reduces the development of branches at the bottom of the stem. These findings contribute to a better understanding of the relationship between planting density and the development of coffee plants (Fig. 13.25).

Table 13.5 Computational plant production for different values of *Sp*

Sp	*c*	N_1 (phytomers on the stem)	N_2 (phytomer on the branch)	Q/DC	D/DC	H
625	0.81	40	100	10	122	91
1250	0.85	40	118	18	150	101
2500	0.89	40	131	30	172	109
5000	0.92	40	140	43	189	115
10,000	0.93	40	145	55	198	120

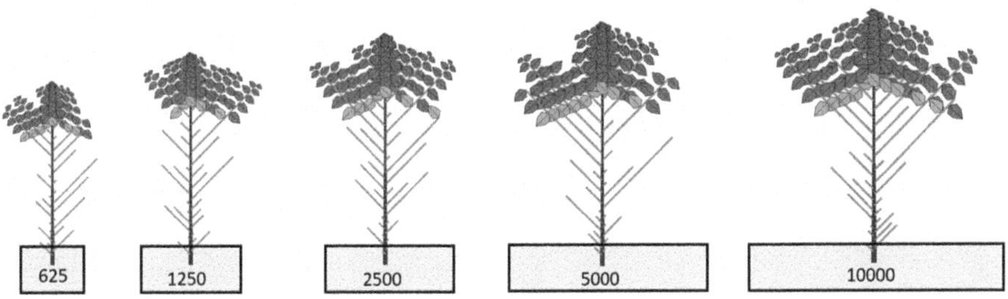

Fig. 13.25 Influence of planting density on branch mortality (Sp = [625, 1250, 2500, 5000, 10,000] cm^2). At the beginning of the growth, the mortality of the branches is strong (base effect); it increases progressively until stabilizing at a value that depends on the density of plantation. The lower the density, the lower the mortality

Table 13.6 Values obtained at 40 DCs for different fruiting rate f depending of Sp in tomato

Sp	f	N_1 (phytomers on the stem)	Q	D	H
625	0.71	40	7.3	29	165
1250	0.77	40	11	30	184
2500	0.82	40	15	31	200
5000	0.89	40	27	33	211
10,000	0.89	40	27	31	229
20,000	0.87	40	22	32	218

13.5.1.8 Fruiting Controlled by Production Surface *Sp*

Kang et al. (2011) made detailed observations on the relationship between tomato fruit abortion and trophic pressure (Q/D), especially under different planting densities. Such an experiment can be reproduced in silico with the GreenLab Loop model.

The planting density expressed by the production area Sp influences fruit abortion that depends on the abortion rate f of fruits. We chose $K_f = 10$, in this example, and the *Bernoulli* parameter is set to 1.

We choose six production areas Sp = [625, 1250, 2500, 5000, 10,000, 20,000] (cm^2). We simulate 30 random plants at the age of 40 DCs. For source–sink parameters, we set: $r = 30$, $\varepsilon = 0.05$, $Q_0 = 1$, and $t_a = t_x = 10$; and organ sinks: $P_a = 1$, $P_i = 0.5$, and $P_f = 2$.

Table 13.6 gives the values obtained at 40 DCs for the number of phytomers produced on the stem (N_1), the synthesized biomass Q, the demand D, the height of the plant H, and the fruiting rate f, for the different values of Sp.

The fruiting rate f increases with the production area Sp. The behavior of the tomato is retrieved qualitatively. However, beyond the value Sp = 10,000 cm^2, the fruiting rate f starts to decrease. Surprisingly, for the value Sp = 20,000 cm^2 (a condition of quasi-free growth), the production Q also decreases. This can be explained by the fact that the demand for fruits becomes too high, and as a result of feedback, both the fruiting rate f and biomass production Q decrease (Fig. 13.26).

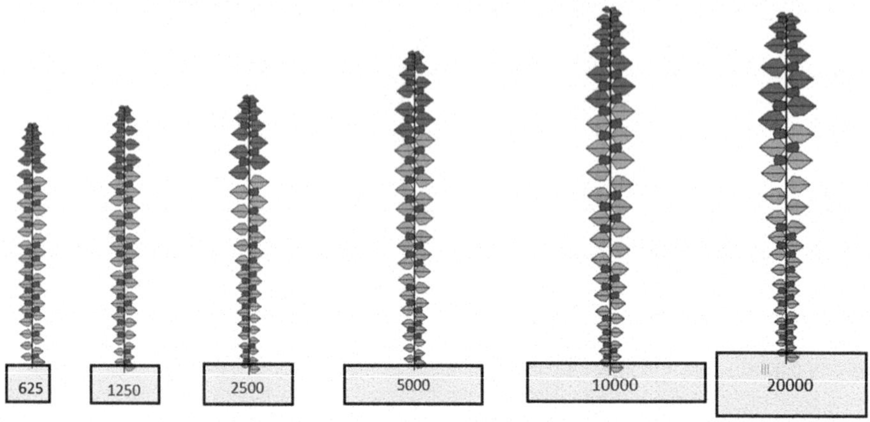

Fig. 13.26 Influence of planting density d ($Sp = 1/d$) on fruiting rate ($Sp = [625, 1250, 2500, 5000, 10{,}000, 20{,}000]$ cm^2). In early growth stages, the fruiting rate f increases progressively and then stabilizes at a value that depends on the plantation density. When Sp becomes large (free growth), there is a slight decrease in both Q and f due to feedback, caused by an increase in demand for fruits

Fig. 13.27 Effect of radiation E on plant development and growth. (**a**) Normalized radiation fixed at $E = 1$. (**b**) Fixed radiation $E = 0.5$. We observe a reduction in the volume of the organs and an increase in the branch mortality

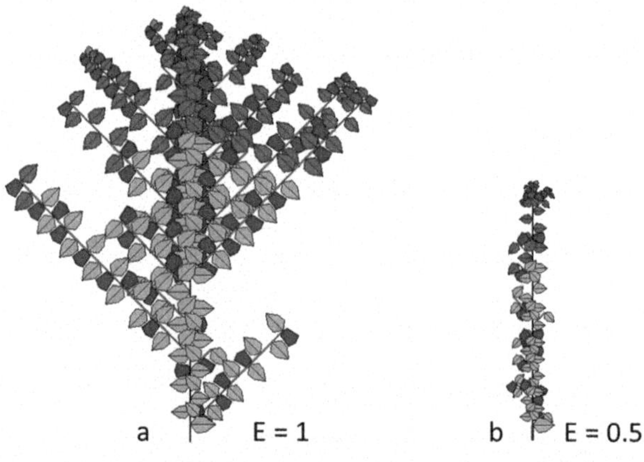

13.5.1.9 Growth Controlled by Photosynthetic Radiation *E*

The parameter E represents the radiation level, and it has a significant effect on growth and trophic pressure. Reduction in E increases meristem mortality and fruit abortion, as shown in Fig. 13.27, using the same parameters as in Fig. 13.24f.

13.5.2 Case of Rhythmic Growth

The stochastic case of trees with rhythmic growth is not covered in this book. This case is studied in de Reffye (2016). The growth behavior is like that of the deterministic case studied in Sect. 13.4 of this chapter.

13.6 Modeling Source–Sink Relationship Functioning

Throughout its history, the GreenLab model has undergone a few changes. While the overall philosophy of a sink–source structural–functional system based on the notion of phytomer has remained the same, the formulation of certain concepts has been modified locally. A good fitting to experimental data is necessary for any model validation, but it is not sufficient: the plant functioning can be approximated by various hypotheses that give equivalent results, making it necessary to make relevant choices that are as close as possible to biologically realistic hypotheses. Some trials have thus been given up or put on standby, while others have been accepted. We believe it is useful to present these attempts here, as they provide an in-depth look at the mechanisms of plant growth and are a source of future improvements to the model.

13.6.1 Continuous GreenLab

The basis of the development engine of the GreenLab model is the discretization of time into development cycles (DC). The notion of DC is a natural approach because the plant produces integer numbers of organs per cohort. This is derived from the Renewal Theory, which allows the transition from continuous to discrete time and treats the functioning of a meristem to a *Bernoulli* process (Chap. 3). The concept of thermal time can also be easily integrated into this approach.

However, in nature, the development of plant architecture is continuous, and the *Bernoulli* process is only an approximation using the Renewal Theory:

- On a practical level, the discretization of time into development cycles (DC) has proven sufficient, as evidenced by the high quality of model fits on cultivated plants (Chaps. 9, 10 and 11). Moreover, the concept of DC makes it possible to establish the Potential Structure, which is the basis of the principle of fitting organic series in the case of stochastic development.
- On the theoretical level, it would be interesting to formulate mathematically the development of plants based on continuous time by exploiting completely the Renewal Theory. Undoubtedly, the mathematical complexity of the system increases. In any case, the numerical resolution of the differential equations resulting from this continuous time formulation requires a discretization: a similar

formalism will thus be possible. The synchronization of events in the plant structure at the level of meristems, concerning branching, abortion, etc., could then be achieved by introducing discrete events in the numerical solution scheme (Le Chevalier et al. 2007).

The basis of the growth engine of the discrete GreenLab model is the growth cycle (GC). In most cases, the chosen GC is merged with the DC. However, it is not a mandatory requirement, and using the rhythm ratio, the time devoted to growth can be subdivided to that devoted to development. For instance, a rhythm ratio of $w = 1/3$, which gives the time series [1 0 0 1 0 0 1 0 0, . . .] where the "1"s represent the cycles at which new phytomers have been created, imposes one DC for three GCs on the meristem. This system allows for a more detailed time scale to follow the effect of climate on growth (see coffee tree in Chap. 10).

One disadvantage of the system is that the values of the sink–source functions depend on the chosen time step, or, alternatively, the chosen rhythm ratio. This problem is troublesome when a generic model is used to simulate different plant species together, as individual rhythm ratios need to be defined to synchronize them. However, it is possible to have continuous growth while keeping a discrete development, which makes the values of the sink–source parameters invariant (Le Chevalier et al. 2007; Li et al. 2009) with respect to the chosen time step. This allows for the growth of a mixture of species to be modeled.

13.6.2 Possible Alternatives Around the Source Functions

The first version of GreenLab (Model 1, Yan et al. 2004) for the computation of photosynthesis is inherited directly from the formalism adopted in the AMAPhydro model (de Reffye et al. 1999). This model incorporated the concept of leaf transpiration with respect to the climatic demand E (chosen to be the potential evapotranspiration, PET), along with water use efficiency (Chap. 5). Model 2 is designed here like the Standard model.

In Model 1, the number of leaves N_i per cohort in cycle i has an individual resistance to transpiration, which is given by $\frac{r_1}{S_i} + r_2$, depending on their individual surface S_i. This resistance decreases with increasing leaf area to reach a minimum value r_2 attributed to the action of the petiole.

The production equation of the hydraulic model (Model 1) can be expressed as:

$$Q = E \cdot \sum\nolimits_{i=1,t} \frac{N_i}{\frac{r_1}{S_i} + r_2} \tag{13.15}$$

Which can be compared to the equation used in the Standard model (Model 2):

Table 13.7 Error values of the two models' differences from the measurements

Plant	Error of model 1	Error of model 2
Cotton	6.26	6.06
Wheat	3.27	3.89
Sunflower	193.4	193.4
Maize	47.7	47.8
Tomato	1.76	1.64

$$Q = \frac{E \cdot Sp}{r_1} \cdot \left(1 - \exp\left(-k\left(\frac{\sum_{i=1,t} N_i \cdot S_i}{Sp}\right)\right)\right) \tag{13.16}$$

Note: In the case of free growth (*Sp* very large and r_2 very small), Eqs. (13.15) and (13.16) are equivalent.

Model 1 in GreenLab gives good results when faced with the organic series of single-stemmed cultivated plants, such as sunflower, maize, tomato, cotton, and wheat. The models' parameters r_1, r_2, and the sink parameters, can be well identified in such cases. If $r_2 > 0$, the production in Model 1 tends toward a limit value, and r_2 plays a role like that of the production area *Sp* in Model 2. It is not possible to distinguish between the two models based on the quality of plant fits. Table 13.7 shows that the error values between the measured observations and the corresponding simulated values are similar, regardless of the model used. Figure 13.28 shows the similar quality of fits obtained with both models in the case of the organic series.

Through the collaboration with agronomists from CIRAD (Centre International de Cooperation en Recherche Agronomique pour le Developpement) (Michael Dingkuhn) and Wageningen University (Ep Heuvelink), GreenLab has evolved toward crop models (Model 2) that use the Beer-Lambert law for modeling light intercepted at different canopy levels. This formalism has the advantage of accounting for planting density by incorporating the production area *Sp* of the individual plant. This allows the transition from the functioning of the single plant to that of the stand. Despite the change in modeling approach, the similar quality of fits obtained using both Model 1 and Model 2 made it an easy decision to replace the former with the latter.

To ensure a source model is well adapted, its parameters must be computed by an inverse method, and the series of biomasses produced at each cycle ($Q_1, Q_2, \ldots, Q_t$) must be correctly estimated.

The main advantage of the GreenLab model, as mentioned earlier, is that it uses an inverse method to compute the photosynthetic output from the organic series, bypassing the need to directly compute light interception by the leaf surface.

Another source model, which is more complex to implement, involves computing light interception using a simulated 3D plant architecture (Chap. 12). Although this approach provides more detailed information and allows to compute the interception of each leaf individually, it provides similar results to the previous models when it comes to evaluating the biomass production at the scale of the whole plant or plant canopy.

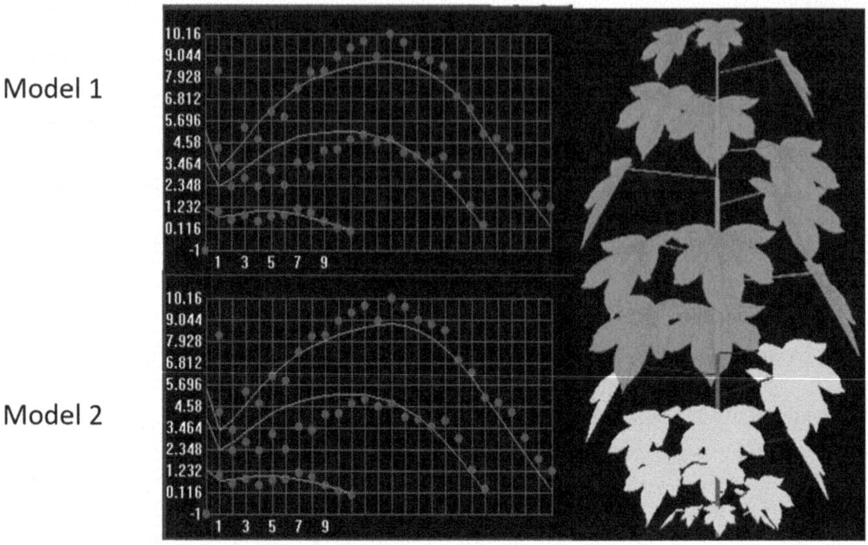

Fig. 13.28 Comparison between the two models of the fittings of the computed (lines) and measured (points) organic series of the internode weights of a single-stemmed cotton plant and visualization of the simulated plant. Both models have equivalent performance (Cornerfit, by Zhan Z-G, CAU)

13.6.3 Organ Functioning

The modeling of source and sink functions relies on the physiological functioning of organs.

13.6.3.1 Modeling the Functioning of the Organs as Sources

It is not only the leaves that are sources of the plant system. The internodes can take part in photosynthesis in their young stages before they get woody. The same goes for the sheaths of grasses (e.g., maize) or the pods of crucifers (e.g., rapeseed). It is essential for agronomist to consider these organs in the model, which is not challenging to implement. However, it is necessary to consider parameters such as shape allometry or the orientation of the organ to estimate the interception of the light by its surface.

For leaves, the surface is deduced from the allometric ratio weight/thickness (specific leaf weight: SLW), which can vary with the age of the plant (Christophe et al. 2008). However, if the leaf area index (LAI) is high (>3), this has little influence on the value of photosynthesis, making the "constant thickness" hypothesis robust.

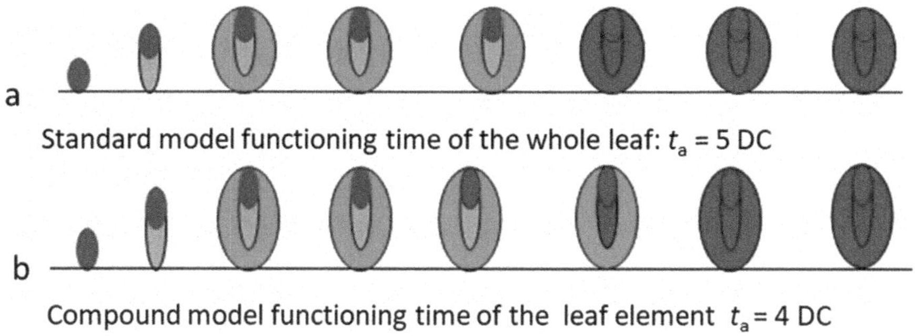

Fig. 13.29 Diagrams of the operation of an expanding leaf over three cycles (DC). (**a**) The leaf ceases to function as a source organ after five DCs. (**b**) The leaf elements operate for four cycles. The compound leaf gradually stops functioning (Color figure online)

The model accounts for the vertical orientation of leaves using the cultural coefficient k in production Eq. (13.3) (another form of Eq. (13.16)).

$$Q = C \cdot \mathrm{PAR} \cdot (1 - \exp(-k \cdot \mathrm{LAI})) \quad (13.17)$$

Photosynthetic active radiation (PAR) is the active part of solar radiation in photosynthesis (mole photon m^{-2} s^{-1}). In agronomy, for each crop, an empirical value of k is proposed and generally adopted by GreenLab. This value corresponds to the orientation of the leaf surface normal with respect to the vertical. The angle is characterized by its cosine. If $k = 1$, the leaf surface is horizontal. The adoption of this value improves the quality of the fittings of the organic series in the grasses whose leaves have an upright orientation, such as rice.

Leaves have a functioning time during which they are sources for the system, with photosynthetic production exceeding consumption due to the maintenance respiration. The end of the functioning time is visible by yellowing, but it is a gradual process that generally spreads from the periphery toward the interior of the organ. Thus, in the gramineous plants, the necrosis of the leaves starts from the top, and in the leaves of the other herbaceous plants it starts from their periphery (Fig. 13.29).

In the GreenLab model, the leaf organ grows during several cycles corresponding to its expansion time, which is less than or equal to the functioning time. Each expansion corresponds to a surface element. It is possible to differentiate these elements by assigning them their own functioning time. It is interesting to compare the Standard model, which assumes that the leaf ceases to function all at once when it exceeds its functioning time, with a compound model, which makes the surface elements stop functioning one by one as their functioning time is exceeded. Despite their differences, both models have similar behaviors.

Thus, the condition for free growth at equilibrium for the Standard model is written (Chap. 5):

$$\left(\frac{t_a+1}{2}\right)\cdot\frac{E}{A}=1.$$

In the progressive leaf senescence model (the so-called compound model), the equation becomes: $(\frac{t_a}{2})\cdot\frac{E}{A}=1$, where $A = r{\cdot}e$; $E = 1$, and t_a is the duration of functioning and expansion of the organs. It is possible to easily adjust a plant from the Standard model to the compound model. As an example, let us consider a monoculture plant with a rhythm ratio $w = 0.75$, corresponding to the development sequence of 1 1 1 0 1 1 1 0, ... In the first model, the leaf functioning time is five DCs, while in the second model, the leaf element functioning time is five DCs. The expansion time of the leaf is also five DCs in both cases.

Using the inverse method, we fit the organic series of the Standard model to the compound model. Table 13.8 demonstrates that all the source–sink parameters of the compound model are affected, i.e., their values must be modified to correctly fit the organic series. Figure 13.30 shows that the biomass productions per cycle are equivalent. The dashed curve of the Standard model has oscillations due to leaf drop, but these nearly disappear in the smoother, solid curve of the compound model. This is because the cessation of functioning in a surface element has less influence than that of a whole leaf.

Both models are equivalent, but the compound model is more complex to program and does not improve the fit quality. Therefore, it has not yet been

Table 13.8 Comparison of the parameters of the Standard model and the compound model

	w	e	t_a	t_x	P	r	Sp	Q_0
Standard model	0.75	0.05	5	5	1	15	1000	1
Compound model	0.75	0.05	5	5	1	27	1604	2.1

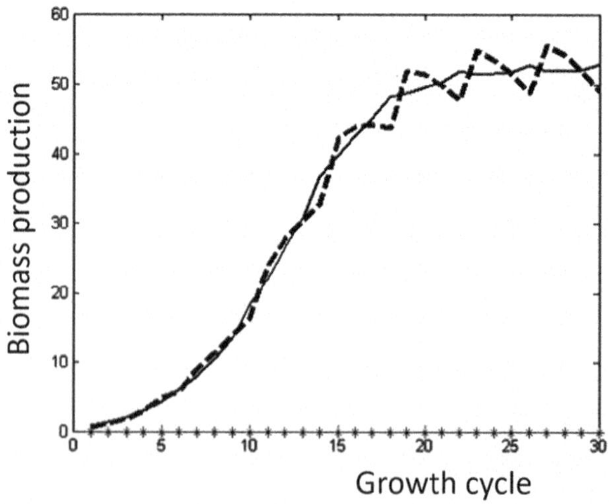

Fig. 13.30 Comparison of biomass production per cycle between the two models, standard (dotted line) and compound (solid line)

implemented in the standard Gloups software. However, determining the functioning time of leaves is a practical problem as it cannot be determined by inverse method due to being strongly correlated with the parameters r and t_a (which is non-identifiable). If t_a increases, the parameter r will also increase to fit the organic series correctly. In practice, the phenological stage of the observed wilting (yellowing) of the leaf is used to determine the functioning time empirically.

13.6.3.2 Reminder on the Formulation of the Descriptive Sinks in the Standard Model

In the GreenLab model, the expansion of an organ is characterized by its duration t_x, in DC, which is dependent on thermal time. In calendar time, the growth of the organ slows down with lower temperatures, as observed. In this modeling formalism, the variation of the sink is independent of the supply. Other crop models, such as Pilote model in Chap. 11, also define the evolution of leaf area index (LAI) only as a function of thermal time.

The method gives good results, but it is not the only one and has its limitations. One issue is the difficulty in determining the parameter t_x, which represents the expansion time of an organ, through observation. Additionally, t_x can vary according to the rank of the phytomer on the stem, especially during the early growth stage (known as the base effect). The first leaves that emerge on the stem are generally smaller and have a shorter expansion time compared to the mature leaves. Hervé Rey (Chap. 9) conducted a long-term growth follow-up and precisely measured the $t_x(r)$ values of the leaf organs on sunflower as a function of the rank r on the stem. Fortunately, the simulation process can empirically fit the evolution of the leaf expansion time along the stem, which typically shows a linear phase of increase starting from a minimal value t_{xmin}, followed by a stabilization at the value t_{xmax}, as well modeled by:

$$t_x(r) = t_{x\,\mathrm{min}} + a \cdot (r - 1) \quad \mathit{if}\ r < r_{\mathrm{max}}$$

and:

$$t_x(r) = t_{x\,\mathrm{max}} \quad \mathit{if}\ r \geq r_{\mathrm{max}} \ \text{with} \quad t_{x\,\mathrm{max}} = t_{x\,\mathrm{min}} + a \cdot (r_{\mathrm{max}} - 1)$$

While experimentation can be a tedious and discouraging task, the inverse method for identifying parameters only works correctly when the $tx(r)$ tables are defined for each organ type. Fortunately, it is possible to estimate the parameters $t_{x\mathrm{min}}$, r_{max}, and $t_{x\mathrm{max}}$ empirically through trial and error, resulting in the best fits of the organ series. This process becomes easier with practice and eliminates the need for precise growth monitoring.

In the GreenLab model, the sink function of an organ that appeared at cycle i when the plant is at age t is defined by the equation (Chap. 5):

$$P_o(t,i) = P \cdot F(t-i+1, t_x) \tag{13.18}$$

We have: $j = t - i + 1$:

$$F(j,t_x) = K\left(\frac{j-0.5}{t_x}\right)^{a-1} \cdot \left(1 - \frac{j-0.5}{t_x}\right)^{b-1} \tag{13.19}$$

The parameter P is the sink strength, while the function F is derived from a Beta law density function. K is a normalized factor that sets the mode of the function F to the value 1, and t_x is the expansion time of the organ. This value can evolve according to the rank of the stem, marking a base effect, as said before. This supply-independent formulation of the sink function, which depends on DC indexed to thermal time, is referred to here as the "descriptive sink". Thus, during its expansion, the sink of an organ evolves according to the value imposed by the function φ. Note that the choice of the Beta law formulation of the expansion function is arbitrary: it is simply well suited, by its form and flexibility, to the representation of a growth phenomenon. Other functions, such as the Gompertz function, often chosen to track fruit growths, give equivalent results.

13.6.4 *Functional Sinks for the Loop Model*

Research conducted at the Chinese Agricultural University (CAU) has led to a complete reconsideration of a formulation of organ expansion, based on more physiological assumptions (Zhang et al. 2009). This new formulation is required to fit the organ series as correctly as the model using beta functions. This is where the Loop model comes into play.

Physiologically, the emergence and growth of an organ are composed of two main phases: cell creation, and expansion of existing cells. During organ emergence, cell creation is the preponderant phenomenon, which gradually decreases as cell expansion becomes more important, leading to the final size of the organ. In the model, we focus on modeling the creation of cells or *cell sets* during the formation of organs in meristems, known as primordia. The volumes of the primordia depend on the *Q*/*D* ratio at their creation. As each cell set can only expand up to a certain maximum size, the maximum weight an organ can reach at the end of its expansion depends directly on the volume of the primordia at its creation.

13.6.4.1 Adaptation of Primordia to Sink Functions

So, each cell set that forms an organ or part of an organ corresponds to a primordium. At each DC, a new cell set can be added to each organ, and the number of sets thus depends on the duration of the phase of cell creation. The number of sets in a leaf is

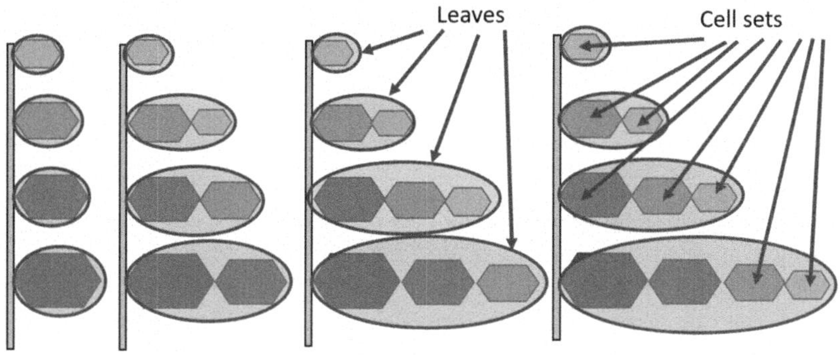

Fig. 13.31 Expansion of an organ (here a leaf) in the organic series of a stem, according to the number of cell sets defined by organ. At each DC, a new set can be created. Each set has a fixed duration of functioning

fixed, and the expansion of the organ is then achieved through the expansion of each cell set that composes a leaf. This expansion occurs sequentially according to the cycle of appearance of the sets (Fig. 13.31). The cell set of leaf organs has a functioning time t_c in DC.

Here it is necessary to set a functioning duration per cell set and not per leaf.

13.6.4.1.1 Case of a Single Cell Set Per Organ

The initial volume $q_o(i, i)$ of the organ created during DC i of a plant at age i can be modeled according to the following relation ($i = t$):

$$q_o(i,i) = C_o \cdot \frac{Q(i-1)}{D(i)} \tag{13.20}$$

C_o is the cell sink that corresponds to a sink parameter linked to a cell set from an organ of type o and physiological age φ (o can be b [leaf blade], p [petiole], e [internode], or f [fruit]). Only one physiological age is considered here.

X_o is the expansion sink, which defines the maximum volume that the cell set can expand to relative to its initial biomass according to the equation:

$$q_{xo}(i) = X_o q_o(i,i) \tag{13.21}$$

An individual maximum volume is assigned to each organ cohort, which depends on the creation cycle. The sink of the cell set created during cycle i, when the plant has age t, can be defined using the following equation:

$$P_o(i,t) = D_o \cdot \left(\frac{q_o(i,t-1)}{q_{xo}(i)}\right)^{a_o} \cdot \left(1 - \frac{q_o(i,t-1)}{q_{xo}(i)}\right)^{b_o} \cdot \Delta t \quad \left(t > i \quad \text{and} \quad \Delta t = 1\right) \tag{13.22}$$

Where $q_o(i, t-1)$ is the volume of the organ at cycle $t-1$, while $q_{xo}(i)$ is its limit volume computed at its creation (Eq. 13.21).

D_o is the sink strength of the organ of type o. Parameters a_o and b_o are corrective factors that can be adjusted as needed (value 1 gives a logistic function, 0 gives a constant sink). However, in the plants studied, no corrections were needed except for one, so we typically set $a_o = 1$ and $b_o = 1$ for the plant analysis.

The shape of the sink is still determined by an algebraic equation, as in the case of descriptive sinks (Beta laws). However, it now takes on a more functional aspect. Firstly, it is no longer regulated solely by thermal time in a fixed way, but also by the biomass variable itself. Secondly, the time of expansion is no longer predetermined, but rather depends on when the limit size is reached. As a result, the biological meaning of the sink function parameters is more clearly defined, and the functions are more closely linked to the underlying biological processes.

At the beginning, the sink is proportional to the biomass of the organ, which ensures exponential growth. As the organ approaches its maximum size, the sink gradually decreases until it is canceled when the organ reaches its limit. This sink formalism is called a "functional sink" because the demand depends here on the relative size of the organ (as modeled by the Loop model), as opposed to the descriptive sink, which is based on the age of the organ in DC (as determined by thermal time). For the analysis presented here, we generally set $a_o = 1$ and $b_o = 1$, except in specified cases.

The growth of an organ cell set on a phytomer of rank i, when the plant has t DC at cycle t, can be expressed as:

$$\Delta q_o(i,t) = D_o\left(\frac{q_o(i,t-1)}{q_{xo}(i)}\right) \cdot \left(1 - \frac{q_o(i,t-1)}{q_{xo}(i)}\right) \cdot \frac{Q(t-1)}{D(t)} \tag{13.23}$$

The biomass of the organ at cycle t is increased relative to cycle $t-1$ according to the equation:

$$q_o(i,t) = q_o(i,t-1) + \Delta q_o(i,t) \tag{13.24}$$

For a single-stemmed plant with definite development (case of one cell set per organ), the demand $D(t)$ of the organic series can be expressed as:

$$D(t) = D_o + \sum_{i=1}^{t-1} D_o \cdot \left(\frac{q_o(i,t-1)}{q_{xo}(i)}\right) \cdot \left(1 - \frac{q_o(i,t-1)}{q_{xo}(i)}\right) \tag{13.25}$$

At each DC, a new cell set is created until the maximum number of cell sets is reached, allowing for the expansion of the organs in place. Notable, the expansion time no longer appears in the equations, as the increases in biomass become infinitely small when the organ tends toward its maximum value. Equation (13.25) can be generalized by introducing the numbers of organs per cohort. To ensure that the system is well adapted for modeling real plants, it is necessary for the parameters C_o, D_o, X_o to be assessed using an inverse method based on the analysis of organic series.

13.6.4.1.2 Case of Several cell Sets Per Organ

We can set a maximum number of cell sets (n_o) per organ type. This means that a new cell set will be added to a given organ during *no* DCs. This process refines the variation of the organ sink. The sink of the organ is equal to the sum of the sinks of its constituent sets. The volume of the organ at its maximum expansion is equal to the sum of the maximum volumes over all the cell sets (Fig. 13.31).

13.6.4.1.3 Behavior of the Model

The study of the Loop model's behavior looks more complicated because of the interaction between the supply Q and demand D. The free growth at equilibrium could not be demonstrated as for the Standard model. However, the growth limited by the production surface Sp is still obviously valid. In the following example, we choose a single cell set per leaf organ, which forms the only organic series of the computational single-stemmed plant (the internodes having a negligible sink as in a rosette). The parameters of the system are specified as follows:

One set/leaf, $t_c = 5$, $n_o = 1$, $r = 30$, $Sp = 5000$, $Q_0 = 1$, $e = 0.05$, $D_o = 50$, and $X_o = 30$. t_c is the functioning duration of a cell set for photosynthesis.

The system generates an organic series that has a theoretical maximum envelope, resulting in. a limit organ weight for each rank on stem (red line in Fig. 13.32). The most recently created leaves are still expanding (green line in Fig. 13.32).

Note that the expansion time t_x of the organ has disappeared. Indeed, the expansion tends toward zero, when the weight of the organ created at the age i approaches the maximum computed value $q_{xo}(i, t)$.

Figure 13.33 shows various types of behaviors. The parameter C_o is set to 1. If D_o is too small, the organs of the plant cannot reach the maximum envelope and the plant collapses under demand, and supply tends toward zero. On the other hand, if D_o is too large, the organs quickly reach their maximum sizes, and growth is continuous, leading to a stabilization of demand. For intermediate values of D_o,

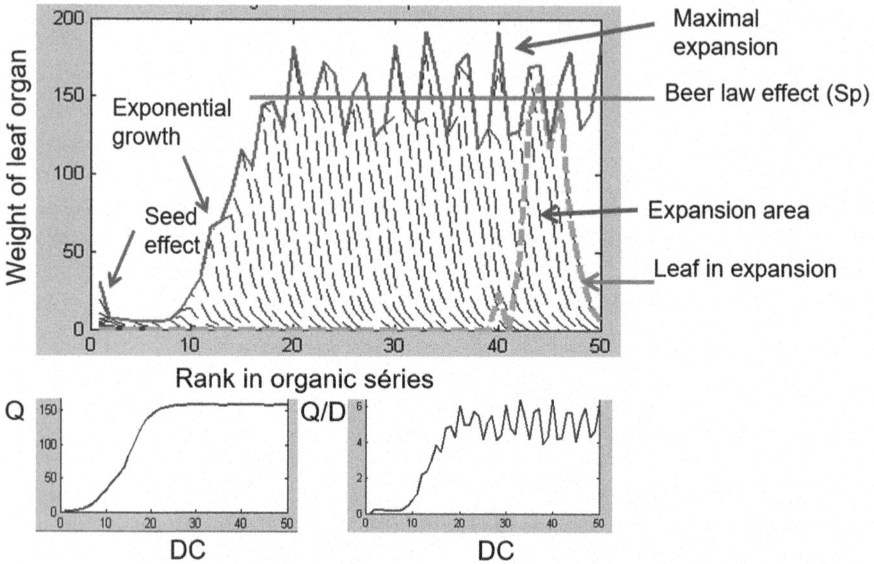

Fig. 13.32 Growth of a plant with functional sinks. There is one cell set per organ. The organic series of the plant is limited by a maximum envelope (red curve), which is generated at the creation of the organ and reaches progressively according to the rank of the organ in the phytomer. The organs of the last phytomers (green curve) are still expanding. Note that the growth is continuous and reaches a limit value because of *Sp*, while the demand looks chaotic (Color figure online)

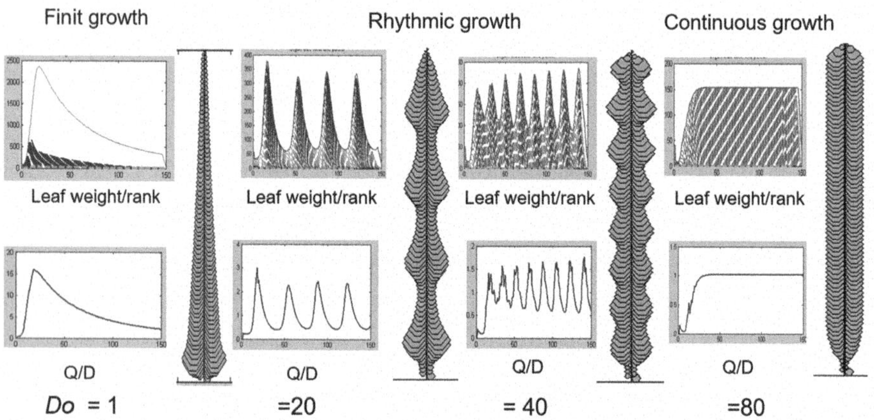

Fig. 13.33 Growth behavior of a single-stemmed plant, dependent on functional sinks. There is one set per organ. The largest organic series is delimited by the red curve. Parameter C_o is set to 1. If D_o is too small, the organic series cannot reach its maximal envelope, causing the growth of the plant to decrease. For $20 < D_o < 80$, the organic series reach their maximum sizes with stable oscillations, resulting in rhythmic growth. For $D_o > 80$, the oscillations disappear, and the plant has continuous growth (Color figure online)

we can see that the limit envelope of the organic series has stable oscillations, and the weights of the organs reach their limit values.

13.6.4.2 Inverse Method for the Computation of Functional Sinks

The parameters of the functional sinks can be computed using the WLSM from the organic series, like how the parameters of the descriptive sinks in the Standard model are computed.

13.7 Application of GreenLab Loop Model to Agronomic Plants

13.7.1 Methodology

In Sects. 13.5 and 13.6, we have proposed simple and arbitrary functions depending on the trophic pressure *Q*/*D* to simulate meristem functioning and organ expansion and to study the behavior of computational plants that involve growth–development interactions. Thus, Eq. (13.14) is used to compute the probabilities associated with meristem functioning, and Eq. (13.22) is used to compute organ expansion. Note that the formulations of organ expansion in the cases are interchangeable, meaning descriptive sinks can be used in the Loop model as well as functional sinks. It has been shown that the parameters of these functions can be computed with the WLSM from simulated plants. However, it is necessary to define a method to find empirical formulations of these functions adapted to real plants. This should be done first for development and second for growth. In the end, the quality of the Loop model fittings should be found comparable to that of the Standard model.

13.7.2 Plant Development with Stochastic Functioning for Meristems: Case of Coffea robusta *(Letort et al. 2020)*

In the first step, the plant crown analysis can give numerical values for meristem functioning during development (*Bernoulli* parameter, branching probability). In the second step, the analysis of the organic series can provide the evolution of demand and biomass produced as a function of the DCs, depending on sink parameters. We thus have the evolution of the trophic pressure *Q*/*D*, as a function of the DCs. From this, empirical functions can be obtained that link the development parameters to the trophic pressure.

13.7.2.1 Results of the Crown Analysis Step

The method of crown analysis, as established in Chap. 3 and used in Chap. 10, was applied to a *Coffea robusta* dataset.

Figure 13.34 presents in a schematic form the topology of eight different trees out of the 15 plants measured, illustrating the variability in their development. Some branches are always missing at the bottom, and the ramification rate increases with plant age. In this first step, this observed variability is assumed to be simply driven by stochastic events with constant or varying probability rates that must be estimated.

Figure 13.35a shows the variability of the number of phytomers of the branches located at rank k below the top of the trunk. The crown analysis method (Chap. 3) provides the values of phytomer *Bernoulli* rates for the trunk ($b_1 = 0.88$) and the lateral axes ($b_2 = 0.93$). This average value is obtained by pooling all branches, regardless of their rank, as well as the rhythm ratio between branches and trunk ($w = 0.9$). Since the values of b_1 and b_2 are similar, they are considered identical in the simulations below.

A bottom-up analysis of the progressive implementation of the branching makes it possible to quantify the evolution of the branching rate at the ranks of the trunk phytomers (Fig. 13.35b). The increase in the probability of branching observed is almost linear, up to rank 14, then it stabilizes at a constant value of 1. Fitting of a power function to these data gives:

$$\begin{aligned} a(x) &= \left(\frac{x}{14}\right)^{1.25} \quad \text{if } x \leq 14 \\ a(x) &= 1 \quad \text{if } x > 14 \end{aligned} \tag{13.26}$$

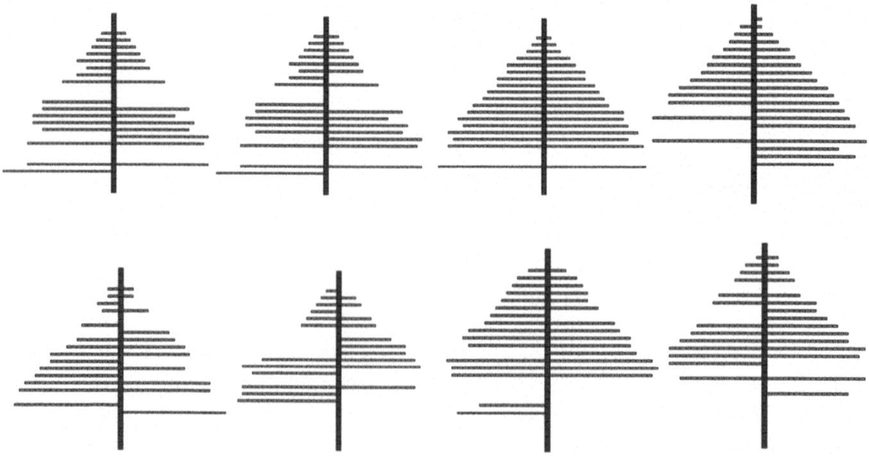

Fig. 13.34 Schematic representation of eight coffee tree crowns among the 15 measured plants

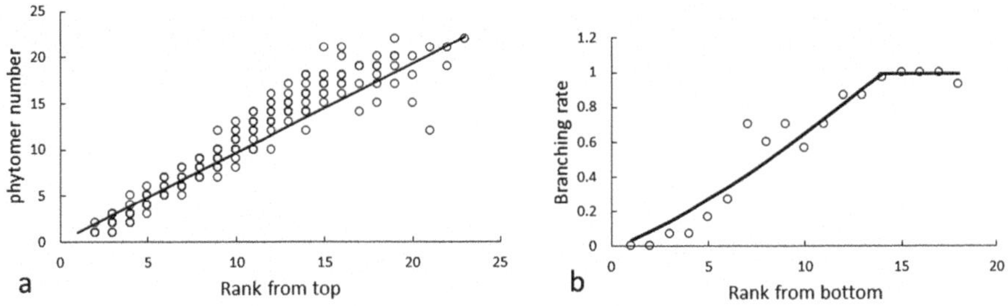

Fig. 13.35 Variability of the number of phytomers and evolution of the branching rate. (**a**) Variability in the number of phytomers of the branches located at rank *k* below the top. (**b**) Evolution of the branching probability from the base of the tree. Lines represent the computed values

The estimated development parameters can now be used to compute the probability of phytomer realization at any position in the potential structure. Based on stochastic development, the average number of phytomers produced at 27 DCs is $N_1 = 24$ for the trunk and $N_2 = 261$ for the branches.

13.7.2.2 Results of the Organic Series Analysis

Some parameters can be directly extracted from observations: the duration of photosynthetic activity of the leaves $t_a = 10$ DCs, and the expansion duration of the leaf and internodes $t_x = 4$ DCs. Average specific leaf weight (SLW), a proxy for leaf thickness, is $e = 0.0095$ g cm^{-2}. In this study, a normalized environment with radiation $E = 1$ at each DC is considered, which assumes average environmental conditions throughout tree growth. This modeling assumption is reasonable as the objective with GreenLab is not a detailed exploration of the environmental influences on plant growth (which would be challenging to decipher for growth in open conditions) but rather the representation of the main dynamics of growth. The values for the hidden parameters are presented in Table 13.9, and the graphs comparing computed to observed data are given in Fig. 13.36. The higher variability of data points for the base of the branches is because the means are computed over a decreasing sample size, as older phytomers are less numerous because branches are aligned from their tips and longer axes are less probable.

13.7.2.3 Reconstruction of the Trophic Dynamics During Tree Growth

Using the estimated parameter values, simulations provide in-depth insights into growth dynamics of coffee trees. By fitting of the model to the observed organic series, we can simulate biomass supply and demand at each growth cycle (DC) (Fig. 13.37). The simulations reveal an initial exponential phase, followed by a linear increase in demand. Biomass produced at each time step exhibits a sigmoidal

Table 13.9 Estimated parameter values (with their estimated standard deviation values)

Symbols	Name	Estimated value	SD	CV%
r	Inverse of resource use efficiency	164.1	13	8
Sp	Production surface	6417	939	15
Q_0	Seed biomass	8.3	2.1	25
P_{a1}	Leaf sink trunk	1	Reference	
P_{a2}	Leaf sink branch	1	0.06	6
P_{i1}	Intern. sink trunk	5.2	3.6	69
P_{i2}	Intern. sink branch	0.77	0.4	52
Pc	Girth growth sink	0.2	0.02	10
α (a)	Sink function parameter 1 for leaf	2.8	1.1	39
β (a)	Sink function parameter 2 for leaf	6	Fixed	
α (i)	Sink function parameter 1 for internode	1.7	0.36	21
β (i)	Sink function parameter 2 for internode	2.5	Fixed	

Note: The reference value for sink values is that of the leaf of the trunk, set to 1. Due to practical non-identifiability, and without loss of generality, only one parameter of the beta density function was estimated. The other one was empirically fixed

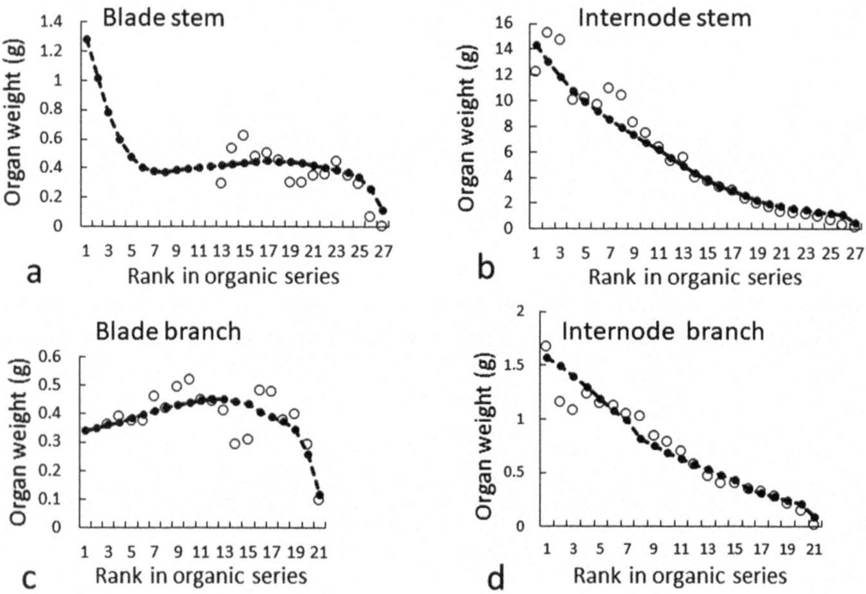

Fig. 13.36 Fitting organic series (o, observed; --, computed) on the trunk (**a**, **b**) and on the branches (**c**, **d**) of *Coffea* tree with Standard model) ($R^2 = 0.95$ on $n = 81$ data). Girth growth is important on the trunk. First leaves on the trunk are missing because they have fallen, but for simulated data, all values are displayed to show the reconstructed weights of the first leaves (leaf weight is displayed even for senescent leaves)

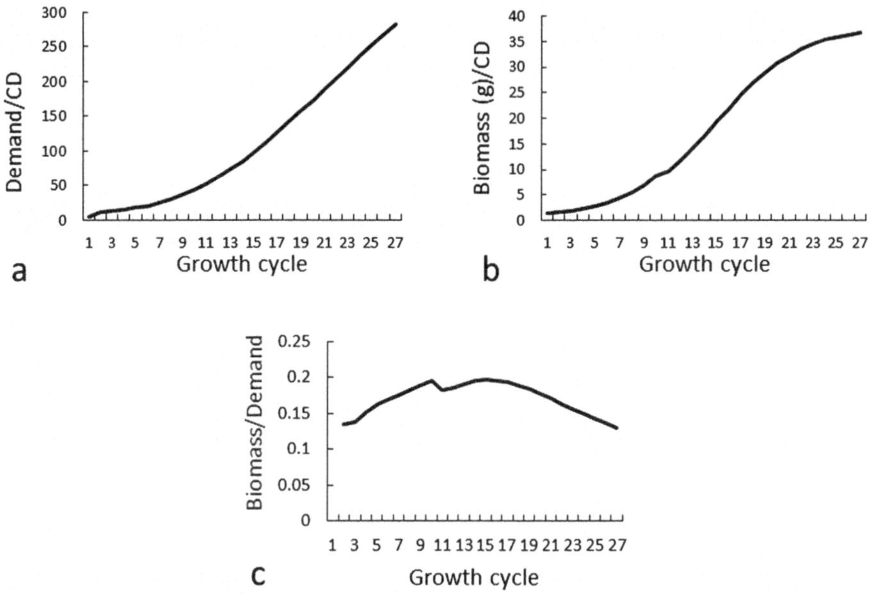

Fig. 13.37 Evolution of the demand (**a**) and biomass production (**b**) during the growth of *Coffea*, as well as their ratio (**c**)

shape, with saturation attributed to self-shadowing of leaves and competition with neighboring trees (production surface $Sp = 6417$ cm^2 from Table 13.9). Note that these coffee trees were only observed for a relatively short period (27 DCs correspond to 2 years). Looking at the ratio of supply to demand, *Q/D*, in Fig. 13.37c, we observe a progressive increase in *Q/D* in the first phase as plant architecture is established, and organ size gradually increases. This is accompanied by the appearance of first branches. In the second phase, there is a subsequent decline in *Q/D* due to the increase in demand while production reaches saturation. It is worthing noting that the irregularity in the *Q/D* curve at DC 10 is an artifact of the model due to the abrupt termination of the oldest leaf's contribution to production. This process is of course smoother (see Sect. 13.6.3 and Fig. 13.30).

13.7.2.4 Influence of Growth on Development Parameters

Results from Figs. 13.35 and 13.37, once combined, allow us to investigate the relationship between internal trophic competition, represented by *Q/D*, and development probabilities. Figure 13.38a shows a clear increasing relationship between the branching rate and *Q/D*. This relationship was fitted by a power function, giving ($x = Q/D$):

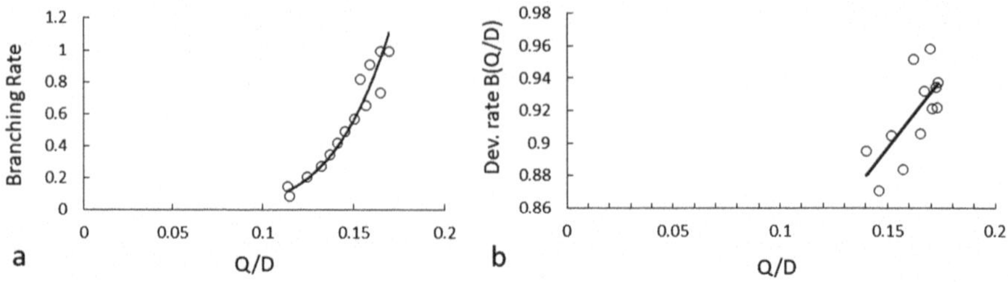

Fig. 13.38 Empirical relationship between *Q/D* and bottom-up branching rate (**a**) and development probability (**b**), respectively

$$a(x) = (23483)^{5.61 \cdot x}, \quad R^2 = 0.95 \tag{13.27}$$

Note that only the data from the establishment phase were included in the regression for the branching rate, as subsequent values consistently had a branching rate of 1, although *Q/D* decreased. A similar analysis was performed for the *Bernoulli* probability of phytomer emission, using the values $b_{2,k}$ estimated from crown analysis at rank K from the top, with the rhythm ratio w kept constant. Only those points corresponding to large enough samples of branches ($n > 15$) were kept, which excluded most basal branches. The points are more dispersed, so the linear regression should be considered as indicating a general trend:

$$b(x) = 1.72 \cdot x + 0.64, \quad R^2 = 0.52 \tag{13.28}$$

13.7.2.5 Recomputation of Demand and Biomass with Trophic Pressure

Equations (13.27) and (13.28) allow us to compute the parameter values for branching and development probabilities as functions of the trophic pressure *Q/D* at each DC. These values were then incorporated into the Loop model, and the resulting complete simulations were compared to previous ones from the Standard model (those without a feedback effect of *Q/D* on the development probabilities) in Fig. 13.39 for the organic series. The agreement between the two models is good for the organic series of leaves and internodes.

13.7.2.6 Discussion

The study of the coffee tree highlights the potential for controlling meristem function during growth through trophic pressure. While the Standard model describes the evolution of the meristem function as a function of DCs, the Loop model explains it as a function of the trophic pressure *Q/D*. The Loop model provides a better insight into plant functioning than the Standard model. However, the Standard model is still

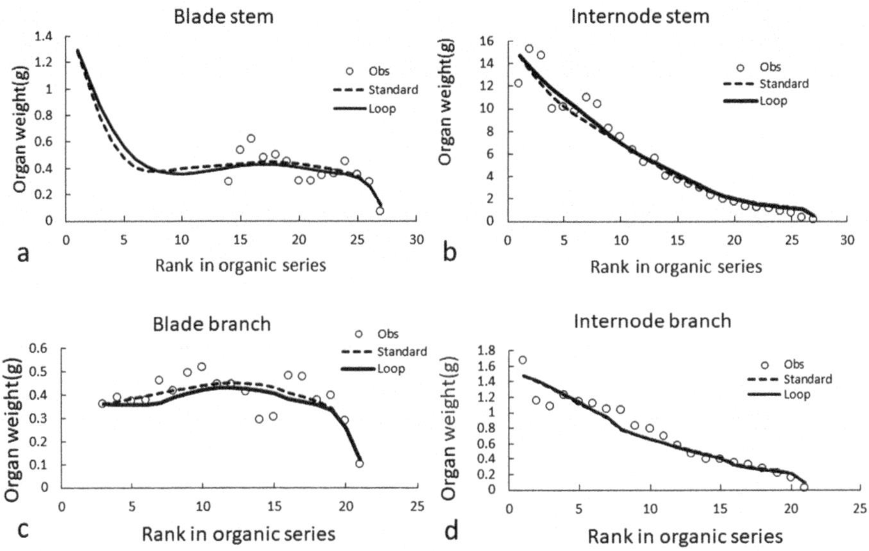

Fig. 13.39 Comparison of organic series calculated by the Standard model without retroaction (dotted lines) and the Loop model, including the regulation of development probabilities by trophic pressure *Q*/*D* (solid lines). Markers of measurements are "o." The differences between both models are hardly visible

necessary for computing the evolution of supply and demand, which are hidden variables. Their numerical values allow us to find empirical functions to link the probabilities of meristem activities to trophic pressure.

13.7.3 Plant Growth with Functional Organ Sinks

The new formulation of the sink functions defined in Sect. 13.6 is used here to fit the organic series already studied with the Standard model (Chaps. 9 and 10), for the single-stemmed plants with non-stochastic meristem functioning. For the Loop model to be used in crop simulation, it is essential that it fits well with the agronomic data from various crops studied using the Standard model (Chaps. 9 and 10). We consider here the deterministic case and single-stemmed plants.

13.7.3.1 Fitting a Computational Plant Produced by Standard Model with the Loop Model

The question is whether the Loop model with functional sinks can correctly approximate the Standard model with Beta law sinks. The answer is affirmative, as showed

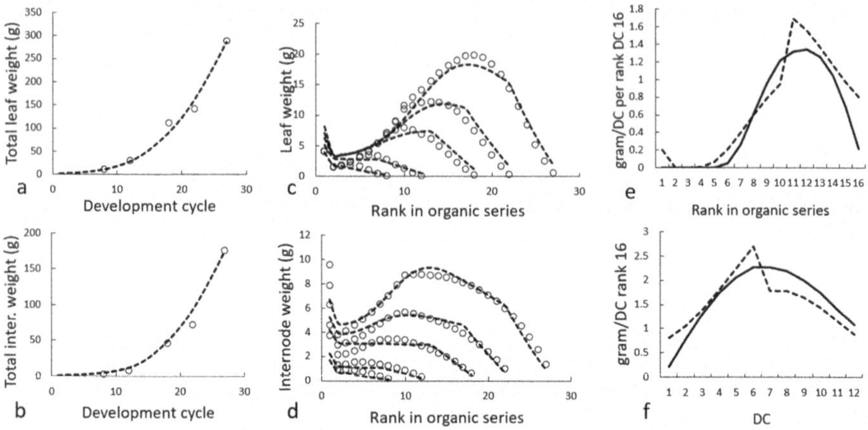

Fig. 13.40 Fitting of organ compartments (**a**, **b**) and corresponding organic series (**c**, **d**) from the Standard model (circles) by the Loop model (dotted lines). The results are good (Gloups software). (**e**) Comparison of biomass fluxes entering each leaf on the stem, at each rank of the organic series when the plant is at DC 16. (**f**) Biomass coming into the leaf at rank 16 when the plant grows from DC 16 to DC 27 (the Standard model shown as solid lines, and the Loop model shown as dotted line)

in Fig. 13.40. We simulate a cotton plant with the Standard model, thus generating a target with its organic series. We then try to fit this simulated target with the Loop model, and all parameters are easily estimated, resulting in good fit. To match this target the best, we need to create six cell sets per organ.

As an example, we compare the biomass fluxes that enter each leaf organ from rank 1 to rank 16 along the stem when the plant is 16 DC aged. We also compare the biomass fluxes that enter the leaf organ of rank 16 to ensure its expansion in both models when plant grows from DC 16 to DC 27 (Fig. 13.40f). The global evolution of the biomass fluxes fits rather well in the two models.

The organic series produced by the simulation with the Standard model is accurately fitted by the Loop model, which provides estimated parameters. This kind of verification is essential for complex systems such as GreenLab model.

This result is important as it demonstrates that both the Loop model and the Standard model are equally capable of calibrating source–sink relationships in crops, despite their differences. In the following, we will study the fitting of the single-stemmed crop plants from Chaps. 9 and 10 using the Loop model. The source–sink parameters of these plants according to the Standard model can be found in Chaps. 9 and 10.

13.7.3.2 Multi-Fitting of Single-Stemmed Cotton Plant by the Loop Model

The data used in this section come from pruning experiments conducted in China on single-stemmed cotton plants (see Chap. 10, Sect. 10.2.1.2, Li Dong, CAU). The organic series of leaves (blades + petioles) and internodes were measured at five developmental stages: [8 12 18 22 27] DCs, with only one plant sampled at each stage. The functioning and expansion times of the organs are estimated to be 20 DCs and are constant (the base effect is negligible). The SLW of the blade is $\varepsilon = 0.024$ cm. As for the previous section, we set $n_a = 6$ and $n_i = 3$ for the number of cell sets for leaf and internode, respectively, that give the best fit for the Loop model.

Parameter C_o is the common sink strength for cell set and is not distinguished by organ type to simplify the optimization procedure. Parameters D_o and X_o are the sink strength of the cell set and expansion rate of cell set belonging to organ *o*, respectively (see Sect. 13.6.4.1.1 of this chapter). The sink strength value for leaf is normalized to $D_b = 1$. Note that for the ring element, there is no expansion, and thus the number of cell sets is 1. D_c corresponds to the sink strength of the ring element.

The fitting error for the Standard model was Error = 23, while for the Loop model, it was Error = 18. Table 13.10 displays the parameter values computed for the Loop model.

The quality of the fitting is quite similar and even better in some cases, with the Loop model compared to the Standard model (Fig. 13.41). However, the sink values cannot be compared as they have different meanings in the two models. Nevertheless, the source parameters *r* and *Sp* are comparable between the two models. For example, values of $r = 87$, $Sp = 7039$ cm^2 for the Standard model are close to the values computed with the Loop model (Table 13.10), which is a satisfying result.

13.7.3.3 Multi-Fitting of Sunflower Plant by the Loop Model

Here we repeat the sunflower data presented in Chap. 9, Sect. 9.5.5. Sunflower is characterized by a large base effect that causes variability in expansion time t_x of organs in 15 ranks on the stem, ranging from 40 DCs to 100 DCs. Fitting the organic series by the Standard model is impossible without considering this variability. In contrast, the Loop model accounts for this variability through functional sinks, where the maximum weight of the organ computed at its creation automatically

Table 13.10 Source–sink parameter values for cotton plant with the Loop model

	r	Sp	Q_o	C_o	D_b	D_p	D_i	D_c	X_b	X_p	X_i
Mean	86	6103	1.51	2.35	1	0.15	0.34	18.0	3.6	2.6	4.2
SD	3.4	708	0.25	0.26	0	0.01	0.06	3.42	0.27	0.20	0.65
CV (%)	4	12	17	11	0	7	16	19	8	8	16

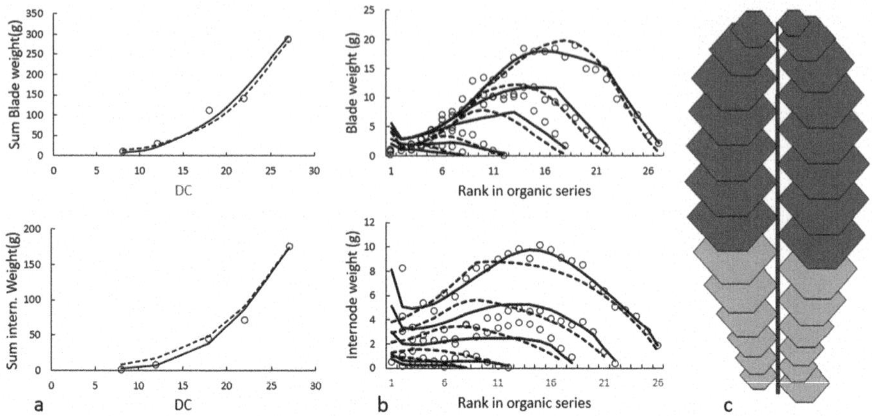

Fig. 13.41 Five-stage fitting [8 12 18 22 27 DCs] of cotton plant. (**a**) Compartments. (**b**) Measures organic series of leaf blades and internodes in a single-stemmed cotton plant (circles) and computed ones by the Loop model (solid lines). The fitting with the Standard model is superimposed (dotted lines). (**c**) The visualization of the computational plant corresponds to the fitting (Data from Li Dong, CAU, and Gloups software)

Table 13.11 Source–sink parameter values for sunflower plant with the Loop model

	r	Sp	Q_0	C_o	D_b	D_p	D_i	D_f	X_b	X_p	X_i	X_f
Values	104.7	1843	0.84	0.09	1	0.46	1.14	1390	172	192	226	3987
SD	5.9	141.7	0.3	0.04	0	0.01	0.04	684	51.5	56.8	65.6	2052
CV (%)	6	8	33	42	0	2	4	49	30	30	29	51

incorporates the effect of Q/D, and the notion of expansion time is not needed. Table 13.10 displays the computed values of 11 parameters for the Loop model.

The functional parameters are well identified from the organic series (Table 13.11), and a value of $n_o = 2$ was empirically set to get the best fitting. The fitting error is Error = 73, which is higher than the error of the Standard model (Error = 40).

For the Standard model, the values for the source parameter are: $r = 144.3$ and $Sp = 2546$, which are close to those of Table 13.11.

Figure 13.42 shows the multi-stage fitting of the organic series (only the 64 and 120 DCs stages are shown). The functional sinks can therefore represent organ growth over long periods of expansion with their variations.

13.7.3.4 Multi-Fitting of Cucumber Plant by the Loop Model

Here we repeat the cucumber data presented in Chap. 9, Sect. 9.5.6. The duration of organ expansion varies with plant development. For leaf blades and petioles, the expansion time t_{xa} varies linearly from 4 DCs to 20 DCs in 4 ranks along the stem. The fruits are developed in positions 3, 4, 5 and 15, 16, 17, which allows the mask to

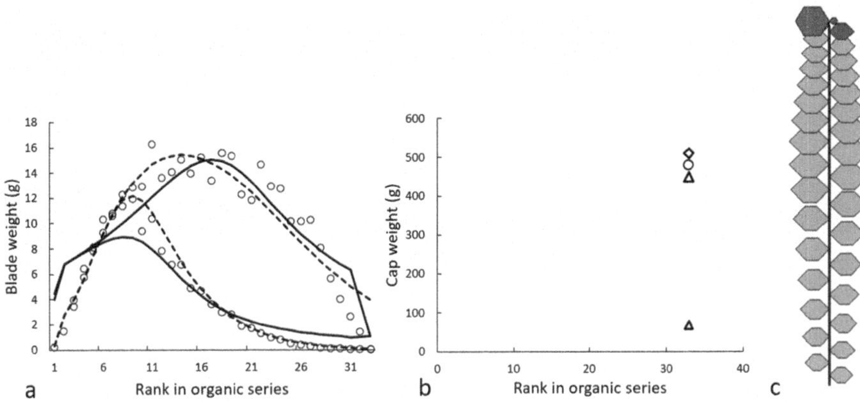

Fig. 13.42 Two-stage fitting [63 120 DCs] of organic series of sunflower. (**a**) Leaf blades. (**b**) Cap in a sunflower plant (circles) by the Loop model (solid lines). Fitting with the Standard model is superimposed (dotted lines). The Standard model is better for fitting leaf organic series mainly for the first rank of phytomers. The cap is well fitted at 120 DCs (Data circle, Standard diamond, Loop. Triangle). (**c**) Visualization of the computational plant corresponds to the fitting for 120 DCs (data from Li Dong, CAU, and fitting by Gloups software)

Table 13.12 Source–sink parameter values for cucumber plant with the Loop model

	r	Sp	Qo	C_0	D_b	D_p	D_i	D_f	X_b	X_p	X_i	X_f
Value	11.1	995	1.5	0.63	1	0.68	0.43	55.0	33.9	35.7	27.9	23.1
SD	0.6	86	0.4	0.05	0	0.04	0.02	4.0	3.0	3.5	2.3	2.4
CV (%)	5.1	9	27	7.46	0	5.51	4.80	7.3	8.8	9.9	8.4	10.4

be calibrated for the fruit series. In other positions, the fruits are aborted at an early stage, with their weight considered negligible.

The functional parameters are well identified from the organic series (Table 13.12). Here, the number of cell sets must be empirically set to the value $n_o = 3$ to get the best fitting. The fitting error is Error = 34, which is close (and even better) to that of the Standard model (Error = 40).

The source parameters for the Standard model were: $r = 13.2$ and $Sp = 1269$, which are similar to those in Table 13.12.

Figure 13.43 shows the multi-stage fitting of the organic series (only the stages of 15 and 31 DCs are shown). The Loop model poorly fits the first ranks of the phytomers, but after the rank six on the stem, the organic series are well fitted.

13.7.3.5 Multi-Fitting on Tomato Plant by the Loop Model ("Zigzag" Phenomenon)

Here we repeat the tomato data presented in Chap. 9, Sect. 9.5.7 (Zhang Baogui, CAU). After an initial vegetative phase, the stem produces sympodial modules of

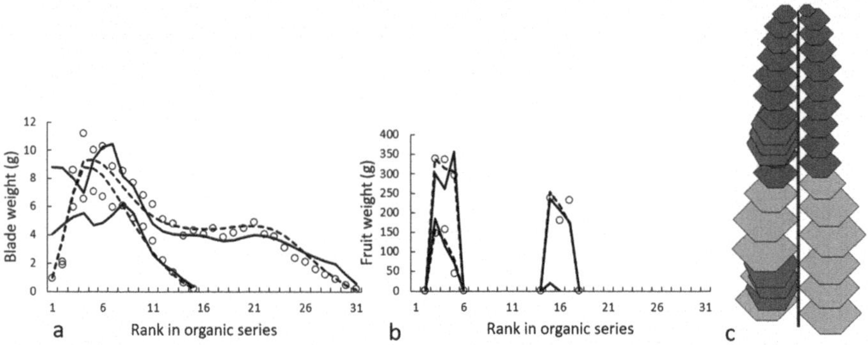

Fig. 13.43 Two-stage fitting [15 31 DCs] of organic series of cucumber. (**a**) Leaf blades. (**b**) Fruits in a cucumber plant (circles) by the Loop model (solid lines). Fitting with the Standard model is superimposed (dotted lines). The Standard model is better for fitting leaf organic series, mainly for the first rank of phytomers. (**c**) The visualization of the computational plant corresponds to the fitting at 31 DCs (Data from Zhang Baogui, CAU, and Gloups software)

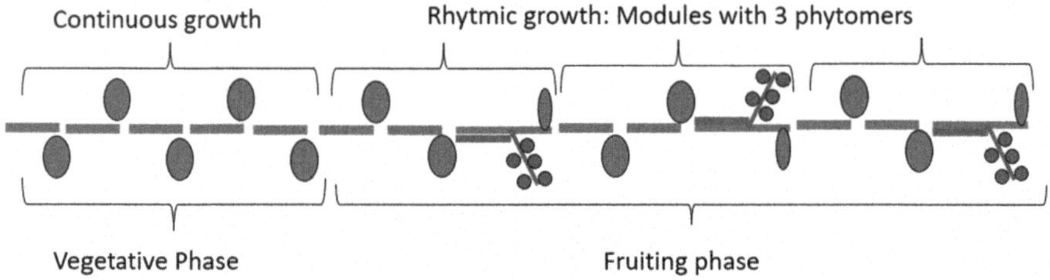

Fig. 13.44 Development scheme of a single-stemmed tomato plant

three phytomers. The first two are vegetative and consist of short internodes and large leaves, while the last one is fruit-bearing and consists of a long internode and a small leaf. It bears a truss of fruits whose petiole is partially fused to the vegetative internode, which artificially increases its weight (Fig. 13.44).

This produces pronounced "zigzags" in the profile of the organic series, including leaves and internodes. Such an effect could be considered as a kind of alternation of physiological sub-ages within a module. Each module reproduces the same pattern: the first two phytomers produce small internodes and large leaf, while the last one produces a large internode, a small leaf, and the truss.

Only the last phytomer of a module bears a truss of fruit. When a new phytomer is created, the cell sets compete for resources. The creation of a truss will automatically reduce the expansion of the leaves created at the same time. This creates two different sizes of leaves and internodes, including the one carrying the truss.

The functional parameters are well identified from the organic series (Table 13.13). Here, the number of cell sets must be set empirically to the value $n_o = 1$ to get the best fitting. The fitting error is Error = 30, which is close to that of the Standard model (Error = 40).

Table 13.13 Source–sink parameter values for tomato plant with Loop model

	r	Sp	Q_0	C_o	D_b	D_p	D_i	D_f	X_b	X_p	X_i	X_f
Value	16.9	2169	1.3	0.21	1	1.10	0.70	28.1	10.1	10.0	10.2	25.7
SD	1.0	223	0.3	0.03	0	0.07	0.05	2.7	1.2	1.3	1.5	3.2
CV (%)	6	10	21	14	0	6	7	9	11	13	14	12

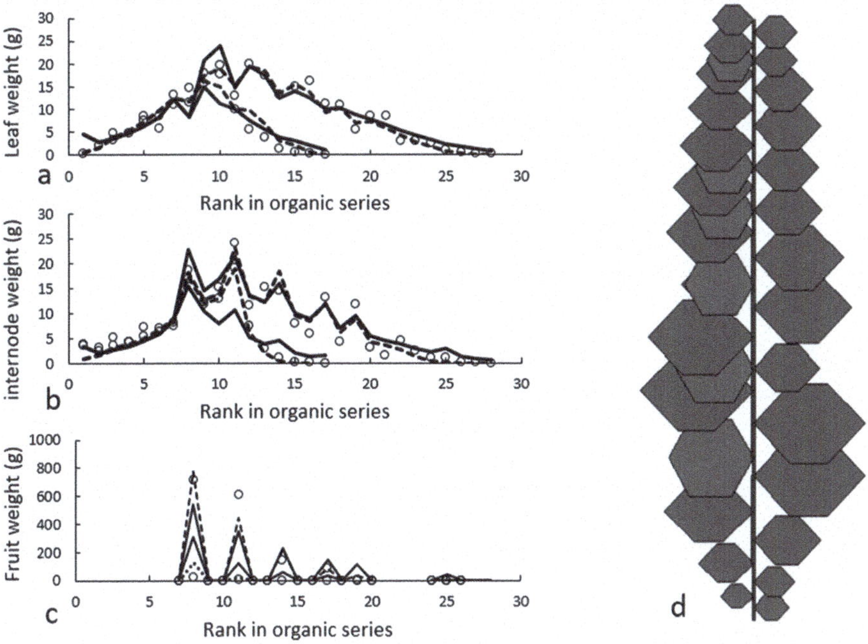

Fig. 13.45 Two-stage fitting [17 28 DCs] of tomato organic series with the functional sinks. Circles represent the observed values; Solid lines represent the values computed by the Loop model; Fitting with the Standard model is superimposed (dotted lines). (**a**) Blades. (**b**) Internodes. (**c**) Trusses. (**d**) Visualization of the simulated plant (Gloups software)

For the Standard model, the values for the source parameters were: $r = 13.4$, $Sp = 2381$, which are close to those of Table 13.13.

The internode weights data had to be corrected (see Fig. 13.44). It was found that a part of the weight of the petiole of the bunch artificially increases the weight of the internode to which it is attached. The corrective term is estimated to be $Zg = 0.03$ (part of the bunch weight that inflates the internode). This correction is applied only to the weight of the bunch and internodes, but the Loop model automatically restored the zigzags on the organic series of blades and petioles.

Note that the Standard model imposes two types of internodes and leaves for the stem, which leads to the zigzags effect. However, this is not the case for the Loop model (Fig. 13.45).

13.7.4 Comparison Between the Performances of the Standard and Loop Models

Table 13.14 shows a comparison of the performance of the Standard and Loop models on the same set of crop data, using the error given by WLSM.

The parameters C_o, D_o, and X_o are well computed separately by the WLSM, making their identification relevant.

The Loop model achieves comparable performance to the Standard model and seems more ecophysiologically relevant. It eliminates the need for variations in expansion time, which complicates the description of the organic series. In the Loop model, the same leaf, internode, and fruit repeat identically along the stem, but their expansion duration can vary along the organic series (a trend particularly noticeable in sunflowers), which is mechanistically generated. This phenomenon is an emergent property of the model, rather than a forcing input. Note that the morphological change of a cotton leaf bottom-up along the stem, for example, which changes from unilobed at the base of the stem to tri- and finally penta-lobed (constitutes heteroblasticity), does not make the organs different from the perspective of their sink function.

The functional sink model is still in its early stages, and therefore should be improved. Its parameters could be used for phenotyping and selection since they are closely related to the physiological functioning of the plant. Moreover, modeling the impact of stress on organ abortion may be easier, as it involves simply reducing the maximum volume of the organ computed at its creation during growth.

13.8 Conclusion

In this chapter, a thorough study of growth–development interactions has been carried out using simple architectural models. The general scheme of the source–sink relationships during the growth of a computational plant (Chap. 5) is well defined and not easily modifiable. But in detail, the source and sink functions can be formulated differently based on different assumptions, which may be comparably effective in fitting the model to the data. However, not all assumptions are equally relevant from an ecophysiological point of view, and determining the best formulation can be challenging in certain cases.

Table 13.14 Accuracy comparison between the two models (error in WLSM)

Plant	Standard model	Loop model
Cotton	23	**18**
Cucumber	40	**30**
Sunflower	**40**	74
Tomato	40	**30**

Regarding development, in Sect. 13.4, a first attempt to link the functioning of meristems to the trophic pressure Q/D is made according to the deterministic mode. The model proposes that the functioning of a meristem is triggered when the Q/D value exceeded a threshold. The model provides a qualitative understanding of the phases of plant development. However, the lack of a stochastic aspect prevents the model from being calibrated on cultivated plants.

In Sect. 13.5, the Loop model incorporates a stochastic aspect where the probabilities of meristem activity depend on Q/D for development, mortality, and branching of the axes, creating a feedback loop between growth and development. The study mainly focused on the free and limited growth modes, whose growth behaviors are already known by reference for the Standard model (Chap. 6). It is important to note that in these versions of the Standard model, the development is forced, even if it can be modified at each stage.

In the Standard model, the mortality of meristems mechanically leads to pruning of the branches, eventually compensating for their renewal. As a result, the demand for plant crown stabilizes.

This phenomenon induces a stabilization of biomass production in free growth if the relation $\frac{(t_a+1)}{2} \cdot \frac{E}{A} = 1$ is verified. The Q/D ratio also stabilizes, leading to a stabilization of the organic series described top-down into living axes. However, in models with feedbacks, where development interacts with growth, it is unclear how this phenomenon would magnify.

These models have a natural boundary for development, which is the potential structure. If all meristems were working, the potential structure would be filled. This remark indicates that the Standard and Loop models share the same potential structure. In the Standard model, the probabilities associated with the creation of each new cohort are predetermined and are programmed by the botanical automaton. In contrast, in the Loop model, these probabilities must be computed at each cycle according to the previous biomass supply and current demand.

The study showed that the analyses done on the Standard models remained valid for the Loop model in the case of continuous and branched growth model. The condition $\frac{(t_a+1)}{2} \cdot \frac{E}{A} = 1$, which leads to the stabilization of the supply in free growth, also holds true for the Loop model.

One of the most interesting results in this study is the ability to determine the functional parameters of the source–sink relationships and the parameters related to the stochastic development of the meristems through inverse method of the organic series. The influence of stochastic development on the profiles of the organic series is strong enough to allow this methodology.

Additionally, introducing fruiting in the plant system enhances the complexity of growth–development interaction. Regulations are established, which allow growth to be maintained by retroaction, thanks to the abortion of fruits and vegetative meristems.

Nevertheless, using homographic functions depending on Q/D for the study of development is an arbitrary choice that does not correspond necessarily to reality, although it qualitatively restores the effect of pruning and branching in growth–

development interactions. In order to efficiently calibrate a cultivated plant, it is necessary to first analyze the plant with the Standard model to obtain the joint evolution of biomass supply and demand. Then, empirical functions can be derived to compute the probabilities of meristem function as a function of *Q*/*D*. This procedure has been successfully applied on the coffee tree (Sect. 13.7.2).

In the Standard model, regarding source–sink relationships, the first formulation of the leaf function was based on the link between transpiration and photosynthesis via the hydraulic resistance of the architecture. However, this approach has been given up since it did not allow the transition from individual plant to stand level, despite the good fit of the source model to the plants.

The source model finally adopted in GreenLab, as presented in Chap. 5, does not particularly improve the fittings to the data, but it places GreenLab in the agronomic framework of the crop models (Chap. 9). It thus allows the transition from the individual plant functioning to the stand functioning and the fine modeling of spatial competition at the individual level thanks to the notion of production surface *Sp*, which is validated in various agronomic experiments where the model is tested (see Chaps. 9, 10, and 11). This method solves the computation of light interception by the plant without the complex and time-consuming use of architecture coupled with ray tracing or radiosity.

The choice between the two kinds of sink function models attached to the Standard model and the Loop model is a challenging task. Both models ensure a proper distribution of synthesized biomass in the plant structure, but from two different points of view.

The first point of view is a descriptive approach based on the evolution of sink strength of an organ during its expansion as a function of thermal time, which agrees with agronomic knowledge. The expansion time of an organ is an important parameter of the Standard model that must be measured or estimated, as it often evolves along an organic series due to the base effect.

The second point of view is a mechanistic approach, which aims to be more explicative and closer to the functioning. When an organ is created, it is initiated by a cell set whose volume depends on its type and *Q*/*D* (via a factor C_o). The maximum volume it can reach depends on an expansion rate X_o, which affects cells specific to its type. The organ stops growing when it reaches its maximum computed volume, which is achieved quickly depending on a sink strength D_o. Thus, sink strength is no longer described by an empirical law that follows its evolution, but by a differential equation (Logistic shape).

The second model eliminated the need for expansion durations as organs stop growing when they reach their maximum computed volume. Thus, the base effect on expansion times is simulated mechanistically without being imposed by a list of expansion times given as input to the simulation.

However, the study of the model's behavior is more complicated for the Loop model and remains to be done in a purely mathematical framework. Despite this drawback, in the long term, it could be beneficial to replace the current descriptive Standard model with the new explanatory Loop model based on the mechanistic approach. This is because it provides a better understanding of plant functioning rather than just describing it. The parameters, which are more representative of biological functions, should prove even more stable and ultimately more suitable for plant selection or optimizing cropping itineraries.

References

Chauvet G-A (2006) A new paradigm for theory in integrative biology—the principle of auto-associative stabilization: biochemical networks and the selection of neuronal groups. J Integr Neurosci 5(3): 381–415

Christophe A, Letort V, Hummel I, Cournède P-H, de Reffye P, Lecœur J (2008) A model-based analysis of the dynamics of carbon balance at the whole-plant level in Arabidopsis thaliana. Funct Plant Biol 35(11): 1147–1162

de Reffye P, Blaise F, Chemouny S, Jaffuel S, Fourcaud T, Houllier F (1999) Calibration of hydraulic growth model on the architecture of cotton plants. Agronomie-Paris 19(3): 265–280

de Reffye P, Marc J, Daniel B, François H (2016) Architecture et croissance des plantes: modélisation et applications. Ed. Quae, Versailles, 684 p. (Synthèses: Quae).

Hallé F, Oldeman R-A-A, Tomlinson P-B (1978) Tropical trees and forests. Springer Verlag, Berlin/Heidelberg/New York. 441 p.

Kang M-Z, Yang L-L, Zhang B-G, de Reffye P (2011) Correlation between dynamic tomato fruit-set and source–sink ratio: a common relationship for different plant densities and seasons? Ann Botany 107(5): 805–815

Le Chevalier V, Jaeger M, Mei X, Cournède P-H (2007) Simulation and visualisation of functional landscapes: effects of the water resource competition between plants. J Comp Sci Technol 22(6): 835–845

Letort V, Sabatier S, Okoma MP, Jaeger M, De Reffye P (2020) Internal trophic pressure, a regulator of plant development? Insights from a stochastic functional–structural plant growth model applied to Coffea trees. Ann Bot 126(4): 687–699

Li Z-P, Le Chevalier V, Cournède P-H (2009) Towards a continuous approach of functional-structural plant growth. In: Li B-G, Jaeger M, Guo Y (eds) Proceedings of plant growth modeling, and their applications (PMA09). IEEE CPS, Beijing, pp 334–340, November 9–13

Mathieu A, Cournède P-H, de Reffye P (2004) A dynamical model of plant growth with full retroaction between organogenesis and photosynthesis. In: CARI'04, 7ème Colloque Africain sur la Recherche en Informatique, Hammamet, 22–25/11/2004

Mathieu A, Cournède P-H, Barthélémy D, de Reffye P (2007a) Condition for the generation of rhythms in a discrete dynamic system. Case of a functional structural plant growth model. In: Fourcaud T, Zhang XP (eds) Second international symposium on plant growth modeling, simulation, visualization and applications, PMA06, 13–17 November 2006, Beijing, P.R. China. IEEE Computer Society, Los Alamos, pp 26–33

Mathieu A, Zhang B-G, Heuvelink E, Liu S-J, Cournède P-H, de Reffye P (2007b) Calibration of fruit cyclic patterns in cucumber plants as a function of source-sink ratio with the GreenLab model. In: FSPM 2007 The 5th international workshop on functional structural plant models, Napier, pp 1–4, 04–05/11/2007

Mathieu A, Cournède P-H, Barthélémy D, de Reffye P (2008) Rhythms and alternating patterns in plants as emergent properties of a model of interaction between development and functioning. Ann Bot 101(8): 1233–1242

Mathieu A, Cournède P-H, Letort V, Barthélémy D, de Reffye P (2009) A dynamic model of plant growth with interactions between development and functional mechanisms to study plant structural plasticity related to trophic competition. Ann Bot 103(8): 1173–1186

Mathieu A, Letort V, Cournède P-H, Zhang B-G, Heuret P, de Reffye P (2012) Oscillations in functional structural plant growth models math. Model Nat Phenom 7(6): 47–66

Nicolini E, Chanson B (1999) La pousse courte, un indicateur du degré de maturation chez le hêtre (*Fagus Sylvatica* L.). Can J Bot 77:1539–1550

Sabatier S, Barthélémy D (2001) Bud structure in relation to shoot morphology and position on the vegetative annual shoots of *Juglans Regia* L. (*Juglandaceae*). Ann Bot 87: 117–123

Yan H-P, Kang M-Z, de Reffye P, Dingkuhn M (2004) A dynamic, architectural plant model simulating resource-dependent growth. Ann Bot 93(5):591–602

Zhang B-G, Kang M-Z, Letort V, Wang X-J, de Reffye P (2009) Comparison between empirical or functional sinks of organs-Application on Tomato plant. In: 2009 third international symposium on plant growth modeling, simulation, visualization and applications. IEEE, pp 191–197

Chapter 14
Computer Plant Simulation Technics: Case of GreenLab Model

Marc Jaeger, Philippe De Reffye, Jing Hua, and Mengzhen Kang

Abstract The GreenLab model emphasizes modeling natural plants according to the physics approach rather than adhering strictly to computer science methodologies. It forms the basis of the Gloups software, which creates detailed two-dimensional (2D) and three-dimensional (3D) representations of plant structures by simulating meristem activities. This chapter delves into the intricacies of the GreenLab model and how it translates into the Gloups software's graphic generation process. The chapter begins with a review of various plant simulation models and then narrows in on the GreenLab model, detailing its parameters and constraints that shape the graphical output. The GreenLab-based computer simulations construct plant topologies using nested substructures like Russian dolls and convert them into architectural models, with photosynthesis and harvest considerations being optional depending on the study's focus. The construction of the geometric structure uses vector operators, which are described in detail. The longevity of such software is contingent on regular updates to reflect technological progress and theoretical evolutions. Despite the need for further refinement for commercial application, the Gloups software's versatility and Matlab compatibility suggest a promising future for its use in plant modeling.

M. Jaeger · P. De Reffye
CIRAD, UMR AMAP, Montpellier, France

AMAP, Univ Montpellier, CIRAD, CNRS, INRAE, IRD, Montpellier, France
e-mail: marc.jaeger@cirad.fr

J. Hua
State Key Laboratory for Multimodal Artificial Intelligence Systems, Institute of Automation, Chinese Academy of Sciences (CASIA), Beijing, China

M. Kang (✉)
State Key Laboratory for Multimodal Artificial Intelligence Systems, Institute of Automation, Chinese Academy of Sciences (CASIA), Beijing, China

School of Artificial Intelligence, University of Chinese Academy of Sciences, Beijing, China
e-mail: mengzhen.kang@ia.ac.cn

P. De Reffye et al. (eds.), *Plant Model Greenlab for Botany and Agronomy*,
https://doi.org/10.1007/978-981-97-7541-5_14

Keywords Miscellaneous Functional-Structural Plant Models (FSMPs) · Topological and geometrical algorithms · GreenLab story

14.1 Introduction

In 1988, at the Siggraph conference, two pioneering papers were presented that focused on the modeling of plant growth.

The first paper (Prusinkiewicz et al. 1988) is the first important communication of the application of L-system-based computer grammars to plant growth. The main goal was to develop an elegant formalism able to handle the complexity of plant architectures. At the beginning, this mono-disciplinary approach in computer science did not integrate botanical or agronomical knowledge, and virtual plants were generated. It was only later that the L-systems began to integrate the necessary concepts to stimulate agronomic plants (see in the Introduction, L-Peach and L-Kiwi).

The second paper (de Reffye et al. 1988) was a multidisciplinary approach to plant growth where knowledge from Botany and Agronomy are integrated in the AMAP (botAny and Modelling of Plant Architecture and vegetation) mathematical model to generate computational plants. What counts here was the modeling of meristems (stochastic process) functioning and the organ (source–sink relationships) functioning at the phytomer scale, based on observations. The computer formalism of the plant construction was secondary as long as it was efficient enough to take in account the experimentally validated mathematical models.

Jean Françon (1990) has discussed this approach:

> The objective of AMAP requires that the modeling be validated in all its details by accurate observations and measurements in the field, which no other method has done. This methodology is that of Physics. This is why I consider that the AMAP model is to be ranked as an experimental validated theory (p. 242).

Jean Françon adds: "The two approaches [L-systems and stochastic simulation] can be merged into one". (p. 243)

L-systems were widely adopted afterwards, especially in the field of computer science. For its part, the GreenLab model remained more concerned with validating the model on natural plants in the sense of physics than with adopting a formalism more in line with the requirements of computer science. The attempts in this direction by the mathematicians of the GreenLab group (Kang et al. 2007) make the model more readable for computer scientists (but not for botanists!) but have not really added any improvement to the model. Finally, the method of Graphical Models was found to be generic and well adapted to model plant architectures in their complexity.

So, the computer simulation of plant architecture based on GreenLab relies on the functioning of meristems, which build the plant topological structure. This structure is then transformed into two-dimensional (2D) or three-dimensional (3D) architecture using geometric operators (Chaps. 2, 3, 4, and 5), resulting in a computational plant. If the goal is only to build the plant architecture without considering photosynthesis and harvest, the source–sink relationships do not need to be factored into the computation. However, for computational plants (Chaps. 5, 6, 7 and 8), photosynthesis and harvest must be considered.

This chapter refers to the specificities of the GreenLab model. The nature of the parameters involved in this model, as in many others, responds to specific constraints. Generating a graphic structure is a complex process, and this imposes certain choices in its implementation.

After an overview of the different types of existing models dedicated to plant simulation, this chapter will focus on the GreenLab model and its implementation in the Gloups software.

14.2 History of Computer Plant Simulation

At the beginning, plant architectural models only considered the branching pattern, and less frequently, the cessation of axis development, also known as mortality [Fischer and Honda 1977; Aono and Kunii 1984; Oppenheimer 1986]).

The elongation of the axes was often viewed as too trivial to model, and the notion of meristem was absent. However, it is crucial to model the growth of single-stemmed plants (i.e., Corner's model), which includes rhythm, *Bernoulli* and viability processes as discussed in Chap. 3, is essential to properly understand plant growth, including photosynthesis. This section provides an overview of the various approaches to computer simulation of plant growth. The following descriptions of these different plant architecture models are directly inspired by the authors themselves, as found in literature.

14.2.1 Pattern Approach

One of the inherent difficulties in procedural or rule-based approaches is the challenge of defining procedures or rules to obtain a specific effect. While inverse methods based on statistical analysis of plant architectures can address this difficulty, they can be difficult for beginners to master. In multimedia areas, graphic designers appreciate the easy and intuitive control over the shape and expression of the objects they manipulate (Zhao and Barbič 2013). This constraint is very strong, as it requires the procedural aspects to be adaptable by an external user and relies on a specific, highly graphical interface (Chen et al. 2008). However, the plants obtained through this approach may only have graphic applications because the development and growth are not modeled, and the botanical inclusion is limited.

14.2.2 Structured Recursion (Bosanac 1990)

ONYX_TREE software uses a structural tree model that employs structured recursion to build a tree image, making the modeling process more intuitive. This approach allows for functional differentiation between the trunk, branches, and twigs, which are the principal elements of tree perception. This differentiation makes it possible to implement a top-down (perceptual) modeling process and strengthens the ability of the model to represent a variety of species. The hierarchical structure of the model is a direct result of the application of structural recursion (Bosanac 1990). The 3D plant library available through ONYX_TREE is impressive. More information can be found on their website: https://www.onyxtree.com/about.html.

14.2.2.1 Rule-Based Approach (Deussen and Lintermann 1997)

The Xfrog model uses a rule-based approach along with traditional geometric modeling techniques, making it easy to generate many branched objects including flowers, bushes, trees, and even non-botanical objects. This is achieved through a set of components that describe the structural and geometric elements of plants, forming a graph that generates the geometry of a specific plant. Users get immediate feedback on what they have created: geometric parameters, tropisms, and free-form deformations can control the overall shape of a plant. The method can handle the complexity of most real plants (Deussen and Lintermann 1997). The 3D plant library provided by Xfrog is also impressive. For more information, see the website: https://www.xfrog.net/.

14.2.3 Formal Grammars and Combinatory Trees

Formal grammars and combinatory trees are used to generate tree structures.

The course: https://perso.liris.cnrs.fr/eric.guerin/glpoly.pdf, introduces the relation between grammar and automaton including:

- Factorization in grammars.
- Specific grammars for 3D object (including plants) generation.

14.2.3.1 Combinatory Trees (Viennot et al. 1989)

The ultimate appearance of a tree depends on the developmental processes leading to its unique forms. To quantify the form of a tree or its branching pattern, a new concept called the "ramification matrix" is introduced. Using this matrix, an algorithm is developed to create a random tree with a given arbitrary stochastic triangular matrix. The geometry of the tree is defined by combinatorial parameters that analyze its form. This method provides powerful control over the final form of the tree, while still allowing for variety and the rendering of the images (Fig. 14.1). The combinatorial analysis involved represents a more detailed approach to a concept previously

Fig. 14.1 Some tree realizations from the combinatorial approach

utilized in hydrogeology in the morphological study of river networks. Recently, the concept of ramification matrix has also been used in physics to investigate fractal ramified patterns (Viennot et al. 1989).

The structure is a binary tree with branches generated through random draws. The image in the top left depicts a perfect binary tree.

14.2.3.2 L-Systems

An L-system, also known as a Lindenmayer system, is a type of rewriting system or formal grammar used in theoretical computer science. It was invented in 1968 by the biologist Aristid Lindenmayer to model the development and proliferation of plants or bacteria. This system consists of an alphabet of symbols that can be used to make a collection of production rules that expand each symbol into some larger string of symbols, an initial "axiom" string from which to begin construction, and a mechanism for translating generated strings into geometric structures. See:

https://en.wikipedia.org/wiki/Formal_grammar.

Originally, L-systems were devised to provide a formal description of the development of such simple multicellular organisms, and to illustrate the neighborhood relationships between plant cells. Later, this system was extended to describe higher plants and complex branching structures (Fig. 14.2). L-systems have been the subject of many implementations, which allow the transition from the virtual plant to the computational plant. They simulate architectural models and "source–sink relationships" (Prusinkiewicz et al. 1996). The L-systems are a major reference for the simulation of virtual plants, thanks to the contribution of Prusinkiewicz. The

Fig. 14.2 Trees generated by L-systems

book *The Algorithmic Beauty of Plants* written by Lindenmayer and Prusinkiewicz in 1990, is a must-read in this field. To learn more, see the website: https://allenpike.com/modeling-plants-with-l-systems/.

14.2.4 Multi Agent System Object Oriented (Eschenbach 2005)

The functional–structural tree growth model ALMIS (Eschenbach 2005) uses the individual-based modeling approach and is implemented in the object-oriented programming language SIMULA. The model uses local specification features (state variables) and functions (processes) at the level of the single plant organs. These elementary units include internodes, leaves, meristems, and roots, and increasing numbers of these units are replicated to form the growing tree. By combining plant physiology and architecture, ALMIS enables the study of how single ecophysiological and structural processes impact whole tree growth and the tree–environment system. The model employs various procedures, including photosynthesis, nutrient uptake, transport, storage, mobilization, respiration, and growth, to describe carbon and nutrient uptake, and matter fluxes between different plant organs.

ALMIS can also simulate the effects of microclimate, self-shading, variable nutrient availability, variable transport dynamics, and branching patterns on tree growth. At higher levels of plant organization, interactions at the organ and sub-organ levels can give rise to emergent properties such as lifetime spectrum of single organs, space filling ("architecture"), and self-thinning of the crown. The model can simulate different growth forms of trees under various conditions, resulting in the emergence of different emergent properties (Fig. 14.3).

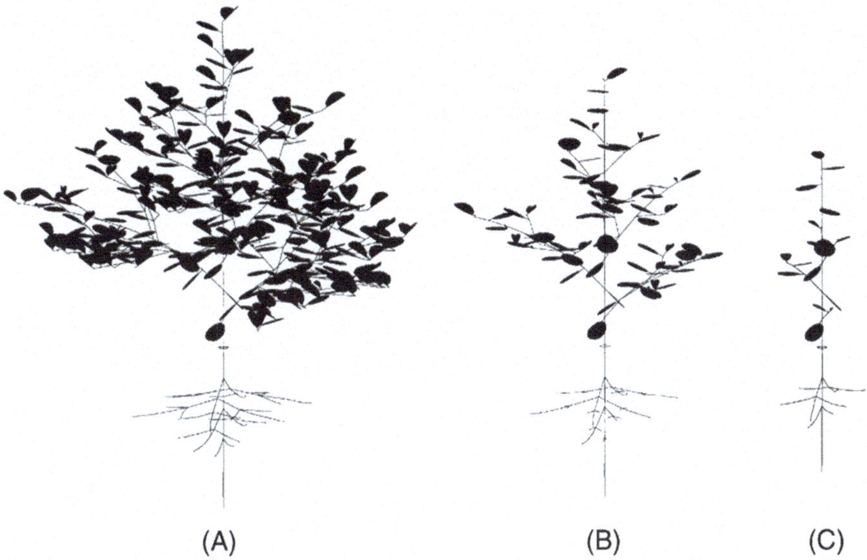

Fig. 14.3 The effects of favorable and unfavorable photosynthetic conditions on modeled tree growth. Each of the seedlings was grown under three different constant microclimatic conditions for 110-time steps: optimal conditions (**a**), low temperature (**b**), and low light (**c**)

14.2.5 Self-Organizing Tree Models (Palubicki et al. 2009)

This method using L-systems is based on the biological hypothesis that the form of a growing tree results from a self-organizing process where the buds and branches compete for light or space, regulated by internal signaling mechanisms. Through simulations of this process, a diverse array of realistic trees and bushes can be generated reliably. Various interactive techniques such as procedural brushes, sketching, and editing operations like pruning and bending of branches can be employed to control the produced forms (Palubicki et al. 2009).

Although this method generates beautiful and realistic tree architectures, it lacks significant botanical and agronomical knowledge (Fig. 14.4).

14.2.6 Graphical Model (Zhao et al. 2001)

A graphical model, also known as a probabilistic graphical model (PGM; see https://en.wikipedia.org/wiki/Graphical_model), is a type of probabilistic model that uses a graph to represent the conditional dependence structure between random variables. These models are commonly used in probability theory, statistics (particularly Bayesian statistics), and machine learning.

Fig. 14.4 Development and tree forms of self-organizing tree model. (**a**) The process of self-organizing tree development. Each terminal segment attempts to create new branches by growing forward. A segment is only added to the structure if it does not collide with any existing or attempted segment. The colors in the figure represent different generation steps. (**b**) Sampled tree forms generated by using the priority model of bud fate control. *Left*: apical control is removed from both main stem and lateral branches early in development. *Center*: apical control is limited to the main stem and removed late in development. *Right*: apical control is removed early in the main stem but persistent in first-order lateral branches

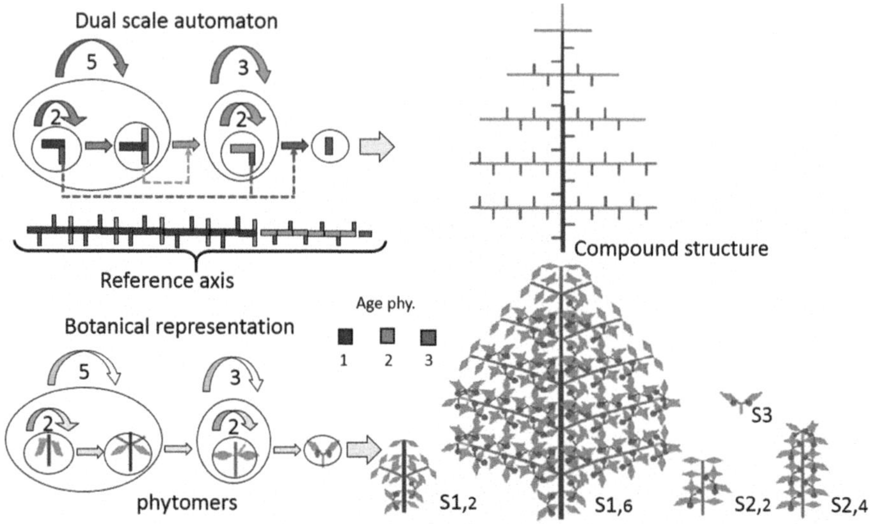

Fig. 14.5 GreenLab organogenesis model. The program of the development of a rhythmic plant using a botanical automaton. The model includes detailed information about the characteristics of the phytomers, including their morphology, repetitions patterns, ramifications, and PAs. The structures and substructures of the plant are generated directly from the botanical automaton, allowing for precise control over the development of the plant (Zhao et al. 2001)

The GreenLab model, which is based on a Botanical Automaton, utilizes a graphical model in the form of a graph (Fig. 14.5) to program the development of the axes based on the morphological description of the phytomer and its functional characteristics such as physiological age (PA), repetition, branching, and mortality. This method simplifies the programming of the development processes significantly (Chap. 5).

14.3 Elements for the Development of Software Tools for Plant Simulation

- Software tools are indispensable for simulating and representing plant models, as they serve diverse purposes, and are designed to achieve associated goals:
  - They serve as a means of communication, facilitating information exchanges (Jaeger and de Reffye 1991).
  - They are useful for internally verifying the validity of the model, which often requires a computer implementation and testing (Kang et al. 2012). The model evaluation requests comparisons with real data.
  - The results of the simulations in the form of tables and graphs allow a better understanding of the data coming from the measurements on the plants (Dauzat et al. 2008).
  - Software tools enable the studies of new cases, capitalizing on knowledge obtained through partnerships, thereby valorizing knowledge and skills (Jaeger and de Reffye 1992; Soler et al. 2003).
  - The implementation of the model presents algorithmic challenges, which often contribute to new advances in the model itself. For instance, the use of instantiations to reduce computational and memory costs has resulted in the formalization of the notion of substructures (Yan et al. 2002; Cournède et al. 2006) (Chaps. 2 and 4).
- Like many digital applications, we can categorize tools between data entry and analysis, processing tools like simulation or calibration, and restitution tools such as property or structure visualizations. The originality of software tools lies mainly in the complexity of the processing tools. These tools must implement the formalism used, often with a growth dynamic, while still being manageable in terms of complexity in both implementation and computation time. This complexity comprises two distinct yet interconnected aspects. The first is related to determine the dimensioning of the implementation of the model, the second is related to the manipulated quantities, which are integrated into a structure that is constructed incrementally by selecting a specific path.

14.4 Simulation Complexity and Dimensioning

The architecture of a modeled plant can be quite complex, not only in terms of the number of organs it comprises but also due to the intricate topological and chronological relationships between them.

This complexity arises from the multidimensional elements involved in constructing the plant structure. When examining a specific state (at a time t) of the formal or computerized system implementing the GreenLab model, the phytomers are indexed according to four main quantities, as outlined in Chap. 2:

Table 14.1 Evolution of the number of phytomers and computation time for a simple theoretical tree

N° simulation	DC	$N\varphi 1$	$N\varphi 2$min	$N\varphi 2$max	*Nb*	*Taxd(ms)*	*Tstr(ms)*	*Tfn(ms)*
1	5	25	80	100	196	7.5	28.5	53.9
2	9	45	240	295	568	12.1	100.9	257.3
3	12	60	335	495	973	15.4	216.7	534.8
4	15	75	520	695	1486	18.6	398.2	1098

Each line represents a stochastic simulation conducted on 70 individuals. The DC indicates the number of development cycles. The growth units consist of six growth tests for two physiological ages (PAs). $N\varphi 1$ denotes the number of phytomers created at PA 1, while $N\varphi 2min$ and $N\varphi 2max$ specify the minimum and maximum numbers of phytomers created at PA 2, respectively. *Nb* represents the total number of phytomers tested, including both created phytomers and pauses. Finally, *Taxd*, *Tstr*, and *Tfn* indicate the computation times (in ms) required to generate the development axes, and the development and growth of this structure, respectively (see also Chap. 8, Sect. 8.6.5)

- The PA
- The chronological age (CA) (and ontogenic age), expressed in number of development cycles (DCs)
- A stochastic index
- The level of reiteration

In certain cases, additional quantities related to the plant structure can be incorporated. For instance, in the case of inflorescences, a topological distance from the seed or from the end of the axis is used (Chap. 7).

These concepts are indexed for each organ compartment, such as internodes, leaf, male fruit, female fruit, and roots. They can be regarded as independent variables. Moreover, irrespective of the topological aspects, the models are dynamic, and the relationships that describe the evolution of the organs are continuous functions, such as the beta law, whose computation is not straightforward.

Due to this complexity, memory resources can become limiting quickly. When all the phytomers have to be accessible for all the limiting dimensions, the potential of high-performance computers becomes saturated. Table 14.1 illustrates this explosion in a simple branched architecture.

14.5 Procedural and Dynamic Aspect: Branched Structures Paths

The formalism of the GreenLab model is suitable for the use of iterative or recursive procedures. The structure carrying the organs is built according to development cycles. However, the internal computer representation of the plant's structure can differ significantly from that of the botanical establishment of the structure.

In computer science, trees are well-known objects and derived from graph theory, consisting of nodes and edges linking them. Each node has a unique link (an edge) to a father node and right and left son nodes (in a binary tree). The root is the only node without a father, while the leaves are nodes without sons. Two typical tree traversals are depth-first and breadth-first. Due to the inherent hierarchy of plant structures induced by PAs, leading to dominant axes (trunk versus branches and twigs), several types of traversals can be introduced.

Implementing structural plant growth models such as GreenLab organogenesis model is challenging because it is necessary to traverse a structure that is built up as we go along. The choice between breadth-first and depth-first paths leads to different simulation approaches and algorithms.

In the breadth-first traversal, the structure is built at all ages of development, and the simulation can stop or extend at any time. All nodes (meristems) are necessary in memory to contribute to the evolution of the structure. If mechanical effects are included (curvature of axes), all phytomers are involved.

On the other hand, in a depth-first traversal, the final age of development must be specified. Traversals are established from the root to the terminal axes, advancing in time. In terms of simulation, this means going back and forth at different dates to build the structure. Different methods can be used to build optimized traversals. Following and prioritizing the ramifications minimize the storage requirements, called depth-first pre-order traversal, which corresponds to the prefix order. Alternatively, duplications can be exploited as much as possible by exploring ramifications but processing only the explored one, called depth-first post-order traversal, which corresponds to the creation of cohorts.

Both types of paths are classic in computer science. Introducing PAs, or more precisely, the order of ramification, leads to specific depth-first traversals called "by rank-first". In this case, the structure is constructed and explored in ascending order: the trunk, the branches coming from the trunk, etc., ending with the ultimate branches. The advantages of depth-first traversals are that it is not necessary to store the entire structure, and it is possible to work on notions of integrated subsystems biologically relevant.

In practice, there is no "optimal" pathway in implementation when examining the advantages and disadvantages of the various traversals. In our developments, the depth-first pre-order is favored for its economical path, while the breadth-first traversal is explored for gene simulation (see below). Graphic applications with the rank-first traversal have been developed, but visualization is carried out a posteriori, with full knowledge of the complexity of the structure to be represented (Fig. 14.6).

The program Gloups uses the List mode to store the phytomers of the simulated structures in a prefixed order.

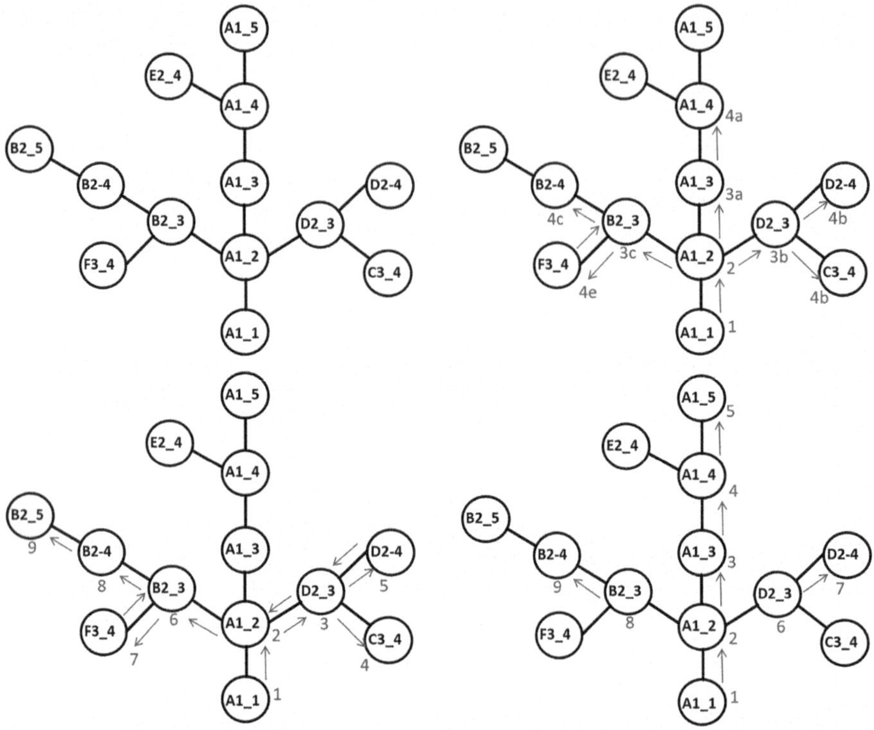

Fig. 14.6 Schematic representation of the width and depth of tree path. *Top left*: a topology example is shown, where the first letter represents an axis, the following number denotes the order of branching (or a PA), and the third number defines the date of appearance in the structure (the ontogenic age). *Top right*: the breadth-first traversal, which reflects the dynamics of the structure's node placement that corresponds to a cohort. The number in red indicates the stage at which the node is traversed or constructed. Nodes can be constructed simultaneously in this path (e.g., 3a, 3b, 3c). *Bottom left*: the depth-first pre-order traversal, which is built with a systematic prior on branching that corresponds to the prefix order. *Bottom right*: the rank-first traversal, which is constructed by systematically favoring the axis on which one is located. The depth paths in this traversal are not related to the temporality of the dynamics of the structure's installation

14.6 Vector Operators for the 3D Representation of Plant

To construct the 3D structure of a plant, we use vector operators that define the geometrical information of each organ. We encode a plant structure from a stack of items (organs), and each organ dimension, orientation, and position are defined by four geometrical parameters (Fig. 14.7).

1. Point of attachment $M(x_1, x_2, x_3)$ defined with three spatial coordinates, which defines the absolute position (origin) of the organ.
2. Principal direction vector: $\overrightarrow{V_1}\,(a_1, a_2, a_3)$, whose norm defines its length.
3. Secondary direction vector: $\overrightarrow{V_2}\,(b_1, b_2, b_3)$ (V_2 is orthogonal to $\overrightarrow{V_1}$), whose norm defines the width.

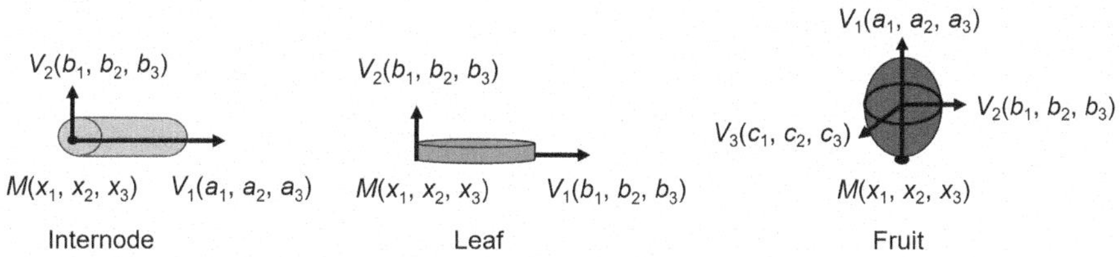

Fig. 14.7 Schematic representation of the organ shapes

4. The last secondary vector V_3 is orthogonal to V_1 and V_2, whose norm is generally the same as V_2.

Note that the size of the organ, which is related to its volume and its allometric rules of development in several directions, is encoded in the V_1, V_2, and V_3 vectors.

Plane defined by $\overrightarrow{V_1}$ and $\overrightarrow{V_2}$ is used to distinguish three main types of organ shapes. The vector $\overrightarrow{V_3}$ is perpendicular to this plane and is used to represent the three-dimensional orientation of each organ.

- Cylinder: internodes
- Blade: leaves
- Ellipsoids: fruits

14.6.1 Vectors Elementary Operations

The construction of the skeleton of a plant can be described using vector algebra. Here are some common vector algebra operations that are useful in this context.

14.6.1.1 Addition of Vectors

Let $\overrightarrow{V_1} = (x_1, x_2, x_3)$ and $\overrightarrow{V_2} = (y_1, y_2, y_3)$, we have:

$$\overrightarrow{Vs} = \overrightarrow{V_1} + \overrightarrow{V_2} = (x_1 + y_1, x_2 + y_2, x_3 + y_3) \tag{14.1}$$

The operation is used to compute translations.

14.6.1.2 Multiplication of Vectors

The scalar product of two vectors, $\overrightarrow{V_1}$ and $\overrightarrow{V_2}$, is defined as a number R. If the angle between the two vectors is denoted as θ, we have:

$$R = \vec{V}_1 \cdot \vec{V}_2 = |V_1| \cdot |V_2| \cdot \cos(\theta) = x_1 \cdot y_1 + x_2 \cdot y_2 + x_3 y_3 \tag{14.2}$$

Here, $|V_1|$ denotes the modulus of $\overrightarrow{V_1}$. The modulus is equal to one when the vector is normalized. The value of the modulus is defined as:

$$|V_1| = \sqrt{x_1.x_1 + x_2.x_2 + x_3.x_3} \tag{14.3}$$

When $\overrightarrow{V_1}$ and $\overrightarrow{V_2}$ are normalized, the scalar product returns the cosine of the angle θ between them.

The vectorial product of two vectors, $\overrightarrow{V_1}$ and $\overrightarrow{V_2}$, is defined as a vector:

$$\overrightarrow{V_3} = \vec{V}_1 \wedge \vec{V}_2 = (x_2 \cdot y_3 - x_3 \cdot y_2, x_3 \cdot y_1 - x_1 \cdot y_3, x_1 \cdot y_2 - x_2 \cdot y_1)$$
$$|\overrightarrow{V_3}| = |V_1| \cdot |V_2| \cdot \sin(\theta) \tag{14.4}$$

The resulting vector $\overrightarrow{V_3}$ is perpendicular to the plane created by $\overrightarrow{V_1}$ and $\overrightarrow{V_2}$. It is necessary to normalize $\overrightarrow{V_3}$ by dividing its coordinates by $\left|\overrightarrow{V_3}\right|$.

14.6.1.3 Rotation of Vectors

We define the general case of rotating a vector $\overrightarrow{V_1}$ by an angle ω around an axis with direction given by $\vec{U}$, where the modulus of $\vec{U}$ is one. The angle between V_1 and $\vec{U}$ is denoted by α, and $\cos(\alpha)$ is equal to the scalar product $\vec{V}_1 \vec{U}$, since both vectors are normalized. As a result of this rotation, a new vector $\overrightarrow{V_2}$ is obtained, as shown in Fig. 14.8.

$$\overrightarrow{V_2} = (1 - \cos(\omega)) \cdot \left(\vec{U} \cdot \overrightarrow{V_1}\right) \cdot \vec{U} + \cos(\omega) \cdot \overrightarrow{V_1} + \sin(\omega) \cdot \left(\vec{U} \wedge \overrightarrow{V_1}\right) \tag{14.5}$$

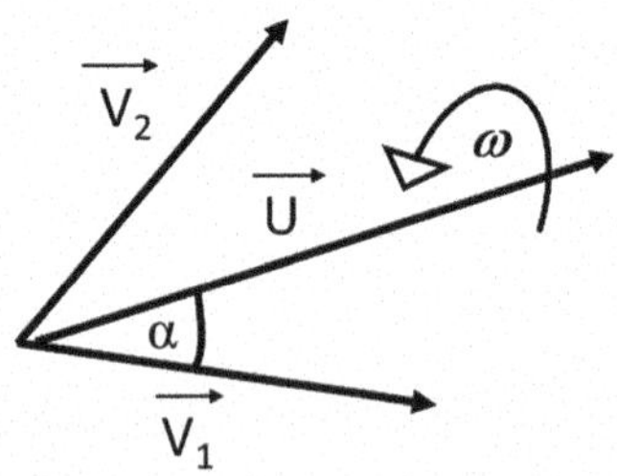

Fig. 14.8 Rotation of a vector around an axis. The vector $\vec{U}$ is selected as the axis of rotation. Vector $\overrightarrow{V_1}$ can transform into $\overrightarrow{V_2}$ by rotating $\overrightarrow{V_1}$ around $\vec{U}$ using angle ω

By using translations and rotations, the positions of all organs on the geometrical structure of the plant can be determined.

Affine matrix operators are of interest implementing these operations.

For instance, we can define $\overrightarrow{V_2}$ as follows: $\overrightarrow{V_2} = R\ \overrightarrow{V_1}$.

Where matrix 3*3 [*R*] is defined as:

$mc(\omega) \cdot a^2 + c(\omega)$	$mc(\omega) \cdot a \cdot b - c \cdot s(\omega)$	$mc(\omega) \cdot a \cdot c + b \cdot s(\omega)$
$Mc(\omega) \cdot a \cdot b + c \cdot s(\omega)$	$Mc(\omega) \cdot b^2 + c(\omega)$	$Mc(\omega) \cdot b \cdot c - a \cdot s(\omega)$
$Mc(\omega) \cdot a \cdot c - b \cdot s(\omega)$	$Mc(\omega) \cdot b \cdot c + a \cdot s(\omega)$	$Mc(\omega) \cdot c^2 + c(\omega)$

with: U as the normed rotation axis: $U = (a, b, c)\ (a^2 + b^2 + c^2 = 1)$

$c(\omega) = \cos(\omega)$

$Mc(\omega) = 1.0 - \cos(\omega)$

$s(\omega) = \sin(\omega)$

14.6.1.4 Geometry of Axes

An axis is built up with a series of cylinders in a single file. In our implementations, the axis is a curve that belongs to elasticity theory and lies within a vertical plane, assuming pure bending. The curve is formed by adding internodes one by one and is composed of vectors $x(i)\ \overrightarrow{V_{(i)}}$ ($\overrightarrow{V_1}$ type), which represents the principal directions (Fig. 14.9). The modulus of $\overrightarrow{V_{(i)}}$ is equal to 1, and $x(i)$ indicates the length of the internode. The angle between two vectors on the curve is computed using elasticity theory or tropism considerations. The normal directions ($\overrightarrow{V_2}$ type) of the internodes lie within a vertical plane. The axis of rotation $\vec{U}$ of vector $\overrightarrow{V_1}$ is defined by the vector product $\left(\overrightarrow{V_1} \wedge \overrightarrow{V_2}\right)$. The length of the axis, *S*, made up with *n* internodes is given by:

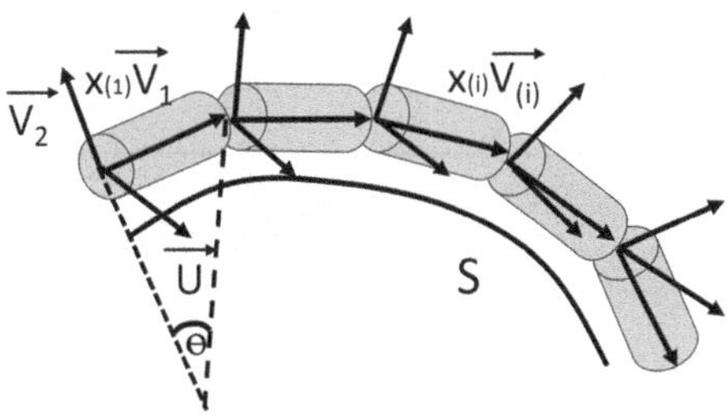

Fig. 14.9 Bending of an axis as a series of cylinder rotations. $\overrightarrow{V_2}$ is in a vertical plane and perpendicular to $\overrightarrow{V_1}$ (the axis of the internode). $\vec{U}$ is in the horizontal plane and equal to $\overrightarrow{V_1} \wedge \overrightarrow{V_2}$. The vector $\overrightarrow{V_1}$ will then rotate by an angle of θ according to the bending law

$$S = \sum_{i=1}^{n} x(i)$$

14.6.1.5 Axis Insertion

The direction vector (vertical) of the first phytomer of the main axis (main stem) is defined as:

$$\overrightarrow{V1} = (0, 0, 1).$$

In Fig. 14.10, to determine the direction of the first branch of an axis for other branches, we need to know the directions of the two previous bearing axes, $\overrightarrow{V_1}$ and $\overrightarrow{V_2}$. Let us call the direction of the first branch $\overrightarrow{V_3}$, which makes an angle θ with $\overrightarrow{V_2}$.

To find $\overrightarrow{V_3}$, we choose an intermediate vector $\vec{A}$ that lies in the plane defined by $\overrightarrow{V_1}$ and $\overrightarrow{V_2}$. Next, we find a rotation axis $\vec{U}$ that is perpendicular to both $\overrightarrow{V_1}$ and $\overrightarrow{V_2}$. This is done using the following operation:

$$\vec{U} = \overrightarrow{V_1} \wedge \overrightarrow{V_2}$$

We can obtain the vector $\vec{A}$ by rotating $\overrightarrow{V_2}$ around $\vec{U}$ with an angle of θ.

$$\vec{A} = (1 - \cos(\theta)) \cdot \left(\vec{U} \cdot \overrightarrow{V_2}\right) \cdot \vec{U} + \cos(\theta) \cdot \overrightarrow{V_2} + \sin(\theta) \cdot \left(\vec{U} \wedge \overrightarrow{V_2}\right) \quad (14.6)$$

Similarly, we can obtain $\overrightarrow{V_3}$ by rotating $\vec{A}$ around $\overrightarrow{V_2}$ by an angle of ω.

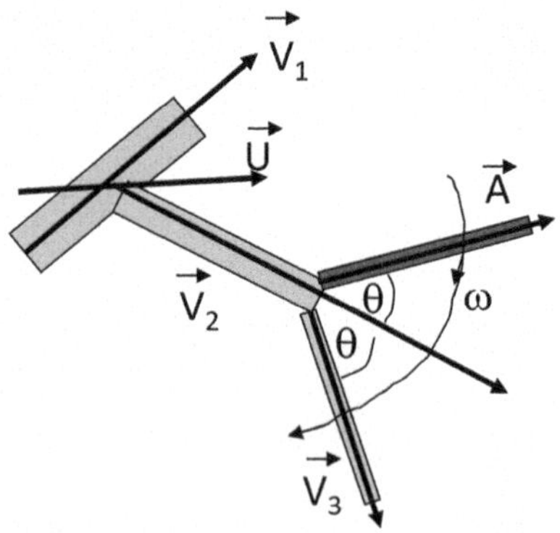

Fig. 14.10 Computation of the insertion direction of an axillary branch

$$\overrightarrow{V_3} = (1 - \cos(\omega)) \cdot \left(\overrightarrow{V_2} \cdot \vec{A}\right) \cdot \overrightarrow{V_2} + \cos(\omega) \cdot \vec{A} + \sin(\omega) \cdot \left(\overrightarrow{V_2} \wedge \vec{A}\right) \quad (14.7)$$

Special cases:

If $\omega = 0$, then $\overrightarrow{V_1}$, $\overrightarrow{V_2}$, and $\overrightarrow{V_3}$ are in the same plane.

If $\omega = \frac{\pi}{2}$, then the axis $\overrightarrow{V_3}$ is perpendicular to the plane defined by $\overrightarrow{V_1}$ and $\overrightarrow{V_2}$, which is suitable for initiating plagiotropic branches on the main axis.

Phyllotaxy can be determined by rotating the vector $\overrightarrow{V_3}$ around the vector $\overrightarrow{V_2}$ and multiplying the phyllotaxis angle ω by the number of internodes from the beginning of the carrier branch. When dealing with a tier of branch, the rotation angle is given by $\omega = \frac{2\pi}{n}$, where n is the number of branches. However, these rules need to be adapted for precisely positioning organs such as leaves and fruits on the internodes.

14.6.2 Modeling Beams' Bending

Plants grow under the constraint of gravity, which contributes to bending the tree architecture. To determine the shape of a bending axis, one needs to consider various factors such as the axis profile (length and diameter evolution), the elasticity of the material, the direction of the beginning of the axis in relation to verticality, and the application points and directions of the forces. The theory of elasticity provides the equations to solve the problem, but plant architecture has certain unique characteristics.

Plants do not follow the usual condition of mass conservation in mechanics, and their growth can result in significant deformation. The woody material of branches is also not homogeneous. While mechanics typically deals with light deformations, the bending of branches can result in high levels of deformation, as seen in weeping-willow. Moreover, as branches grow in length and diameter, they can encounter complex and varying stresses that affect their shape and deformation. Each year, a new layer of wood is deposited along the branch length, which support no constraint, but adds weight to the branch.

The curves of a growing branch bending under its own weight are more complex than those of the pure flexion of an axis supporting forces along its length. As living beings, plants have evolved mechanisms to resist the force of gravity, including the production of reaction wood, which can invert the bending of branches and help the tips of stems reach verticality. This living phenomenon is known as orthotropy. To simulate the effect of gravity on plant architecture during growth accurately, one must take into account all these considerations.

An initial approach to simulating the deformation of plants under the force of gravity was developed by de Reffye in 1974 on coffee trees. This approach dealt with high deformation of beams with conic profiles under a force application and proposed a simulation of orthotropy. The model was simplified because each branch bends independently and the woody internal material was assumed to be homogeneous. Despite this simplification, the model produced good visual results, and by empirically adjusting the mechanics parameters, such as the Young's elasticity modulus, the simulated shape of the plant could look faithful. This model was efficient and suitable for numerous applications, including landscaping.

The complete problem of simulating the equilibrium of a growing tree under its own weight was solved by Fourcaud (1995) (see Chap. 12), and the simulation of the internal allocation of new wood inside the tree structure at each growth cycle was accounted for to compute the internal constraints correctly. This method was much more advanced than the previous approach with a high level of mechanics knowledge, making it possible to set feedbacks between growth and gravity.

Here we will describe the simple approach of beam bending as described by de Reffye (1979), which is based on well-known notions in mechanics. We define:

- E as the Young's elasticity modulus.
- I as the quadratic moment of the beam section. For circular section, the value of I is $I = \pi \cdot \frac{R^4}{4}$, where R is the radius of the section.
- r as the bending radius at a position on the beam. If the beam does not bend, r is infinite.

By definition, the bending at a position on the beam is: $\frac{1}{r} = \frac{d\theta}{ds}$, where θ is the angle of bending and s is the length of the beam.

The theory of elasticity models the relationship between the moment of the force M, which makes the beam bend, and the bending angle θ at a position on the beam. This relationship can be expressed as:

$$M = E \cdot I \cdot \frac{d\theta}{ds} \tag{14.8}$$

Suppose a vertical force F is applied to the tip of a cylindrical beam of length h, which is set at an angle θ_0 with respect to the vertical. In this case, the value of I is constant.

Figure 14.11 shows a visual description of the beam bending.

The vertical force applied on the tip of the beam is decomposed in two forces: one that ensures pure flexion and is perpendicular to the beam, and the other that is parallel and ensures compression. The deformation results in an angle ω from the original direction and a shifting e of the tip.

Using the notations of Fig. 14.11, we can write the fundamental equation as:

$$E \cdot I \cdot \frac{d\theta}{ds} = F \cdot \cos(\theta_0) \cdot (e - y) + F \cdot \sin(\theta_0) \cdot (h - x) \tag{14.9}$$

We also have: $dx = ds \cdot \cos(\theta)$ and $dy = ds \cdot \sin(\theta)$.

Partial differentiation of Eq. (14.9) gives:

Fig. 14.11 Description of the beam bending

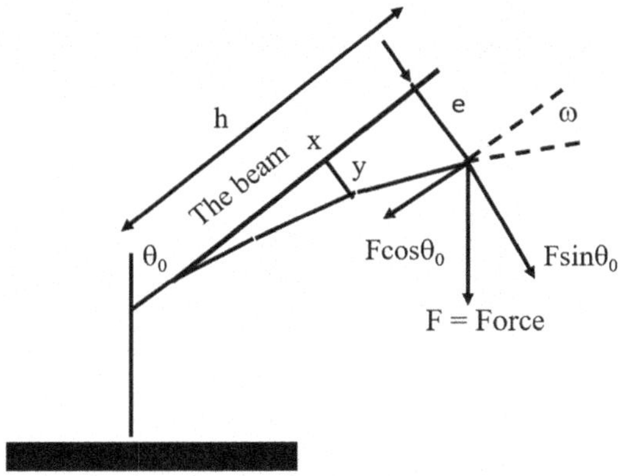

$$E \cdot I \cdot \frac{d^2\theta}{ds^2} = F \cdot \cos(\theta_0) \cdot (-\sin(\theta)) + F \cdot \sin(\theta_0) \cdot (-\cos(\theta)) \tag{14.10}$$

We can multiply both sides of Eq. (14.10) by $\frac{d\theta}{ds}$ and integrate the expression with θ varying from 0 to θf, giving:

$$\frac{1}{2} \cdot E \cdot I \cdot \left(\frac{d\theta}{ds}\right)^2 = F \cdot \cos(\theta_0) \cdot (\cos(\theta) - \cos(\theta f)) + F \cdot \sin(\theta_0) \cdot (\sin(\theta) - \sin(\theta f)) \tag{14.11}$$

If we define $K^2 = 2 \cdot \frac{F}{E \cdot I}$, we can write:

$$d\theta = K \cdot \sqrt{\cos(\theta + \theta_0) - \cos(\theta f + \theta_0)} \cdot ds \tag{14.12}$$

The shape of the curve resulting from the flexion of the beam can be obtained by integrating:

$$\int ds = h = \int_0^{\theta f} \frac{d\theta}{K \cdot \sqrt{\cos(\theta + \theta_0) - \cos(\theta f + \theta_0)}} \tag{14.13}$$

When θ reaches the value θf, the sum of ds has the value h, which is the length of the beam. This allows us to numerically solve Eq. (14.13) and find the value of θf. Starting with a θf that is too small will result in $\int ds < h$, while a θf that is too large will result in $\int ds > h$. A dichotomic search can be used to quickly find the solution for θf.

We can compute the coordinates X_s and Y_s at each point with position s on the beam as follows:

$$
\begin{aligned}
X_s &= \int_0^s \cos(\theta_0 + \theta_s) \cdot ds \\
Y_s &= \int_0^s \sin(\theta_0 + \theta_s) \cdot ds \\
\theta_s &= \int_{x=1}^s K \cdot \sqrt{\cos(\theta_x + \theta_0) - \cos(\theta f + \theta_0))} \cdot ds
\end{aligned}
\tag{14.14}
$$

This system can be easily simulated numerically if we know θf. We choose ds small enough and start with $\theta = 0$. With several iterations, we can draw the curve of the flexion of the beam step by step. At the step x, we have:

$$
\begin{aligned}
d\theta_x &= K \cdot \sqrt{\cos(\theta_x + \theta_0) - \cos(\theta f + \theta_0)} \cdot ds \\
\theta_x &= \sum_1^x d\theta_i \\
X_s &= \sum_1^x \cos(\theta_0 + \theta_X) \cdot ds \\
Y_s &= \sum_1^x \sin(\theta_0 + \theta_X) \cdot ds
\end{aligned}
\tag{14.15}
$$

This method provides a quick and good approximation for bending.

Case of a Cone Shape for the Beam

Let $\alpha = \frac{1}{H}(H > h)$ be the conicity of the beam. At the beginning, the radius is R_0. At a distance, the radius is:$R_s = R_0 \cdot (1 - \frac{s}{H})$

So, the quadratic moment of the section at s is: $I_s = \pi \frac{R_s^4}{4}$

So, the value of the previous parameter $K_0 = \sqrt{2 \cdot \frac{F}{E \cdot I}}$ becomes:

$$
K_s = \sqrt{2 \cdot \frac{F}{E \cdot \pi \cdot \frac{R_0^4 \cdot (1 - \frac{s}{H})^4}{4}}} = \frac{K_0}{(1 - \frac{s}{H})^2}
$$

Equation 14.13 becomes:

$$
K_0 \frac{ds}{(1 - \frac{s}{H})^2} = \frac{d\theta}{\sqrt{\cos(\theta + \theta_0) - \cos(\theta f + \theta_0)}}
\tag{14.16}
$$

Integrating this equation gives the following expression for the first term:

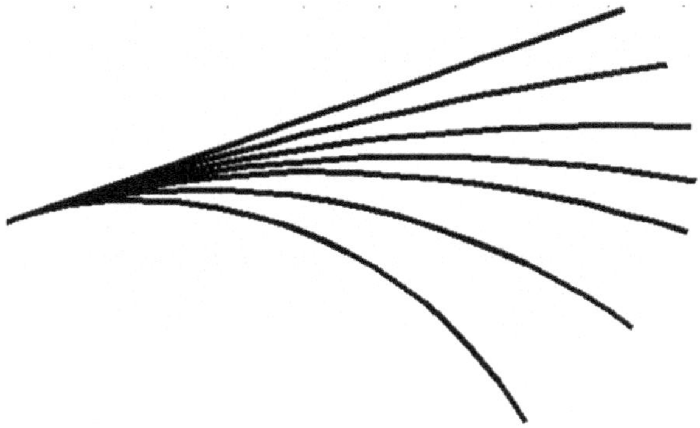

Fig. 14.12 Progressive bending of a beam due to the application of a force at its end and the decrease of Young's modulus

$$\left[\frac{H}{1-\frac{s}{H}}\right]_0^h = \frac{H}{1-\frac{h}{H}} - H = \frac{h}{1-\frac{h}{H}} \tag{14.17}$$

Let us define $\alpha = h/H$ as the conicity. It must be less than 1. When $\alpha = 0$, we find a shape of a cylinder. Therefore, Eq. (14.13) can be rewritten as follows:

$$\frac{h}{1-\alpha} = \int_0^{\theta f} \frac{d\theta}{K \cdot \sqrt{\cos(\theta + \theta_0) - \cos(\theta f + \theta_0)}} \tag{14.18}$$

Using this approach, we can numerically determine the value of θf as given by Eq. (14.13). Figure 14.12 shows an example of beam bending under the influence of an applied force at the top of the beam, illustrating how the degree of bending varies with the Young's modulus.

14.6.3 Various Geometrical Applications Concerning Angles on Plant Architecture

The computation of angles in plant architecture pertains to various aspects such as the angles of insertion, phyllotaxis, subsidence, bending, and straightening of the branches (Fig. 14.13).

14.7 History of the Successive Implementations of the GreenLab Model

The implementation of a structural–functional model such as GreenLab is a complex task that involves developing multiple tools within the same framework. Before using the model for a specific application, it is necessary to create the appropriate

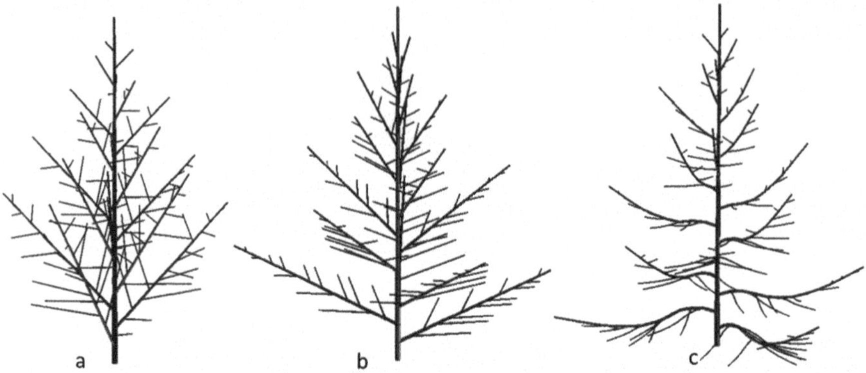

Fig. 14.13 Computation of angles in plant architecture under different factors. (**a**) Phyllotaxis effect. (**b**) Subsiding and plagiotropic effects. (**c**) Bending and righting effect

parameter sets. Similarly, once the simulations are complete, users need to be able to examine the results both qualitatively (such as through visualizing structures) and quantitatively (such as through analyzing curves).

14.7.1 Generations of GreenLab Software Implementations

The initial software development focused on the generic simulation of plant architecture development, utilizing botanical automaton (visual plant software). The second software focused on growth simulation of single-stemmed plants (cornerfit software). The subsequent software evolutions (GreenScilab, Digiplante, QinYuan, Gloups) synthesized the features of the first two.

Since its origin in 1998, the development of the GreenLab formalism has been primarily based on scientific matrix computing environments, and more particularly on the Matlab (R) environment from MathWorks and similar open-source environments such as Scilab (http://www.scilab.org/) or Octave (https://www.gnu.org/software/octave). Currently, three implementations are available under these scientific environments.

- GreenScilab is a free tool designed for education and training, featuring test datasets and focusing on deterministic modeling. This free tool is based on the Scilab environment, and it can be downloaded from the GreenLab course webpage: (http://greenlab.cirad.fr/GLUVED/html/P3_Tools/Tool_GreenScilab.html).
- StemGL is another free tool designed for education and training, featuring test datasets and dedicated to single-stemmed plants. This free tool operates under both Matlab and Octave environments. The Octave environment and documentation for StemGL can be accessed via the GreenLab StemGL project webpage: http://greenlab.cirad.fr/StemGL/.

- Gloups is the most advanced implementation, but access to the tool requires a shared research project with specific terms of use. This version runs under both Matlab and Octave environment, and enables the applications of both deterministic and stochastic models.

These implementations include a set of tools for parameterization, simulation, calibration (inverse method), and optimization. However, the specificity of scientific programming environments such as Matlab or Scilab does not allow for strong scalability in terms of memory space and execution speed, which prohibits simulations of large numbers of phytomers that an adult tree may present. To address this, versions designed in general-purpose languages such as C, C++, and java have been developed. These include:

- The Digiplante tool, piloted by the Digiplante team at École centrale Supélec, which will give birth to the dgpSDK development environment (see below) that enables, among other things, the simulation and visualization of 3D scenes from the GreenLab model.
- The QingYuan simulator, piloted by the Institute of Automation of the Chinese Academy of Sciences (CASIA) (Hua and Kang 2011; Hua et al. 2011).
- The Gloups operator in java under Xplo, piloted by the Amap unit (Feng et al. 2012; Griffon and de Coligny 2012).

14.7.1.1 Cornerfit (Zhan et al. 2003)

Cornerfit software is deterministic and was used to compute the source–sink parameters of single-stemmed plants such as maize, sunflower, tomato, wheat, and beet using the Weighted Least Squares Method (WLSM) (Fig. 14.14). It was also utilized to model their growth as depicted in Figs. 13.20 and 13.28 in the Chap. 13.

14.7.1.2 VisualPlant (Zhao et al. 2001)

VisualPlant (Zhao et al. 2001) belongs to the family of Graphical Models and is a stochastic software used solely for continuous or rhythmic plant development. It is a remake version of the reference axis, modeled by a botanical automaton, and is capable of easily drawing inflorescence models and architectural models of trees (Fig. 14.15).

14.7.1.3 GreenScilab (Kang et al. 2007)

GreenScilab, a tool developed by Kang Mengzhen-Zhen and Qi Rui (Kang et al. 2007) at the Institute of Automation, Chinese Academy of Sciences (CASIA), is the first complete GreenLab software that includes a user manual, an interface for

Fig. 14.14 Three-dimensional (3D) plant simulation with the Cornerfit software (Sunflower, maize, tomato) (Zhan et al. 2003)

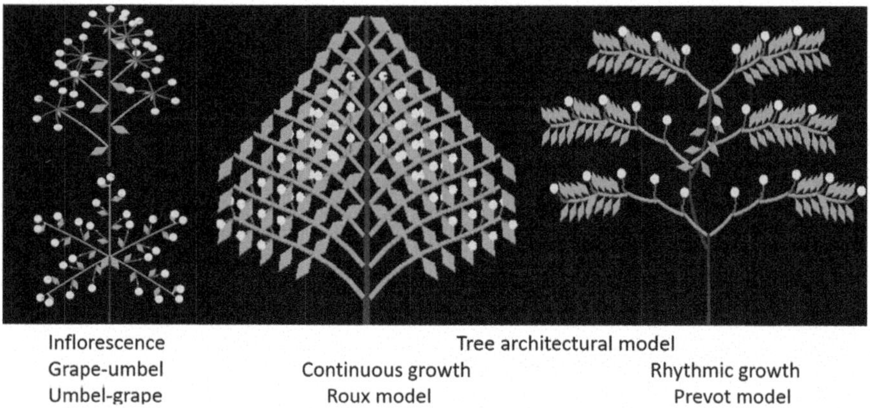

Fig. 14.15 Two-dimensional (2D) plant simulations with the VisualPlant software (inflorescences, trees) (Zhao et al. 2001)

parameter input, a source–sink solver for organic series, and 3D output simulations (Fig. 14.16).

GreenScilab implements structural stochastic development, continuous and rhythmic growth, and sink–source production simulation. It takes advantage of substructure factorization for production and simulation and is particularly effective for plants with simple structures. The tool offers calibrated datasets on plants of agronomic interest based on field measurements. Developed in the SciLab environment, GreenScilab is available from: http://greenlab.cirad.fr/GLUVED/html/P3_Tools/Tool_GreenScilab.html, and includes the Scilab 4.3 environment. Although

Fig. 14.16 Three-dimensional (3D) plant simulations with the GreenScilab software (Kang et al. 2006). (**a**) Gramineous. (**b**) Flowering plant. (**c–e**) Gingko tree (profile, top views, single branch)

still operational, it is no longer maintained due to compatibility issues with Scilab versions 4.5 and above.

14.7.1.4 Digiplante, dgpSDK, and PyGMAlion

The Digiplante team has developed two software packages centered on the use of the GreenLab model: the Digiplante tool and the dgpSDK tool.

Digiplante, created by Paul-Henry Cournède, is a C++ software with a graphical interface that allows for simulation, calibration, and visualization of the model (Fig. 14.17). This tool is available on Windows. After a few years, some components of this software were decoupled to create the dgpSDK (Digiplante Software Development Kit), which is a set of libraries and multi-platform applications, also developed in C++, that implement the GreenLab model.

Initially developed for individual plant simulation, dgpSDK now allows for simulating the growth of stands on large scenes and considers their interactions, such as the *Sp* parameter of the GreenLab model. These tools, initially based on the GreenLab model, provide improved control of the modeling work chain, including modeling, sensitivity analysis, calibration, and uncertainty analysis. They have also led to the creation of PyGMAlion, a tool dedicated to the study of all stochastic discrete dynamic systems (Cournède et al. 2013).

Fig. 14.17 Plant species simulated and calibrated with Digiplante tools. (**a**) Beetroot. (**b**) *Arabidopsis*. (**c**) Wheat. (**d**) Rice. (**e**) Maize. (**f**) Sunflower. (**g**) Chrysanthemum. (**h**) Cucumber. (**i**) Tomato. (**j**) Sweet pepper. (**k**) Coffee. (**l**) Pine

While the notion of growth unit (GU) is still rudimentary and the stochastic mode is not implemented, these applications are fast, reliable, and efficient with inverse methods and visualization. They also manage growth on development feedback through thresholds indexed on Q/D in a deterministic framework.

14.7.1.5 QingYuan

QingYuan is an autonomous tool developed by the research engineer Hua Jing (CASIA) (Hua and Kang 2011; Hua et al. 2011), is built using the C++ language in the Qt environment (http://www.qt.io/), and implements advanced stochastic structural features. Unlike Digiplante, Qingyuan offers interactive 3D visualization and pruning features through a user-friendly graphical user interface (GUI). However, at present, the inverse method is not available yet.

QinYuan is currently the most powerful software available to build large tree architectures from botanical automaton and source–sink relationships (Fig. 14.18). It is capable of simulating architectural models with the detail of the growth units, including pre-formation, neo-formation, and polycyclism, and takes into account all the necessary characteristics for accurate simulation.

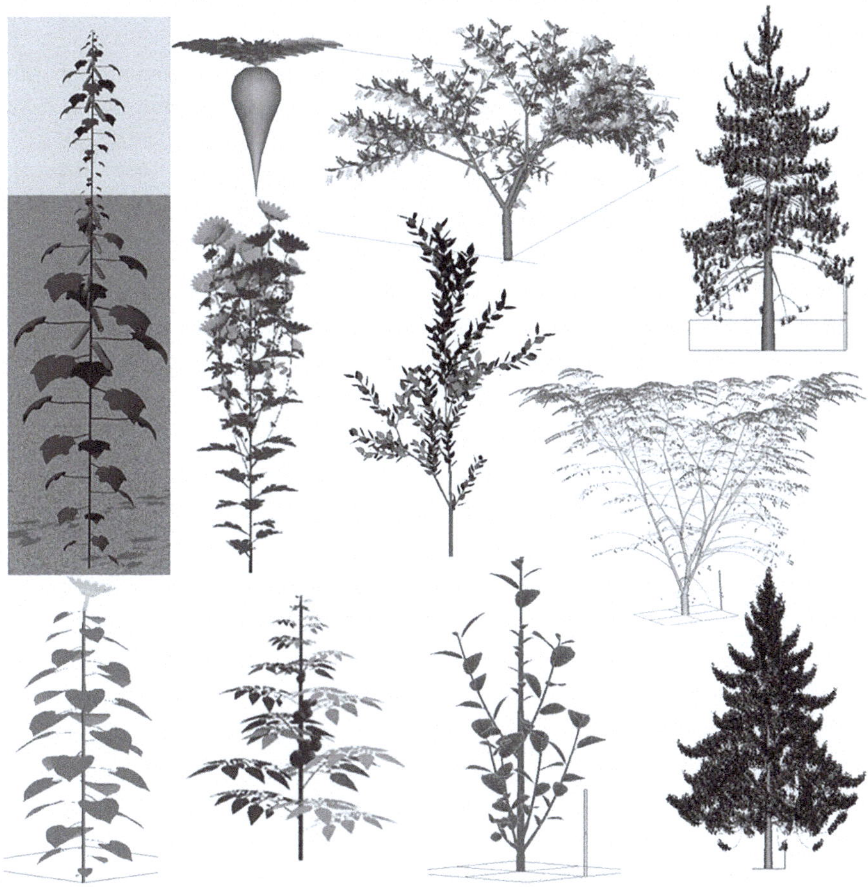

Fig. 14.18 Some simulated output generated from the QingYuan simulator

14.8 GreenLab Operator and Universal Plant Simulator (Gloups)

Gloups is the latest generation of tools based on the GreenLab methodology, developed by Philippe de Reffye in the CIRAD (International Cooperation of Agronomy Research and Development Centre) Amap unit. The software is coded in the Matlab environment and is compatible with all systems supporting Matlab 9.0 and above. Gloups is the most complete tool that implements GreenLab formalisms and is illustrated in many of the images produced in this book.

As a research prototype, Gloups implements the recent advancements in modeling and allows for the calibration of parameters using the weighed least square evaluation based on experimental data. The application is accessible to the scientific community via research partnerships. The application and the principles of its implementation are detailed below.

Gloups covers both the botanical development aspect and agronomic growth aspects, and its inverse methods (analysis of plant crowns and organic series) have been validated on numerous cultivated plants. In terms of computational performance, Gloups benefits from the acceleration of the Matrix mode (see Chap. 3), which preserves the concept of organic series while bypassing the topological structure. Additionally, the PhytoFacto mode (see Chap. 10) transforms growth units into meta-phytomers, allowing the simulation of large tree structures. Gloups also has an interface and 2D/3D visualization capabilities. A free version in Octave is available with reduced performance, and a single-stemmed version (StemGL) is also available. Furthermore, a Gloups prototype with growth feedback on development driven by the *Q*/*D* ratio has been developed (see Chap. 13).

This following section provides a detailed description of the various tools used in applications that utilize the latest GreenLab formalism, including the principles of the simulator and the calibration program.

14.8.1 Organization of Gloups: Data Assimilations and Structure Representations

The GreenLab implementation model consists of two main modules, namely Gloups_Dev and Gloups_Gro. Gloups_Dev is a software tool used for simulating plant development and analyzing plant crowns, while Gloups_Gro is a software tool used for simulating plant growth and analyzing plant organic series.

14.8.1.1 Gloups_Dev Softwares and Data Assimilation for Development

The first module, Gloups_Devs, is responsible for simulating a random collection of plant crowns and creating a target file that contains their description. The age of the crowns is distributed according to a normal distribution and the parameters of the development (a, b, c, w) are given. The simulation results in the display of the tree crowns where the living branches are green, and dead branches are gray. The description of the crowns is recorded in a target file. Each stem has a different number of phytomers, and at each rank under the top, the presence or absence of branches is noted. An absent branch is scored 0, a living branch is scored by its number of phytomers, and a dead branch is scored -1 (there is no need to count its number of phytomers). Phytomers that are older than the age of the stem under study are scored -111, and the maximum number of phytomers is that of the largest stem.

The second module, Gloups_Deva, estimates the parameters of development from a target file filled according to the Crown Analysis format and principles (Chap. 3). The data assimilation is done by means of the target describing the crowns

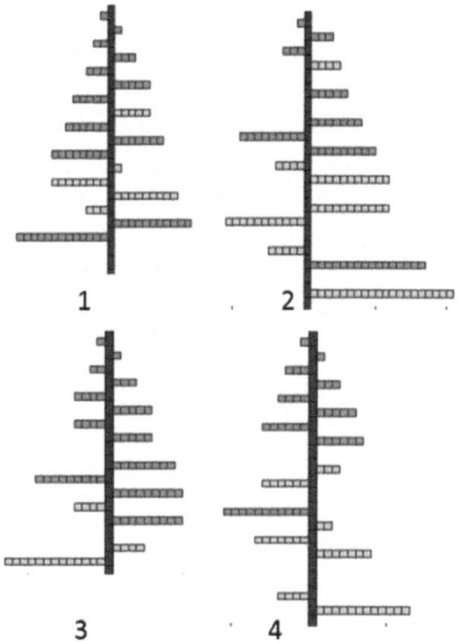

Rank from Top	Phytomer number/branch Plant1	Plant 2	Plant 3	Plant 4
1	1	1	1	1
2	1	3	1	1
3	2	3	2	3
4	3	-1	3	3
5	3	0	4	4
6	5	5	5	5
7	5	0	4	6
8	-1	7	5	6
9	6	9	0	0
10	7	9	8	-1
11	8	-1	9	-1
12	-1	-1	9	0
13	-1	0	-1	11
14	-1	-1	9	-1
15	-1	-1	0	-1
16	11	0	-1	-1
17	13	-1	-1	0
18	0	16	-1	0
19	0	0	-111	-1
20	-1	-1	-111	-1

Fig. 14.19 Plant crown analysis. Four simulated stochastic plants, each with their targets. The *Bernoulli* parameter used is $b = 0.8$ for stems and branches, $a = 0.9$ for branching rate, and $c = 0.95$ for viability rate. The rhythmic ratio is $w = 0.7$. In the structure representation, green branches represent living branches while gray ones represent dead ones. The gaps in the structure indicate the absence of branching. A table beside the structure provides details on the coding of the four crowns (Color figure online)

of the measured trees. The stems of these crowns do not have the same length in number of phytomers. The largest one is used as reference. The analysis of the crowns is done in two directions, top-down and bottom-up, where the quadruplet a, b, c, w functioning parameters of the plant development are estimated from the target file.

A first analysis is done top-down. The purpose of this analysis is to calculate the probabilities associated with the functioning of the meristems (Bernoulli process), the rhythm ratios, the branching rate, as well as the evolution of the branch mortality (there is no mortality on the main stem). At each rank under the top, the presence or absence of branches is noted, and the scoring system for branches is the same as in Gloups_Devs (Fig. 14.19).

A second analysis is done bottom-up to measure the establishment of the branching pattern, starting from the first rank that is not coded -111.

Figure 14.19 shows four random plant crowns simulated with the parameters $w = 0.7$ for the rhythm ratio, $b = 0.8$ for the development rate of stems and branches, $a = 0.9$ for the branching rate, and $c = 0.95$ for the viability rate. The coding file of these crowns is a target that prepares them for data assimilation by the software Gloups_Deva (see Sect. 3.3.1.6.2 in Chap. 3)

The format of the targets produced from the measurements is detailed here (see Chaps. 3, 5, and 6).

14.8.1.2 Gloups_Gro Softwares and Data Assimilation for Growth

The first module, Gloups_GroS, is a plant growth simulator that uses the development parameters obtained from Gloups_Dev to stochastically simulate the structure, biomass production, and partitioning of a set of random plants at a given age. This involves constructing the plant topology, its 2D or 3D geometry, and the organic series extracted from the simulated plant architecture (Fig. 14.20). These organic series only represent the living axes and implicitly contain the history of source–sink relations during the plant's growth. They generate a target file for data assimilation using the Gloups_GroA software.

To illustrate, we consider the same bud functioning parameters (Development) as in the previous section. The organic series are computed top-down along the ranks of the axes, and the averages of organ weights within phytomer ranks are only extracted on living branches. All branches belong to a single organic series once sorted top-down (Chap. 6). Due to the *Bernoulli* process and the rhythm ratio, the series may be incomplete. Hence, none of the plants simulated in Fig. 14.20 have produced 20 phytomers on the stem, and the virtual node of rank 20 is represented by −0.0001. For the branches, there are many virtual internodes due to the rate ratio w.

In this example, for growth, the leaves' lifespan is five DCs, and branches with all yellowed leaves are considered dead. The values of the main functional parameters are $r = 20$, $Sp = 1{,}000{,}000$, and $\varepsilon = 0.05$. The sinks are constant, with a leaf sink

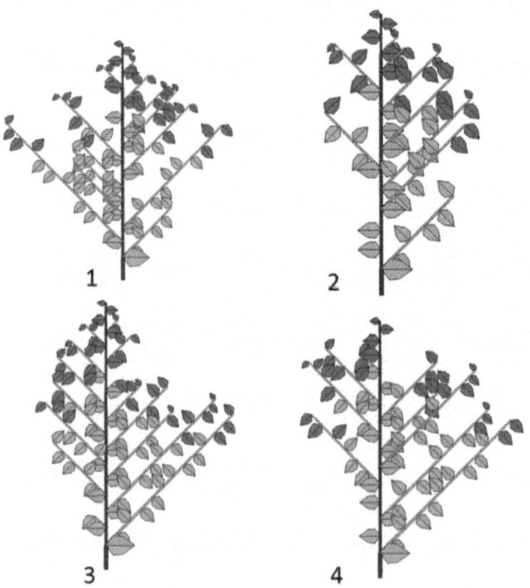

Rank from Top	Stem (g)		Branch (g)	
	Leaf	Internode	Leaf	Internode
20	-0.0001	-0.0001	-0.0001	-0.0001
19	0.66	1.54	-0.0001	-0.0001
18	0.65	1.53	-0.0001	-0.0001
17	0.49	1.15	-0.0001	-0.0001
16	0.41	0.95	-0.0001	-0.0001
15	0.39	0.89	-0.0001	-0.0001
14	0.35	0.78	-0.0001	-0.0001
13	0.31	0.66	0.23	0.63
12	0.30	0.58	0.23	0.58
11	0.30	0.57	0.23	0.55
10	0.29	0.51	0.23	0.51
9	0.29	0.48	0.23	0.47
8	0.30	0.46	0.22	0.42
7	0.31	0.44	0.23	0.40
6	0.32	0.42	0.24	0.37
5	0.33	0.42	0.25	0.35
4	0.31	0.37	0.26	0.34
3	0.25	0.29	0.23	0.28
2	0.17	0.20	0.16	0.19
1	0.09	0.10	0.08	0.09

Fig. 14.20 Analysis of organic series of four stochastic simulated plants with their corresponding target files. Development parameter: $b = 0.8$ for stems and branches, $a = 0$, and $c = 0.95$; Growth parameters: $r = 20$, $Sp = 1000000$, and $E = 0.05$. For physiological age 1 (PA1, stem), $P_{b1} = 1$ for leaf sink and $P_{i1} = 0.5$ for internode sink. For physiological age 2 (PA 2, branch), $P_{b2} = 1$ for leaf sink and $P_{i2} = 0.5$ for internode sink. The ring sink is $P_c = 0.5$. The table describes the organic series (weight of leaves and internodes) along the stems and branches

$P_{b1} = 1$ and an internode sink $P_{i1} = 0.5$ for PA 1 (stem). For PA 2 (branch), the leaf sink $P_{b2} = 1$ and the internode sink $P_{i2} = 0.5$. Finally, the ring sink is $P_c = 0.5$.

The second module, Gloups_GroA, serves as a source–sink solver for computing the functioning parameters of organs based on the organic series measured on the plant. The estimation of model parameters uses the algorithms of the WLSM (Chap. 9). It uses the concept of potential structure defined in Chap. 3. The data required for parameter identification are provided by a target file generated by the simulator Gloups_GroS for various ages (multi-fitting) corresponding to observed and measured data. The target files contain the extracted organic series and compartment of organs. Measured data from field experiments must follow the format of target files built by Gloups_GrowS.

It should be noted that Gloups_GroS and Gloups_GroA share multiple internal functions. Both tools are developed in the same context and programming environment under the Matlab environment. Many functions are common and shared for simulation and estimation.

In the processing, modules, an environment with specific tools for inputs and outputs of the processing, are typically found. Input consists of simple text files, composing parameter files that can be filled in by spreadsheets, generic editors, or specific editors. Graphical user interfaces (GUIs) for managing parameter and processing variables can also be designed from the simulation environment using for instance the forms available in the Matlab programing environment's GUIDE (Graphical User Interface Development Environment) library.

Multiple outputs and results consist in curves and tabs (see Chaps. 5, 6, 7, 8, 9, 10, and 11). Data and graphics can be saved in the application environment or exported to text files in CSV (Comma-Separated Values) format for further use in spreadsheet programs or editors. Spreadsheets, input or output evaluation tools, enable additional classic statistical analyses and the generation of personified graphical representations of curves. The files used are the standard CSV format in ASCII (American Standard Code for Information Interchange) to allow for a wide use of these data by various tools.

Gloups provides 2D and 3D plant architecture figures in Matlab Windows. Another coding of the structure (with its geometry) is specific and available in formats (.lig and .opf) for use with visualization tools such as LigDraw, a tool for schematic or realistic visualization of a simulated plant. Its functionalities are described in the representation/visualization chapter. This format also allows exploration or even redefinition of the geometry of the structure with the Xplo tool. Xplo (Griffon and De Coligny 2014) is a tool for exploring and editing the structure (see below). It allows for the definition and editing of geometrical attributes to the phytomers described in the structure loaded in input.

14.8.2 Simulation Tool and Its Principles

From a technical point of view, the simulation follows a well-defined sequence:

- Load the parameter file and, optionally, the mask file for Gloups_GroS and Gloups_GroA. In the case of running WLSM, also load the target file for Gloups_GroA.
- Compute the development of structures.
- Compute the functioning (expansion of organs).
- Compute the outputs.

14.8.2.1 Loading Parameters and Reading Data

The parameters describing a plant are stored in a text file using a simple coding system. The original coding of each parameter is represented by two lines of text: the first line specifies the nature of the parameter, while the second line indicates its value. Multiple values can be associated with a single index, which typically corresponds to a PA. The data in the file include three types of parameters: those to be simulated (Param), those that can be calibrated, and the output options of the simulation.

14.8.2.2 Simulation of Development and Construction of Structures in Gloups

The development processes depend on meristem properties, such as pauses, mortality, and ramification, indexed by CA and PA differentiated by architectural analysis.

In this construction, the topology of the structure, including filiation relations inherent in the ramifications (including reiterations), is separate from its geometry, which include mechanical effects, angles of insertion, phyllotaxis, and element dimensions.

The GreenLab model does not require construction of the topological structure or geometric construction when using the matrix mode. In this case only cohorts of organs and organic series are built up. A topological (or geometrical) construction is necessary using the List mode for tasks such as a 3D representation.

14.8.3 Implicit and Explicit Structure, Topology, and Geometry

14.8.3.1 Implicit Structure

An implicit structure lacks geometrical and topological properties, making it impossible to analyze and represent the underlying branched system. Instead, the structure is defined by counting organs according to their cohorts, indexed by PA, CA, and age of appearance (ontogenic age). The cohorts are represented by a matrix *M(o, φ, ag)* that indicates the type of organ (o), PA (φ), CA, and age (the cycle) of appearance (ag).

In the case of an implicit structure, also called (structural) construction in Matrix mode, the simulation/construction of the structure is simplified to tracking the evolution of the number of organs of each cohort at each growth cycle.

Additionally, the implicit structure generates automatically organic series that are orthogonal to cohorts. The plant is described as a lot of separated pieces that are the axes extracted from the implicit structure. In most cases, cohorts and organic series generated by the Matrix mode are sufficient to fulfil the computation of the source–sink functioning with the WLSM. However, the system is not suitable for inflorescences of herbaceous plants that lack cohorts for organ expansion (see Chap. 7).

14.8.3.2 Semi-Explicit Structure: Implicit Geometry and Explicit Topology

The definition of a matrix is useful but limited when it comes to accounting for structural relationships that depend on the branching order, or more generally on the path of the structure. This means that the notion of cohort for organ expansion, as seen in the case of many inflorescences, cannot be captured. To overcome this limitation, it is necessary to define topological relationships and construct an explicit representation of the branched system, such as a tree in the computer sense. This involves choosing a type of path (see above) or constructing a representation of the path as a list of phytomers during the simulation. This semi-explicit structure is referred to as a structural construction in List mode.

14.8.3.3 Explicit Structure

An explicit structure that includes both topology and geometry allow for three-dimensional representation of the plant architecture, considering the geometrical sizes and deployment (in the plane or space). This explicit structure completes the construction in List mode by assigning geometric parameters such as dimensions, positioning, and orientation to each phytomer in the list. The plant architecture is then obtained.

The information regarding these geometrical attributes can either be computed during the construction of the list (simultaneously with the topology), or it can be computed later. In the GreenLab model implementations, the latter approach is predominant. This means that the geometrical computations are shifted and fully performed at the end of the simulation when the final representation of the construction in List mode is generated (Fig. 14.21).

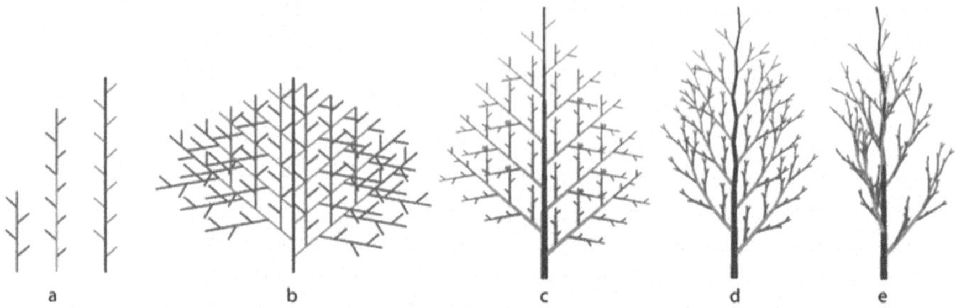

Fig. 14.21 From implicit to explicit structures (Jaeger, CIRAD). In this example, we are examining the structure of a simulated plant with three physiological ages (PAs) over 12 growth cycles. The implicit structure is represented by a matrix of 12 × 3 elements, which indicates the numbers of phytomers by PA. This matrix does not have a graphical representation. (**a**) The introduction of growth filiation relations provided by cohorts allows for the representation of axes, as shown in their final stage of growth (**a**). (**b**) In the List mode construction, the entire topology is explicit, which makes a simple geometrical representation possible in the plane. This is achieved by assigning each phytomer a unit length, each branch an angle (in this case, 45 degrees), and a color to indicate the PA. (**c**) The three figures on the right (**c, d, e**) illustrate explicit representations of the plant architecture. In each figure, various attributes have been gradually integrated for each phytomer present in the List mode. For example, in (**c**), the dimensions (length/diameters); in (**d**), the angles of insertions and deviations (straightening); and in (**e**), the angles of phyllotaxis (the geometric plunge inscribed in space rather than being limited to the plane)

14.8.4 Construction of the Structure: Principles

The principle of building structures involves constructing a main structure by dividing it into smaller substructures in a recursive way (see Chap. 2, Sect. 2.2.9; and Chap. 3, Sect. 3.5.4.2). We give more detail here.

14.8.4.1 Concept of Development Axis

The development axis of a given PA can be described as the topological arrangement of the leafy axis on the plant.

To compute the development axes of all PAs for all CAs, the process starts by determining the physiologically oldest development axis, which is also the physiologically oldest substructure that does not branch. Lateral substructures are then added in a top-down manner on the carrier axis, with increasing chronological age. A new substructure is created with the same PA as the carrier axes, but with a lower PA than the ones of the lateral substructures attached to it. This is repeated for all the CAs of the carrier axis, resulting in a collection of substructures for all the chronological ages. The process continues recursively until the main structure is obtained for all the CAs (see also Chaps. 2, 3, and 4).

For instance, let us consider a plant with continuous development until six DCs, where the rhythm ratios for PAs 1, 2, and 3 are $w = 1$, 0.75, and 0.5, respectively

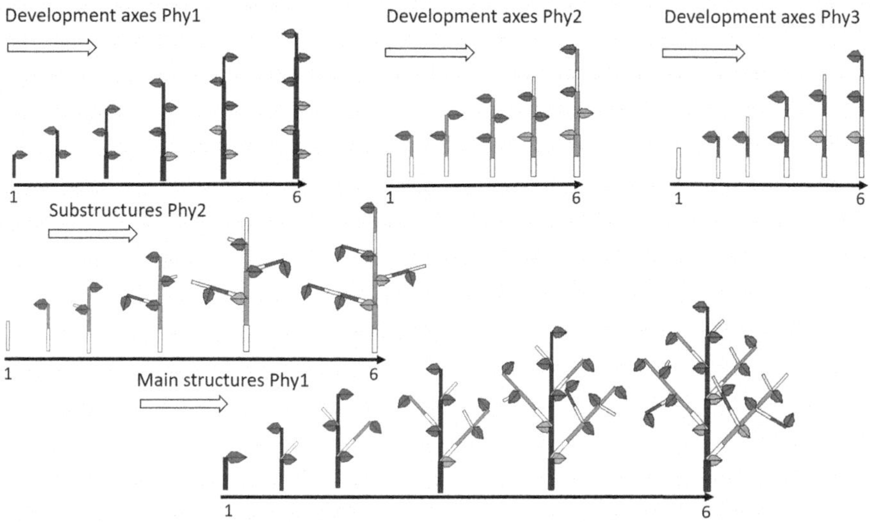

Fig. 14.22 Development of a continuous structure made of three physiological ages (PAs) (representation of the structure in chronological mode obtained with these three development axes for six DCs)

(Fig. 14.22). Using this information, the development axes and the substructures for all PAs can be computed following the process outlined above.

14.8.4.2 Geometric Representation of a Development Axis

In terms of modeling, a development axis for a given PA is defined as a table where the column index corresponds to the CA in the cycle, and the row indices correspond to the PAs with attributes specific to the state of the phytomer (phytomer entity, or pause entity), the number of leaves, etc. The development axis can be continuous or modular if the development is rhythmic.

The graphical or geometrical representation of a development axis can be easily obtained by specifying geometrical attributes to the phytomers making up the axis.

Thus, in the examples illustrated above, each phytomer is associated with a graphic symbol (a rectangle) whose color specifies the PA. Other attributes and/or symbols can be used to characterize breaks or the presence of other organs, such as leaves.

In this case, the following attributes are used:

- Each state, corresponding to a phytomer entity or a pause entity, is represented by a rectangle of unit length;
- The color is gray (or white) for a pause entity or colored, blue, green, red, yellow, and purple, for a phytomer of PAs 1, 2, 3, 4, and 5, respectively.

- The presence of organs carried by each phytomer is represented by symbols, such as the axillary meristems, which are represented by a rectangle whose colors correspond to the PA of the branch.

Any axis of a simulated plant is made up of a succession of development axes if the meristems undergo transitions of PAs. The initialization of this axis can be done with the addition of ad hoc initial pauses at the base, if branching is delayed.

Axes of development of the same PA and different CAs are copies of the initial axis of development. Theses copies of the initial axis are bounded, according to the lowest value of the number of cycles defined among the following criteria:

- By the meeting of the end of the development axis (corresponding, for example, to a transition in a higher PA, or to a determined growth).
- By the age of the simulated plant.
- By a death of the terminal meristem.

When there is a transition, i.e., an evolution of the PA along the axis, the representation of the axis is obtained by concatenating the representations of the development axes of the various PAs concerned, with the same constraints of shifting and demarcation of the copies.

Figure 14.23 shows how the GUs are coded in the development axes.

14.8.4.3 From the Development Axes to the Structure

14.8.4.3.1 Representations of the Construction of Structures: Topological and Chronological Modes

Using a coding that describes the branching attributes of the phytomers and the processes of duplicating the representations of the development axes, schematic planar representations of the simulated structures can be easily obtained.

All that is required is to define the branching angles between the phytomers and the axes derived from the axillaries. We have chosen to define a fixed angle per PA (20, 45, 90 degrees for PA 2, 3, and 4, respectively). The structures can be simulated and displayed with or without pauses.

In the first case, the pauses entities are displayed in the same way as the phytomer entities, which defines the chronological mode (in short, chrono mode). This mode makes it possible to materialize each phytomer, real or void, for each growth cycle, allowing for easy tracking of the chronology of phytomer establishment in the structure. This representation mode is particularly useful in the study and modeling phases.

On the other hand, by suppressing the pauses, the representation of the simulated structure presents a scheme close to the real one. This mode of representation is called topological mode (in short, topo mode), and the detailed flow of time is no longer visible in the simulated structure. This mode is the default one when building

Phy 4 phyto	Inter-node	Bud Phy	leaf	fruit
6	0	0	0	0
5	0	0	0	0
4	0	0	0	0
3	0	0	0	0
2	0	0	0	0
1	1	0	1	1

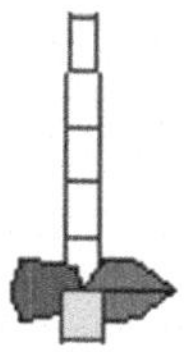

Phy 3 phyto	Inter-node	Bud Phy	leaf	fruit
6	0	0	0	0
5	0	0	0	0
4	0	0	0	0
3	0	0	0	0
2	1	1;4	1	0
1	1	1;4	1	0

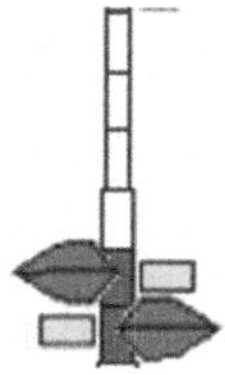

Phy 2 phyto	Inter-node	Bud Phy	leaf	fruit
6	0	0	0	0
5	0	0	0	0
4	0	0	0	0
3	1	1;3	1	0
2	1	1;4	1	0
1	1	1;4	1	0

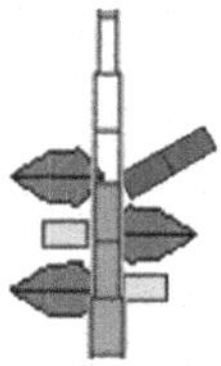

Phy 1 phyto	Inter-node	Bud Phy	leaf	fruit
6	0	0	0	0
5	1	1;2	1	0
4	1	1;3	1	0
3	1	1;3	1	0
2	1	1;4	1	0
1	1	1;4	1	0

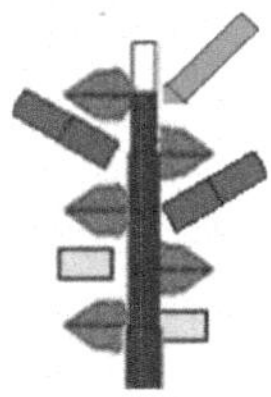

Fig. 14.23 Coding the modules (growth units) of the development axes in a rhythmic structure with four physiological ages (PAs). For each phytomer, the internodes are 1 or 0 (pause); axillary buds are described by their number and PA, while leaves and fruits are described by their number. Each growth unit (GU) is described by an array, and concatenation of the arrays builds a development axis

a realistic three-dimensional structure. Both modes are illustrated below. We use in Fig. 14.24 the modules depending of PA and created in Fig. 14.23.

Figure 14.24 shows the construction of the modular development axes (corresponding to Fig. 14.23), which are created by concatenating GUs and their respective tables.

14.8.4.3.2 Case of Stochastic Simulations

The construction of structures in GreenLab simulations, whether deterministic or stochastic, relies on the use of development axes as the basis (Chap. 5). In the case of stochastic simulations, a collection of development axes is created with a limited number of repetitions per CA and PA. This differs from a usual stochastic simulation, which generate each axis randomly using Monte Carlo method (pseudo-random mode). The collection of development axes used in GreenLab simulation produces a quasi-random plant, as described in Chap. 3 (de Reffye and Hu 2003).

The principle of copying the development axes remains the same in the process of representing the stochastic structure. Just as in the deterministic case, each phytomer produced can carry a certain number of axillary meristems. To each PA corresponds a collection of random, development axes in number *Nrep*φ. For each meristem we associate a number that corresponds to a repetition of the axes. It is a draw with

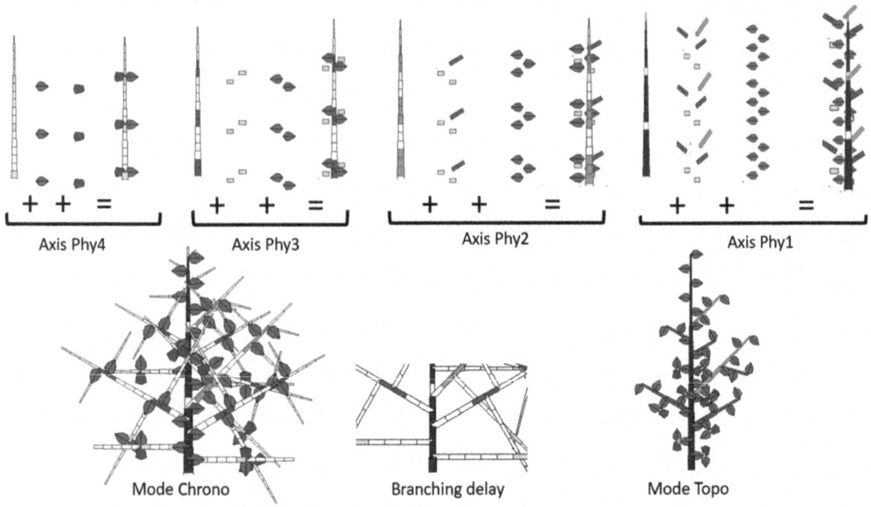

Fig. 14.24 Three representations of a rhythmic growing plant. The structure is reconstructed here at the final age (3 macrocycles = 3*6 = 18 microcycles) from the representations of the development axes illustrated above. *Left* and *right*: representation of the structure in chronological and topological modes. The white phytomers are pause entities. The central figure illustrates pauses that synchronize the development of the axillary branches according to their rank in the carrier GU. The diagrams above show the phases of construction for the development axes. The four physiological ages (PAs) are represented

remise because we can draw the same number twice. The greater the number of repetitions is, the closer the quasi-random mode is to the pseudo-random mode.

14.8.4.3.3 Generation of Stochastic Development Axes

In the deterministic case, a single developmental axis is defined for each PA. In the stochastic case, $Nrep\varphi$ stochastic development axes are created for each PA φ, based on a classical Monte Carlo simulation scheme:

- An identifying number (from 1 to $Nrep\varphi$) is assigned to the development axis being created.
- Seed numbers (or random generators) are then initialized, allowing for distinct random sequences to be generated for each development axis.
- During the development simulation, a *Bernoulli* test is carried out for each appearance of a phytomer. As in the deterministic case, the result is coded in the development axis (0 if there is a pause, 1 if a phytomer appears).
- Viability is then estimated. If viability is not maintained, the axis dies, and a death marker (−1 in this case) is substituted for the status indicator (0 or 1) in the coding. This marker is recorded for all future cycles.

- If the current phytomer is successfully produced, the ramifications are then evaluated. For each branching, a random number is compared to the probability of branching. If branching occurs, the delay in the deployment of the lateral axis needs to be evaluated. This can include a number of cycles imposed by a rhythmic growth and a number of cycles drawn from a random law. In general, these are two additional attributes to be coded in the development axis.

In practice, experimentation suggests that $Nrep\varphi$ values between 10 and 20 give acceptable results. Once the stochastic development axes have been created, the construction of the structure follows the same principle as the deterministic case, by randomly drawing from the collection.

14.8.4.3.4 Construction of Quasi-Random Stochastic Structures

The advantage of the quasi-random method is its computation time, which is proportional to the number of repetitions multiplied by the square of the CA for the first simulated plant, and not to the number of phytomers established. For the other plants, a deterministic approach is used, because the first plant has already created the stochastic collection of development axes.

As an example (Fig. 14.25), let us consider a collection of plants with a CA of five DCs and three PAs. The development axes are simulated at five DCs, and the collection consists of two development random axes per physiological age. Therefore, the computational plant only contains two types of axes per physiological age, and the lengths of these axes are limited by their position and the CA of the structure. The truncation is done bottom-up.

14.8.4.3.5 Performance of the Computational, Quasi-Random, and Pseudo-Random Modes

This paragraph completes the study made in Chap. 8 in Sect. 8.6.5 where the List modes and the PhytoFacto modes were compared in the case of rhythmic trees. The study now focuses on a plant with four PAs, with continuous and indefinite development of axes, and meristems that function with the same probability *b*. The branching orders correspond to the PAs, and the simulation covers 40 DCs of development.

There are three modes of approach for the computation:

- The computational mode (corresponding to the potential structure), which directly gives the theoretical values of the means and variances of the numbers of phytomers through mathematical formulas (see Chap. 6). The computation time is a function of the square of the age of the structure.
- The quasi-random mode, which is based on the factorization of substructures. This mode simulates limited collections of development axes and random structures for each CA and PA starting from the highest, and assembles them by

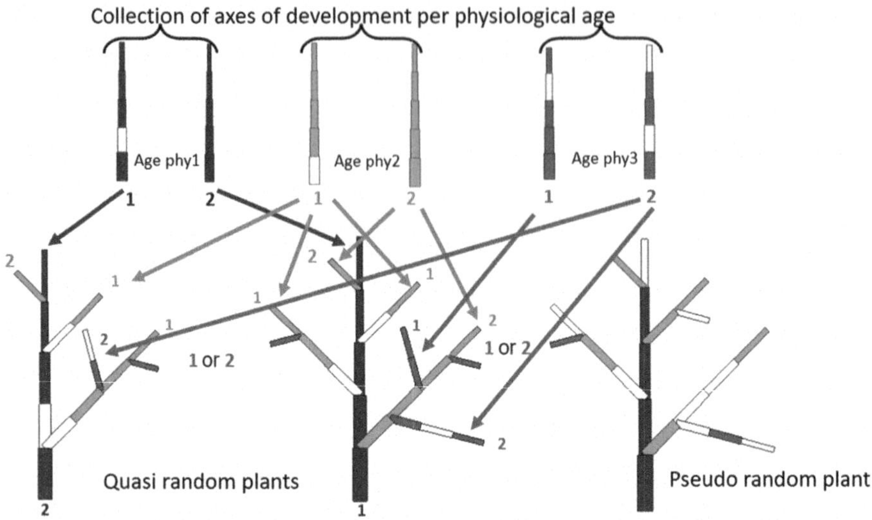

Fig. 14.25 Stochastic axes of development and construction of stochastic structures. In this example, three physiological ages (PAs) are used, and only two random repetitions per PA. The stochastic aspects are related to the development with the presence of random pauses ($b = 0.8$). The collection of development axes is displayed at the top of figure, and at the *bottom left* are two quasi-random plants on which the numbers of the axes of the collection are displayed. The same number appears several times in the structure due to the random draw. On the *bottom right* is a pure pseudo-random plant in which the axes are drawn using the Monte Carlo method

randomly selecting them. This mode corresponds to a random draw (a same substructure can be drawn several times), and the variability obtained is more limited than in a pure Monte Carlo process and depends on the number of repetitions of the structures in the collections. The computation time is proportional to the number of repetitions multiplied by the square of the age of the structure.

- The pseudo-random mode, which corresponds to a classical stochastic Monte Carlo simulation (in prefixed mode, corresponding to List mode). The computation time is proportional to the number of phytomers produced, which can grow exponentially.

To study the performance of the model in these three modes, b is set to 0.8, and the CA t is 40 DCs. The plant produces a number of phytomers:

$X_t^{\varphi} = \binom{t}{\varphi} \cdot b^{\varphi}$ per PA φ at CA t. The coefficients $\binom{t}{\varphi}$ are those of the Newton binom. To ensure a robust analysis, 50 pseudo-random trees and 50 structures for the quasi-random trees are chosen for replication. Three simulations at age ten DC are shown in the chronological mode of this tree in Fig. 14.26.

Table of Fig. 14.27 shows that the three modes (computational, quasi-random, pseudo-random) produced similar productions of phytomers, but the computation times varies significantly.

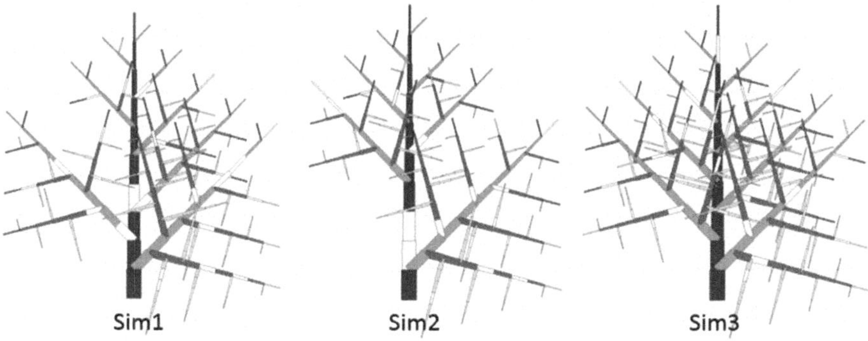

Fig. 14.26 Three random simulations in chronological mode (10 DCs) of a shrub with four physiological ages (PAs) and a development probability $b = 0.8$ for all meristems. The theoretical number of phytomers produced at PA φ and chronological age (CA) t is: $X_t^{\varphi} = \binom{t}{\varphi} \cdot b^{\varphi}$. To evaluate the performance of the computation of means and variances, the plant is tested at 40 DCs

	org numb	Comput. time		
		theo	quasi	pseudo
age DC				
5	17	0.01	0.04	0.06
10	184	0.02	0.16	0.35
15	871	0.03	0.36	1.31
20	2705	0.05	0.63	3.3
25	6571	0.083	0.99	6.7
30	13606	0.12	1.43	12.2
35	25206	0.17	1.96	20.8
40	43023	0.24	2.57	33.4
Total organ nb		43023	42904	43875

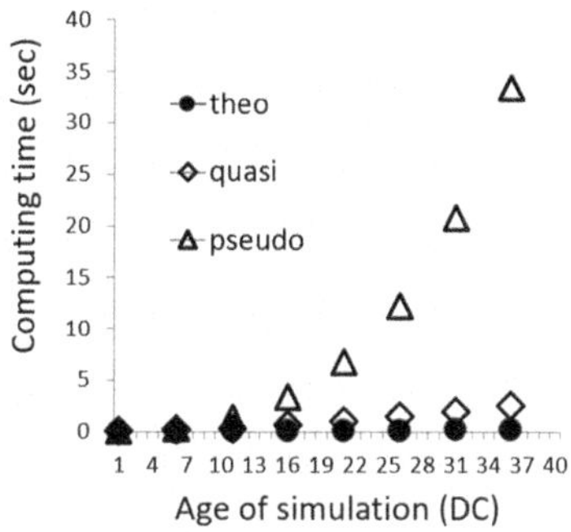

Fig. 14.27 Evolution of the computational time (seconds) of phytomer production according to the three modes (computational, quasi-random, and pseudo-random)

At 40 DCs, the theoretical computation time is 0.24 s. The quasi-random mode took 10 times longer, and the pseudo-random mode (Monte Carlo) took 150 times longer. Additionally, the computation time for the quasi-random mode increases as expected in a quadratic manner, while the pseudo-random mode increased in a bi-quadratic manner, i.e., proportionally to the number of organs computed for four PAs.

This example clearly demonstrates the necessity for reasonable computation times, which may require abandoning List mode if the representation of plant architecture is unnecessary. This is the case for the WLSM, which can only consider cohorts and organic series to compute source–sink parameters.

14.8.4.3.6 Implicit and Explicit Reconstructions of the Structure

Constructing and representing plant structures is a computationally and memory-intensive process, due to the copies of the various development axes involved.

The computation of structures involves assembling substructures on the development axes through a double recursive loop over CAs and PAs (see Chap. 3, Fig. 3.39). This can be accomplished by either concatenating arrays of phytomers (Matrix mode) or concatenating lists of phytomers (List mode).

In Matrix mode, at the end of the computation, a collection of arrays (arranged as Russian dolls) describes the number of phytomers produced at all CAs for all PAs, and repetitions. In List mode, a collection of substructures is created in the same way, with the topology described phytomer by phytomer in prefix order. It is possible to extract or visualize any substructure by knowing the CAs, PAs, and repetition number.

14.8.4.3.7 Matrix Mode

If plant representation is unnecessary, the construction process can be limited to quantifying the number of organs per cohorts and simulating the development axes, which also provide the organic series. The resulting plant is known in separate pieces. The matrix mode of structure construction is fast and implicit, making it useful for the WLSM.

The following example illustrates matrix mode coding. The bottom index ($t = 5$) indicates the age (in DC) of the plant. The top index indicates the number of stochastic repeat (*Rp*). The first line of the matrix provides the number of phytomers of the main structure aged five DCs, corresponding to PAs 1 (the trunk), 2 (the main branches), and 3 (the twigs), i.e., [15 20 30]. The second line of the matrix provides the number of phytomers of the branch aged five DCs, corresponding to PAs 2 (the main branches) and 3 (the twigs), i.e., [0 10 10]. The third line of the matrix provides the production of phytomers of the twigs aged five DCs, corresponding to PA 3 (the twigs), i.e., [0 0 5].

$$[S]_5^1 = \begin{bmatrix} 15 & 20 & 30 \\ 0 & 10 & 10 \\ 0 & 0 & 5 \end{bmatrix}.$$

The number of generated structures is $t*Rp$, and represents all the cohorts of organ produced for any CA, PA, and repetition.

14.8.4.3.8 List Mode

To represent a plant, two options are available: 2D to visualize plant topology or 3D for computer graphics. Both methods require exploring a list to reconstruct the plant geometry. The list is an array with C columns and L rows. The number of columns C is equal to the number of phytomers within the list and each row corresponds to a phytomer's position in prefix order within the topological structure. The number of rows equals to the number of descriptors needed to reconstruct the plant topology in either 2D or 3D.

The Gloups software primarily uses the following descriptors to depict a phytomer at position i in the list in prefix order:

- *EM(i)* $\in \{0, 1\}$ indicates if the phytomer is void or real.
- *MXM(i)* $\in \{-1, 1\}$ indicates if the phytomer is dead or alive.
- *PPM(i)* $\in \{1, 2, \ldots, n\}$ indicates the PA.
- *KXM(i)* $\in \{1, 2, \ldots, n\}$ indicates the order of branching.
- *XXM(i)* $\in \{1, 2, \ldots, n\}$ indicates the position of the phytomer on the axis.
- *FXM(i)* $\in \{1, 2, \ldots, n\}$ indicates the number of phytomers from current position to plant base.

Figure 14.28 shows an example of a plant with 20 phytomers ($NUM = 20$). At position five ($NUM(5) = 5$), there is a pause ($EM(5) = 0$), and the phytomer is dead ($MXM(5) = -1$, black color). The physiological age (PA) is three ($PPM(5) = 3$), and the branching order is three ($KXM(5) = 3$). The position on the branch is three ($XXM(5) = 3$), and the path from the position to the bottom is five ($FXM(5) = 5$). At position 20 ($NUM(20) = 20$), the phytomer is real ($EM(20) = 1$, blue color). The PA is one ($PPM(20) = 1$) and the branching order is one ($KXM(20) = 1$). The position on the axis is five ($XXM(20) = 5$) and the path from the position to the bottom is five ($FXM(20) = 5$).

By following the phytomer number, we can determine whether we are on the same axis or if a new branch is created. This is represented by the relation $KXM(i) = KXM(i-1)$ for the same axis *or* $KXM(i) > KXM(i-1)$ for a new branch.

If $FXM(i) = C$, it indicates that the end of the axis has been reached due to factors such as the plant age.

Additionally, there are descriptors such as *NXM* (the number of the axis in a whorl), and the number of organs type *O* in a phytomer according to *PPM*(i), etc.

List

Reading the list

NUM	1	2	3	4	5	6	7	8	9	10	11	12	13	14	15	16	17	18	19	20
EM	1	1	0	1	0	1	0	0	1	1	1	0	0	0	1	0	1	1	0	1
MXM	1	1	1	1	-1	1	-1	-1	1	1	1	-1	-1	-1	1	1	1	1	1	1
PPM	1	2	3	3	3	2	3	3	2	2	1	2	2	2	1	2	2	1	2	1
KXM	1	2	3	3	3	2	3	3	2	2	1	2	2	2	1	2	2	1	2	1
XXM	1	1	1	2	3	2	1	2	3	4	2	1	2	3	3	1	2	4	1	5
FXM	1	2	3	4	5	3	4	5	4	5	2	3	4	5	3	4	5	4	5	5

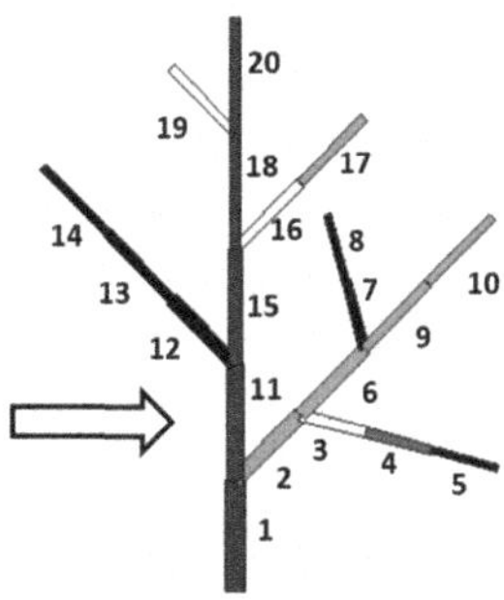

Fig. 14.28 Descriptors in the list to follow the plant topology in the figure

14.8.4.4 Structure: Bi- or Three-Dimensional Reconstruction

The explicit construction of the structure, as defined in the previous section, is limited to a topological representation. A bi- or three-dimensional representation additionally requires the definition of geometric attributes to locate each phytomer in space, and to define its orientations and sizes.

The definition of these parameters, particularly the position and orientation, must respect the continuity of the plant, specifically the organization in axes. There are two classic stages in the computations:

- At the beginning of the axis, the position and orientation of the first phytomer define the orientation and insertion of the axis.
- Along an axis, the sequence of states defines the sequence of internodes, whose position is built up end to end, integrating local deviations of orientation if necessary.

The vector tools for the construction of the architecture have been explained in the previous Sect. 14.6, and the skeleton of the plant is built from this vector computation. The positions of organs in space depends on a series of translations and rotations.

The planar representation is reserved for the display of the topological structure, which does not need to consider biomass. Organs have predefined dimensions in 2D, and their colors depend on the PA and organ type. Branching angles are also defined, and the plant can be represented in either Topo mode or Chrono mode.

In the 3D representation, which include biomass production and allocation, the organ shape depends on their volume and structural allometries (Chap. 5). In addition to branching angles, there are angles that control phyllotaxis, subsiding, and bending.

14.8.4.4.1 Insertion and Orientation of Axes

To compute the position and orientation of the first phytomer at the beginning of the axis, the carrier phytomer is used as a reference point, and angles relative to it are considered:

- The position of the axis insertion, determined by the end of the carrier phytomer (its orientation along the main direction plus its origin).
- The orientation of the first phytomer of the axis, defined with respect to the main direction of the carrier phytomer, adjusted by an insertion angle.
- The arrangement along (around) the carrier axis, defined by a phyllotaxis angle. In the case of a whorl, an angle corresponding to the number of the axillary meristem in the whorl must be added.

In all cases, the parameters involved (i.e., the angle of insertion and the angle of phyllotaxis) depend on the PA of the phytomer connected for insertion, and the carrier for phyllotaxis. Insertion and orientation may also be influenced by botanical topological criteria to simulate the behavior of plagiotropic axes, for example.

14.8.4.4.2 Axis Continuation

The procedure for determining orientations along an axis is similar, but differs in the computation of orientations. The insertion point of a phytomer is defined by the extremity of its successor, while its orientation is determined by a deviation from the predecessor, according to two origins: one mechanical, and the other behavioral. The orientation evolution can be viewed as a deformation along the axis that results from a succession of local deformations.

In our tools, two computation steps are required to simulate the effects of mechanics. Firstly, the length of the axis is estimated, and then, according to specific parameters such as Young's modulus and conicity, a global deformation (a bending angle) is computed over the whole axis, which is considered as a recessed beam. Then, for each phytomer, the local deformations corresponding to a bending share are computed. Additionally, other local deformations are added, controlled by a parameter and/or by the current orientation of the phytomer, such as a point of inflection, torsion, and curvature.

In most structural–functional models, the geometrical aspects are integrated upstream in the simulations. However, in our approach, the reconstruction of the structure and its placement in 3D geometrical space is not required. The 3D reconstruction is thus implemented as post-processing. In the Gloups implementation, this 3D reconstruction is limited to strict topological aspects:

- The list describing the structure (as described in the previous section) is exported, and the geometrical aspects are then generated for visualization.
- The user can then define the geometrical parameters (angles, mechanical aspects) interactively. Similarly, it is possible to associate surface property attributes of the organs according to their age, and thus obtain realistic quality images.

Figure 14.29 shows the outputs of the model in both implicit and explicit modes. For implicit mode, it is necessary and sufficient for the computation of the parameters (the WLSM). However, the explicit mode is required for visualizations. The potential structure showing all possible phytomers should be avoided due to its size (Fig. 14.29d), and, depending on the needs, a planar representation of the simulated structure may be more readable (Fig. 14.29e). For computer graphics images, a 3D representation of the architecture is necessary (Fig. 14.29f).

14.8.4.5 Results and Their Exploitation in Gloups

Chapters 9, 10, and 11 present the results of the plant computation with Gloups software in tables, graphs, and figures.

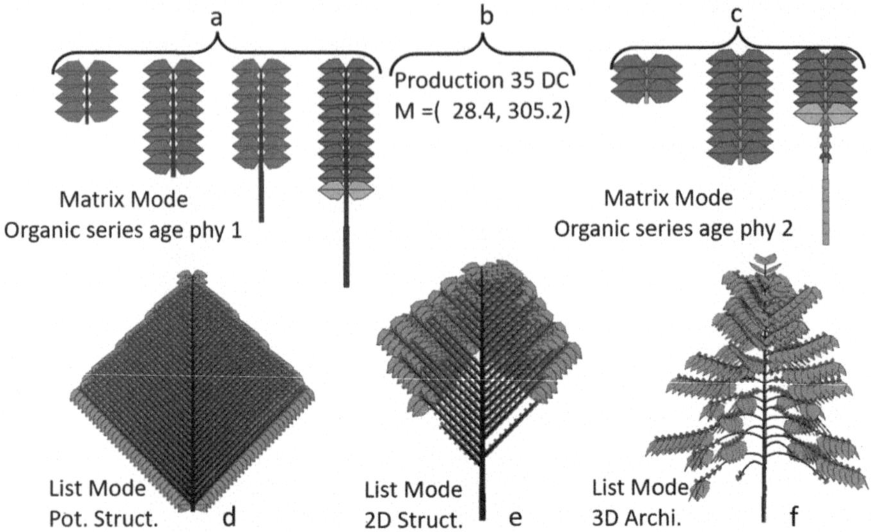

Fig. 14.29 Simulations of a coffee tree with the Gloups software. *Upper* figures: matrix mode (implicit). (**b**) The output of the number of phytomers per chronological age (CA) and per physiological age (PA) (1 and 2) stored in a matrix M at 35 DCs. The organic series can be visualized for all CA and PA. (**a**) Trunk at CA [5 10 20 25 DCs]. (**c**) Branches at CA [5 10 20 DCs]. *Lower* figures: List mode (explicit). The plant is drawn. (**d**) Planar representation of the potential structure at 35 DC. (**e**) 2D stochastic simulation of the coffee tree at 35 DCs. (**f**) 3D simulation of the 3D architecture of the same coffee tree at 35 DCs (Gloups software)

At the end of a simulation, various outputs are computed and attached to each cohort (in matrix mode) or each phytomer (in List mode), and this is done for each random realization if the stochastic mode is chosen. To save memory in List mode, the user selects a limited set of observed data beforehand.

The characteristic quantities of the production are usually represented by curves or graphs in a synthetic way.

14.8.4.5.1 Systematic Outputs

The Gloups software generates many outputs at the end of a simulation. Systematic outputs are:

- The number of phytomers per PA with their attached organs; the total number of phytomers (in mean and variance in the stochastic case) for the structure.
- The value of demand and biomass production per cycle for all organs.
- The computation times allocated to the three tasks: structure, functioning, and restitution of variables.

- The presentation of these output variables (generally indexed to the growth cycle) is typically done through a generic curve representation function that can display one or more curves on a graph and save the related data.

14.8.4.5.2 Generic Results Presentation Function

This feature allows for the display of multiple graphs, including the comparison of magnitudes with different PAs. It also enables the user to save data for each graph to disk in a clear file format (csv) that can be used by spreadsheet programs.

The generic output function requires the following inputs for a specific variable:

- A table of values for the variable according to two indices.
- Presentation methods (colors, markers).
- Texts for legends.
- A file name for saving the graph containers (values, modalities, and texts), if desired.

The first index (main) is typically assigned to the growth cycle, while the second index (optional and secondary) allows for the display of other curves for different PAs or observation periods.

14.8.4.5.3 Optional Outputs

The plot function is used to display both systematic outputs and those selected by the user. To facilitate the exploration of simulation results, various classes of outputs (indicative list below) are available. The following outputs have been predefined:

- Number of phytomers produced by CA and PA during plant development, with the options available in the stochastic case.
- Total biomass production of organ compartments by CA and PA during plant development, with the option for min, max, mean, and standard deviation values available in the stochastic case.
- Graphics of the evolution of biomass demand and production during growth.
- Graphs of sink functions for different types of organs.
- Display of organ biomass along the organic series.

14.8.4.5.4 Specific Outputs

In addition to the predefined outputs, there are specific outputs that represent geometrical aspects. These include a 2D or 3D view of the plant structure or a section of an axis with its ring boundaries (see Chaps. 8 and 10).

- Axis diameters. Section and trunk profile.

The representation of the rings inside the stems that characterize secondary growth of woody plants is done by the software Gloups. The trunk profile can also be displayed. Note that the matrix mode also computes the organic series and thus the stem profile without using the structure computation in List mode, which can save time and prevent time-consuming calculations.

- Geometric representation.

To define a 2D or 3D geometry, it is necessary to consider the distribution, orientation, and size of the organs. The size of the organs is determined by allometry factors based on biomass. In cases where biomass is not computed, default values are assigned using constant scaling factors by PA for the various organs (leaves, fruit, internodes). For growth in diameter, we use an empirical polynomial relationship according to the number of cycles. The position of the organs in relation to each other is determined by the lengths of the internodes and the topological relationships between them. This requires specific data that characterize the orientation during branching, including the angles of phyllotaxis and branching.

In practice, phyllotaxis is ignored to obtain planar representations that develop the structure flat, which is useful for understanding.

The 3D representation is aimed at obtaining a realistic geometric structure of the plant; finer parameters are needed to compute the orientation of the organs in space.

When the structure is explicitly computed, the user can represent, visualize, and explore it through exports to two tools: Ligdraw and Xplo.

14.9 History of the Computer Simulation of Plant Architectures Developed within the Amap Unit

Since its creation in the late 1980s, software development has been a prominent component of the Amap unit. The laboratory specializes in developing plant growth models based on architectural concepts developed by Francis Hallé (1978) and his colleagues (Barthélémy and Caraglio 2007).

These models have led to a generation of software tools (the Amap range), which corresponds to the successive implementation of different modeling paradigms, including stochastic models of the trajectory of meristems, discrete probabilistic automata, resistive models, hierarchical discrete automata, etc.

The Amap range of software now covers a much broader spectrum concerning multiple domains. A non-exhaustive list of software is accessible from the research unit's website at the following link: https://amap.cirad.fr/en/logiciels.php

14.9.1 Pioneering Tools

The first prototype for simulating plant architecture was developed by Philippe de Reffye during his thesis in the late 1970s (de Reffye 1979). It was implemented on a Hewlett Packard HP9825A computer with a proprietary system in HPL (Harmony Programming language). The tool produced outputs in the form of tables of numbers and a two-dimensional plot file, representing the simulated structure on an external plotter. The code structure is compact, integrating both simulation and representation. The simulation of the model built the structure in prefix order, at a given age in advance. The plotting was carried out in the same time on a single-color plotter compatible with the computer, the HP9862. Although the structure was computed in three-dimensional space, the representation of the organs was fixed to two-dimensional textual forms: the internode is materialized by the tracing of a rectangle, the leaf by the tracing of two brackets, and the fruit by the tracing of a star (Fig. 14.30).

This pioneering tool marked the beginning of a long line of software development. The evolution of these tools reflects advancements in modeling and information technology, both in hardware and software aspects (operating systems, programming languages).

The next generation of simulators was developed on Bull Mini6 and Data General Minicomputers, running on proprietary systems. This simulator was implemented in Fortran, which was the scientific language of the time and is still widely used today (Fig. 14.31).

This work, which dates to the mid-1980s, focuses on the evolution of the implementation of the model, beginning with improvements in representation and visualization. Initially, geometric computations were carried out in three dimensions using predefined shapes such as leaves, flowers, and branch tips, replacing graphic symbols. A Benson pen plotter was used, allowing for four distinct colors, and resulting in the first color sketches (Fig. 14.32). The unsuitable plotters used for drawing plant architectures have now been replaced by new graphic screens (Figs. 14.31 and 14.33).

Subsequently, the geometric outputs were deferred, with representation becoming three-dimensional and disconnected from the simulation process. These geometric outputs allowed for the first interactive 3D visualizations of simulated plants using an Evans and Sutherland PS300 vector machine from the University of Strasbourg.

Researchers at CIRAD are equipped with a Tektronix 4107 raster graphic screen, which allows for 16-color visualizations. In this Fortran version, the simulation tool becomes generic, enabling the construction of trees, flowers, crops, palm trees, and inflorescences, using the model at different levels with different parameters. The botanical notion of total and partial reiteration is implemented, allowing for the construction of trees with master branches.

The software can then branch according to distinct sets of parameters, thus creating structures of different natures. For instance, large herbs can be built by

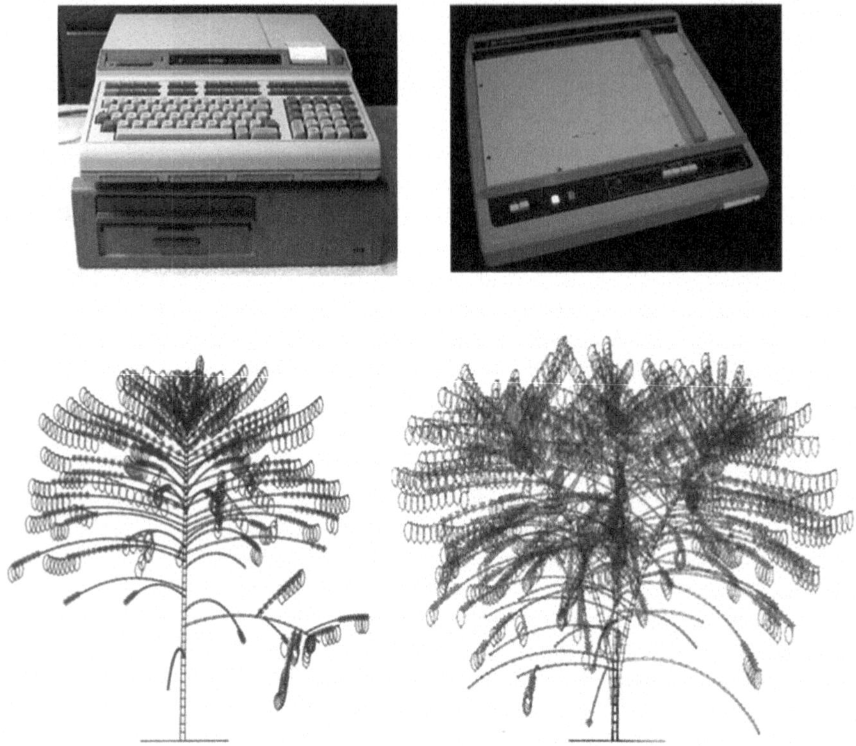

Fig. 14.30 *Top*: The HP9862 and its plotter. *Bottom*: Two simulated coffee trees generated on these devices (de Reffye 1979)

Fig. 14.31 Dashboard of a Bull Mini6 CII (1984) and first images of trees on a semi-graphic screen connected to a General Data computer (1985)

exploiting three levels: one underground, building a rhizome, a second for the main supporting structure, and finally, a last one composing the floral pieces.

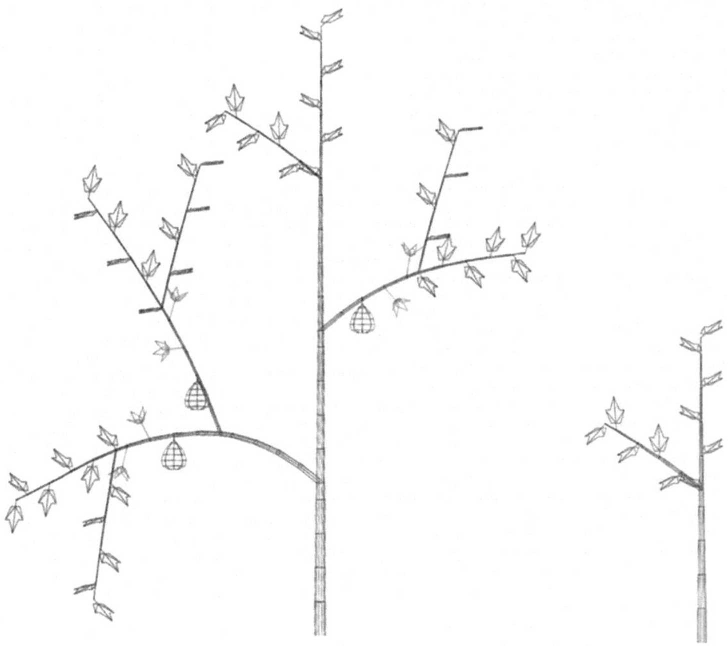

Fig. 14.32 Benson plotter four-color drawings (Jaeger, CIRAD, 1984)

Similarly, the vine can be described using sets of parameters for the vine stock, branches, and bunches. The graphical representation, initially limited to the Benson plot, has been expanded to include perspective, surface representation, and illumination (Fig. 14.34).

Furthermore, the simulation tool is complemented by a parameter input tool. The generation of the geometry is based on the definition of a 3D skeleton, which is then dressed and visualized in the visualization tools. This approach allows the creation of the first scenes of virtual landscapes (Fig. 14.35).

14.9.2 The Graphic Era and the First Technological Transfers

The advent of workstations (CAD: Computer-Aided Design) such as the Silicon Graphics stations enabled the Amap model to be made available to various user communities (Fig. 14.36). By rewriting the computation codes in C language, within a relatively standard environment, namely the Unix/Irix system, the simulator implementing the Amap model could be distributed and used as a 3D object generator (de Reffye et al. 1988; Jaeger and de Reffye 1992). The development of tools for capturing and editing landscapes enables the Amap laboratory to provide

Fig. 14.33 Vector representations of a coconut tree, a zoomed-in simulation of a Rauh model, and photographs of a screen from an Evans and Sutherland station (PS 300) at the IBMC of the University of Strasbourg (Jaeger, CIRAD, University of Strasbourg, 1985)

Fig. 14.34 Representations of flowerbeds and cotton trees (de Reffye, Jaeger, CIRAD, 1987–1988). *Right*: Screenshot of a simulation representation of a young cotton tree, with leaves represented by solid polygons whose color depends on their orientation in relation to the direction of the light. *Left*: Flowerbed computed in a single simulation with three structures: the rhizome (not shown), the leafy axes, and the flowers

Fig. 14.35 First urban and agricultural landscape scenes (Jaeger, CIRAD, 1987)

one of the first virtual planning solutions to image professionals, urban architects, landscape designers, and land managers (Saito et al. 1993).

In the early 1990s, this generation of simulators facilitated technology transfers and audio-visual applications, including the development of the Amap range. Today, Bionatics SA still markets this generation due to the flexible construction of the structures and a large library of plant parameters.

14.9.3 Bionatics SA's Tools Range

In the early 1990s, the plant simulation tool developed in the 1980s in C language (ported to Linux and Windows environments) was commercialized by JMG Graphics. Later, the distributor became Bionatics SA (https://mybionatics.com/), which released integrated versions of the tool, known as "plug-ins" in various applications. The tool allows the user to choose a plant from a library and a simulation age, and then access controls for simulation parameters or output properties, depending on the context.

To integrate the tool into professional applications requiring high-quality visual rendering, specific interfaces were created to allow control of input and output parameters while prospecting the host application's ergonomics, principles, and functionalities.

In architecture (AutoCAD) or image synthesis (Maya), the simulation tool becomes a generator of polygonal 3D objects. The users can control the visual quality and geometric complexity of a simulated plant through specific outputs, and choose the season, modulate the density of branches or foliage, and select different random prints in the simulation parameters. Fine visual properties of colors, bark, or leaf textures can also redefine or ignore to use those of the tool hosting the simulator (Fig. 14.37).

Chapter 12 describes LandSim3D, a professional application for representing the layout of large spaces. One of its features allows users to easily define a plant species with a random age and seed interval. Additionally, LandSim3D includes an adaptative multiresolution 3D output that adjusts the level of detail based on the

Fig. 14.36 The era of graphic workstations, dedicated to CAD; these machines allow realistic representations of simulated plants and scenes integrating numerous plant elements, right down to the landscape

viewing distance of the plant. The host application automatically computes date and season variations, and provides options to adjust the accuracy of the geometric outputs.

14.9.4 The Genesis of the Plant Simulation Software Amap and GreenLab

The image synthesis market has driven significant advancements in simulation software, leading to revisions and updates of the original simulation Amap software to integrate evolutions concerning both structural and functional aspects. This has been achieved through two major steps: firstly, simulation with Amap range software; and secondly, computation with the GreenLab range software. Figures 14.38 and 14.39 depict the evolution of this history in two different forms.

Fig. 14.37 Screenshot of the Bionatics RealNat® tool (Jaeger, CIRAD)

14.9.4.1 Amap Software Generation: Structural Models of Development

In Chap. 3, the AmapSim software was introduced, which implemented the PA to botanically simulate the development of tree architectures from the young to the senile stage. AmapSim (Rey et al. 1998; Barczi et al. 2008) also implemented the notion of reference axis (de Reffye et al. 1991) and modified the organization of data, making it more suitable for dynamic memory management using the C language.

Making simulation tools modular, portable, and reliable allows the construction of specialized applications for professional use. In an exploratory research context, integrating simulation tool into general-purpose software platforms enables comparison with other approaches, utilization of analytical tools for examining results, data sharing, and even coupling with other models upstream or downstream of the simulation structure function of the individual plant.

Jean François Barczi developed the AmapSim software (Chap. 3) according to the rules mentioned above, under the Qt environment, making it easy to port to different operating systems. The latest version, the AmapSim II (Barczi et al. 2008), comes with a collection of open procedures that facilitate dynamic communication

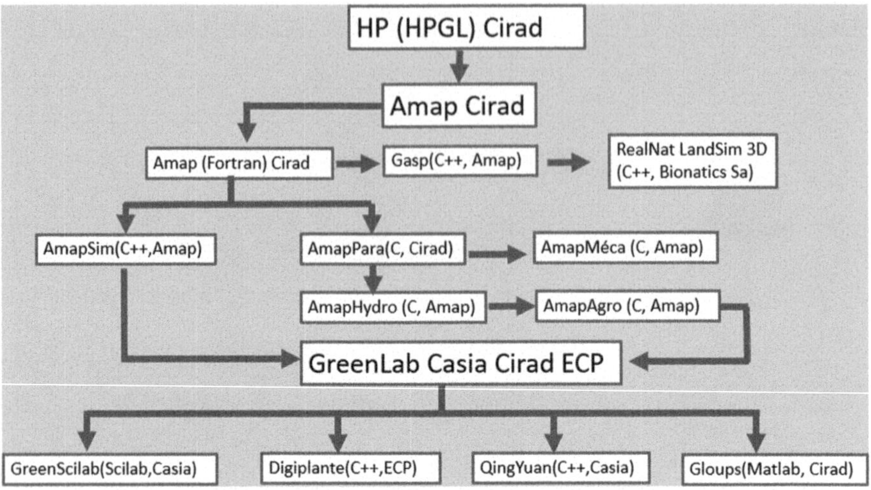

Fig. 14.38 Diagram of the genesis of the simulation software coming from the prototype HPGL to Amap range and finally GreenLab range

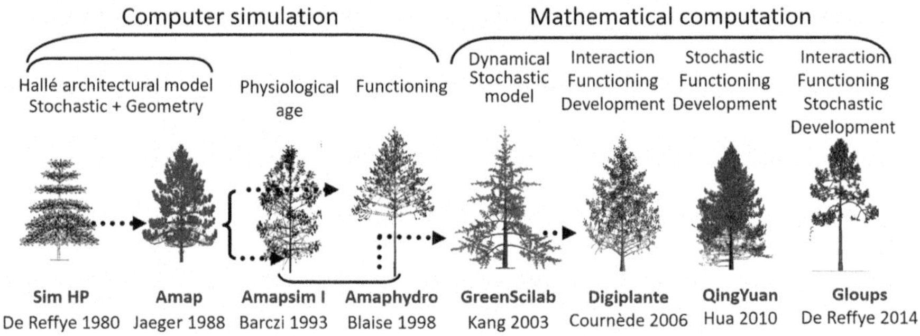

Fig. 14.39 Examples of simulations of plant architecture along the different generations of AMAP_ GrenLab softwares

between the simulation kernel and external applications. This enables modification of simulation behavior, as well as extraction or association of additional data during the simulation. The software package includes parameter editing, simulation, visualization and control tools, reference databases, technical documentation, and examples of using external applications.

Chapter 8 describes the use of AmapPara software for implementing breadth-first tree construction traversal to simulate spatialized growth computations of multiple plants in a voxel space. This software considers the topology and geometry of plants and discretizes the space into voxels. It also simulates spatial competition and physical discomfort between neighboring plants (Blaise and de Reffye 1994; Blaise et al. 1996). AmapMéca, an extension of AmapPara, introduces mechanical elements to simulate feedback processes in stem curvatures and the appearance of tension–compression wood within the rings (Fourcaud 1995; Fourcaud et al. 1998).

AmapAgro, another evolution of AmapPara, calculates biomass production cycle by cycle and distributes assimilates produced by leaves to the different compartments using the branching system as a transport. In this prototype tool, the construction of the structure follows by inheritance from AmapPara using breadth-first tree construction traversal. This evolution marks the first attempt to introduce functioning into model implementations, which was previously only structural (Fourcaud et al. 1997). In parallel, AmapAgro also simulates the size of the organs and diameters of axes (the secondary growth of the trees) as functions of the leaf area carried with significant memory costs and computations.

AmapAgro also inherits the ability to simultaneously analyze both intra-plant and inter-plant functions, such as light interception and global water resource sharing. By further developing these functions, the software enables the exploration of competitive phenomena. This innovative tool allows for the representation of how stand density and technical itineraries (thinning, pruning) affect the aerial development of the tree and wood production, including the diameter of trunks and branches and the distribution of rings in the trunk. AmapAgro serves as a prototype for GreenLab software, which builds upon its capabilities.

14.9.4.2 GreenLab Software Generation: Functional–Structural Models of Growth

The Amap software relies solely on simulation for its philosophy, leading to extended computation times and limiting its practical applications to visualization and forgetting the utilization of applied mathematics methods such as inverse methods, differential statistics, and optimization (Chaps. 9, 11, and 12).

To overcome these computational bottlenecks, GreenLab software was developed with the same botanical and ecophysiological concepts but with a focus on computation over simulation.

The botanical automaton replaces the reference axis, resulting in the plant architecture being factored into nested recursive substructures. This leads to a considerable reduction in computation time, with the exponential increase according to the age of the plant limits to quadratic complexity (Chap. 10).

Moreover, the concept of potential structure (Chap. 3) replaces costly Monte Carlo simulations with the exact calculation of the means and variances of stochastic structures at a given age. This development allowed the computation of parameters using WLSM (Chap. 9).

14.9.5 *Peripheral Developments*

Chapter 12 discusses several computer developments that have been focused on simulating plants. While the focus is on growth simulation, there are three other important areas of development that are briefly mentioned. These include the visualization of plants in computer graphics images, the coding of measured architectures (AMAPmod), and the calculation of light interception using 3D models of plants. Although these areas are outside the strict scope of growth simulation, they are still significant and worthy of mention.

14.9.5.1 Computer Developments for the Visualization of Plants and Landscapes

To create 3D models of synthesized objects, computer-generated image is applied to a mesh of polygons using classical visual rendering techniques such as local illumination, texture mapping, up to global illumination (shadow maps, ray-tracing, radiosity, etc.). Marc Jaeger and his team have adapted these classical techniques to visualize plants and landscapes (Jaeger 2010; Jaeger et al. 2010).

However, visualizing trees can be challenging due to the large number of polygons required. Therefore, algorithms have been developed to display only the essential polygons needed based on the distance from the observer without compromising the rendering quality using multiresolution plant descriptions and adaptative geometrical meshing, so called level of details (Zhang et al. 2017; Guo et al. 2020). These methods are useful for displaying large plant scenes in real time.

14.9.5.2 Encoding of Measurements Taken on the Architecture of the Plants

Christophe Godin developed AmapMod software (Godin et al. 1997) to create and explore databases of plant architectures, and to perform statistical analysis of data samples extracted from these databases. The software is based on two main foundations: a representation of the tree structure using a multiscale graph, and a specialized language called MTG (C; Godin et al. 1999) for the topological coding of such graphs.

14.9.5.3 The Computation of the Radiative Transfer in the Canopies

Jean Dauzat developed the Archimed software (Dauzat et al. 2008; Vezy et al. 2020) for fine simulation of the radiative transfer, photosynthesis, and transpiration of plant formations at the stand scale.

14.10 Conclusion

The Amap and GreenLab models, which have been developed for over 40 years, demonstrate the crucial role that computers play in making mathematical models functional. Although the equations for these models were theoretically available at the beginning of the twentieth century in the field of applied mathematics and physics, their application was only possible with the aid of computers. The development of the Amap and GreenLab models was closely tied to the evolution of computer technology, as depicted in Figs. 14.38 and 14.39. Initially, in 1970, plotters were used for visualizing the plant since computer screens did not exist (Fig. 14.30). However, with the advent of screens in the 1980s, along with advancements in software generations (Fortran, C, etc.) and processors with graphic cards, computer-generated images became possible, enabling the representation of plants and landscapes. Moreover, specific algorithms, such as structure factorization, have contributed to the simulation of large tree architectures by accelerating computation, in addition to the increased computer performance.

When considering plant growth simulation software, it is important to factor in their durability and development time, which often involves writing tens of thousands of lines of code and can coincide with thesis topics lasting at least three years. However, these programs can become obsolete over time due to turnover in compilers, such as Fortran, C, C++, or Java, and changes in operating systems (versions of Windows, Unix). For instance, AMAP software developed before the year 2010 may no longer function due to unavailability of compatible computers.

Furthermore, as the model evolves, it may be necessary to rewrite everything to accommodate new theoretical advances, which can take several years. For example, the transition from architectural models to the reference axis (AmapSim), and from development to growth (GreenLab). It is worth noting that only the first Amap software, which was translated into C and industrialized by Bionatics, has been perpetuated.

Due to the great genericity of the Gloups software, which can be used to simulate various types of plants such as grasses, plants, bushes, and trees, and the durability of the Matlab platform, it is possible to expect that the GreenLab model will remain stable and be able to model many new plants without major issues. However, it should be noted that the current prototype version of the software is only suitable for research purposes and requires further development before it can be industrialized. This includes translation into a more powerful language and fixing any existing limitations or issues.

References

Aono M, Kunii T-L (1984) Botanical tree image generation. IEEE Comput Graph Appl 4(5):10–33

Barczi J-F, Rey H, Caraglio Y, de Reffye P, Barthélémy D, Dong Q, Fourcaud T (2008) AmapSim: an integrative whole-plant architecture simulator based on botanical knowledge. Ann Bot 101(8):1125–1138

Barthélémy D, Caraglio Y (2007) Plant architecture: a dynamic, multilevel and comprehensive approach to plant form, structure and ontogeny. Ann Bot 99(3):375–407

Blaise F, de Reffye P (1994) Simulation de la croissance des arbres et influence du milieu: le logiciel AMAPpara. In: Tankoano J (ed) CARI'94, 2ème colloque africain sur la recherche en informatique, Ouagadougou, Burkina Faso. Orstom Éditions, Paris, pp 61–75

Blaise F, Houllier F, de Reffye P (1996) Simulation of tree architecture and growth in a forest stand: Amappara software. In: Nepveu G (ed) Proceedings of connection between silviculture and wood quality through modelling approaches and simulation softwares, 1st IUFRO WP S5.01-04 workshop. Hook, Suède, pp 46–55

Bosanac B (1990) Tree: a knowledge-based system for generating images of trees. Harvard University, Massachusetts Hall, Cambridge

Chen X, Neubert B, Xu Y-Q, Deussen O, Kang S-B (2008) Sketch-based tree modeling using markov random field. In: ACM SIGGRAPH Asia 2008 papers, pp 1–9

Cournède P-H, Kang M-Z, Mathieu A, Barczi J-F, Yan Y-P, de Reffye P (2006) Structural factorization of plants to compute their functional and architectural growth. Simul Trans Soc Model Simul Int 82(7):427–438

Cournède P-H, Chen Y, Wu Q, Baey C, Bayol B (2013) Development and evaluation of plant growth models: methodology and implementation in the Pygmalion platform. Math Modell Nat Phenom 8(4):112–130

Dauzat J, Clouvel P, Luquet D, Martin P (2008) Using virtual plants to analyse the light-foraging efficiency of a low-density cotton crop. Ann Bot 101(8):1153–1166

de Reffye P (1979) Modélisation de l'architecture des arbres par des processus stochastiques. Simulation spatiale des modèles tropicaux sous l'effet de la pesanteur. In: Application au Coffea robusta. Thèse de doctorat. Université Paris Sud Orsay. 195 p

de Reffye P, Hu B-G (2003) Relevant qualitative and quantitative choices for building an efficient dynamic plant growth model: GreenLab case. In: Hu BG, Jaeger M (eds) Plant growth modelling and applications (PMA03), Proceedings of the 2003' international symposium on plant growth modeling, simulation, visualization and their applications. Tsinghua University Press, Springer, pp 87–107

de Reffye P, Jaeger M, Edelin C, Françon J, Puech C (1988) Plant models faithful to botanical structure and development. Computer Graphics, Siggraph 1988 22(4):151–158

de Reffye P, Dinouard P, Barthélémy D (1991) Modélisation et simulation de l'architecture de l'Orme du Japon *Zelkova serrata* (Thunb.) *Makino* (Ulmaceae): la notion d'axe de référence. Comptes Rendus du 2e Colloque international sur l'arbre, Montpellier, 10–15 Septembre 1990. Naturalia Monspeliensia, Hors-Série:251–266

Deussen O, Lintermann B (1997) A modeling method and user interface for creating plants. In: Graphics interface '97. Morgan Kaufmann, San Francisco, pp 189–197

Eschenbach C (2005) Emergent properties modelled with the functional structural tree growth model ALMIS: computer experiments on resource gain and use. Ecol Model 186(4):470–488

Feng L, de Reffye P, Dreyfus P, Auclair D (2012) Connecting an architectural plant model to a forest stand dynamics model—application to Austrian black pine stand visualization. Ann For Sci 69(2):245–255

Fischer J. B., Honda H., Computer simulation of branching pattern and geometry in Terminalia (Combretaceae), a tropical tree, 1977

Fourcaud T (1995) Analyse du comportement mécanique d'une plante en croissance par la méthode des éléments finis. Thèse de 3e cycle en mathématiques appliquées. Université de Bordeaux, 148 p

Fourcaud T, Blaise F, Barthélémy D, Houllier F, de Reffye P (1997) A physiological approach for tree growth modelling in the software AMAPpara. In: Nepveu G (ed) Proceedings of 2nd workshop connection between sivliculture and wood quality through modelling approaches and simulation software (WP S5.01.04), 25-30/08/1996, Berg-en Dal, Afrique du Sud, pp 321–334

Fourcaud T, Blaise F, de Reffye P, Lac P (1998) Une approche fonctionnelle pour la simulation de la croissance d'arbres fruitiers: le logiciel AmapPara. In: Architecture et modélisation en arboriculture fruitière, Montpellier, 5-6/03/1998, pp 307–316

Françon J (1990) Sur la modélisation informatique de l'architecture et du développement des végétaux. 2ème Colloque International "L'Arbre", Institut de Botanique, Montpellier, France

Godin C, Costes É, Caraglio Y (1997) Exploring plant topological structure with the AMAPmod software: an outline. Silva Fenn 31(3):355–366

Godin C, Costes E, Sinoquet H (1999) A method for describing plant architecture which integrates topology and geometry. Ann Bot 84(3):343–357

Griffon S, de Coligny F (2012) Amapstudio: a software suite for plants architecture modelling. In: Proceedings of plant growth modeling, simulation, visualization and applications (PMA12), Shanghaï, 31 Oct – 3 Nov 2012. IEEE Press, pp 141–147

Griffon S, De Coligny F (2014) AMAPstudio: an editing and simulation software suite for plants architecture modelling. Ecol Model 290:3–10

Guo J, Xu S, Yan D-M, Cheng Z, Jaeger M, Zhang X (2020) Realistic procedural plant modeling from multiple view images. IEEE Trans Vis Comput Graph 26(2):1372–1384. https://doi.org/10.1109/TVCG.2018.2869784

Hallé F, Oldeman R-A-A, Tomlinson P-B (1978) Tropical trees and forests. Springer Verlag, Berlin/Heidelberg/New York. 441 p

Hua J, Kang M-Z (2011) Functional tree models reacting to the environment. ACM SIGGRAPH 2011 posters. ACM, Vancouver

Hua J, Kang M-Z, de Reffye P (2011) An interactive plant pruning system based on GreenLab model: implementation and case study. In: 2011 IEEE international conference on computer science and automation engineering (CSAE), vol 4. IEEE, pp 185–188

Jaeger M (2010) Représenter et visualiser les objets naturels biologiques. De la plante au paysage. HdR. Discipline Informatique. Université Blaise Pascal, Clermont-Ferrand II. 151 p

Jaeger M, de Reffye P (1991) Le logiciel Amap: un outil de simulation et de représentation des végétaux. In: de Reffye P, Jaeger M (eds) L'informatique scientifique dans l'enseignement de la biologie et de la géologie au lycée: actes du colloque. Institut National de la Recherche Pédagogique, Paris, pp 243–248. (Colloque Technologies Nouvelles et Education)

Jaeger M, de Reffye P (1992) Basic concepts of computer simulation of plant growth. J Biosci 17(3):275–291

Jaeger M, Sun R, Jia J, Le Chevalier V (2010) Efficient virtual plant data structure for visualization and animation. In: Proceedings of IADIS international conferences computer graphics, visualization, computer vision and image processing 2010, MCCSIS 2010 Freiburg, Germany, July 27–29, ISBN 978-972-8939-22-9, 2010. IADIS, pp 65–75

Kang M-Z, Qi R, de Reffye P, Hu B-G (2006) GreenScilab: a toolbox simulating plant growth in the Scilab environment. In: 8th Middle Eastern simulation multiconference, Alexandria, pp 174–178

Kang M-Z, Cournède P-H, Quadrat J-P, de Reffye P (2007) A stochastic language for plant topology. In: Fourcaud T, Zhang XP (eds) Proceedings. Second International Symposium on plant growth modeling, simulation, visualization and applications, PMA06, 13-17 november 2006, Beijing, P.R. China. Los Alamitos. IEEE Computer Society, pp 34–39

Kang M-Z, Heuvelink E, Carvalho S-M, de Reffye P (2012) A virtual plant that responds to the environment like a real one: the case for chrysanthemum. New Phytol 195(2):384–395

Oppenheimer P-E (1986) Real time design and animation of fractal plants and trees. ACM SIGGRAPH Comput Graph 20(4):55–64

Palubicki W, Horel K, Longay S, Runions A, Lane B, Měch R, Prusinkiewicz P (2009) Self-organizing tree models for image synthesis. ACM Trans Graph 28(3):1–10

Prusinkiewicz P, Lindenmayer A, Hanan J (1988) Development models of herbaceous plants for computer imagery purposes. In: Proceedings of the 15th annual conference on computer graphics and interactive techniques, pp 141–150

Prusinkiewicz P, Hammel M, Hanan J, Mech R (1996) L-systems: from the theory to visual models of plants. In: Proceedings of the 2nd CSIRO symposium on computational challenges in life sciences, vol 3. Citeseer, pp 1–32

Rey H, Barczi J-F, Caraglio Y, de Reffye P (1998) AMAPsim: un outil de construction et simulation de plantes numériques tridimensionnelles: principes et applications. In: Architecture et modélisation en arboriculture fruitière, 11ème colloque sur les recherches fruitières, Montpellier, France 5-6/03/1998. INRA-Ctifl, pp 296–306

Saito K, Kumagai Y, Honjo T, Ishida Y, Lecoustre R, de Reffye P (1993) Photo-realistic forest landscape simulation—application of GIS and plant modeling technique. In: Proceedings of NICOGRAPH'93. Tokyo, Japan, NICOGRAPH'93, 11/1993, pp 226–236

Soler C, Sillion F-X, Blaise F, de Reffye P (2003) An efficient instantiation algorithm for simulating radiant energy transfer in plant models. ACM Trans Graph 22(2):204–233

Vezy R, Perez R, Grand F, Dauzat J (2020) Light exchanges in discrete directions as an alternative to raytracing and radiosity. In: FSPM 2020: towards computable plants. 9th international conference on functional-structural plant models

Viennot X, Eyrolles G, Janey N, Arques D (1989) Combinatorial analysis of ramified patterns and computer imagery of trees. In: Proceedings of the 16th annual conference on computer graphics and interactive techniques, pp 31–40

Vos J, Marcelis L-F-M, Visser P-H-B, Struik P-C, Evers J-B (eds) (2007) Functional structural plant modelling in crop production. Springer. Printed in the Netherlands, pp 43–52. https://doi.org/10.1007/1-4020-6034-3_4

Yan H-P, Barczi J-F, de Reffye P, Hu B-G, Jaeger M, Leroux J (2002) Fast algorithms of plant computation based on substructure instances. J WSCG 10(1–3):145–153

Zhan Z-G, de Reffye P, Houllier F, Hu B-G (2003) Fitting a functional-structural growth model with plant architectural data. In: Hu B-G, Jaeger M (eds) Proceedings of 1st international symposium on plant growth modeling, simulation, visualization and applications (PMA03). Springer/Tsinghua University Press, Beijing, pp 108–117

Zhang X, Bao G, Meng W, Jaeger M, Li H, Deussen O, Chen B (2017) Tree branch level of detail models for forest navigation. Comput Graphics Forum 36(8):402–417. https://doi.org/10.1111/cgf.13088

Zhao Y, Barbič J (2013) Interactive authoring of simulation-ready plants. ACM Trans Graph 32(4): 1–12

Zhao X, de Reffye P, Xiong F-L, Zhan Z-G (2001) Dual-scale automaton model for virtual plant development. Chin J Comput 24(6):608–617

Conclusion

At the end of this book, it is necessary to review the formulation of the GreenLab model with the botanical and agronomic concepts used and their adaptations necessary to make it work. The aim here is to highlight the strengths of the model in its validation on cultivated plants, its genericity, its forecasting capabilities, and its potential applications in agronomy.

When we look for information on plant development and growth in academic literature, we find that data at the level of the whole plant (the level that essentially concerns the present work) is poorly represented. First and foremost, plant architecture is ignored, with a focus instead on organ-specific function (cell multiplication in meristems, leaf photosynthesis, etc.). A great deal of attention is paid to growth factors, including the roles of temperature, light, and hormones. Growth is typically examined from a nutritional perspective, with the role of water or light use efficiency. However, our understanding is limited regarding the quantity of matter synthesized and its distribution within the plant. While the factors influencing growth are studied, the impact of plant architecture on growth is often neglected.

The existing academic understanding of plant growth is not sufficient to predict how much plant matter enters a plant at a given developmental stage and in a given environment, as well as how it is distributed, which is the main challenge in agronomy. This question can only be addressed by taking into account plant architecture and modeling source-sink relationships, which is what this book is all about.

Reminder on the Plant Models Used in Agronomy

Process-based models (PBMs) aim to predict harvests. Only the stand is concerned and the individual plant is ignored. Production is known only at the level of plant compartments (leaves, stems, fruits, roots). It is quantified per unit area (m^2, hectare)

P. De Reffye et al. (eds.), *Plant Model Greenlab for Botany and Agronomy*,
https://doi.org/10.1007/978-981-97-7541-5

and uses synthetic variables such as the leaf area index (LAI) and the harvest index (HI). These models, which take into account the environment factors (light, temperature, irrigation, soil type), are well suited to monitor the evolution of stands and make harvest forecasts. They integrate phenological data, such as flowering (e.g., Pilot model), which depends on the sum of the temperatures. Moreover, data collection is simple: LAI is estimated by instruments (Licor) in the canopy, and HI is calculated as the ratio between the biomass of the useful compartment and the total biomass of the stand. This global approach has its limitations in managing abortion-related stress and in predicting HI. The latter can vary greatly due to source–sink relationships whose genetic parameters are important components of yield. Crop models are therefore less effective for plant selection and stress management.

Virtual plants aim to represent plant architecture either by computer simulation or by digitizing architectures using lasers (Lidar). Simulated virtual plants contain little botanical knowledge and their beauty is mainly due to the talent of the designers and the capabilities of the software used in computer graphics. Those obtained by digitization faithfully reproduce the architecture of the plants, but they do not contain any botanical information on growth and development (as the growth units (GUs) are not identified). They are virtual and static models. They can also be used in agronomy to compute light interception. Their limitation is that they cannot integrate the dynamics of growth through feedback.

Functional–Structural Plant Models (FSPMs), the subject of this book, aim to faithfully reproduce plant functioning in order to predict plant growth by taking the environment into account. In this context, it is more the topology of the plant structure (relations between organs) than its geometry (architecture) that is concerned. It is a question of modeling the functioning of meristems to compute the number of phytomers set up and the functioning of the source–sink relationships of organs to compute the biomass production and distribution in the plant at each stage of growth. FSPMs use environmental variables in a manner similar to crop models, but they can interact with the architecture of the plant. Light is intercepted by the leaves and the produced biomass is distributed step by step through the plant to ensure the creation and expansion of organs according to their sink strength. A computer simulation grows the plant by feedback of growth on development. In terms of simulation time, the computation of light interception and biomass distribution is a bottleneck in computational plants.

Although FSPMs, through simulation, provide information about plant functioning, it has not yet been proven that they can provide applications in agronomy. The numerous parameters identified in these models are not computable and are empirically based. There is no validation in the models by adjusting the computed values to the measured data, by an inverse method, which optimizes the hidden parameters.

The Advantages of the GreenLab Model

From a conceptual and operational perspective, the FSPM model "GreenLab" has several strengths.

On the Conceptual Level

Since the level of observation of crop models is the field, subdivided into plant compartments (leaves, stems, fruits, roots), the GreenLab level is the plant subdivided into organic series. In both cases, the organs are either grouped into compartments or considered individually. By grouping the organic series into compartments, GreenLab can effectively revert to a crop model. This explains why most of the functionalities of the crop models are compatible with the GreenLab model, which increases its interest and distinguishes it from other FSPMs.

A characteristic of this model, compared to other computational plant models, is its restriction to the phytomer as a spatial unit. The model does not attempt to model structures and processes at a scale of resolution finer than that of a single plant organ (leaves, internodes, buds). The cellular levels, metabolic networks, and molecular genetics are excluded from the scope of this modeling approach. We do not look for a causal origins of plant functions, such as the action of hormones. The variability induced by these different phenomena cannot be represented in a causal way but can be characterized and reproduced in a stochastic way.

It is legitimate to explore how far we can go with a model strictly limited to the phytomer. This restriction is justified by the desire to be oriented toward agronomic applications and to build models whose parameters are all measurable, which is far from being the case in general for FSPMs. If the model is descriptive at the phytomer level, it becomes explanatory for the functioning at the whole plant level. The search for causality at a lower scale is not our objective.

On the Operational Level

The model is a guide for analyzing plant architecture and designing data assimilation. When data are collected outside the context of a model, they often prove to be inappropriate or insufficient: "Without theory, there is no valid measurement". The model is thus an indispensable tool for the development of the experimental method, in particular thanks to the numerous tests that can be carried out on the virtual data in order to develop the measurement protocols.

Botanical Aspects of Development

The concept of physiological age (PA) subdivides the plant structure into categories of axes that support the organic series (Chap. 2). One advantage is that the number of physiological ages is small (< 6) and does not depend on the architectural pattern and its reiterations. The PA decomposes the plant architecture into nested substructures in the manner of Russian dolls, which allows an efficient factorization of the topological structure.

The organic series are orthogonal to the phytomer cohorts and easily identifiable. They are found repeated many times at multiple levels in the plant and are the basis for sampling and data assimilation for estimating hidden parameters. Missing data due to organ abortions are no longer a problem for crown analysis and organic series.

Ecophysiological Aspects of Growth

The GreenLab model operates in development cycles (DCs) (Chap. 5). This feature allows the plant's architecture to maintain relative independence from environmental variables when they vary within reasonable limits. Plant development is more or less rapid depending on the temperature, but ultimately the same plant is obtained. The thermal time is discretized in DCs, which correspond to the average duration of formation of a phytomer by a meristem. Modeling the meristem functioning by analyzing the crown does not require following the development of the plant in real-time; instead, the dynamics of development are reconstructed a *posteriori*.

Although organ expansion is a continuous phenomenon, the model discretizes it into a growth cycles (GCs) whose duration is equal to the DC, which makes it possible to easily combine growth and development through feedback. Modeling the source–sink relationships of the organs by analyzing the organic series does not require monitoring the growth of the plant.

The model uses organ weight as a predictable and observable variable. The weights can be measured either in fresh matter or in dry matter. The sinks obtained will only be corrected by a factor corresponding to the hydration rates of the organs. Initially, only net photosynthesis is considered. The costs of the assimilates necessary for the production of different types of organs, according to their water or oil content, as well as the computation of respiration, are calculations to be made a *posteriori*, if necessary, by the users of the model. The methods to be implemented are the same as those used in the PBMs, are adaptable, and can, therefore, be easily added to the GreenLab model.

A particularity of the model is the consideration of the seed for initiating growth (Q_0) and the planting density (Sp) (Chap. 5) rarely used in PBMs and FSPMs. Moreover, the step-by-step distribution of biomass in the structure, common to the FSPMs, is replaced by its allocation to each cycle in all organs of the plant according to their sink strengths. This is possible thanks to the conservation of the concept of common pool, which is relatively well established for agronomic plants and which can be extended to woody plants, because it is compatible with secondary growth according to Pressler's law (Chap. 8). The biomass allocation is computed by matrix operators which greatly reduce the computation time.

Mathematical Aspects of the Modeling

Stochastic Processes and Development

The natural variability of plant architecture is taken into account through the notion of development cycle and the use of stochastic processes to control meristem functioning. The renewal theory (Chap. 3) transforms the law of time intervals between the emission of two phytomers by the same meristem into a *Bernoulli* process that assigns a phytomer production rate (parameter b) to each DC. This transformation preserves the mean and variance of the number of phytomers of axes of the same age and even, with a good approximation, their distribution. Thus, the rhythm rate (ratio w), the viability rate (parameter c), and branching rate (parameter a) are used. The parameters (w, a, b, c) can be computed by the crown analysis.

The GreenLab model has obvious affinities with Markov processes. It uses the visual form of graphical models of these processes. The botanical automaton (Chap. 4) proceeds by repeating phytomers of the same nature, grouping them into growth units (GUs) and then into substructures to finally create a potential structure. It ensures the transitions in the PA either by continuation or by branching considering the stochastic aspects. The development operators that compute the production of cohorts of phytomers according to the cycles are deduced from the botanical automaton. They implement the factorization of the substructures in matrix form that greatly reduces the computation time.

The notion of potential structure (Chap. 3) is a consequence of the *Bernoulli* process. It is a chronological structure which expresses all the possibilities of development. Each phytomer of this structure corresponds a probability of existence that can be computed with the quadruplet (w, a, b, c).

There are two uses of the potential structure:

- The first is the computation of the plant demand (D) which corresponds to the sum of the sinks P_o^φ multiplied by their probability of existence, (Eq. 5.8 of Chap. 5). The number of organs N_o^φ is computed by the botanical automaton according to a convolution. The demand equation is written as:

$$D(t) = \sum\nolimits_{o,\varphi} \left(\sum\nolimits_{i=1}^{t} N_o^\varphi(t-i+1) \cdot P_o^\varphi(i) \right) \tag{A.1}$$

The demand is a hidden parameter that cannot be measured directly and is only accessible through computation. It plays an important role in the biomass production.

- The second is the adjustment by the model of the observed organic series by the model. This is done by a mathematical transformation of the organic time series of the potential structure into organic topological series. Indeed, the observed organic series contain variability that requires data assimilation processing (Chap. 6).

The botanical automaton, which is based on the notion of physiological age, potentially simulates all architectural models in their anatomical and stochastic components. This gives it a good genericity for the computation of the development.

Computation of Photosynthesis

The growth in the model is computed by recurrence from the seed. The demands $D(j)$ from $D(1)$ to $D(t)$ is known and also the produced biomass $Q(j)$ from $Q(0)$ to $Q(t-1)$. The photosynthetic production is computed by the Eq. 5.2 in Chap. 5:

$$Q(t)=\frac{E\cdot Sp}{r}\cdot\left(1-\exp\left(-\frac{k}{\varepsilon\cdot Sp}\cdot\sum_{i=t-t_a+1}^{\max(t,t_a)}\sum_{\varphi=1}^{mx\varphi}N_a^{\varphi}(i)\cdot\sum_{j=i}^{t}\frac{p_a^{\varphi}(j-i+1)\cdot Q(j-1)}{D(j)}\right)\right) \tag{A.2}$$

This equation is the detailed expression of biomass production in the classical crop model (Eq. A.3). The number of functional leaves of the DC i cohort, $Na(i)$ is computed by the botanical automaton. The leaf area is computed as the sum of individual leaf area deduced from the allometric area–weight relationship which gives the leaf thickness ε. Beer Lambert's law, which describes light interception by leaf area, is adapted for individual plants by computing the projected surface of the leaf area as a function of planting density characterized by the parameter Sp which is a solution of Eqs. A.2 and A.3.

$$Q=\frac{E\cdot Sp}{r}\cdot\left(1-\exp\left(-k\cdot\frac{Sf}{Sp}\right)\right) \tag{A.3}$$

The Sf/Sp ratio is an optimized individual LAI. This formulation has the advantage of not depending on the planting arrangement. Indeed, Beer's law does not apply well at the stand scale when the plants are small or grouped in rows, which leaves many empty spaces.

The formulation of plant growth by the GreenLab model is similar to that of the crop models with which it is compatible, but it goes down to the level of detail of the phytomer. A crop model can be considered to be a projection of the GreenLab model, which is a synthesis between architectural and crop models.

The expansion q_o of an organ "o" that appeared in cycle $t - i$, in cycle t of the plant's growth, depends on the value of its sink $p_o(i)$, its age i in DC, and the ratio of the synthesized biomass $Q(t-1)$ in the previous cycle to the demand $D(t)$ in the current cycle (Eq. A.4).

$$q_o(t-i+1,t)=p_o(i)\frac{Q(t-1)}{D(t)} \tag{A.4}$$

The volume of an organ is the cumulative result of its expansions, and its shape depends on the allometries. The Eqs. A.1–A.4 compute the expansion state of organs of the phytomers in all locations of the time-series structures.

The *Q/D* ratio is a common factor in growth equations. It is interpreted as trophic pressure. It is evident that there is a link between plant growth (initially exponential) and the *Q/D* ratio. Indeed, the appearance of branching, the size of GUs, the mortality of meristems, the volumes, and the abortions of organs are well correlated statistically with this ratio. However, does this correlation imply a functional causality? Chapter 13 shows that this hypothesis is credible. From a causal point of view, thresholds of this ratio could trigger hormonal actions, etc.

Various Applications of GreenLab as a Dynamic Model

The GreenLab model belongs to the class of dynamic models. These have a mathematical library of applications that can be used by the model:

- The implementation of inverse methods for the calibration of parameters (here the Weighted Least Square Method (WLSM))
- Optimization of parameters under certain conditions with applications in genetics through the definition of ideotypes.
- Optimal control of environmental variables (irrigation, inputs, etc.) to optimize the harvest.

Computer Aspects of the Simulation

The simulation of plant growth with the Gloups software is done in two ways with the GreenLab model. Two operating modes are associated: the list mode and the matrix mode.

The demathematized form consists of a multiformalized computer simulation. The Monte-Carlo method manages the stochastic aspect of this simulation. The computer simulation is fragmented into submodels that simulate the functioning of the basic entities composing the phytomers (meristems, organs). The results of the simulation are computational plants described by their topological structures (planar graphs) and by their architecture if we add geometrical operators (dimensions, branching angles, phyllotaxis, etc.) which place the organs in space. The computer simulation provides datasets, which are a pure realization of the model. These datasets allows the development of algorithms and procedures related to the import of data and inverse methods. This is a crucial step before comparing the model with measurements on real plants, as it determines how well the model can only be adapted to actual plant growth data.

The remathematized form consists of a posteriori in building a purely computational formalism of the model. It is certainly possible to simulate a large number of plants in order to understand their variability, but the study of the model's behavior using only simulation is quickly made prohibitive by the heaviness of the computations. In the case of the GreenLab model, this mathematization is possible thanks to two considerations derived from the operation of the botanical automaton:

- The concept of the potential structure (Chap. 3) which implicitly contains all the computer simulations in chronological mode.
- The factorization of the development under the dependence of the PA which allows to consider a plant as an assembly and a nesting in Russian dolls of substructures. The computation time no longer depends on the number of phytomers to be set up, but only on the age of the plant.

Thus, even in the stochastic case, at each DC, the demand and the produced biomass are obtained directly by computation, as well as the theoretical means and variances of the numbers and biomasses of the organs.

In the list mode, the whole plant is simulated in detail. The structure of the plant is known explicitly. The phytomers are marked in a prefixed way. The computation time is obviously dependent on the number of simulated phytomers (Chap. 3) and can become consequent as well as the memory of the computer that stores the plant data.

In the matrix mode, only the organ cohorts and the organic series (which are orthogonal to them) are simulated. The structure of the plant is only implicitly known. The computation time only depends on the age of the plant. This efficient procedure is adapted to the estimation of the parameters.

Applications of the GreenLab Model

Genericity of the Model

A plant growth model must be validated on a dataset to be applicable, but it is also essential that the model has a good genericity in order to be easily adaptable from one plant species to another. This is the case of the GreenLab model.

All the developments of the architectural models come from the same botanical automaton of which they are particular realizations. This is because the botanical notions of meristem, phytomer, GU, and PA are common to all plants.

Photosynthesis is the engine of growth which is the same for all plants. The notions of thermal time, water and light use efficiency (WUE, LUE), common pool, and source–sink relationships are the components of this functioning.

All plants benefit from the same development botanical automaton and the same growth engine.

The Gloups software has been specifically developed on these bases in order to be generic and to simulate correctly, according to the basic botanical and physiological knowledge, for herbaceous, shrubs, and trees.

Validation of the Model

The validation is considered as acquired when there is a method of estimating the parameters, which optimally fit the model with a set of data measured on the plant and satisfies the statistical tests. It is confirmed whether the model has the capacity to predict the behavior in targeted experiments.

Data Assimilation

The measurements made on the plants are generic and the methodology applicable to all architectural models. The observable variables, and thus the associated dataset, must be chosen in such a way that all the parameters of the system have a certain influence. Without this, one runs into obvious problems of non-identifiability. The estimation procedure makes each measurement correspond to a numerical value that can be computed by the model.

Calibration of the Parameters on the Data

The crown analysis (Chap. 3) provides the parameters of the meristems functioning and the analysis of the organic series (Chap. 5) offers insights into the source–sink relationships. The fits of the crown and organic series on the thirty or so plants studied of all types (Chaps. 3, 9, 10, 11) are all of satisfactory quality and validate the operational quality of the GreenLab model.

Experiments on the Model

The experiments on the model allow to test its validity on its predictive capacity.

- The parameters of a plant, once computed from a particular experiment, should correctly predict its growth in other cropping systems. This case is verified for beetroot, maize, and chrysanthemum (Chap. 11).

- In some cases, the hidden parameters can be estimated directly. For example, in limited growth, the parameter *Sp* representing the planting density, which is known can also be computed by the WLSM (Chap. 5). This fact was found on cotton, tomato, wheat, corn, beet, maple, etc. Growth monitoring of tomato and sweet pepper fruits empirically provides the same evolution of sinks as computed by the WLSM.
- The effect of plant size is well simulated by the model on coffee and cotton plants (Chap. 5), as well as the effect of seed (Chap. 5).
- The model correctly recomputes the source–sink relationships even in the absence of measurement on an organic series, as the sink action of the organs remains imprinted in the plant demand (beet, cucumber, sunflower in Chap. 9).

Capitalization of Experimental Data

The capitalization of measurements made on plants to develop the AMAP-GreenLab models forms an important database. This database, which contains data on crowns and organic series, was acquired over many years through various long-term research projects. The database (crowns, organic series) thus constituted in architecture and plant growth remains a useful reference that can be used in the framework of any structural–functional model. If the models can evolve, they will always have to be confronted with the reality of the measurements to be validated!

The software developed to implement these models of plant growth and architecture requires skills in computer development, particularly in computing and imaging for 3D reconstruction. Software development requires several years, and the question of maintenance is a concern.

The most generic and complete software implementation simulating plant growth and architecture according to the GreenLab model is the Matlab-based software Gloups developed at CIRAD. It is a good starting point for other attempts.

Finally, it should be remembered that only plants that have been the subject of agronomic experiments under controlled growing conditions can be correctly modeled. On the other hand, the virgin forest seems to be an unattainable target because of the complexity of the interactions of varieties, individual chronological ages, and the variability of the space-time environment around the plants. For the time being, a successful simulation of a plant stand in a controlled environment can only concern the case of monospecific even-aged plantations (Chap. 11), which is already of great interest for agriculture. In the medium term, it is reasonable to aim at the application of these models to associated crops.

To the End

At the end of this book, I hope that the readers will see the plants under a new ecological angle. They will have understood the difference between the computational plants in agronomy, which aim at the plant production of the fields, and the virtual plants of computer graphics which aim at an aesthetic representation of the landscapes. The former is concerned with the functioning of agronomic plantations, while the latter involves the visualization of plant scenes that rely on the talent of computer graphics, without any real connection to the botanical content.

I still hope that the agronomists will consider more the whole plant and not only the stand, in order to capitalize on all the information that can be provided at this scale on plant functioning. The parameters that control meristem functioning and source–sink relationships must be used to improve the efficiency of crop models, plant breeding methods (QTL), and the optimization of cropping systems (density, size, irrigation).

The quality of the model's fits to the tree crowns and organic series shows the amazing adequacy of GreenLab to correctly model and simulate plant growth. It is a mathematical beauty of a different order and, in my opinion, surpasses the beauty of computer-generated images of plants, which ultimately rely only on sophisticated visual rendering.

Systems Intelligence for Plant Science: From GreenLab to PlantVerse

It is my great pleasure and honor to be invited by Professor Philippe De Reffye to write this postscript during my visit to Montpellier, France from May 23 to May 27, 2024. This is my second visit to Philippe's house, and 12 years have passed since my first visit at the end of August 2012, when I attended the European Conference on Artificial Intelligence (ECAI 2012) in Montpellier.

The purpose of my visit this time is twofold, the first is to record his Alfred North Whitehead Laureate Lecture for IEEE/CAA Journal of Automatica Sinica, and the second, the main one, is to talk with Philippe about the potential ways to revitalize, extend, and deepen the past research of his GreenLab and his unique approach to computer plant simulation and computational plant science. After three full days of video recording, casual discussion, site seeing, and visiting his office and test fields at Europe's oldest botanic garden, as well as his past graduates, colleagues, and lab facilities in the French Agricultural Research Centre for International Development (CIRAD), I have a much deep understanding and profound appreciation of Philippe's work and impact to our society and our sustainability. I told myself that we must continue his work and his team effort. In addition, we must start to think and appreciate the significance of his contribution to plant science from a new perspective of social science, as revealed in part by Professor Franck Varenne's book "From Models to Simulations".

P. De Reffye et al. (eds.), *Plant Model Greenlab for Botany and Agronomy*,
https://doi.org/10.1007/978-981-97-7541-5

A photograph of Fei-Yue, Philippe, Marc, Thierry and Hervé, taken at the CIRAD

The first time I learned Philippe's work was over 20 years ago when I was still a professor in the USA and was in the process of preparing for the establishment of a Sino-USA joint research center at the Institute of Automation, Chinese Academy of Sciences (CASIA) in the late 1990s. At that time CASIA had begun to host the Sino-French Laboratory in Computer Science, Automation, and Applied Mathematics (LIAMA), where Philippe worked first as Principal Investigator and then as its French Co-Director in Beijing with his Chinese colleagues and students. Many years later, as CASIA's vice president for research, education, and international cooperation, I was assigned to oversee the operation of LIAMA and learned more about Philippe and his team's plant modeling work. I had witnessed the launch of the International Symposium of Plant Modeling, Simulation, Visualization and Application (PMA), initiated by Philippe's Team and the Late Professor Hu Bao-Gang, one of my two best friends at CASIA at that time, and the growth of GreenLab and cPlant during the last two decades.

Under the recommendation of Bao-Gang, Dr.Kang Meng-Zhen, the major Chinese principal investigator in the GreenLab Project and Philippe's PostDoc, jointed my team 15 years ago, initiated our research program on parallel plants and systems agriculture, as well as blockchain and metaverse-based smart agronomic and ecological systems, especially DeCASA and AgriVerse projects, and in the process, provided me a unique opportunity to understand and appreciate Philippe's thought and work deeper and better. With Philippe's help, we established Institute of Smart Agriculture at Qingdao Academy of Intelligent Industries (QAII) in 2014, and is planning to continue its activities internationally as a new DeSci Institute using smart contracts and metaverses, i.e., a Decentralized Autonomous Organization with Decentralized Autonomous Operation (DAO) for Decentralized Autonomous Science (DeSci) in plant, agricultural, and ecological systems.

This book summarizes the scientific achievements during the long-lasting and ongoing Sino-French cooperation since 1998, including the underlying botanical concepts, the knowledge about the organization of plant architectures, the modeling of plant structure, the modeling of plant growth, as well as the modeling of inflorescence, the secondary growth of trees. A special chapter is dedicated to the model validation through real plants, including herbaceous and woody plants, making the model valuable for agriculture. Additionally, the work is embodied by elegant visualization of plants, making us happy and enjoyable, the original purpose and intrinsic nature of our research and development.

Philippe should be extremely proud of himself for his lifetime achievement of over half a century in plant modeling, which deeply touched me in our talks during my second visit to him. Early in the 1970s, he pioneered research on three-dimensional simulation of plants with coffee plants in Africa. With his background in agronomy and mathematics, his deep understanding of science, as well as his love of classical music and painting, Philippe established a "GreenLab School" of plant modeling and applications, which distinguished itself with generic models, mathematical description, and inverse modeling. With his devoted and sincere effort in Beijing over two decades, he won the Friendship Prize in 2006 from the Chinese Central Government. Currently, hundreds of scientific papers on GreenLab have been published by researchers around the world, and software systems for GreenLab have been developed and licensed. So far, the model has been successfully applied in many areas, including controlled environment agriculture, the simulation of closed ecological systems, the training of researchers and farmers, and various other innovative deployments.

Fei-Yue and Philippe at Montpellier

I believe the work by Philippe and his team would be extremely useful in "growing" or constructing artificial plants or digital twins of actual plants. Their plant models will be the scientific foundation of various foundation models, i.e., large language models for plants and plant science, as well as the technical foundation for decision intelligence in the management and control of parallel agriculture and ecological systems.

Now the important and urgent issue before us is how to continue Philippe's effort, especially his work in GreenLab, and make GreenLab School stronger, more useful, and last forever as a school of thought in plant science and agricultural engineering.

My childhood experience offers me a unique perspective to address this issue: I grew up in a hop experimental station in 1960s, helped my father in his trials and tests in making China's own hops and building its first hop field, first for Tsingdao Brewery and then for the entire beer industry in China. This makes me deeply passionate and understand the effort and difficulty in growing plants, its long periods, tedious labors, and anxious expectations; I hope we could have a computer model for computer simulations of hop seeding, root cutting, branch trimming, and other growth caring problems at that time, then we would have much more time for normal family activities in my childhood.

This fortunate family experience also made growing flowers and plants my favorite pastime in my life, but in the spring 1995, I decided to make modeling plants and controlled environment agriculture (CEA) a new field of my professional research and started to work with Biosphere 2 and CEA Station in Tucson, Arizona, USA, for computational experiments of plant growth with linguistic dynamic systems (LDS), i.e., modeling-based experience and knowledge expressed with data and words. In my robotics and automation class SIE 585/485, I tried to use LDS aided with computer vision for modeling and monitoring plant growth, and introduced the first web camera for CEA at the University of Arizona (UA) in the late 1990s. However, after a few initiatives, I did not follow up due to my other interests and more urgent projects, and fortunately, my colleagues and friends, such as Dr. Guanghui Li in Biosphere 2 and others in UA, continued and made huge contributions in plant modeling and smart plants research, such as iPlants project.

Yes, this was the main reason that my thinking and attitude towards Philippe's work are different from other people when I learned his effort at CASIA in the late 1990s. When I was told that plant modeling was too far out the traditional scope of CASIA's research, I expressed my strong support to CASIA's leadership and believed it would be an extremely important and valuable addition to CASIA's future portfolio, strategic significance to China's development. In the meantime, Philippe's research at CASIA motivated me to launch my own research on Parallel Plants, Parallel Seeds, Parallel Agriculture, Parallel Ecosystems, CASA, Agriculture 5.0, and later aPlants (Artificial Plants), bPlants (Blockchained Plants), cPlants (Computational Plants), AgriVerse, and other ventures, where Meng-Zhen, Xiu-Juan, Jing, and many others led and made significant contributions. Since 2002, parallel agriculture and parallel ecology have been listed as one of four strategic important fields in CAS' Key Laboratory of Complex Systems and Intelligence Science and The State Key Laboratory for Management and Control of Complex Systems.

How to move forward? Personally and first of all, I would like to add education as an important function of GreenLab, and move GreenLab out of professional research labs and into schools and households so that pupils and ordinary people can grow their real plants with the help of GreenLab, providing them with deep and easy understanding of plant biology, physiology, botany, ecology, and other related knowledge, as well as enjoyment and appreciation in the miracle and beauty of the nature, especially our plants.

For this purpose, we must transform Philippe's GreenLab to PlantVerse by integrating AI and metaverse technology into plant science and implementing them through cloud and edge intelligent systems, so virtual plants and real plants can grow in parallel and feedback each other in real-time and closed-loop fashion. To protect individuals' privacy and intellectual as well as other rights, we also need to introduce blockchain intelligence to "chain" artificial/actual plants for safety and security, smart contacts to govern their behaviors for ordered interactions, DAOs to manage their production and utilization through decentralized autonomous organizations and operations, and DeSci to conduct plant science research and deployment by decentralized autonomous science. This would open a new world of opportunities for innovations in GreenLab, smart agriculture, and sustainable ecology, and in the end, make Philippe's GreenLab a foundation model and/or an infrastructure model to "grow" Systems Intelligence for Sustainability Industry in our future smart societies.

All said, my only message at this point to Philippe is

$$\text{SEE}: \ \text{SCE}^{+} \ \text{H}_3\text{O in Your GreenLab of Life}$$

That is, Sleeping well, Eating well, Exercising well; Slow, Casual, Enjoying, Elegant, Easy, ……; Happy, Healthy, and Helping Organized, Operated, and Optimized.

Fei-Yue Wang
The State Specially Appointed Expert
and Founding Director

The State Key Laboratory for
Management and Control of
Complex Systems
Chinese Academy of Sciences
Beijing, China
June 18, 2024

Fei-Yue, Didier and Philippe at the test field in Montpellier Botanical Garden

From left to right: Daniel Barthélémy, François Houllier, Philippe De Reffye
Photo by De Reffye, taken in 2006

From left to right: Marc Jaeger, Philippe De Reffye, Bao-Gang Hu.
Photo by De Reffye, taken in 2006

Véronique Letort and Mengzhen Kang.
Photo by Jaeger, taken in 2016

From left to right: Véronique Letort, Ep Heuvelink, Yuntao Ma.
Photo by De Reffye, taken in 2007

From left to right: Philippe De Reffye, Amélie Mathieu, Paul-Henry Cournède.
Photo by De Reffye, taken in 2001

From left to right: Bao-Gui Zhang, Philippe De Reffye,Yan Guo.
Photo by De Reffye, taken in 2000

Jean Pierre Quadrat and Hongping Yan.
Photo by De Reffye, taken in 2002

From left to right: Zhigang Zhan, Bao-Gang Hu, Philippe De Reffye, Xing Zhao.
Photo by De Reffye, taken in 2002

From left to right: Xiujuan Wang, Jing Hua, Philippe De Reffye, Mengzhen Kang.
Photo by Wang, taken in 2013

Philippe De Reffye and Beda Innocent Adji.
Photo by De Reffye, taken in 2021

The manufacturer's authorised representative in the EU is Springer Nature Customer Service Centre GmbH, Europaplatz 3, 69115 Heidelberg, Germany. If you have any concerns regarding our products, please contact ProductSafety@springernature.com

Printed and bound by CPI Group (UK) Ltd, Croydon, CR0 4YY
30/09/2026
02225298-0020